SYSTEMATIC LAB EXPERIMENTS IN ORGANIC CHEMISTRY

SECOND EDITION

Arun Sethi
Professor
Department of Chemistry
Lucknow University
Lucknow, Uttar Pradesh

Publishing Globally

NEW AGE INTERNATIONAL (P) LIMITED, PUBLISHERS

LONDON • NEW DELHI • NAIROBI

Bangalore • Chennai • Cochin • Guwahati • Hyderabad • Kolkata • Lucknow • Mumbai

Visit us at **www.newagepublishers.com**

Published by New Age International (P) Ltd., Publishers
First Edition: 2003
Second Edition: 2010
Reprint: 2020

GLOBAL OFFICES

- **New Delhi** **NEW AGE INTERNATIONAL (P) LIMITED, PUBLISHERS**
7/30 A, Daryaganj, New Delhi-110002, (INDIA)
Tel.: (011) 23253771, 23253472, **Telefax:** 23267437, 43551305
E-mail: contactus@newagepublishers.com • Visit us at www.newagepublishers.com
- **London** **NEW AGE INTERNATIONAL (UK) LTD.**
27 Old Gloucester Street, London, WC1N 3AX, UK
E-mail: info@newacademicscience.co.uk • Visit us at www.newacademicscience.co.uk
- **Nairobi** **NEW AGE GOLDEN (EAST AFRICA) LTD.**
Ground Floor, Westlands Arcade, Chiromo Road (Next to Naivas Supermarket)
Westlands, Nairobi, KENYA, **Tel.:** 00-254-713848772, 00-254-725700286
E-mail: kenya@newagepublishers.com

BRANCHES

- **Bangalore** 37/10, 8th Cross (Near Hanuman Temple), Azad Nagar, Chamarajpet, Bangalore- 560 018
Tel.: (080) 26756823, **Telefax:** 26756820, **E-mail: bangalore@newagepublishers.com**
- **Chennai** 26, Damodaran Street, T. Nagar, Chennai-600 017, **Tel.:** (044) 24353401
Telefax: 24351463, **E-mail: chennai@newagepublishers.com**
- **Cochin** CC-39/1016, Carrier Station Road, Ernakulam South, Cochin-682 016
Tel.: (0484) 2377303, **Telefax:** 4051304, **E-mail: cochin@newagepublishers.com**
- **Guwahati** Hemsen Complex, Mohd. Shah Road, Paltan Bazar, Near Starline Hotel
Guwahati-781 008, **Tel.:** (0361) 2513881, **Telefax:** 2543669
E-mail: guwahati@newagepublishers.com
- **Hyderabad** 105, 1st Floor, Madhiray Kaveri Tower, 3-2-19, Azam Jahi Road, Near Kumar Theater
Nimboliadda Kachiguda, Hyderabad-500 027, **Tel.:** (040) 24652456, **Telefax:** 24652457
E-mail: hyderabad@newagepublishers.com
- **Kolkata** RDB Chambers (Formerly Lotus Cinema) 106A, 1st Floor, S N Banerjee Road
Kolkata-700 014, **Tel.:** (033) 22273773, **Telefax:** 22275247
E-mail: kolkata@newagepublishers.com
- **Lucknow** 16-A, Jopling Road, Lucknow-226 001, **Tel.:** (0522) 2209578, 4045297, **Telefax:** 2204098
E-mail: lucknow@newagepublishers.com
- **Mumbai** 142C, Victor House, Ground Floor, N.M. Joshi Marg, Lower Parel, Mumbai-400 013
Tel.: (022) 24927869, **Telefax:** 24915415, **E-mail: mumbai@newagepublishers.com**
- **New Delhi** 22, Golden House, Daryaganj, New Delhi-110 002, **Tel.:** (011) 23262368, 23262370
Telefax: 43551305, **E-mail: sales@newagepublishers.com**

ISBN: 978-81-224-2828-5
C-19-05-12058

Printed in India at Ajit Printing Press, Delhi.
Typeset at In-house, Delhi.

NEW AGE INTERNATIONAL (P) LIMITED, PUBLISHERS
7/30 A, Daryaganj, New Delhi - 110002
Visit us at **www.newagepublishers.com**
(CIN: U74899DL1966PTC004618)

DEDICATED

TO

MY PARENTS

PREFACE TO THE SECOND EDITION

It is an honour to present the second edition of my book 'Practical Organic Chemistry'. As a result of the feedback from different teachers/academicians and researchers of different Universities and Research Institutes across the country, this book now features several improvements in comparison to the earlier edition. The major changes made in the book are outlined below:

- New methods of preparation like Grignard synthesis, alkylation etc.
- Synthesis of Drug and Drug Intermediates to meet the requirement of pharmaceutical industry and to prepare the student as future medicinal chemist.
- Synthesis of Polymers and Soaps to make the student acquainted with different industrial preparations.
- Estimation of iodine value, COD, DO and BOD to make the student aware of the different methods of estimation and their usefulness in daily life.

It is hoped that the second edition will assist the teachers, researchers and students in a better way to attain and develop better skill in the field of practical organic/synthetic/medicinal chemistry. The author would appreciate for any constructive suggestions being made for further improvement of the book. In the end, the author would like to thank the entire staff of the New Age International (P) Limited for their constant cooperation.

Arun Sethi

PREFACE TO THE FIRST EDITION

In view of the commercial importance and widespread academic research interest in organic chemistry, the author felt that a book dealing with the practical aspect of organic chemistry will help to familiarize the students with the identification, preparation, isolation and estimation procedures involved in organic chemistry. Perhaps the most important application of this book is to organise the material in a systematic manner and to explain to the students the basic principle of a number of reactions, followed by illustrations in the form of experiments.

Identification of organic compounds still continues to be one of the basic research in organic chemistry. A thorough understanding of the functional group, directions for the preparation of some novel derivatives besides some more common derivatives have been consolidated in Section I.

A comprehensive account of the reaction of various functional groups essentially from the view of laboratory practice, use of old as well as new techniques and modern application of older methods of synthesis have been included in Section II of this book. The experiments have been carefully selected keeping in view the time available for a normal practical class and also the resources available with different colleges and universities.

The study of organic chemistry has its origin in the study of natural products. Certain important drugs are still either exclusively natural products or these natural products are being used as starting material for the synthesis of such drugs. Thus the study of natural products should form a part of our curriculum. Section III of this book provides a survey of some important classes of natural products and goes on to describe a fair number of these including their isolation.

Section IV is devoted to the simple methods of analytical techniques involved in the estimation of elements and some simple functional groups.

The author would like to express his gratitude to his teacher late Prof. M.P. Khare whose teachings have always been helpful to the author while undertaking this project. Additional acknowledgement is also rendered to Prof. V.K. Pandey who has always been helpful in giving advice about the possible mechanism of some experiments.

There is no sufficient way to thank my wife, family and research students for their help and understanding during this project. The author would appreciate if any constructive suggestions are made for further improvement of this book.

Thanks are also due to publishers of New Age International (P) Limited for their constant cooperation.

Arun Sethi

CONTENTS

SECTION II
SYSTEMATIC ORGANIC PREPARATIONS

1

Simple Laboratory Techniques and Apparatus

In the chemical laboratory various types of glass apparatus like test tubes, beakers, erlenmeyer flask, round bottom flask, flat bottom flask, condensors, separating funnels etc are used for performing various types of experiments.

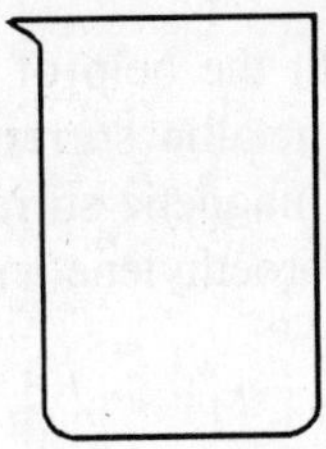

Beaker

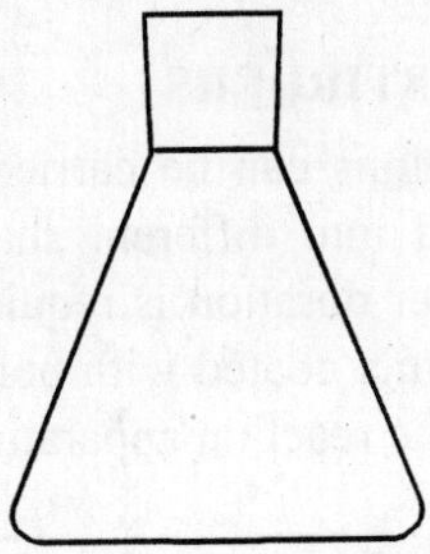

Erlenmeyar flask

Round bottom flask

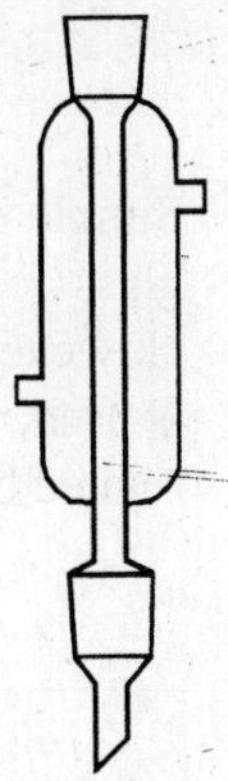

Condensor

APPARATUS FOR ORGANIC REACTIONS

The most common apparatus required for carrying out various reactions consists of a round bottom flask or erlenmeyer flask fitted with a reflux condensor. However, when several operations are to be carried out simultaneously; like introduction of gas, introduction of liquid, stirring, cooling etc, then two or three necked flask form the standard apparatus for the preparative organic synthesis. Anhydrous conditions during the course of reaction can be achieved by attaching a drying tube (a long tube filled with $CaCl_2$) at the outlet of condensor or by adding molecular sieve to the reaction medium.

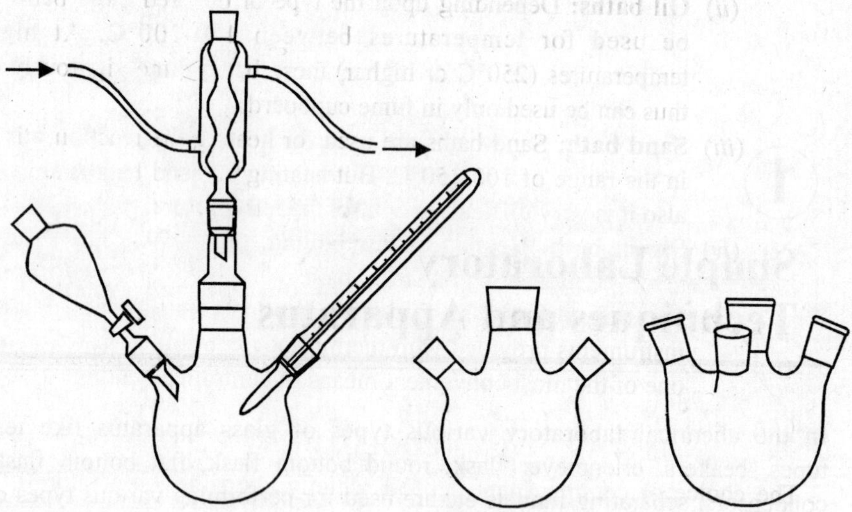

Three necked round bottom flask

Four necked round bottom flask

STIRRERS

Simple stirring of organic reactions can be carried out with the help of a glass rod or glass rod moulded into different shapes or metallic stirrers. However when stirring for longer duration is required then magnetic stirrer is used, which consists of a magnet coated with polytetrafluoroethylene and rotated by means of a motor in the reaction apparatus.

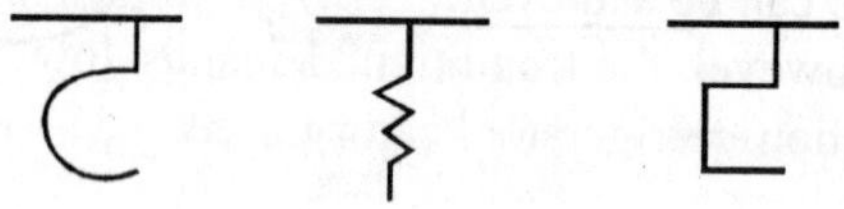

HEATING

The apparatus containing the reaction mixture can be heated with gas, steam or electricity. The choice of heat source, however, depends upon the desired temperature and rate of heating.

Higher temperatures can be achieved by heating directly on the free flame of the burner or on electric hot plates. As it is very difficult to regulate the temperature, hence in some cases, direct heating often leads to local overheating.

Heating can thus be carried out by the use of various heating baths :-

(*i*) **Water baths:** Can be used for heating the reaction mixtures upto 100°C.

(*ii*) **Oil baths:** Depending upon the type of oil used these baths can be used for temperatures between 150-200°C. At higher temperatures (250°C or higher) these baths fume vigorously and thus can be used only in fume cupboards.

(*iii*) **Sand bath:** Sand baths are used for heating the reaction mixture in the range of 100-150°C. But heating on sand bath is slow and also it is very difficult to control the temperature.

(*iv*) **Steam bath:** Heating of low boiling inflammable liquid can be achieved by heating on steam bath.

(*v*) **Heating mantle and electric hot plates:** These electrical instruments provided with thermostatic control have proved to be one of the most convenient means of controlled heating.

COOLING

The choice of cooling agent depends chiefly upon the temperature of the cooling and the amount of heat to be removed during the cooling process.

(*a*) Because of its cheapness, easy availability and high heat capacity water is the best and the most widely used cooling agent.

(*b*) Crushed ice containing little water can however be used for faster and better heat transfer process.

(*c*) Crushed ice mixed with one third its weight with coarse salt (called ice bath) is most commonly used for cooling the reaction mixture upto –5°C. By adding solid carbon dioxide (dry ice) to acetone or other suitable solvent temperature between –50°C to –70°C can be achieved.

(*d*) If however the condition demands low temperature for longer duration, refrigerators can be used.

DRYING

A good drying agent is one that should not react with the organic reactant/ product, should possess not only good drying intensity but also high drying capacity. Thus those drying agents which can take up greater amount of water while its drying intensity remains intact have greater drying capacity. Some common drying agents along with application are given below:

Drying agent	Application for
P_2O_5	Placed in desiccator and drying pistol for drying of hydrocarbon, halogenated hydrocarbon and acid solutions
Silica gel	Placed in desiccator
Molecular sieve	Organic solvents.

Sulphuric acid	Placed in desiccator and also for the drying of neutral and acid gases.
Sodium	Drying of ethers, hydrocarbons and tertiary amines
Na_2SO_4, $CaSO_4$, $MgSO_4$	Esters, ethers or solution of unknown sensitive materials
KOH, NaOH	Amines, ethers and hydrocarbons
K_2CO_3	Amines
$CaCl_2$	Placed in desiccator and also for the drying of hydrocarbon, ethers, alkenes and neutral gases.

FILTRATION

Solid particles can be separated from the liquid by the process of decantation or filtration. By the decantation process complete separation is however not possible. Filtration involves the passing of the reaction mixture through a funnel containing a filter paper. For rapid and relatively large amount of reaction product, buchner funnel with a suction flask connected to a filter pump or vacuum pump can be used for the filtration process. The process involves cutting the filter paper into the size of buchner funnel, moistened with solvent and then pressed on to the filter plate. The reaction mixture is then poured on to it. Suction is applied. The solid product is then pressed with flat glass stopper until no more mother liquor passes over.

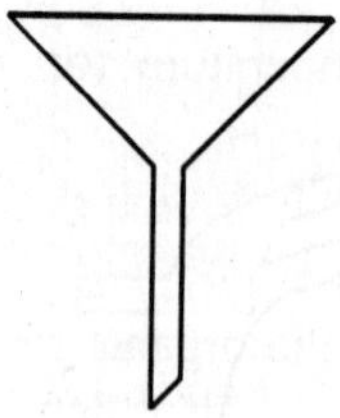

Funnel

Buchner funnel

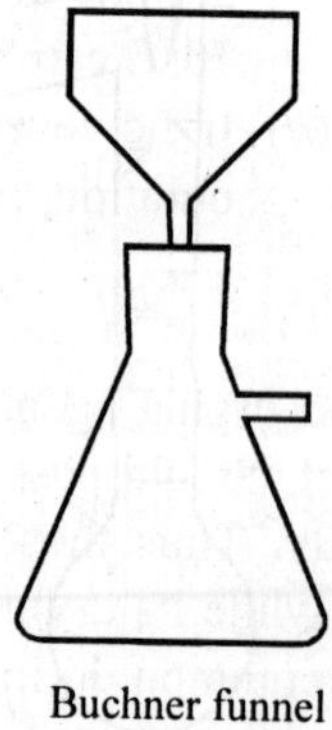

Buchner funnel with suction flask

DISTILLATION

Separation and purification of the liquids can be achieved by the process of distillation. The method consists of boiling the liquid by the supply of heat. Vapours evolved condense in the condensor as distillate.

Round bottom flask is normally used as the distillation vessel. The flask and the condensor are connected by means of distillation head. The thermometer is then so placed, such that its mercury bulb is at a short distance below the side tube. The lower end of the condensor is then connected to the second round bottom flask by means of an adaptor which then acts as a receiver.

When one of the component of the mixture is volatile then the separation can be carried out by the process of steam distillation.

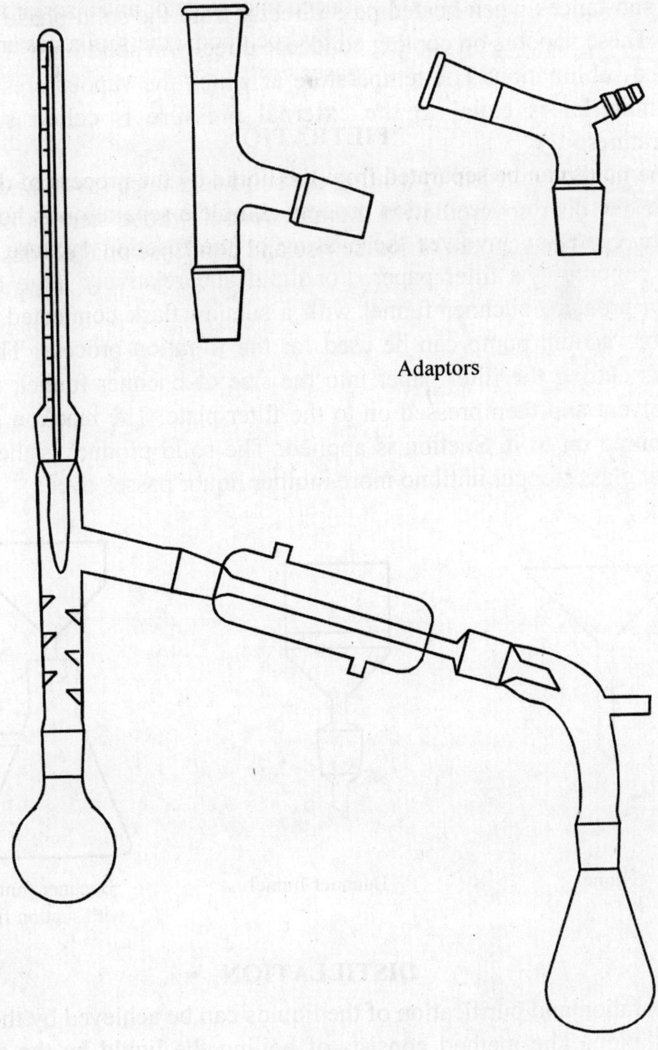

Adaptors

Distillation Unit

CRYSTALLIZATION

Crystallization is one of the important method for the purification of the reaction product. The method consists of heating the crude product with **suitable** solvent (substance sparingly soluble in the solvent in cold and highly soluble on warming) and then filtered hot to remove any insoluble material. Upon cooling the substance generally in pure form crystallises out

SUBLIMATION

Some substances when heated pass directly from the solid phase to gaseous phase. These vapours on cooling condense directly in solid form. This process is called sublimation. The temperature at which the vapour pressure of the solid material is equal to the external pressure is called sublimation temperature.

The method consists of heating the solid substance say iodine, placed in a porcelain dish covered with a small funnel whose stem is closed with cotton wool. The vapours of iodine rise and condense on the sides of funnel.

2

Chromatography

Organic compounds are either isolated from natural source or they are synthesized. The product obtained after isolation or from a chemical reaction is a mixture of number of compounds. As a result, the physical constants of the desired compound are not sharp and the exact structure of the compound cannot be elucidated. There are several physical methods besides the chemical methods by which the desired compound can be obtained in a high state of purity like fractional crystallization, distillation, steam distillation and sublimation . These methods suffer from two disadvantages – firstly, these methods work satisfactorily only on large scale and secondly, these methods fail if the crude product is a mixture of two or more than two components which have quite similar solubilities in a range of solvents or their melting point or boiling points are quite close together. These methods are now complimented by different chromatographic techniques which not only ward off the above disadvantages but also works efficiently on micro scale separation.

Chromatography literally means colour writing .It is a separation technique working on the principle of differential distribution of components of a mixture between the mobile phase and the stationary phase. As the individual components of the mixture move through the packed bed of the stationary phase at different rates with the aid of fluid mobile phase , separation occurs due to the difference in the rates of migration of the individual components. The least hindered component elutes first whereas the highly hindered component elutes the last.

Mixture

↓

Individual Components

↓

Detection of Individual Components

All chromatographic techniques work on the principle of adsorption or partition.

(i) When the stationary phase is a solid adsorbent like silica or alumina and the mobile phase is a liquid , the process is usually called liquid- solid chromatography or adsorption chromatography. This form of chromatography depends upon the adsorption of solute molecules (mixture of components) on the hydroxyl group sites on silica or alumina. Included into this category of chromatography are Thin Layer Chromatography (TLC) and Column Chromatography.

(ii) When stationary phase is a liquid strongly adsorbed on the surface of the solid support whereas the mobile phase is a liquid, the solutes are now distributed between the stationary liquid phase and mobile phase, the process is usually called liquid-liquid or partition chromatography. Included into this category is Paper Chromatography .

(A) THIN LAYER CHROMATOGRAPHY

Thin Layer Chromatography has become an indispensable part of the organic research. It not only helps in the separation procedure but it also helps in checking the course of chemical reaction.

Principle: The chromatographic process in thin layer chromatography involves the spreading of the thin layer of adsorbent (stationary phase) on a suitable support which may be either a glass plate, an aluminium foil or a plastic sheet. But more commonly glass plates are used in laboratory. Sample is applied and when the solvent has evaporated off the plate is placed vertically in a TLC chamber containing the mobile phase. As soon as the mobile phase front reaches the necessary distance, the plate is taken out , dried in air and then submitted for detection. The position of each component on the plate is expressed by its R_f value

$$R_f = \frac{\text{distance travelled by the component}}{\text{distance travelled by the solvent}}$$

The basic procedure and essential requirement for running the Thin Layer Chromatography is as follows:

(i) Choice of Stationary Phase (Adsorbent): The most commonly used stationary phase in thin layer chromatography are silica gel, alumina and powdered cellulose. All these adsorbents contain binder which is usually plaster of paris and which holds the adsorbent firmly on to the plate.

(ii) Preparation of thin layers on glass plates: The solid stationary phase on the solid support (glass plate) can be prepared by any of the following method.

(a) **Spreading:** This is the most widely used method for obtaining thin uniform layer of adsorbent on the glass plate. In this technique several strips of thoroughly cleaned and dried surgical glass plates are glued on to the solid support. A slurry (in water or solvent) of coating material, which is usually silica gel, is then poured onto the first plate and spread over the other plates with the help of glass rod. The layer may be made smooth by drawing the clean glass rod over the surface again before the slurry gets hardened. The coated plates are air dried for 30-40 minutes, and then processed to remove surplus adsorbent from the edges.

However a more uniform layer of adsorbent can be obtained by using TLC spreader (as shown below)

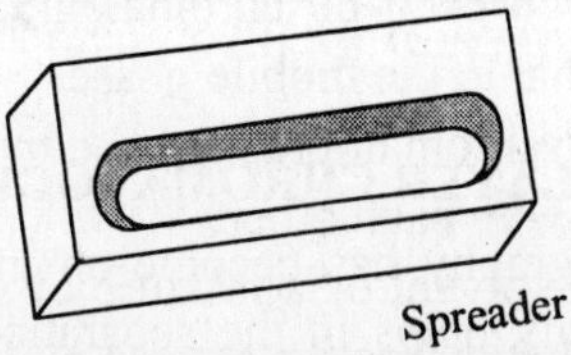

Spreader

(b) **Pouring:** In this technique the slurry of silica is usually made in organic solvent and then evenly distributed on glass plates by slightly tipping the plate to and fro without touching the edges. However it is not possible to obtain thin and uniform layer with this technique.

(c) **Dipping:** In this method the slurry is prepared by shaking the silica with organic solvent (chiefly carbon tetrachloride or chloroform) in a stoppered bottle. Two thoroughly cleaned, dried and sandwiched plates are dipped into the slurry, carefully withdrawn after 15 seconds and then allowed to drain. The two slides are then carefully separated, placed horizontally and then allowed to dry for 15-20 minutes. Surplus adsorbent adhering to the edges is then scrapped off the the help of flat spatulla .

(iii) Loading of Sample: Carefully remove any surplus adsorbent from the back and edges of the plate with the help of flat spatulla. Prepare the sample solution by dissolving 10mg of compound in few drops of suitable solvent. Dip the capillary (or micropipette) tube into the sample solution. Now apply the sample solution on to the plate by lightly touching the capillary to the adsorbent. Remove the capillary tube and allow the solvent to evaporate. The spotting of sample should be done in the middle of the width and about .5- 1 cm from the bottom of the plate.

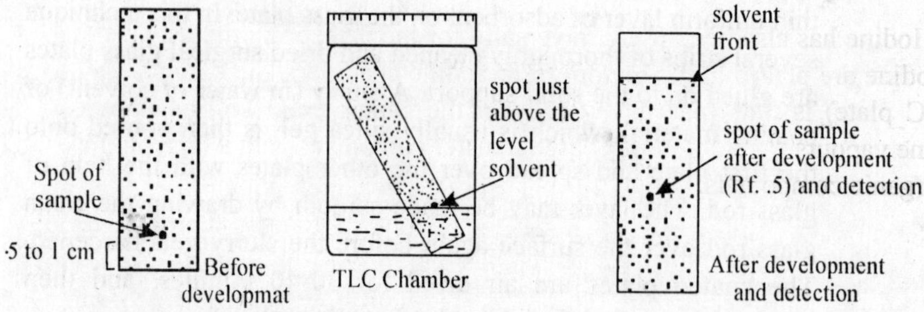

(iv) Choice of Mobile Phase: For the choice of mobile phase the following rules should be observed.

(a) The sample which is to be chromatographed should at least be moderately soluble in the mobile phase.

(b) The sample to be chromatographed and the solvent (mobile phase) should not react with each other.

(c) The best chosen solvent or a mixture of solvents for running the TLC is the one that makes the sample move half way up the plate.

(d) The mobile phase can be a single solvent or a combination of solvents and much depends upon the nature of the sample that is to be chromatographed. For eg, if the given sample is polar then a polar solvent would produce greater migration and better separation takes place in its presence.

(v) Detection: Detection of separated compounds of the sample on the chromatogram also known as visualisation is the important and final step in the TLC. The methods of detection may be classified into two categories

(a) **Physical Methods:** The developed plate is first observed in daylight and if there are any coloured spots they can be marked. Then it may be observed under UV light to detect any fluorescent spots.

(b) **Chemical Methods:** In these methods the compound on the chromatogram are sprayed with a suitable reagent giving a coloured or a fluorescent spot. The spraying can be carried out with a suitable sprayer (usually using compressed air) and whole area of the chromatogram should be sprayed evenly. After spraying the chromatogram may be heated in a drying oven. Some universal spraying agents are 50% aqueous solution of sulphuric acid, sulphuric acid mixed with a suitable aldehyde, sodium or potassium dichromate and potassium permanganate.

Iodine has also been used universally for the detection process. Crystals of iodine are placed at the bottom of an iodine chamber. The chromatogram (TLC plate) is then placed vertically in the chamber for several minutes. Iodine vapours form coloured charge transfer complexes with a large number of organic compounds.

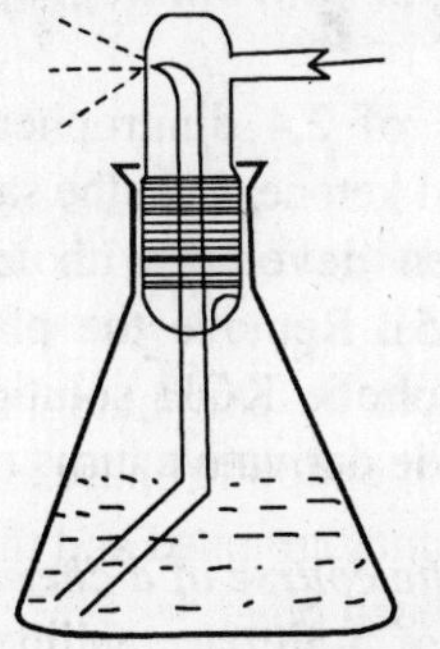

Sprayer

Experiment 1 : Thin Layer Chromatography of Spinach Leaves

Grind 5-10gm of finely chopped spinach leaves and place it in a 200ml beaker. Add sufficient 50% methanol so as to cover the leaves. Shake the beaker thoroughly, stir with a glass rod and then remove the upper liquid layer. Extract the residual portion with n hexane (3 × 50ml). Dry the combined hexane extract over sodium sulphate. Filter and concentrate the filtrate.

Apply the spot of the sample on the TLC plate and then develop with a suitable solvent (hexane-acetone-carbon tetrachloride or toluene:methanol) in the TLC chamber. Several coloured spots will be detected on the TLC plate which would include carotenes followed by chlorophyll a and b and xanthophylls. Calculate their R_f value.

Experiment 2: Thin Layer Chromatography of Amino Acids

Apply spots of glycine, aspartic acid on the TLC plate. Allow the solvent to evaporate and then develop with water : ethanol : acetic acid (1: 5 : 0.1) in the TLC chamber. Remove the plate from the chamber, dry and spray with ninhydrin solution. Heat the plate at 100-110°C for few minutes. Glycine will appear as pink spot whereas aspartic acid as violet. Calculate the R_f value.

Experiment 3: Thin Layer Chromatography of Sugars

Activate the TLC plate by keeping it in an oven at 100-110°C for 1hr. Spot the sample of glucose, sucrose, rhamnose and fructose and then develop the plate with n-butanol:acetic acid : ether : water (4.5 :3: 1.5 : .5) Remove the plate from the chamber, dry and then spray with anisaldehyde reagent. Heat the plate at 100°C for few minutes. Glucose will appear as blue spot, fructose-violet, sucrose- violet and rhamnose-green. Calculate their R_f value.

Experiment 4 : Thin Layer Chromatography of 2,4 dinitrophenyl hydrazones

Carry out the synthesis of 2,4 dinitrophenyl hydrazones of acetone, acetaldehyde and methyl ethyl ketone. Spot the sample of the three derivatives on the TLC plate and then develop with toluene: pet ether (2:3) or n-heptane:benzene (3.5 : 6.5). Remove the plate from chamber, dry and spray with 5% aqueous alcoholic KOH solution. Ketone derivative turns black brown whereas aldehyde derivative turns red brown.

Experiment 5: To monitor the course of a chemical reaction

Carry out the synthesis of p-nitroacetanilide from acetanilide. After the addition of nitrating reagent is complete ,spot the sample of acetanilide (on left) and reaction mixture (on right) after 30 minutes of the start of the reaction. Develop the plate in Hexane : ethyl acetate (4:1) and then spray with 50% sulphuric acid (heat at 100°C for few minutes)or 40% H_2SO_4: $K_2Cr_2O_7$. Appearance of starting material (acetanilide) along with the new product (p-nitro acetanilide) in the reaction mixture indicates that the reaction is still not complete. Allow the reaction to proceed and again spot the same samples after 60 minutes. Absence of acetanilide in the reaction mixture indicates the completion of the reaction.

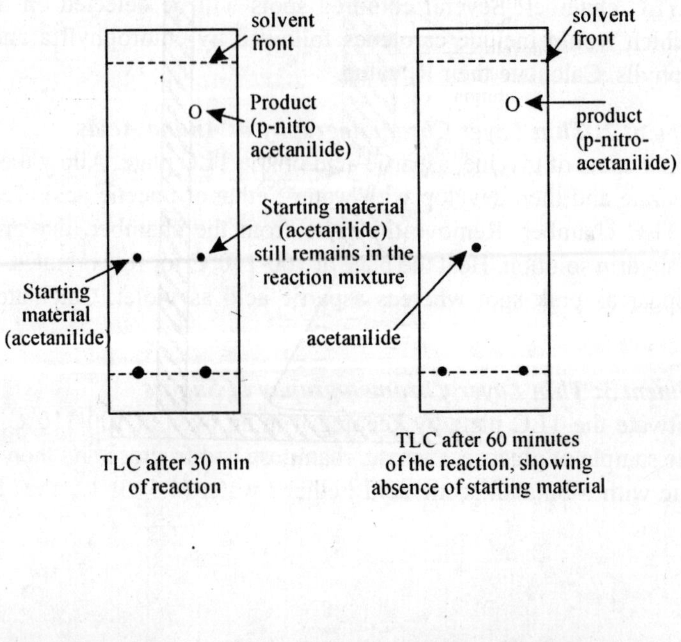

(B) COLUMN CHROMATOGRAPHY

Principle and Procedure: Column chromatography works on the same principle as that of TLC except that it differs in the manner of execution. In TLC the stationary phase is laid down on a flat solid support whereas in column chromatography the stationary phase (adsorbent) is supported in a cylindrical tube (column) which is clamped in a vertical position. Before the adsorbent is added press a small piece of cotton/ glass wool down the bottom of the column and then half fill the column with the solvent (mobile phase) that is to be used for the separation. Next appropriate amount of adsorbent should be added taking care that there are no air bubbles within the packed column (tap the column during the addition of the adsorbent). After the column is packed, apply concentrated sample solution to the top bed of the adsorbent in the column. Start the column by opening the stopcork from the bottom and simultaneously adding the solvent (mobile phase) from the top. As the mobile phase moves down the column under gravitational pull several bands are developed. The most highly adsorbed component forms the topmost band whereas the least adsorbed component forms the lower most band. Continuous passage of the solvent from the top eventually results in the emergence of individual components of the mixture from the bottom of the column. This process is known as elution.

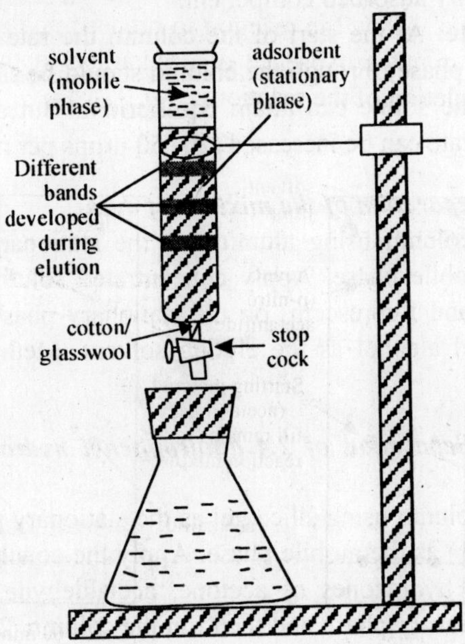

(i) Choice of Adsorbent: For the choice of adsorbent following rules should be observed.

(a) The adsorbent should be colourless and not soluble in the mobile phase.

(b) The sample to be chromatographed and the adsorbent should not react with each other.

The most common adsorbent that are frequently used are silica gel and alumina, but material such as cellulose , starch, charcoal, clay and sugar have also been used.

(ii) Elution: In the elution process the function of the solvent (mobile phase) is to differentially remove the solute molecules from the surface of adsorbent. Polarity is thus an important factor in the adsorption chromatography. A polar component is more strongly held by the stationary phase whereas the non polar component easily moves down with the mobile phase. It is therefore best to start with solvent of low polarity and gradually increase its polarity so as to elute more strongly adsorbed components. It is always not possible to carry out chromatographic separation with a single solvent. In such cases solvent mixtures can be used starting with the weak eluent. This will result in the elution of most weakly adsorbed component. The strength of the eluting solvent is then gradually increased by adding small proportion of a stronger eluent to the weak one. This will wash down the more powerfully adsorbed component.

(iii) Flow rate: At the start of the column the rate of flow of solvent (eluent or mobile phase) through the column should be slow (5-10 drops per minute) so that the solute can attain equilibrium. But after collecting 4-5 fraction the flow rate can be increased (20- 50 drops per minute)

Experiment 1 : Separation of the mixture of dyes

Prepare the column using alumina as the stationary phase and ethyl alcohol as the mobile phase. Apply concentrated solution of dye mixture (methylene blue and fluorescein) on the stationary phase and then run the column with ethyl alcohol as the eluting solvent. Methylene blue will be eluted first.

Experiment 2 : Separation of 2,4 dinitrophenyl hydrazones of carbonyl compounds

Prepare the column using silica gel as the stationary phase and hexane : ethyl acetate (19:1) as the mobile phase. Apply the concentrated solution of 2,4 dinitrophenyl hydrazones of acetone, acetaldehyde and methyl ethyl ketone on the stationary phase and then run the column. 2, 4 DNP of methyl ethyl ketone will be eluted first. Gradually increase the strength of the solvent by increasing the proportion of ethyl acetate. 2, 4 DNP of acetone will be eluted next and then finally 2, 4 DNP of acetaldehyde.

(C) PAPER CHROMATOGRAPHY

Paper partition chromatography is a technique which is similar to TLC and in which analysis/separation is carried out by the flow of solvent on specially designed filter paper. The separation being effected due to the differences in their partition coefficient.

Principle: The chromatographic process starts by applying the spot of the sample on the paper at a suitable distance from the upper or the lower end of the paper strip.The chromatogram is then placed in the chromatographic chamber which should be saturated both with the stationary phase and the mobile phase. The chromatogram (paper) should be either hung from the trough which is placed at the top of the chamber containing the mobile phase,the solvent travels down by capillary action as well as by gravitational pull (descending chromatography), or the chromatogram is hung from the top of the chamber so that its lower end is immersed in the solvent (mobile phase) at the bottom of the chamber. The solvent travels up the chromatogram by capillary action (ascending chromatography) . As soon as the mobile phase front has reached the necessary distance the chromatogram is taken out of the chamber, dried and submitted for detection. The position of each component on the paper (chromatogram) is expressed by its R_f value.

$$R_f = \frac{\text{distance travelled by the component}}{\text{distance travelled by the solvent}}$$

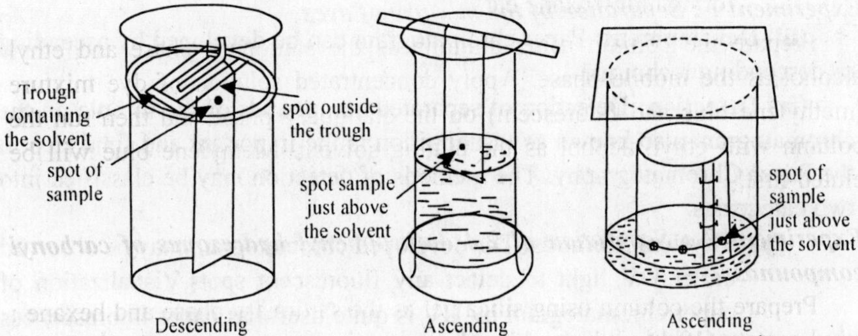

Descending Chromatography   Ascending Chromatography   Ascending Chromatography

(i) Choice of Paper: The choice of paper depends upon the flow rate and thickness of paper. Whatman No 1 is generally used for medium flow rate, Whatman No 4 is used for fast flow rate whereas Whatman No 20 is used for slow flow rate paper chromatography.

(ii) Choice of Solvent: The success of chromatographic separation depends upon the correct choice of solvent system (stationary as well as mobile phase). In principle the sample to be chromatographed should be more soluble in stationary phase than in the mobile phase.

(a) *System using water as stationary phase:* The basic material of all paper is cellulose. The free hydroxyl group of the cellulose gets easily hydrated by association with water. Water is thus one of the best solvent that can be used as stationary phase. The mobile phase which is an organic solvent can be used in a mixture of water (methyl or ethyl alcohol) . However if the organic solvent is immiscible with water (any higher alcohol) then it should be first saturated with water. This type of solvent system is well suited for compounds which are soluble in water as carbohydrates and polyhydric compounds. However if weak acids, bases or their salts are chromatographed then strong acids or bases should be added to the mobile phase. Thus some more common combination for such types of system are ethyl alcohol : ammonia : water, n butanol : acetic acid : water.

(b) *System using polar organic solvent as stationary phase:* Propylene glycol, formamide, dimethyl formamide and ethylene glycol are used as stationary phase. A weakly polar organic solvent which is immiscible or slightly miscible with stationary phase is used as mobile phase. Some more common combination for this type of system are formamide-hexane, formamide –cyclohexane, dimethyl formamide-hexane etc.

(iii) Development: Paper chromatogram can be developed by ascending or descending technique.

(iv) Detection: Detection of separated compounds of the sample on the chromatogram also known as visualisation is the important and final step in the Paper Chromatography. The methods of detection may be classified into two categories.

(a) Physical Methods: The developed chromatogram may be observed under UV light to detect any fluorescent spots.Visualization of paper chromatogram by UV is quite useful as paper fluorescences dimly in dark room and a number of organic compounds will quench that fluorescence.

(b) Chemical Methods: In these methods the compound on the chromatogram are sprayed with a suitable reagent giving a coloured or a fluorescent spot. The spraying can be carried out with a suitable sprayer (usually using compressed air) and whole area of the chromatogram should be sprayed evenly. After spraying the chromatogram may be heated in a drying oven. Some suitable spraying agents used by different categories of compounds are as follows :-

Organic acids	bromocresol green or bromo phenol blue solution in ethanol
Amines	Ninhydrin solution, iodine solution
Amino acids	Ninhydrin solution
2, 4 DNP of carbonyl compounds	10% KOH solution
Hydroxy and keto acids	Ammonia and Nessler's solution
3, 5 dinitrobenzoates of alcohols	5% KOH solution.

Experiment No 1: Paper chromatography of 3,5 dinitrobenzoates (of alcohol)

Synthesize 3,5 dinitrobenzoates of methanol and ethanol. Spot the samples on paper and develop in 50% dimethyl formamide/ cyclohexane by ascending technique. Dry the paper and view under UV light. Alcohol derivatives would appear as violet spots. Calculate their R_f value.

Experiment No 2: Paper chromatography of amino acids

Apply spots of glycine, aspartic acid , proline, phenylalanine and glutamic acid on the paper. Allow the solvent to evaporate and then develop with n-butanol : acetic acid : water (12 : 3 : 5) by ascending technique. Remove the paper from the chamber, dry and spray with ninhydrin solution.Heat the chromatogram in an oven at 100-105°C for few minutes Glycine, aspartic acid , phenylalanine and glutamic acid will appear as blue spot whereas proline as yellow. Calculate their R_f value.

Experiment No 3: Paper chromatography of sugars.

Spot the samples of glucose, sucrose, rhamnose and fructose on two separate papers. Allow the solvent to evaporate and then develop the papers with n butanol : pyridine : water (6 : 4 : 3) by ascending or descending technique. Remove the papers from the chambers, dry and then spray with anisaldehyde reagent or aniline hydrogen phthalate. Heat the chromtograms at 100-105°C for few minutes. Coloured spots will appear on the paper chromatogram. Calculate their R_f values in both the paper chromatograms.

SPRAY REAGENTS

1. Anisaldehyde- Sulphuric acid

Add 2 ml of concentrated sulphuric acid to a solution of 1ml of anisaldehyde in 100ml acetic acid.

2. Aniline hydrogen phthalate

Dissolve 1.86 gm of aniline and 3.32gm of phthalic acid in 200ml of water saturated n-butanol.

3. Ninhydrin

Dissolve .6 gm of ninhydrin in 200ml of ethanol (or n-butanol) and then add 6ml of acetic acid to it.

3

Seperation of Organic Mixtures

The separation of mixtures of organic compounds can be carried out by a number of physical methods like distillation, chromatography (paper, column and gas phase chromatography) and electrophoresis. But only those methods that take into account the chemical nature of the compound like the difference in the polarities, difference in the acidic and basic strength etc are discussed here. The organic mixtures can be separated into its constituent components by the following solvents in order of preference

(*i*) Water
(*ii*) 10 % sodium bicarbonate solution
(*iii*) 10% sodium hydroxide solution
(*iv*) dilute hydrochloric acid.

Separation based on salt formation

A general method developed for the separation of the organic mixtures is based on the concept of converting one of the component into salt which being polar becomes soluble in water. This basic principle can be explained by taking into account a number of simple examples.

The separation of aniline from toluene can be effected by treatment with dilute hydrochloric acid. Though aniline is soluble in toluene and insoluble in water, on treatment with dilute hydrochloric acid, aniline goes into the aqueous layer in the form of its salt –aniline hydrochloride. Similarly benzoic acid can be separated from benzaldehyde by shaking the mixture with 10% sodium bicarbonate solution. The benzoic acid is transformed into its salt, sodium benzoate, whose highly polar character makes its soluble in water.

Compound differing in the acidic strength can also be separated from one another by this method of extraction. Strong acids form salts with sodium bicarbonate solution whereas the weaker acids do not react with this weak base. Thus, if a mixture of benzoic acid and 1-naphthol is treated with 10% sodium bicarbonate solution, the acid passes into the aqueous layer as sodium benzoate, leaving the weakly acidic phenolic component 1-naphthol behind.

Separation based on treatment with non polar solvent(ether)

Hydrogen bonding between solute molecules and solvent is an important factor that controls solubility. Compounds possesing functional group that facilitate the hydrogen-bond formation process are soluble in polar solvents and insoluble in non polar solvent. Non polar and slightly polar substances will in general dissolve in ether. Mannitol and benzoic acid can very effectively be separated by ether. Mannitol being highly polar will not dissolve in ether whereas benzoic acid will immediately dissolve. Similarly glucose can be separated from benzophenone by adding ether. While benzophenone dissolves, glucose remains insoluble in ether.

(1) One liquid, two solids

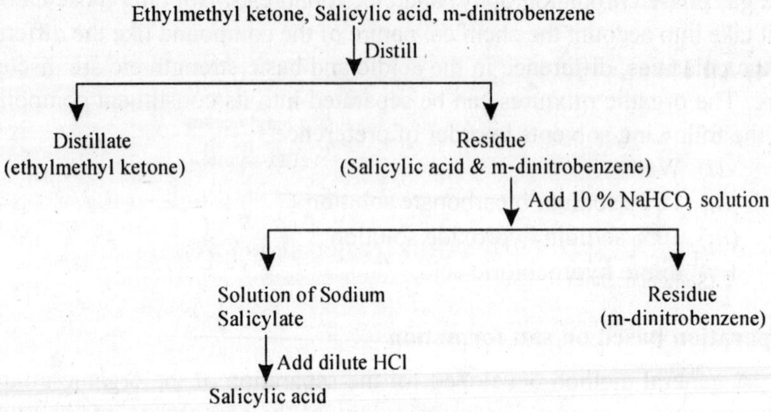

(2) Two liquids, one solid

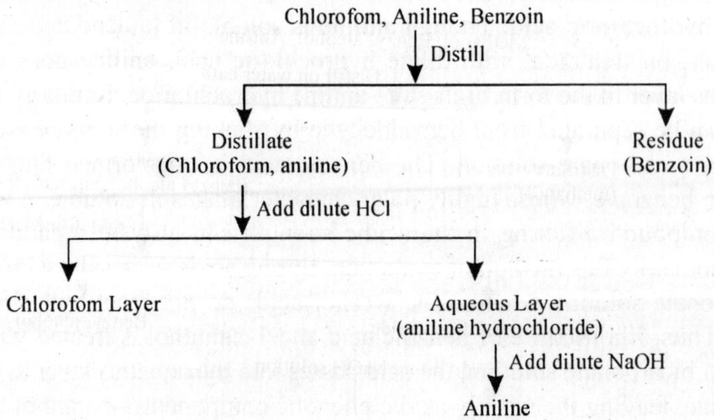

(3) One liquid, two solids (without distillation)

Phenol, Aniline, Benzophenone

- Add dilute HCl
  - Solution of aniline hydrochloride
    - Add dilute NaOH solution
      - Aniline
  - Residue (Phenol, Benzophenone)
    - Add 10% of NaOH solution
      - Solution of Sodium Phenoxide
        - Add dilute HCl
          - Phenol
      - Residue (Benzophenone)

(4) All Three Liquids

Acetic acid, n-butyl alcohol, n-butyl acetate

- Add Na_2CO_3 solution
  - Lower Layer (Sodium acetate)
    - Add dilute HCl
      - acetic acid
  - Upper Layer (n-butyl alcohol, n-butyl acetate)
    - Add $CaCl_2$ solution
      - Lower Layer (solution of n-butyl alcohol)
      - Residue (n-butyl acetate)

(5) All Three Liquids

Methanol, Benzyl alcohol, Aniline

- Distill on water bath
  - Distillate (methanol)
  - Residue (benzyl alcohol, aniline)
    - Add dilute HCl
      - Salt of aniline hydrochloride
        - Add dilute NaOH solution
          - Aniline
      - Benzyl alcohol

(6) All Three Solids

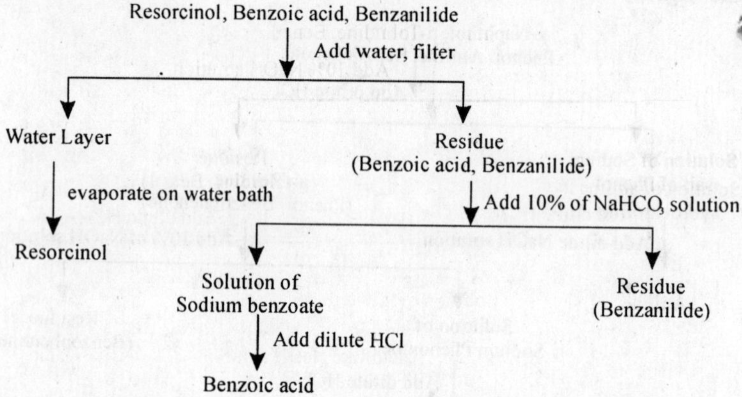

(7) All Three Solids

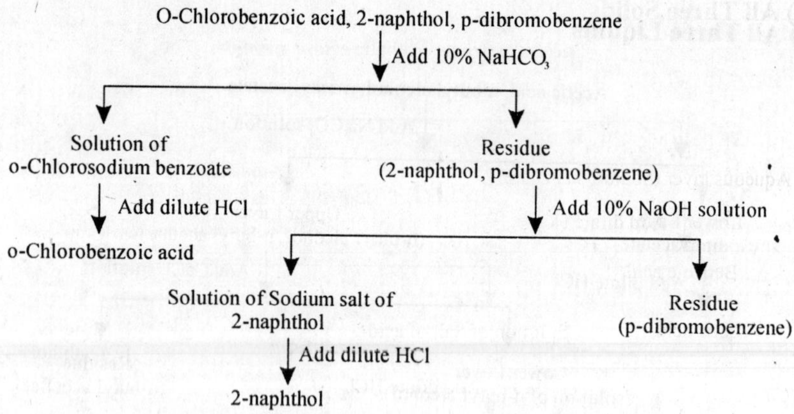

(8) All Three Solids

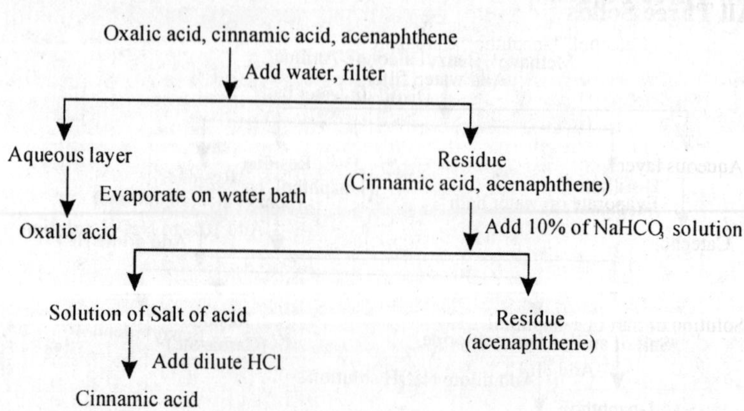

(9) All Three Solids

1-Naphthol, p-Toluidine, Benzil

↓ Add 10% NaOH solution

- **Solution of Sodium salt of Phenol**
  - ↓ Add dilute HCl
  - 1-Naphthol
- **Residue (p-toluidine, Benzil)**
  - ↓ Add dilute HCl
  - **Solution of Salt of amine**
    - ↓ Add dilute NaOH solution
    - p-toluidine
  - **Residue (Benzil)**

(10) All Three Solids

Sodium benzoate, Phthalic acid, 2-naphthol

↓ Add water, filter

- **Aqueous layer (Sodium benzoate)**
  - ↓ Add dilute HCl
  - Benzoic acid
- **Residue (Phthalic acid, 2-naphthol)**
  - ↓ Add 10% $NaHCO_3$
  - **Solution of Salt of acid**
    - ↓ Add dilute HCl
    - Phthalic acid
  - **Residue (2-naphthol)**

(11) All Three Solids

Catechol, 1-naphthol, Carbazole

↓ Add water, filter

- **Aqueous layer**
  - ↓ Evaporate on water bath
  - Catechol
- **Residue (1-naphthol, carbazole)**
  - ↓ Add 10% of $NaOH_3$ solution
  - **Solution of Salt of 1-naphthol**
    - ↓ Add dil HCl
    - 1-naphthol
  - **Residue (Carbazole)**

(12) All Three Solids

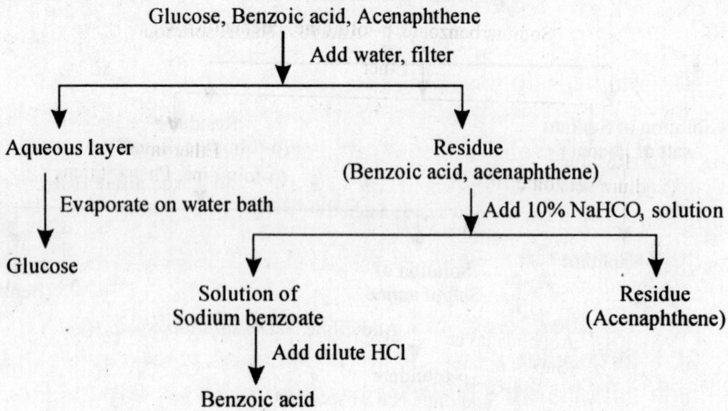

(13) Solvent Separation

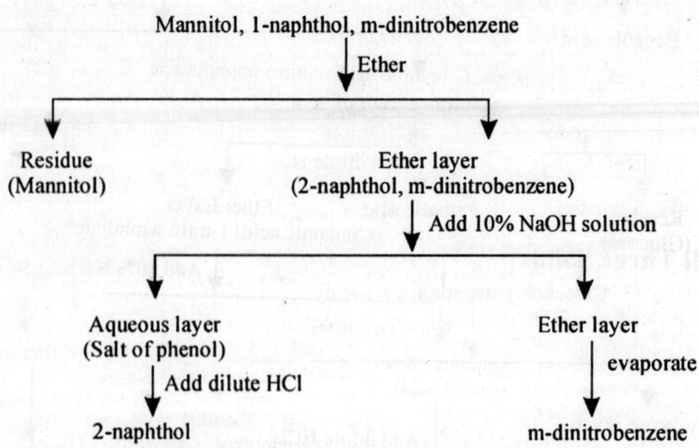

(14)

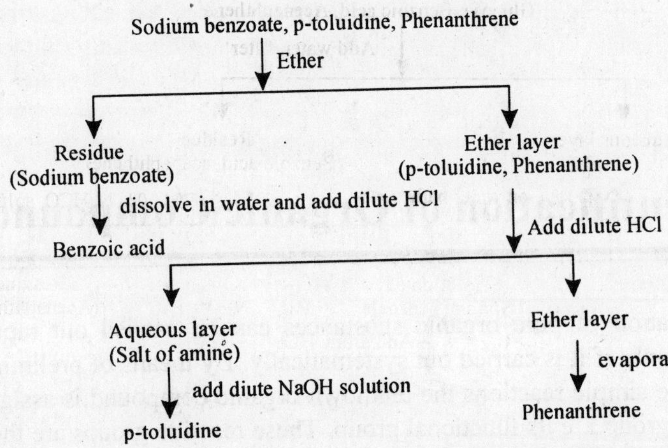
Sodium benzoate, p-toluidine, Phenanthrene
Ether
Residue
(Sodium benzoate)
dissolve in water and add dilute HCl
Benzoic acid
Ether layer
(p-toluidine, Phenanthrene)
Add dilute HCl
Aqueous layer
(Salt of amine)
add diute NaOH solution
p-toluidine
Ether layer
evaporate
Phenanthrene

(15)

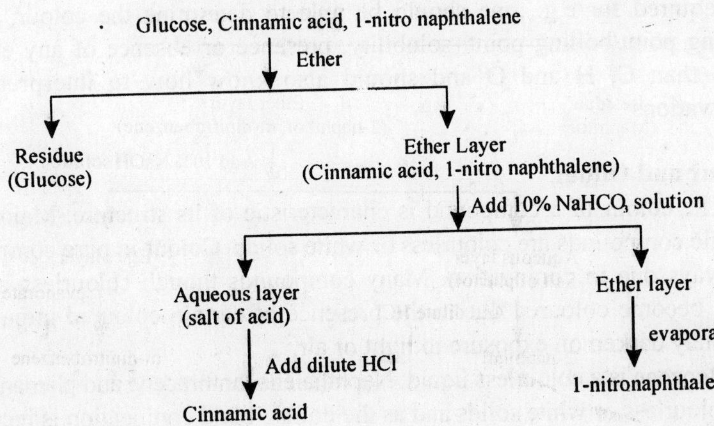
Glucose, Cinnamic acid, 1-nitro naphthalene
Ether
Residue
(Glucose)
Ether Layer
(Cinnamic acid, 1-nitro naphthalene)
Add 10% NaHCO3 solution
Aqueous layer
(salt of acid)
Add dilute HCl
Cinnamic acid
Ether layer
evaporate
1-nitronaphthalene

4

Identification of Organic Compounds

Identification of pure organic substances can be carried out rapidly and reliably only if it is carried out systematically. By means of preliminary test and some simple reactions the unknown organic compound is assigned to a specific group i. e its functional group. These reactive groups are then made to react with suitable reagents to form new crystalline compounds or derivatives. By comparing the melting point of the unknown as well as its derivative with those provided in the literature the identity can very well be established.

PRELIMINARY TESTS

For the qualitative characterisation of an organic compound certain techniques are required for e.g., one should be able to determine the colour, odour, melting point/boiling point, solubility, presence or absence of any element other than C, H and O and should also know how to interpret these observations

Colour and Odour

The colour of a compound is characteristic of its structure. Majority of organic compounds are colourless or white solids. Colour in pure compounds is always due to conjugation. Many compounds though colourless in pure form, become coloured due to the presence of certain coloured impurity or they may darken on exposure to light or air.

Benzene is a colourless liquid. Naphthalene, anthracene and phenanthrene are colourless or white solids and as the double bond conjugation is increased further the compound becomes coloured as in the case of tetracene which is yellow orange solid. Aniline is a colourless liquid whereas azo dyes are coloured because there is an extension of conjugation between two aromatic rings through a nitrogen-nitrogen double bond. Acetone is a colourless liquid whereas its 2, 4 dinitrophenyl hydrazone is yellow. This is again attributed due to extension of conjugation. Anthracene is colourless, on oxidation it gives anthraquinone which is yellow in colour. Anthraquinone has two carbonyl groups interposed between two aromatic rings giving a p-quinoid structure.

Some examples of typical coloured compounds are:-

o-nitro aniline	orange
p-nitro aniline	pale yellow
m-nitro aniline	golden yellow
p-nitro toluene	lemon
resorcinol	light pink
1-naphthol	light pink
2-naphthol	faint violet
picric acid	yellow
azobenzene	orange

Although a vast majority of organic compounds are odourless but still a number of them possess characteristic odour and this information may sometimes be valuable in determing the compounds identity. The odour is more pronounced in lower members of the class (low molecular weight) since they are more volatile. For e.g., a strong smell of bitter almonds may help you in identifying your compound to be either nitrobenzene or benzaldehyde or benzonitrile. If the compound has a strong smell of cinnamon then it is quite possible one of the unknown compound may be either cinnamic acid or cinnamaldehyde or cinnamonitrile.

Characteristic odour of some of the organic compounds are :-

Fruity	esters
Fish like	aliphatic and aromatic amines
Mouse like	acetamide
Vanilla	vanillin, anisaldehyde

Combustion and Ignition Test

(*i*) Ignition test: Place few milligrams of the solid on a metallic spatula or in a porcelain dish and ignite it directly on the non luminous flame, first slowly and then strongly. Note the nature of flame

Aromatic compounds	burns with a smoky or sooty flame
Aliphatic compounds	burns with a non smoky or non sooty flame
Compounds containing oxygen	burns with a bluish flame
Aromatic halogen compounds	burns with a smoky or sooty flame
Carbohydrates	burns with a non smoky flame with a characteristic odour

In all above ignition tests there is a rapid disappearance of carbon and tarry products which is an indication that the compound is rich in carbon, hydrogen or other hetro atom. However the persistence of carbon is an indication of the presence of heat resistance material which shields the carbon from complete combustion. The given compound may therefore be analysed for its existence as a salt.

***(ii)* Le Rosen Test for Aromatic Compounds:** Aromatic hydrocarbon that are insoluble in concentrated H_2SO_4 and have a free para position or if a hydroxyl group is present at this position react with formaldehyde and concentrated H_2SO_4 to produce red, violet or green colouration. Sulphuric acid being a good oxidant first brings about the condensation of aromatic compound with formaldehyde forming diarylmethylene compound. Oxidation of the diarylmethylene compound results in the formation of a coloured compound having p-quinoidal structure.

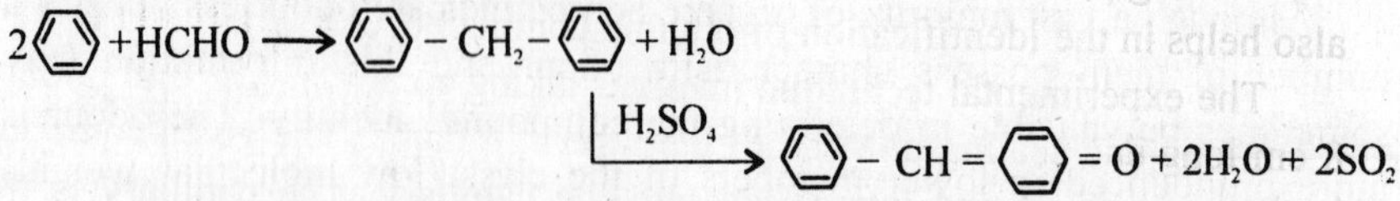

In a porcelain dish dissolve 60mg of compound in 2ml of hexane or cyclohexane or chloroform and add 4-5 drops of formaldehyde –sulphuric acid reagent. A positive response is signalled by the appearance of a colour.

Reagent:- Add 5-6 drops of 37% formaldehyde solution to 5 ml of concentrated H_2SO_4. The reagent should be freshly prepared

Anthracene	yellow green	Naphthalene	green
Toluene	red	fluorene	green
Diphenyl	blue green	benzene	red
Phenanthrene	green	hydroquinone	black
2 Naphthol	brown	pyrogallol	red
Phenol	red violet	resorcinol	red
Cinnamic acid	brick red	mandelic acid	violet red
Salicylic acid	red	benzaldehyde	red
Benzyl alcohol	red	carbazole	violet red
Methyl salicylate	red		

***(iii)* Chloroform-Aluminium Chloride test:** Heat strongly 200 mg of $AlCl_3$ in a test tube so that it sublimes and sticks to the walls of the test tube. Now run down 2ml solution of compound in $CHCl_3$ through the sides of test tube and note the colour produced by the contact of solution with $AlCl_3$.

Non aromatic compounds do not produce any colour, however non-aromatic compounds containing bromine and iodine produce yellow and violet colouration respectively.

Monocyclic aromatic compounds develop yellow-orange to red colour, bicyclic aromatic compounds develop blue to purple colour whereas complex aromatic compounds produce green colour.

Note: This test is applicable for compound which are insoluble in sulphuric acid

Determination of Melting and Boiling Point

***(i)* Melting Point:** Melting point is that range of temperature at which the sample compound changes from solid phase to liquid phase. A pure crystalline compound however has a sharp melting point range (difference in temperature at which the crystal first collapse and the temperature at which it completely becomes liquid) which should not exceed 0.5°C. Presence of impurities not only produce a marked increase in the melting point range but it also cause substance to melt at a temperature lower than the melting point of pure substance.

Melting point determination not only helps in determing the purity but it also helps in the identification of organic compound.

The experimental technique involves taking a capillary tube of about 6-8 cm long and sealing one end of the tube in a flame. The given compound is then powdered and introduced into the capillary. The capillary is then held vertically and tapped causing the powder to settle to the bottom. The capillary is then fastened to the thermometer taking care that the lower portion is close to the mercury bulb. The thermometer along with capillary is then placed in a long tube containing colourless mineral oil or conc H_2SO_4. The tube in turn is placed in a flat round bottom flask which is three-fourth filled with conc H_2SO_4. The melting point apparatus is then heated and the temperature at which the compound starts melting and at which it is entirely liquid are noted.

***(ii)* Boiling point:** Boiling point of a liquid is ascertained by the process of distillation. During the distillation process the first few millilitres of the distillate may contain water or more volatile impurities but the second fraction will consist of the pure liquid.

The liquid whose boiling point is to be determined is placed in a distilling flask along with a small piece of unglazed porous porcelain. The distillation flask is attached to water condensor which in turn is connected to the receiver. The thermometer is then inserted from the top of the flask with the mercury bulb just below the side arm. Heat the flask on an asbestos centered wire gauze. As soon as the boiling commences the flame must be adjusted so that the distillate is collected at the rate of one or two drops per second. During the distillation the temperature will first rise rapidly until it is near the boiling point of the liquid and then slowly and finally will remain constant. This constant temperature is the boiling point of the liquid. Towards the end the temperature may again rise due to superheating.

The boiling point of iquids which have high boiling point or which get decomposed, should be ascertained at reduced pressure.

Test for Acidic and Basic Character

The acidic or basic character or an organic compound is due to its ability to accept or deliver hydrogen ions (protons) or closely related process leading to the formation of salts. All basic compounds are derivatives of ammonia, hydroxylamine or hydrazine. The basic strength of these nitrogen derivatives depend on the molecular remainder group bound to nitrogen atom. For e.g. when NH_2 and NH are bound to alkyl radicals, they are strongly basic whereas when they are attached to aryl group they are weakly basic and if these groups are directly linked to carbonyl group then the product is neutral.

All acidic compounds have the ability to split off hydrogen atom as hydrogen ion or these hydrogen ions can easily be replaced by equal number of metal equivalent to form salts.

Test for acidic character

(*i*) An aqueous or alcoholic solution of the compound turns blue litmus red.

(*ii*) To 10mg aqueous or alcoholic solution of compound in a dry test tube add 3-4 drops of 2% solution of KI and 3-4 drops of 4% potassium iodate. Heat the test tube on boiling water for 1 minute. Cool and add 3-4 drops of starch solution. If the solution turns blue or violet then the compound is acidic

Test for basic character

(*i*) An aqueous or alcoholic solution of the compound turns red litmus paper blue.

(*ii*) To few milligrams or 2-3 drops of test solution add 2 drops of nickel sulphate-dimethyl glyoxime reagent. A red precipitate of Ni-dimethyl glyoxime indicates the test sample to be basic.

Preparation of reagent: To 1ml of nickel sulphate solution (7.7mg in 1ml) add 1ml of dimethyl glyoxime solution (8.1mg in 1 ml alcohol) and shake the mixture thoroughly. Filter if necessary and use the clear filtrate for the test.

Acidic	Basic	Neutral
Carboxylic acid, Phenols, Sulphonic Acid	Primary and Secondary amines	Aldehydes and ketones, esters, alcohols, amides, hydrocarbon, nitro-compounds, ethers, carbohydrates

Test for unsaturation

Addition and oxidative reactions are used to detect unsaturated linkages. However no single reaction gives sufficient information that can be used to identify the presence of double bond. Therefore both the test given below should be performed before reaching to any conclusion regarding the nature of substance under examination.

(1) Bromine in carbon tetrachloride test: Bromine undergoes addition reaction with nearly all active unsaturated compounds like alkenes and alkynes. The decolourisation of solution takes place during the course of reaction. However the discharge of bromine colour accompanied by the liberation of HBr gas indicates substitution and not addition. Such reaction is given by phenols, aromatic amines, aldehydes, ketones and some compounds containing active methylene group.

$$CH_2=CH-CH_3 \xrightarrow{Br_2/CCl_4} CH_2Br\text{-}CHBr\text{-}CH_3$$

In a test tube dissolve 100mg of compound in 2ml CCl_4 and add dropwise with shaking 5% solution of bromine in CCl_4. If decolourisation is observed, the compound may be unsaturated. However if more than 1ml bromine solution is used and fumes of HBr acid are formed then the test may not be considered to be positive.

(2) Potassium permanganate test (Bayer's test): Unsaturated compounds cause disappearance of the purple colour of potassium permanganate and formation of brown coloured precipitate of manganese hydroxide. With water soluble substances this reaction is carried out in water whereas with those insoluble in water, benzene or acetone solvent is used. The test is based on the fact that permanganate ion oxidise alkene and alkyne type of compounds to glycols. However permanganate ion also oxidises phenols and aldehydes. Thus result of permagnate ion test together with the bromine test should be used to detect unsaturated linkages in the compound under study.

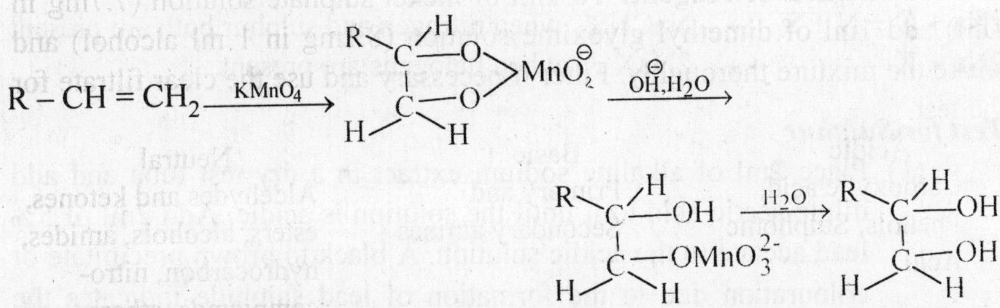

In a test tube dissolve 100 mg of compound in 5ml of water or acetone. Add 2% aqueous permanganate solution drop by drop and shake the test tube vigorously. Disappearance of purple colour with the formation of brown solution indicates that the compound may be unsaturated.

(3) Tetranitromethane test: Dissolve 100 mg of compound in 1-2 ml of chloroform and add 2-3 drops of tetranitromethane. Generation of yellow to red colour indicates the presence of olefinic double bond.

Test for Additional Elements

Tests for the presence of elements in an unknown organic compound are carried out after its oxidative or reductive mineralization. Although methods are available for the identification of carbon and hydrogen, but it is assumed that all organic compounds generally contain carbon and hydrogen hence tests for the detection of additonal elements like nitrogen, halogens and sulphur are carried out.

The most effective method of mineralization is the sodium fusion method also called the Lassaigne's method. Organic compounds on fusion with alkali metals are first converted into ions –nitrogen in presence of carbon is converted to cyanide ion, sulphur and halogens are converted to sulphide and halide ion respectively.

Fusion with sodium (Lassaigne's Test)

Place 20-25mg of the solid or 1ml of liquid sample in an ignition tube. Cut a small piece of sodium metal, dry it with filter paper so as to remove the adhering liquid and then slide it along the side of the ignition tube to its bottom. Add few milligrams of compound more so that the sodium is now sandwiched between the organic sample. Heat the lower end of the tube first gently until the sodium has melted and then heat vigorously until the glass becomes red hot. Immerse the red hot tube into 10 ml of distilled water contained in a small porcelain dish. Break the ignition tube with glass rod. Boil the mixture for 2-3 minutes and filter. The colourless filtrate called sodium extract can be used for the detection of additional elements.

$2Na + S \rightarrow Na_2S$ when sulphur is present

$Na + C + N \rightarrow NaCN$ when nitrogen is present

$2Na + C + N + S \rightarrow NaCNS$ when nitrogen and sulphur both are present

$2Na + X \rightarrow 2NaX$ when halogens are present.

Test for Sulphur

(1) Place 2ml of alkaline sodium extract in a dry test tube and add dilute acetic acid to it until the solution is acidic. Add 2ml of 5% lead acetate to the acidic solution. A black to brown precipitate or colouration due to the formation of lead sulphide indicates the presence of sulphur.

$$Na_2S + Pb(CH_3COO)_2 \rightarrow PbS \text{ (black)} + 2CH_3COONa$$

(2) Place 1m of sodium extract in a dry test tube and add 3-4 drops of freshly prepared 1% sodium nitroprusside solution to it. Appearance of pink to purple colouration indicates the presence of sulphur.

$$Na_2S + Na_2[Fe(CN)_5NO] \rightarrow \underset{\text{Sodium sulphonitro prusside}}{Na_4[Fe(CN)_5NOS]}$$

Test for Nitrogen

(1) In a test tube place 1ml of sodium extract, add 1ml of freshly prepared ferrous sulphate solution. A dark green or greenish grey precipitate of ferrous hydroxide is formed. Boil the mixture gently for one minute, cool and acidify by adding dilute H_2SO_4. Appearance of prussian blue colour or pecipitate indicates nitrogen.

$$FeSO_4 + 2NaOH \rightarrow Fe(OH)_2 + Na_2SO_4$$
$$Fe(OH)_2 + 2NaCN \rightarrow Fe(CN)_2 + 2NaOH$$
$$Fe(CN)_2 + 4NaCN \rightarrow Na_4[Fe(CN)_6]$$
$$FeSO_4 + 6NaCN \rightarrow Na_4[Fe(CN)_6] + Na_2SO_4 \text{(final reaction)}$$

In the presence of sulphuric acid ferrous sulphate changes to ferric sulphate. The product sodium ferrocyanide reacts with ferric sulphate to give prussian blue colour of ferric ferro cyanide.

$$3Na_4[Fe(CN)_6] + 2Fe_2(SO_4)_3 \rightarrow Fe_4[Fe(CN)_6]_3 + 6Na_2SO_4$$

(2) Acidify 1.5 ml of sodium fusion extract with glacial acetic acid, add 1-2 drops of freshly prepared 1% solution of benzidine in 50% acetic acid and then stir the solution. Add 1 drop of 1% $CuSO_4$ solution. A blue coloured solution or precipitate indicates the presence of nitrogen.

Test for nitrogen and sulphur present together

To 2ml of sodium fusion extract add dilute HCl until acidic and then add 3-4 drops of freshly prepared $FeCl_3$ solution. Appearnce of red or blood red colouration due to formation of ferric thiocyanate indicates the presence of both nitrogen and sulphur.

$$3NaCNS + FeCl_3 \rightarrow Fe(CNS)_3 + 3NaCl$$

Test for Halogens

If nitrogen and/ or sulphur are present then addition of silver nitrate to acidified sodium extract will precipitate silver cyanide and /or silver sulphide along with silver halide. The cyanide and sulphide ions can be removed in the following manner.

Place 5ml of sodium fusion filtrate in a porcelain dish and add 4ml of glacial acetic acid or concentrated nitric acid to it. Boil the mixture until the volulme is reduced to 1ml or almost to dryness, so as to drive off hydrocyanic acid and hydrogen sulphide formed during the reaction.

$$NaCN + CH_3COOH \rightarrow HCN + CH_3COONa$$
$$Na_2S + 2CH_3COOH \rightarrow H_2S + 2CH_3COONa$$
$$NaCN + HNO_3 \rightarrow HCN + NaNO_3$$
$$Na_2S + 2HNO_3 \rightarrow H_2S + 2NaNO_3$$

Cool the nitogen and sulphur free sodium extract and then first add 2ml dilute HNO_3 to it followed by 2ml of silver nitrate solution. If the precipitate is obtained halogen may be present. Then observe the colour and solubility of the precipitate as follows.

NOTE :- If both nitrogen and sulphur are absent then halogens can be detected by using fresh sodium extract

(1) Chlorine: A white precipitate soluble in ammonium hydroxide and reprecipitation on adding dilute nitric acid indicates the presence of chlorine.

$$NaCl + AgNO_3 \rightarrow AgCl + NaNO_3$$
$$AgCl + 2NH_4OH \rightarrow Ag(NH_3)_2Cl + 2H_2O$$
$$Ag(NH_3)_2Cl + 2HNO_3 \rightarrow AgCl + 2NH_4NO_3$$

(2) Bromine: A pale yellow precipitate sparingly soluble in ammonium hydroxide and reappears on adding dilute nitric acid indicates the presence of bromine

$$NaBr + AgNO_3 \rightarrow AgBr \quad NaNO_3$$
$$AgBr + NH_4OH \rightarrow Ag(NH_3)_2Br \quad 2H_2O$$

(3) Iodine: (*i*) A yellow precipitate insoluble in ammonium hydroxide indicates the presence of iodine.

$$NaI + AgNO_3 \rightarrow AgI \quad NaNO_3$$

(*ii*) To 1ml of sodium fusion extract add glacial acetic acid so that it is just acidic. Now add 1ml of carbon tetrachloride followed by drop by drop addition of sodium nitrite solution with constant shaking (1ml). A purple or violet colour in the organic layer indicates iodine.

$$2NaI + 2NaNO_2 + 4CH_3COOH \rightarrow I_2 + 2NO + 4CH_3COONa + 2H_2O$$

(4) Fluorine: Acidify 2ml of sodium fusion extract with glacial acetic acid, concentrate the solution to half its volume. Cool and place one drop of the solution on zirconium-alizarin test paper. A yellow colour on red paper confirms the presence of fluorine.

Zirconium-alizarin paper:-Dip dry filter paper in 5% zirconium nitrate solution in 5% HCl. Drain the filter paper and dip it in 2% aqueous solution of sodium alizarin sulphonate. Wash the violet red colour paper with water until the wash water is colourless and then dry in air.

Test for bromine and iodine

Acidify 2ml of sodium fusion extract with either 2ml dilute nitric acid or sulphuric acid and then heat to boiling. Cool, add 2ml of chloroform or carbon tetrachloride followed by drop by drop addition of chlorine water with shaking. If organic layer becomes violet then iodine is present and if the organic phase becomes brown or reddish brown then bromine is present.

$$2NaI + Cl_2 \rightarrow 2NaCl + I_2 \text{ violet}$$
$$2NaBr + Cl_2 \rightarrow 2NaCl + Br_2 \text{ brown}$$

Beilstein test for halogens

Make one end of the copper wire in the form of a loop and heat it on burner till the flame is no longer coloured. Cool the wire at room temperature and dip the loop in a little of the test sample, heat it on the flame. A green flame indicates the presence of halogens.

Solubility

Solubility is a phenomenon which is dependent on the intermolecular forces exhibited by both the solute and the solvent. For one substance to be dissolved in another, bonds in each must be broken and new bonds must be formed between the molecule or ions of the solute with molecules or ions of the solvent. A substance is said to be soluble if 30mg of it dissloves in about 1ml of the solvent at room temperature. The study of solubility behaviour in various liquids like water, 5% sodium bicarbonate, 5% sodium hydroxide, 5% hydrochloric acid and cold concentrated sulphuric acid usually gives some useful information about the type of functional group that may be possibly present in the unknown molecule. For e.g. cinnamic acid is insoluble in polar solvent water but on treatment with a base like sodium hydroxide it is converted into a salt which readily dissolves in water. Thus the solubility in a base points to the presence of an acidic functional group.

(*i*) Solubility in water: Compounds are first tested for their solubility in water which is polar and thus a poor solvent for all types of hydrocarbon. Introduction of halogens into the hydrocarbon molecule however does not alter the solubility behaviour. Salts which are extremely polar and some low molecular weight acid and bases are soluble in water. Some neutral compounds like carbohydrates are also soluble in water. Few amines are also soluble due to their ability to form hydrogen bonded complex with water molecule.

(*ii*) Solubillity in dilute sodium bicarbonate and dilute sodium hydroxide: Sodium bicarbonate and sodium hydroxide are basically used to detect water insoluble acidic substances. The aqueous sodium hydroxide is called the detecting solvent whereas aqueous sodium bicarbonate is called the subclassifying solvent as it serves to distinguish between stronger and weaker acid.

Carboxylic acids which have Ka's (acids diassociation constant) in the range of $10^{-3}$to 10^{-5} are stronger acids than phenols which have Ka's in the range of 10^{-9} to 10^{-10}. Therefore phenols can react with sodium hydroxide solution but not with sodium bicarbonate solution. But phenols having two or more nitro groups are soluble in sodium bicarbonate as well as sodium hydroxide.

Carboxylic acids, sulphinic acid and sulphonic acid are soluble in dilute sodium bicarbonate. Some negatively substituted phenols like 2, 4 dinitrophenol, 2, 4, 6-trinitrophenol and 2, 4, 6-tribromophenol are highly acidic and thus they are also soluble in dilute sodium bicarbonate. Aromatic carboxylic acid and diaryl carboxylic acids substituted with one amino group are also soluble in sodium bicarbonate.

Introduction of a nitro group in a hydrocarbon confers sufficient acidity therefore the aci form of nitroethane dissolves in dilute sodium hydroxide. The imides, p-nitro acetanilide and sulfonamides also dissolves in dilute sodium hydroxide.

***(iii)* Solubility in dilute hydrochloric acid:** Compounds that contain a basic nitrogen atom are soluble in dilute hydrochloric acid. Falling into this category are amines whose basicity is due to the presence of an unshared pair of electron on nitrogen atom. The strength of these basic amines depend on the availability of this electron pair for reaction with acid. Since this electron pair is easily available in case of p-amines they are readily soluble in dilute hydrochloric acid whereas p-aromatic amines have diminished basicity but they are still soluble. But when two or three aryl groups are attached to nitrogen the basic strength is diminished to such an extent that such compounds (secondary and tertiary aromatic amines) do not form salts with dilute hydrochloric acid and hence do not dissolve.

Certain disubstituted amides of the type $RCONR_2$ which have high molecular weight are soluble in dilute hydrochloric acid.

***(iv)* Solubility in concentrated sulphuric acid:** Oxygen containing compounds such as alcohols, ethers, esters, aldehydes and ketones are weak bases due to the presence of unshared pair of electron on oxygen atom. They react with sulphuric acid to form oxonium salts which in turn are soluble in excess sulphuric acid. Unsaturated hydrocarbons like alkenes undergo addition reaction with sulphuric acid to form alkyl hydrogen sulphates which through hydrogen bonding becomes soluble in excess sulphuric acid.

Although the majority of aromatic hydrocarbons are not soluble but some meta derivatives of polyalkyl substituted aromatic compounds on sulphuric acid treatment undergo sulphonation and the sulphonic acid thus produced are soluble

DETERMINATION OF FUNCTIONAL GROUP AND THEIR DERIVATION

Compounds containing C and H only

Hydrocarbons

Hydrocarbons are compounds that contain only carbon and hydrogen and have no reactive functional group. A hydrocarbon containing carbon-carbon double bond (alkene) or carbon-carbon triple bond (alkyne) is called unsaturated hydrocarbon while those without double or triple bond are said to be saturated. Hydrocarbons with benzene rings are called aromatic hydrocarbons while the others are classified as aliphatic.

Identification tests:

(1) **Reaction with sulphuric acid:** To 4ml of 20% fuming sulphuric acid add 1ml of hydrocarbon and shake vigorously Only aromatic hydrocarbon dissolve completely

(2) Unsaturated hydrocarbons can be detected by their reaction with bromine or potassium permanganate

(*a*) *Bromine test*: To a solution of 400mg of compound in 4ml CCl_4 add few drops of 2% solution of bromine in CCl_4. Decolourisation indicates that the compound may be an unsaturated hydrocarbon. However if more than 1ml bromine solution is used and fumes of HBr acid are formed then the test may not be considered to be positive.

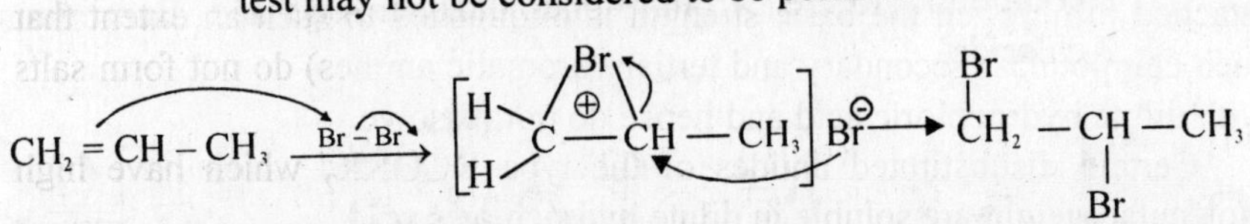

(*b*) *Potassium permanganate test*: Dissolve 400mg of compound in 4ml water or 4ml acetone and add dropwise 2% aqueous solution of potassium permanganate. If the colour of permanganate disappears, then the given compound may be an unsaturated hydrocarbon

Note: The test is negative if more then 5-6 drops of the reagent are used.

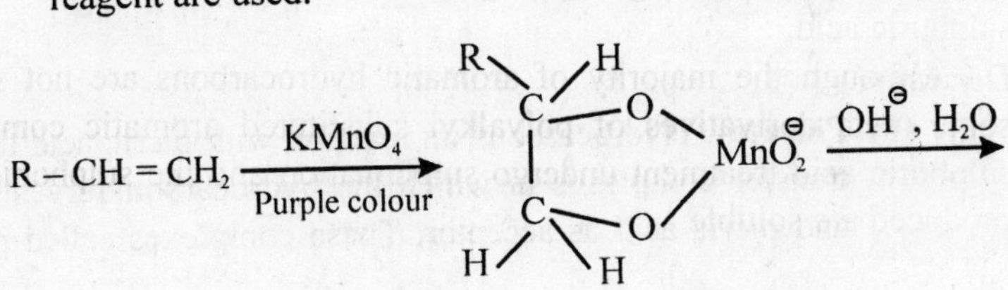

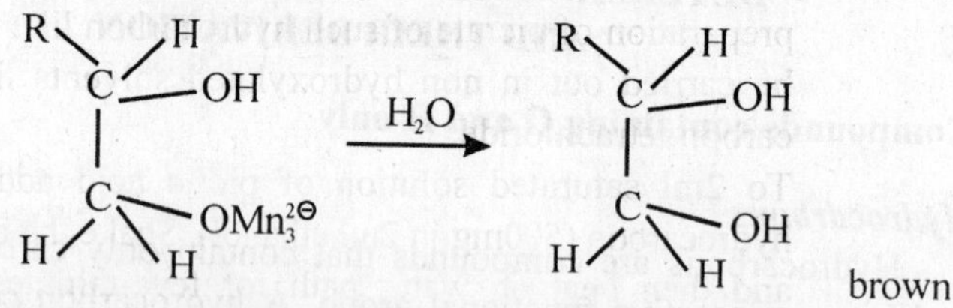

(3) **Reaction with formaldehyde –sulphuric acid reagent:** In a dry test tube place 50mg of hydrocarbon and add to it 3ml of formaldehyde-sulphuric acid reagent. Development of red-brown or brown colour is considered to be a positive test. (Unsaturated cyclic hydrocarbon, aromatic hydrocarbon and their derivatives and polycyclic hydrocarbon respond to this test)

Formaldehyde-Sulphuric and Reagent: To 5ml of 30% formaldehyde add 0.24ml of concentrated H_2SO_4.

(4) **Reaction with Aluminium chloride and azoxybenzene:** To 4ml of benzene add 4-5 crystals of azoxybenzene and about 200mg of anhydrous aluminium chloride. If no colour develops then warm the mixture for few minutes. Development of deep orange to dark red colour or precipitate indicates the hydrocarbon to be aromatic or its halogen derivative. Condensed polynuclear hydrocarbon produce brown colour.

The colour is produced due to the formation of an addition compound formed from p-phenylazobenzene and aluminium chloride.

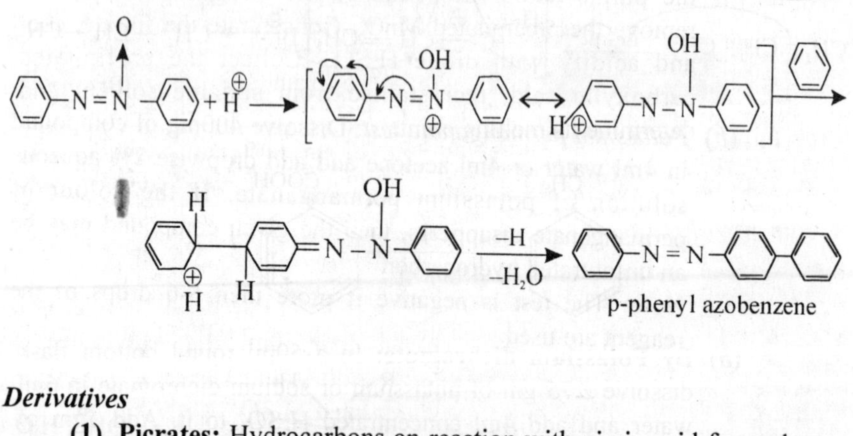

Derivatives

(1) **Picrates:** Hydrocarbons on reaction with picric acid form charge transfer complexes in which the hydrocarbon serves as a donor and picric acid as acceptor. These complexes called picrates are

usually yellow or orange coloured. Some picrates are usually unstable and are decomposed by hydroxylic solvents. So the preparation of picrate of such hydrocarbon like anthracene should be carried out in non hydroxylated solvents like chloroform or carbon tetrachloride.

To 2ml saturated solution of picric acid add 2ml solution of hydrocarbon (500mg in 2ml ethanol). Shake the mixture thoroughly and then heat on water bath for few minutes. Cool, filter the precipitated coloured solid, recrystallise and determine its melting point.

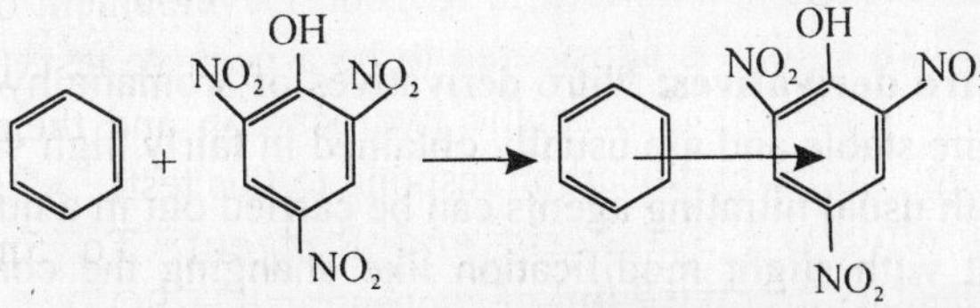

(2) Oxidation of side chain: The side chain alkyl group attached to aromatic ring for eg., toluene is oxidised to benzaldehyde and further to benzoic acid. The mechanism by which this oxidation occurs is however complex. Various intermediates like the chromate and permanganate esters are involved but it also appears to involve an intermediate benzyl cation.

(*a*) **By Potassium permanganate:** In a 100ml round bottom flask dissolve 3gm of $KMnO_4$ in 10ml of 5% NaOH solution. Now add 1gm of unknown sample and reflux for 2-3hrs or until the purple colour of permanganate disappears. Filter and remove the precipitated MnO_2. Concentrate the filtrate, cool and acidify with dilute H_2SO_4. Collect the precipitated carboxylic acid, recrystallise from suitable solvent and determine its melting point.

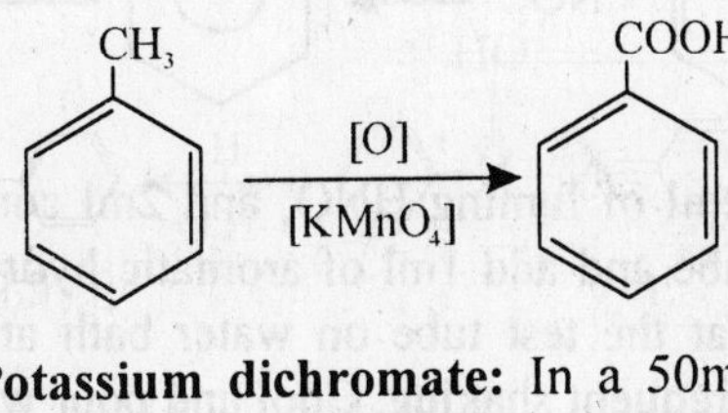

(*b*) **By Potassium dichromate:** In a 50ml round bottom flask dissolve 2.75 gm of potassium or sodium dichromate in 6ml water and add 4ml concentrated H_2SO_4 to it. Add 1gm of unknown hydrocarbon and boil the resulting mixture for 30-

40 min. Cool the mixture to room temperature, add 10ml water to it and filter. Wash the precipitated carboxylic acid with cold water, recrystallise from suitable solvent and determine its melting point.

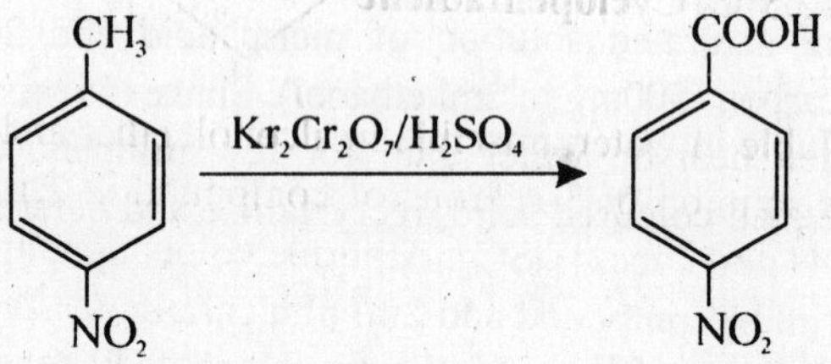

(3) Nitro derivatives: Nitro derivatives of aromatic hydrocarbon are quite stable and are usually obtained in fairly high yield. Nitration with usual nitrating agents can be carried out in a number of ways but with slight modification like changing the concentration of nitrating mixture, temperature and reaction time.

(*a*) Place 6ml of nitrating mixture (3ml each of concentrated HNO_3 and concentrated H_2SO_4) contained in a dry test tube in an ice bath. Slowly add drop by drop 1ml of hydrocarbon(benzene or toluene) to the cooled nitrating mixture. After the addition is complete heat the mixture on water bath for 5-10min. Cool and pour with stirring into ice water. Filter the precipitated solid, wash with water, recrystallise and determine its melting point.

$$HNO_3 + H^+(H_2SO_4) \longrightarrow H-\overset{\oplus}{\underset{H}{O}}-NO_2 \longrightarrow H_2O + NO_2^{\oplus}$$

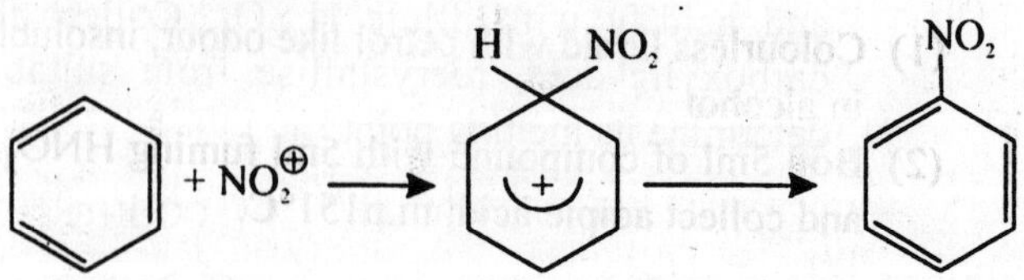

(*b*) Mix 2ml of fuming HNO_3 and 2ml concentrated H_2SO_4 in a test tube and add 1ml of aromatic hydrocarbon (p-xylene) to it. Heat the test tube on water bath at 85-95°C for 30 min with frequent shaking. Cool and pour with stirring in a 100ml beaker containing 20gm crushed ice. Filter the precipitated poly nitro derivative, recrystallise from alcohol and determine its melting point.

Aliphatic Liquid

b.p 42.5°C **Cyclopentadiene** 

(1) Insoluble in water, miscible in alcohol, ether and benzene.
(2) To a solution of 3-4 drops of compound in 1ml $CHCl_3$ and 1ml AcOH add conc. H_2SO_4. A purple coloured solution is obtained.
(3) To 1ml of compound add 2ml of alcoholic benzoquinone solution, shake thoroughly and collect cyclopentadienebenzoquinone, m.p 76°C.
(4) Add 1ml of compound to a cooled suspension of 500mg of maleic anhydride in 5ml benzene (exothermic reaction). Cool and collect cis-3, 6 endomethylene tetrahydrophthalic anhydride, m.p 165°C.

b.p 69°C **n-Hexane** $CH_3CH_2CH_2CH_2CH_2CH_3$

(1) Colourless liquid with petrol like odour, highly inflammable. Insoluble in water, soluble in alcohol and ether

b.p 81°C **Cyclohexane** (hexahydrobenzene) 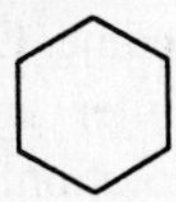

(1) Colourless liquid with petrol like odour, insoluble in water, soluble in alcohol
(2) Boil 5ml of compound with 5ml fuming HNO_3. Cool, pour in ice and collect adipic acid, m.p151°C.

b.p 83°C **Cyclohexene** (Tetrahydrobenzene) 

(1) Colourless liquid, insoluble in water, soluble in alcohol.
(2) Heat with stirring 15ml cyclohexene, 42gm KI, 45gm orthophosphoric acid at 80°C for 3hrs. Cool and add 40ml water

to the reaction mixture. Extract with ether. Evaporate off the ether and collect iodocyclohexane, b.p 48-49°C.

(3) To an ice cold mixture of 10.5ml benzene and 1ml of conc H_2SO_4 add dropwise with stirring 5ml of cyclohexene (temp 5-10°C). Continue stirring for additional one hour. Remove the hydrocarbon layer, cool in ice, wash with conc H_2SO_4, water, NaOH and finally with water. Distill and collect the fraction distilling at b.p 238-243°C. This is pure cyclohexylbenzene.

(4) To an ice cold solution of 4ml cyclohexene, 6ml CCl_4 and 0.5ml absolute ethanol (temp below-5°C) add dropwise with stirring a solution of 1.5ml bromine in 3ml CCl_4. After the addition is complete distil off the solvent on water bath and then distil the residual portion on oil bath. Dibromocyclohexane distills at 99-103°C.

(5) To a mixture of formic acid and 1.5ml 30% hydrogen peroxide add dropwise with stirring 1.5ml cyclohexene (temp 40-50°C). Keep the resultant mixture at 40°C for an hour and then keep overnight. Add 3ml of sodium hydroxide solution (0.8gm in 3ml water). Warm the alkali solution to 40°C and add 10ml ethyl acetate. Remove the ethyl acetate and repeat the process twice. Combine the organic layers and distill off the ethyl acetate. Cool and collect trans cyclohexane 1, 2 diol, m.p 102-103°C.

(6) To 10ml of cyclohexene add 40ml ether, 2ml water and 12gm of mercuric oxide. Cool and add in portion with stirring 25gm of iodine. Continue stirring until the colour of iodine has disappeared. Filter and to the filtrate add potassium iodide and bisulphite solution. Evaporate off the solvent, cool the residue and collect 2-Iodocyclohexanol, m.p 42°C.

b.p101°C **Methylcyclohexane** (hexahydrotoluene)

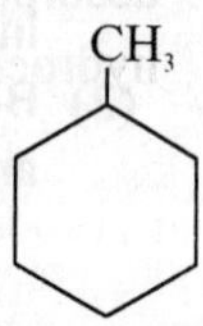

(1) Colourless liquid, soluble in benzyl alcohol.

b.p 156°C α **Pinene**

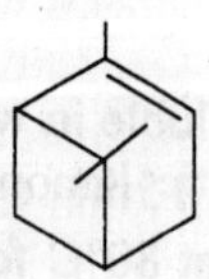

(1) Colourless transparent liquid with a characteristic penetrating odour

(2) Cool 2ml of compound to 0°C and add 2-3ml of bromine saturated with carbon dioxide gas. Collect the precipitated 2, 6-dibromocamphane, recrystallise from chloroform, m.p 169°C.

(3) Cool a mixture of 3.5ml pinene, 5ml amyl nitrite and 8ml glacial acetic acid. Add with stirring a mixture of 2.75ml of conc HCl and 2.75ml of glacial acetic acid. Crystals of nitrosochloride separate out. Dissolve it in excess of 80% piperidine and then warm. After the reaction has subsided add cold water to it. Nitrolpiperidine separates out, m.p 118-119°C.

b.p 177° C **d-limonene** 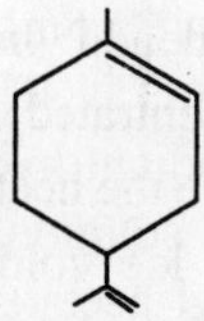

(1) Colourless oily liquid with characteristic lemon odour.

(2) Dissolve 2 ml of compound in 2ml ethylacetate and 3ml ether. Add dropwise 13ml of bromine solution (3ml bromine in 10ml ether). Shake thoroughly, cool and collect d-1, 2, 8, 9 tetrabromo-p-menthane, m.p104°C.

(3) To a solution of 5ml of compound in 5ml glacial acetic acid pass dry HCl gas. Shake the flask frequently in order to facilitate the absorption of gas. Cool and collect the precipitated limonene bis hydrochloride, m.p 50-51°C.

b.p 178°C **Dipentene** (d, l limonene) 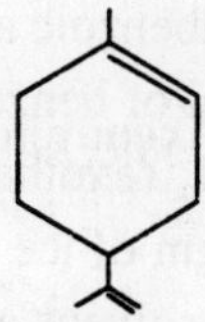

(1) Oily liquid with characteristic lemon odour, insoluble in water, miscible with alcohol.

(2) Dissolve 1ml of compund in 3ml cold alcohol and 3ml cold ether. Add with stirring 2.2ml of bromine. After the addition is complete evaporate off the solvent and collect dipentene tetrabromide, recrystallise from ether, m.p125°C.

Aromatic Liquid

b.p 80°C **Benzene** 

(1) Colourless liquid with characteristic aromatic odour, insoluble in water, soluble in organic solvents
(2) To 2ml of compound add 4ml of nitrating mixture (2ml conc HNO_3 and 2ml concentrated H_2SO_4). Heat the mixture until it boils and then continue the heating process for further 2-3 minutes. Cool, pour in 25ml ice cold water with stirring. Collect the precipitated m-dinitrobenzene, recrystallise from aqueous alcohol, m.p 90°C.
(3) To 10ml fuming H_2SO_4 add in portion 5.5ml benzene(temp 30-50°C)After the addition is complete stir it and then pour in 10ml water, cool and neutralise by adding $NaHCO_3$ and 10gm NaCl. Heat the solution, filter hot, cool the filtrate and collect sodium benzene sulphonate.
(4) To 5gm phthalic anhydride and 20ml benzene add in portion 10gm anhydrous $AlCl_3$. After the addition is complete first warm and then reflux the reaction mixture. Cool and add 50gm crushed ice. Add about 7-8ml conc HCl or until the solution is clear. Distil off the unreacted benzene, cool the residue and collect o-benzoylbenzoic acid, m.p 124°C
(5) To 15ml of benzene add 13gm of chloroacetyl chloride and 20gm of $AlCl_3$ (exothermic reaction). After the reaction has subsided add 20gm of ice and conc HCl. Separate the benzene layer. Distil off the solvent and cool the residue. Crystals of ω chloroacetophenone separates out, m.p 59°C.
(6) Picrate (colourless), m.p 84°C.

b.p 110°C **Toluene** (methylbenzene) 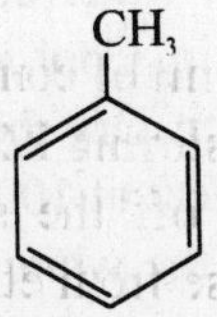

(1) Colourless liquid, odour similar to benzene, insoluble in water, miscible with organic solvents.

(2) Heat 1ml of compound together with few milligrams of MnO_2, $CuSO_4$ and 1ml conc H_2SO_4. Smell of bitter almond due to formation of benzaldehyde is observed.

(3) Heat a mixture of 7ml toluene and 3.3ml conc H_2SO_4in an oil bath at 110-120°C. When toluene layer has disappeared, cool and pour in ice water. Neutralise with 3gm $NaHCO_3$. Heat the solution to boiling, saturate with NaCl and filter hot. Cool the filtrate and collect sodium toluene p-sulphonate.

(4) To 5gm phthalic anhydride and 15ml toluene add in one portion 10gm anhydrous $AlCl_3$. The mixture becomes warm, keep at this tempeature for 10 hrs and then distil off the excess toluene. To the aqueous layer add $NaHCO_3$ until alkaline. Pass steam into this mixture for 5hrs and filter. Acidify the filtrate and collect o-(p-toluyl) benzoic acid, m.p 138°C.

(5) Dissolve 2ml of compound in 3ml fuming HNO_3 and then add immediately 3ml fuming H_2SO_4 to it. After 2-3 minutes pour the mixture in 10ml ice cold water, stir and cool further. Collect the precipitated 2, 4 dinitro toluene, recrystallise from hot aqueous alcohol, m.p 70°C.

(6) Reflux 5ml of compound with 10ml aqueous $KMnO_4$ for 8hrs. Filter hot and concentrate the filtrate. Acidify with dil HCl and collect the precipitated benzoic acid, m.p 121°C

(7) Picrate (light yellow), m.p 88°C.

b.p136°C **Ethyl benzene** (Phenyl ethane) 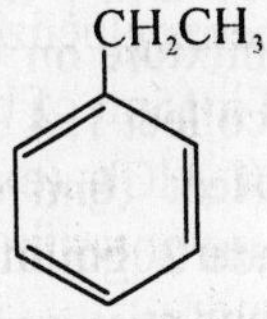

(1) Colourless liquid, insoluble in water, miscible with organic solvents.

(2) Heat on water bath 2ml of compound with 8ml conc H_2SO_4 for 15-20min. Cool, pour in saturated NaCl solution, shake thoroughly and filter. Dry the white solid and then mix with equal amount of PCl_5. Heat at 100°C for 10min. Cool and pour in 10ml ice water. Decant the white solid, add 10-12ml NH_4OH to it and boil. Add 20ml water, boil and filter hot. On cooling p-ethyl benzene-sulfonamide m.p 109°C, is precipitated.

(3) To cold 7ml of ethylbenzene add dropwise a mixture of 3.75ml conc fuming HNO_3 and 3.4ml conc H_2SO_4. Heat the mixture at 100°C for 4hrs. Keep at room temperature for 2hrs, cool and remove the oily layer. Distil and collect the o-nitro ethyl benzene at b.p 135°C/37mmHg. (b.p. 223°C), while the residue is p-nitro ethyl benzene, b.p. 241°C

(4) Picrate (light yellow), m.p 97°C.

b.p 138°C **p-Xylene (1, 4 dimethyl benzene)** 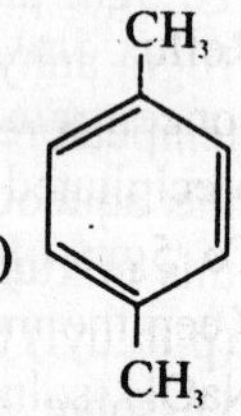

(1) Colourless liquid with characteristic benzene like odour, insoluble in water, soluble in alcohol and ether.

(2) Reflux 5ml of compound with 15ml aqueous $KMnO_4$ for 8hrs. Filter hot, concentrate the filtrate. Acidify with dil HCl and collect the precipitated terephthalic acid (which sublimes readily at 300°C)

(3) To a mixture of 2ml fuming HNO_3 and 4ml conc H_2SO_4 add 1.5ml of compound. Boil the mixture for 2-3 min. Cool and pour in ice with stirring, 2, 3, 5 trinitro-p-xylene, m.p 139°C, precipitates out.

(4) To 2ml p-xylene add with stirring 3.5ml conc H_2SO_4. Heat the mixture on water bath for 10-15min. Cool the mixture in ice and collect 1, 4 dimethyl benzene sulphonic acid, m.p 82°C.

(5) Heat 10ml of compound in an oil bath at 140-160°C. When it starts boiling add dropwise 22.5ml bromine. Continue heating and stirring for 6hrs. Cool and dissolve in 50ml warm chloroform. Cool the chloroform solution and recover tetrabromo-p-xylene, recrystallise from chloroform, m.p 165-169°C.

(6) Picrate (lemon yellow), m.p 91°C.

b.p 139°C **m-Xylene** (1, 3 dimethyl benzene)

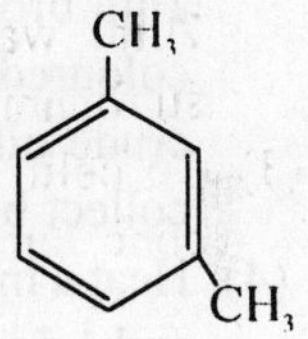

(1) Colourless liquid with characteristic benzene like odour, insoluble in water, soluble in alcohol and ether.

(2) To a mixture of 2ml fuming HNO_3 and 4ml conc H_2SO_4 add 1.5ml of compound. Boil the mixture for 2-3 min. Cool, pour in ice with stirring, 2, 4, 6 trinitro-m-xylene, m.p 181°C, precipitates out.

(3) Reflux 5ml of compound with 15ml aqueous $KMnO_4$. Filter hot, concentrate the filtrate. Acidify with dil HCl and collect the precipitated isophthalic acid, m.p 348°C.

(4) To 5gm of compound add dropwise with stirring 5ml of Br_2. Keep the mixture at room temperature for 2hrs and then stir with NaOH solution. Filter the precipitated solid, wash with water, recrystallise and collect 4, 6 dibromo-m-xylene, m.p 69°C.

(5) To 40gm of compound add 14gm $AlCl_3$. Cool to 0°C and then add in portion with stirring 7gm of p-benzoquinone. Continue stirring for 2hrs and then keep overnight in a refrigerator. Pour in 15ml ice water containing 10ml conc HCl. Collect the precipitated 2, 5 di-m-xylylquinol, recrystallise from benzene, m.p 188-189°C

(6) Picrate (lemon yellow), m.p 91°C.

b.p141°C **Phenylacetylene** (Phenyl ethyne)

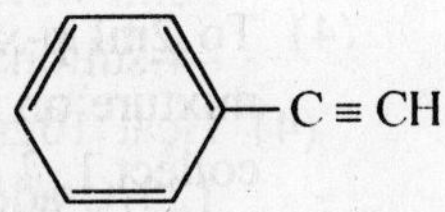

(1) Colourless liquid with penetrating and persistent odour, mildly lachrymatory

(2) Reflux together saturated solution of 2.75gm of cupric acetate monohydrate in 10ml of 1:1 volume pyridine-methanol mixture and 2ml phenylacetylene for 1hr. Cool, pour dropwise with stirring the mixture in an ice cooled 30ml of 18N H_2SO_4. Extract with

ether. Distil off the ether and collect diphenylacetylene, m.p 87-88°C.

(3) To ice cold NaOH solution (3gm in 10ml water) add with stirring 1ml bromine. Now add 2ml phenyl acetylene to the yellow coloured solution and shake mechanically for 60hrs. Separate the crude oil from aqueous phase. Distil under reduced pressure and collect bromoethyne, b.p 40-41°C.

(4) Heat a mixture of 37gm triethylorthoformate, 27ml phenylacetylene and 1.5gm zinc iodide. Distil off ethanol formed during the reaction (at 135°C). Now raise temperature to 200°C and reflux. Cool and filter. Carry out the distillation process and collect phenyl-propargylaldehyde diethylacetal, b.p 99-100°C.

b.p 144°C **o-Xylene** (1, 2-dimethyl benzene) 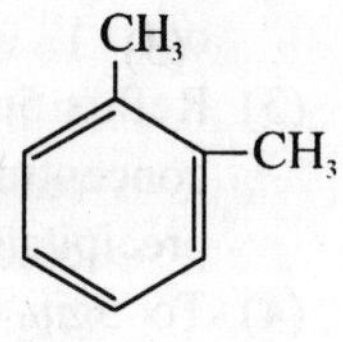

(1) Colourless liquid, insoluble in water, soluble in nearly all organic solvents.

(2) Reflux 5ml of compound with 15ml aqueous $KMnO_4$. Filter hot and concentrate the filtrate. Acidify with dil HCl, collect the precipitated phthalic acid, m.p 195°C.

(3) Heat on water bath 2ml of compound with 8ml conc H_2SO_4 for 15-20min. Cool, pour in saturated NaCl solution, shake thoroughly and filter. Dry the white solid and then mix with equal amount of PCl_5. Heat at 100°C for 10min. Cool and pour in 10ml ice water. Decant the white solid, add 10-12ml NH_4OH to it and boil. Add 20ml water, boil and filter hot. On cooling 1, 2 dimethyl benzene –4-sulfonamide m.p 144°C, is precipitated.

(4) Heat 10gm of o-xylene on oil bath. When temperature reaches 125°C add dropwise 12ml of bromine. After addition is complete continue heating at 125-150°C for 30min. Cool, pour in 10ml pet-ether. Cool further and keep overnight in a refrigerator. Collect the precipitated o-xylylene dibromide, m.p 89-94°C.

(5) Reflux a mixture of 16ml water, 8ml conc HNO_3 and 4ml o-xylene on oil bath at 140-155°C for 50-55hrs. Pour the reaction mixture on crushed ice. Collect the solid product, dissolve in

10ml 10% NaOH solution. Heat on steam bath, filter hot and add 3ml conc HCl. Collect o-toluic acid and recrystallise, m.p 94-101°C.

(6) Picrate (lemon yellow), m.p 88°C.

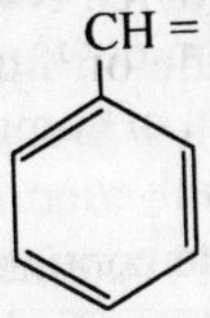

b.p146°C **Styrene** (Phenyl ethylene)

(1) Colourless liquid having odour similar to benzene, sparingly soluble in water, easily soluble in alcohol and ether.

(2) Heat 1ml of compound with acidic $KMnO_4$ solution. Smell of bitter almond due to formation of benzaldehyde is produced.

(3) To a clear solution of 2.73gm of N-Phenylhydroxylamine in 4ml ethanol(heat 40-60°C) add 3ml benzaldehyde (exothermic reaction). Keep the reaction mixture overnight in dark. Collect colourless crystals of N-α diphenylnitrone, m.p111-113°C. To 4gm N-α diphenylnitrone add 5ml styrene and heat it to 60°C for 4hrs. Cool, pour in 10ml pet ether, isoxazoline, m.p 96-98°C, precipitates out.

(4) Reflux 67.5gm of 37% formalin, 4.8gm of H_2SO_4 and 35ml of styrene for 7hrs. Cool and add 50ml benzene. Separate the two layers and extract the aqueous layer with benzene. Mix the organic layers and distil off the benzene. Distil the residual liquid under reduced pressure and collect 4-phenyl m-dioxane.

(5) To a solution of 22gm of perbenzoic acid in 200ml $CHCl_3$ add 15gm of styrene. Keep the solution at 0°C for 24hrs. Add NaOH solution and filter. Distil off the solvent and collect styrene oxide, b.p 188-192°C from residue.

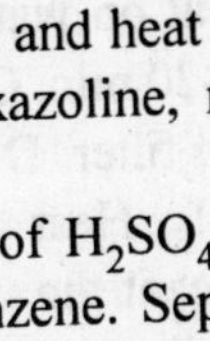

b.p 152°C **Cumene** (Isopropylbenzene)

(1) Colourless liquid, insoluble in water, soluble in alcohol and ether.

(2) Heat on water bath 2ml of compound with 8ml conc H_2SO_4 for 15-20min. Cool, pour in saturated NaCl solution, shake thoroughly and filter. Dry the white solid and then mix with equal amount of PCl_5. Heat at 100°C for 10min. Cool and pour in 10ml ice water. Decant the white solid, add 10-12ml NH_4OH to it and boil. Add 20ml water, boil and filter hot. On cooling 1-isopropyl benzene sulfonamide-4, m.p 106°C is precipitated.

(3) To 10ml of compound add dropwise a mixture of 7ml conc HNO_3 and 11ml conc H_2SO_4 (temp below 4°C). After the addition is complete continue stirring for 1hr. Pour in 20ml ice water. Isolate with ether. Distil and collect 2-nitro cumene at b.p 115-120°C/15mmHg and 4-nitrocumene at 128-132 °C/15mmHg.

b.p 159°C **n-Propylbenzene**

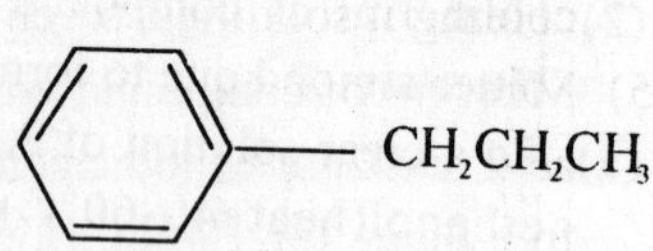

(1) Colourless liquid, soluble in organic solvents.

(2) Heat on water bath 2ml of compound with 8ml conc H_2SO_4 for 15-20 min. Cool, pour in saturated NaCl solution, shake thoroughly and filter. Dry the white solid and then mix with equal amount of PCl_5. Heat at 100°C for 10min. Cool, pour in 10ml ice water. Decant the white solid, add 10-12ml NH_4OH to it and boil. Add 20ml water, boil and filter hot. On cooling 1, n-propylbenzene-sulfonamide-4, m.p 110°C, is precipitated.

(3) Picrate (yellow), m.p 103°C

b.p 164°C **Mesitylene** (1, 3, 5 trimethyl benzene)

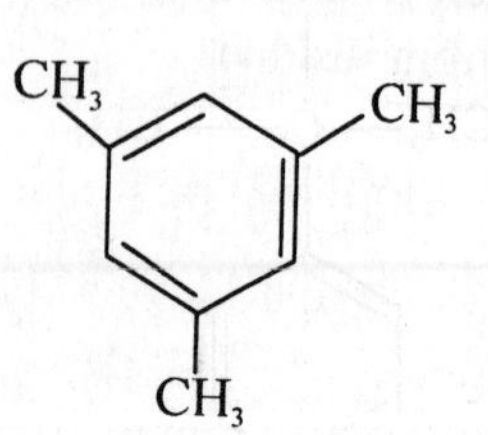

(1) Colourless strongly refractive liquid, insoluble in water, soluble in organic solvents.

(2) To a mixture of 2ml fuming HNO_3 and 4ml conc H_2SO_4 add

1.5ml of compound. Boil the mixture for 2-3 min. Cool and pour in ice with stirring, 2, 4, 6 trinitro-mesitylene, m.p 230°C, precipitates out.

(3) To 2ml of compound add 4ml conc H_2SO_4 and shake thoroughly. Pour this yellow liquid in 8ml conc HCl (temp below 10°C). Pour the resulting mixture in ice and collect 1, 3, 5 trimethylbenzene sulphonic acid, m.p 78°C.

(4) To a mixture of 4ml mesitylene, 20ml conc HCl and 4ml of 37% formaldehyde solution pass HCl gas and heat on water bath at 55°C. After 2hrs add 4ml formaldehyde, continue passing HCl and heating process for 5hrs. Cool, extract with benzene. Distil off benzene and fraction distilling at 130-140°C is collected. On cooling it solidifies, α^2 chloroisodurene, melts at 38°C.

(5) Mix 5ml of compound with 5gm of iodine, 15ml of glacial acetic acid and 2gm of iodic acid in 5ml water. Heat the mixture over a free flame for 15min. Cool, add 15ml cold water to it. Add 10ml dil NaOH solution and shake thoroughly. Extract with ether. Distil off ether and cool the residue, iodomesitylene crystallises out, m.p 30°C.

(6) Add 5gm of fuming nitric acid to a solution of 5gm of mesitylene in 10ml glacial acetic acid. Reflux the mixture for 1hr. Cool and pour with stirring in 50ml ice cold water. Filter the precipitated nitro mesitylene, m.p 47°C.

b.p 169°C **t-Butyl benzene** (2-methyl-2-phenyl propane)

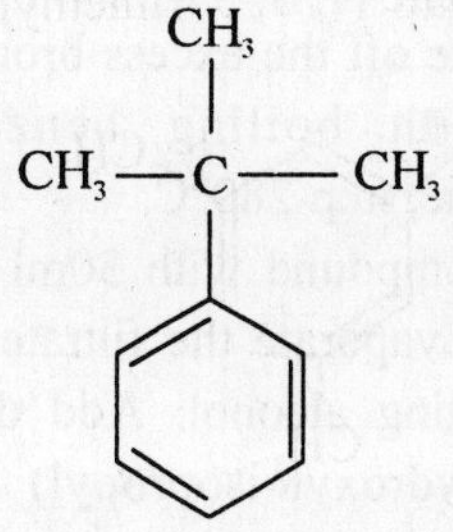

(1) Colourless liquid, insoluble in water, soluble in alcohol

(2) 2, 4 dinitro-t-butylbenzene ($HNO_3 + H_2SO_4$, 60°C for 5hrs)

b.p 169°C **Pseudocumene** (1, 2, 4 trimethylbenzene)

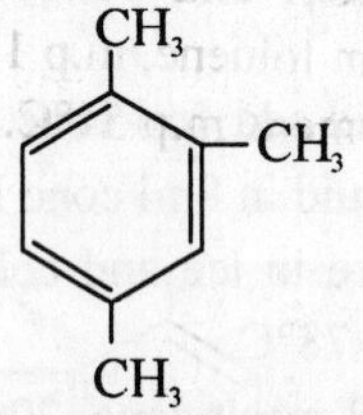

(1) Colourless liquid, insoluble in water, soluble in alcohol and ether.
(2) To a mixture of 2ml fuming HNO_3 and 4ml conc H_2SO_4 add 1.5ml of compound. Boil the mixture for 2-3 min. Cool, pour in ice with stirring, 3, 5, 6 trinitropseudocumene, m.p 185°C precipitates out.
(3) Picrate (yellow), m.p97°C. (4) 3, 5, 6 tribromo- 1, 2, 4 trimethylbenzene m.p 229°C.

b.p 177°C **p-Cymene** (4-isopropyl-1-methyl benzene)

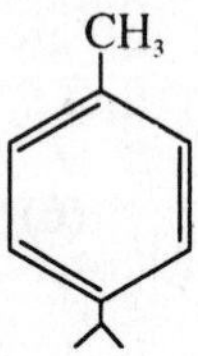

(1) Oily liquid with eucalyptus like odour, insoluble in water, soluble in alcohol and ether
(2) To 3ml compound add 7-8 ml ice cooled bromine to which 1gm Al has been added. Keep the mixture at room temperature for sometime and shake occassionally. Pour the mixture in a porcelain dish and evaporate off the excess bromine on water bath. Extract the residue with boiling benzene. Cool and collect pentabromotoluene, m.p 280°C.
(3) Reflux 3ml of compound with 30ml aqueous $KMnO_4$ for 6hrs. Cool and filter. Evaporate the filtrate to dryness and extract the residue with boiling alcohol. Add dil H_2SO_4 to the alcoholic solution, 4 (α-hydroxyl isopropyl) benzoic acid, m.p 157°C, separates out.
(4) Boil gently 5ml of p-cymene with 30ml conc HNO_3 for 8hrs. Cool and filter. Dissolve the solid in 10ml NaOH and add 200mg

Zn to it. Distil the mixture and filter. To the filtrate add 10ml boiling HCl. Cool and collect the precipitated p-toluic acid, recrystallise from toluene, m.p 176-177°C.

(5) 2, 6 dinitro p-cymene m.p 54°C.

b.p 182°C **Indene** 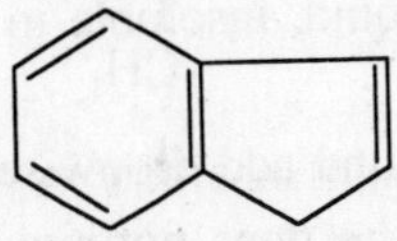

(1) Colourless water like liquid turns yellow on standing, polymerises quite readily even in dark. Insoluble in water, soluble in most organic solvents.

(2) To a mixture of 16ml formic acid and 2.8ml H_2O_2 add dropwise with stirring 2.5ml indene (temp 30-40°C) Stir the reaction mixture for 7hrs. Remove the excess formic acid by distillation. Pour the residue in 40ml 10% H_2SO_4 and steam distil. The distillate 2-indanone is cooled which soon solidifies, m.p 57-58°C.

(3) To a solution of $K_2Cr_2O_7$ (25gm in 350ml water) and 72ml conç H_2SO_4 heated to 65°C add 8ml indene. Heat the mixture on water bath (65°C) for 2hrs. Cool in ice bath for 5hrs and collect homophthallic acid, recrystallise from benzene, m.p 180-185°C.

(4) Picrate (golden yellow) m.p 98°C (5) 1, 2-dibromoindane, m.p 32°C.

b.p 204°C **Prehnitene** (1, 2, 3, 4-tetramethyl benzene)

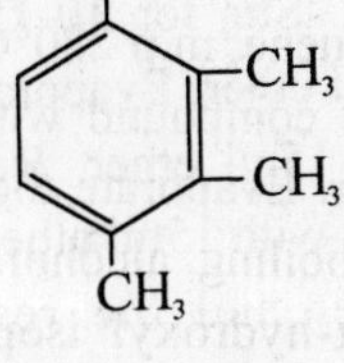

(1) Colourless liquid, insoluble in water, soluble in alcohol.

(2) 5, 6 dibromo prehnitene m.p, 208°C (3) 5, 6 dinitro prehnitene, m.p 176°C (4) Picrate, m.p 89°C.

b.p 244°C **1-Methyl naphthalene**

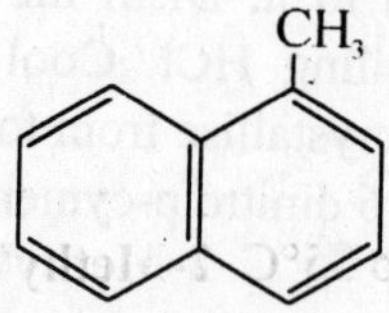

(1) Colourless oily liquid, insoluble in water, soluble in alcohol and ether.

(2) To 5ml of compound add dropwise 7ml conc. HNO_3 taking care that the temperature does not rise above 5°C. After addition is complete remove the ice bath, shake the mixture thoroughly, keep at room temperature for sometime and recover 4-nitro-1-methyl naphthalene,. m.p 72°C.

(3) Picrate (lemon yellow) m.p 141°C.

b.p 264°C m.p 26°C **Diphenylmethane** (Benzylbenzene)

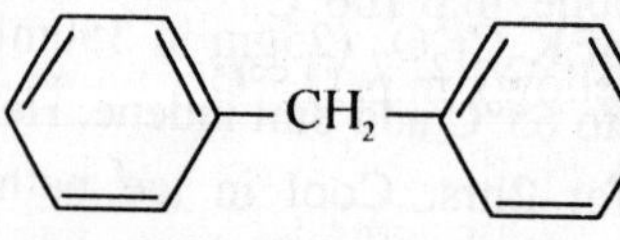

(1) Liquid with orange like odour, insoluble in water, easily soluble in alcohol and ether.

(2) To 1ml of compound add few milligrams of $AlCl_3$. A yellow orange coloured solution is obtained.

(3) To 30ml ammonia solution add a small piece of Na followed by few crystals of ferric nitrate hydrate (10mg) and 1.2gm of Na. After all Na is converted to amide 8.4gm of diphenyl –methane in 5ml ether is added. Stir for 10-15min and then add 6.5gm of butylbromide in 5ml ether. Evaporate off the ammonia, add 15ml water followed by 15ml ether. Remove the ethereal layer and repeat the extraction with 15ml ether. Combine the ethereal layers, distil off the ether and the residual liquid is pure 1, 1-diphenylpentane, b.p 138-139°C.

(4) 2, 4, 2', 4' tetranitro diphenylmethane, m.p 172°C.

Aromatic Solid

m.p 35°C **2-Methyl Naphthalene**
b.p. 241°C

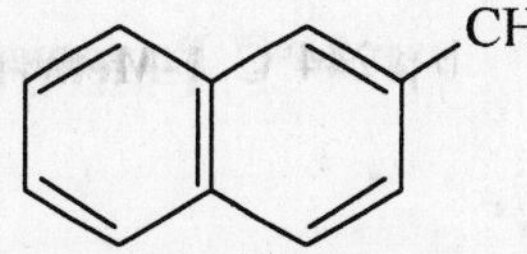

(1) Colourless solid, insoluble in water, soluble in nitromethane and ether.

(2) Dissolve 1.5gm of compound in 6ml carbon tetrachloride, add 1.8gm of N-bromosuccinimide. Reflux the mixture for 16hrs. Filter off the succinimide and concentrate the filtrate. Cool in ice, 2-(bromomethyl) naphthalene precipitates out, m.p 56°C.

(3) Stir 2gm of compound with 1.2gm chromic anhydride and 10ml acetic acid. Keeping the temperature between 15-20°C, continue stirring for further 40-60 min at 40°C. Cool and recover 2-methyl 1, 4 naphthaquinone, m.p 106°C.

(4) Picrate (lemon yellow), m.p 115°C.

m.p 52°C **Bibenzyl** (1, 2-diphenyl ethane, sym diphenyl ethane)

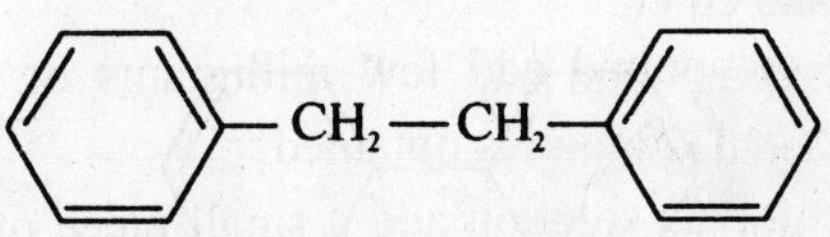

(1) Colourless crystalline solid, insoluble in water, soluble in alcohol and ether.

(2) To 500mg of compound and 1.5gm $K_2Cr_2O_7$ add dropwise 4-5ml conc H_2SO_4. Reflux the mixture, cool and pour in ice, collect benzoic acid, m.p 121°C.

(3) To 500mg of compound add 4-5ml conc HNO_3, heat the mixture at 70-80°C for 6hrs. Cool and collect yellow precipitated 4, 4' dinitrobibenzyl, recrystallise from alcohol, m.p 180°C.

m.p 54°C **Pentamethylbenzene** 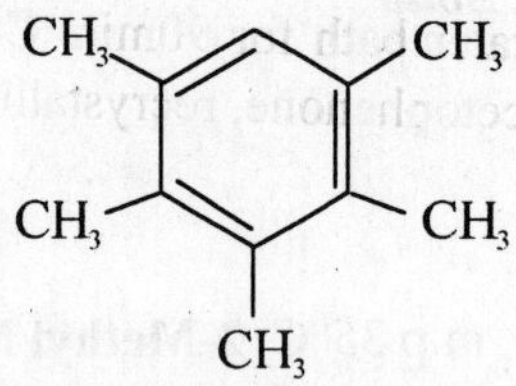

(1) Insoluble in water, soluble in alcohol and benzene
(2) 6-bromo 1, 2, 3, 4, 5-pentamethylbenzene, m.p 160°C (3) Picrate, m.p 131°C

m.p 56°C **o-Diphenylbenzene** 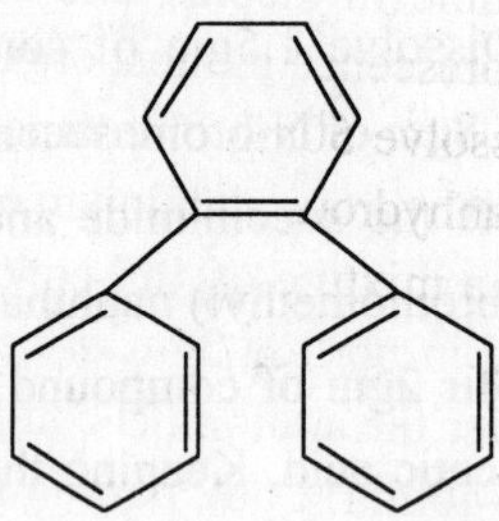

(1) Colourless crystalline compound, soluble in chloroform and acetone.
(2) On oxidation with CrO_3 in AcOH, o-phenyl benzoic acid is obtained, m.p 110°C

m.p70°C **Diphenyl** (Biphenyl, phenyl benzene)

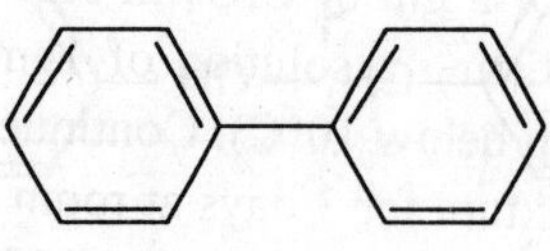

(1) White crystalline solid, insoluble in water, soluble in methanol, ethanol, ether and benzene.
(2) To 1.54gm of compound add 5 ml acetic acid and 1.5ml of bromine. Reflux the reaction mixture for 3hrs. Cool and collect 4, 4'-dibromobiphenyl, m.p 164°C
(3) To 500mg of compound add 5ml acetic acid and 1ml fuming nitric acid. Heat the mixture for an hour at 80-95°C. Cool and collect 4-nitrobiphenyl, recrystallise from alcohol, m.p 114°C.
(4) To a mixture of 3gm biphenyl, 6gm anhydrous $AlCl_3$ and 14ml CS_2 add 2ml of acetic anhydride. Reflux the reaction mixture on

water bath for 90min. Cool and collect the precipitated p-phenyl acetophenone, recrystallise from ethanol, m.p 120°C.

m.p80°C **Naphthalene** 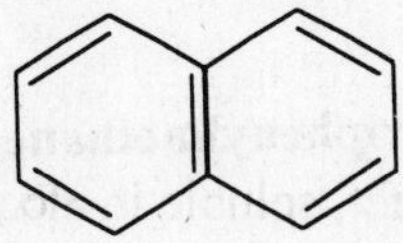

(1) White crystalline solid with characteristic odour, insoluble in water, soluble in alcohol and ether. Sublimes readily and gives violet fluorescence in most organic solvents.

(2) Dissolve 50mg of compound in 2ml chloroform and add a pinch of anhydrous aluminium chloride. A deep green colour is produced.

(3) To a mixture of 4ml conc nitric acid and 4ml conc sulphuric acid add in portion 5gm of naphthalene with stirring (temp 30-40°C). After the addition is complete heat it on water bath at 50-60°C for 30-40min. Cool, pour in ice and collect 1-nitronaphthalene, purify by sublimation, m.p 61°C.

(4) Heat 2gm of naphthalene until it melts and then add dropwise with stirring 2ml of conc sulphuric acid and maintain the temp at 160°C. Stir the mixture and pour in 50ml ice cold water. Neutralise by adding 4gm sodium bicarbonate. Heat the solution to boiling, saturate with sodium chloride and cool. Collect crystals of sodium naphthalene-2-sulphonate.

(5) Cool a solution of 2 gm of CrO_3 in 3ml acetic acid to 0°C. Add dropwise with stirring a solution of 1gm of compound in 10ml acetic acid. (temp below 10°C). Continue stirring for 8-10hrs and then keep the mixture for 3 days at room temperature. Pour in ice and collect 1, 4-naphthaquinone, m.p 125°C.

(6) To 50ml NaOH solution add 3.2gm naphthalene. Heat the solution to boiling and add in portion 150ml of $KMnO_4$ solution (21gm in 150ml water), after the addition is complete continue boiling for 45min, cool and filter. Acidify the filtrate, concentrate, cool and filter. Neutralise the filtrate with 15ml NaOH and 5gm sodium sulphite. Evaporate to dryness. Extract the residue with benzene. Evaporate off the benzene and dissolve the residue in hot water and filter. Cool the filtrate and collect Phthalaldehydic acid, m.p 96°C.

m.p 93°C **Triphenylmethane** (tritan)

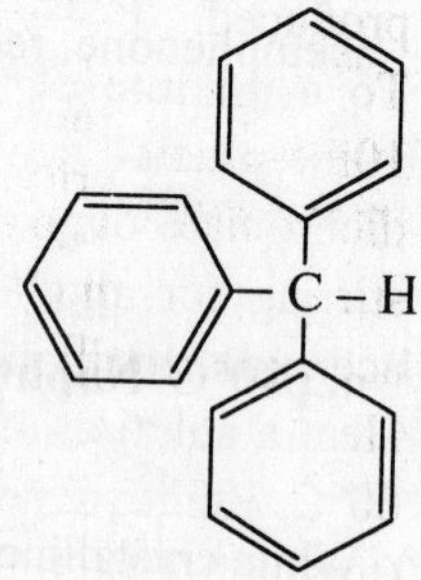

(1) Colourless crystalline solid, sparingly soluble in alcohol, easily soluble in ether and chloroform.
(2) Dissolve 50mg of compound in 2ml CCl_4 and add a pinch of anhydrous aluminium chloride. A yellow coloured solution is obtained.
(3) To few milligrams of compound add conc H_2SO_4. A yellow coloured solution is obtained after keeping for sometime.
(4) To 500mg of compound add 10ml fuming HNO_3 and shake. Pour in 10ml water. Recover the trinitro derivative, dissolve in 40ml hot glacial acetic acid and add a pinch of zinc dust, a red coloured solution will be obtained. Continue to add Zn dust until the red colour is discharged. Decant and add few milligrams of PbO_2 to the solution. An intense fuchsin -red coloured solution is produced.
(5) Reflux 2gm of compound with 10ml conc HNO_3. Cool, pour in cold water and recover triphenylcarbinol, m.p 162°C.
(6) Tris-(4-nitrophenyl) methane, m.p 212°C.

m.p 95°C **Acenaphthene**

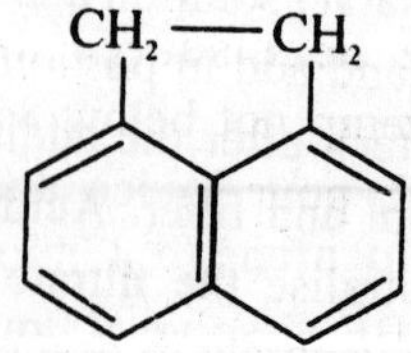

(1) Colourless crystalline solid, sparingly soluble in alcohol, easily soluble in ether and benzene.
(2) Dissolve 50mg of compound in 2ml chloroform and add a pinch

of anhydrous aluminium chloride. A greenish blue colour is produced.

(3) To a mixture of 2gm acenaphthene, 100 gm ceric acetate and 10ml glacial acetic acid add with stirring 7.5gm of sodium dichromate over a period of 1hr. (temp below 40°C). Continue stirring for another 5-8hrs. Dilute with water, cool and collect acenaphthaquinone, m.p 260°C.

(4) Heat a solution of 2gm compound in 11ml glacial acetic acid to 60°C. Remove the heat source and add 8gm of red lead in portion (temp 60-70°C). Pour in ice water. Extract the acetate with ether. Distil off the ether and dissolve the residual liquid in 5ml methanol and add 5ml of NaOH solution(500mg in 5ml water). Reflux for 2hrs and then cool. Collect the yellow acenaphthenol, recrystallise from benzene, m.p 144-145°C.

(5) Picrate (orange), m.p 161°C.

m.p 100°C **Phenanthrene** 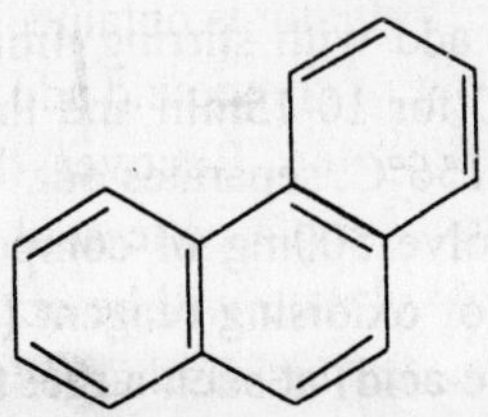

(1) Colourless crystalline solid which sublimes readily, insoluble in water, soluble in alcohol, ether and benzene.

(2) Dissolve 100mg of compound in 5ml benzene. A light blue fluorescence is observed.

(3) Heat sulphuric acid solution (5ml in 10ml water) on water bath and add 500mg of compound followed by 3gm of potassium dichromate(in parts). When the temperature reaches 100-110°C remove the heat source and continue adding potassium dichromate(temp not below 85°C). Then heat on water bath for 30min. Cool, pour in ice and collect phenanthraquinone, m.p 205°C.

(4) In a 50ml RBF dissolve 500mg of compound in 5ml CCl_4. Keep the flask in ice bath. Shift the assembly in sunlight and add 1.5-2 ml bromine. Phenanthrene 9, 10 dibromo, m.p 98°C, separates out.

(5) Dissolve 4.5gm of phenanthrene in 50ml glacial acetic acid and warm to 90°C. Add in portion 17ml of H_2O_2. Heat the mixture on water bath for 3-4hrs. Reduce the volume by distillation and cool. Diphenic acid precipitates out, m.p 228°C.
(6) Picrate (orange yellow) m.p143°C.

m.p 114°C **Fluorene** (Diphenylenemethane)

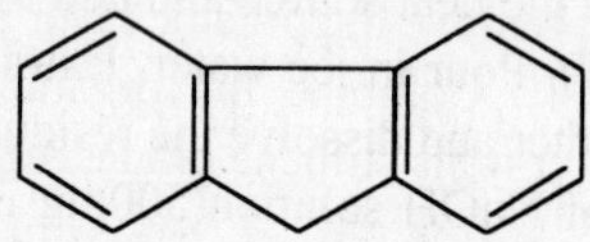

(1) White crystalline solid which sublimes readily, insoluble in water, sparingly soluble in alcohol, easily soluble in benzene and ether.
(2) To few milligrams of compound add conc H_2SO_4 and heat. A blue coloured solution is obtained.
(3) Dissolve 3gm of compound in 25ml acetic acid(heat to 50°C) and then add with stirring 4ml conc HNO_3. Heat the mixture at 60-80°C for 10-15min and then pour in ice water, 2 nitro-fluorene, m.p 156°C, separates out.
(4) Dissolve 500mg of compound in 10ml acetic acid. Add slowly 5ml of oxidising reagent (1.5gm CrO_3 in 1.5ml water and 3.5ml acetic acid) at such a rate that the mixture refluxes. After addtion is complete reflux further for 10min. Cool, pour in ice and collect fluorenone, m.p 83°C.
(5) Heat a mixture of 2.3gm of Na in 45ml absolute methanol and 11.3gm fluorene at 220°C for 16hrs. Cool, dilute with water and then neutralise with conc HCl. Extract with benzene, distil off the benzene and recrystallise the crude 9-methyl fluorene from methanol, m.p 45°C.
(6) Picrate (red brown), m.p 80°C.

m.p 124°C **Stilbene** (trans 1, 2-diphenyl ethylene)

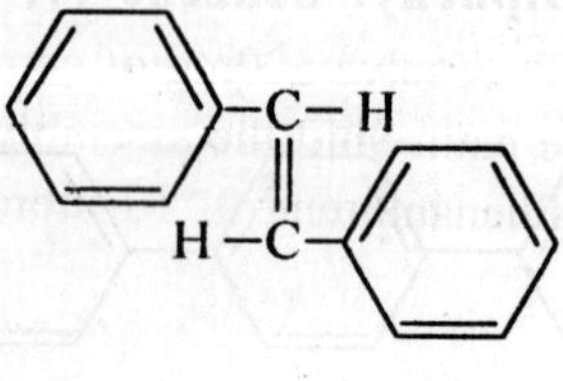

(1) Crystalline solid, insoluble in water, spaingly soluble in alcohol, easily soluble in benzene and ether. Sublimes.
(2) Heat few milligrams of compound with $KMnO_4$ and dil H_2SO_4. Smell of bitter almond due to the formation of benzladehyde is observed.
(3) Cool a solution of 2.7gm of stilbene in 27ml CH_2Cl_2 to 20°C. Add a solution of 4ml of peracetic acid in acetic acid containing 0.25gm of sodium acetate and stir for 15hrs (temp below 35°C). Pour in ice water, remove the organic layer and extract the aqueous layer with CH_2Cl_2. Combine the organic layers, distil off the solvent and recover trans stilbene oxide, recrystallise from hexane, m.p 68-69°C.
(4) To 2.5gm of stilbene in 35ml ether add dropwise with stirring 2.1ml of bromine. Continue stirring for 30min, cool and collect stilbene dibromide, m.p 235°C. To the cooled KOH solution (4.5gm in 8ml absolute ethanol) add in portion the dibromide and then reflux the mixture on oil bath for 15-20hrs. Cool, pour in ice and collect diphenylacetylene, recrystallise from ethanol, m.p 60-61°C.
(5) Picrate (red brown), m.p 91°C (unstable)

m.p 165°C **Hexamethylbenzene**

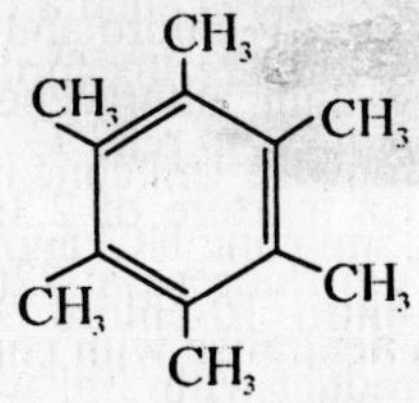

(1) Crystalline compound, soluble in hot alcohol and benzene. Sublimes
(2) Picrate, m.p 170°C.

m.p 213°C **p-Diphenyl benzene** (1, 4 diphenyl benzene, Terphenyl)

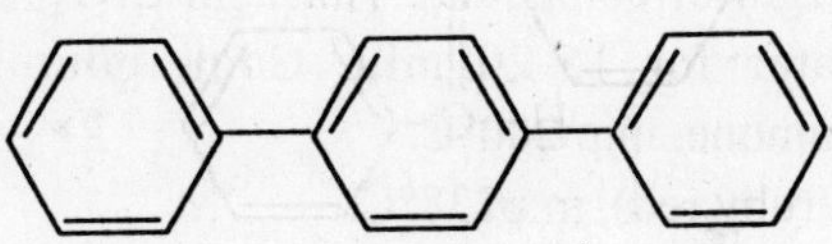

(1) White crystalline solid, insoluble in water, sparingly soluble in alcohol and ether, easily soluble in benzene and nitrobenzene. Sublimes.

(2) Reflux 1.5gm of compound with 7ml benzoyl chloride and 4gm $AlCl_3$ at 100°C for 2hrs. Cool and collect 4, 4-dibenzoylterphenyl, m.p 294°C

m.p 217°C **Anthracene** 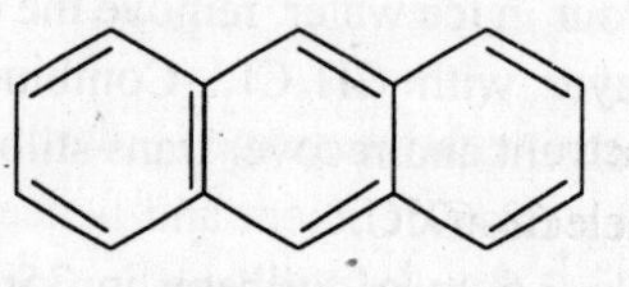

(1) Colourless crystalline solid, insoluble in water, sparingly soluble in alcohol, easily soluble in benzene, ether and chloroform. Sublimes.

(2) Reflux 1gm anthracene, 2.8gm maleic anhydride and 20ml benzene on water bath for 3hrs. On cooling anthracene 9, 10 endo α, β succinic anhydride, m.p 263°C, separates out.

(3) Reflux a mixture of 2gm anthracene with 3gm anhydrous cupric chloride and 50ml CCl_4 for 18-24hrs. Cool and filter. Evaporate the filtrate to dryness. Recrystallise the residue from pet ether and collect 9-chloro anthracene, m.p104-106°C.

(4) To 2gm anthracene add 5ml glacial acetic acid, cool and add dropwise 1ml conc HNO_3, stir for 30 min. Slowly add with stirring 2.5ml conc HCl and 2.5ml glacial AcOH. Cool and collect yellow 9-nitro 10-chloro 9, 10-dihydroanthracene. Triturate the solid product with 3ml warm 10% NaOH solution. Collect the orange coloured 9-nitro anthracene, recrystallise from glacial acetic acid, m.p 145-146°C.

(5) Dissolve 1gm of anthracene in 7ml benzene and add 2.5ml acetylchloride to it. Cool to 0°C, add in portion 1.5gm anhydrous $AlCl_3$. (temp below0°C). Stir the mixture for 30min. Filter the red coloured complex, pour in ice containing 10ml conc HCl. Collect 9-acetyl anthracene, recrystallise from ethanol, m.p 75-76°C.

(6) Reflux 1gm of compound with 2gm CrO_3, 10ml acetic acid and 2ml water for 15-20 min. Cool, pour in ice and collect anthraquinone, m.p 280°C.

(7) Picrate (ruby red), m.p138°C.

Compounds containing C, H and oxygen

Carbonyl Compounds (Aldehydes and ketones)

The carbonyl group can be characterized as the functional group containing a carbon and an oxygen with a double bond between them. The carbonyl carbon is sp^2 hyberdized with bond angle of 120°.

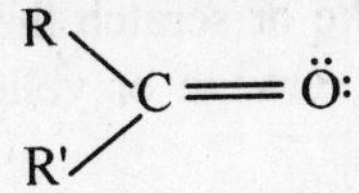

The electronegative oxygen polarizes the carbon-oxygen bond, rendering the carbon electron deficient and hence subject to nucleophilic substitution. Attack on the sp^2 hybridized carbon occur via π electron cloud lying above or below the plane. Thus many reaction of carbonyl groups are acid catalysed. Under acidic conditions the carbon of the protonated carbonyl group is more electrophilic and hence can even react with weak nucleophile.

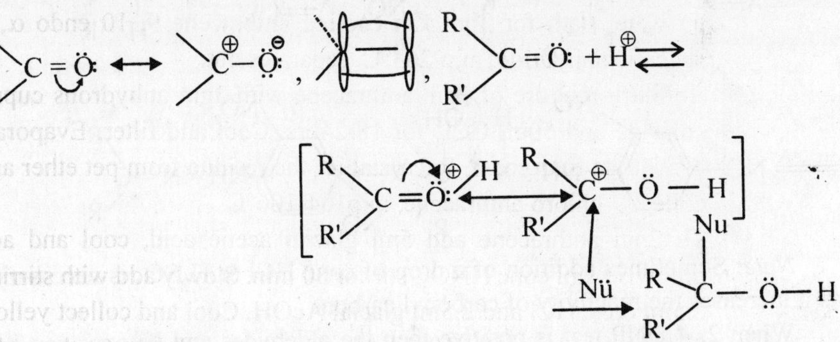

The general structure by which an enormous variety of carbonyl compounds are identified is as follows

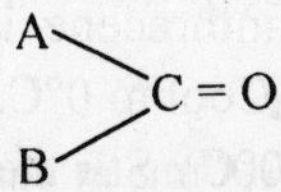

If A is an organic group ie., an alkyl or aryl group and B is hydrogen then the compound is an aldehyde. If B is also alkyl or aryl then the compound is a ketone. Since a hydrogen is bonded to the carbonyl in an aldehyde, they are sterically less hindered than ketones and thus react more rapidly in comparison to the ketones.

Identification Tests

1. Test by interaction with 2, 4 dinitrophenyl hydrazine(2, 4 DNP): Place 2-3 ml of a saturated solution of 2, 4 dinitrophenyl hydrazine (Brady's reagent) in a test tube and to this dark red orange solution add 1 ml of liquid sample or 30-40 mg of solid sample. Stopper the test tube and shake vigorously. If no precipitate separates out, heat the test tube on water bath for 1-2 minutes and again shake or scratch the sides of test tube with glass rod. A precipitate usually red, orange or yellow indicates the presence of carbonyl group

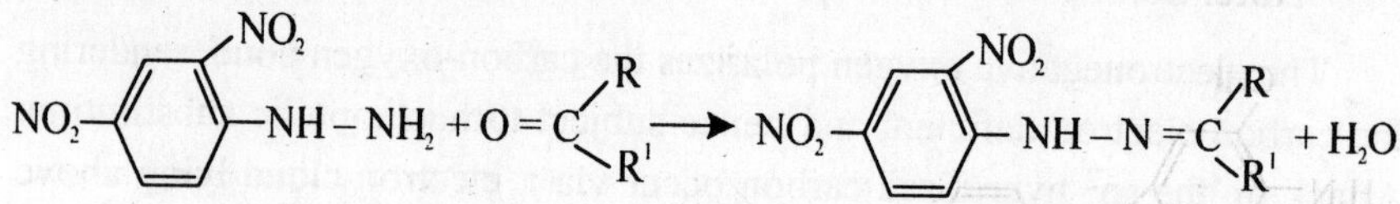

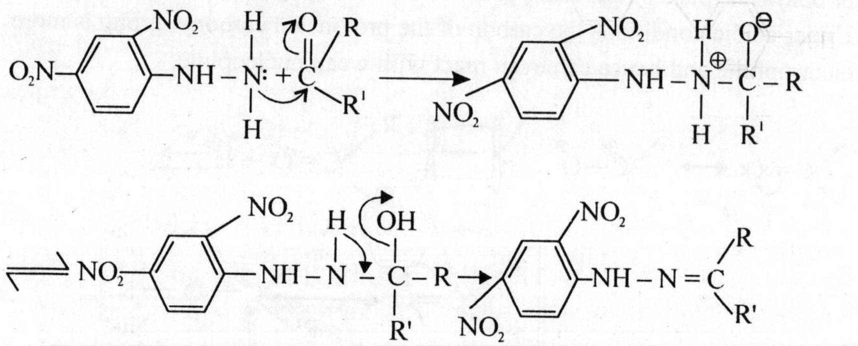

Note: Sometimes addition of a drop of conc HCl or H_2SO_4 is necessary as it increases the reactivity of carbonyl carbon.

When 2, 4 DNP test is positive then the aldehydes and ketones can be distinguished from one another by their potential towards oxidation. Ketones do not oxidise as readily as aldehydes because carbon-carbon bond cleavage would be required whereas aldehydes can be oxidised quite rapidly to carboxylic acids which then resist further oxidation

Classification test for Aldehydes

1. Tollens Test: Add 1ml (100mg in case of solid) or 2ml of ethanolic solution of the sample to 2ml ammonical silver nitrate solution (Tollens reagent) in a clean dry test tube. Shake for 1 minute and then warm the solution on a water bath for 5 minutes. Formation of silver mirror or separation of dark brown to black precipitate indicates the presence of aldehyde

$$AgNO_3 + NaOH \rightarrow AgOH + NaNO_3$$

$$AgOH + NH_4OH \rightarrow 2Ag[(NH_3)_2^+]OH^-$$

$$RCHO + 2Ag[(NH_3)_2^+]OH^- \rightarrow RCOO^-NH_4^+ + 2Ag\downarrow + 3NH_3\uparrow + H_2O$$

Preparation of Tollens reagent:- Mix 1 ml of aqueous silver nitrate solution (1gm $AgNO_3$ in 10 ml water) and 1ml aqueous sodium hydroxide solution (1gm NaOH in 10 ml water) in a test tube. A white precipitate is obtained. Add aqueous ammonia solution dropwise until the white precipitate just dissolves.

2. Schiff's Test: To 2ml of the alcoholic or aqueous solution of the test sample add 2ml of Schiff's reagent and shake. Appearance of red to violet-purple colouration indicates the presence of aldehyde.

Note: Some aldehydes like vanillin give a negative test

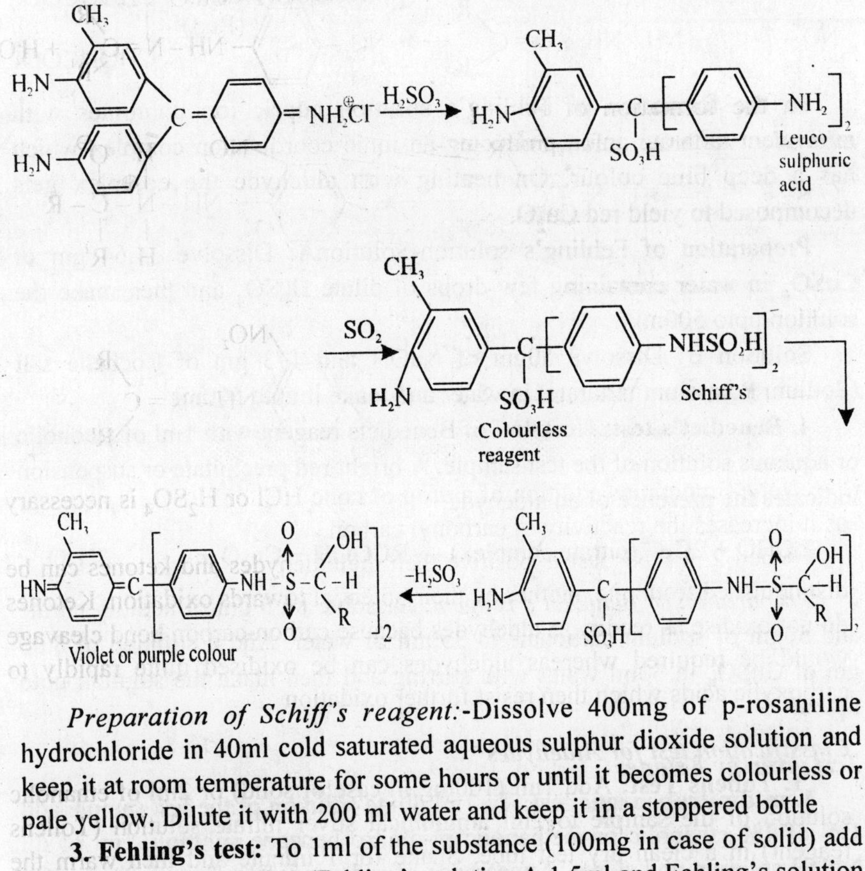

Preparation of Schiff's reagent:-Dissolve 400mg of p-rosaniline hydrochloride in 40ml cold saturated aqueous sulphur dioxide solution and keep it at room temperature for some hours or until it becomes colourless or pale yellow. Dilute it with 200 ml water and keep it in a stoppered bottle

3. Fehling's test: To 1ml of the substance (100mg in case of solid) add 3ml of Fehling's solution (Fehling's solution A 1.5ml and Fehling's solution B 1.5ml mixed in equal amount). Heat the solution for 3-4 minutes on water bath. Formation of bright red precipitate indicates the presence of aldehydes

Note: Aliphatic aldehyde reduce Fehling's solution rapidly in comparison to aromatic aldehydes.

$$CuSO_4 + 2NaOH \rightarrow Cu(OH)_2 + Na_2SO_4$$

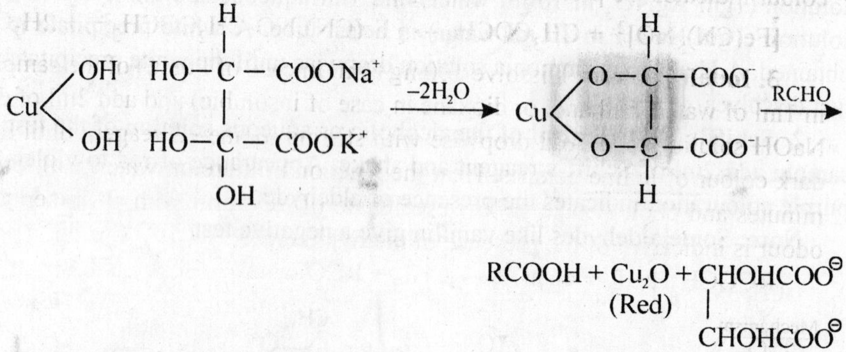

In the formation of Fehling's solution cupric ion combines with tetravalent tartarate anion producing an ionic coordination complex which has a deep blue colour. On heating with aldehyde the complex gets decomposed to yield red Cu_2O.

Preparation of Fehling's solution-SolutionA: Dissolve 34.64 gm of $CuSO_4$ in water containing few drops of dilute H_2SO_4 and then make the solution upto 500ml

Solution B: Dissolve 50gm of NaOH and 173 gm of Rochelle salt (Sodium Potassium tartarate) in water and make it upto 500ml

4. Benedict's test: Heat 4ml of Benedicts reagent with 1ml of alcoholic or aqueous solution of the test sample. A bright red precipitate or suspension indicates the presence of an aldehyde.

$$RCHO + 2Cu^{++}\text{(citrate complex)} \rightarrow RCOOH + Cu_2O + 2H_2O$$

Preparation of Benedict's reagent: Dissolve 86.5 gm of sodium citrate and 50gm of sodium carbonate in 350ml of water. Add a solution of 8.65 gm of $CuSO_4$ in 50ml water with stirring and then make the solution upto 500ml

Test given by Ketones

1. m-dinitrobenzene test: To 1ml dilute solution of the test sample add 1ml of 1% ethanolic or methanolic m-dinitro benzene solution, 1ml of dil NaOH solution and shake. Appearance of red or violet colour indicates the presence of methyl ketones.

2. Sodium nitroprusside test: To 1ml aqueous or alcoholic solution of the test sample add 1ml of 5% sodium nitroprusside solution and few drops of NaOH solution. After few minutes when a slight colour has developed add 2-4 drops of glacial acetic acid. Change of wine red colour to blue colour indicates the presence of methyl ketones

$$[Fe(CN)_5NO]^{2-} + CH_3COCH_3 \rightarrow [Fe(CN)_5NO\text{-}CH_2COCH_3]^{2-} + 2H_2O$$

3. Iodoform test:. Dissolve 100mg (1ml in case of liquid) of the sample in 1ml of water(methanol or dioxane in case of insoluble) and add 2ml of dil NaOH solution to it. Add dropwise with stirring iodine-KI reagent until the dark colour of iodine persists. Heat the reaction mixture on water bath for 5 minutes and cool. A pale yellow precipitate of iodoform with characteristic odour is indicative of the presence of methyl ketones

$$RCOCH_3 + 3I_2 + 4NaOH \rightarrow CHI_3 + RCOO^-Na^+ + 3NaI + 3H_2O$$

Mechanism

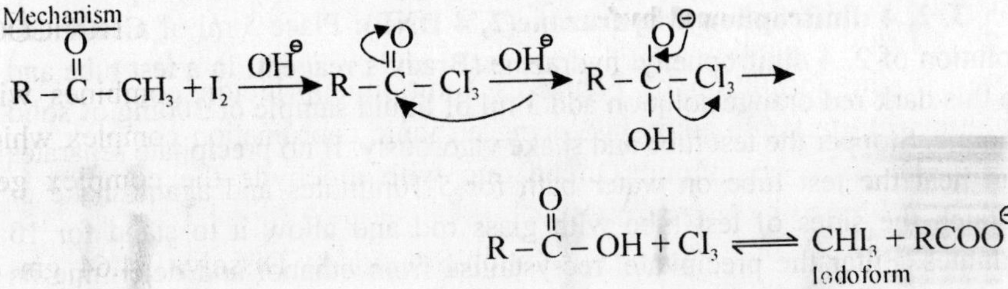

Derivatives

1. Bisulphite addition compound: Neutral solution of aldehydes and aliphatic methyl ketones react with sodium bisulphite to form well crystallized water soluble products known as aldehyde bisulphite or ketone bisulphite.

To 5ml saturated solution of sodium metabisulphite add 2ml of compound and shake thoroughly. The temperature of mixture will rise, cool and allow precipitation to take place. Filter the solid, wash with ethanol and dry.

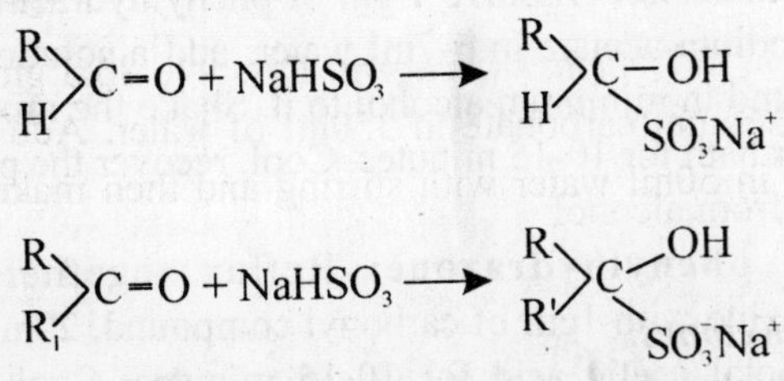

Preparation of reagent. To saturated aqueous sodium metabisulphite solution add 70% of its volume of ethanol and then a little water to produce a clear solution.

2. Oximes: (*a*) For water soluble aldehydes and ketones: In a test tube place 1 gm of hydroxalamine hydrochloride, 2gm of fused sodium acetate, 500mg of given carbonyl compound (5ml in case of liquid) and then add 5ml water to it. Heat the reaction mixture on water bath for 15-20 minutes. Cool and if necessary scratch the sides of test tube to induce precipitation. Filter the precipitated solid, recrystallise from ethanol and determine its melting point.

(*b*) For water insoluble aldehydes and ketones: To 1gm of carbonyl compound in 50ml round bottom flask add 1gm of hydroxylamine hydrochloride, 10ml ethanol and 1ml pyridine. Reflux the reaction mixture on a steam bath for 30-60 minutes. Pour the reaction mixture in a porcelain dish and evaporate most of the liquid. Scratch the dish with spatulla and grind the residue with 3ml of water. Filter and recrystallise the oxime with methanol or ethanol and determine its melting point.

3. 2, 4 dinitrophenyl hydrazone(2, 4 DNP): Place 5 ml of a saturated solution of 2, 4 dinitrophenyl hydrazine (Brady's reagent) in a test tube and to this dark red orange solution add 3 ml of liquid sample or 300mg of solid sample. Stopper the test tube and shake vigorously. If no precipitate separates out heat the test tube on water bath for 5-10minutes and again shake or scratch the sides of test tube with glass rod and allow it to stand for 10 minutes. Filter the precipitate, recrystallise from ethanol and determine its melting point.

4. Semicarbazone: In a 50ml conical flask place 500mg of carbonyl compound, 1gm of semicarbazide hydrochloride and a solution of 1.5 gm of sodium acetate in 6ml water. Shake the mixture vigorously. Semicarbazone of some carbonyl compounds usually separates out immediately on stirring. If there is no precipitate heat the mixture gently on steam bath for 10-15 minutes and then cool in ice water. Filter the precipated semicarbazone derivative and recrystallise from methanol, ethanol or ethanol-water mixture

5. Phenylhydrazone: Dissolve 1 gm of phenylhydrazine hydrochloride and 1.5 gm of sodium acetate in 6-7ml water, add a solution of 500 mg of carbonyl compound in minimum alcohol to it. Shake the mixture thoroughly and heat on water bath for 10-15 minutes. Cool, recover the phenylhydrazone and recrystallise from alcohol

6. p-Nitro phenylhydrazone: Reflux together 1 gm of p-nitrophenylhydrazine with 1gm of carbonyl compound, 20ml of ethanol and 4-5 drops of glacial acetic acid for 10-15 minutes. Cool, recover the p-nitrophenylhydrazone and recrystallise from alcohol.

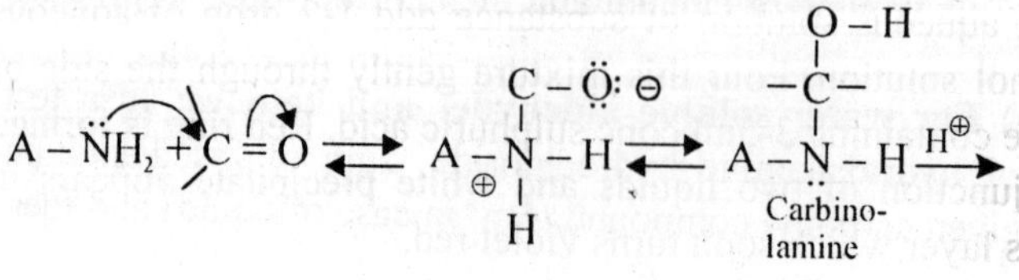

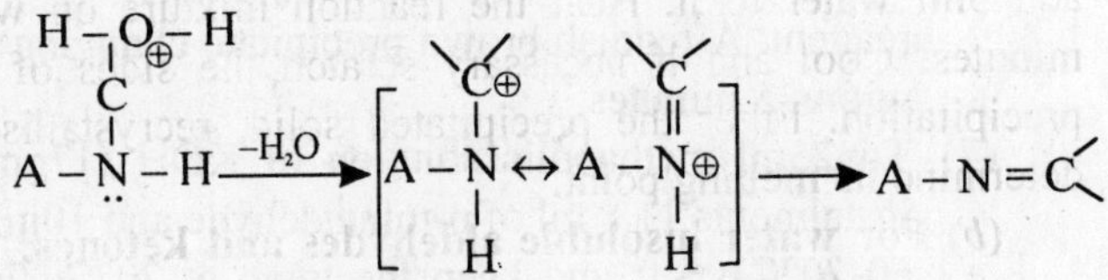

A =	Reagent	Product
A = OH	hydroxylamine	oxime
= NH_2	hydrazine	hydrazone
= $C_6H_5NH_2$	Phenyl hydrazine	Phenyl hydrazone
= $p\text{-}NO_2C_6H_4NH_2$	p-nitrophenyl hydrazine	p-nitrophenyl hydrazone
= NH_2CONH_2	Semicarbazide	Semicarbazone

7. Benzilidene derivatives(given by compounds containing CH_2CO group): Dissolve 2gm of carbonyl compound and 4-5 ml benzaldehyde in 15 ml of rectified spirit, add 2ml of 2N NaOH solution to it. Shake the mixture thoroughly for 5-10min and then keep it at room temperature for an hour. The benzilidene derivative usually precipitates out on scratching the test tube with glass rod. Filter the precipitated derivative, recrystallise from alcohol and determine its melting point.

$$R-COCH_2R^1 + C_6H_5CHO \rightarrow R.CO\,C(=CHC_6H_5)R^1 + H_2O$$

Aldehydes

Aliphatic Liquid

Gas **Formaldehyde** (Methanal) HCHO

(1) Colourless gas at room temperature with an irritating odour, whereas the commercial aqueous solution called formalin, b.p 99°C, (40% aqueous solution) is soluble in alcohol and ether.

(2) Reduces Fehling's solution.

(3) Reduces Tollens reagent.

(4) To 1ml of substance add 50mg of pyrogallol and 1ml of conc sulphuric acid. A white precipitate gradually changing to red is obtained within 5 minutes.

(5) To 1ml aqueous solution of substance add 1-2 drop of aqueous resorcinol solution. Pour this mixture gently through the side of test tube containing 3-5ml conc sulphuric acid. Red ring is formed at the junction of two liquids and white precipitate appears in aqueous layer which soon turns violet red.

(6) To 2ml aqueous solution of compound add 2-3 drops of Nessler's reagent. A reddish brown precipitate changing to grey is obtained within 5 minutes.

(7) To 25ml methanolic solution of KOH (17gm in 25ml) add in instalments 11.8 ml of p-tolualdehyde and 10ml of formalin (temp 60-70°C). Stir and keep the temp at 70°C. Distil off methanol, add water and cool. Extract the aqueous layer with toluene, distil off toluene and collect p-methyl benzyl alcohol, m.p 61°C.

(8) To 4ml of compound add excess of liquid ammonia until the smell of ammonia persists. Evaporate on water bath. A solid residue of hexamethylene tetramine is obtained.

(9) Formaldehyde 2, 4 dinitro phenylhydrazone, m.p 167°C.

(10) Formaldehyde semicarbazone, m.p 169°C (d)

(11) Formaldehyde p-nitrophenylhydrazone, m.p 182°C.

b.p 20°C **Acetaldehyde** (Ethanal) CH_3CHO

(1) Colourless liquid, miscible with water, alcohol and ether, concentrated liquid has disagreeable suffocating odour however dilute aqueous solution has apple like odour. Pure liquid boils at room temperature hence its aqueous solution is used, b.p 98°C.

(2) Reduces Fehling's solution.

(3) Reduces Tollens reagent.

(4) To 1ml aqueous solution of the compound add 1ml sodium nitroprusside solution and a drop of piperidine. A blue coloured solution is obtained.

(5) Warm 2ml aqueous solution of compound with 7ml dilute sodium hydroxide solution. The solution first turns yellow and then separates out as yellow orange resin with penetrating odour.

(6) To 2ml dilute aqueous solution of compound add Nessler's reagent. A yellow precipitate is obtained.

(7) Cool a mixture of 5ml acetaldehyde, 9gm malonic acid, 12ml dry pyridine and 1ml piperidine for 24 hours. Warm the mixture on water bath until evolution of carbon dioxide ceases and then cool. Collect crystals of crotonic acid, m.p 72°C.

(8) To 8gm of compound add 40ml Ac_20 and 1 drop of conc H_2SO_4. The solution becomes warm. Cool and after 30 min add 2 more drops of conc H_2SO_4. After 12hrs pour the mixture in ice water. Extract with ether. Distil and collect ethylidene diacetate at b.p 168°C.

(9) Acetaldehyde semicarbazone, m.p 163°C
(10) Acetaldehyde phenylhydrazone, m.p 100°C.
(11) Acetaldehyde p-nitro phenylhydrazone, m.p 128°C
(12) Acetaldoxime, m.p 47°C.
(13) Acetaldehyde 2, 4 dinitro phenylhydrazone, m.p 168°C
(14) Iodoform, m.p 119°C.

b.p 42°C **Methylal** (Formaldehyde dimethyl acetal) $CH_2(OCH_3)_2$

(1) Liquid with alcoholic odour, miscible with water.
(2) Respond to Schiff's test on warming with one drop of mineral acid.
(3) Reflux 5ml of compound with dil H_2SO_4. Distill, collect the distillate and test for the presence of formaldehyde.

b.p 49°C **Propionaldehyde** (Propanal) CH_3CH_2CHO

(1) Colourless liquid having pungent odour, miscible with water.
(2) Reduces Fehling's solution
(3) Reduces Tollens reagent
(4) Gives crystalline bisulphite addition compound with sodium metabisulphite.
(5) Add with stirring 5ml of compound to 2ml of NaOH solution and then cool in ice. Isolate 2-methyl pent-2-enal with ether, distil and collect fraction boiling at 136-137°C/760mmHg.
(6) Propionaldehyde –2, 4 dinitro phenylhydrazone, m.p 155°C
(7) Propionaldehyde semicarbazone, m.p 154°C
(8) Propionaldehyde p-nitro phenylhydrazone, m.p 124°C.

b.p 52°C **Acrolein** (Propenal, acrylaldehyde) CH_2=CHCHO

(1) Colourless liquid having pungent odour, soluble in water. Powerful lachrymator
(2) Reduces Fehling's solution.
(3) Reduces Tollens reagent.
(4) To 2ml aqueous solution of the compound add alkaline potassium permanganate solution. Purple colour of permanganate disappears.
(5) To 1ml of compound add equal volume of 3% hydrogen peroxide, after 1minute add 6ml of conc hydrochloric acid and 6ml of

phlorogucinol. Shake the mixture thoroughly, the acid layer will become deep red.

(6) Acrolein –2, 4 dinitro phenylhydrazone, m.p 165°C.
(7) Acrolein semicarbazone, m.p 171°C.
(8) Acrolein p-nitro phenylhydrazone, m.p 151°C.

b.p 64°C **Isobutyraldehyde** (2-methyl propanal) $(CH_3)_2CHCHO$

(1) Colourless liquid, is soluble in water, soluble in alcohol.
(2) Gives crystalline bisulphite addition compound with sodium metabisulphite.
(3) Isobutyraldehyde-2, 4 dinitro phenylhydrazone, m.p 187°C.
(4) Isobutyraldehyde semicarbazone, m.p 125°C
(5) Isobutyraldehyde p-nitro phenylhydrazone, m.p 130°C.

b.p 64°C **Dimethyl acetal** (Acetaldehyde dimethyl acetal) $CH_3CH(OCH_3)_2$

(1) Colourless liquid sparingly soluble in water, soluble in alcohol and ether.
(2) Responds to Schiff's test on warming with one drop of mineral acid.
(3) Reflux 5ml of compound with dil H_2SO_4. Distil, collect the distillate and test for the presence of acetaldehyde.

b.p 75°C **Butyraldehyde** (n-butanal) $CH_3CH_2CH_2CHO$

(1) Colourless liquid, sparingly soluble in water, soluble in alcohol and ether.
(2) Gives crystalline bisulphite addition compound with sodium metabisulphite
(3) Add 18ml of compound to a mixture of 15ml acetone and 32ml of 15% NaOH solution. The mixture is allowed to reflux and then stirred for 2 days. Separate the oily layer and extract the aqueous layer with ether. Distill the combined oily and ethereal layer. Collect pure cis n-butylidene acetone at b.p 68-70°C /12mmHg
(4) Heat 10ml NaOH solution to 80°C and add immediately 2.6 ml of butyraldehyde. Reflux the mixture for 1hr. Cool and separate the

organic layer and fractionally distil. Collect pure 2-ethyl hex-2-enal at 66-67°C/25mmHg.

(5) Butyraldehyde p-nitro phenylhydrazone, m.p 91°C.
(6) Butyraldehyde semicarbazone, m.p 104°C.
(7) Butyraldehyde 2, 4 dinitro phenylhydrazone, m.p 123°C.

b.p 75°C **Trimethylacetaldehyde** (Pivalaldehyde) $(CH_3)_3CCHO$

(1) Liquid with characteristic odour.
(2) Gives crystalline bisulphite addition compound with sodium metabisulphite.
(3) Reduces Tollens reagent.
(4) Trimethylacetaldehyde 2, 4 dinitrophenylhydrazone m.p 210°C
(5) Trimethylacetaldehyde p-nitrophenylhydrazone, m.p 119°C
(6) Trimethylacetaldehyde semicarbazone, m.p 190°C
(7) Trimethylacetaldoxime, m.p 41°C.

b.p 89°C **Ethylal** (Formaldehyde diethyl acetal) $CH_2(OC_2H_5)_2$

(1) Sparingly soluble in water.
(2) Responds to Schiff's test on warming with one drop of mineral acid.
(3) Reflux 5ml of compound with dil H_2SO_4. Distil, collect the distillate and test for the presence of formaldehyde.

b.p 92°C **Isovaleraldehyde** (3-methyl butanal –1) $(CH_3)_2CHCH_2CHO$

(1) Colourless liquid with sweet apple like odour, slightly soluble in water, soluble in alcohol and ether.
(2) Gives crystalline bisulphite addition compound with sodium metabisulphite.
(3) Isovaleraldehyde 2, 4 dinitro phenylhydrazone, m.p 123°C.
(4) Isovaleraldoxime, m.p 49°C.
(5) Isovaleraldehyde p-nitro phenylhydrazone, m.p 110°C
(6) Isovaleraldehyde semicarbazone, m.p 132°C.

b.p 103°C **Acetal** (1:1 diethoxy ethane, acetaldehyde diethyl acetal) $CH_3CH(OC_2H_5)_2$

(1) Colourless liquid, sparingly soluble in water, soluble in alcohol.
(2) Reflux 5ml of compound with dil H_2SO_4. Distil, collect the distillate and test for the presence of acetaldehyde.

b.p 104°C **Crotonaldehyde** (2 Butenal) $CH_3CH{=}CHCHO$

(1) Colourless liquid having characteristic pungent odour, fairly soluble in water. Lachrymator
(2) Reduces Tollens reagent.
(3) Gives crystalline bisulphite addition compound with sodium metabisulphite.
(4) To 4ml aqueous solution of compound add 2ml sodium nitroprusside solution and 1-2 drops of sodium hydroxide solution. Deep wine red coloured solution is produced.
(5) Reflux together 5ml of crotonaldehyde with 6gm of malonic acid and 6ml of pyridine on water bath for 3 hrs. Cool, acidify with dil sulphuric acid and collect sorbic acid, m.p 134°C
(6) Crotonaldehyde p-nitro phenylhydrazone, m.p 185°C
(7) Crotonaldehyde, 2, 4 dinitro phenylhydrazone, m.p 190°C
(8) Crotonaldehyde semicarbazone, m.p 201°C.
(9) Crotonaldoxime, m.p 119°C.

b.p 103°C **n-Valeraldehyde** (Pentanal) $CH_3CH_2CH_2CH_2CHO$

(1) Colourless mobile liquid with penetrating odour, sparingly soluble in water, soluble in alcohol and ether.
(2) To 2ml aqueous suspension of the compound add 3-4 drops of sodium nitroprusside and 1ml of sodium hydroxide. A violet red coloured solution is obtained.
(3) Valeraldehyde 2, 4 dinitrophenyl hydrazone, m.p 107°C.
(4) Valeraldoxime, m.p 52°C
(5) Valeraldehyde p-nitrophenylhydrazone, m.p 74°C

b.p 124°C **Paraldehyde** $(CH_3CHO)_3$

(1) Colourless liquid with disagrecable odour, sparingly soluble in water, soluble in most organic solvents.
(2) Heat 6gm of compound with 10ml of acetic anhydride at 125°C.

Add 10gm of diethylmalonate in portion at an interval of 30min. Continue refluxing for 4hrs. Distil and collect diethyl ethylidenemalonate, b.p 104-106°C/10mmHg.

(3) Heat a mixture of 4.2ml of paraldehyde, 6ml of 28% aqueous ammonium hydroxide and 100mg of ammonium acetate at 230°C for 60 min. Cool and separate the organic layer. Extract the aqueous layer with chloroform, Combine the two layers and distil. Collect 5-ethyl 2-methylpyridine at b.p 174°C.

(4) Heat 25gm of compound with 5-6 drops of conc H_2SO_4 on water bath at 50°C. Distill and collect acetaldehyde at b.p 21°C.

b.p 131°C **Caproaldehyde** (Hexanal) $CH_3CH_2CH_2CH_2CH_2CHO$

(1) Colourless mobile liquid with penetrating odour, insoluble in water.
(2) Gives crystalline bisulphite addition compound with sodium metabisulphite.
(3) Caproaldoxime m.p 51°C (4) Caproaldehyde semicarbazone, m.p 106°C (5) Caproaldehyde 2, 4 dinitrophenylhydrazone, m.p 104°C.

b.p 143°C **Tetrahydrofurfural** (Tetrahydrofuran –2 aldehyde)

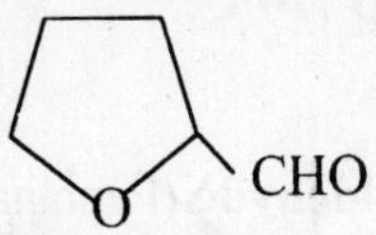

(1) Colourless viscous liquid, soluble in organic solvent
(2) Reduces Fehling's solution.
(3) Reduces Tollens reagent.
(4) Tetrahydrofurfural semicarbazone, m.p 166°C

b.p 155°C **n-Heptaldehyde** (Heptanal) $CH_3CH_2CH_2CH_2CH_2CH_2CHO$

(1) Colourless oily liquid with penetrating odour, slightly soluble in water, soluble in ether. Hygroscopic
(2) Gives crystalline bisulphite addition compound with sodium metabisulphite.
(3) Reflux 28gm of compound with 57ml Ac_2O and 5gm K_2CO_3 in an oil bath at 155-160°C for 1hr. Cool, wash with water and 5%

Na_2CO_3 solution. Distil and collect Hept-1ene-1-ol acetate at b.p 88-90°C/17mmHg.

(4) To a suspension of 30gm of silver oxide in 50ml of 10% aqueous alkali warmed to 95°C add dropwise with stirring 12gm of heptanal. Continue stirring at this temperture for 6hrs. Cool to 70°C, add 60ml of conc HNO_3 and continue stirring for another hour. Remove the oily layer, distil and collect n-heptanoic acid at b.p 223°C.

(5) n-Heptaldehyde semicarbazone, m.p 109°C.

(6) n-Heptaldehyde p nitro phenylhydrazone, m.p 73°C

(7) n-Heptaldehyde 2, 4 dinitro phenylhydrazone, m.p 108°C.

b.p 163.4°C **n-Caprylaldehyde** (Octanal)
$CH_3CH_2CH_2CH_2CH_2CH_2CH_2CHO$

(1) Colourless liquid with fruity odour, soluble in 70% alcohol.

(2) Gives crystalline bisulphite addition compound with sodium metabisulphite.

(3) n-Caprylaldehyde semicarbazone, m.p 98°C

(4) n-Caprylaldoxime, m.p 60°C

(5) n-Caprylaldehyde 2, 4 dinitro phenylhydrazone, m.p106°C.

(6) n-Caprylaldehyde p nitrophenylhydrazone, m.p 80°C

b.p 206°C **d-Citronellal**

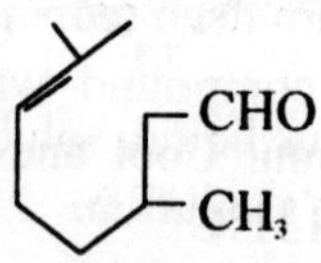

(1) Optically active liquid with strong geranium odour, insoluble in water, soluble in alcohol and ether.

(2) Gives crystalline bisulphite addition compound with sodium metabisulphite.

(3) d-Citronellal semicarbazone, m.p 84°C (4) d-Citronellal 2, 4 dinitrophenyl hydrazone, m.p 78°C

b.p 229°C **Citral** (geranial)

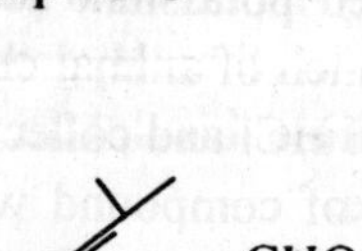

(1) Colourless optically active liquid with strong lemon oil like odour, turns yellow on exposure to sunlight. Insoluble in water, soluble in 5 volumes 60% alcohol.
(2) Reduces Fehling's solution.
(3) Heat 2ml of compound with chromium trioxide and sulphuric acid. A sweet smell due to the formation of acetone is observed.
(4) Citral semicarbazone [Geranial semicarbazone, m.p164°C, Neral semicarbazone, m.p132°C]
(5) Citral 2, 4 dinitrophenyl hydrazone [Geranial 2, 4 DNP, m.p 110°C, Neral 2, 4 DNP, m.p 96°C]

Aromatic liquids

b.p 161°C **Furfural** (2-furfuraldehyde)

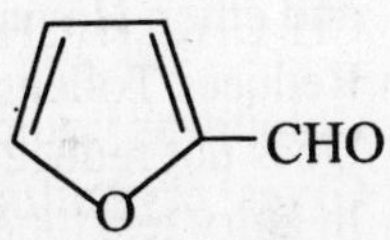

(1) Colourless liquid which darkens in air, soluble in alcohol and ether, odour similar to that of bitter almond.
(2) Reduces Fehling's solution.
(3) Reduces Tollens reagent.
(4) Boil aqueous solution of the substance and treat the vapours with paper moistened with equal amount of aniline and acetic acid. The paper will turn deep red.
(5) Reflux 5ml of compound with acidic potassium dichromate solution for 30min. Cool and pour the solution in ice, collect furoic acid, m.p 131°C
(6) To 13ml furfural and 120ml water add 25ml acetone. Cool the mixture to 10°C and add solution of 1gm NaOH in 2ml water. Stir the mixture for 4hrs and then add 14ml dil H_2SO_4. Remove the organic layer, distil under reduced pressure and collect furfurylideneacetone, m.p 39°C.
(7) Reflux together 4.2ml of furfural, 7.1ml of acetic anhydride and 4.9gm of fused potassium acetate for 4hrs(exothermic rection). Cool, add a pinch of animal charcoal, boil and filter hot. Cool the filtrate, add dil HCl and collect furylacrylic acid, m.p138-139°C
(8) Reflux 20gm of compound with 40ml water, 15ml alcohol and 2gm KCN for 30-40min. Cool, acidify with acetic acid. Furoin separates out, recrystallise from methyl alcohol, m.p 135°C.

(9) Furfural 2, 4-dinitrophenylhydrazone, m.p 200 °C(213)
(10) Furfuraldoxime, m.p 91°C
(11) Furfural semicarbazone, m.p 203°C
(12) Furfural phenylhydrazone, m.p 97°C
(13) Furfural p-nitro phenylhydrazone, m.p154°C.

b.p 179°C **Benzaldehyde** (Phenyl methanal)

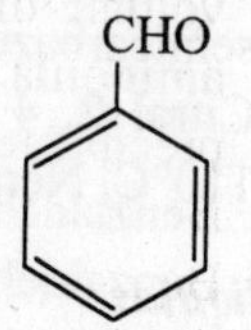

(1) Colourless liquid, sparingly soluble in water, soluble in alcohol and ether. Having a characteristic odour of bitter almonds.
(2) Reduces Tollens reagent
(3) Does not reduce Fehling's solution
(4) In a dry test tube mix 2-3drops of substance with equal amount of melted phenol and conc H_2SO_4 respectively. Then add 4-5ml of dil NaOH to the resultant solution. A violet red coloured solution will be produced.
(5) Heat 2ml of substance with alkaline $KMnO_4$ solution. Cool, acidify with dil HCl and collect Benzoic acid, m.p121°C.
(6) To 1ml of benzladehyde add equal amount of dimethyl aniline, a pinch of fused $ZnCl_2$ and heat. A deep green coloured precipate is obtained.
(7) To 2ml aqueous solution of compound add a pinch of 2-naphthol and 5-10ml water. Filter and pour the filtrate on to the surface of 3ml of conc H_2SO_4. A violet red coloured ring is formed at the junction of two layers.
(8) Heat 3ml benzaldehyde with 3.5gm m-nitroaniline in 15ml ethanol on water bath for 20min, add water until cloudiness appears and cool. Collect the prcipitated Schiffs base, m.p 71-72°C. Dissolve the base in methanol, heat on water bath at 40°C, add 1gm $NaBH_4$ and then reflux for15min. Cool and collect N Benzyl-m-nitroaniline, m.p106-107°C.
(9) Cool a mixture of 2.25gm of NaOH, 25ml water and 10ml ethanol. Stir and add 5ml acetophenone and 4.5ml benzaldehyde (temp below 25°C). Stir for 3hrs, keep overnight at 0°C and then collect benzylidene -acetophenone, m.p 56-57°C.

(10) Add 10gm benzaldehyde to 6gm KCN and 5gm of NH_4Cl dissolved in minimum amount of water. Add methyl alcohol to get a clear solution. After 3hrs extract with ether. Evaporate off the solvent. Dissolve the residue in 10% HCl and cool. Crystals of hydrochloride of amino nitrile are precipitated. Dissolve the salt in fuming HCl, keep overnight and then dilute with equal volume of water. Heat the mixture to boiling, cool, dilute with ammonia solution. Crystals of phenylaminoacetic acid are precipitated, m.p 255°C.

(11) Benzaldehyde phenylhydrazone, m.p 158°C
(12) Benzaldehyde semicarbazone, m.p 224°C
(13) Benzaldehyde 2, 4 DNP, m.p 237 °C
(14) Benzaldoxime (β), m.p 132°C
(15) Benzaldehyde p-nitro phenylhydrazone, m.p 192°C.
(16) m-nitro benzaldehyde, m.p 58°C.

b.p 195°C **Phenylacetaldehyde** (Phenyl ethanal, α Toluic aldehyde)

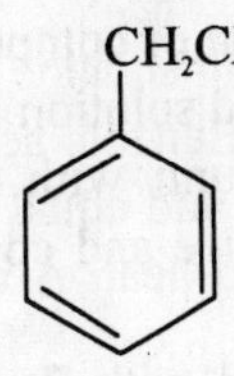

(1) Colourless liquid with hyacinths like odour. Sparingly soluble in water, soluble in ether.
(2) Gives crystalline bisulphite addition compound with sodium metabisulphite.
(3) Reflux 5ml of compound with alkaline potassium permanganate solution. Cool, acidify with dilute hydrochloric acid and collect benzoic acid, m.p121°C
(4) Add 10ml of 5% methyl alcoholic solution of sodium methoxide to a cooled mixture of 30gm of phenylacetaldehyde and 26.5 gm of benzaldehyde. Neutralise the mixture with acetic acid and keep overnight. Distil and collect the straw coloured oily liquid at b.p 196-201°C/16mmHg which on cooling soon solidifies. Recrystallise the crude α phenylcinnamaldehyde from alcohol, m.p 95°C.

(5) To 5ml of compound add dilute sulphuric acid and leave the mixture for several days. A polymer triphenyl paraldehyde is formed, m.p 155°C.
(6) Phenylacetaldehyde phenylhydrazone, m.p 63°C
(7) Phenylacetaldehyde semicarbazone, m.p156°C
(8) Phenylacetaldehyde 2, 4-dinitro phenylhydrazone, m.p 121°C
(9) Phenylacetaldoxime, m.p 99°C.

b.p 197°C **Salicylaldehyde** (o-hydroxy benzaldehyde, o-formyl phenol)

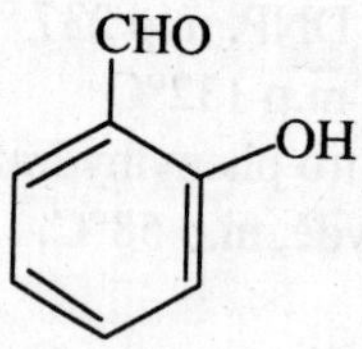

(1) Oily liquid with faint aromatic odour, sparingly soluble in water, easily soluble in alcohol and ether
(2) Reduces Tollens reagent.
(3) To 2ml aqueous solution of compound add neutral ferric chloride solution. Violet coloured solution is obtained.
(4) Reflux 5ml of compound with 10ml of acetic anhydride for 4 hours. Cool, pour in ice and collect Salicylaldehyde triacetate, m.p 102°C.
(5) Heat 5ml of compound with 7ml of aniline at 100°C for few minutes. Cool and collect yellow crystals of Salicylaldehyde anil, m.p 50°C.
(6) Heat 2ml of compound with 5ml of alkaline $KMnO_4$ for 20-30 minutes. Cool, add dil HCl and collect Salicylic acid, m.p 158°C.
(7) To a mixture of 5ml salicylaldehyde, 4.5 gm chloroacetic acid and 35ml water add with stirring a solution of 3.8gm of NaOH in 75ml water. Reflux for 2-3 hrs. Acidify with dil HCl. Steam distil to remove unreacted salicylaldehyde. Cool and collect o-formyl phenoxyacetic acid, m.p132°C.
(8) Reflux 5ml of salicylaldehyde with 4ml triethylamine and 10ml Ac_2O for 12hrs. Steam distil the mixture. To the residue add solid $NaHCO_3$ until alkaline. Cool and filter. Collect the solid coumarin m.p 69°C. Acidify the filtrate with HCl and collect acetoxy cinnamic acid, m.p153-154°C.

(9) Reflux 20gm of compound with 30gm Ac_2O, 32.5gm anhydrous Na_2CO_3 and 0.5gm iodine at 120°C for 2hrs and then at 180-190°C for 5hrs. Cool, pour in ice and extract with ether. Add sodium bisulphite solution to the ethereal layer and filter. Evaporate off the ether and collect coumarin, recrystallise from pet-ether, m.p 67°C.

(10) Salicylaldehyde p nitrophenylhydrazone, m.p 228°C.

(11) Salicylaldehyde semicarbazone, m.p 232°C (d).

(12) Salicylaldoxime, m.p 63°C.

(13) Salicylaldehyde 2, 4 dinitro phenylhydrazone, m.p 252°C(d)

b.p 199°C **m-tolualdehyde** (m-methyl benzaldehyde)

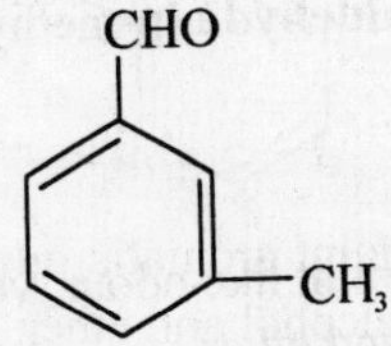

(1) Colourless liquid, very sparingly soluble in water, soluble in alcohol and ether.

(2) Gives crystalline bisulphite addition compound with sodium metabisulphite.

(3) m-tolualdoxime, m.p 60°C.

(4) m-tolualdehyde p nitro phenylhydrazone, m.p 157°C.

(5) m-tolualdehyde phenylhydrazone, m.p 91°C

(6) m-tolualdehyde semicarbazone, m.p 224°C

(7) m-tolualdehyde 2, 4 dinitrophenylhydrazone, m.p194 °C.

(8) m-toluic acid, m.p 110°C.

b.p 200°C **o-tolualdehyde** (o-methyl benzaldehyde)

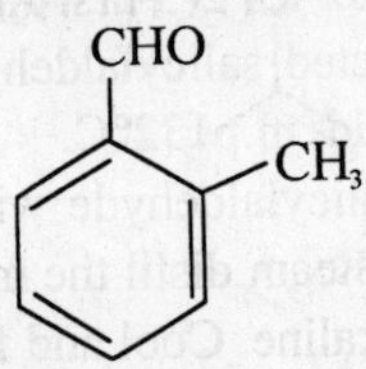

(1) Liquid with odour similar to that of bitter almond, very sparingly soluble in water, soluble in alcohol and ether.

(2) Gives crystalline bisulphite addition compound with sodium metabisulphite.
(3) Reduces Tollens reagent.
(4) o-tolualdehyde 2, 4 dinitrophenylhydrazone, m.p 194°C
(5) o-tolualdehyde semicarbazone, m.p 212°C.
(6) o-tolualdehyde p nitrophenylhydrazone, m.p 222°C.
(7) o-tolualdehyde phenylhydrazone, m.p 105°C.
(8) o-toluic acid, m.p 104°C

b.p 205°C **p-tolualdehyde** (p-methyl benzaldehyde)

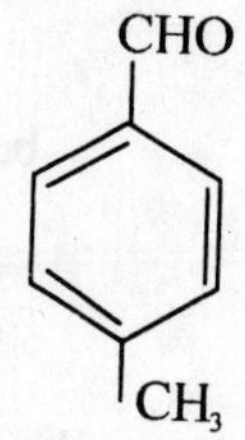

(1) Liquid with peppermint like odour, very sparingly soluble in water, soluble in alcohol and ether.
(2) Gives crystalline bisulphite addition compound with sodium metabisulphite.
(3) p-tolualdoxime, m.p 80°C.
(4) p-tolualdehyde 2, 4 dinitrophenylhydrazone, m.p 234°C
(5) p-tolualdehyde semicarbazone, m.p 234°C.
(6) p-tolualdehyde p nitrophenylhydrazone, m.p 200°C.
(7) p-tolualdehyde phenylhydrazone, m.p 112°C.
(8) p-toluic acid, m.p 178°C

b.p 224°C **Hydrocinnamaldehyde** (3-phenyl propanal, benzyl acetaldehyde)

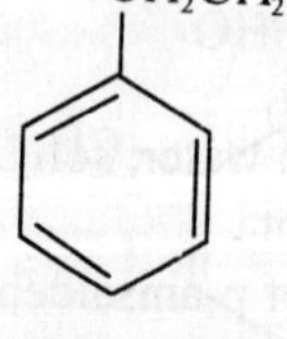

(1) Pale yellow coloured liquid with hyacinth like odour, insoluble in water, soluble in 70% alcohol.

(2) Gives crystalline bisulphite addition compound with sodium metabisulphite.
(3) Hydrocinnamaldehyde 2, 4 dinitrophenylhydrazone, m.p149°C.
(4) Hydrocinnamaldoxime, m.p 94°C
(5) Hydrocinnamaldehyde p-nitrophenylhydrazone, m.p 123°C.
(6) Hydrocinnamaldehyde semicarbazone, m.p 127°C
(7) hydrocinnamic acid, m.p 48°C.

b.p 143°C (50mm) **m-Anisaldehyde** (m-methoxy benzaldehyde)

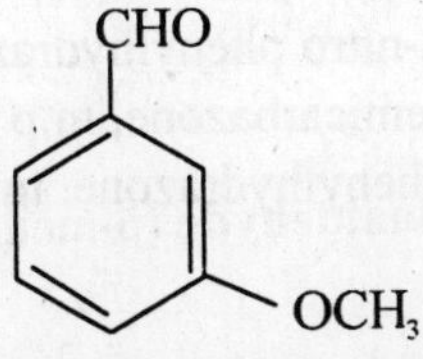

(1) Pale yellow coloured liquid, insoluble in water, soluble in organic solvents
(2) Reflux together 10ml of m-anisaldehyde with 15gm of malonic acid, 35ml of pyridine and 0.5ml of piperidine on water bath for 1hr. Cool, pour in water containing HCl. Filter and collect the precipitated m-methoxy cinnamic acid, m.p117°C.
(3) m-Anisaldehyde p-nitro phenylhydrazone, m.p171°C
(4) m-Anisaldoxime, m.p 40°C.

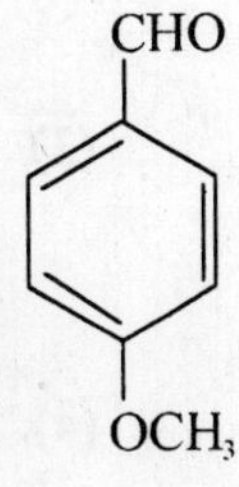

b.p 248°C **p-Anisaldehyde** (p-methoxy benzaldehyde)

(1) Oily liquid, insoluble in water, soluble in alcohol and ether
(2) Reduces Tollens reagent
(3) Reflux together 10ml of p-anisaldehyde with 15gm malonic acid, 25ml pyridine and 0.5ml piperidine for 1 hr on water bath. A rapid evolution of CO_2 takes place. Boil for another 5minutes.

Cool, pour in ice water containing HCl. Filter and collect p-methoxy cinnamic acid, m.p172°C.

(4) Heat 10ml of $KMnO_4$ solution with 4ml of compound until the colour of $KMnO_4$ persists. Cool, acidify with dil HCl and collect anisic acid, m.p 184°C.

(5) Reflux 5gm of compound with 8ml alcohol, 4ml water and 1gm of KCN for 2hrs. Add further 1gm of KCN and continue refluxing for another 1hr. Cool, filter the precipitated anisoin and recrystallise from alcohol, m.p 113°C.

(6) p-Anisaldehyde 2, 4 dinitro phenylhydrazone, m.p 253°C (d)

(7) p-Anisaldoxime (β), m.p 133°C.

(8) p-Anisaldehyde p-nitro phenylhydrazone, m.p161°C

(9) p-Anisaldehyde semicarbazone, m.p 210°C.

(10) p-Anisaldehyde phenylhydrazone, m.p 120°C

b.p 251°C **Cinnamaldehyde** (3-phenyl propenal)

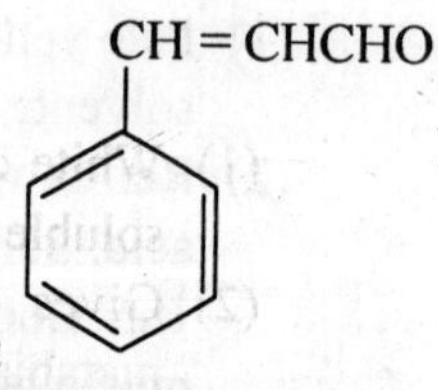

(1) Colourless oily liquid with odour of cinnamon, insoluble in water and pet-ether, soluble in alcohol and ether.

(2) Cool a mixture of 4ml of cinnamaldehyde and 16.7ml of acetic acid, add dropwise with stirring 2ml of bromine followed by 2.3gm of K_2CO_3. Reflux the mixture for 30minutes, cool and pour in ice. Collect α bromo cinnamaldehyde, m.p 73°C.

(3) Cool a mixture of 5ml of cinnamaldehyde and 22.5ml of Ac_2O. Add dropwise with stirring 2ml of conc HNO_3 (temp below 50°C). Keep the mixture at room temperature for 2 days, cool and then add HCl. Collect α nitro cinnamaldehyde, m.p 125-127°C.

(4) Reflux 5ml of compound with alkaline $KMnO_4$. Cool, acidify with dil HCl and collect benzoic acid, m.p 121°C.

(5) Reflux together 6ml of aldehyde with 3gm of succinic acid, 5.6gm of lead oxide and 7.2ml of Ac_2O for 2hrs. Cool, filter and collect tetraene. To the filtrate add few ml of Ac_2O and boil for 2hrs. Cool and collect the second crop. Recrystallise from chlorofrom and collect 1, 8-diphenyl octa 1, 3, 5, 7-tetraene, m.p 232°C.

(6) To 1ml of compound add 2ml of Ac_2O, 1-2 drops of conc H_2SO_4 and shake. The solution becomes warm and on cooling cinnamal diacetate is precipated, m.p 85°C

(7) Cinnamaldehyde 2, 4 dinitrophenylhydrazone, m.p 255°C (d)

(8) Cinnamaldoxime, m.p 139°C.

(9) Cinnamaldehyde p nitrophenylhydrazone, m.p195°C

(10) Cinnamaldehyde semicarbazone, m.p 216°C.

(11) Cinnamaldehyde phenylhydrazone, m.p 168°C.

Solid Aldehydes

m.p 37°C **Piperonal** (3, 4-methylene dioxy benzaldehyde)
b.p 263°C

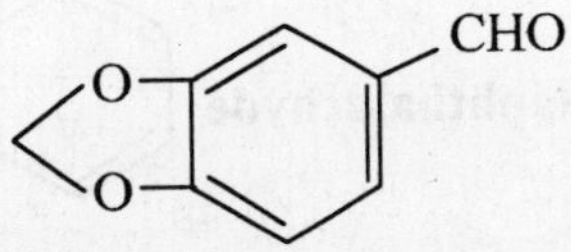

(1) White crystalline solid with heliotrope odour, insoluble in water, soluble in alcohol and ether.

(2) Gives crystalline bisulphite addition compound with sodium metabisulphite.

(3) Dissolve 5gm of compound and 7.5gm of malonic acid in a mixture of 15ml pyridine and 0.5ml of piperidine. Reflux on water bath for 60min. Cool, pour in water containing conc HCl. Filter the precipitated 3, 4 methylenedioxycinnamic acid, recrystallise from glacial acetic acid, m.p 238°C.

(4) Warm 100gm of compound with 1ml of conc nitric acid. Cool and collect 6 nitro piperonal, m.p 95°C

(5) Piperonal 2, 4 dinitrophenyl hydrazone, m.p 265°C (d)

(6) Piperonal semicarbazone, m.p 230°C.

(7) Piperonal p nitophenylhydrazone, m.p 200°C.

(8) Piperonal phenylhydrazone, m.p 102°C.

(9) Piperonaldoxime, m.p 110°C.

m.p 56°C **p-Homosalicylaldehyde** (2-hydroxy-5-methyl benzaldehyde)

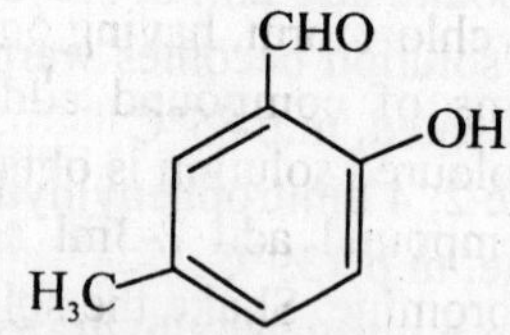

(1) Crystalline compound, sparingly soluble in water, easily soluble in alcohol, ether and chloroform
(2) To few milligrams of compound add neutral aqueous $FeCl_3$ solution. Deep blue coloured solution is obtained.
(3) p-Homosalicylaldehyde phenylhydrazone, m.p 149°C
(4) p-Homosalicylaldoxime, m.p 105°C.

m.p 60°C **2-Naphthaldehyde**

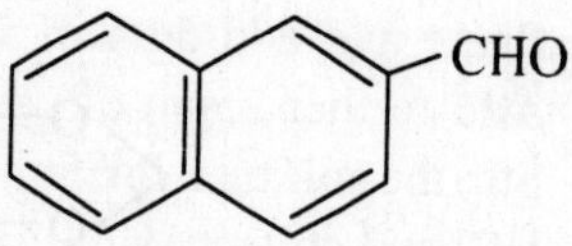

(1) White crystalline solid, insoluble in water, soluble in alcohol and ether
(2) Reduces Tollens reagent.
(3) Gives crystalline bisulphite addition compound with sodium metabisulphite.
(4) To 400mg of compound add 10ml alkaline $KMnO_4$ and heat for 15-20minutes. Cool, acidify with dil HCl and collect 2-naphthoic acid, m.p 184°C.
(5) 2-Naphthaldehyde 2, 4 dinitrophenylhydrazone, m.p 270°C.
(6) 2-Naphthaldoxime, m.p 156°C.
(7) 2-Naphthaldehyde p-nitrophenylhydrazone, m.p 230°C
(8) 2-Naphthaldehyde semicarbazone, m.p 245°C.
(9) 2-Naphthaldehyde phenylhydrazone, m.p 206°C.

m.p 81°C **Vanillin** (4-hydroxy-3-methoxy benzaldehyde)

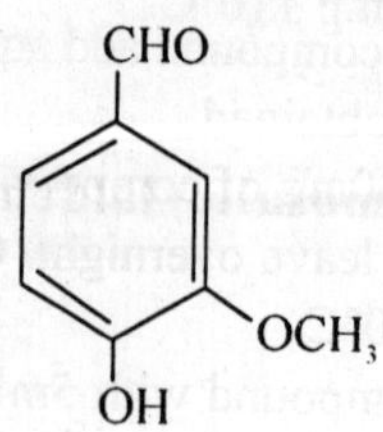

(1) Colourless solid, sparingly soluble in water, easily soluble in alcohol, ether and chloroform, having odour of vanilla. Sublimes.
(2) To few milligrams of compound add neutral aqueous $FeCl_3$ solution. Violet coloured solution is obtained
(3) To 500mg of compound add 2-3ml acetic acid and then add dropwise 1ml of bromine. Shake the solution thoroughly, pour in ice. Collect 5-bromo vanillin, m.p 164°C.
(4) Dissolve 100mg of compound in 5ml water, add 2-3drops of conc HCl and 7-8 drops of 10% $FeCl_3$ solution Boil the reaction mixture for 2minutes. Cool and collect dehydrovanillin, m.p 304°C
(5) Shake thoroughly 500mg of compound with 5ml NaOH and 2ml benzoyl chloride. Cool and collect vanillin monobenzoate, m.p 78°C
(6) Dissolve 30gm of KOH in 10ml water, heat to 160°C, remove the flame and add 5gm of vanillin. Temperature will rise to 195°C. Add further 25gm of vanillin in parts so as to maintain the temp. Stir the mixture for 5min, add 50ml cold water, acidify with dil HCl and collect Vanillic acid, m.p 206-208°C.
(7) Vanillin 2, 4 dinitrophenylhydrazone, m.p 271°C
(8) Vanillin phenylhydrazone, m.p 105°C.
(9) Vanillin p-nitrophenylhydrazone, m.p 229°C
(10) Vanillin semicarbazone, m.p 230°C
(11) Vanillin oxime, m.p 117°C

m.p 100-102°C **m-Hydroxy benzaldehyde** (m-formyl phenol)

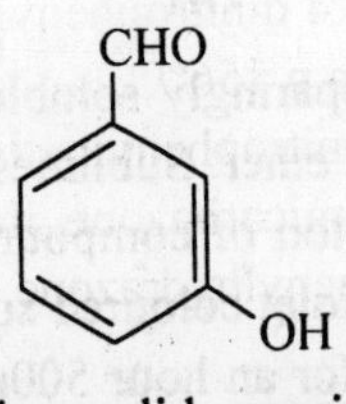

(1) Colourless crystalline solid, sparingly soluble in water, easily soluble in alcohol, ether and benzene
(2) To few miligrams of compound add aqueous $FeCl_3$ solution. Violet coloured solution is obtained
(3) Shake thoroughly 200mg of compound with 2ml aniline for 2 hrs at 35-40°C and then leave overnight. Cool and collect m-hydroxy benzalaniline, m.p 90°C.
(4) Reflux 500mg of compound with 5ml of Ac_2O for 3-4 hrs. Cool,

pour the mixture in ice and collect m-acetoxybenzaldiacetate, m.p 76°C.

(5) Shake thoroughly 200mg with 10ml of NaOH and 10ml of benzoyl chloride until the smell of benzoyl chloride persist. Cool and recover m-benzoxy benzaldehyde, m.p38°C.

(6) To 5gm of compound in 16ml 2N NaOH solution add dropwise with stirring 3.3ml of dimethyl sulphate (temp 40-45°C). Add 9ml of NaOH solution and 1.6ml of dimethyl sulphate (temp 50°C). Continue stirring for 30min, cool, isolate with ether, distil off ether and collect m-methoxy benzaldehyde as pale yellow liquid at b.p 143°C (50mmHg).

(7) m-Hydroxy benzaldehyde 2, 4 dinitrophenylhydrazone, m.p 260°C (d) (8) m-Hydroxy benzaldehyde p-nitrophenylhydrazone, m.p 222°C (9) m-Hydroxy benzaldehyde phenylhydrazone, m.p 130°C. (10) m-Hydroxy benzaldehyde semicarbazone, m.p 199°C. (11) m-Hydroxy benzaldoxime, m.p 88°C.

m.p 117°C **p-Hydroxy benzaldehyde** (p-formyl phenol)

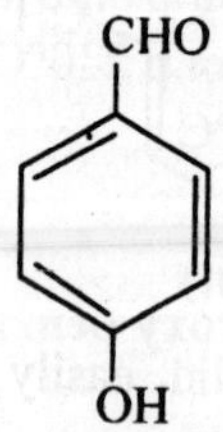

(1) Pink coloured solid, sparingly soluble in water, easily soluble in hot water, alcohol and ether. Sublimes

(2) To 1ml aqueous soluiton of compound add few drops of neutral $FeCl_3$ solution. Pale violet coloured solution is obtained.

(3) Reflux at 175-200°C for an hour 500mg of compound with 10ml aqueous KOH solution. Cool, acidify with HCl and collect p-hydroxy benzoic acid, m.p 210°C.

(4) Dissolve 200mg of compound in 2ml glacial acetic acid and add in portion 1ml bromine solution (bromine in acetic acid). Shake thoroughly, pour in ice and collect 3, 5 dibromo 4-hydroxy benzaldehyde, m.p 179°C.

(5) To 200mg compound add 600mg fused sodium acetate and 2ml

acetic anhydride. Heat at 170-180°C for 10 hrs. Cool and reflux the precipitated solid (acetate) with excess alkali for an hour. Cool, acidify the filtrate with HCl and collect p-hydroxy cinnamic acid, m.p 206-207°C.

(6) Heat on water bath 6.1gm of compound with 9gm of malonic acid, 0.5ml pyridine and 1-2 drop of piperidine for 4hrs. Cool, wash the precipitated solid with dil HCl and dissolve in methyl alcohol. On cooling, yellow crystals of p-hydroxycinnamanilide separate out, m.p 208°C

(7) p-Hydroxy benzaldehyde 2, 4 dinitrophenylhydrazone, m.p 280°C (d).

(8) p-Hydroxy benzaldoxime, m.p 72°C.

(9) p-Hydroxy benzaldehyde phenylhydrazone, m.p 178°C.

(10) p-Hydroxy benzaldehyde p-nitro phenylhydrazone, m.p 266°C.

(11) p-Hydroxy benzaldehyde semicarbazone, m.p 224°C.

m.p 136°C **β-Resorcylaldehyde** (2, 4-dihydroxy benzaldehyde)

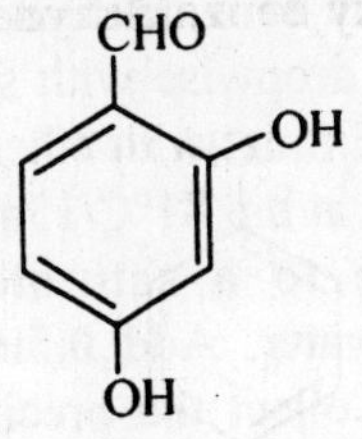

(1) Colourless crystalline solid, easily soluble in water, alcohol and ether

(2) To few milligrams of compound add neutral aqueous $FeCl_3$ solution. Deep blue coloured solution is obtained.

(3) Dissolve 2.5gm of compound and 2.5gm acetophenone in 12ml glacial acetic acid. Pass HCl gas into the cooled solution. After few hours yellow crystals of 7-hydroxy-2-phenyl-benzopyrylium chloride separate out. Keep the mixture overnight, filter and wash with acetic acid. Recrystallise from dil HCl.

(4) 2, 4-dihydroxy benzaldehyde 2, 4 dinitrophenylhydrazone, m.p 286°C (d)

(5) 2, 4-dihydroxy benzaldehyde phenylhydrazone, m.p 159°C (6) 2, 4 dihydroxy benzaldoxime, m.p 191°C.

KETONES

Liquid Aliphatic ketones

b.p 56°C **Acetone** (dimethyl ketone, Propanone) CH_3COCH_3

(1) Colourless pleasant smelling liquid, miscible with water, ethanol and ether

(2) To 1ml of compound add 2-3 drops alkaline sodium nitroprusside solution. Orange coloured solution which later changes to yellow is obtained.

(3) Gives crystalline bisulphite addition compound with sodium metabisulphite.

(4) To 1ml of acetone add 100mg of m-dinitrobenzene and excess of sodium hydroxide solution. A violet coloured solution is obtained.

(4) To a mixture of 5ml of compound and 15ml sodium hydroxide solution add iodine reagent drop by drop till the solution is yellow. Heat on water bath for 2-3 min, cool and collect iodoform, m.p 119°C.

(5) Add 15gm of acetone to a solution of 15gm of KCN in 50ml water. Cool and add dropwise with stirring 45ml of 30% H_2SO_4 (temp below 20°C). Extract with ether. Distil off ether and collect Acetone cyanohydrin at b.p 81°C/15mmHg. [b.p 120°C (d)]

(6) Add 20ml acetone to a saturated solution of 50gm of cyanoacetamide in water. Add 0.5ml piperidine and keep the solution for 24hrs. Collect the precipitated solid, dissolve it in 40ml conc H_2SO_4 and 15ml water. Reflux the mixture for 5hrs. Cool, dilute with water and add ammonium sulphate. Extract with ether. Evaporation of the ether yields ββ-dimethylglutaric acid, m.p 101°C.

(7) Heat a mixture of 2ml acetone, 4ml benzaldehyde, 3ml alcohol and 4ml dil NaOH on water bath for 2-3 min. Cool and collect dibenzalacetone, m.p 112°C.

(8) Acetone 2, 4 dinitrophenylhydrazone, m.p 126°C. (9) Acetone p-nitro phenylhydrazone, m.p 148°C.

(10) Acetone semicarbazone, m.p 188°C (11) Acetoxime, m.p 59°C.

b.p 80°C **Ethyl methyl ketone** (Butanone-2) $CH_3CH_2COCH_3$

(1) Colourless pleasant smelling liquid, miscible with water, alcohol and ether.

(2) To 1ml of compound add 100mg of m-dinitrobenzene and excess of sodium hydroxide solution. A violet coloured solution is obtained.

(3) Add alkaline sodium nitroprusside solution to 1ml of compound. A wine red coloured solution is obtained.

(4) To a mixture of 5ml of compound and 15ml sodium hydroxide solution add iodine solution drop by drop till the solution is yellow. Heat on water bath for 2-3 min, cool and collect iodoform, m.p 119°C.

(5) Add with stirring 20ml of compound to a mixture of 5.3gm of Na wire and 90ml ethyl acetate cooled to 0°C. Keep the mixture for 12hrs and then heat on water bath for 90min. Cool and pour in ice. Remove the ethereal layer. Neutralise the aqueous layer with acetic acid and then add the required N/8 cupric acetate solution. Collect the blue copper propionyl acetone, m.p 198°C. Decompose the copper salt with H_2SO_4 in presence of ether. Distil the ethereal solution and collect propionylacetone at b.p 158°C.

(7) Ethyl methyl ketone 2, 4-dinitrophenylhydrazone, m.p 115°C. (7) Ethyl methyl ketone semicarbazone, m.p 136°C.

(8) Ethyl methyl ketone p-nitrophenylhydrazone, m.p 128°C.

b.p 88°C **Biacetyl** (Dimethyl glyoxal, 2, 3-Butane dione) $CH_3COCOCH_3$

(1) Yellowish green coloured liquid, soluble in water, alcohol and ether

(2) Biacetyl 2, 4 dinitrophenylhydrazone(di), m.p 315°C

(3) Biacetyl bis semicarbazone, m.p 279°C.

(4) Biacetyl p-nitrophenylhydrazone, m.p 230°C.

(5) Biacetyl bis phenylhydrazone, m.p 243°C.

b.p 93°C **Isopropyl Methyl ketone** (3-methyl butan-2 one) $(CH_3)_2CHCOCH_3$

(1) Colourless liquid, sparingly soluble in water, easily soluble in alcohol and ether.

(2) Add alkaline sodium nitroprusside solution to 1ml of compound. A red coloured solution is obtained.

(3) To a mixture of 5ml of compound and 15ml sodium hydroxide

solution add iodine-KI reagent drop by drop till the solution is yellow. Heat on water bath for 2-3 min, cool and collect iodoform, m.p 119°C.

(4) Stir 25gm of compound with 150ml of diethyl carbonate and cool to 0°C. Add slowly a suspension of 15gm of sodium methoxide in diethyl carbonate and leave at room temperature for 2hrs. Distil off the alcohol. Cool and acidify the reaction mixture with acetic acid. Separate the organic layer and distil. Collect 4 methylpent-3-one-anoic acid at b.p 75-81°C/10mmHg.

(5) Isopropyl Methyl ketone 2, 4-dinitrophenylhydrazone, m.p 117°C.

(6) Isopropyl Methyl ketone p-nitro phenylhydrazone, m.p 109°C

(7) Isopropyl Methyl ketone semicarbazone, m.p 112°C.

(8) Isopropyl Methyl ketoxime, m.p 109°C.

b.p 102°C **Diethyl ketone** (Pentan-3-one) $CH_3CH_2COCH_2CH_3$

(1) Colourless liquid, sparingly soluble in water, miscible with alcohol and ether.

(2) Add 1ml of ethylbromide to 6gm Mg turning covered with ether (exothermic reaction). After the reaction has subsided add a mixture of 20ml diethyl ketone, 25ml ethyl bromide and equal volume of ether. Heat the mixture under reflux for 2-3 hrs. Add 20gm crushed ice and dil H_2SO_4. Remove the ethereal layer and distil. Collect triethyl carbinol at b.p 135-141°C.

(3) Diethyl ketone 2, 4-dinitrophenylhydrazone, m.p 156°C

(4) Diethyl ketone semicarbazone, m.p 139°C.

(5) Diethyl ketone p-nitrophenylhydrazone, m.p 144°C.

(6) Diethyl ketoxime, m.p 69°C.

b.p 102°C **Methyl n-Propyl ketone** (Pentan-2-one)

$CH_3COCH_2CH_2CH_3$

(1) Colourless liquid, very slightly soluble in water, miscible with alcohol and ether.

(2) Gives crystalline bisulphite addition compound with sodium metabisulphite.

(3) Add alkaline sodium nitroprusside solution to 1ml of compound. A red coloured solution is obtained

(4) To a mixture of 5ml of compound and 15ml sodium hydroxide

solution add iodine-KI reagent drop by drop till the solution is yellow. Heat on water bath for 2-3 min, cool and collect iodoform, m.p 119°C.

(5) Methyl n-Propyl ketone 2, 4 dinitrophenylhydrazone, m.p 142°C.

(6) Methyl n-Propyl ketone semicarbazone, m.p 110°C. (7) Methyl n-Propyl ketone p-nitro phenylhydrazone, m.p 117°C. (8) Methyl n-Propyl ketoxime, m.p 58°C.

b.p 106°C **Pinacolone** (t-butyl methyl ketone, 3, 3-dimethyl-2-butanone) $(CH_3)_3CCOCH_3$

(1) Colourless oily liquid with characteristic peppermint like odour, soluble in alcohol and ether.

(2) Add alkaline sodium nitroprusside solution to 1ml of compound. A red coloured solution is obtained

(3) Cool NaOH solution to 0°C (1.6gm in 14ml water) and add 1ml bromine with stirring (temp below 10°C). Again cool to 0°C and add 1gm pinacolone and continue stirring for 3hrs. Steam distill to remove bromoform. Cool the reaction mixture and add 2ml conc H_2SO_4. Distil and collect trimethyl acetic acid which on cooling soon solidifies, m.p 34-35°C.

(4) Pinacolone 2, 4 dinitrophenylhydrazone, m.p 125°C

(5) Pinacolone semicarbazone, m.p 157°C.

(6) Pinacolone oxime, m.p 78°C (7) Benzylidene, m.p 41°C.

b.p 119°C **Isobutyl methyl ketone** (4-methyl-2-Pentanone) $(CH_3)_2CHCH_2COCH_3$

(1) Colourless liquid, with camphor like odour, insoluble in water, soluble in alcohol and ether.

(2) Gives crystalline bisulphite addition compound with sodium metabisulphite

(3) Add alkaline sodium nitroprusside solution to 1ml of compound. A red coloured solution is obtained

(4) To a mixture of 5ml of compound and 15ml sodium hydroxide solution add iodine-KI reagent drop by drop till the solution is yellow. Heat on water bath for 2-3 min, cool and collect iodoform, m.p 119°C.

(5) Isobutyl methyl ketone 2, 4 dinitrophenylhydrazone, m.p 95°C.

(6) Isobutyl methyl ketone semicarbazone, m.p 132°C. (7) Isobutyl methyl ketone p-nitrophenylhydrazone, m.p 79°C. (8) Isobutyl methyl ketoxime, m.p 58°C.

b.p 127°C **n-Butyl methyl ketone** (Hexanone-2)

$CH_3CH_2CH_2CH_2COCH_3$

(1) Colourless liquid, soluble in alcohol and ether, very slightly soluble in water.
(2) Gives crystalline bisulphite addition compound with sodium metabisulphite
(3) n-Butyl methyl ketone 2, 4-dinitrophenylhydrazone, m.p 106°C
(4) n-Butyl methyl ketone semicarbazone, m.p 122°C (5) n-Butyl methyl ketone p-nitrophenylhydrazone, m.p 88°C.

b.p 131°C **Mesityl Oxide** (4 methyl pent-3-en-2-one)

$(CH_3)_2C{=}CHCOCH_3$

(1) Colourless oily liquid with peppermint like odour, sparingly soluble in water, easily soluble alcohol and ether
(2) To few ml of compound add dil H_2SO_4, pleasant odour due to formation of acetone is observed.
(3) Gives crystalline bisulphite addition compound with sodium metabisulphite
(4) To 1ml of ketone add bromine solution. Deep brown colour of bromine disappears.
(5) To 1.2gm of sodium add 20ml absolute ethanol and stir. After the sodium has dissolved add with stirring 8.5gm of diethyl malonate and 5gm of mesityl oxide. Reflux for 2hrs. Add a solution of 6gm KOH in 30ml water and reflux again on water bath for 6hrs. Acidify with dil HCl and distill off excess alcohol. Cool and recover dimedone, m.p 147-148°C.
(6) To ice cold 11ml of Ac_2O add with stirring 3.5ml of perchloric acid (temp below 8°C). Cool and again add with stirring 5ml of mesityl oxide followed by 7ml of Ac_2O. Heat the mixture on water bath at 50-70°C for 15-20 minutes. Cool and collect 2, 4, 6 trimethylpyrylium perchlorate, m.p 240°C (d).
(7) To a mixture of 5ml of compound and 15ml sodium hydroxide solution add iodine –KI reagent drop by drop till the solution is

yellow. Heat on water bath for 2-3 min, cool and collect iodoform, m.p 119°C.

(8) Mesityl Oxide 2, 4-dinitrophenylhydrazone, m.p 203°C.

(9) Mesityl Oxide semicarbazone, m.p 164°C.

(10) Mesityl Oxide p-nitrophenylhydrazone, m.p 134°C.

(11) Mesityl Oxide oxime (β), m.p 49°C.

(12) Mesityl oxide phenylhydrazone, m.p 142°C.

b.p 131°C **Cyclopentanone** 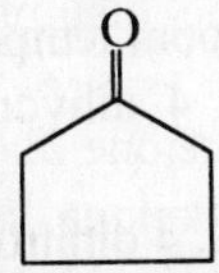

(1) Colourless liquid with peppermint like odour, soluble in water, miscible with alcohol and ether.

(2) Gives crystalline bisulphite addition compound with sodium metabisulphite

(3) To 2ml of cyclopentanone add 4ml of benzaldehyde, 3ml of alcohol and 4ml of NaOH. Heat the mixture on water bath for 2-3min. Cool and filter the yellow precipitate of 1, 3 dibenzalcyclo-pentanoné-2, m.p 189°C.

(4) Cyclopentanone 2, 4-dinitrophenylhydrazone, m.p 146°C.

(5) Cyclopentanone semicarbazone, m.p 206°C

(6) Cyclopentanone p-nitrophenylhydrazone, m.p 154°C.

(7) Cyclopentanone oxime, m.p 57°C.

(8) Cyclopentanone phenylhydrazone, m.p 55°C.

b.p 139°C **Acetyl acetone** (Pentane 2, 4-dione)

$CH_3COCH_2COCH_3$

(1) Colourless liquid miscible with water, alcohol and ether, odour similar to acetone.

(2) To 3ml aqueous solution of compound add 3-4 drops of freshly prepared $FeCl_3$ solution, an orange red coloured solution is produced.

(3) To ice cold solution of 3.2 gm of hydrazine hydrate in 20ml dil NaOH add with stirring 3ml of acetyl acetone (temp below 15°C). Stir the mixture at 15°C for 1hr. Cool and collect the precipitated 3, 5 dimethyl pyrazole, m.p 107-108°C.

(4) Heat a mixture of 5 gm $CaSO_4$, 2.5 gm p-toluidine and 3ml of acetyl acetone on steam bath for 1hr. Cool and extract with 25ml of ether. Evaporate off the ether and recover 2, 4, 6 trimethyl quinoline, m.p 67-69°C.

(5) Reflux 2.5ml of compound with 2.25gm of cyanoacetamide, 18ml ethanol and 1.2ml piperidine for 30 min. Cool and filter the precipitated 3-cyano-4, 6 dimethyl –2-pyridine, recrystallise from alcohol, m.p 299-300°C.

(6) To a solution of 4gm of ammonium carbonate in 40ml water add 5ml acetyl acetone, 2ml ethanol and 1.2ml acetaldehyde. Keep the mixture at room temperature for 4 days. Filter the precipitated 3, 5 diacetyl-1, 4 dihydro-2, 4, 6 trimethyl pyridine, m.p 152-153°C.

(7) Acetyl acetone 2, 4 dinitrophenylhydrazone, m.p 209°C. (8) Acetyl acetone dioxime, m.p 149°C.

b.p 148°C **Acetoin** (Acetyl methyl carbinol, 3-hydroxy butan-2-one)
m.p 15°C $CH_3CHOHCOCH_3$

(1) Hygroscopic liquid, soluble in water and alcohol, insoluble in ether.

(2) Reduces Tollens reagent.

(3) Reduces Fehling's solution.

(4) Gives crystalline bisulphite addition compound with sodium metabisulphite

Biacetyl bis 2, 4-dinitrophenylhydrozone, m.p 318°C

Acetoin semicarbazone, m.p 185°C

Biacetyl phenylosazone, m.p 243°C

(8) Iodoform, m.p 119°C.

b.p 151°C **n-Amyl methyl ketone** (Heptan-2 one)

$CH_3(CH_2)_4COCH_3$

(1) Colourless liquid, soluble in alcohol and ether.

(2) Gives crystalline bisulphite addition compound with sodium metabisulphite

(3) n-Amyl methyl ketone 2, 4-dinitrophenylhydrazone, m.p 89°C.

(4) n-Amyl methyl ketone semicarbazone, m.p 123°C (4) n-Amyl methyl ketone phenylhydrazone, m.p 207°C.

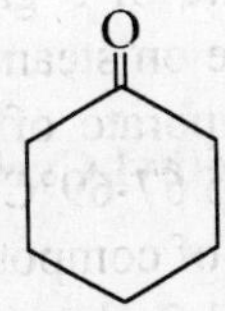

b.p 155°C **Cyclohexanone**

(1) Colourless liquid, slightly soluble in water, soluble in alcohol and ether.

(2) Heat 5ml of compound with 10ml of dil HNO_3 for 5-10 minutes. Cool and collect adipic acid, m.p 150°C.

(3) To 2ml of compound add 2ml alcohol, 5ml benzaldehyde and 2ml of dil NaOH solution. Shake the mixture thoroughly and keep it at room temperature for 5-10min. Cool and collect yellow crystals of 1, 3 dibenzalcyclohexanone-2, m.p 118°C.

(4) To a mixture of 30ml benzene, 2.8gm KOH and 3.8 gm CaC_2add with stirring and dropwise 5ml of ketone. Continue stirring for 2-4hrs. Cool, add dil HCl and collect 1, 1'ethylene-bis cyclohexanol, m.p 110°C.

(5) Heat under reflux 5ml of compound with 10ml glacial acetic acid. Add dropwise 5ml of phenylhydrazine and continue reflux for 1hr. Cool and collect 1, 2, 3, 4 tetrahydrocarbazole, m.p 116-117°C.

(6) Reflux 5ml cyclohexanone with 6.5ml morpholine and 100mg p-toluenesulphonic acid dissolved in 20ml benzene for 3hrs. Carefully evaporate off benzene and morpholine and distil the residue under reduced pressure and collect 1-morpholino-cyclohexene at b.p 121-122 °C/12mmHg.

(7) To a solution of 3ml of compound in 11ml conc HCl add 2gm of sodium azide. Keep the mixture at room temperature for several hours and then evaporate under reduced pressure. Dissolve the residue in water and make it alkaline with NaOH solution. Extract with chloroform. Evaporate off the solvent and crystallise the precipitated caprolactam from benzene, m.p 64°C.

(8) Cyclohexanone 2, 4-dinitrophenylhydrazone, m.p 160°C.

(9) Cyclohexanone semicarbazone, m.p 166°C.

(10) Cyclohexanone p-nitrophenylhydrazone, m.p 146°C.

(11) Cyclohexanone phenylhydrazone, m.p 81°C.

(12) Cyclohexanone oxime, m.p 91°C.

b.p 163°C **2-Methyl Cyclohexanone**

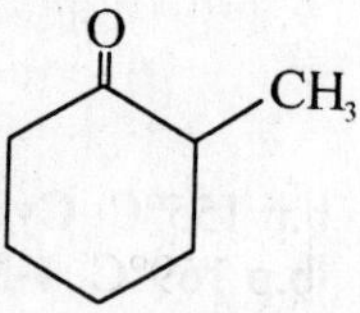

(1) Liquid, easily soluble in alcohol and ether, insoluble in water.
(2) Gives crystalline bisulphite addition compound with sodium metabisulphite
(3) Heat 7.5gm of compound with 10ml benzaldehyde, 40ml water, 25ml alcohol and NaOH on steam bath for 3hrs. Extract with ether. Distil of ether and collect 2-benzylidene-6-methylcycohexanone at b.p 155-160°C /4mmHg which on cooling soon solidifies, m.p 61-62°C.
(4) 2-Methyl Cyclohexanone 2, 4 dinitrophenylhydrazone, m.p 136°C.
(5) 2-Methyl Cyclohexanone oxime, m.p 43°C
(6) 2-Methyl Cyclohexanone semicarbazone, m.p 197°C.
(7) 2-Methyl Cyclohexanone p-nitrophenylhydrazone, m.p 132°C.

b.p 166°C **Diacetone Alcohol** (4-hydroxy 4-methyl Pentan-2-one)
$CH_3COCH_2C(OH)(CH_3)_2$

(1) Colourless liquid, miscible with water, alcohol and ether.
(2) Reduces Fehling's solution.
(3) To 5ml of compound add dil NaOH solution and heat. Sweet smell due to formation of acetone is observed.
(4) Distil 50ml diacetone alcohol containing 50mg of iodine through a fractionating column. Collect mesityl oxide at b.p 127-131°C.
(5) Diacetone Alcohol oxime, m.p 58°C.
(6) Diacetone 2, 4-dinitrophenylhydrazone, m.p 202°C.

b.p 168°C **Diisobutyl ketone** (2, 6 dimethylheptanone-4)
$[(CH_3)_2CHCH_2]_2CO$

(1) Oily liquid, insoluble in water, miscible with alcohol and ether.
(2) Diisobutyl ketone 2, 4 dinitrophenylhydrazone, m.p 66°C.
(3) Diisobutyl ketone semicarbazone, m.p 122°C.

b.p 169°C **4-Methylcyclohexanone** 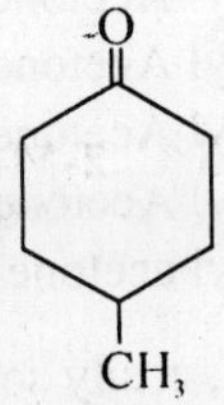

(1) Insoluble in water, soluble in alcohol and ether.
(2) 4-Methylcyclohexanone p-nitrophenyl hydrazone, m.p 128°C
(3) 4-Methylcyclohexanone semicarbazone, m.p 199°C
(4) 4-Methylcyclohexanone phenylhydrazone, m.p 110°C
(5) 4-Methylcyclohexanone oxime, m.p 39°C.

b.p 173°C **n-Hexyl Methyl ketone** (Octanone –2)

$CH_3(CH_2)_5COCH_3$

(1) Colourless liquid, insoluble in water, miscible with alcohol and ether.
(2) Gives crystalline bisulphite addition compound with sodium metabisulphite.
(3) n-Hexyl Methyl ketone 2, 4 dinitrophenylhydrazone, m.p 58°C.
(4) n-Hexyl Methyl ketone semicarbazone, m.p 123°C. (5) n-Hexyl Methyl ketone p-nitrophenylhydrazone, m.p 93°C.

b.p 194°C **Acetonyl Acetone** (Hexane –2, 5-dione)

$CH_3COCH_2CH_2COCH_3$

(1) Colourless liquid, miscible with water, alcohol and ether.
(2) Reflux 5.4ml of compound with 10gm of ammonium carbonate at 100°C for 90min. Cool and add 5ml water and again heat for 30min. Remove the upper organic layer and extract the aqueous layer with chloroform. Combine the two organic layers, distil and collect 2, 5 dimethylpyrrole at b.p 165-168°C.
(3) Heat 2.5ml of compound with 5.6gm of phosphorous pentasulphide. When the exothermic reaction has started remove the heat source and add in portion 10ml of acetonyl acetone such that the mixture boils. Finally heat the mixture under reflux for 1hr. Cool, extract with ether, distil off ether and collect the foul smelling 2, 5 dimethylthiophen, b.p 134-136°C.

(4) Acetonyl Acetone 2, 4-dinitrophenylhydrazone (di), m.p 257°C
(5) Acetonyl Acetone dioxime, m.p 138°C.
(6) Acetonyl Acetone bis-phenylhydrazone, m.p 120°C.
(7) Acetonyl Acetone semicarbazone(di), m.p 223°C
(8) Acetonyl acetone p-nitrophenylhydrazone (di), m.p 210°C.

b.p 207°C **(-) Menthone**

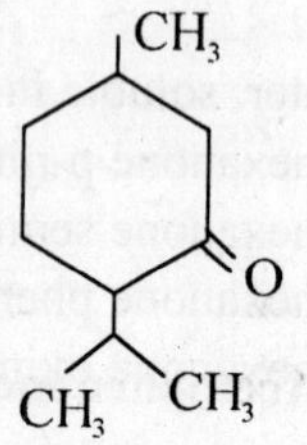

(1) Colourless liquid with peppermint like odour, slightly slouble in water, easily soluble in alcohol and ether
(2) (-) Menthone 2, 4-dinitrophenylhydrazone, m.p 146°C.
(3) (-) Menthone semicarbazone, m.p 184°C.
(4) (-) Menthone phenylhydrazone, m.p 53°C.
(5) (-) Menthone oxime, m.p 59°C.

b.p 215°C **Isophorone** (1, 1, 3-trimethylcyclohexene-3-one-5)

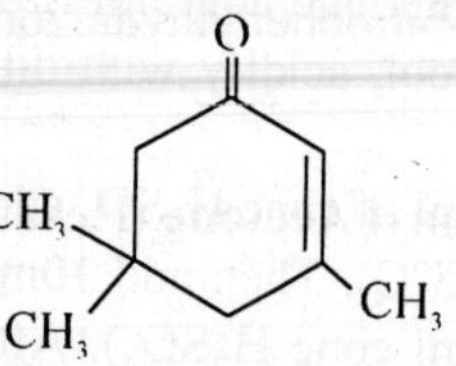

(1) Liquid with charcteristic peppermint like odour, insoluble in water.
(2) Isophorone semicarbazone, m.p 199°C. (3) Isophorone phenylhydrazone, m.p 68°C. (4) Isophorone oxime, m.p 76°C.

b.p 230°C **(+)Carvone**

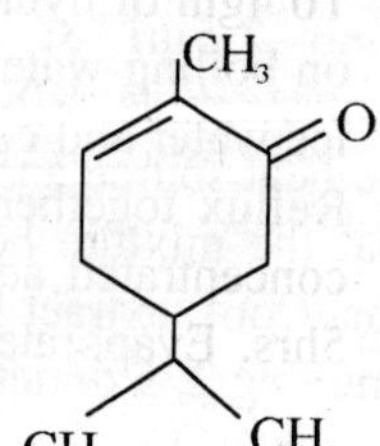

(1) Oily liquid with characteristic spearmint odour, very slightly soluble in water, miscible with alcohol and ether.

(2) (+)Carvone 2, 4 dinitrophenylhydrazone, m.p 191°C (3) (+) Carvone semicarbazone, m.p 162°C (4) (+) Carvone p-nitrophenylhydrazone, m.p 175°C (5) (+) Carvone phenylhydrazone, m.p 110°C (6) (+) Carvone oxime, m.p 73°C.

Aromatic liquids

b.p 202°C **Acetophenone**

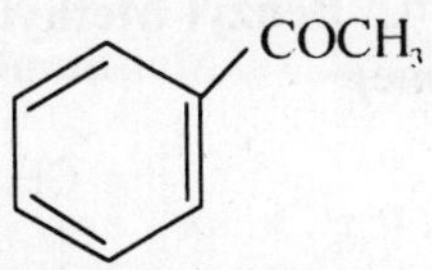

(Acetyl benzene,
Phenyl methyl ketone)

(1) Colourless pleasant smelling hypnotic liquid, insoluble in water, easily soluble in alcohol, ether, chloroform and benzene.

(2) To 2ml aqueous solution of ketone add 3-4 drops of sodium nitroprusside solution and 3-4 drops of dil NaOH. A wine red coloured solution which turns blue on acidification with acetic acid is obtained.

(3) To 5ml of compound add 5ml KOH and 5ml $KMnO_4$. Reflux for 15-20 min. Cool, acidify with dil HCl and collect benzoic acid, m.p 121°C.

(4) To cooled 15ml of conc H_2SO_4 add dropwise 5ml of acetophenone (temp below 5°C). Then add 10ml of nitrating mixture(4ml conc HNO_3 and 6ml conc H_2SO_4). After addition is complete stir the mixture for 10 min, pour in ice and collect m-nitro acetophenone, m.p 76-78°C.

(5) To 5ml acetophenone add 4.2 gm of phenylhydrazine, 15ml ethanol and few drops of glacial acetic acid. Heat the mixture on water bath for few minutes, cool and collect the hydrazone, m.p 105°C. To 4gm of hydrazone add 25gm of polyphosphoric acid and heat on boiling water for 10-15min(exothermic reaction). Add 100ml ice water and collect 2-phenyl indole, m.p 188-189°C.

(6) Reflux together 12.5ml acetophenone with 18gm sulphur, 25ml concentrated aqueous ammonia and 15ml pyridine at 160°C for 5hrs. Evaporate off the reaction mixture and extract the residue

with boiling water. Concentrate and cool, phenylacetamide separate out, m.p 157-158°C.

(7) Acetophenone 2, 4-dinitrophenylhydrazone, m.p 249°C.
(8) Acetophenone semicarbazone, m.p 198°C.
(9) Acetophenone p nitrophenylhydrazone, m.p 184°C.
(10) Acetophenone phenylhydrazone, m.p 105°C.
(11) Acetophenone oxime, m.p 59°C.
(12) Iodoform, m.p 119°C

b.p 216°C **Benzyl Methyl ketone** (Phenyl acetone, 1-phenyl-2-propanone)

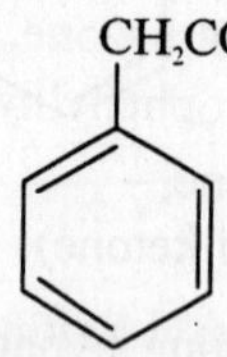

(1) Colourless liquid, insoluble in water, soluble in alcohol and ether.
(2) Gives crystalline bisulphite addition compound with sodium metabisulphite.
(3) Benzyl Methyl ketone 2, 4-dinitrophenylhydrazone, m.p 156°C.
(4) Benzyl Methyl ketoxime, m.p 70°C.
(5) Benzyl Methyl ketone p-nitrophenylhydrazone, m.p 145°C.
(6) Benzyl Methyl ketone semicarbazone, m.p 199°C.
(7) Benzyl Methyl ketone phenyhydrazone, m.p 87°C. (8) Iodoform, m.p 119°C

b.p 218°C, m.p 21°C **Ethyl phenyl ketone** (Propiophenone)

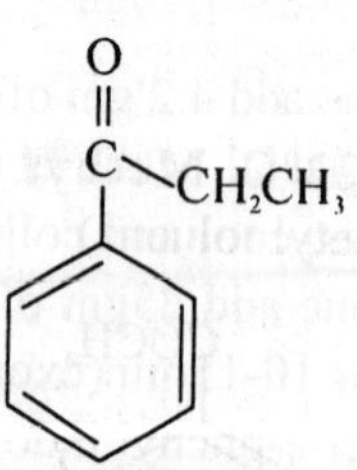

(1) Colourless liquid, insoluble in water, soluble in alcohol and ether.
(2) To a solution of of 5.2ml of compound in 25ml ether pass HCl gas and simultaneously add in portion 4.6gm of isopently nitrite.

After addition is complete continue passing HCl for another 15min and then keep overnight. Pour with stirring into 10% of 50ml of cold NaOH solution. Retain the aqueous layer and extract the ethereal layer with dil NaOH solution. Combine the extracts with aqueous layer, cool and acidify with conc HCl. Collect the precipitated isonitrosopropiophenone, crystallise from benzene, m.p 115°C.

(3) Reflux 10gm of compound with 10gm of KOH, 75ml triethylene glycol and 7.5ml of 85% hydrazine hydrate for 4hrs. Extract with ether. Distil and collect n-propylbenzene at b.p 162-163°C.
(4) Ethyl phenyl ketone 2, 4-dinitrophenyl hydrazone, m.p 191°C.
(5) Ethyl phenyl ketoxime, m.p 53°C.
(6) Ethyl phenyl ketone semicarbazone, m.p 178°C.
(7) Ethyl phenyl ketone p-nitrophenylhydrazone, m.p 147°C.
(8) Iodoform, m.p 119°C.

b.p 218°C **Isopropyl Phenyl ketone** (Isobutyrophenone)

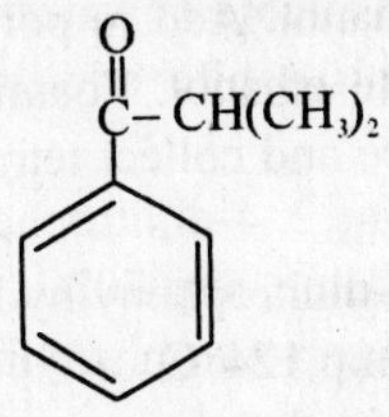

(1) Colourless liquid, insoluble in water, soluble in alcohol and ether.
(2) Isopropyl Phenyl ketone 2, 4-dinitrophenylhydrazone, m.p 163°C
(3) Isopropyl Phenyl ketone semicarbazone, m.p 181°C
(4) Isopropyl Phenyl ketone phenylhydrazone, m.p 73°C

Solid Ketones

Aromatic

m.p 28°C, b.p 222°C **Methyl p-tolyl ketone** (p-Methyl Acetophenone, p-acetyl toluene)

(1) Colourless crystalline solid, insoluble in water, soluble in alcohol and ether.
(2) Methyl p-tolyl 2, 4 dinitrophenylhydrazone, m.p 258°C (3) Methyl p-tolyl semicarbazone, m.p 205°C
(4) Methyl p-tolyl phenylhydrazone, m.p 95°C (5) Methyl p-tolyl p-nitrophenylhydrazone, m.p 198°C.
(6) Methyl p-tolyl ketoxime, m.p 88°C

m.p 34°C, b.p 330°C **Dibenzyl ketone** (Diphenyl acetone, 1, 3 diphenyl 2-propanone)

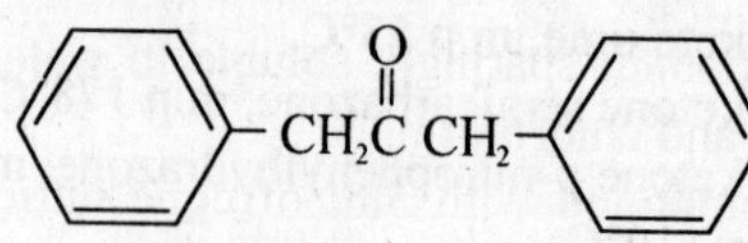

(1) Colourless crystalline solid, insoluble in water, soluble in alcohol and ether.
(2) Heat on water bath a mixture of 2.1gm of compound and 2.1gm of benzil in 15ml ethanol. Add in portion a solution of 0.3gm of KOH in 3ml absolute ethanol. Foaming takes place. Reflux for 15-20 min, cool in ice and collect tetraphenyl cyclopentadienone, m.p 217-220°C.
(3) Dibenzyl ketone 2, 4-dinitrophenylhydrazone, m.p 100°C.
(4) Dibenzyl ketoxime, m.p 124°C.
(5) Dibenzyl ketone semicarbazone, m.p 146°C. (6) Dibenzyl ketone phenylhydrazone, m.p 129°C.

m.p 38°C, b.p 258°C **p-Methoxy Acetophenone** (p-acetyl anisole)

(1) Crystalline solid, slightly soluble in water, soluble in alcohol and ether.
(2) p-Methoxy Acetophenone 2, 4-dinitrophenylhydrazone, m.p 220°C.

(3) p-Methoxy Acetophenone oxime, m.p 87°C.
(4) p-Methoxy Acetophenone p-nitrophenylhydrazone, m.p195°C.
(5) p-Methoxy Acetophenone phenylhydrazone, m.p 142°C.
(6) p-Methoxy Acetophenone semicarbazone, m.p 196°C.

m.p 41°C **1-Indanone** (α-hydrindrone)

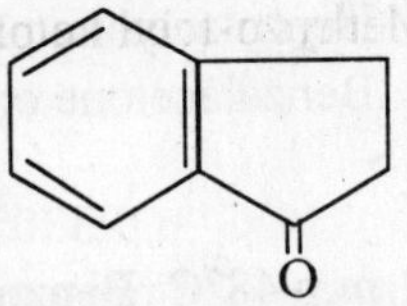

(1) Crystalline compound, sparingly soluble in water, easily souble in alcohol, ether and chloroform.
(2) Heat 1gm of compound with 5ml of conc nitric acid on water bath. Cool and pour in ice and collect phthalic acid, m.p 206-208°C (d).
(3) 1-Indanone 2, 4-dinitrophenylhydrazone, m.p 258°C.
(4) 1-Indanone semicarbazone, m.p 233°C.
(5) 1-Indanone p-nitrophenylhydrazone, m.p 234°C.
(6) 1-Indanone oxime, m.p 145°C.

m.p 42°C, b.p 262°C **Benzalacetone** (Benzylidene acetone, 4-Phenyl but-3-ene-2-one)

$CH = CHCOCH_3$

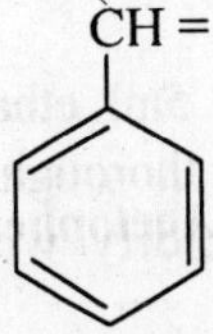

(1) Yellow coloured solid, insoluble in water, easily soluble in alcohol, ether and chloroform.
(2) Dissolve few milligrams of compound in conc sulphuric acid, orange red coloured solution is obtained.
(3) Heat few milligrams of compound with acidic $KMnO_4$. Smell of bitter almond is observed due to formation of benzaldehyde.
(4) Dissolve 1gm of compund in 3ml chloroform and cool. Add with stirring 1 ml of bromine in 3ml chloroform and shake thoroughly till deep brown colour of bromine disappears. Cool, pour in ice water and collect benzalacetone dibromide, m.p 125°C.

(5) To a cooled solution of 1gm of NaOH in 10ml water and 8ml alcohol add 2ml benzaldehyde and 0.75gm of benzalacetone. Stir for 60min, cool and collect dibenzalacetone, m.p 112°C.
(6) Benzalacetone 2, 4-dinitrophenylhydrazone, m.p 227°C.
(7) Benzalacetone semicarbazone, m.p 186°C.
(8) Benzalacetone p-nitrophenylhydrazone, m.p 167°C.
(9) Benzalacetone phenylhydrazone, m.p 157°C.
(10) Benzalacetone oxime, m.p 115°C.

m.p 48°C **Benzophenone** (diphenyl ketone)

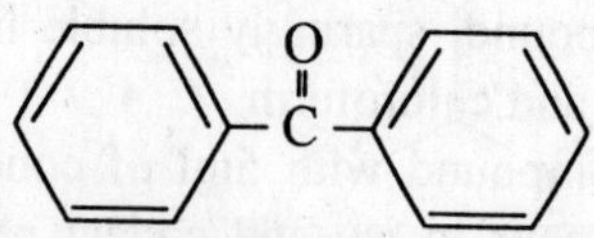

(1) Colourless crystalline pleasant smelling solid, insoluble in water, soluble in alcohol and ether.
(2) Dissolve few milligrams of compound in conc sulphuric acid, yellow coloured solution is obtained.
(3) Fuse 50mg of compound with a small piece of sodium metal, deep blue colour is generated.
(4) Dissolve 1.5gm of benzophenone in 8ml of isopropyl alcohol containing a drop of glacial acetic acid. Keep the reaction mixture in direct sunlight. Benzopinacol precipitates out in 2-3 days, m.p 190°C.
(5) To 500mg of compound add 5ml ethanol, 500mg NaOH and 500mg Zn. Shake the mixture thoroughly and keep it at RT for 3 hrs. Pour in ice water, acidify with dil HCl and collect benzhydrol, m.p 68°C.
(6) To ice cold solution of 2.5gm of benzophenone in 13ml alcohol pass H_2S and HCl gas simultaneously for 3hrs. Then pass only the H_2S gas for 20hrs with continued ice cooling. The solution turns intense violet and thiobenzophenone is precipitated, m.p 54°C.
(7) Heat 9gm of compound with 10gm bromoacetic acid, 4gm Zn in 45ml benzene. As soon as the reaction commenses remove the heat source (take care to prevent the reaction to become too violent). After the reaction has subsided reflux for 45min and add ice cold H_2SO_4 to it. Separate the benzene layer, evaporate off

the solvent and collect ester of β-hydroxy β, β-diphenyl propionic acid, recrystallise from alcohol, m.p 87°C.

(8) Benzophenone 2, 4-dinitrophenylhydrazone, m.p 238°C.
(9) Benzophenone semicarbazone, m.p 164°C.
(10) Benzophenone p-nitrophenylhydrazone, m.p 154°C.
(11) Benzophenone phenylhydrazone, m.p 137°C.
(12) Benzophenone oxime, m.p 142°C.

m.p 55°C **Methyl 2-naphthyl ketone** (2-acetyl naphthalene)

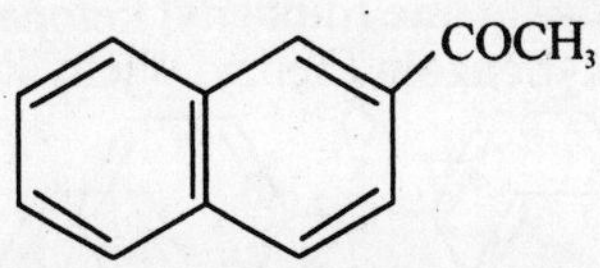

(1) Crystalline compound, sparingly soluble in water, easily soluble hot alcohol and ether
(2) Heat 1gm of compound with 5ml of dil nitric acid on water bath. Cool, pour in ice and collect 2-naphthoic acid, m.p 185°C.
(3) Methyl 2-naphthyl ketone 2, 4-dinitrophenylhydrazone, m.p 262°C(d). (4) Methyl 2-naphthyl ketone semicarbazone, m.p 235°C. (5) Methyl 2-naphthyl ketone phenylhydrazone, m.p 171°C. (6) Methyl 2-naphthyl ketoxime, m.p 145°C. (7) Picrate, m.p 85°C.

m.p 55-57°C **Benzalacetophenone** (Chalcone)

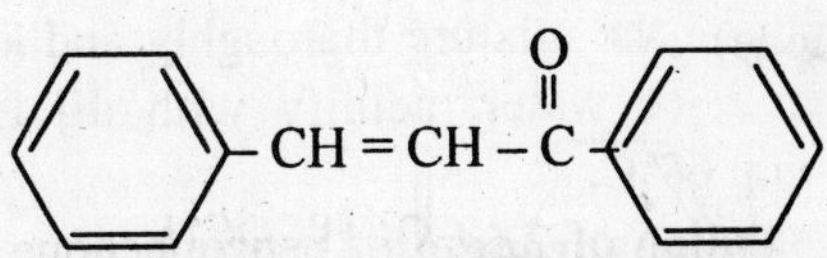

(1) Yellow coloured solid, insoluble in water, sparingly soluble in alcohol, easily soluble in ether and chloroform
(2) Dissolve few milligrams of compound in conc sulphuric acid, intense yellow coloured solution is obtained
(3) Dissolve 100mg of compound in carbon tetrachloride and add 0.5ml of bromine dissolved in carbon tetrachloride. Shake thoroughly and pour in ice water. Crystals of benzalacetophenone dibromide separate out, m.p 158°C.
(4) Cool a mixture of 17ml benzene and 1.6gm of $AlCl_3$ to 10°C.

Add a solution of 1.2gm of compound in 3ml benzene. Remove the ice bath and stir the mixture for 30minutes. Decant the benzene layer in beaker containing 1ml HCl and 10ml water. Steam distil the resultant solution to remove benzene. Cool the residual oily portion and collect ββ-diphenyl propiophenone, m.p 91-92°C.

(5) Benzalacetophenone 2, 4 dinitrophenylhydrazone, m.p 244°C(d)
(6) Benzalacetophenone semicarbazone, m.p 168°C(180)
(7) Benzalacetophenone phenylhydrazone, m.p 118°C.

m.p 60°C **Deoxybenzoin** (benzyl phenylketone)

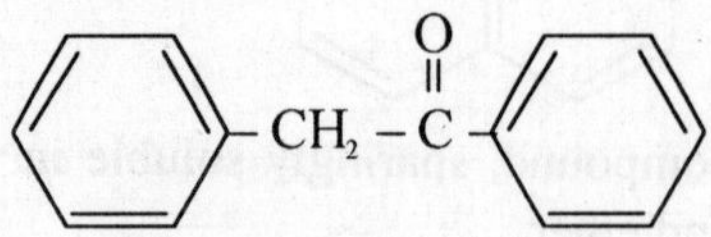

(1) Crystalline compound, insoluble in water, soluble in alcohol and ether
(2) Deoxybenzoin 2, 4-dinitrophenylhydrazone, m.p 204°C
(3) Deoxybenzoin semicarbazone, m.p148°C.
(4) Deoxybenzoin p-nitrophenylhydrazone, m.p163°C.
(5) Deoxybenzoin phenylhydrazone, m.p 116°C.
(6) Deoxybenzoin oxime, m.p 98°C. (7) Benzylidene, m.p 102°C.

m.p 61°C **Benzoylacetone** (Methyl phenacyl ketone, 1-phenyl 1, 3-butanedione)

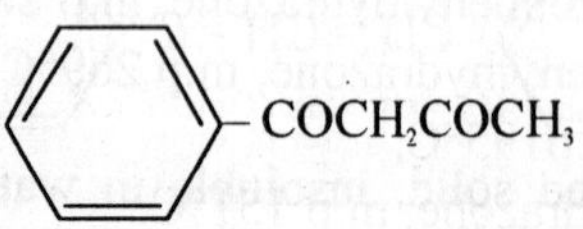

(1) Crystalline compound with persistent penetrating odour, soluble in alcohol and ether. Exhibits keto-enol tautomerism.
(2) To few milligrams of compound add $FeCl_3$ solution. An intense red coloured solution is obtained.
(3) Benzoyl acetone 2, 4-dinitrophenylhydrazone, m.p 151°C
(4) Benzoyl acetone p-nitrophenylhydrazone, m.p 100°C
(5) Iodoform, m.p 119°C.

m.p 68°C **Cinnamalacetone**

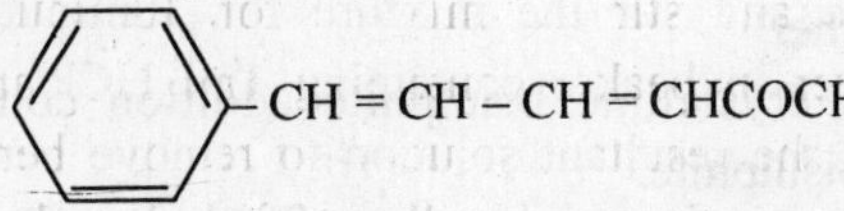

(1) Crystalline compound, insoluble in water, soluble in alcohol, ether and benzene
(2) Dissolve few milligrams of compound in conc sulphuric acid, yellow coloured solution is obtained.
(3) Cinnamalacetone 2, 4-dinitrophenylhydrazone, m.p 220°C.
(4) Cinnamalacetone semicarbazone, m.p 186°C.
(5) Cinnamalacetone phenylhydrazone, m.p 180°C.
(6) Cinnamalacetone oxime, m.p 152°C.

m.p 84°C **Fluorenone** (Diphenylene ketone)

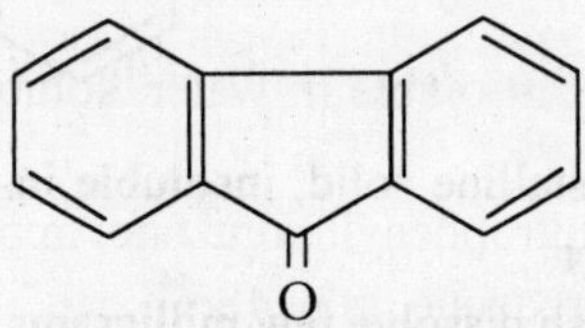

(1) Bright yellow coloured solid, insoluble in water, soluble in alcohol.
(2) Dissolve few milligrams of compound in conc sulphuric acid. A deep reddish violet coloured solution is obtained
(3) Reflux 1.5gm of fluorenone with 22ml of fuming HNO_3 and 20ml of conc H_2SO_4 for 9hrs. Keep the reaction mixture overnight and then pour in ice. Collect light yellow crystals of 2, 4, 5, 7 tetranitrofluorenone, m.p 253-254°C
(4) Fluorenone 2, 4-dinitrophenylhydrazone, m.p 283°C.
(5) Fluorenone p-nitrophenylhydrazone, m.p 269°C.
(6) Fluorenone oxime, m.p195°C.
(7) Fluorenone phenylhydrazone, m.p 151°C.

m.p 84°C α **hydroxy acetophenone** (phenacyl alcohol)

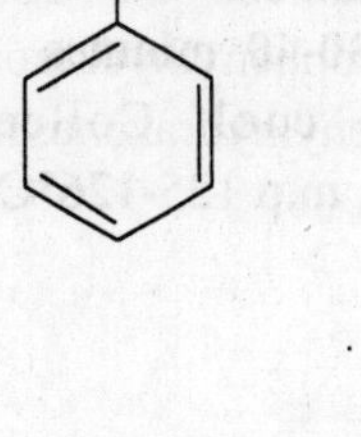

(1) Colourless crystalline compound, soluble in hot water, alcohol and ether
(2) Gives crystalline bisulphite addition compound with sodium metabisulphite.
(3) Reduces Tollens reagent.
(4) Reflux 500mg of compound with 5ml Ac_2O at 100°C. Cool, pour in ice and collect phenacyl acetate, m.p 49°C.
(5) α hydroxy acetophenone semicarbazone, m.p 146°C. (6) α hydroxy acetophenone oxime, m.p 70°C
(7) α hydroxy acetophenone phenylhydrazone, m.p 112°C.

m.p 95°C **Benzil** (Bibenzoyl)

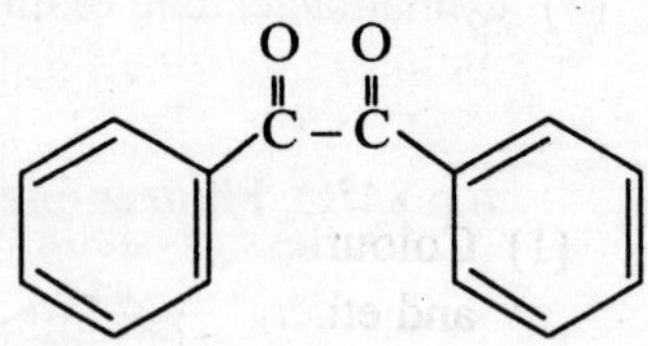

(1) Pale yellow crystalline solid, insoluble in water, easily soluble in alcohol and ether
(2) In a porcelain dish dissolve few milligrams of compound in alcohol and add a pellet of KOH to it. Heat the dish on water bath. A purple red coloured solution is produced.
(3) To 3.5 gm of compound add 3-4ml of KOH solution(3.5gm in 7ml water) and 5ml of alcohol. Boil the solution for 10-15minutes. Cool and acidify with dil H_2SO_4. Intense red colouration is observed and benzilic acid is precipitated, m.p 150°C.
(4) Reflux 200mg of compound with alkaline $KMnO_4$ for 10 min. Cool, acidify with dil HCl and collect white crystals of benzoic acid, m.p121°C.
(5) Reflux together 2.5gm of benzil with 1.5gm of urea, 7.5ml of 30% NaOH solution and 25ml of ethanol for 2hrs. Cool, pour in 50ml water and filter. Acidify the filtrate with conc HCl and collect 5, 5 diphenyl hydantoin, m.p 297-298°C.
(6) To a warm solution of 1gm of benzil in 5ml ethanol add a solution of 0.5gm of o-phenylene diamine in 5ml ethanol. Warm the mixture on water bath for 30-40 minutes. Add water until cloudiness persists and then cool. Collect the precipitated 2, 3 diphenylquinoxaline, m.p 125-126°C.

(7) Benzil 2, 4 dinitrophenylhydrazone(di), m.p189°C. (8) Benzil bis-semicarbazone, m.p 243°C.

(9) Benzil bis p-nitrophenylhydrazone, m.p 290°C (10) Benzil bis phenylhydrazone, m.p 235°C. (11) Benzil (di) oxime, m.p 237°C.

m.p 107°C **p-hydroxy Acetophenone** (p-acetyl phenol)

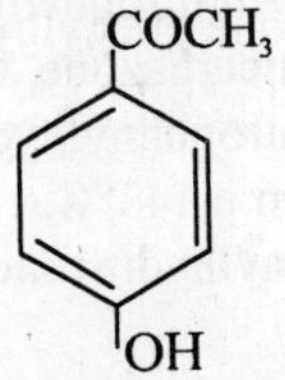

(1) Colourless crystalline solid, insoluble in water, soluble in alcohol and ether.

(2) To alcoholic solution of compound add 1-2 drops of neutral $FeCl_3$ solution. Red to violet coloured solution is obtained.

(3) Heat 500mg of compound with 5ml Ac_2O and 200mg of sodium acetate. Cool pour in ice water with stirring and collect p-acetoxy acetophenone, m.p 58°C.

(4) Shake thoroughly for 15-20 minutes, 500mg of compound with 10ml of NaOH solution and 6ml of benzoyl chloride. Cool and collect the benzoyl derivative, m.p 135°C.

(5) Reflux 5gm of compound with 10gm amalgamated zinc and 50ml dil HCl until the evolution of hydrogen commences. Add few drops of conc HCl during this period. Steam distil, saturate the distillate with potassium carbonate and isolate with ether. Distil off ether, cool the residue and collect p-ethylphenol, m.p 45°C.

(6) p-hydroxy Acetophenone 2, 4-dinitrophenylhydrazone, m.p 261°C.

(7) p-hydroxy Acetophenone oxime, m.p 145°C.

(8) p-hydroxy Acetophenone semicarbazone, m.p 199°C.

(9) p-hydroxy Acetophenone phenylhydrazone, m.p 151°C.

m.p 112°C **Dibenzalacetone** (1, 5 diphenylpenta-1, 4-dien-3-one)

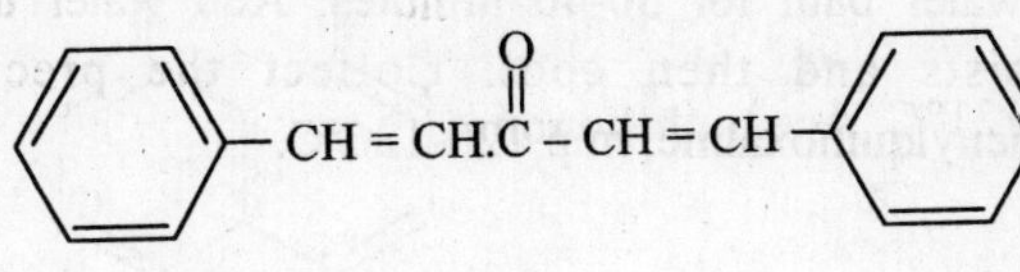

(1) Pale yellow coloured solid, sparingly soluble in alcohol and ether, easily soluble in chloroform and acetone
(2) Dissolve few milligrams of compound in conc sulphuric acid, a red coloured solution is obtained.
(3) Dissolve 100mg of compound in chloroform and add a solution of bromine to it (1.5ml Br_2 in 2ml $CHCl_3$). Shake thoroughly and pour in ice water. Crystals of dibenzalacetone tetrabromide, m.p 211°C separate out.
(4) Dibenzalacetone 2, 4-dinitrophenylhydrazone, m.p 180°C.
(5) Dibenzalacetone semicarbazone, m.p 190°C.
(6) Dibenzalacetone p-nitrophenylhydrazone, m.p 173°C.
(7) Dibenzalacetoxime, m.p 143°C.
(8) Dibenzalacetone phenylhydrazone, m.p 153°C.
(9) Picrate, m.p 114°C.

m.p 113°C **Anisoin** (p,p'-dimethoxy benzoin)

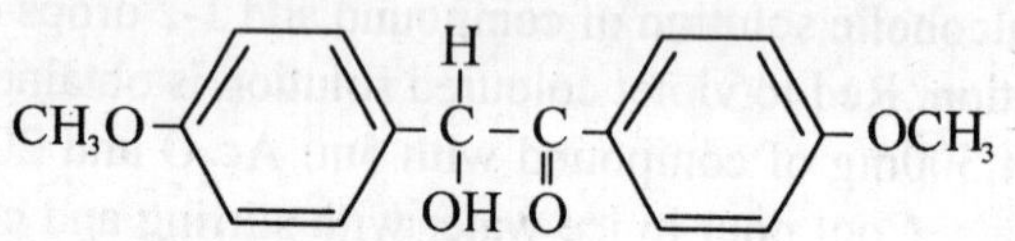

(1) Crystalline solid insoluble in water, soluble in hot alcohol, slightly soluble in ether
(2) Reduces Fehling's solution.
(3) Dissolve few milligrams of compound in conc H_2SO_4. A pale green coloured solution is obtained which changes to purple red on warming.
(4) Add Fehling's solution to a solution of 5gm of anisoin in 25ml of 70% boiling alcohol till the solution remains permanently blue. Add 20ml water, filter the precipitated cuprous oxide and anisil. Extract with benzene, concentrate the benzene layer, cool and collect crystals of anisil, m.p 133°C.
(5) Anisoin semicarbazone, m.p 185°C
(6) Anisoin acetate, m.p 95°C.

m.p 121°C **Acenaphthenone**

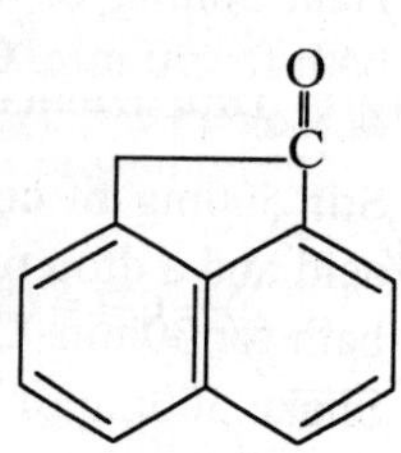

(1) Colourless crystalline compound, easily soluble in alcohol, chloroform and benzene.

(2) Acenaphthenone phenylhydrazone, m.p 90°C. (3) Acenaphthenone oxime, m.p 183°C.

m.p 134°C **p-hydroxy benzophenone** (p-benzoyl phenol)

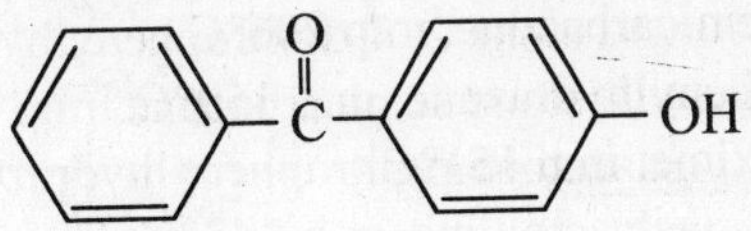

(1) Crystalline compound, sparingly soluble in water, easily soluble in alcohol and ether

(2) p-hydroxy benzophenone 2, 4 dinitrophenylhydrazone, m.p 242°C.

(3) p-hydroxy benzophenone semicarbazone, m.p 194°C. (4) p-hydroxy benzophenone phenylhydrazone, m.p 144°C. (5) p-acetoxy benzophenone, m.p 81°C (6) p-benzoxy benzophenone, m.p 114°C.

m.p 137°C **Benzoin** (benzyl phenyl carbinol)

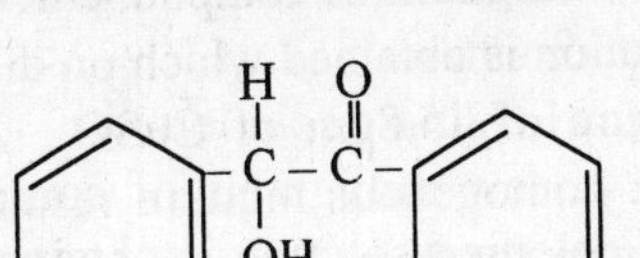

(1) Pale yellow solid, insoluble in water, sparingly soluble in ether, easily soluble in alcohol

(2) Reduces Tollens reagent

(3) Reduces Fehling's solution.

(4) Heat few milligrams of compound with dil NaOH solution. A red violet colouration is produced.

(5) Heat 500mg of compound with 3-4ml of conc HNO_3 on water bath for 90 min. Cool, pour in ice with stirring and collect benzil, m.p 95°C.

(6) Stir 500mg of compound with 5ml Ac_2O, 5ml of glacial acetic acid and a drop of conc sulphuric acid. Heat the mixture on steam bath for 30min. Cool and collect benzoin acetate, m.p 83°C.

(7) Shake 500mg of compound with 10ml NaOH and 6-7ml benzoyl

chloride till the smell of benzoyl chloride persists. Cool and collect benzoin benzoate, m.p 125°C.

(8) Reflux a mixture of 2.12gm benzoin, 1.10 gm urea and 8ml of glacial acetic acid for 6-7 hrs. Cool, keep it at RT for 3hrs and filter. Stir the solid material with 10ml ether and filter. Dissolve by heating the solid material in glacial acetic acid, pour in ice and collect diphenylglyoxalone, m.p 330-335°C.

(9) Benzoin 2, 4 dinitrophenylhydrazone, m.p 234°C.

(10) Benzoin semicarbazone, m.p 206°C.

(11) Benzoin phenylhydrazone, m.p 159°C.

(12) Benzoin oxime, m.p 151°C.

m.p 134-137°C (d) **Furoin** 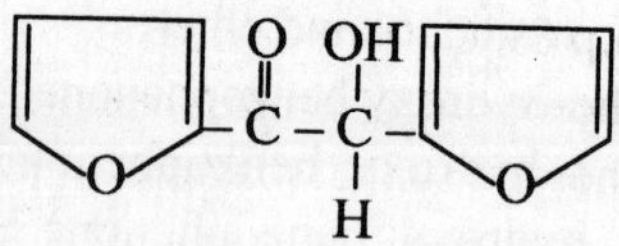

(1) Colourless crystalline compound, soluble in warm alcohol, ether and toluene

(2) Dissolve few milligrams of compound in conc sulphuric acid, a deep blue -green coloured solution is obtained.

(2) Dissolve few milligrams of compound in dil NaOH, a bluish green coloured solution is obtained which on dilution changes to green.

(3) Heat a mixture of 15.8gm of $CuSO_4$, 21gm pyridine and 9ml water until a homogenous mixture results. Add 6gm furoin and heat the mixture for 2hrs. Pour in 100ml water and filter. Wash the black precipitate with water. Extract with methyl alcohol. Concentrate the alcohol, cool and collect furil, m.p 166°C.

(5) Furoin 2, 4 dinitrophenylhydrazone, m.p 216°C

(6) Furoin phenylhydrazone, m.p 81°C.

(7) Furoin oxime, m.p 161°C.

(8) Furoin acetate, m.p 76°C

(9) Furoin benzoate, m.p 93°C.

m.p 165°C **Furil** 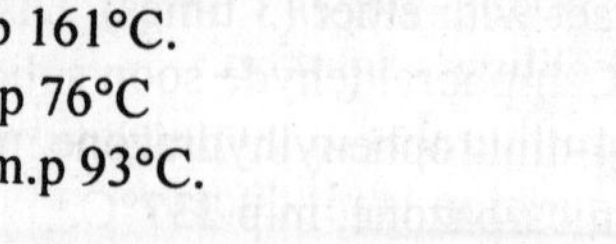

(1) Yellow crystalline solid insoluble in water, soluble in alcohol and ether

(2) Furil 2, 4 dinitrophenylhydrazone, m.p 215°C (3) Furil p-nitrophenylhydrazone, m.p 199°C (4) Furildioxime (α), m.p 166°C

Aliphatic solid ketones

m.p 28°C, b.p 198°C **Phorone** (diisopropylideneacetone) $(CH_3)_2C{=}CHCOCH{=}C(CH_3)_2$

(1) Yellow green coloured solid with disagreeable odour, soluble in alcohol and ether
(2) To 500mg of compound in 5ml CS_2 add dropwise with stirring 3ml of bromine. Continue stirring for 15min, distil off the solvent and collect the phorone tetrabromide, recrystallise from alcohol, m.p 89°C.
(3) Phorone 2, 4-dinitrophenylhydrazone, m.p 112°C
(4) Phorone semicarbazone, m.p 221°C

m.p 176-177°C **Camphor** 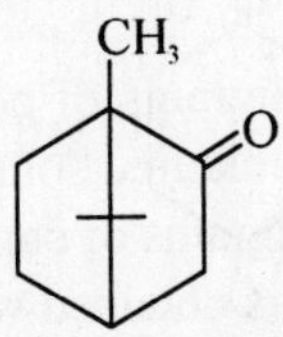

(1) White crystalline optically active solid with charcteristic odour, insoluble in water, easily soluble in alcohol, ether and acetone. Sublimes at 204°C.
(2) Dissolve 9 gm of camphor in toluene and add 0.6gm of sodium wire to it. Cool the mixture and add 1.9gm of ethylformate to it. Keep in refrigerator for 24hrs. Pour in ice water with vigorous stirring. Extract with ether (3 times). Distill off the ether, cool the residual oil, Camphoraldehyde soon solidifies, m.p 81°C.
(3) Camphor 2, 4-dinitrophenylhydrazone, m.p 177°C.
(4) Camphor semicarbazone, m.p 237°C.
(5) Camphor p-nitrophenylhydrazone, m.p 217°C.
(6) Camphor phenylhydrazone, m.p 233°C.
(7) Camphor oxime, m.p 118°C. (8) Benzilidene, m.p 98°C.

(ii) *Quinones*

Quinones are unsaturated cyclic diketones which are easily identified by their colour and their sensitivity to alkali. With sulphuric acid quinones form highly coloured onium compounds. They are generally insoluble in water and have a quite penetrating odour.

Identification tests

(*i*) **Test with alkali:** Place 200mg of compound in a test tube and add 5ml of alkali solution to it. Development of colour from brown to black indicates quinones.

(*ii*) **Test with potassium iodide:** Dissolve 200mg of compound in 5ml of alcohol. Add 10ml of 10% aqueous potassium iodide solution to a mixture of 5ml alcohol and 5ml conc sulphuric acid. Now introduce alcoholic quinone solution to the acidic KI solution. The solution turns pale yellow or brown due to the liberation of iodine. Add few drops of starch to this coloured solution. The solution turns blue or violet. (Anthraquinone however do not respond to this test)

$$KI + H_2SO_4 \rightarrow KHSO_4 + HI$$

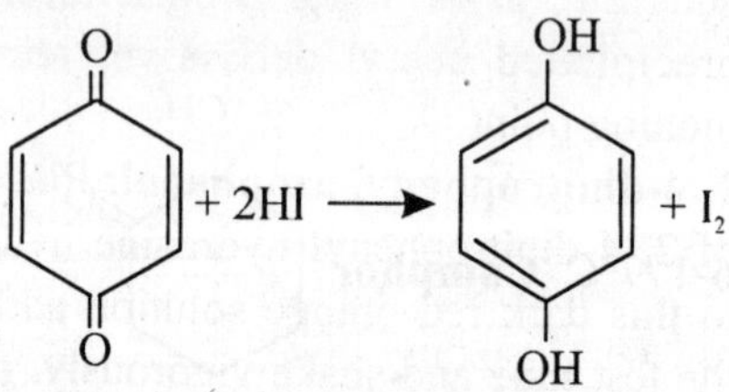

Derivatives

(*i*) **Reduction**: Shake thoroughly 250mg of compound with 5ml dil HCl and a pinch of zinc dust until the colour of quinone disappears. Filter and neutralise the filtrate with sodium hydrogen carbonate. Isolate the dihydric phenolic component with ether. Evaporate off the ether, recrystallise the product and determine its melting point.

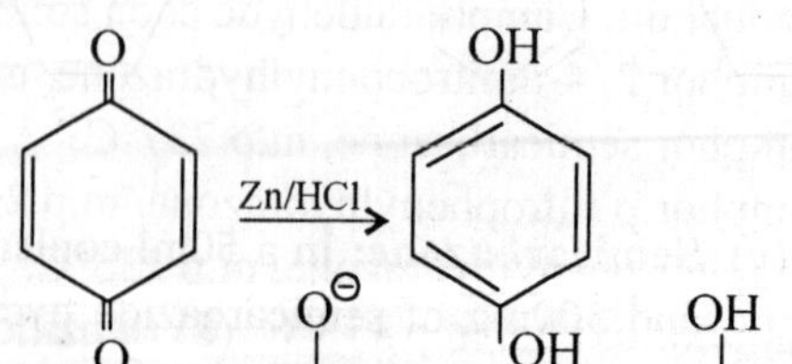

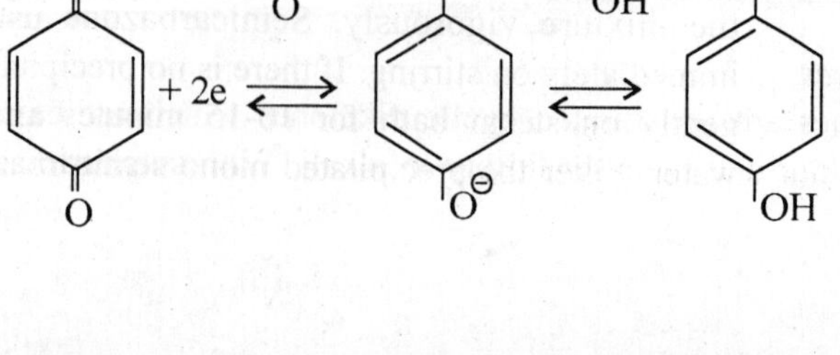

(*ii*) **Quinoxaline derivative**: Dissolve 500mg of compound in 3ml alcohol and add 3ml of alcoholic o-phenylene diamine solution (500mg dissolved in 3ml alcohol). Heat the mixture on water bath for 15-20minutes. Cool, pour the mixture in ice and collect the precipitated solid, recrystallise and determine its melting point.

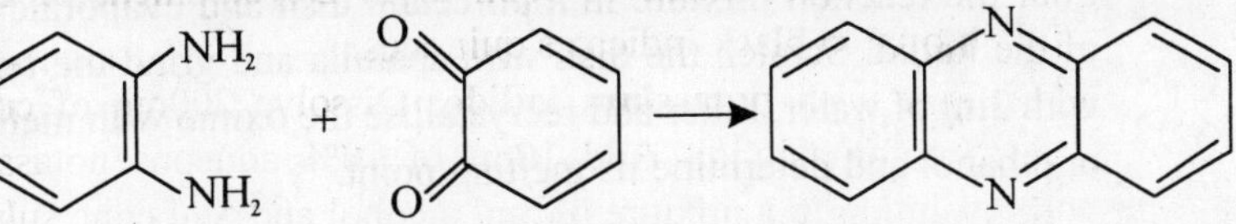

(*iii*) **Reductive acetylation:** Warm a mixture of 1gm of compound, 5ml acetic anhydride, 1gm zinc and 250 mg of sodium acetate until the colour of quinone has disappeared. Warm the colourless mixture for another 2-3minutes and then add 5ml of glacial acetic acid to it. Boil this acidic solution to dissolve the precipitated solid. Decant the hot solution, heat the filtrate to boiling, cool and add few ml of water so that turbidity occurs. Cool, filter the precipitated acetyl derivative, recrystallise and determine its melting point.

(*iv*) **2, 4-dinitrophenyl azophenol:** Place 5 ml of a saturated solution of 2, 4-dinitrophenyl hydrazine hydrochloride in a test tube and to this dark red orange solution add 300mg of quinone. Stopper the test tube and shake vigorously. If no precipitate separates out heat the test tube on water bath for 5-10minutes and again shake or scratch the sides of test tube with glass rod and allow it to stand for 10 minutes. Filter the precipitate, recrystallise from ethanol and determine its melting point.

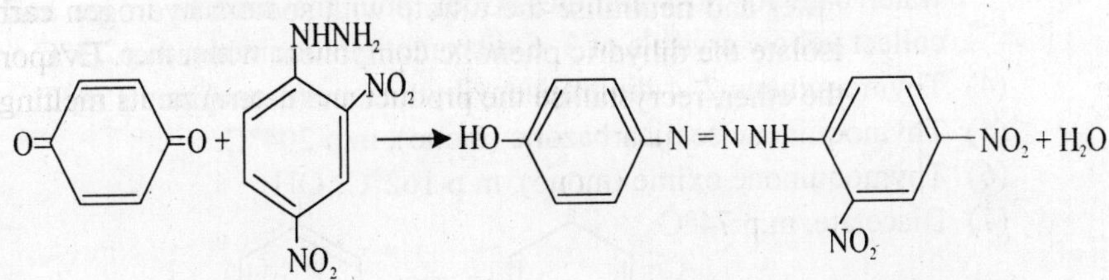

(*v*) **Semicarbazone:** In a 50ml conical flask place 500mg of quinone and 500mg of semicarbazide hydrochloride in 6ml water. Shake the mixture vigorously. Semicarbazone usually separate out immediately on stirring. If there is no precipitate, heat the mixture gently on steam bath for 10-15 minutes and then cool in ice water. Filter the precipitated mono semicarbazone derivative and recrystallise from methanol, ethanol or ethanol-water mixture or hot water. Bis semicarbazone however remains undissolved

(*vi*) **Oxime:** To 1gm of quinone in 50ml round bottom flask add 1gm of hydroxylamine hydrochloride, 10ml ethanol and 1ml pyridine. Reflux the reaction mixture on a steam bath for 30-60 minutes. Pour the reaction mixture in a porcelain dish and evaporate most of the liquid. Scratch the dish with spatulla and grind the residue with 3ml of water. Filter and recrystallise the oxime with methanol or ethanol and determine its melting point.

Solid

m.p 45°C **Thymoquinone** (2-methyl-5-isopropyl-benzoquinone-1, 4)

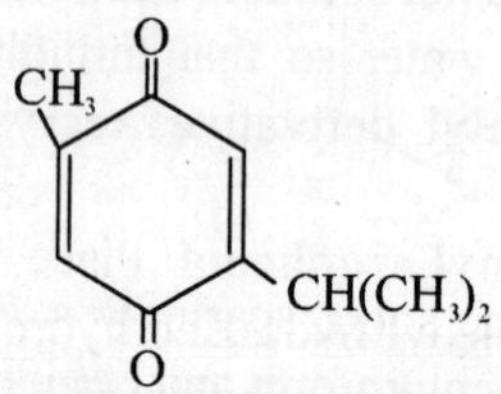

(1) Yellow coloured solid with smell like thymol, sparingly soluble in water, easily soluble in alcohol, ether, chloroform and benzene
(2) Heat few milligram of compound with dil alkali. A dark coloured solution is obtained.
(3) Heat 200mg of compound with 5ml water and 0.75ml bromine on water bath for 30-40minutes. Cool, pour the mixture in ice and collect yellow crystals of 3, 6-dibromothymoquinone, m.p 73°C.
(4) Thymoquinone 2, 4-dinitrophenylhydrazone (mono) m.p 179°C.
(5) Thymoquinone semicarbazone (mono), m.p 204°C.
(6) Thymoquinone oxime (mono), m.p 162°C.
(7) Diacetate, m.p 74°C

m.p 69°C **p-Toluquinone** (2-methyl 1, 4-benzoquinone)

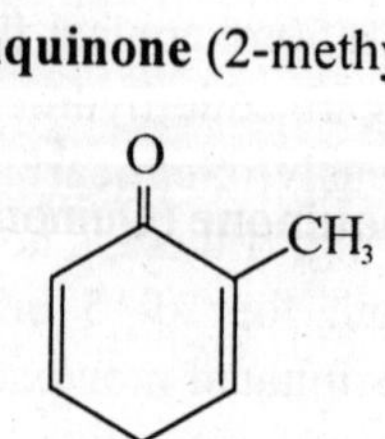

(1) Golden yellow coloured solid, sparingly soluble in water, easily soluble in alcohol and ether. Sublimes

(2) To few milligrams of compound add dil NaOH solution. A brown-red coloured solution is obtained.

(3) Heat at 50-60°C, 500mg of compound with 6ml of acetic anhydride containing 2-3 drops of conc H_2SO_4. Cool and keep the mixture overnight in a refrigerator, collect crystals of 2, 4, 5-triacetoxy toluene, m.p 114°C.

(4) p-Toluquinone 2, 4-dinitrophenylhydrazone (di) m.p 269°C. (5) p-Toluquinone semicarbazone (mono) m.p 179°C. (6) p-Toluquinone oxime (mono) m.p 135°C. (7) Diacetate m.p 52°C

m.p 106°C **2-methyl 1, 4-Naphthaquinone** (menadione)

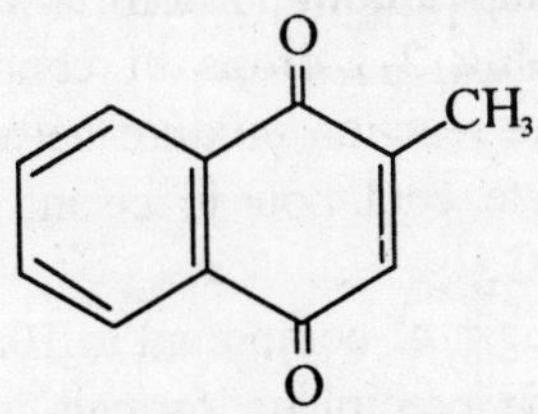

(1) Yellow crystalline solid, sparingly soluble in water and alcohol, easily soluble in chloroform and carbon tetrachloride

(2) To few milligrams of compound add dil NaOH solution. A dark brown coloured solution is obtained.

(3) To few milligrams of compound add conc H_2SO_4. A red coloured solution is obtained.

(4) Reflux 300mg of compound with 3ml glacial acetic acid, 3ml Ac_2O, 100mg of fused sodium acetate and a pinch of zinc dust. Cool, pour the mixture in crushed ice and collect 1, 4-diacetoxy 2-methyl naphthalene, m.p113°C.

(5) 2-methyl 1, 4-Naphthaquinone semicarbazone, m.p 247°C

(6) 2-methyl 1, 4-Naphthaquinone oxime, m.p 167°C.

m.p 115°C **p-Benzoquinone** (Quinone)

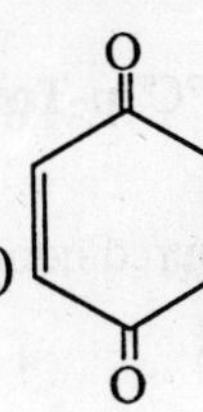

(1) Golden yellow crystalline solid with characteristic pungent odour similar to chlorine, sparingly soluble in water, easily soluble in alcohol and ether. Sublimes readily.
(2) Reduces Tollens reagent
(3) To few milligrams of compound add aqueous KI solution, a drop of conc H_2SO_4 and shake thoroughly. A deep brown coloured solution is generated due to liberation of iodine.
(4) Suspend 5gm of quinone in water and pass SO_2 gas until the decolourisation has occurred. Extract with ether. Evaporate off the solvent, recrystallise the residue from sulphorous acid and collect quinol, m.p 169°C.
(5) Dissolve 2gm of quinone and 2gm of hydroquinone separately in minimum amount of water. Mix the two solutions and warm. Cool, collect the precipitated Quinhydrone, m.p 171°C.
(6) Add 1.1gm of compound in portions to a stirred soloution of 5ml of Ac_2O containing 2-3 drops of conc H_2SO_4. Maintain the temperature of the reaction mixture between 40-60°C. After the addition is complete, cool, pour in ice and collect 1, 2, 4-triacetoxy benzene, m.p 97°C.
(7) To a solution of 5gm of compound in 10ml $CHCl_3$ pass dry HCl gas till the brown precipitate formed is converted into white. Cool, filter the precipitated 2-chloroquinol, recrystallise from $CHCl_3$, m.p 106°C.
(8) Reflux 2.5gm of compound with 15ml ethyl alcohol and 3gm of fused zinc chloride for 24hrs. Cool, filter the precipitated 2, 5-diethoxybenzoquinone, recrystallise from alcohol, m.p 183°C
(9) p-Benzoquinone 2, 4-dinitrophenylhydrazone (mono), m.p 186°C.
(10) p-Benzoquinone semicarbazone (mono), m.p 165°C.
(11) p-Benzoquinone dioxime, m.p 140°C.

m.p 116 °C (d) **1, 2 Naphthaquinone** (β-Naphthaquinone)

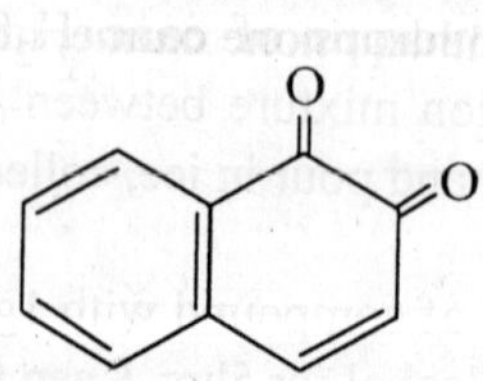

(1) Red coloured odourless solid, insoluble in water, soluble in alcohol and ether

(2) To few milligrams of compound add dil NaOH solution. A yellow coloured solution is obtained

(3) To few milligrams of compound add conc H_2SO_4. A green coloured solution is obtained.

(4) Add in portion with stirring 500mg of compound to a well stirred mixture of 5ml Ac_2O containing 2-3 drops of conc H_2SO_4. Maintain temperature between 40-45°C. After addition is complete keep the mixture at room temperature for sometime and then pour in ice, collect 1, 2, 4-triacetoxy naphthalene, m.p 135°C.

(5) 1, 2-Naphthaquinone semicarbazone, m.p 184°C.

(6) 1, 2-Naphthaquinone p-nitrophenylhydrazone, m.p 235°C.

(7) 1, 2-Naphthaquinone phenylhydrazone, m.p 138°C.

(8) 1, 2-Naphthaquinone oxime, m.p 162°C.

m.p 125°C **1, 4 Naphthaquinone**

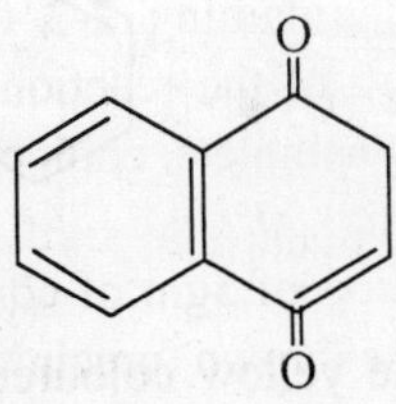

(1) Yellow coloured solid with pronounced odour, insoluble in water and alcohol, easily soluble in ether, benzene and chloroform. Sublimes readily

(2) To few milligrams of compound add dil NaOH solution. A reddish brown coloured solution is obtained

(3) Boil 100mg of compound with 2ml aniline in 2ml alcohol for 2-3 minutes. Cool, dilute the reaction mixture with 10ml water and then add 2ml acetic acid to it. Shake thoroughly and filter the precipitated crystals of 2-anilinonaphthoquinone 1, 4, m.p190°C.

(4) Add 1.1gm of compound in portions to a stirred soloution of 5ml of Ac_2O containing 2-3 drops of conc H_2SO_4. Maintain the temperature of the reaction mixture between 40-60°C. After the addition is complete cool and pour in ice, collect 1, 2, 4-triacetoxy naphthalene, m.p135°C.

(5) Reflux a mixture of 1gm of compound with 1gm of 2, 3-dimethyl butadiene -1, 3 and 5ml alcohol for 5hrs. Keep the reaction mixture in refrigerator for 10-12hrs, recover the addition product, m.p 147-149°C.

(6) Heat in a sealed tube 6gm of compound with 10ml alcohol containing 2-3gm of butadiene at 100°C for 3hrs. Cool and recrystallise the precipitated tetrahydroanthraquinone from alcohol, m.p 105-106°C.
(7) 1, 4-Naphthaquinone 2, 4-dinitrophenylhydrazone (mono), m.p 278°C.
(8) 1, 4-Naphthaquinone semicarbazone (mono) m.p 247 °C (d).
(9) 1, 4-Naphthaquinone oxime (mono) m.p 198°C.

m.p 170°C **Benzanthrone** [7H-Benz (de)anthracen-7-one]

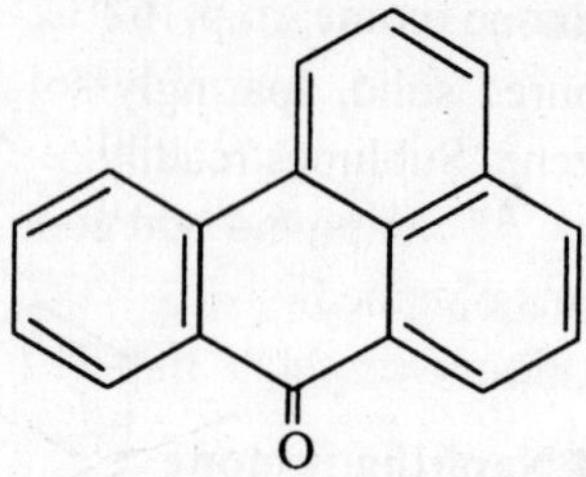

(1) Pale yellow coloured solid, insoluble in cold alcohol and dilute alkali.
(2) Dissolve few milligrams of compound in conc H_2SO_4. A reddish brown solution with deep orange fluorescence is obtained.

m.p 171°C **Quinhydrone**

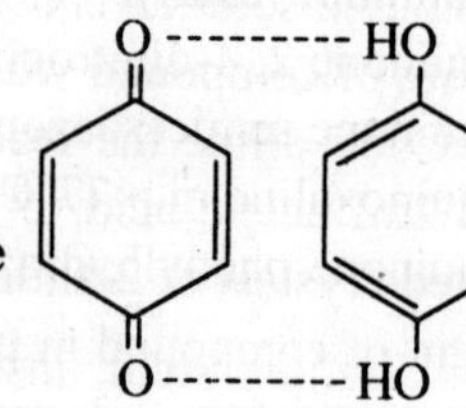

(1) Dark green coloured solid which appears red brown by transmitted light, soluble in alcohol and ether.
(2) Reduces Tollens reagent.
(3) Dissolve few milligrams of compound in alcohol. A yellow coloured solution is obtained.
(4) Suspend 5gm of compound in water and pass SO_2 gas until the decolourisation has occurred. Extract with ether. Evaporate off

the solvent, recrystallise the residue from sulphorous acid and collect quinol, m.p169°C.

m.p 177°C **2-methyl anthraquinone**

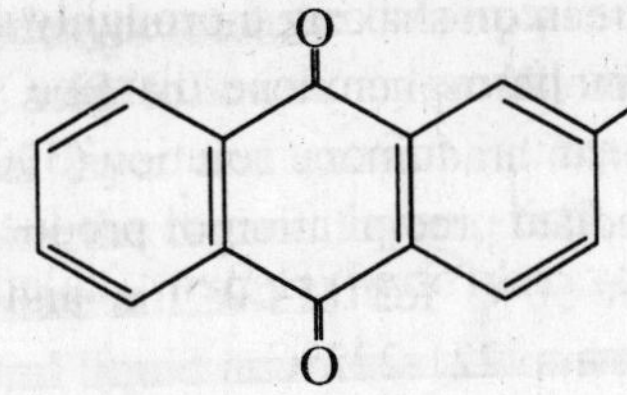

(1) Pale yellow coloured solid, sparingly soluble in alcohol, soluble in ether and benzene. Sublimes readily
(2) To few milligrams of compound add conc H_2SO_4. A pale yellow coloured solution is obtained.
(3) 2-methyl anthraquinone diacetate m.p 217°C.

m.p 199°C **Camphorquinone** 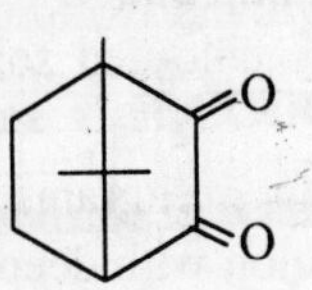

(1) Yellow crystalline compound, insoluble in water, easily soluble alcohol. Sublimes readily.
(2) Camphorquinone 2, 4-dinitrophenylhydrazone (di) m.p 190°C.
(3) Camphorquinone semicarbazone (mono) m.p 236°C.
(4) Camphorquinoxaline m.p 77°C.
(5) Camphorquinone phenylhydrazone m.p 183°C.

m.p 205°C **Phenanthraquinone** (9, 10-dihydrophenanthrene -9, 10 dione)

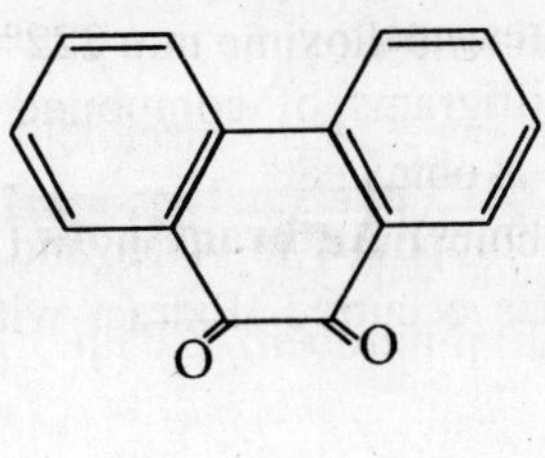

(1) Yellow-Orange coloured odourless solid, insoluble in water, soluble in alcohol and benzene,
(2) To few milligrams of compound add conc H_2SO_4. A green coloured solution is obtained.
(3) Boil for a minute 50mg of compound with 10ml dil NaOH and 400mg of zinc dust. Filter the solution while still hot. The filtrate which is intense green on shaking thoroughly slowly turn yellow.
(4) Add 10ml of ethanolic o-phenylene diamine solution (150mg in 10ml) to a hot phenanthraquinone solution (300mg in 10ml glacial acetic acid). Immediate precipitation of product takes place. Keep the temperature at 50°C for 15-20 min and recover 1, 2:3, 4 dibenzophenazine, m.p 221-222°C.
(5) Heat 4gm of compound with 100ml of 2N NaOH solution on water bath for 1½ hr. Neutralise the excess alkali and filter. Concentrate the filtrate, acidify and cool. Crystals of 9-hydroxyfluorene-9-carboxylic acid separate out.
(6) Phenanthraquinone p-nitrophenylhydrazone (mono) m.p 245°C.
(7) Phenanthraquinone phenylhydrazone (mono) m.p164°C.
(8) Phenanthraquinone (mono) oxime, m.p 158°C.
(9) Diacetate, m.p 202°C

m.p 261°C **Acenaphthenequinone**

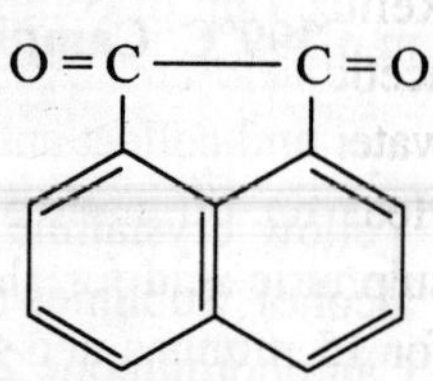

(1) Yellow coloured solid, insoluble in water, sparingly soluble in alcohol, soluble in toluene. Sublimes
(2) Reflux 500mg of compound dissolved in 10ml glacial acetic acid with 300mg of o-phenylene diamine for 3hrs. Cool, add cold water and collect α, α naphthaquinoxaline, m.p 234°C.
(3) Acenaphthenequinone semicrbazone (mono) m.p 192°C.
(4) Acenaphthenequinone phenylhydrazone (mono) m.p 179°C.
(5) Acenaphthenequinone dioxime m.p 222°C.

m.p 286°C (Sublimes) **Anthraquinone** (9, 10-dihydroanthracene-9, 10 dione)

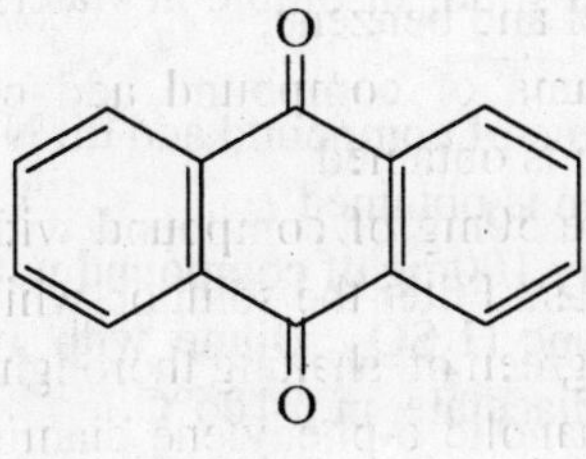

(1) Pale yellow crystalline solid, insoluble in water, difficult soluble in alcohol, ether and benzene, soluble in toluene.
(2) Boil for a minute 50mg of compound with 10ml dil NaOH and 400mg of zinc dust. Filter the solution while still hot. The filtrate which is deep red on shaking thoroughly slowly turn yellow.
(3) Boil 200mg of compound with 5ml Ac_2O, 200mg of sodium acetate and 200mg of zinc dust. Filter the solution while still hot, cool and pour in ice water. Collect colourless crystals of anthraquinone diacetate, m.p 260°C.
(4) Reflux 2gm of compound with 6gm zinc and 100ml glacial acetic acid. Add few drops of conc HCl at a time till no colouration occurs. After 15 minutes cool and pour the reaction mixture in dil HCl and collect anthranol, m.p 170°C (d).
(5) Reflux 1gm of compound with 1gm of granulated tin, 8ml glacial acetic acid and 2.5ml conc HCl for 2hrs. Cool, dilute with cold water and collect anthrone, m.p 155°C.
(6) Heat on oil bath 5gm of anthraquinone with 3ml of fuming sulphuric acid for 1hr. Cool, pour in cold water. Heat the mixture for 15 minutes and filter. To the filtrate add NaOH solution and recover anthraquinone 2-sulfonate.
(7) Anthraquinone (mono) oxime, m.p 224°C.

m.p 290°C **Alizarin** (1, 2-dihydroxy anthraquinone)

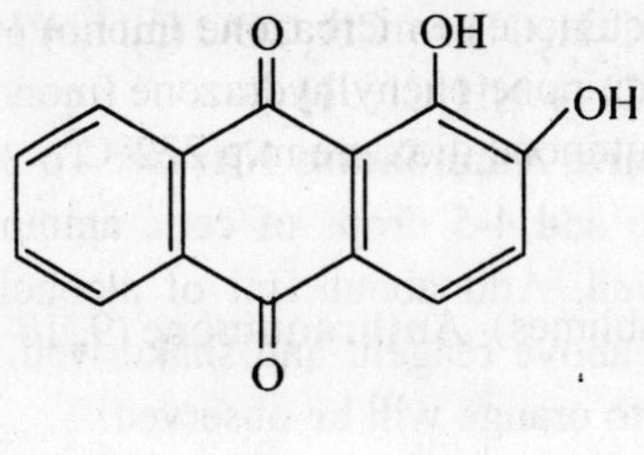

(1) Orange coloured solid, insoluble in water, soluble in alcohol and ether.

(2) To few milligrams of compound add dil NaOH solution. A purple coloured solution is obtained

(3) Shake thoroughly 100mg of compound with 3ml Ac_2O containing 2-3 drops of conc H_2SO_4. Dilute with 10ml of cold water and collect alizarin diacetate, m.p 186°C.

(iii) *Alcohols*

Alcohols are regarded as alkyl derivatives of water. They are amphoteric and usually act as a weak acid as well as a weak base. Over the years alcohols are being used as sterilizers, analgesic and also as a tranquilizers.

H-O-H	R-O-H
Water	Alcohol

Identification Tests

(*i*) **Test with sodium metal:** In a dry watch glass/beaker cover add a small piece of sodium metal to 1ml of alcohol. Effervescence takes place which increases with time due to evolution of hydrogen.

$$2ROH + 2Na \rightarrow 2RO^{-}Na^{+} + H_2$$

(*ii*) (*a*) **Acetyl Chloride Test:** In a dry test tube mix 1ml of acetyl chloride with 2ml of alcohol. Shake the mixture slowly for 5 minutes. Add about 2ml of cold water to decompose excess of acetyl chloride and then neutralise the aqueous layer with solid sodium bicarbonate. Now carefully smell the reaction mixture, if the smell is pleasant and fruity then ester formation has taken place.

$$CH_3COCl + C_2H_5OH \rightarrow CH_3COO\,C_2H_5 + HCl$$

(*b*) In a dry test tube mix 1ml of acetyl chloride with 2ml of alcohol. Shake the mixture slowly for 5 minutes. Bring a glass rod dipped in conc ammonia solution at the mouth of the test tube. White fumes due to formation of ammonium chloride are produced.

$$CH_3COCl + C_2H_5OH \rightarrow CH_3COO\,C_2H_5 + HCl$$

$$NH_4OH + HCl \rightarrow NH_4Cl + H_2O$$

(*iii*) **Test with Ceric Ammonium Nitrate:** To 3ml of distilled water in a test tube add 4-5 drops of ceric ammonium nitrate reagent and shake well. Add about 1ml of alcohol (100mg in case of solid) to the above reagent and shake well. A change of colour from yellow to orange will be observed.

(Note:-Phenols give green to brown precipitate)

$(NH_4)_2 [Ce(NO_3)_6] + 2R^1OH \rightarrow [Ce(NO_3)_4 (ROH)_2] + 2NH_4NO_3$

Ceric Ammonium Nitrate Reagent: Dissolve 5gm of Ceric Ammonium Nitrate in 12.5 ml of 2N HNO_3 (warm if necessary)

(*iv*) **Chromic acid test**: To 2ml conc H_2SO_4 add 10ml saturated solution of potassium dichromate and 4ml of alcohol, shake the reaction mixture vigorously and note the change.

(*a*) A colour change from orange to opaque blue green indicates the presence of primary or secondary alcohol.

(*b*) In case of tertiary alcohols no change will occur.

$$K_2Cr_2O_7 + H_2O + 2H_2SO_4 \longrightarrow 2H-O-\overset{\overset{O}{\|}}{\underset{\underset{O}{\|}}{Cr}}-O-H + 2HSO_4^-$$

$$R-CH_2OH + HO-\overset{\overset{O}{\|}}{\underset{\underset{O}{\|}}{Cr}}-OH \longrightarrow ROOH + Cr^{3+} \text{ (blue)}$$

Orange

$$R-CHOHR^1 + H_2CrO_4 \longrightarrow R-\overset{\overset{O}{\|}}{C}-R^1 + Cr^3 + \text{(blue)}$$

$$R-\overset{\overset{OH}{|}}{\underset{\underset{R''}{|}}{C}}-R' + H_2CrO_4 \longrightarrow \text{No change in colour}$$

(*v*) **Oxidation with Agulhon's reagent:** To 5ml of unknown sample add 10ml of freshly prepared Solution A and 0.5ml of Solution B. Shake the resultant mixture thoroughly. Appearance of blue colour indicates the presence of alcohol.

Solution A :-5ml of conc HNO_3 mixed with glacial acetic acid.

Solution B :- 15% Potassium dichromate solution.

(*vi*) **Deniges's test for t-alcohol:** To 1ml of alcohol add 3ml of deniges's reagent and boil for 2-3 minutes. Appearance of yellow colour which changes from yellowish to reddish precipitate indicates the presence of t-alcohol. Primary and secondary alcohol however give only white suspension.

Deniges's reagent :- Dissolve 2.5gm mercuric oxide and 10ml conc H_2SO_4 in 50ml distilled water.

(*vii*) **Boric acid test for polyhydric alcohol** :-To 2ml 1% solution of borax add 3-4 drops of phenolphthalein, a pink colour develops.

Add 2ml polyhydric alcohol(200mg in case of solid) to the pink coloured solution. The pink colour disappears but it reappears on heating and as the solution becomes cool it again vanishes.

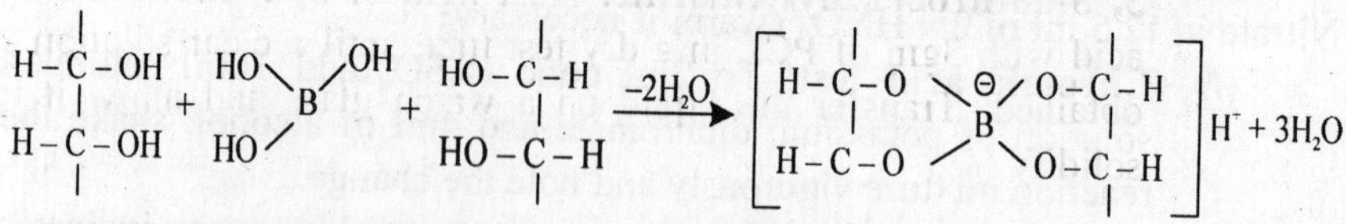

(*viii*) **Lucas test for differentiation between primary secondary and tertiary alcohol.**

To 3ml alcohol add 15ml lucas reagent, shake thoroughly, keep aside the reaction mixture and observe the changes.

(a) Two phase separate immediately due to formation of tertiary chloride indicating the alcohol to be tertiary.

(b) The clear solution becomes turbid and a distinct layer is visible after 5 minutes indicating a secondary alcohol.

(c) The aqueous layer remains clear indicating it to be primary alcohol.

Lucas Reagent: Dissolve 6.8gm anhydrous zinc chloride in 5.5ml conc HCl (cool if necessary to avoid loss of HCl)

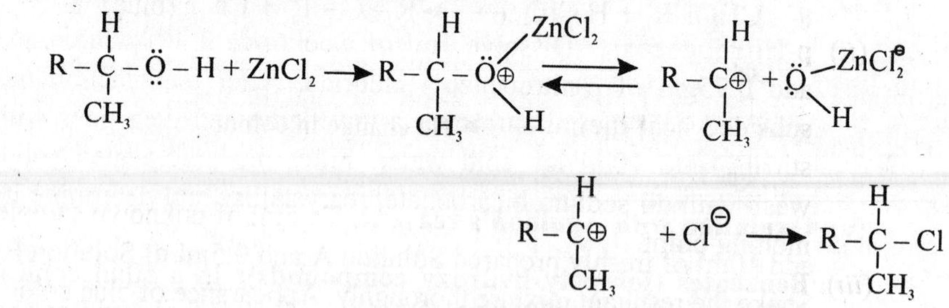

Derivatives

(*i*) **3, 5-dinitrobenzoate [Method A]:** In a test tube heat 2ml alcohol with 3, 5-dinitrobenzoyl chloride (prepared from 2gm 3, 5-dinitrobenzoic acid and 3gm PCl_5) on water bath for 10-15 minutes. Cool the mixture and add 10ml of dilute sodium bicarbonate solution. Filter off the solid ester, wash first with sodium bicarbonate and then with water. Recrystallise from a suitable solvent and determine its melting point.

Method B : In a 50ml conical flask place 2ml of alcohol, 4 gm of 3, 5-dinitrobenzoyl chloride and 4ml of pyridine. Heat the mixture on steam bath for 15-20 minutes. Cool and pour the reaction mixture in ice containing 15ml of 10% HCl. Add 15ml ether, shake the mixture thoroughly. Isolate the ethereal layer, evaporate off the solvent and recrystallise the residue from a suitable solvent.

3, 5-dinitrobenzoyl chloride: Heat 2gm of 3, 5-dinitrobenzoic acid with 3gm of PCl_5 in a dry test tube until a clear solution is obtained. Transfer the liquid on a watch glass and allow it to solidify

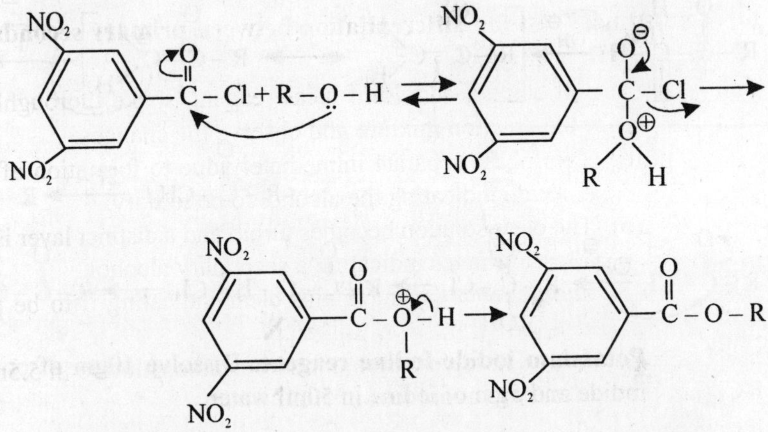

(*ii*) **p-Nitrobenzoate:** Dissolve 1ml of alcohol in 3ml pyridine and add 0.75gm of p-nitrobenzoyl chloride. After the reaction has subsided heat the mixture over a low flame and then pour with stirring into 10ml ice cold water. Filter the precipitated solid, wash with dil sodium bicarbonate, recrystallise and determine its melting point.

(*iii*) **Benzoates (for poly hydroxy compounds):** In a 50ml conical flask place 1ml alcohol (500mg in case of solid) 5ml pyridine and 5ml benzoyl chloride. Shake the flask vigorously and reflux for 30-45min. Cool and add 10ml saturated sodium bicarbonate solution. Filter the precipitated solid, wash with water and recrystallise from ethanol

(*iv*) **Iodoform:** given by alcohols containing CH_3-CH(OH)-R and R-CH(OH)-CH_2-CH(OH)-R grouping

Dissolve 500mg (3-5ml in case of liquid) of the sample in 5ml water (methanol or dioxan in case of insoluble) and then add 10ml dil NaOH to it. Add dropwise with stirring Iodine-potassium iodide reagent until the dark colour of iodine persists. Keep the mixture at room temperature for 4-5 minutes and allow precipitation to take place. If however no iodoform separates, heat the reaction mixture on water bath for 2-3 minutes and add few drops of iodine reagent. Cool and collect the yellow precipitate of iodoform, m.p 119°C.

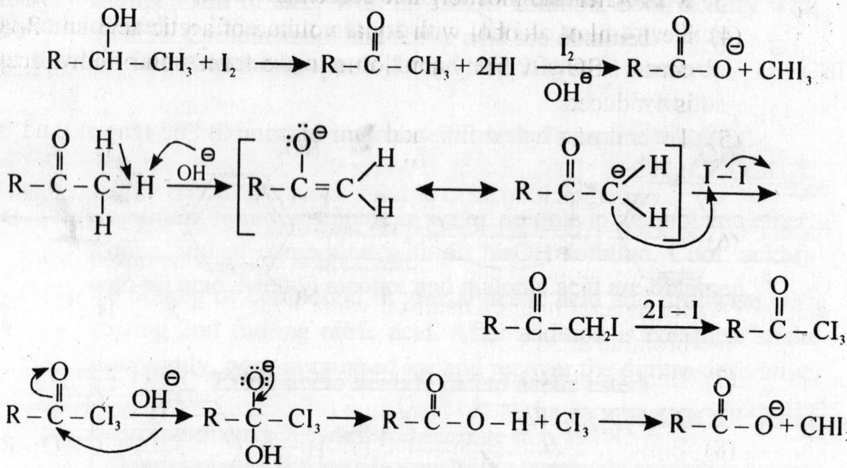

Potassium iodide-Iodine reagent: Dissolve 10gm of potassium iodide and 5gm of iodine in 50ml water

Aliphatic Liquid

b.p 65°C **Methyl Alcohol** (Carbinol, Methanol) CH_3OH

(1) Colourless inflammable liquid, miscible with water and other organic solvents

(2) Heat a small piece of copper wire and drop it in a test tube containing 2-3 ml of alcohol. A pungent odour due to formation of formaldehyde is produced.

(3) Heat 2ml of alcohol with 50mg of salicylic acid and add 2-3 drops of conc H_2SO_4. Smell of oil of winter green due to formation of methyl salicylate is produced.

(4) To 10ml alcohol add 2gm phosphorous and then add with stirring 10gm of iodine, continue shaking for an hour and then keep at room temperature for 3hrs. Carry out the distillation at 50-60°C and collect the distillate in a receiver containing ice cooled water. To the distillate add caustic soda, remove the lower alkyl halide layer. Distill and collect methyl iodide at b.p 43°C.

(5) Methyl 3, 5-dinitrobenzoate, m.p 109°C. (6) Methyl p-nitrobenzoate, m.p 96°C.

b.p 78°C **Ethyl Alcohol** (Ethanol) CH_3CH_2OH

(3) Colourless inflammable liquid with a spirit like odour, miscible with water and most organic solvents.

(4) Heat 4ml of alcohol with equal volume of acetic acid and 2ml of conc H_2SO_4. A fruity smell due to the formation of ethyl acetate is produced.

(5) To 2ml of alcohol first add 1ml of conc iodine reagent and then add 5ml NaOH solution. Warm the solution on water for 2 minutes and cool. Yellow crystals of iodoform, m.p 119°C are produced.

(6) To 10ml alcohol add 2gm of phosphorous, cool and add dropwise 4ml of bromine and then keep the mixture at room temperature for several hours. Distill on water bath at 50°C and collect ethyl bromide at 38°C.

(7) Heat 2ml of alcohol with acidic potassium dichromate solution. A pungent smell due to formation of acetaldehyde is generated.

(8) Ethyl 3, 5-dinitrobenzoate, m.p 93°C. (7) Ethyl p-nitrobenzoate, m.p 57°C.

b.p 82°C **Isopropyl Alcohol** (Isopropanol, Propanol-2) $(CH_3)_2CHOH$

(1) Colourless liquid, miscible with water and most organic solvents.

(2) Heat 1ml of alcohol with 3ml of potassium dichromate solution and 0.5ml of conc H_2SO_4. A pleasant sweet odour due to formation of acetone is generated.

(3) To 2ml of compound add few milligrams of resorcinol and 2-3 drops of conc H_2SO_4. An amber coloured ring is produced.

(4) In a test tube place 2ml of test sample and 2ml of iodine reagent. Now add 6-7ml dil NaOH solution. Heat on water bath for 2min, cool and collect iodoform, m.p 119°C.

(5) Isopropyl 3, 5-dinitrobenzoate, m.p 122°C.

(6) Isopropyl p-nitrobenzoate, m.p 110°C.

b.p 83°C **t-Butyl Alcohol** (Trimethyl carbinol) $(CH_3)_3COH$

(1) Colourless liquid, miscible with water and most organic solvents.

(2) To 2ml of compound add few milligrams of resorcinol and 2-3 drops of conc H_2SO_4. A red coloured solution is produced.

(3) To ice cold 10ml conc H_2SO_4add slowly 6gm urea (temp below

20°C). Now add dropwise 19ml t-butyl alcohol (temp 20-25°C). Stir for 30-40min, keep overnight and then pour in ice. Make the solution alkaline by adding NaOH, cool and filter. Dissolve the solid in water, warm and cool. Collect crystals of t-butyl urea, m.p 180-182°C.

(4) t-Butyl 3, 5-dinitrobenzoate, m.p 142°C (5) t-Butyl p-nitrobenzoate, m.p 116°C.

b.p 97°C **n-Propyl Alcohol** (Propanol) $CH_3CH_2CH_2OH$

(1) Colourless liquid, miscible with water, alcohol and ether
(2) Heat 2ml of alcohol with acidic potassium dichromate solution. A pungent smell due to formation of propionaldehyde is observed. Bring filter paper moistened with ammonical silver nitrate solution near the mouth of test tube, a black stain is produced.
(3) To 2ml of compound add few milligrams of resorcinol and 2-3 drops of conc H_2SO_4. An amber coloured ring is produced.
(4) n-Propyl 3, 5-dinitrobenzoate, m.p 74°C
(5) n-Propyl p-nitrobenzoate, m.p 35°C.

b.p 97°C **Allyl Alcohol** (prop-2-en-1-ol) $CH_2{=}CHCH_2OH$

(1) Colourless liquid with pungent irritating mustard like odour, miscible with water and ether, possess lachramatory property.
(2) To 2ml of compound add few milligrams of resorcinol and 2-3 drops of conc H_2SO_4. A brown coloured ring is produced.
(3) Heat 2ml of alcohol with acidic potassium dichromate solution. A pungent odour even more than the parent compound due to the formation of acrolein is produced.
(4) To 2ml of compound add 1ml of bromine water. Colour of bromine disappears.
(5) Allyl 3, 5-dinitrobenzoate, m.p 50°C.
(6) Allyl p-nitrobenzoate, m.p 29°C.

b.p 100°C **s-Butyl Alcohol** (butan-2-ol, ethyl methyl carbinol) $CH_3CH_2CH(OH)CH_3$

(1) Colourless liquid, soluble in water, miscible with alcohol and ether.
(2) To 2ml of compound add few milligrams of resorcinol and 2-3

drops of conc H_2SO_4. A pale lemon-yellow coloured ring is produced.

(2) To 2ml of alcohol add acidic potassium dichromate solution. A pleasant sweet odour due to formation of ethyl methyl ketone is produced.

(3) In a test tube place 2ml of test sample and 2ml of iodine reagent. Add 6-7 ml dil NaOH solution and heat on water bath for 2-3min, cool and collect iodoform, m.p 119°C.

(4) In 5ml alcohol pass HCl gas till there is an increase in weight by 15-20gm. Mix the acidic solution with 10ml glacial acetic acid and reflux for 10hrs. Cool, dilute with water and isolate s-butyl acetate with ether (2 × 10ml). Distill off the ether and collect the ester at b.p 110-112°C.

(5) Add a solution of 10ml of alcohol in 25ml ether to a solution of ethyl magnesium bromide (from 14gm ethyl bromide, 3.2gm Mg and 50ml ether). Now add a solution of 13gm of methyl ester of methylacrylic acid in 15ml ether and reflux the mixture for 10-12hrs. Cool and pour in 150ml dil HCl. Separate the ethereal layer and extract the aqueous layer with ether. Combine the two layers and distill. Collect s-butyl methacrylate at b.p 70-71°C/ 50mmHg

(6) s-Butyl 3, 5-dinitrobenzoate, m.p 76°C. (8) s-Butyl p-nitrobenzoate, m.p 26°C.

b.p 102°C **t-Amyl Alcohol** (2-methyl butan-2-ol, Dimethylethyl carbinol) $(CH_3)_2C(OH)CH_2CH_3$

(1) Colourless liquid, miscible with alcohol and ether.

(2) Heat 1ml of alcohol with 3ml of potassium dichromate solution and 0.5ml of conc H_2SO_4. A pleasant sweet odour due to formation of acetone is generated.

(3) t-Amyl 3, 5-dinitrobenzoate, m.p 117°C. (4) t-Amyl p-nitrobenzoate, m.p 85°C.

b.p 108°C **Isobutyl Alcohol** (2-methylpropan-1-ol) $(CH_3)_2CHCH_2OH$

(1) Colourless liquid, soluble in water, miscible with alcohol and ether.

(2) To 2ml of compound add few milligrams of resorcinol and 2-3 drops of conc H_2SO_4. An amber coloured ring is produced.
(3) Add in portion a solution a 14gm of potassium permanganate in 200ml water to a mixture of 10ml of alcohol and 15ml water containing 1.5gm sodium carbonate. Cool the mixture and slowly allow it to attain room temperature. After 12 hrs filter and concentrate the filtrate. Cool, add 15ml ether and then acidify with dil H_2SO_4. Separate the ethereal layer and extract the aqueous layer with ether. Combine the two organic layers and distill. Collect isobutyric acid at b.p 155°C.
(4) Isobutyl 3, 5-dinitrobenzoate, m.p 88°C. (5) Isobutyl p-nitrobenzoate, m.p 68°C.

b.p 116°C **sym s-Amyl Alcohol** (Pentan-3-ol) $CH_3CH_2CH(OH)CH_2CH_3$

(1) Colourless liquid sparingly soluble in water, soluble in alcohol and ether
(2) s-Amyl 3, 5-dinitrobenzoate, m.p 100°C
(3) s-Amyl p-nitrobenzoate, m.p 17°C

b.p 118°C **n-Butyl Alcohol** (Butan-1-ol) $CH_3CH_2CH_2CH_2OH$

(1) Colourless liquid, sparingly soluble in water, miscible with alcohol and ether.
(2) Reflux a mixture of 5ml alcohol with 6ml glacial acetic acid and a drop of conc H_2SO_4 for 3-5hrs. Pour the mixture in a separating funnel containing water. Remove the upper layer and wash first with sodium bicarbonate and then with water. Distill on wire gauze and collect Butyl acetate at b.p 124-125°C.
(3) To ice cold sodium nitrite solution (3.8gm in 15ml water) add ice cold n-butyl alcohol solution(1ml water + 1.5ml conc H_2SO_4 + 5ml n-butyl alcohol)The resulting mixture is allowed to stand in ice bath until it separates into layers. The lower aqueous layer is removed and n-butyl nitrite is recovered at b.p 76-77°C.
(4) Add 27gm anhydrous zinc chloride to 19gm conc HCl and to this solution add 10ml of alcohol. Reflux the mixture for 4hrs. Remove the upper layer, mix with equal volume of conc H_2SO_4 and heat for 30 min. Distill and collect n-butyl chloride at b.p 76-78°C.

(5) n-Butyl 3, 5-dinitrobenzoate, m.p 63°C. (6) n-Butyl p-nitrobenzoate, m.p 36°C.

b.p 119°C **s-Amyl Alcohol** (Pentan-2-ol, methyl n-propyl carbinol) $CH_3CH(OH)CH_2CH_2CH_3$

(1) Colourless liquid, soluble in slightly warm water, miscible with alcohol and ether.
(2) To 2ml of alcohol add acidic potassium dichromate solution. A pleasant sweet odour due to formation of methyl n-propyl ketone is produced.
(3) In a test tube place 2ml of test sample and 2ml of iodine reagent. Now add 6-7 ml dil NaOH solution. Heat on water bath for 2min, cool and collect iodoform, m.p 119°C.
(4) s-Amyl 3, 5-dinitrobenzoate, m.p 62°C.

b.p 128°C **Active amyl alcohol** (2-Methyl butanol-1) $CH_3CH_2CH(CH_3)CH_2OH$

(1) Optically active liquid, sparingly soluble in water, miscible with alcohol and ether.
(2) Active amyl 3, 5-dinitrobenzoate, m.p 70°C

b.p 132°C **Isoamyl Alcohol** (3-methyl butan-1-ol, isopentyl alcohol) $CH_3CH(CH_3)CH_2CH_2OH$

(1) Colourless liquid having irritating odour, insoluble in water, miscible with alcohol and ether.
(2) Stir a mixture of 20gm of NaOH, 40ml of alcohol, 300ml ether and 25ml CCl_4 for 30min. Add 36gm of CS_2 drop by drop (temp below 30°C). Continue stirring for 3hrs and then add 75gm of methyl iodide. Reflux the mixture for 6hrs. Distill and collect methyl xanthate at b.p 101-103°C/10mmHg
(3) Isoamyl 3, 5-dinitrobenzoate, m.p 62°C.
(4) Isoamyl p-nitrobenzoate, m.p 21°C.

b.p 138°C **n-Amyl Alcohol** (Pentan-1-ol) $CH_3CH_2CH_2CH_2CH_2OH$

(1) Colourless liquid, sparingly soluble in water, miscible with alcohol and ether.

(2) n-Amyl 3, 5-dinitrobenzoate, m.p 46°C.

b.p 139°C **Hexan-2-ol** (n-Butyl methyl carbinol)

$CH_3CH_2CH_2CH_2CHOHCH_3$

(1) Colourless liquid, very slightly soluble in water, soluble in alcohol and ether.
(2) To 10ml alcohol add 15ml HBr and 3.5ml conc H_2SO_4. Distill the reaction mixture and collect 2-bromohexane at 140-144°C.
(3) In a test tube place 2ml of test sample and 2ml of iodine reagent. Add dil NaOH solution and heat on water bath for 2min, cool and collect iodoform, m.p 119°C
(4) n-Butyl Methyl Carbinyl 3, 5-dinitrobenzoate, m.p 38°C.
(5) n-Butyl Methyl Carbinyl p-nitrobenzoate, m.p 40°C.

b.p 140°C **Cyclopentanol** 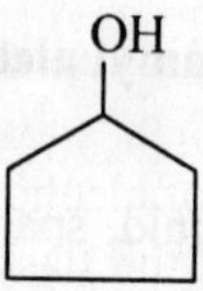

(1) Colourless liquid, sparingly soluble in water, easily soluble in alcohol.
(2) Heat 5ml of compound with 10ml dil HNO_3 until the evolution of oxides of nitrogen has stopped. Cool, pour in ice and collect glutaric acid, m.p 97°C.
(3) To 10ml alcohol add 14.2ml HBr and 3.5ml conc H_2SO_4. Distill the reaction mixture and collect cyclopentyl bromide, b.p 135-136°C.
(4) Cyclopentyl 3, 5-dinitrobenzoate, m.p 115°C.
(5) Cyclopentyl p-nitrobenzoate, m.p 62°C

b.p 157°C **n-Hexyl Alcohol** (Hexan-1-ol)

$CH_3CH_2CH_2CH_2CH_2CH_2OH$

(1) Colourless liquid, insoluble in water, soluble in alcohol, miscible with ether.
(2) Reflux 10ml of alcohol with 0.5gm red phosphorous and slowly add dropwise 2ml of bromine. When all bromine has been added

continue refluxing for 30min. Carry out the downward distillation and collect 1-bromo hexane at 154-156°C.

(3) n-Hexyl 3, 5-dinitrobenzoate, m.p 61°C.

b.p 160°C **s-Heptyl alcohol** (Heptan-2-ol, n-Amyl methyl carbinol)

$CH_3CH_2CH_2CH_2CH_2CH(OH)CH_3$

(1) Colourless liquid, insoluble in water, soluble in alcohol and ether.
(2) s-Heptyl 3, 5-dinitrobenzoate, m.p 49°C.

b.p 161°C **Cyclohexanol** (hexahydrophenol)

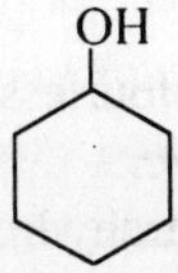

(1) Colourless hygroscopic liquid, insoluble in water, easily soluble alcohol and ether.
(2) Reflux together 5ml of alcohol with 10ml of conc HNO_3 on water bath for 4-5hrs. Cool the flask and collect crystals of adipic acid, m.p 152°C.
(3) Pass dry HCl gas to 7.5gm of cyclohexanol so that its weight is increased by 2-3gm. Add 13.5gm of glacial acetic acid and reflux for 14hrs. Cool, remove the upper layer and wash first with sodium bicarbonate and then with water. Distill on wire gauze and collect cyclohexyl acetate at b.p 168-172°C.
(4) Reflux 10gm of cyclohexanol with 25ml of conc HCl and 8gm of anhydrous $CaCl_2$ for 10hrs. Cool and separate the upper layer, wash with $NaHCO_3$ and distill. Recover chlorocyclohexane at b.p 140-142°C.
(5) Cyclohexyl 3, 5-dinitrobenzoate, m.p 113°C.
(6) Cyclohexyl p-nitrobenzoate, m.p 50°C.

b.p 176°C **n-Heptyl Alcohol** (Heptanol-1)

$CH_3CH_2CH_2CH_2CH_2CH_2CH_2OH$

(1) Colourless liquid, insoluble in water, miscible with alcohol and ether.
(2) Reflux with stirring 10ml of alcohol with 500mg of red

phosphorous, add dropwise 2ml bromine. When all bromine has been added continue refluxing for further 30 min. Carry out downward distillation and collect 1-bromo heptane, b.p 180°C.

(3) n-Heptyl 3, 5-dinitrobenzoate, m.p 48°C.

b.p 177°C **Tetrahydrofurfuryl alcohol** (tetrahydro 2-furan carbinol)

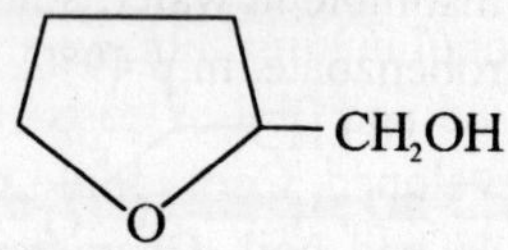

(1) Colourless, odourless liquid, miscible with water, alcohol and ether.

(2) Tetrahydrofurfuryl 3, 5-dinitrobenzoate, m.p 84°C.

(3) Tetrahydrofurfuryl p-nitrobenzoate, m.p 48°C.

b.p 178°C **s-Octyl Alcohol** (Octan-2-ol, n-hexyl methyl carbinol)

$CH_3CH_2CH_2CH_2CH_2CH_2CH(OH)CH_3$

(1) Colourless liquid, insoluble in water, soluble in alcohol and ether.

(2) s-Octyl 3, 5-dinitrobenzoate, m.p 32°C

(3) s-Octyl p-nitrobenzoate, m.p 28°C.

b.p 189°C **Propylene glycol** (Propan 1, 2 diol, 1, 2 dihydroxy propane)

CH_3
|
CHOH
|
CH_2OH

(1) Viscous liquid with sweet taste, miscible with water and alcohol, soluble in ether.

(2) Heat 2ml of compound with HI, the solution turns yellow.

(3) Propylene glycol (di) p-nitrobenzoate, m.p 127°C.

(4) Propylene glycol 3, 5-dinitrobenzoate (di), m.p 147°C

b.p 198°C **Ethylene glycol** (Ethane –1, 2 diol, Glycol) $\begin{array}{c} CH_2OH \\ | \\ CH_2OH \end{array}$

(1) Colourless viscous hygroscopic liquid with sweet taste, miscible with water and alcohol but insoluble in ether and benzene.
(2) To 2ml of compound add 2ml of conc H_2SO_4 and heat until the brown colour is developed. Cool, dilute with 6ml of water, make alkaline with NaOH and boil. Characteristic smell of aldehyde resin is obtained.
(3) Dissolve 1.5ml of compound in 7.5ml of Na_2CO_3 solution and add 75mg of $KMnO_4$ to it. Boil for 2 minutes and filter. If the filtrate is pink or purple add H_2O_2 to destroy the colour. Acidify the filtrate with glacial acetic acid, heat to boiling and then add few drops of $CaCl_2$. A white precipitate of calcium oxalate is obtained.
(4) To 3ml of compound add 6ml benzoyl chloride. Now add NaOH solution and shake thoroughly until the smell of benzoyl chloride persists. Cool and collect the dibenzoate derivative, m.p 73°C.
(5) Ethylene glycol (di) 3, 5-dinitrobenzoate, m.p 169°C. (6) Ethylene glycol (di) p-nitrobenzoate, m.p 140°C

b.p 198°C **Linalool** (l linalyl alcohol)

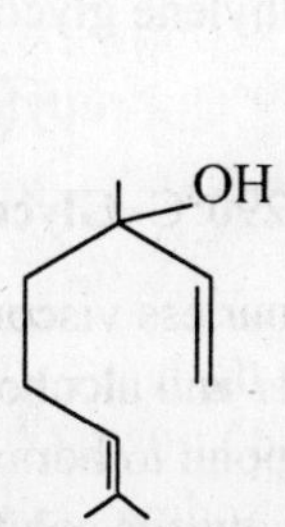

(1) Colourless optically active liquid with a pleasant odour, very slightly soluble in water, miscible with alcohol and ether.
(2) Reflux 10ml of compound with chromic acid. Cool and recover citral a (geranial), b.p 229-230°C.
(3) Reflux 5ml of compound with $KMnO_4$, followed by CrO_3 and conc H_2SO_4. A sweet smell due to the formation of acetone and levulinic acid is obtained.

(4) L Linalyl p-nitrobenzoate, m.p 70°C

b.p 229°C **Geraniol** 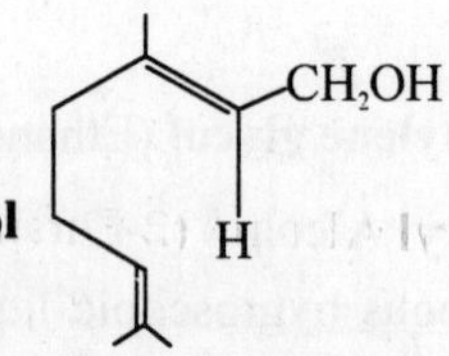

(1) Colourless liquid with smell of rose and geranium, insoluble in water, easily soluble alcohol and ether.
(2) To 2ml of compound add 3ml of bromine solution (bromine in CCl_4) and shake thoroughly. Cool and collect geraniol tetrabromide, m.p 71°C.
(3) Geranyl 3, 5-dinitrobenzoate, m.p 63°C.
(4) Geranyl p-nitrobenzoate, m.p 35°C.

b.p 245°C **Diethylene glycol** (Di-2-hydroxy ethyl ether)

$HOCH_2CH_2OCH_2CH_2OH$

(1) Colourless, odourless viscous liquid, miscible with water and nearly all organic solvents.
(2) Diethylene glycol bis-(3, 5-dinitrobenzoate), m.p 149°C.
(3) Diethylene glycol bis-(p-nitrobenzoate), m.p 149°C

b.p 290°C **Glycerol** (Propan-1, 2, 3-triol)

(1) Colourless viscous hygroscopic liquid with sweet taste, soluble in water and alcohol, insoluble in ether.
(2) Respond to borax –phenolphthalein test
(3) To aqueous solution of compound (3 drops in 2ml water) add 7-8 drops of 10% pyragallol and 4ml conc H_2SO_4. Boil for ½ minute, cool and add 15ml alcohol. A purplish red coloured solution is obtained.
(4) Heat 1ml of compound with few crystals of $KHSO_4$ in a dry test tube. A pungent irritating smell of acrolein is obtained.
(5) To 2ml of compound add 6ml benzoyl chloride and 20ml dil NaOH solution. Shake the mixture thoroughly until the smell of

benzoyl chloride disappears. Cool and collect glyceryl tribenzoate, m.p 72°C.

(6) Glyceryl (tri) p-nitrobenzoate, m.p 188°C.

Aromatic Liquids

b.p 171°C **Furfuryl Alcohol** (2-Furan Carbinol)

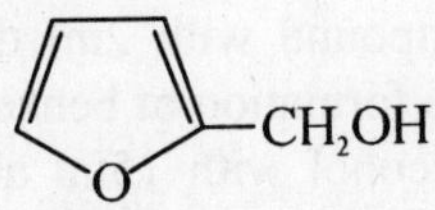

(1) Colourless liquid, miscible with water, alcohol and ether
(2) Reduces Tollens reagent.
(3) Heat 2ml of compound in a test tube and bring a pine splinter soaked in conc HCl at the mouth of test tube. It turns green.
(4) Dissolve 3.8gm thiourea in 5ml water and 4ml conc HCl. Then add dropwise 5ml furfuryl alcohol (exothermic reaction). After the addition is complete keep the green solution for 12hrs at RT. Now add 5ml NaOH solution (4gm in 5ml water). A heavy brown oily layer will separate. Steam distill the mixture and collect 2-furfuryl mercaptan, b.p 155°C.
(5) Furfuryl 3, 5-dinitrobenzoate, m.p 80°C.
(6) Furfuryl p-nitrobenzoate, m.p 76°C.

b.p 204°C **Phenyl ethyl alcohol** (methyl phenyl carbinol)

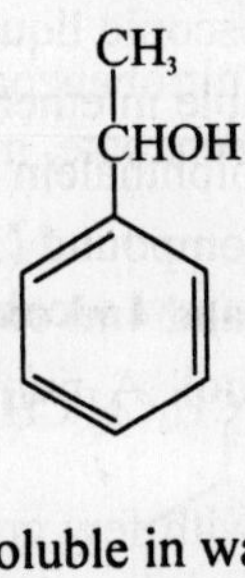

(1) Colourless liquid, insoluble in water, soluble in alcohol and ether
(2) Methyl phenyl carbinyl 3, 5-dinitrobenzoate, m.p 93°C.
(3) Methyl phenyl carbinyl p-nitrobenzoate, m.p 43°C.

b.p 205°C **Benzyl alcohol** (Phenyl methanol, phenyl carbinol)

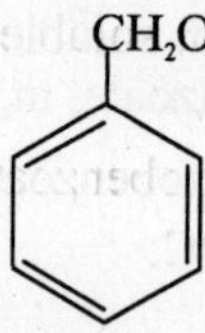

(1) Colourless liquid with faint aromatic odour, soluble in alcohol and ether, insoluble in pet ether.
(2) Heat 1ml of compound with 2ml dil HNO_3. A smell of bitter almond due to the formation of benzaldehyde is observed.
(3) Reflux 5ml of alcohol with 15ml alkaline $KMnO_4$. Cool, filter and acidify the filtrate with dil HCl. Crystals of benzoic acid separate out, m.p 121°C.
(4) Benzyl 3, 5-dinitrobenzoate, m.p 112°C.
(5) Benzyl p-nitrobenzoate, m.p 85°C.

b.p 221°C **2-Phenyl ethyl alcohol** (2-phenyl ethan-1ol, Benzyl carbinol)

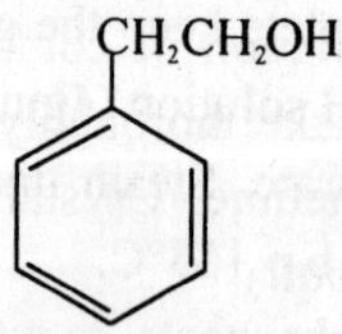

(1) Colourless liquid with faint rose like odour, soluble in water, easily soluble in alcohol and ether
(2) Heat 2ml of alcohol with 100mg anhydrous oxalic acid for 2-3 minutes, cool, first add 2ml water and then 2ml ethanol and then heat again. On cooling crystals of di –(2-phenyl ethyl) oxalate, m.p 51°C separate out.
(3) 2-Phenyl ethyl 3, 5-dinitrobenzoate, m.p 108°C.
(4) 2-Phenylethyl p-nitrobenzoate, m.p 63°C.

b.p 235°C **Hydrocinnmyl alcohol** (3-phenyl-1-propanol)

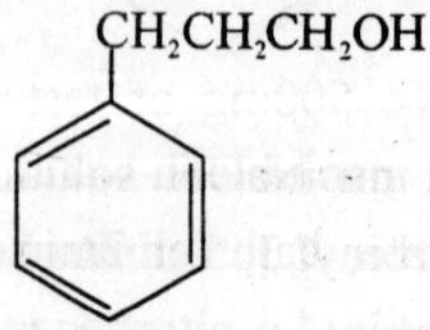

(1) Colourless viscous liquid, soluble in water, miscible with alcohol and ether.

(2) Hydrocinnamyl 3, 5-dinitrobenzoate, m.p 92°C. (3) Hydrocinnamyl p-nitrobenzoate, m.p 46°C.

Aliphatic Solid

m.p 35°C, b.p 220°C α-**Terpineol**

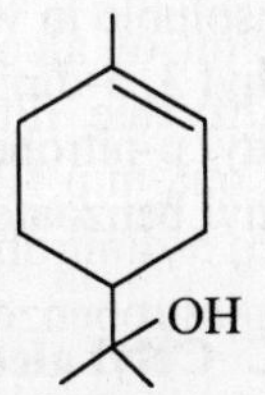

(1) Colourless optically active compound with lilac like odour, insoluble in water, easily soluble alcohol and ether

(2) To 3ml ethanolic solution of compound add 1.5ml of bromine drop by drop and shake thoroughly. Leave the solution at room temperature for sometime. Crystals of dipentene tetrabromide, m.p 124°C seperate out.

(3) α-Terpinyl 3, 5-dinitrobenzoate, m.p 79°C

(4) α-Terpinyl p-nitrobenzoate, m.p 139°C.

m.p 42°C **Pinacol** (Tetramethylene glycol, 2, 3-dimethyl butane-2, 3-diol) $(CH_3)_2C(OH)C(OH)(CH_3)_2$

(1) White cyrstalline solid, sparingly soluble in water, easily soluble alcohol and ether

(2) Boil few milligrams of compound with conc H_2SO_4. A peppermint like odour is produced due to formation of methyl butyl ketone (pinacolone).

(3) In a test tube place 500mg of test sample and 4ml of iodine reagent. Now add dil NaOH solution until the solution turns yellow. Heat on water bath for 2min, cool and collect iodoform, m.p 119°C

(4) Acetate (di), m.p 65°C.

m.p 42°C, b.p 215°C **(-) Menthol**

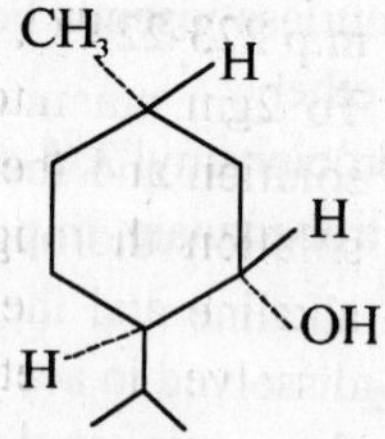

(1) Colourless crystalline optically active solid with peppermint like odour, insoluble in water, soluble in alcohol and ether.
(2) (-) Menthyl 3, 5-dinitrobenzoate, m.p 153°C
(3) (-) Menthyl p-nitrobenzoate, m.p 61°C.
(4) (-) Menthyl benzoate, m.p 54°C.

m.p 50°C **Cetyl alcohol** (Hexadecanol-1) $CH_3(CH_2)_{14}CH_2OH$

(1) Crystalline solid, insoluble in water, soluble in ether.
(2) Cetyl 3, 5-dinitrobenzoate, m.p 66°C (3) Cetyl p-nitrobenzoate, m.p 52°C

m.p 110°C **D-sorbitol** (1, 2, 3, 4, 5, 6-hexane hexol)
$CH_2OH(CHOH)_4\ CH_2OH$

(1) Crystalline solid with sweet taste, soluble in water but insoluble in alcohol
(2) Reflux 1gm of compound with 6-7ml Ac_2O and 500mg of fused $ZnCl_2$ for 2hrs. Cool, pour in ice and stir vigorously with glass rod. Hexa acetyl d-sorbitol separates out, m.p 99°C.
(3) Shake thoroughly 1gm of compound with 6-7ml benzoyl chloride and 15ml of 20% NaOH solution until the smell of bnezoyl chloride perisists. Cool and recover hexabenzoyl d-sorbitol, m.p 216-217° C.

m.p 166°C **D-Mannitol** $CH_2OH(CHOH)_4\ CH_2OH$

(1) White crystalline solid with sweet taste, soluble in water, insoluble in alcohol and ether. Sublimes slowly above its melting point.
(2) To 1ml of $CuSO_4$ solution add NH_4OH solution. Blue precipitate is obtained which dissolves in excess NH_4OH solution. Dilute the solution with 2ml water, add 50mg of compound to it and shake. A blue precipitate is obtained.
(3) To 1gm of compound add 5ml benzaldehyde and 1gm P_2O_5. Heat

on water bath for 5 minutes, cool and collect tribenzal –d-mannitol, m.p 223-224°C.

(4) To 2gm mannitol in 10ml water add 500mg $FeSO_4$ cold water solution and then gradually add 7-8 ml H_2O_2 solution. Cool the solution thoroughly, add Na_2CO_3 solution so that it becomes alkaline and then filter. To the filtrate add 1ml phenylhydrazine dissolved in acetic acid and keep the solution for sometime. Collect the precipitated mannose phenylhydrazone, m.p 198°C.

(5) Heat 200mg of compound with 5ml Ac_2O, 100mg fused sodium acetate containing 2-3drops of pyridine. Cool and pour the reaction mixture in ice with stirring. Mannitol hexa acetate separate out, m.p 126°C.

(6) Mannitol hexa benzoate, m.p 148°C.

m.p 208°C **D Borneol**

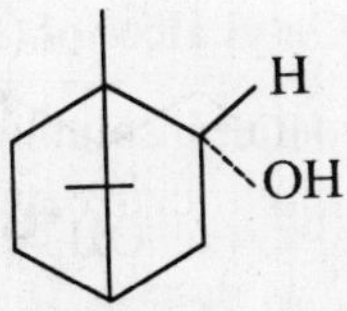

(1) White solid with odour similar to camphor, sparingly soluble in water, easily soluble alcohol and ether. Sublimes readily.

(2) Dissolve 3.75 gm of compound in 10ml ether and cool to 0°C. Add with stirring 25ml chromic acid solution to the cold vigorously stirred borneol solution. Continue stirring for 5minutes. Remove the ether layer and extract the aqueous layer with ether (2 × 15ml). Combine the ether layers. Distill off the ether and collect (-) camphor, m.p 159-164°C.

(3) D-borneol 3, 5-dinitrobenzoate, m.p 154°C.

(4) D-borneol p-nitrobenzoate, m.p 137°C.

m.p 225°C **meso Inositol** (hexahydroxy cyclohexane)

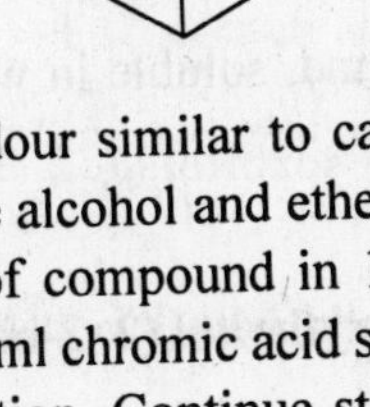

(1) White crystalline solid with a sweet taste, soluble in water, insoluble in alcohol and ether.

(2) Reduces Tollens reagent.

(3) To 20mg of compound in a porcelain dish add few drops of conc HNO_3. Evaporate it to dryness and then add few drops of NH_4OH and equal amount of $CaCl_2$ to the residue. A rose red colouration results.

(4) Heat 500mg of compound with 10ml Ac_2O and 200mg of $ZnCl_2$. Cool and collect hexacetylinositol, m.p 212°C.

(5) Hexabenzoylinositol, m.p 258°C

m.p 234°C **d-Quercitol** (1, 2, 3, 4, 5 pentahydroxy cyclohexane)

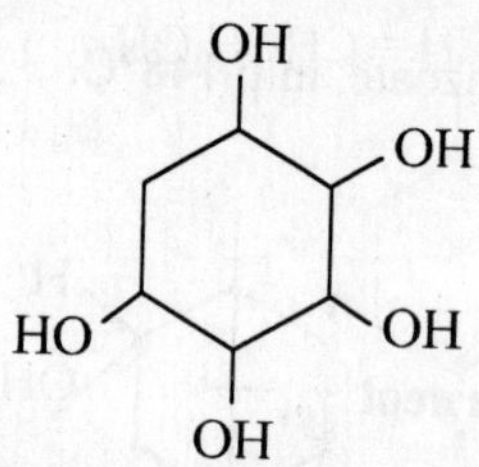

(1) Crystalline compound, soluble in water, difficult soluble alcohol, insoluble in ether.

m.p 261°C **Pentaerythritol** (2, 2-bishydroxy methyl-1, 3-propane diol)

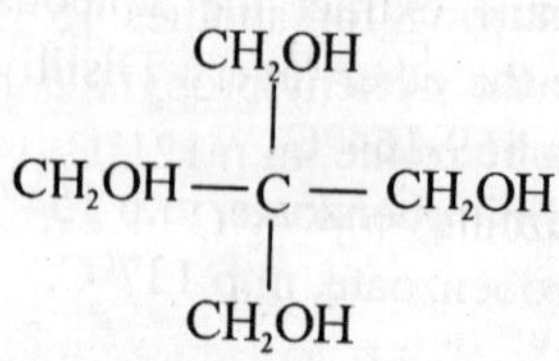

(1) Crystalline compound, soluble in water, slightly soluble in alcohol, insoluble in ether. Sublimes.

(2) Reflux for 2hrs 500mg of compound with 7-8ml acetic anhydride containing few crystals of $ZnCl_2$. Cool, pour in ice water with stirring and collect pentaerythrityl tetraacetate, m.p 83°C.

(3) To 5gm of compound add dropwise with heating on steam bath 7ml of phosphorous tribromide. When the addition is complete heat the mixture on oil bath for 20hrs. Cool, pour in ice and collect Pentaerythrityl bromide, recrystallise from alcohol, m.p

163°C.

(4) Reflux on steam bath 5gm of compound with 10gm NaI and 30ml of ethylmethyl ketone for 48hrs. Distill off the excess solvent. Wash the residue with water and collect Pentaerythrityl iodide, recrystallise from hot alcohol, m.p 233°C.

(5) Pentaerythrityl tetrabenzoate, m.p 99°C.

Aromatic Solid

m.p 33°C, b.p 257°C **Cinnamyl alcohol** (3-phenyl-2-propene-1-ol)

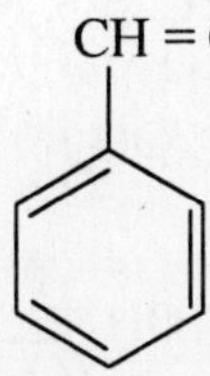

(1) Crystalline compound, slightly soluble in water, soluble in alcohol and ether.

(2) Dissolve 500mg of compound in glacial acetic acid and add 3ml of HBr. Shake the mixture thoroughly for 3hrs. Cool and pour in ice. Collect the precipitated cinnamyl bromide, m.p 28°C.

(3) Dissolve 500mg of compound in 5ml chloroform and then cool the mixture in ice bath. Add dropwise with stirring 1.5ml of bromine. Continue stirring for another 15 minutes. Pour in ice and collect β, γ-dibromo γ-phenylpropyl alcohol, m.p 74°C.

(4) Cinnamyl 3, 5-dinitrobenzoate, m.p 121°C

(5) Cinnamyl p-nitrobenzoate, m.p 78°C

m.p 69°C **Benzhydrol** (diphenyl carbinol, diphenyl methanol)

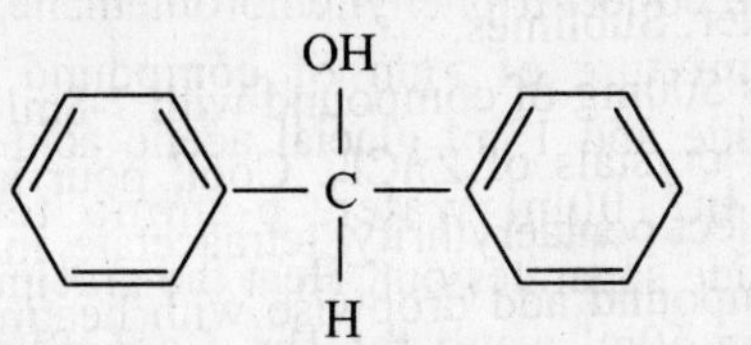

(1) Colourless crystalline solid, slightly soluble in water, easily soluble in alcohol, ether and chloroform.

(2) To few milligrams of compound add 4-5 drops of conc H_2SO_4. A

deep red colouration is produced.

(3) Heat a mixture of 3.6gm of ethylene chlorohydrin, 0.5ml of conc H_2SO_4 and 5ml benzene. Add a solution of 5.5gm of compound in 13ml benzene to it. Heat the mixture under reflux for 4hrs. Cool and add 10ml benzene to the reaction mixture. Remove the benzene layer, distill off the benzene, distill the residual liquid under reduced pressure and collect benzhydryl chloroethyl ether at 144-148°C/1.0mmHgwhich on cooling soon solidifies to a hard white mass, m.p 28°C.

(4) Benzhydrol 3, 5-dinitrobenzoate, m.p 141°C

(5) Benzhydrol p-nitrobenzoate, m.p 131°C.

(6) Benzhydrol benzoate, m.p 88°C.

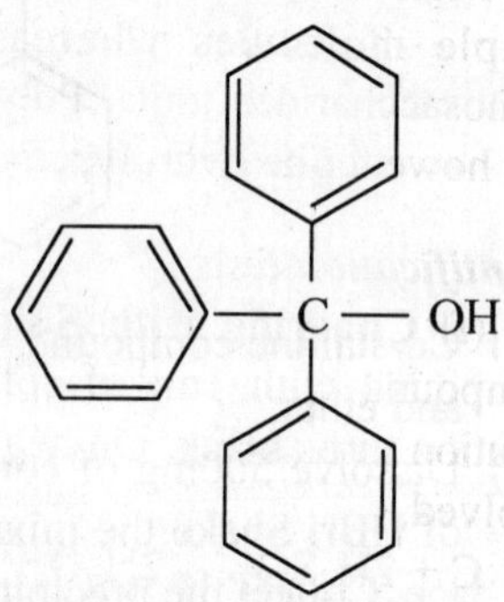

m.p 164°C **Triphenyl carbinol**

(1) Colourless crystalline compound, insoluble in water, soluble in alcohol and ether.

(2) To few milligrams of compound add 4-5 drops of conc H_2SO_4. An intense yellow colouration is produced.

(3) Heat on steam bath 10gm of compound with 5ml benzene and then add 2ml acetyl chloride. When all triphenyl carbinol has disappeared add further 6ml of acetyl chloride and continue heating for another 30 min. Cool and add 5ml pet ether. Keep in ice bath for 2hrs and collect triphenyl chloromethane, m.p 110-112°C.

(4) Reflux a mixture of 5gm of compound with 5.1gm aniline hydrochloride and 13ml glacial acetic acid for 3hrs. While still hot pour in 100ml water, p-amino tetraphenyl methane hydrochloride separates out. Heat the precipitated solid with 2gm of NaOH in 50ml water for 1hr. Cool, filter the precipitated p-amino tetraphenyl methane, recrystallise from toluene, m.p 249-250°C.

(5) Heat 2gm of compound with 15ml alcohol and 15ml conc H_2SO_4

at 70-80°C. Cool and collect triphenyl methane, 92°C.

(6) To a cooled solution of 1gm of triphenylcarbinol in 12ml acetic anhydride add dropwise with stirring 1ml of 70% perchloric acid. Continue stirring the mixture in cooling bath for 60min. Filter the red coloured crystals of triphenylmethyl perchlorate, m.p 142-143°C

(7) Triphenyl carbinyl benzoate, m.p 167°C.

Carbohydrates

Carbohydrates are polyhydroxyaldehydes, polyhydroxyketones or compounds that are easily hydrolyzed to them. They are insoluble in ether but freely soluble in water.

Monosaccharides are carbohydrates which cannot be hydrolyzed to simple molecules, whereas disaccharides can be hydrolyzed to two monosaccharides units. Polysaccharides or biopolymers on the other hand can however be hydrolyzed to a number of monosaccharide units.

Identification tests

(*i*) **Charring with Sulphuric acid:** In a dry test tube heat 0.2gm of compound with 1ml of conc H_2SO_4. Immediate charring occurs and the solution turns black. On heating the test tube further CO_2 and SO_2 gases are evolved.

$$C + 2H_2SO_4 \rightarrow 2H_2O + CO_2 + 2SO_2$$

(*ii*) **Molisch test:** In a dry test tube dissolve 40mg of unknown compound in 2ml water and add 4-5 drops of 10% solution of 1-naphthol in ethanol (molisch reagent). Hold the test tube at an angle and slowly add 2ml conc H_2SO_4 along the sides of test tube. The development of a reddish violet ring at the junction of the two liquids indicate the presence of carbohydrate.

Pentoses on treatment with sulphuric acid usually give furfural whereas hexoses produce 5-hydroxymethyl furfural. This colour development is probably due to the formation of condensation product between 1-naphthol and furfural or its hydroxymethyl derivative.

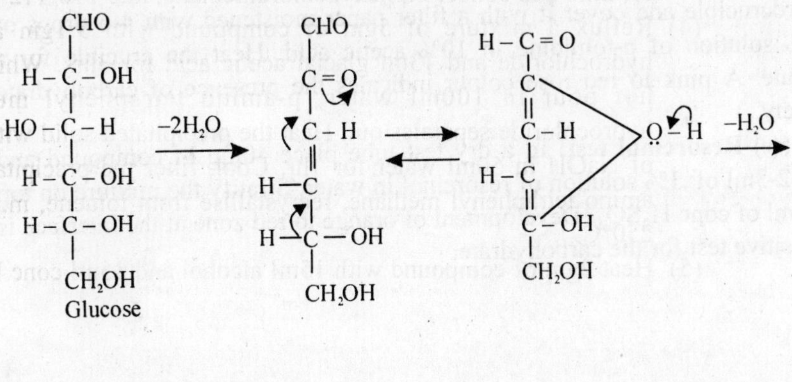

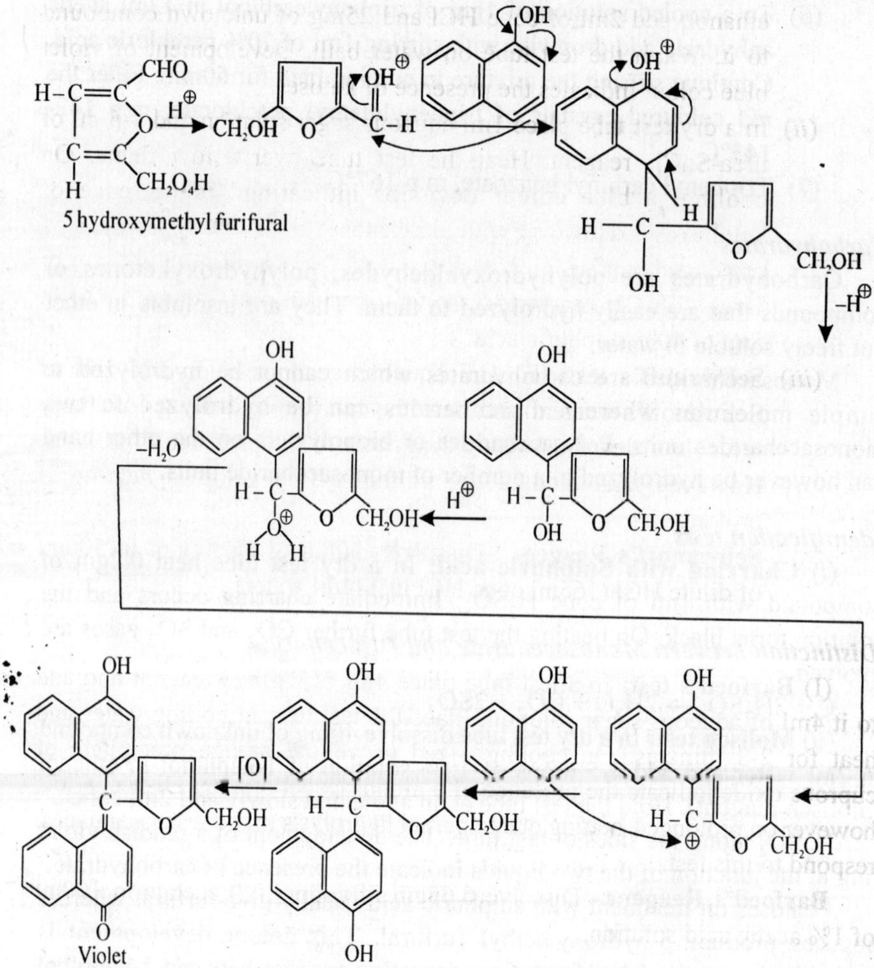

(*iii*) **p-toluidine acetate test:** Place 20-30mg of substance in a microcrucible and cover it with a filter paper moistened with 4-6 drops of 10% solution of p-toluidine in 10% acetic acid. Heat the crucible for a minute. A pink to red rose colour indicates the presence of carbohydrate moiety.

(*iv*) **Resorcinol test**: In a dry test tube place 40mg of compound and add 2-3ml of .1% solution of resorcinol in water. Stratify the mixture on top of 4ml of conc H_2SO_4. Development of orange to red zone at the interface is a positive test for the carbohydrate.

Distinction between Ketoses and Aldoses

(*i*) In a dry test tube place 2ml of 2% diphenylamine solution in ethanol, add 2ml of conc HCl and 25mg of unknown compound to it. Warm the test tube on water bath. Development of violet blue colour indicates the presence of ketoses.

(*ii*) In a dry test tube place 1ml aqueous sugar solution and 4-5 ml of urea-$SnCl_2$ reagent. Heat the test tube over a low flame. On cooling, a blue colour develops indicating the presence of ketohexoses. However with aldohexoses a red colour develops on prolonged heating.

Urea - $SnCl_2$ reagent: dissolve 2gm urea and 200mg $SnCl_2$ in 5ml of 40% sulphuric acid.

(*iii*) **Seliwanoff's test:** To 1ml aqueous sugar solution add 1ml of Seliwanoff's reagent and heat the mixture to boiling. A cherry red colour develops within 2 minutes in the presence of ketoses. However with aldoses the colour will develop if heating is prolonged.

Seliwanoff's Reagent :-Dissolve 250mg of resorcinol in 500ml of dilute HCl (166ml conc HCl in 334ml water)

Distinction between Monosaccharide and Disaccharide

(I) **Barfoed's test:** In a test tube place 4ml of barfoed reagent and add to it 4ml of aqueous sugar solution. Place the test tube in boiling water and heat for 2-3 minutes. On cooling a red or yellow orange precipitate of cuprous oxide indicate the presence of a monosaccharide unit. Disaccharide, however on prolonged heating due to partial hydrolysis to monosaccharaiades, respond to this test.

Barfoed's Reagent:- Dissolve 6.65gm of copper (*II*) acetate in 100ml of 1% acetic acid solution.

Distinction between Reducing and Non-reducing Sugars

(*i*) **Fehling's test :** To 4ml of warm Fehling's solution (2ml of Fehling's A + 2ml of Fehling's B solution) add 2ml of aqueous carbohydrate solution (200mg in 4ml water). Heat the mixture to boiling for 2-3 minutes. A yellow or red precipitate of cuprous oxide indicates the presence of reducing sugar.

Preparation of **Fehling's solution-<u>SolutionA</u>**: Dissolve 34.64 gm of $CuSO_4$ in water containing few drops of dilute H_2SO_4 and then make the solution upto 500ml

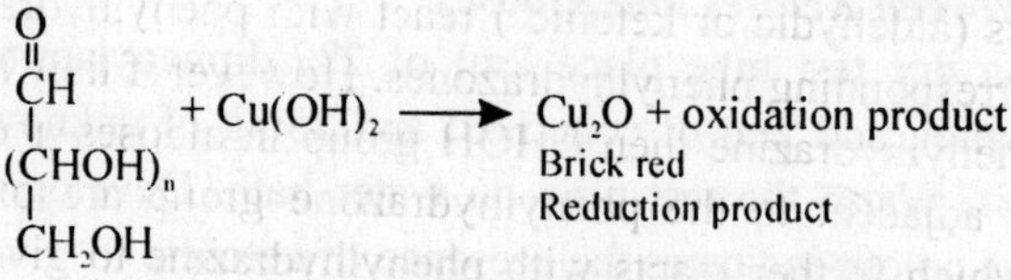

Solution B:-Dissolve 50gm of NaOH and 173 gm of Rochelle salt (Sodium Potassium tartarate) in water and make it upto 500ml.

(ii) **Benedicts test:** Place 4ml of benedicts reagent in a test tube and add to it 30-40mg of unknown sample or 2ml of its aqueous solution. Heat the test tube in boiling water for 3 minutes. Appearance of a red precipitate indicate the presence of a reducing sugar. However in case of non reducing sugar the solution remains clear.

Preparation of **Benedicts reagent**: Dissolve 43.25 gm of Sodium citrate and 25gm of sodium carbonate in 175ml of water. Add a solution of 4.32 gm of $CuSO_4$ in 25ml water with stirring and then make the solution upto 250ml

Derivatives

(i) **Osazones:** In a test tube dissolve 400mg of the carbohydrate in 8ml water, add 800mg of phenylhydrazine hydrochloride and 1.2gm of sodium acetate to it. Place the test tube in boiling water. Note the time of immersion and the separation of osazone. Cool the test tube and filter the precipitated product. The precipitated osazone may be white, yellow or yellow orange in colour.

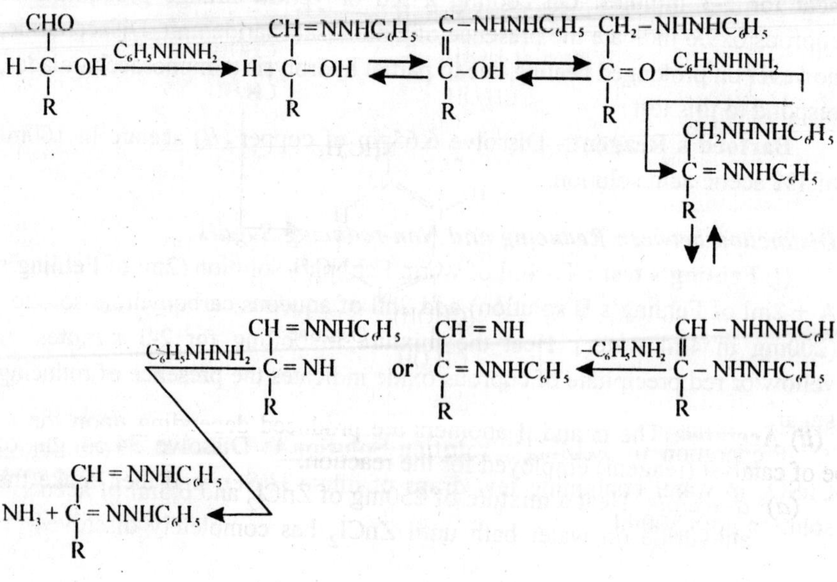

Carbohydrates (aldehydic or ketonic) react with phenylhydrazine (one mole) forming corresponding phenylhydrazones. However if they are heated with excess of phenylhydrazine then CHOH group in aldoses and CH_2OH group in ketoses adjacent to the phenylhydrazone group are oxidised to carbonyl group which further reacts with phenylhydrazine to give osazone or di phenylhydrazone.

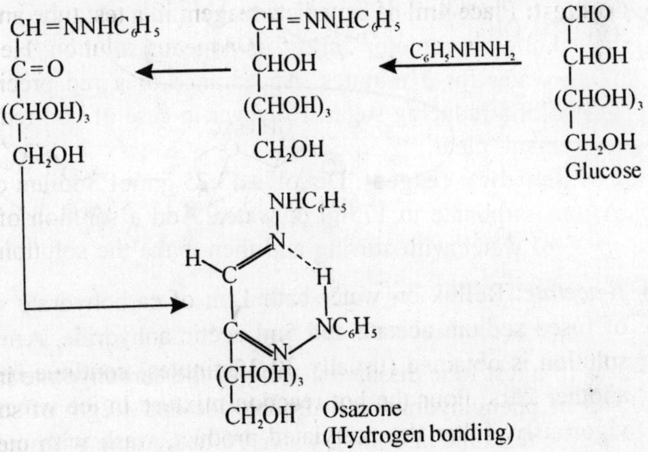

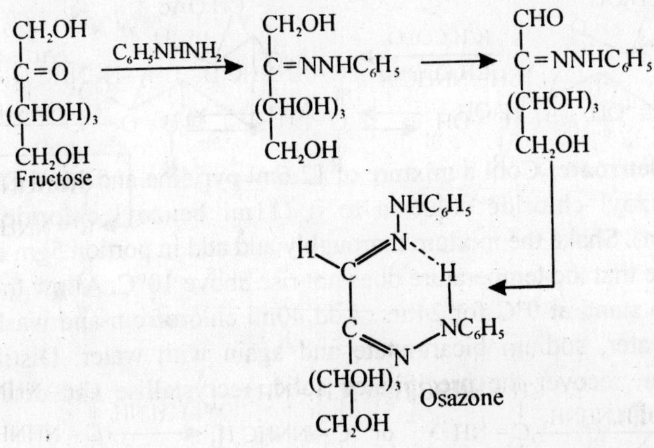

***(ii)* Acetates:** The α and β anomers are produced depending upon the type of catalyst (reagent) employed for the reaction.

(*a*) *α-acetate:* Heat a mixture of 250mg of $ZnCl_2$ and 6.5ml of acetic anhydride on water bath until $ZnCl_2$ has completely dissolved.

Slowly add 1gm of glucose and heat the resultant mixture on water bath for 45-60 minutes. Cool, pour the mixture in ice and stir vigorously. Filter the precipitated solid, recrystallise from ethanol and determine its melting point.

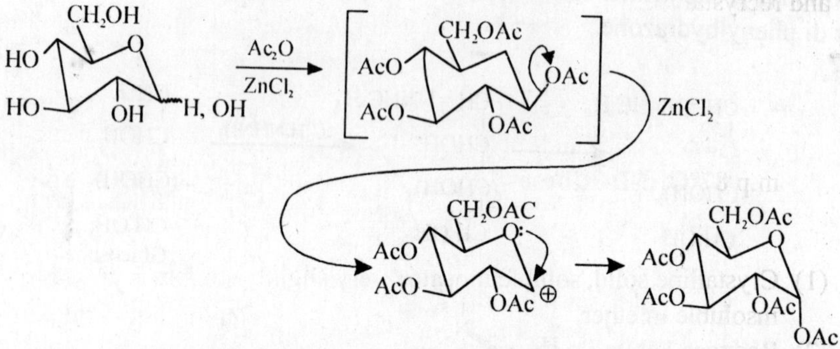

(*b*) *β-acetate*: Reflux on water bath 1gm of carbohydrate with 1gm of fused sodium acetate and 5ml acetic anhydride. After a clear solution is obtained (usually 30-35minutes) continue heating for another 2hrs. Pour the hot reaction mixture in ice water and stir vigorously. Filter the acetylated product, wash with cold water, recrystallise from ethanol and determine its melting point.

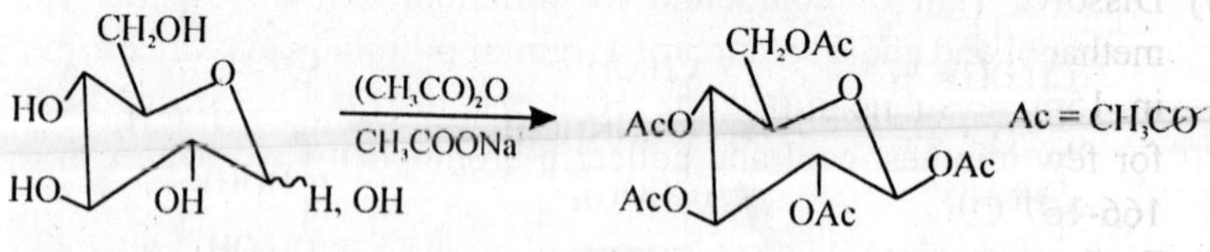

(*iii*) **Benzoate:** Cool a mixture of 12.6ml pyridine and 11ml $CHCl_3$, add 22ml benzoyl chloride solution to it (11ml benzoyl chloride in 11ml chloroform). Shake the mixture thoroughly and add in portion 5gm of glucose taking care that the temperature does not rise above 10°C. Allow the reaction mixture to stand at 0°C for 24hrs. Add 40ml chlorofrom and wash with dil H_2SO_4, water, sodium bicarbonate and again with water. Distill off the chloroform, recover the precipitated solid, recrystallise and determine its melting point.

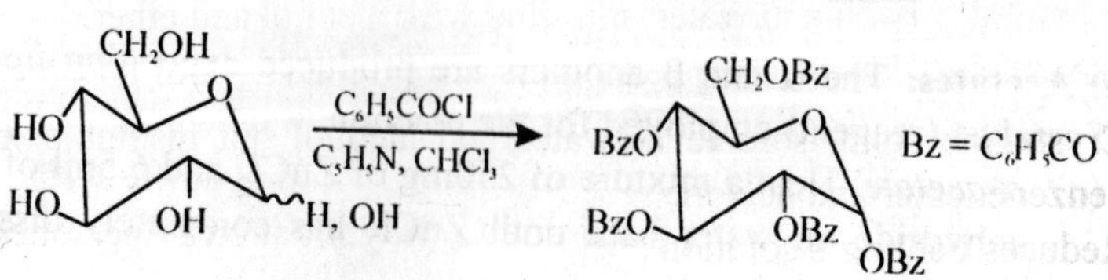

(*iv*) **Oxidation:** In a dry test tube place 1gm of carbohydrate and add 4ml of conc HNO_3 to it. Heat the reaction mixture on water bath until a syrupy liquid is obtained. Add 10ml water to it and shake. Allow the reaction mixture to cool. Filter the precipitated meso acid (mucic acid), wash with water and recrystallise.

m.p 87°C β-**D-Ribose**

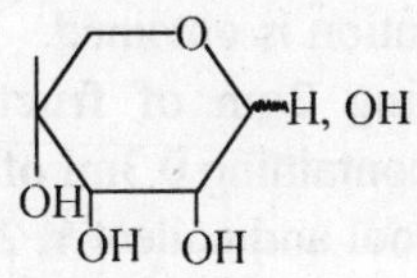

(1) Crystalline solid, soluble in water, very slightly soluble in alcohol, insoluble in ether
(2) Reduces Fehling's solution
(3) Reduces Barfoed's reagent.
(4) Dissolve 1gm of compound in 14ml water and add .3N H_2SO_4 so that the pH becomes 4. Now add with stirring 11ml aniline solution (1ml in 10ml alcohol). Continue stirring for 20 minutes, keep in refrigerator and collect crystals of D-ribopyranosylamine, m.p 125-127°C.
(5) Dissolve 1gm of compound in minimum amount of absolute methanol and add a solution of 1.1gm of p-bromophenylhydrazine in 15ml absolute methanol. Heat the mixture on steam bath for few minutes, cool and collect p-bromophenylhydrazone, m.p 166-167°C.
(6) D-ribose phenylosazone, m.p 163°C.
(7) D-ribose tetraacetate (β), m.p 109-110°C

m.p 103-105°C **D-Fructose** (Laevulose)

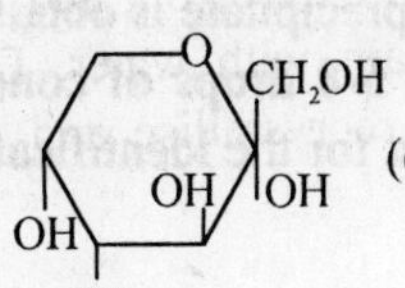

(α - D - pyranose form)

(1) Crystalline solid, soluble in water and alcohol but insoluble in benzene, ether and acetone.
(2) Reduces Fehling's solution

(3) Reduces Barfoed's reagent.
(4) To 500mg of substance add few milligrams of resorcinol and 0.5ml of conc HCl. On heating wine red colour is obtained.
(5) To 500mg of compound add 1ml ammonium molybdate solution and heat. A blue coloured solution is obtained.
(6) In a porcelain dish scatter 50mg of compound and add 4-5 drops of 2N NaOH solution followed by 50 mg of solid alkali. A blood red coloured solution is obtained.
(7) Add with stirring 2gm of fructose to 4.5ml of ice cooled cyclohexanone containing 0.3ml of conc H_2SO_4. Continue stirring for 30minutes, cool and collect 1, 2:4, 5-di-O-cyclohexylidene-D-fructopyranose, recrystallise from n-heptane, m.p 145-156°C.
(8) Fructose pentacetate (α), m.p 70°C.
(9) Fructose pentacetate (β), m.p 108°C.
(10) Fructose pentabenzoate, m.p 79°C.
(11) Phenylosazone, m.p 204 °C (time of formation 2-5 minutes)

m.p 118°C, hydrated m.p 78°C **Raffinose** (melitose)

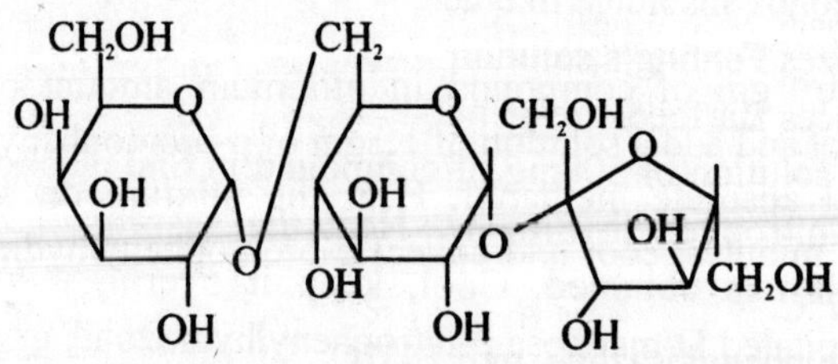

(1) Trisaccharide, fairly soluble in water, insoluble in alcohol.
(2) Does not reduce Fehling's solution.
(3) To 4ml aqueous solution of the compound add 1gm lead acetate and heat the contents to boiling. Dilute the reaction mixture with ammonium hydroxide. A white precipitate is obtained.
(4) To 10ml raffinose solution add 5-8 drops of conc HCl and boil gently. Cool and perform the test for the identification of glucose, galactcse and fructose.
(5) Hendecaacetate, m.p 99°C.

m.p 126°C, hydrated m.p 105°C **L-Rhamnose** (6-deoxy-L-Mannose)

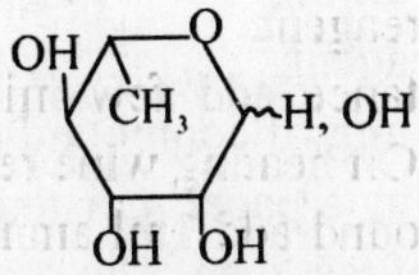

(1) Crystalline solid with sweet taste, slightly soluble in water.
(2) Reduces Fehling's solution.
(3) L-Rhamnose tetraacetate (β), m.p 98-99°C.
(4) L-Rhamnosephenylosazone, m.p 182-185°C (9-10 min)
(5) L-Rhamnose phenylhydrazone, m.p 159°C
(6) L-Rhamnose 2, 4-dinitrophenylhydrazone, m.p 164-165°C

m.p 132°C **D-Mannose** 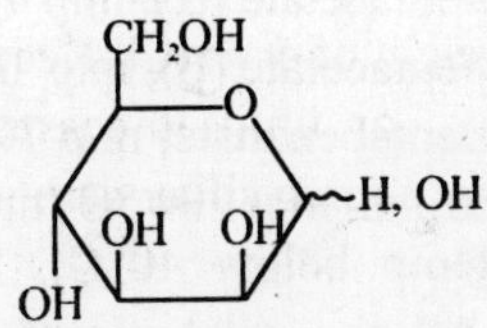

(1) White solid with sweet taste, soluble in water, sparingly soluble in alcohol, insoluble in ether.
(2) Reduces Fehling's solution
(3) Reduces Barfoed's reagent.
(4) To a solution of 500mg of compound in 6ml alcohol add 500mg of p-nitrophenylhydrazine. Heat the suspension until a clear solution is obtained. Cool, keep it overnight and filter the precipitated D-mannose p-nitrophenylhydrazone, m.p 201-202°C.
(5) D-Mannose pentacetate (α), m.p 64°C. (6) D-Mannose pentacetate (β), m.p 117°C. (7) Phenylhydrazone, m.p 199°C (within 30 sec) (8) Oxime, m.p 176-184°C

m.p 145°C **D-Xylose** 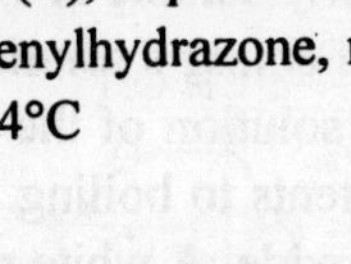

(α-pyranose form)

(1) Crystalline solid with sweet taste, soluble in water and alcohol, insoluble in ether

(2) Reduces Fehling's solution.
(3) Reduces Barfoed's reagent.
(4) In a test tube place 50mg of compound and add 3ml dil HCl to it. Heat the test tube, bring a paper moistened with aniline and acetic acid near the mouth of test tube. The paper turns red.
(5) To 5gm of xylose in 6ml conc HCl add 6ml ethanethiol and shake thoroughly. Cool and add lead carbonate until neutral and then filter. Remove excess lead by passing H_2S gas and filter. Evaporate the filtrate and dissolve the syrupy solution in 3.5ml pyridine and 7ml Ac_2O, cool to 0°C and then leave overnight. Pour in ice and leave for 4 days. Collect crystals of D-xylosediethyl dithioacetal tetraacetate, m.p 46-48°C.
(6) Dissolve 6gm of dry xylose in 120gm of methyl alcohol containing 1% of its weight of dry HCl. Keep the solution at room temperature for 5-7 days or until it ceases to reduce Fehling's solution. Neutralise by adding silver carbonate and evaporate under reduced pressure (temp below 40°C). Extract with hot ethyl acetate. Evaporate off the solvent under reduced pressure and collect methylxylofuranoside, b.p 161°C/.03mmHg
(7) D-xylose tetraacetate (α), m.p 59°C
(8) D-xylose tetraacetate (β), m.p 126°C.
(9) D-xylose phenylosazone, m.p 163°C (10-15 min)
(10) D-xylose tetrabenzoate (α), m.p 119-121°C

m.p (α) 146°C (β) 150°C, hydrated m.p 85°C **D-Glucose** (Dextrose, Pento hexanal)

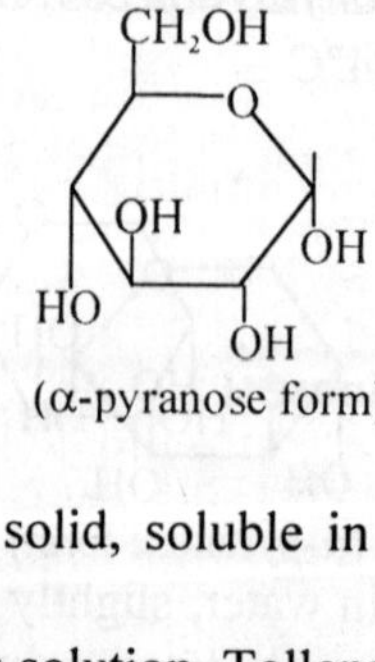

(α-pyranose form)

(1) White crystalline solid, soluble in water, insoluble in ether and alcohol
(2) Reduces Fehling's solution, Tollens and Benedict's reagent.
(3) Reduces Barfoed's reagent.

(4) Dissolve 1gm of substance in 3ml water and add 1ml lead acetate followed by excess of NH_4OH. A white precipitate is obtained which turns pink on heating for few minutes.
(5) To 8ml ice cooled Ac_2O add 1-2 drops of 6% perchloric acid. Now add in portion 2gm D-glucose (temp 30-40°C). Cool and add 700mg of red phosphorous followed by dropwise addition of 1.1ml bromine. Continue stirring for 2hrs (temp below 20°C). Add 5ml water and 15ml methylene chloride and filter. Add the organic layer to 10ml ice cold $NaHCO_3$. Distill off the solvent and collect α-acetobromo D-glucose, recrystallise from pet ether, m.p 88-89°C.
(6) Add with stirring 4.5gm of D-glucose to 10ml of ice cooled cyclohexanone containing 1.3ml of conc H_2SO_4. Keep the mixture overnight, add 8ml of n-heptane and boil on water bath. Remove the n-heptane and repeat the process twice. Combine the organic layer, concentrate and keep in refrigerator. Collect the crystalline 1, 2:5, 6 di-O-cyclohexylidene- α-D-glucofuranose, m.p 131-132°C.
(7) Reflux 20gm of anhydrous glucose with 40gm of dry methyl alcohol containing 2-3gm HCl for 4½ hr. Add a pinch of animal charcoal and filter. Cool the filtrate in ice bath and leave for 30 min. Fine crystals of α-methylglucosides separate out, m.p 165°C (recrystallise from methyl alcohol). Filter, concentrate the filtrate and cool, β-methyl glucoside separate out, m.p 110°C
(8) D-glucose pentaacetate (α), m.p 114°C.
(9) D-glucose pentaacetate (β), m.p 135°C.
(10) D-glucose pentabenzoate, m.p 179°C.
(11) D-glucose phenylosazone, m.p 205°C (5-10 minute)

m.p 159°C **L-Arabinose**

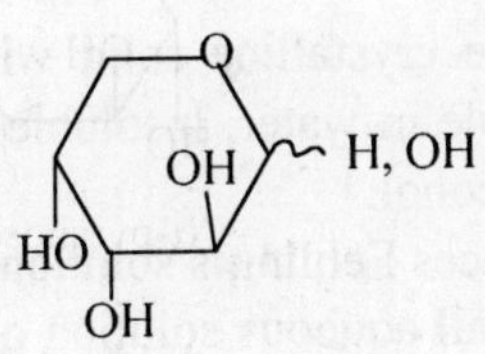

(1) White solid soluble in water, slightly soluble in alcohol, insoluble in ether.
(2) Reduces Fehling's solution.
(3) Reduces Barfoed's reagent.

(4) Heat 200mg of compound with a equal amount of phloroglucinol and conc HCl. A red precipitate is obtained.
(5) Dissolve 1gm of compound in 10ml water and add a solution of 2gm of p-bromophenylhydrazine in 20ml water and 7ml 50% acetic acid. Shake thoroughly and keep the mixture overnight. Filter the prcipitated L-arabinose p-bromophenylhydrazone, recrystallise from alcohol, m.p 167-168°C.
(6) To 10ml ice cold Ac_2O add 1-2 drops of 60% perchloric acid. Now add in portion 2.5gm L-arabinose (temp 30-40°C). Cool and add 700mg of red phosphorous followed by dropwise addition of 1.5ml bromine (temp below 0°C). Continue stirring for 2hrs and then add 5ml water. Now add 15ml methylene chloride and shake. Remove the organic layer and wash with 10ml $NaHCO_3$. Evaporate off the solvent and collect 2, 3, 4-tri-O-acetyl-β-L-arabinopyranosyl bromide, recrystallise from benzene, m.p 136-138°C.
(7) L-arabinose tetraacetate (α), m.p 96°C
(8) L-arabinose tetraacetate (β), m.p 86°C.
(9) L-arabinose phenylosazone, m.p 166°C (15-30 min).
(10) L-arabinose tetrabenzoate (α), m.p 164-165°C.
(11) L-arabinose tetrabenzoate (β), 157-158°C

m.p 165°C, hydrated m.p 102-110°C **Maltose** (4-O-α-D-Glucopyranosyl-β-D-Glucopyranose)

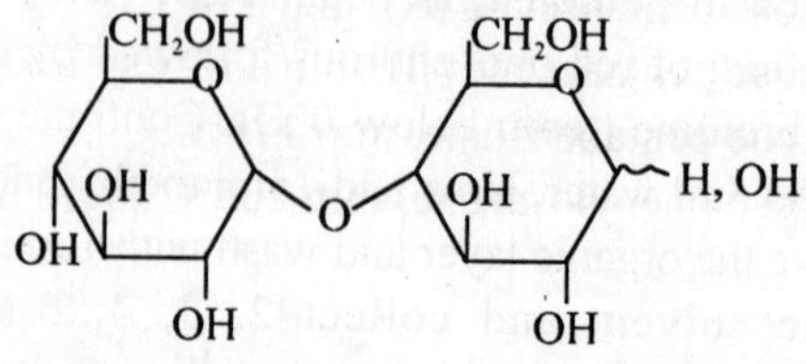

(1) White crystalline solid with slight odour and sweet taste, very soluble in water, insoluble in benzene and ether. Slightly soluble in alcohol.
(2) Reduces Fehling's solution.
(3) To 4ml aqueous solution of the compound add 1gm lead acetate, heat the contents to boiling. Dilute the reaction mixture with ammonium hydroxide. A white precipitate is obtained.
(4) Maltose octacetate (α), m.p 125°C.
(5) Maltose octacetate (β), m.p 159°C.

(6) Maltose phenylosazone, m.p 206°C (time of formation 1hr).

m.p 168°C, hydrated 120°C **D-Galactose** 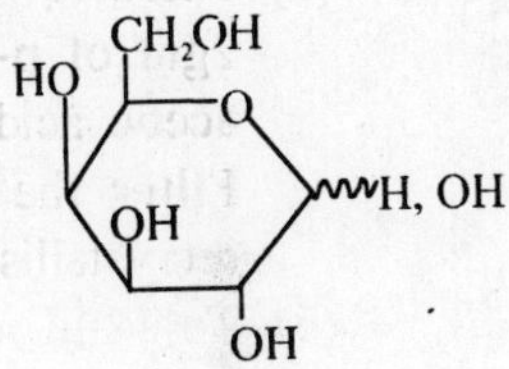

(1) White crystalline solid, soluble in water, sparingly soluble in alcohol, insoluble in ether.
(2) Reduces Fehling's solution.
(3) Reduces Barfoed's reagent.
(4) In a dry test tube place 1gm of carbohydrate and add 4ml of 50% HNO_3 to it. Heat the reaction mixture on water bath until a syrupy liquid is obtained. Add 10ml water to it and shake. Allow the reaction mixture to cool. Filter the precipitated mucic acid, wash with water and recrystallise, m.p 214°C (d).
(5) To 5gm of galactose add 35ml methanolic HCl solution. Reflux for 14hrs, cool and add 10ml water. Add lead carbonate until neutral and filter. Evaporate off the solvent and triturate the brown syrup with absolute ethanol, cool in ice. Collect methyl-D-galactopyranoside and crystallise from water, m.p 109-110°C.
(6) To 8ml ice cold Ac_2O add 1-2 drops of 60% perchloric acid. Now add in portion 2gm D-galactose (temp 30-40°C). Cool and add 700mg of red phosphorous followed by dropwise addition of 1.1ml bromine (temp below 0°C). Continue stirring for 2hrs and then add 5ml water. Now add 15ml methylene chloride and shake. Remove the organic layer and wash with 10ml $NaHCO_3$. Evaporate off the solvent and collect 2, 3, 4, 6-tetra-O-acetyl-α-D-galactoyranosyl bromide, recrystallise from pet ether, m.p 84-85°C.
(7) D-galactopentaacetate (α), m.p 96°C.
(8) D-galactopentaacetate (β), m.p 142°C.
(9) D-galactose phenylosazone, m.p 196°C (20-30 min).

m.p 186°C (d), hydrated 169°C **Sucrose** (β-D fructofuranosyl-α-D glucopyranoside)

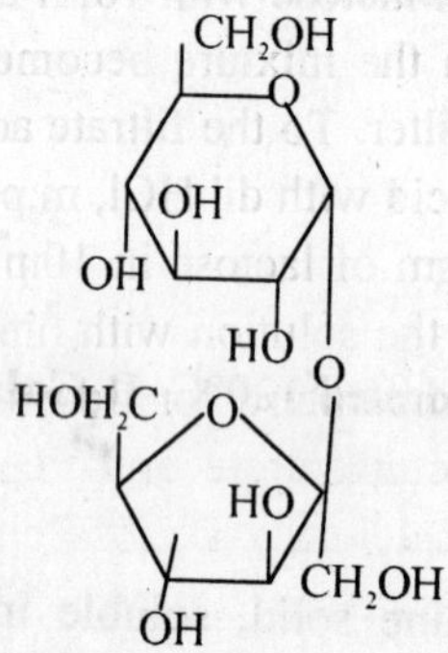

(1) White crystalline solid, soluble in water, slightly soluble in alcohol, insoluble in ether, benzene and chloroform.
(2) Does not reduce Fehling's solution or barfoed reagent.
(3) To 1ml aqueous solution of substance add a little $CuSO_4$ and 1ml NaOH. A blue coloured solution is obtained.
(4) To 1ml aqueous solution of substance add a little cobalt nitrate and 1ml NaOH. A violet coloured solution is obtained.
(5) Sucrose octaacetate (α), m.p 69°C.
(6) Sucrose octaacetate (β), m.p 60-63°C

m.p 219°C, hydrated m.p 202°C **Lactose** (4-O—β-D-galactopyranosyl-D-glucopyranose)

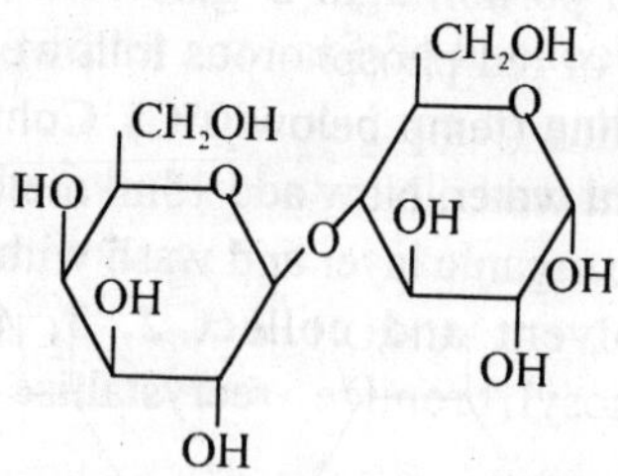

(1) White crystalline solid, readily soluble in water, insoluble in ether and alcohol
(2) Reduces Fehling's solution.
(3) Does not reduce barfoed reagent.
(4) In a test tube place 4ml aqueous solution of compound, add 1ml of lead acetate followed by NH_4OH. Heat gently for a minute. A pink colouration is produced.

(5) Mix 1gm of lactose with 10ml aqueous nitric acid, heat on water bath. When the mixture becomes thick and pasty cool, add 5ml water and filter. To the filtrate add dil NaOH and reprecipitate D-galactaric acid with dil HCl, m.p 212-213°C.

(6) Dissolve 1gm of lactose in 10ml water, add 300 mg of $Ca(OH)_2$ to it. Flush the solution with nitrogen, stopper the flask and keep at room temperature for 3days. Heat on water bath for 10hrs, filter hot, concentrate and keep in refrigerator. Collect white crystals of calcium α-D isosaccharin.

(7) Lactose octaacetate (α), m.p 152°C.

(8) Lactose octaacetate (β), m.p 90°C.

(9) Lactose phenylosazone, m.p 200°C (1½ hr).

(10) Lactose octabenzoate (β), m.p 140-142°C

Starch (Dextrin)

(1) White tasteless powder, insoluble in water, alcohol and ether, soluble in hot water. A polysaccharide.

(2) Does not reduce Fehling's solution or Tollens reagent.

(3) Heat few milligrams of compound with conc H_2SO_4. Sulphur dioxide and carbon dioxide gases are evolved with charring.

(4) Treat 1gm of compound with water and make it in the form of paste, pour into 40ml boiling water. To 5ml cold portion of this solution add iodine solution (iodine dissolved in potassium iodide solution); a blue colouration is produced.

(5) To 10ml starch solution add 5-8 drops of conc HCl and boil gently. Cool and perform the tests of glucose.

(6) Heat 100mg of compound with 5ml dil H_2SO_4. Cool, add iodine solution, a red colour due to the formation of dextrin is produced.

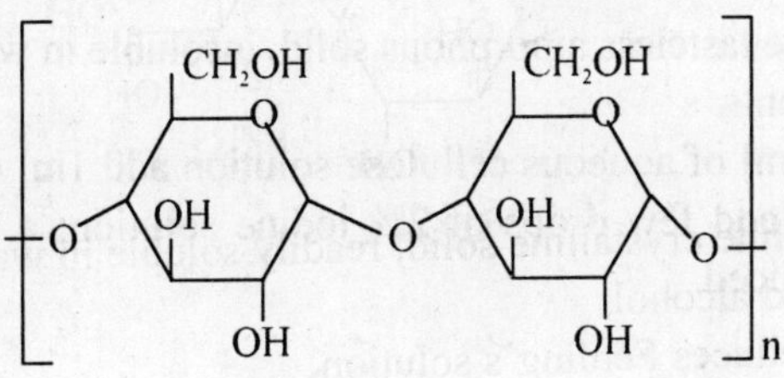

Inulin

(1) White, odourless, tasteless powder, sparingly soluble in water, soluble in hot water. A polysaccharide.

(2) To 10ml solution of inulin add 5-8 drops of conc HCl, boil gently. Cool and perform the tests of fructose.
(3) Does not reduce Fehling's solution.
(4) No colour with iodine.

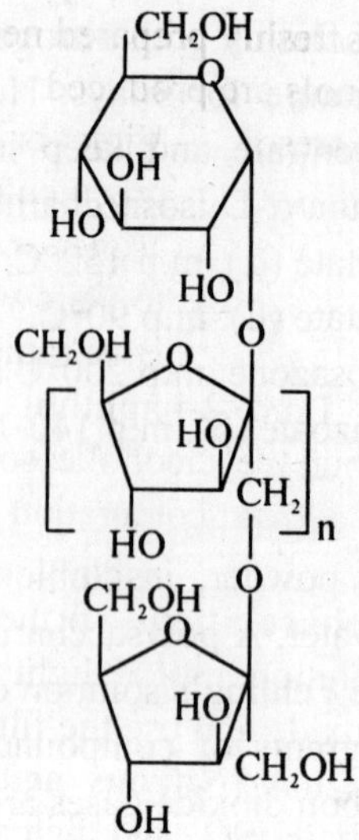

Cellulose

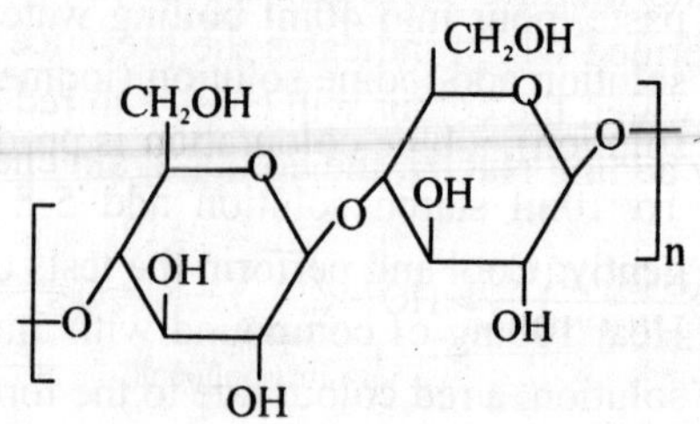

(1) White tasteless amorphous solid, insoluble in water and all organic solvents.
(2) To 1ml of aqueous cellulose solution add 1ml of conc H_2SO_4 and then add few drops of 2% iodine solution; a blue colouration is produced.

(*v*) *Phenols*

Compounds having hydroxyl group directly attached to the aromatic ring are called phenols. They have distinctive odour and are usually colourless crystalline compound which however become coloured due to their oxidation. They are however more acidic than alcohols and thus easily soluble in alkali to form phenolates.

Identification Tests

***(i)* Litmus test:** Place a drop of aqueous or alcoholic solution on blue litmus paper. Conversion of blue litmus to red indicates the compound to be acidic.

***(ii)* Ferric Chloride test:** To 2ml aqueous or alcoholic solution of compound add 5-6 drops of freshly prepared neutral ferric chloride solution. Coloured complexes of phenols are produced

Phenol, resorcinol, o, p cresol	Violet or blue colouration
Catechol	green colouration
Hydroquinone	Violet or transient blue which changes to yellow on excess addtion of $FeCl_3$
Pyragallol	Blue rapidly changing to red.

Aqueous solution of 1and 2-naphthol however do not give any characteristic colouration but the alcoholic solution of 1-naphthol and 2-naphthol give violet-blue and green colouration respectively due to formation of dinaphthols.

Neutral Ferric chloride solution: To the ferric chloride solution add dilute sodium hydroxide solution until a slight turbidity is formed. Filter off the precipitated ferric hydroxide and use the filtrate for the test.

***(iii)* Libermann Reaction (Nitrous acid test):** Dissolve 100mg of unknown phenol in 2ml conc H_2SO_4 and then add 40mg of sodium nitrite to it. Shake the mixture thoroughly and then warm. A blue, green or violet colour is formed which indicates the presence of phenol. Dilute the solution with 10ml water, the colour will change to red or blue-red. Make the solution alkaline by adding NaOH, the colour again changes to blue or bluish green.

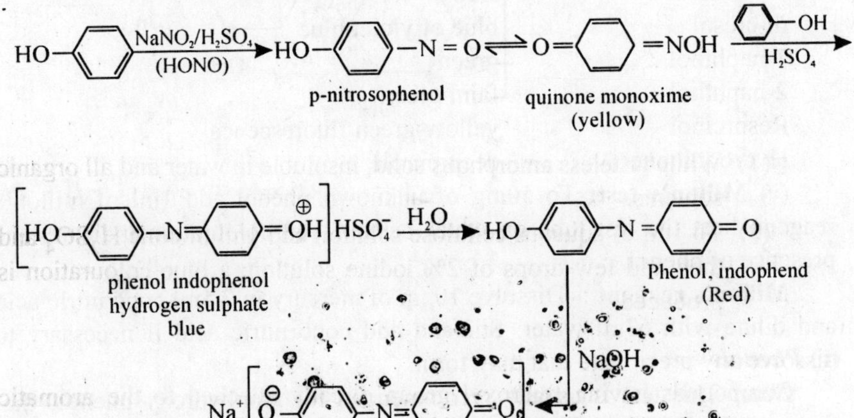

Note :- This test is given by phenols which contain a free para position

***(iv)* Phthalein test:** In a test tube place 200mg of unknown phenol and 200mg of phthalic anhydride. Add 2-3 drops of conc H_2SO_4 and heat the test tube for 3-4 minutes in an oil bath previously heated to 160°C. Cool and pour the fused material in a beaker containing 5ml dil NaOH solution and then add equal volume of water to it. Characteristic colours are produced which can be viewed against a white background. If the solution exhibits fluoresecence view it against a black background.

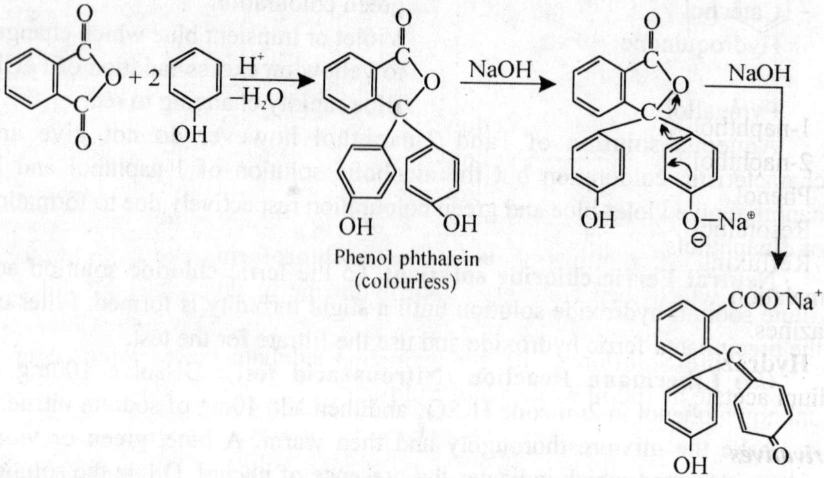

Phenol	Reddish pink (phenolphthalein)
o-cresol	Red
m-cresol	blue or violet blue
1-naphthol	green
2-naphthol	faint green
Resorcinol	yellow-green fluorescence
Hydroquinone	deep purple

***(v)* Millon's test:** To 50mg of unknown phenol add 1ml of millon's reagent, heat the test tube on water bath. A red colouration indicates the presence of phenol.

Millon's reagent :- Dissolve 10gm of mercury in 7.5ml conc nitric acid and dilute with 63ml water. Stir and add conc nitric acid if necessary to dissolve any precipitate that may form.

Note: This test is usually recommended for monhydroxy phenols that has at least one ortho position free.

(vi) Dissolve 50mg of compound in 1ml NaOH solution. Transfer 5-10 drops of this solution in a crucible and evaporate it to dryness. Add 20 drops of chloroform to the crucible and evaporate it off. Repeat the process twice each time using 20 drops of chloroform. Take the residue in a drop of 6N acetic acid and a drop of hydrazine solution. Transfer a drop of this solution on filter paper. Yellow to orange fluorescence under UV light indicates the presence of phenol.

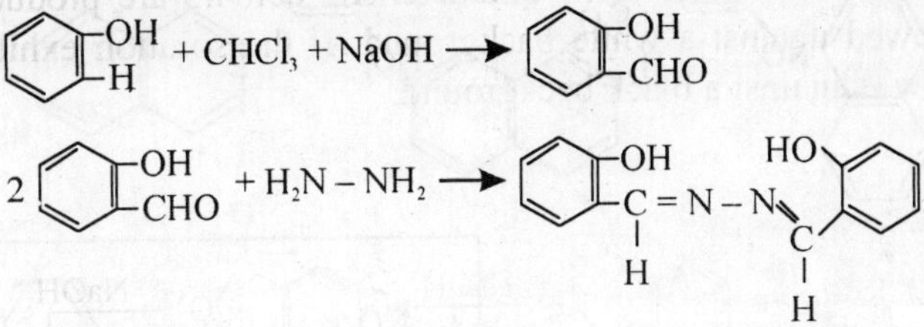

1-naphthol	Yellow
2-naphthol	Yellow
Phenol	Yellow
Resorcinol	Yellow

Refluxing of alkaline phenolate with chloroform results in ortho formalation which then condenses with hydrazine to give water insoluble aldazines.

Hydrazine Solution :- 10 gm hydrazine sulphate boiled with 10gm sodium acetate in 100ml water. Cool and filter.

Derivatives

(1) **3, 5-dinitrobenzoates and p-nitrobenzoate:** In a 100ml round bottom flask fitted with a reflux condensor place 100mg of phenol, 2ml pyridine and 250mg of 3, 5 dinitrobenzoyl chloride/p-nitro benzoyl chloride. Reflux the mixture for 30-35 min. Cool, pour the reaction mixture into 20ml dil HCl. Decant the aqueous liquid and stir the precipitated solid or oily layer with 10ml of dil NaOH solution. Filter the solid ester, wash with water, recrystallise from methanol, ethanol or cyclohexane and determine its melting point.

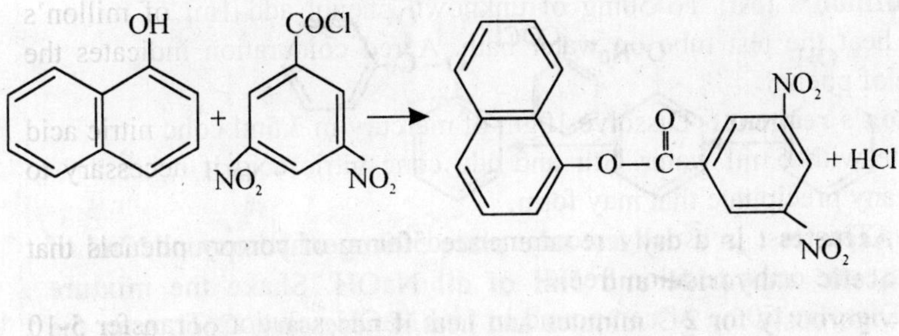

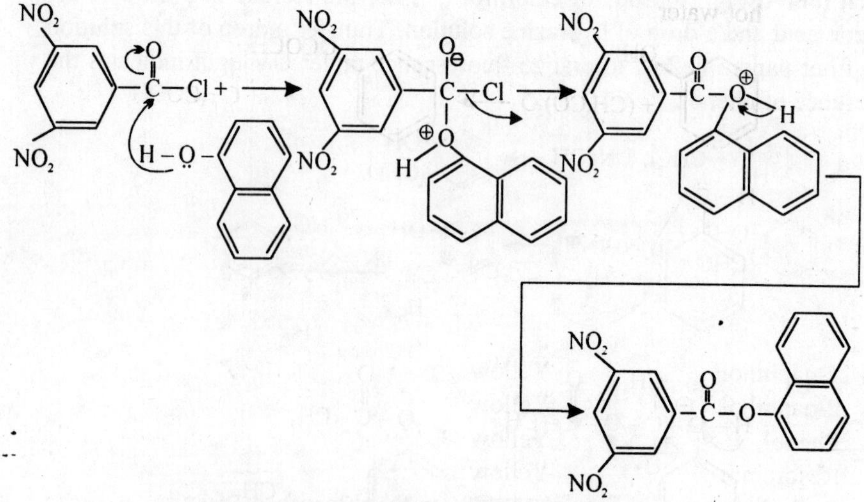

3, 5-dinitrobenzoyl chloride :- Heat 2gm of 3, 5-dinitrobenzoic acid with 3gm of PCl_5 in a dry test tube until a clear solution is obtained. Transfer the liquid on a watch glass and allow it to solidify

p-nitrobenzoyl chloride :- Heat 2gm of p-nitrobenzoic acid with 3gm of PCl_5 in a dry test tube until a clear solution is obtained. Transfer the liquid on a watch glass and allow it to solidify

(2) **Benzoates (Schotten –Baumann method):** Dissolve 500mg of phenol in 10ml of 5% NaOH solution in a conical flask and then add 3ml of benzoyl chloride to it. Stopper the flask and shake vigorously for 15minutes or until the smell of benzoyl chloride persists. Cool, filter the precipitated ester derivative, wash with water and recrystallise.

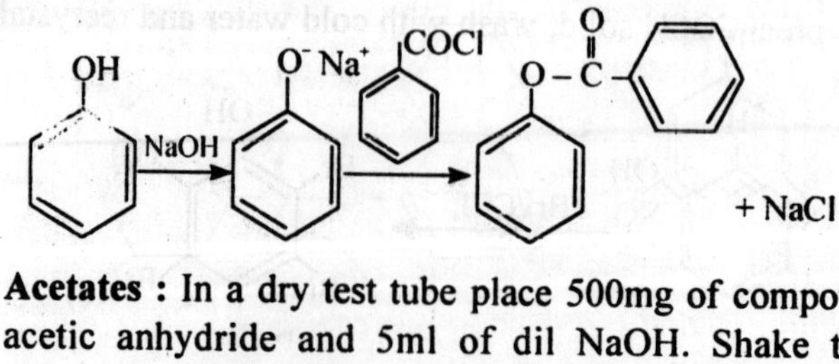

(3) **Acetates :** In a dry test tube place 500mg of compound, 2ml of acetic anhydride and 5ml of dil NaOH. Shake the mixture vigorously for 2-3 minutes and heat if necessary. Cool, pour the reaction mixture with stirring into 20ml of ice water. Filter the

crude product, wash with water and recrystallise from ethanol or hot water.

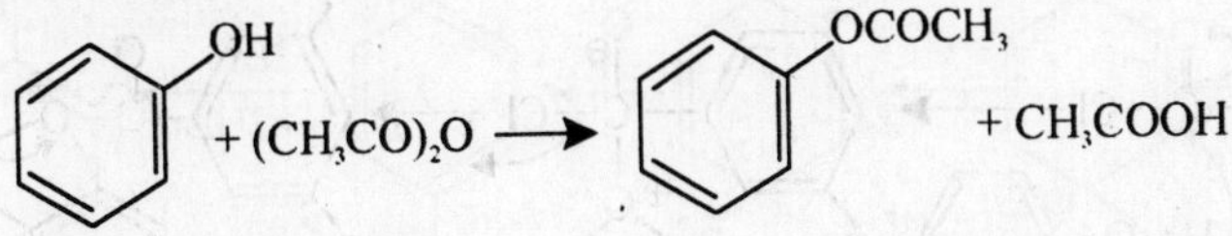

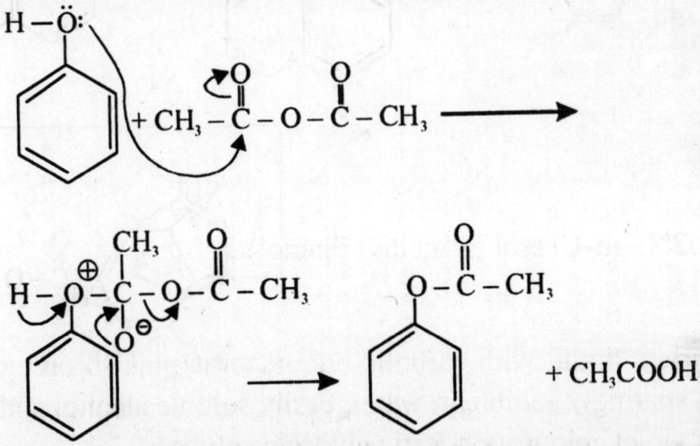

(4) **Bromo derivative (Method A)** In a 50ml stoppered conical flask dissolve 500mg of compound in 7ml glacial acetic acid/5ml CCl_4. Add dropwise with stirring 8ml of bromine solution (2ml bromine in 6ml glacial acetic acid/6ml CCl_4). After the addition is complete continue stirring for 10 minutes and then allow the mixture to stand at room temperature for 10-15 minutes. Pour the mixture in ice water, filter the bromo derivative, wash with water and recrystallise from methanol, ethanol or aqueous ethanol.

Method B: In a 50ml stoppered conical flask dissolve 500mg of compound in 3ml ethanol and then add with stirring bromine soloution (1.5ml bromine to a solution of 7.5gm of KBr in 25ml water). Shake thoroughly, keep the reaction mixture at room temperature for 5-10 minutes and then pour in ice water. Filter the precipitated solid, wash with cold water and recrystallise.

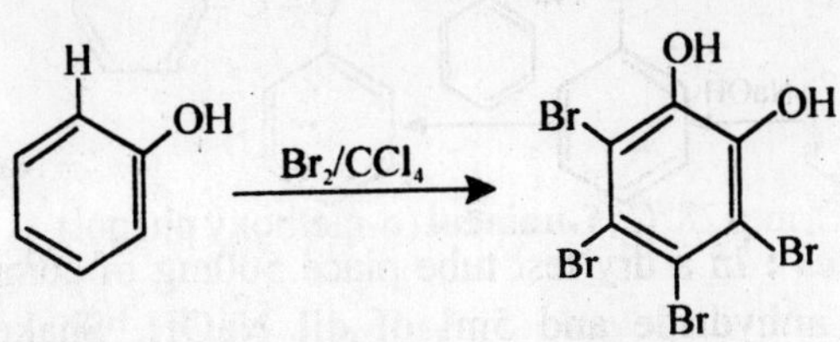

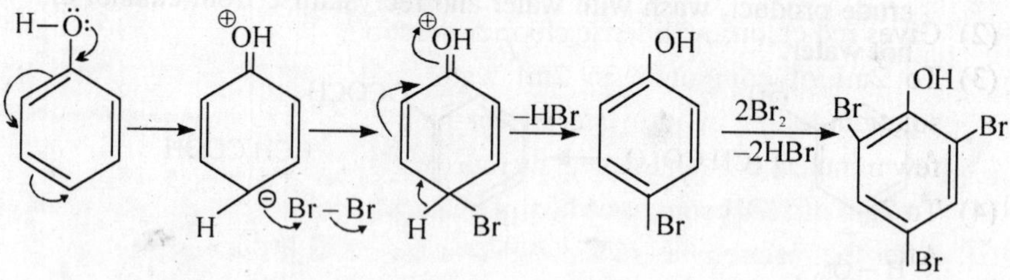

Liquid Phenols

b.p 202°C **m-Cresol** (3-methyl Phenol) 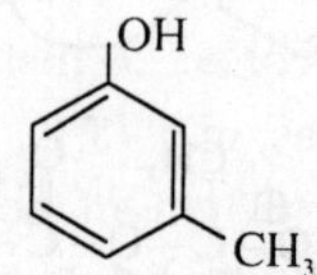

(1) Colourless liquid with carbolic odour, turns pinkish on exposure to air, sparingly soluble in water, easily soluble alcohol and ether
(2) Gives violet colour with ferric chloride solution.
(3) Responds to libermann test to give blue or green colour.
(4) Responds to phthalein test to give blue colour
(5) To 1ml compound add 1ml conc H_2SO_4 and then pour the resulting mixture into a 50ml conical flask containing 3ml nitrating mixture (1.5ml conc H_2SO_4 +1.5ml conc HNO_3) Heat the mixture on water bath for 8-10 min and then pour in ice with stirring. Collect the precipated 2, 4, 6-trinitro-3-methylphenol, m.p 106°C.
(6) To 3ml alcoholic solution of compound add 3ml alcoholic picric acid solution. Shake well and heat if necessary. On cooling cryst: 's of picrate, m.p 88°C separate out.
(7) m-tolyl benzoate, m.p 55°C.
(8) m-tolyl p-nitrobenzoate, m.p 90°C.
(9) m-tolyl 3, 5-dinitrobenzoate, m.p 165°C.
(10) 2, 4, 6-tribromo-3-methyl phenol, m.p 84°C.

b.p 205°C, m.p 28°C **Guaiacol** (o-methoxy phenol) 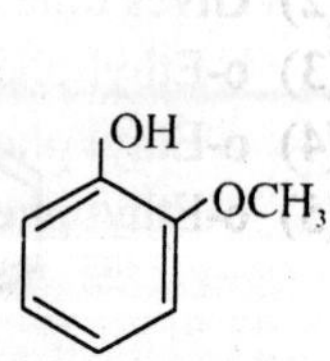

(1) Colourless liquid with characteristic aromatic odour, sparingly soluble in water, easily soluble in most organic solvents.

(2) Gives red colour with ferric chloride solution.
(3) To 2ml of compound in 2ml water add 5ml of picric acid solution(400mg in 5ml water). Heat the mixture on water bath for few minutes, cool and collect picrate (orange), m.p 87°C.
(4) To 2gm of KOH solution add 5ml guaiacol (exothermic reaction). After the reaction has subsided heat in an oil bath for 3hrs. Cool, add a pinch of Cu powder, 5ml of bromobenzene and 1 ml of guaiacol. Heat at 200°C for 2hrs. Cool and isolate the product with ether (2 × 10ml). Combine the ethereal layer and steam distill, collect 2-methoxy diphenyl ether which on cooling soon solidifies, m.p 78°C.
(5) Reflux 6.3gm of compound with 6.6 gm allyl bromide, 7gm potassium carbonate and 10ml acetone for 8hrs. Cool and dilute with 20ml water. Extract with ether. Distill and collect guaiacol allyl ether at b.p 111-113°C/12mmHg
(6) o-methoxy phenyl benzoate, m.p 57°C.
(7) o-methoxy phenyl p-nitrobenzoate, m.p 93°C.
(8) o-methoxy phenyl 3, 5-dinitrobenzoate, m.p 141°C.
(9) 4, 5, 6-tribromoguaiacol, m.p 116°C.

b.p 207°C **o-Ethyl phenol** (o-hydroxy ethyl benzene, Phlorol)

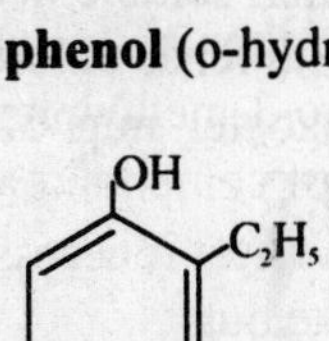

(1) Colourless liquid sparingly soluble in water, miscible with alcohol and ether.
(2) Gives blue colour with ferric chloride solution.
(3) o-Ethyl phenyl benzoate, m.p 39°C.
(4) o-Ethyl phenyl p-nitrobenzoate, m.p 56°C
(5) o-Ethyl phenyl 3, 5-dinitrobenzoate, m.p 108°C.

b.p 211°C, m.p 26°C **2, 4-dimethyl Phenol** (4-hydroxy-1, 3-dimethyl benzene, 1, 2, 4-xylenol)

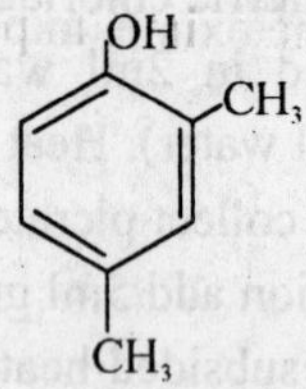

(1) Colourless liquid sparingly soluble in water, soluble in alcohol and ether.
(2) Gives blue-violet colour with ferric chloride solution.
(3) 2, 4-dimethyl phenyl benzoate, m.p 37°C.
(4) 2, 4-dimethyl phenyl p-nitrobenzoate, m.p 105°C.
(5) 2, 4-dimethyl phenyl 3, 5-dinitrobenzoate, m.p 165°C.

b.p 214°C **m-Ethyl phenol** (m-hydroxy ethyl benzene)

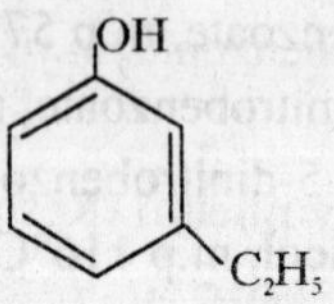

(1) Slightly soluble in water, soluble in alcohol and ether.
(2) Gives violet colour with ferric chloride solution.
(3) m-Ethyl phenyl benzoate, m.p 52°C. (4) m-Ethyl phenyl p-nitrobenzoate, m.p 68°C

b.p 218°C, m.p 28°C **o-Hydroxy acetophenone** (o-acetyl Phenol)

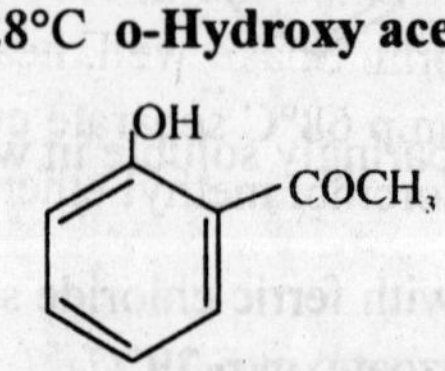

(1) Oily liquid, insoluble in water, soluble in alcohol and ether.
(2) Gives reddish-violet colour with ferric chloride solution.
(3) To 1 ml of compound add dil NaOH solution. A yellow coloured solution is obtained.
(4) o-Benzoxy acetophenone, m.p 87°C.
(5) o-acetoxy acetophenone, m.p 89°C
(6) o-Hydroxy acetophenone semicarbazone, m.p 210°C.

(7) o-Hydroxy acetophenone phenylhydrazone, m.p 110°C.
(8) o-Hydroxy acetophenone oxime, m.p 117°C.

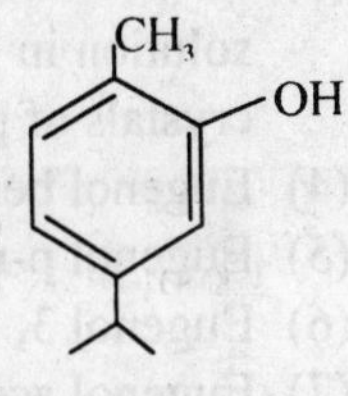

b.p 237°C **Carvacrol** (2-hydroxy p-cymene)

(1) Viscous oily liquid very slightly soluble in water, soluble in alcohol and ether.
(2) Gives transient green colour with ferric chloride solution.
(3) Dissolve 2ml of compound in 5ml alcohol saturated with HCl gas. Cool the solution to 0°C and add $NaNO_2$ solution. Stir and recover 4-nitroso carvacrol, m.p 153°C.
(4) Carvacryl p-nitrobenzoate, m.p 51°C.
(5) Carvacryl 3, 5-dinitrobenzoate, m.p 78°C.

b.p 244°C **3-methoxy phenol** (Resorcinol monomethyl ether)

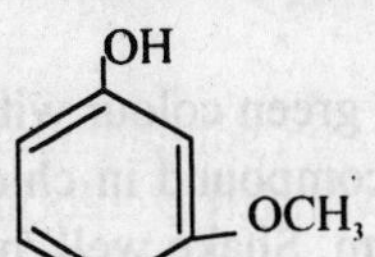

(1) Sparingly soluble in water, miscible with alcohol and ether.
(2) Gives pale violet colour with ferric chloride solution.
(3) To 3ml solution of compound in chloroform add 3ml picric acid solution in chloroform. Shake well, heat if necessary. On cooling crystals of picrate, m.p 68°C separate out.
(4) 2, 4, 6-tribromo resorcinol methyl ether, m.p 104°C.

b.p 253°C **Eugenol** (4-allyl-2-methoxy phenol)

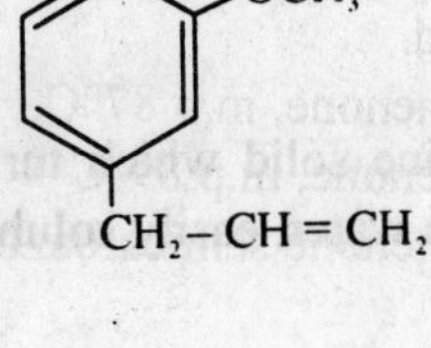

(1) Colourless oily liquid with odour of cloves, sparingly soluble in water, easily soluble alcohol and ether.
(2) Gives blue colour with ferric chloride solution.
(3) To 3ml solution of compound in chloroform add 3ml picric acid solution in chloroform. Shake well, heat if necessary. On cooling crystals of picrate(red brown), m.p 63°C separate out.
(4) Eugenol benzoate, m.p 70°C
(5) Eugenol p-nitrobenzoate, m.p 81°C.
(6) Eugenol 3, 5-dinitrobenzoate, m.p 131°C.
(7) Eugenol acetate, m.p 29°C.

b.p 268°C **Isoeugenol** (2 –methoxy-4-propenylphenol)

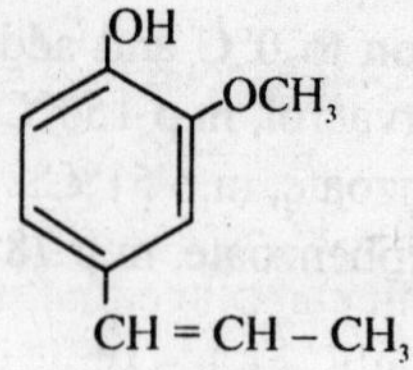

(1) Pale yellow coloured oily liquid, insoluble in water, soluble in alcohol and ether
(2) Gives transient olive green colour with ferric chloride solution.
(3) To 3ml solution of compound in chloroform add 3ml picric acid solution in chloroform. Shake well, heat if necessary. On cooling crystals of picrate (red), m.p 47°C separate out.
(4) Isoeugenol benzoate, m.p 103°C.
(5) Isoeugenol p-nitrobenzoate, m.p 109°C.
(6) Isoeugenol 3, 5-dinitrobenzoate, m.p 158°C.
(7) Isoeugenol acetate (trans), m.p 79°C.

Solid Phenols

m.p 31°C, b.p 191°C **o-Cresol** (2-methyl phenol)

(1) Colourless crystalline solid which turns red on exposure to air, sparingly soluble in water, easily soluble in alcohol and ether.

(2) Gives violet colour with ferric chloride solution.
(3) Responds to libermann test to give green or blue colouration
(4) Responds to phthalein test to give pink or red colouration
(5) To 3ml alcoholic solution of compound add 3ml alcoholic picric acid solution. Shake well, heat if necessary. On cooling yellow crystals of picrate, m.p 88°C. separate out
(6) o-Tolyl p-nitrobenzoate, m.p 94°C
(7) o-Tolyl 3, 5-dinitrobenzoate, m.p 138°C
(8) 4, 6-dibromo 2-methyl phenol, m.p 56°C.

m.p 34°C, b.p 202°C **p-Cresol** (4-methyl phenol)

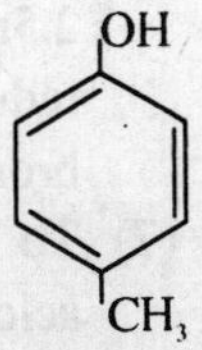

(1) Colourless solid with carbolic odour, turns pinkish on exposure to light and air, sparingly soluble in water, easily soluble in alcohol and ether.
(2) Gives blue colour with ferric chloride solution.
(3) Responds to liberman test to give red coloured solution.
(4) 2, 6-dibromo 4-methyl phenol, m.p 49°C
(5) p-Tolyl benzoate, m.p 70°C
(6) p-Tolyl p-nitrobenzoate, m.p 98°C.
(7) p-Tolyl 3, 5-dinitrobenzoate, m.p 188°C

m.p 41°C, b.p 182°C **Phenol** (Carbolic acid)

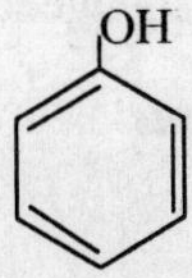

(1) Crystalline compound, sparingly soluble in water, easily soluble alcohol and ether.
(2) Gives violet colour with ferric chloride solution.
(3) Responds to libermann test to give blue colouration.
(3) Responds to phthalein test to give reddish pink colour
(4) Dissolve 3gm of phenol in 15ml of 10% NaOH solution. Add

10ml of 40% formaldehyde solution and keep the mixture at room temperature for 6 days. Neutralise with dil HCl and extract with ether. Evaporate off the ether and recover the o-and p-hydroxy benzyl alcohol. The ortho isomer is more soluble in benzene than para. Ortho isomer, m.p 82°C, Para isomer, m.p 112°C.

(5) To a solution of 200 mg of compound in 2ml conc H_2SO_4 add 4ml of nitrating mixture (2ml conc H_2SO_4 + 2ml conc HNO_3). Heat the mixture on water bath for 10-15 min. Cool, pour in ice and collect pale yellow crystals of picric acid, m.p. 122°C.

(6) Cool a mixture of 4.7 gm of phenol in CS_2 in ice bath (below 5°C). Slowly add dropwise and with stirring 2.5ml of bromine in 2.5ml CS_2. After the addition is complete distill the mixture under reduced pressure and collect o and p isomer but on cooling p-bromo phenol solidifies, m.p 63°C.

(7) To 3ml alcoholic solution of compound add 3ml alcoholic picric acid solution. Shake well, heat if necessary. On cooling yellow crystals of picrate, m.p 83°C separate out.

(8) Reflux a mixture of 1.18gm of succinic acid, 1.88gm phenol and 2ml phosphorous oxychloride on steam bath for 90min. Add 6ml benzene and reflux for further 60min. Decant off benzene. Extract the syrupy layer with hot benzene (2 × 15ml). Concentrate the combine benzene layer and cool, diphenylsiccinate precipitate out, m.p 120-121°C.

(9) Phenyl benzoate, m.p 68°C. (10) Phenyl p-nitrobenzoate, m.p 126°C. (11) Phenyl 3, 5-dinitrobenzoate, m.p 146°C.

m.p 46°C, b.p 219°C **p-ethyl phenol** (p-hydroxy ethyl benzene)

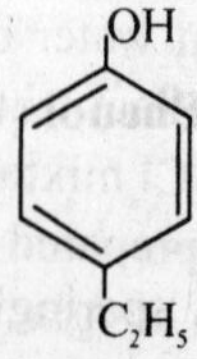

(1) Crystalline compound, slightly soluble in water, soluble in alcohol and ether.

(2) Gives deep blue colour with ferric chloride solution.

(3) p-Ethyl phenyl benzoate, m.p 60°C. (4) p-Ethyl phenyl p-

nitrobenzoate, m.p 81°C. (5) p-Ethyl phenyl 3, 5-dinitrobenzoate, m.p 132°C.

m.p 49°C, b.p 212°C **2, 6-dimethyl phenol** (1, 2, 6-xylenol)

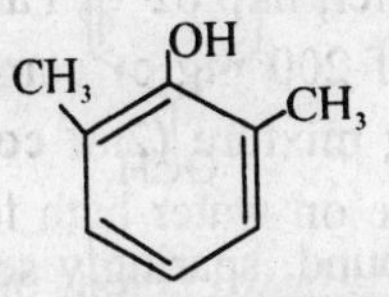

(1) Colourless crystalline compound, soluble in hot water and alcohol.
(2) 2, 6-dimethyl phenyl 3, 5-dinitrobenzoate, m.p 159°C. (3) 3, 4, 5-tribromo-2, 6-dimethyl phenol, m.p 201°C. (4) Picrate (orange-yellow), m.p 53°C.

m.p 51.5°C **Thymol** (3-methyl-6 isopropyl phenol, 3-hydroxy-4 isopropyl toluene)

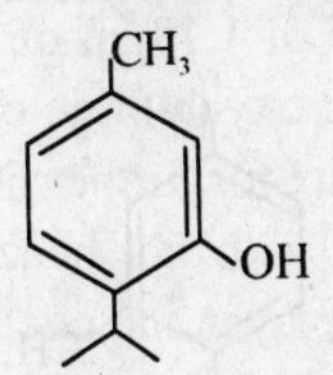

(1) Colourless crystalline solid, soluble in alcohol, ether, chloroform and benzene, characteristic odour of thyme (ajwain)
(2) Gives transient green colour with alcoholic ferric chloride solution.
(3) Responds to phthalein test to give intense blue colour.
(4) To a solution of 200 mg of compound in 2ml conc H_2SO_4 add 4ml of nitrating mixture(2ml conc H_2SO_4 + 2ml conc HNO_3). Heat the mixture on water bath for 10-15min. Cool, pour in ice with stirring and collect trinitro thymol, recrystallise from water containing a little HCl mixture (20+8+1), m.p 108-110°C
(5) Pass chlorine gas (prepared from permanganate and HCl) into a solution of thymol in glacial acetic acid. Cool and collect 6-chlorothymol, recrystallise from pet ether, m.p 58°C.
(6) Thymol benzoate, m.p 33°C
(7) Thymol p-nitrobenzoate, m.p 70°C
(8) Thymol 3, 5-dinitro benzoate, m.p 103°C.

m.p 53°C **4-methoxy phenol** (p-hydroxy anisole, hydroquinone monomethyl ether)

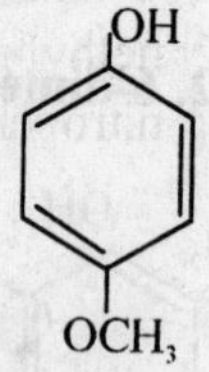

(1) Crystalline compound, sparingly soluble in water, easily soluble benzene, chloroform and ether.
(2) To 3ml solution of compound in chloroform add 3ml picric acid solution in chloroform. Shake well, heat if necessary. On cooling crystals of picrate(orange-yellow), m.p 44°C separate out.
(3) 4-methoxy phenyl benzoate, m.p 87°C. (4) 4-methoxy phenyl acetate, m.p 32°C.

m.p 63°C, b.p 225°C **3, 4-dimethyl phenol** (1, 3, 4-Xylenol)

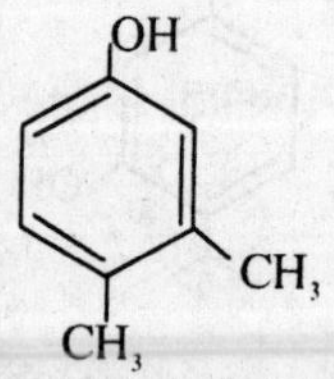

(1) Crystalline compound, soluble in water and alcohol, miscible with ether.
(2) Gives green colour with ferric chloride solution.
(3) 3, 4-dimethyl phenyl benzoate, m.p 58°C.
(4) 3, 4-dimethyl phenyl 3, 5-dinitrobenzoate, m.p 182°C.
(5) Picrate (yellow), m.p 83°C.

m.p 64-68°C **3, 5-dimethyl phenol** (1, 3, 5-xylenol)

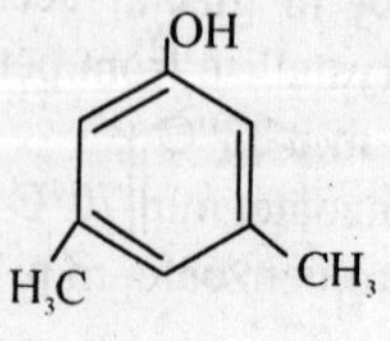

(1) Crystalline compound sparingly soluble in water, soluble in alcohol and benzene.
(2) 3, 5-dimethyl phenyl 3, 5-dinitrobenzoate, m.p 195°C.
(3) 2, 4, 6-tribromo 3, 5-dimethyl phenol, m.p 166°C
(4) 3, 5-dimethyl phenyl p-nitrobenzoate, m.p 109°C

m.p 68-71°C **Mesitol** (2, 4, 6-trimethyl phenol)

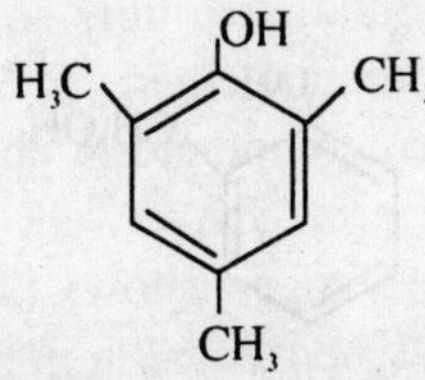

(1) Colourless compound, sparingly soluble in water, easily soluble in alcohol and ether. Sublimes readily
(2) Mesityl benzoate, m.p 62°C

m.p 72°C **Pseudocumenol** (2, 4, 5-trimethyl phenol)

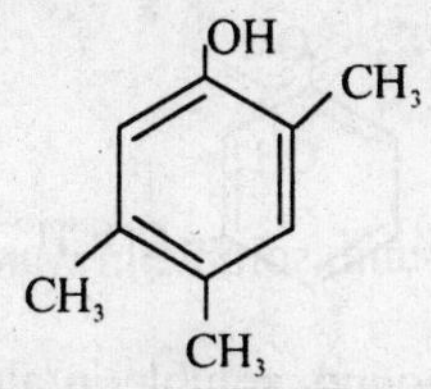

(1) Crystalline solid, insoluble in water, soluble in alcohol and ether
(2) Pseudocumyl benzoate, m.p 63°C (3) Pseudocumyl acetate, m.p 34°C.

m.p 75°C **2, 5-dimethyl phenol** (1, 2, 5-Xylenol)

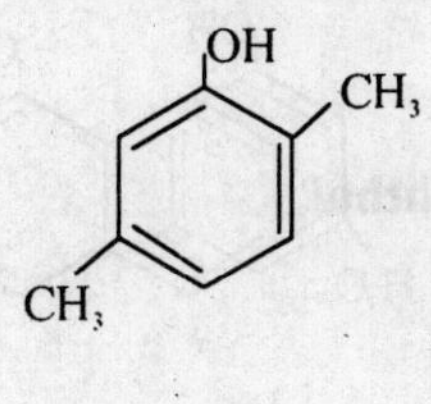

(1) Crystalline compound, volatile with steam, slightly soluble in water, soluble in alcohol, benzene and ether.
(2) 2, 5-dimethyl phenyl benzoate, m.p 61°C
(3) 2, 5-dimethyl phenyl p-nitrobenzoate, m.p 87°C.
(4) 2, 5-dimethyl phenyl 3, 5-dinitrobenzoate, m.p 137°C.
(5) Picrate, m.p 81°C.

m.p 87°C **o-Hydroxy benzyl alcohol** (Saligenin)

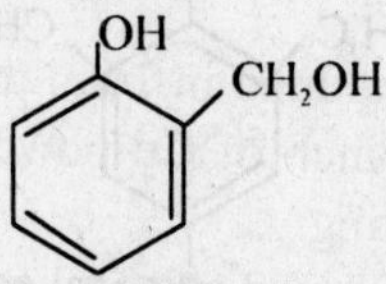

(1) Crystalline compound, soluble in hot water, alcohol and ether. Sublimes readily.
(2) Gives red- violet colour with alcoholic ferric chloride solution.
(3) o-Benzoxybenzyl benzoate, m.p 51°C.

m.p 87-89°C **3-hydroxy Acetophenone** (3-acetyl phenol)

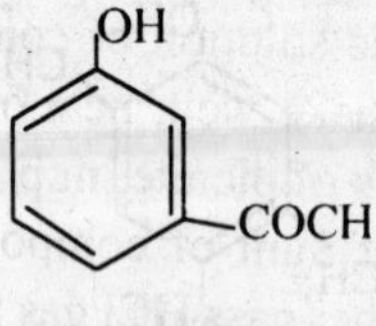

(1) Crystallline compound, soluble in alcohol, ether, chloroform and benzene
(2) To few milligram of compound add conc H_2SO_4. A deep yellow coloured solution is obtained.
(3) 3-hydroxy acetophenone semicarbazone, m.p 194°C.
(4) 2, 4, 6-tribromo-3 hydroxy acetophenone, m.p 127°C.

m.p 96°C **1-Naphthol**

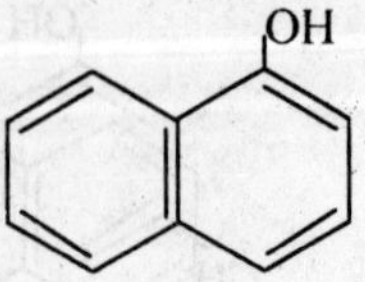

(1) Crystalline solid, sparingly soluble hot water, easily soluble benzene, chloroform, alcohol and ether.
(2) To 1gm of compound in minimum quantity of boiling water, $FeCl_3$ solution is added on cooling until the precipitate is bright reddish violet. Filter the solid, boil once with water, twice with benzene and collect di –1-naphthol, m.p 300°C.
(3) To 100mg of compound in 15ml dil NaOH solution add 7-10 drops of $CHCl_3$, heat on water bath for few seconds. A blue coloured solution is obtained which slowly changes to bluish green.
(4) To 50mg of compound in 10ml dil NaOH solution add 3-4 drops of CCl_4 and a pinch of Cu powder. A blue coloured solution is obtained on heating.
(5) To 50mg of compound add 5ml each of iodine and NaOH solution and shake thoroughly. A violet coloured solution is obtained.
(6) Dissolve 5.4 gm of compound in 50ml liquid ammonia. Add 1.04gm of Li metal in small pieces. Continue stirring for 20 min after addition is complete and then add dropwise 10ml of absolute ethanol. Remove ammonia and add 50ml water. Extract with ether (2 × 10ml)and add conc HCl to the resultant solution. Extract with ether (2 × 10ml). Evaporate off the ether and collect 5, 8 dihydro 1-naphthol, m.p 69-72°C.
(7) To 3ml alcoholic solution of compound add 3ml alcoholic picric acid solution. Shake well, heat on water bath. On cooling orange coloured crystals of picrate, m.p 189°C separate out.
(8) To a mixture of 3gm of compound, 3gm of $ZnCl_2$ and 2gm of HCN in 15ml ether pass HCl gas for 2 ½ hr. Decant the precipitated 4-hydroxynaphthaldehyde, recrystallise from hot water, m.p 181°C.
(9) 1-Naphthyl benzoate, m.p 56°C. (10) 1-Naphthyl p-nitrobenzoate, m.p 143°C (11) 1-Naphthyl 3, 5-dinitrobenzoate, m.p 217°C (12) 1-Naphthyl acetate, m.p 49°C (13) 2, 4 dibromo-1-Naphthol, m.p 105°C

m.p 103-104°C **1, 2-dihydroxy naphthalene**

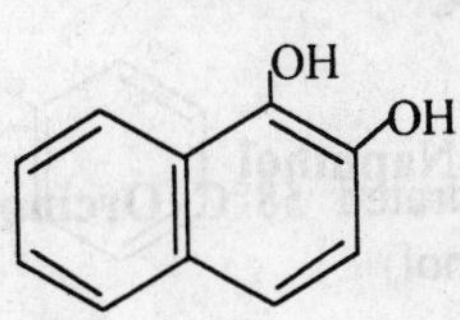

(1) Colourless crystalline solid, slightly soluble in water, soluble in alcohol and ether.
(2) To 1ml aqueous solution of compound add 2ml NaOH solution. A yellow coloured solution is obtained which slowly turns green.
(3) 1, 2 diacetoxy naphthalene m.p 106°C.

m.p 104-106°C **Catechol** (1, 2-dihydroxy benzene)

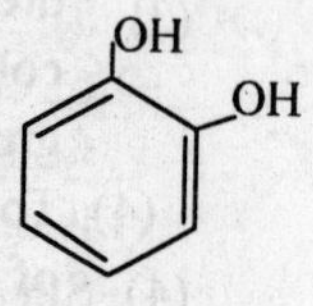

(1) Colourless crystalline solid, soluble in water, alcohol and ether.
(2) Gives green colour with ferric chloride solution.
(3) Reduces Fehling's solution and Tollens reagent.
(4) To aqueous solution of compound add equal volume of lead acetate solution. A white precipitate is obtained which is soluble in acetic acid.
(5) Moisten 5.5gm of catechol with 1ml water and then add dropwise with stirring 6.6ml PCl_3. Keep the mixture at room temperature for 1 hr and then heat on water bath for 2hrs. Distill under reduced pressure and collect o-phenylene phosphorochloridite which soon solidifies, m.p 30°C.
(6) Cool a solution of 10gm of compound in 20ml methyl alcohol containing 25gm dimethyl sulphate to –5°C. Add a solution of 15gm of caustic soda in 35ml water(exothermic reaction). After 15 minutes dilute with 10ml water. Extract with ether, distill off ether and collect veratrole at b.p 205°C, m.p 15°C.
(7) Catechol dibenzoate, m.p 84°C.
(8) Catechol di p-nitrobenzoate, m.p 169°C.
(9) Catechol di (3, 5-dinitrobenzoate), m.p 152°C.
(10) Catechol diacetate, m.p 65°C.
(11) tetra bromocatechol, m.p 192°C.
(12) picrate m.p 122°C.

m.p 107°C, hydrated 58°C **Orcinol** (3, 5-dihydroxy toluene, 5-methyl resorcinol)

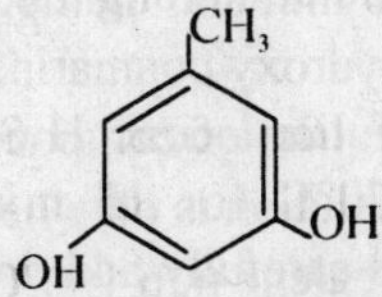

(1) Colourless crystalline solid, soluble in water, alcohol and ether.
(2) On warming reduces Tollens reagent.
(3) To 50mg of compound in 5ml dil NaOH solution add 5-7 drops of $CHCl_3$. An orange red coloured solution is obtained. Add 30ml water to the coloured solution, an intense violet green fluorescence is observed.
(4) Stir a mixture of 12.4gm of potassium carbonate, 41ml acetone and 4.2gm of orcinol. Add dropwise 7.1ml of dimethyl sulphate. Mixture becomes warm and begins to reflux. After the reaction has subsided reflux for 4hrs. Distill off acetone. Add 5ml aqueous ammonia, stir and heat for 10min. Remove the organic layer and extract the aqueous layer with ether. Combine the organic layers, distill off ether and collect orcinol dimethyl ether at b.p 134-135°C/40mmHg.
(5) Orcinol dibenzoate m.p 88°C.
(6) Orcinol di p-nitrobenzoate m.p 214°C.
(7) Orcinol di (3, 5-dinitrobenzoate) m.p 190°C.
(8) 2, 4, 6-tribromo orcinol m.p 104°C.
(9) Picrate (orange-red) m.p 92°C.

m.p 110°C **Resorcinol** (1, 3-dihydroxy benzene)

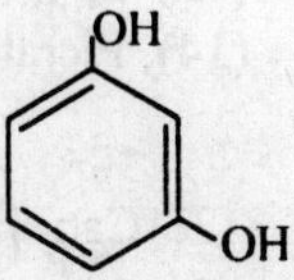

(1) Pinkish crystalline solid, soluble in water, alcohol and ether, insoluble in chloroform.
(2) Gives blue –violet colour with ferric chloride solution.
(3) Responds to phthalein test to give yellow –green fluorescence.
(4) To 2ml aqueous solution of compound add equal volume of NaOH soluion followed by 2-4 drops of $CHCl_3$. On warming a red coloured solution with green fluorescence is obtained.
(5) Add 16gm of polyphosphoric acid to a solution of 1.1 gm of resorcinol in 1.3 gm ethyl aceto acetate. Stir and heat the mixture

at 70-80°C for 15-20 min and then pour in ice water, collect pale yellow 4-methyl 7-hydroxy coumarin, m.p 178°C.

(6) Warm a mixture of 10ml conc H_2SO_4 and 1gm resorcinol on water bath at 60-70°C for 15 minutes. Cool the slurry of disulphonic acid and carefully add dropwise cold mixture of 1ml conc HNO_3 and 1.2ml conc H_2SO_4 such that temp does not rise above 20°C. Pour the mixture in ice and collect 2-nitro resorcinol, m.p 76°C.

(7) Dissolve 3gm fused $ZnCl_2$ in 3ml hot glacial acetic acid and add 2gm resorcinol to it. Heat the mixture to 140-150°C, remove the heat source and then allow the mixture to complete itself. Cool and add 20ml dil HCl, stir and filter. Dissolve the dark brown solid in caustic soda solution and filter. Acidify the filtrate with dil HCl and collect resacetophenone, m.p 142°C (d).

(8) Heat 4gm of resorcinol with 2gm potassium carbonate and 40ml water on water bath for 4hrs. Heat the mixture on free flame for 30min and pass CO_2 gas. Cool, 2, 4 dihydroxy benzoic acid separate out, recrystallise from hot water, m.p 216-217°C.

(9) To a solution of 5gm of compound in 15ml ether add in portion 3ml of anydrous HCN. Pass HCl into the reaction mixture. Aldimine hydrochloride separate out. Keep the mixture at room temperature for few hours, decant the ether and then add boiling water to it. Cool and collect resorcylaldehyde, m.p 134°C.

(10) Resorcinol dibenzoate m.p 117°C.

(11) Resorcinol di p-nitrobenzoate m.p 182°C

(12) Resorcinol di (3, 5-dinitrobenzoate m.p 201°C.

(13) 2, 4, 6-tribromo resorcinol m.p 111°C.

(14) Picrate (orange-yellow) m.p 90°C.

m.p 122°C **2-Naphthol** (2-hydroxy naphthalene)

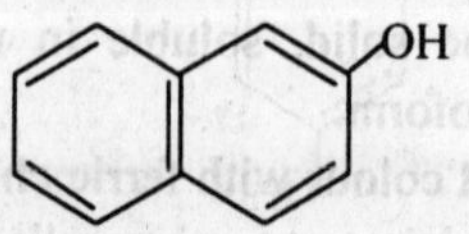

(1) Pink coloured solid, insoluble in water, soluble in alcohol, ether and chloroform.

(2) To 50mg of compound in 5ml dil NaOH solution add 5-7 drops of $CHCl_3$, heat it on water bath for few seconds. A blue coloured solution is obtained.

(3) Dissolve 2 gm of compound in 10ml ether and then gradually add anhydrous $FeCl_3$. Reflux on water bath and then distill off the ether. Add $CaCO_3$, water followed by excess of caustic soda to the residue and filter. To the filtrate add dil H_2SO_4 and collect di-2 naphthol, m.p 218°C.

(4) Heat a mixture of 5gm of compound, 15ml ethanol, 10gm NaOH and 21ml water to 70°C. Add dropwise 5ml $CHCl_3$. After the reaction starts remove the water bath and continue adding $CHCl_3$. After addition is complete continue stirring for 1hr. Distill off $CHCl_3$ and alcohol and add dil HCl to the residue until acidic. Extract with ether, distill off ether and collect 2 hydroxy 1-naphthaldehyde, m.p 80°C.

(5) Dissolve 5gm of compound in a warm solution of 1.2gm of NaOH in 60ml water. Cool to 0°C and add 2.5gm of sodium nitrite. Add dropwise 5ml dil H_2SO_4 (temp below 0°C). Continue stirring for 30min, filter the precipitated nitroso 2-naphthol, m.p 106°C.

(6) Heat 10gm of compound with 15ml of 40% aqueous solution of ammonium sulphite and 10ml of 20% aqueous ammonia solution at 100-150°C for 8hrs. Cool, filter the precipitated 2-naphthylamine, wash with NaOH, m.p 112°C.

(7) 2-Naphthyl benzoate m.p 107°C. (8) 2-Naphthyl p-nitrobenzoate m.p 169°C. (9) 2-Naphthyl 3, 5-dinitrobenzoate m.p 210°C. (10) 6-bromo 2-naphthol m.p 127°C. (11) 2-naphthyl acetate m.p 70°C. (12) Picrate (yellow orange) m.p 156°C.

m.p 124°C **p-Toluhydroquinone** (2, 5-dihydroxy toluene, 2-methyl hydroquinone)

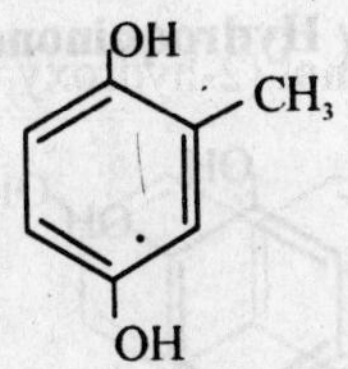

(1) Crystalline compound, soluble in water, alcohol and ether. Sublimes.

(2) Reduces Fehling's solution and Tollens reagent.

(3) To aqueous solution of compound add ammonia solution. A red coloured solution which shows orange fluorescence is obtained.

(4) 2-methyl hydroquinone dibenzoate m.p 119°C. (5) 2-methyl hydroquinone diacetate m.p 49°C.

m.p 133°C **Pyrogallol** (1, 2, 3-trihydroxy benzene)

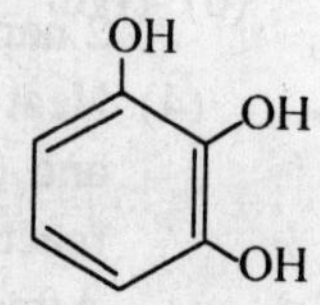

(1) White crystalline solid which sublimes undecomposed. Sparingly soluble in benzene, chloroform, easily soluble water, alcohol and ether. Turns black on exposure to air.
(2) Gives orange-yellow colour with ferric chloride solution.
(3) Reduces Tollens reagent.
(4) To 4ml water first add 10drops of 1% aqueous solution of compound, 2-3 drops of glycerol and 4ml of conc H_2SO_4, boil for 20-30 sec. A violet red colouration against white background is obtained.
(5) Heat 100mg of compound with 2ml Ac_2O and 150mg of fused $ZnCl_2$ for 40-45 minutes at 140-150°C. Cool, pour in ice, collect gallacetophenone, m.p 172°C.
(6) Pyrogallol tribenzoate m.p 90°C.
(7) Pyrogallol tri p-nitrobenzoate m.p 230°C.
(8) Pyrogallol tri 3, 5 dinitrobenzoate m.p 205°C.
(9) Picrate (lemon-yellow) m.p 128°C. (10) Pyrogallol triacetate m.p 165°C

m.p 140°C **Hydroxy Hydroquinone** (1, 2, 4-trihydroxy benzene)

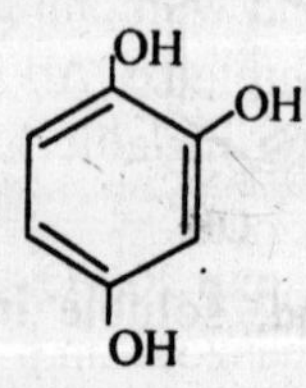

(1) Crystalline compound, insoluble in benzene and chloroform, soluble water, alcohol and ether.
(2) Gives transient green colour with ferric chloride solution.

(3) Responds to phthalein test to give green fluorescence.
(4) To few milligrams of compound add conc H_2SO_4. A green coloured solution slowly changing to violet is obtained.
(5) Hydroxy hydroquinone tribenzoate m.p 120°C.
(6) Hydroxy hydroquinone triacetate m.p 96°C.

m.p 165°C **p-xenol** (4-Hydroxy Biphenyl, p-phenyl phenol)

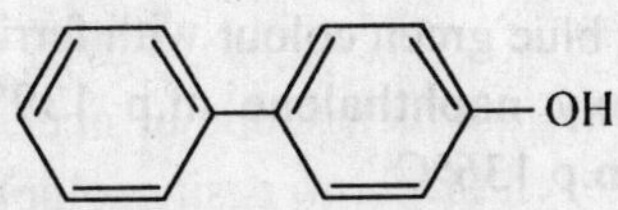

(1) Crystalline compound, soluble in alcohol, ether and chloroform.
(2) p-Xenyl benzoate m.p 149°C.
(3) p-xenyl acetate m.p 87°C.
(4) 3, 5 dibromo 4-hydroxy biphenyl m.p 91°C

m.p 171°C **Hydroquinone** (Quinol, 1, 4-dihydroxy benzene)

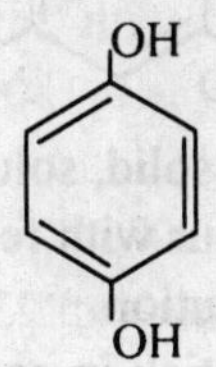

(1) White crystalline solid, soluble in water, alcohol and ether. Sublimes undecomposed.
(2) Gives yellow colour with ferric chloride solution.
(3) Reduces Fehling's solution in cold and Tollens reagent on warming.
(4) Warm 100mg of compound with 2ml conc H_2SO_4. Cool and add a pinch of potassium dichromate. An immediate precipitation of dark green needles having metallic lustre of quinhydrone, m.p 171°C take place.
(5) Hydroquinone dibenzoate m.p 205°C.
(6) Hydroquinone di- p-nitrobenzoate m.p 258°C.
(7) Hydroquinone di 3, 5-dinitrobenzoate m.p 317°C
(8) Hydroquinone diacetate m.p 123°C.
(9) Picrate (yellow) m.p 117°C.

m.p 187°C (d) **2, 7-dihydroxy naphthalene**

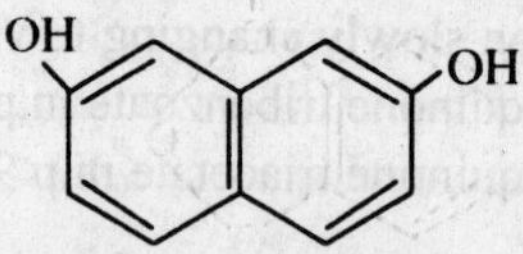

(1) Crystalline compound, slightly soluble in water, soluble in alcohol and ether. Sublimes with some decomposition.
(2) Gives blue or blue green colour with ferric chloride solution
(3) 2, 7-dibenzoxy naphthalene m.p 139°C. (4) 2, 7-diacetoxy naphthalene m.p 136°C.

m.p 219°C, hydrated m.p 108-109°C **Phloroglucinol** (1, 3, 5-trihydroxy benzene)

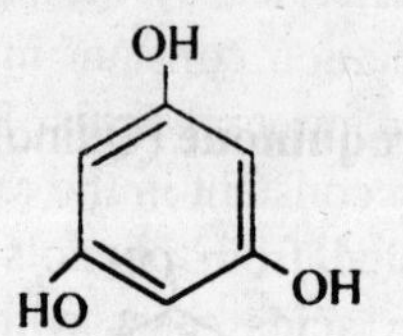

(1) Colourless crystalline solid, soluble in water and ether.
(2) Gives blue-violet colour with ferric chloride solution.
(3) Reduces Fehling's solution.
(4) Soak wooden match stick in conc HCl and add to it 2-3 drops of aqueous solution of substance. The match stick turns red.
(5) Dissolve 200mg of compound in 2ml conc H_2SO_4 and pour with stirring this solution into 4ml cold nitrating mixture (2ml each of conc HNO_3 and conc H_2SO_4). Keep the mixture at room temperature for 5-10 min and then pour in ice. Collect yellow crystals of 2, 4, 6-trinitrophloroglucinol, m.p 166°C.
(6) Phloroglucinol tribenzoate m.p 174°C
(7) Phloroglucinol tri-p-nitrobenzoate m.p 283°C.
(8) Phloroglucinol tri 3, 5-dinitrobenzoate m.p 162°C
(9) Phloroglucinol triacetate m.p 104°C.
(10) 2, 4, 6-tribromophloroglucinol m.p 151°C.

m.p 258-62°C **Phenolphthalein** (4, 4' Dihydroxytriphenyl-methanecarboxylic acid –2)

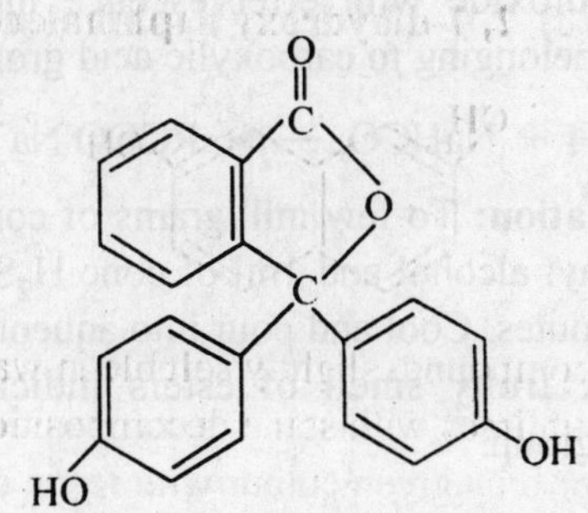

(1) Crystalline compound, insoluble in water.
(2) Diacetylphenolphthalein m.p 143°C
(3) Dibenzoylphenolphthalein m.p 169°C.

(*vi*) *Carboxylic Acids*

Organic acids are characterised by the presence of a carboxyl group – COOH. They have disassociation constant in the range of 10^{-6} to 10^{-5} and are much weaker acids than mineral acids. Phenols which are also acidic however have disassociation constant in the range of 10^{-10}. Carboxylic acids can however be differentiated from phenols in respect of their solubility behaviour. Most carboxylic acids react with weak base such as sodium bicarbonate to generate carbon dioxide and sodium carboxylate anion which is soluble in aqueous solution. Phenols are however too weak to decarboxylate the bicarbonate and the proton tend to remain on phenol and thus insoluble in bicarbonate solution.

Identification tests

(*i*) **Litmus Paper test:** Dissolve 50mg of compound in minimum amount of water or alcohol. Put a drop of this solution on the blue litmus paper. Conversion of blue litmus to red indicates the unknown compound to be acidic.

(*ii*) **Iodate-Iodide test:** To 20mg of unknown compound first add 4-5 drops of 2% solution of potassium iodide and then add 4-5 drops of 4% solution of potassium iodate. Heat the mixture on boiling water for 1minute, cool and then add 3-4 drops of freshly prepared 0.1% solution of starch. Development of blue colour indicates the unknown compound to be acidic.

$$5I^{\ominus} + IO_3^{\ominus} + 6H^{\oplus} \longrightarrow 3I_2 + 3H_2O$$

(*iii*) **Sodium Bicarbonate test:** Add few milligrams of unknown compound to 1ml of 5% sodium bicarbonate solution. Evolution of carbon dioxide with effervescence indicates the unknown compound belonging to carboxylic acid group.

$$RCOOH + NaHCO_3 \longrightarrow RCOO^-Na^+ + CO_2$$

(*iv*) **Ester formation:** To few milligrams of compound add 1-2ml of absolute ethyl alcohol and 1ml of conc H_2SO_4. Heat the solution for 5-10 minutes. Cool and pour into aqueous sodium bicarbonate solution. A fruity smell of esters indicates the presence of carboxylic group.

Derivatives

(*i*) **Amide:** In a dry test tube place 200mg of carboxylic acid and add to it 400mg of PCl_5. Reflux gently till a clear solution is obtained. Cool and add 4-5 ml of concentrated ammonia solution to this acid chloride. A vigorous reaction sets in. After the reaction has subsided heat on water bath for 4-5 minutes. Cool and filter the precipitated solid, wash with water, recrystallise and determine its melting point.

$$RCOOH + PCl_5 \rightarrow RCOCl + POCl_3 + HCl$$
$$RCOCl + NH_4OH \rightarrow RCONH_2 + HCl + H_2O$$

Mechanism

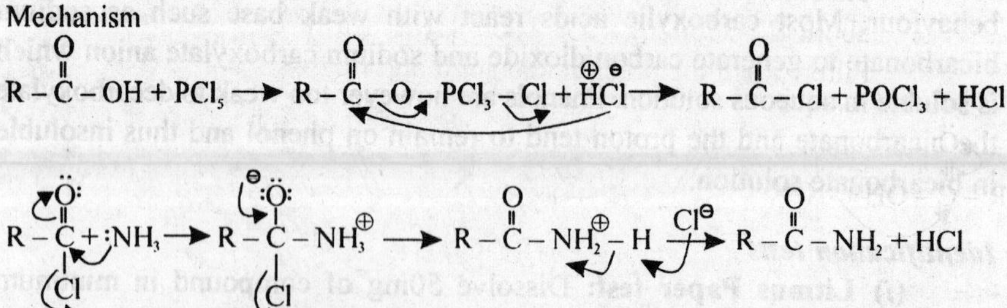

(*ii*) **Anilide (Method A):** In a dry test tube place 200mg of carboxylic acid and add to it 400mg of PCl_5. Reflux gently till a clear solution is obtained. Dissolve the crude acid chloride in 5ml ether and add a solution of 3-5ml of aniline in 15ml ether until the smell of acid chloride has disappeared. Shake the mixture thoroughly for 10min. Cool and add 5ml of dil HCl so as to remove excess of aniline as its salt. Remove the ethereal layer, evaporate off the solvent and recrystallise the anilide from ethanol, benzene or pet ether.

Method B: Heat 2gm of compound with 4gm of aniline in an oil bath at 150°C for 2 hrs. Cool the reaction mixture and add about

20-30ml of 10% HCl. Heat nearly to boiling and cool. Filter, wash the precipitated solid and recrystallise

$RCOOH + PCl_5 \rightarrow RCOCl + POCl_3 + HCl$

$RCOCl + C_6H_5NH_2 \rightarrow RCONHC_6H_5 + HCl$

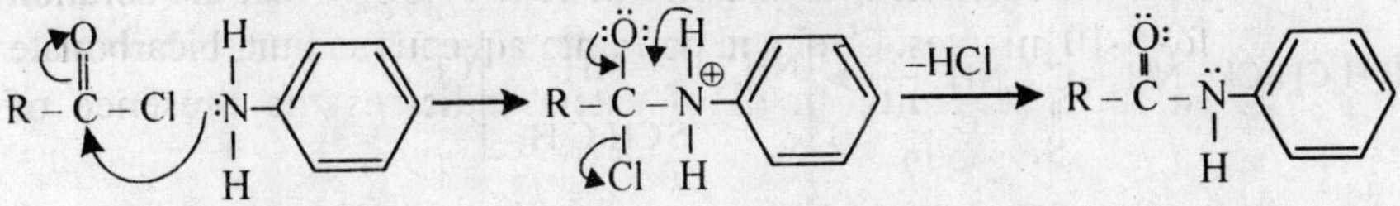

(*iii*) **p-toluidide: (Method A)** - In a dry test tube place 200mg of carboxylic acid and add to it 400mg of PCl_5. Reflux gently till a clear solution is obtained. Dissolve the acid chloride in 5ml ether and add a solution of 3-5ml of p-toluidine in 15ml ether until the smell of acid chloride has disappeared. Shake the mixture thoroughly for 10min. Cool and add 5ml of dil HCl so as to remove excess of amine as its salt. Remove the ethereal layer, evaporate off the solvent and recrystallise the p-toluidide from ethanol, benzene or pet ether.

Method B: Heat 2gm of compound with 4gm of p-toluidine in an oil bath at 150°C for 2 hrs. Cool the reaction mixture and add about 20-30ml of 10% HCl. Heat nearly to boiling and cool. Filter, wash the precipitated solid and recrystallise.

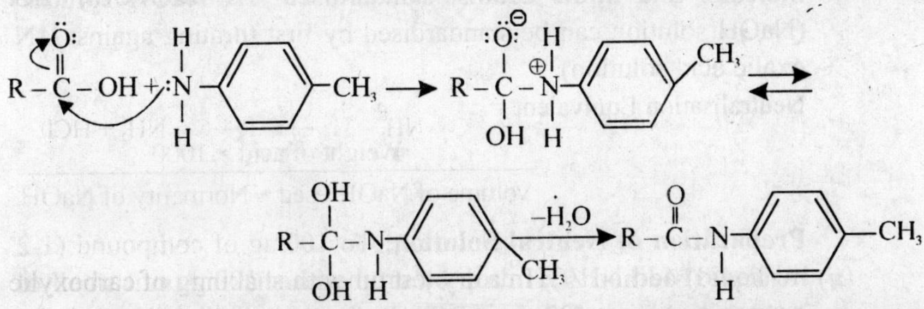

(*iv*) **S Benzyl isothiouronium salt:** Dissolves 500mg of organic acid in 10ml water and warm if necessary. Add 2-3 drops of phenolphthalein indicator and then neutralise with 1N NaOH solution until a pink colour is obtained. Now add 2-3 drops of .1N HCl so that the solution attains a faint pink colouration. Now add 10ml of S-Benzyl isothiouronium chloride solution (2gm in 10ml water) and cool the resulting mixture in ice until precipitation is complete. Collect the crude derivative, recrystallise and determine its melting point.

Preparation of S Benzyl isothiouronium chloride :- Dissolve 15gm of thiourea in 40ml water and add 27ml of ethanol and 25.4 ml of benzyl chloride. Reflux the mixture on water bath for 45 minutes. Cool and shake the mixture vigorously. Collect S Benzyl isothiouronium chloride and recrystallise from HCl, m.p 175°C.

$$C_6H_5CH_2Cl + NH_2-\underset{\underset{S}{\|}}{C}-NH_2 \longrightarrow \left[NH_2-\underset{\underset{SCH_2C_6H_5}{|}}{C}=\overset{\oplus}{N}H_2\right]Cl^-$$

$$\left[C_6H_5CH_2-SC\begin{matrix}\nearrow \overset{\oplus}{N}H_3\\ \searrow NH_2\end{matrix}\right]Cl^{\ominus} + RCOO^-Na^+ \longrightarrow \left[C_6H_5CH_2-SC\begin{matrix}\nearrow \overset{\oplus}{N}H_2\\ \searrow NH_2\end{matrix}\right]^+ RCOO^- + NaCl$$

(*v*) **Anhydrides:** Heat 300mg of acid with 2ml acetic anhydride for 20-30 minutes. Distill off the excess reagent. Crystallise the cyclic anhydride from chloroform or benzene and determine its melting point.

(*vi*) **Neutralisation Equivalent (NE)**: Neutralisation equivalent is the weight of an organic acid in grams needed to neutralise one gram equivalent of alkali.

Weigh about 0.2gm of dry finely powdered organic acid and dissolve it in 50ml distilled water. (if the acid is insoluble in water dissolve in 5ml ethanol). Add 2 drops phenolphthalein indicator and titrate against standardised .1N NaOH solution (NaOH solution can be standardised by first titrating against .1N oxalic acid solution).

Neutralisation Equivalent =

$$\frac{\text{Weight of acid} \times 1000}{\text{Volume of NaOH used} \times \text{Normality of NaOH}}$$

Preparation of Neutral Solution: To 100mg of compound (1-2 ml liquid) add NH_4OH drop by drop with shaking until smell of ammonia persists. Dilute it with water and boil to remove excess ammonia.

Aliphatic Carboxylic Acids

Liquid

b.p 101°C **Formic acid** (Methanoic acid) HCOOH

(1) Colourless liquid having pungent odour, miscible with water, alcohol and ether.

(2) Reduces Tollens reagent.
(3) Heat 2ml of acid with 2ml of conc H_2SO_4 and bring a burning piece of paper near the mouth of test tube. Carbon monoxide evolved during the reaction burns with a blue flame.
(4) Heat 2ml neutral solution of the acid with 1-2ml mercuric chloride solution. A white or grey precipitate of mercurous chloride is obtained.
(5) To 1ml of compound add 1ml of absolute alcohol and few drops of conc H_2SO_4, heat the solution. A fruity smell due to the formation of ethyl formate is obtained.
(6) Add 2ml neutral $FeCl_3$ solution to 2ml neutral aqueous solution of the acid. A wine red colour changing to brown precipitate is obtained.
(7) To 2ml of the acid add a little $NaHCO_3$, warm and add sodium nitroprusside solution. A greenish blue colour is produced.
(8) Reflux together 3ml of formic acid with 20ml of methyl acrylate, 6 gm of hydroquinone and a drop of conc H_2SO_4 on oil bath for 2-3hrs. Distill off methyl formate and collect acrylic acid b.p 140°C from the residual portion.
(9) Formanilide m.p 50°C
(10) Formo-p-toluidide m.p 53°C
(11) S Benzyl thiouronium formate m.p 146°C. (12) NE 46

b.p 118°C **Acetic acid** (Ethanoic Acid) CH_3COOH

(1) Colourless hygroscopic liquid with characteristic odour of vinegar, miscible with water, alcohol and ether.
(2) To 1ml neutral aqueous solution of the acid add few drops of neutral $FeCl_3$ solution. A red colour due to the formation of ferric acetate is obtained.
(3) To 2ml of organic acid add 2ml absolute alcohol and few drops of conc H_2SO_4. Heat the resultant solution. A fruity smell due to the formation of ethyl acetate is obtained.
(4) To a solution of 15ml acetic acid and 1.5gm red phosphorous add dropwise 10ml bromine (cool if the reaction becomes vigorous). Warm the mixture until colour of bromine persists. Cool and distill off bromoacetyl bromide under reduced pressure. Add water to the distillate whereby the mixture solidifies to a white crystalline mass. Collect colourless crystals of bromoacetic acid, m.p 51°C.
(5) Acetamide m.p 82°C [Amide]

(6) Acetanilide m.p 114°C [Anilide]
(7) Acet-p-toluidide m.p 148°C.
(8) S Benzyl thiouronium acetate m.p134°C. (9) NE 60

b.p 141°C **Acrylic acid** (propenoic acid) CH_2= CHCOOH

(1) Colourless liquid with smell similar to acetic acid, miscible with water, alcohol and ether.
(2) Reduces $KMnO_4$ solution
(3) Acrylamide m.p 84°C
(4) Acrylanilide m.p 104°C.
(5) Acrylo-p-toluidide m.p141°C.
(6) NE 72

b.p 141°C **Propionic acid** (Propanoic acid) CH_3CH_2COOH

(1) Colouless liquid with pungent odour, miscible with water, alcohol and ether.
(2) To 1ml neutral solution of acid add 1ml neutral $FeCl_3$ solution and 1ml amyl alcohol. Shake it thoroughly. Change of red colour from aqueous phase to alcoholic phase takes place.
(3) Propionamide m.p 79°C.
(4) Propionanilide m.p104°C.
(5) Propio-p-toluidide 124°C.
(6) S Benzyl thiouronium propionate m.p148°C. (7) NE 74

b.p 155°C **Isobutyric acid** (2-methyl propanoic acid) $(CH_3)_2CHCOOH$

(1) Colourless liquid with unpleasant odour, sparingly soluble in water, easily soluble alcohol and ether.
(2) Isobutyramide m.p 129°C.
(3) Isobutyranilide m.p 105°C.
(4) Isobutyro-p-toluidide m.p 104°C.
(5) S Benzyl thiouronium isobutyrate m.p 143°C.
(6) NE 88

b.p 163°C **n-Butyric acid** (Butanoic acid) $CH_3CH_2CH_2COOH$

(1) Colourless oily liquid with unpleasant smell of rancid butter, miscible with water, alcohol and ether.

(2) To 2ml of organic acid add 2ml of absolute alcohol and few drops of conc H_2SO_4. Heat the resultant solution. A sweet smell of pine apple due to the formation of ethyl butyrate is obtained.

(3) n-Butyramide m.p 114°C

(4) n-Butyranilide m.p 96°C.

(5) n-Butyro-p-toluidide m.p 72°C.

(6) S-Benzyl thiouronium n-butyrate m.p 146°C.

(7) NE 88

b.p 168-169°C (d) **Isocrotonic acid** (cis-buten-2-oic acid-1)

$CH_3CH{=}CHCOOH$

(1) Liquid with sharp odour, soluble in water and alcohol.

(2) Reduces $KMnO_4$ solution

(3) Isocrotonamide m.p102°C

(4) Isocrotonanilide m.p 102°C.

(5) Isocroto-p-toluidide m.p 132°C. (6) NE 86

b.p 176°C **Isovaleric acid** (3-methyl butanoic acid)

$(CH_3)_2CHCH_2COOH$

(1) Colourless liquid with decayed cheese like odour, miscible with alcohol and ether.

(2) To 10gm isovaleric acid is added 5ml Br_2 and .5ml PCl_3. The mixture is heated at 70-80°C for 15-20hrs. The crude bromoacid is then distilled under reduced pressure and fraction boiling at 110-125°(15mm) is collected. 20ml of NH_4OH is added to the bromoacid and the mixture is kept at room temperature for a week. Ammonia is removed and solution is concentrated to thick paste. Disssolve in hot water and filter hot. Add 2ml ethanol to the filtrate and leave overnight. Collect crystals of Valine, m.p 280-282°C (d).

(3) Isovaleramide m.p135°C.

(4) Isovaleranilide m.p 110°C

(5) Isovalero-p-toluidide m.p 106°C.

(6) S Benzyl thiouronium isovalerate m.p 153°C.

(7) NE 102

b.p 187°C **n-Valeric acid** (Pentanoic acid –1) $CH_3(CH_2)_3COOH$

(1) Colourless liquid with decayed cheese like odour, miscible with alcohol and ether

(2) Heat 11ml of valeric acid on water bath and add in portion 9ml of thionyl chloride to it. Reflux for 30 min. Distill and collect valeryl chloride at 125-127°C.

(3) Reflux a mixture of 6ml of valeric acid, 7ml of propanol, 10ml of benzene and 1.2ml H_2SO_4 for 36 hrs. Pour in ice water. Extract with ether. Distill and collect propyl valerate at b.p 163-164°C.

(4) n-Valeramide m.p104°C.

(5) n-Valeranilide m.p 63°C.

(6) n-Valero-p-toluidide m.p 74°C

(7) NE 102

(8) S Benzyl thiouronium valerate m.p 156°C

b.p 205°C **Hexanoic acid** (n-Caproic acid) $CH_3(CH_2)_4COOH$

(1) Oily liquid with unpleasant odour, steam volatile, slighltly soluble in water, soluble in alcohol and ether.

(2) To ice cooled 12.6ml acid pass 2-3gm of keten. Carefully distill the reaction mixture on oil bath. First collect acetone, keten and acetic acid and then raise the temperature of an oil bath to 220°C and maintain at this temperature for 3hrs. Now distill the residue under reduced pressure and collect hexanoic anhydride at 118-121°C (6mm). [b.p 241-243°C (d)]

(3) Reflux 20gm of acid with 10ml of bromine and .3gm of phosphorous oxychloride at 60-70°C. Slowly raise the temperature of the reaction mixture to 100°C. When all bromine has reacted, distill under reduced pressure and collect α bromocaproic acid at b.p 135-140°C/15mmHg [b.p 240°C]

(4) n-Hexanoamide m.p 100°C. (4) n-Hexanoanilide m.p 95°C.

(5) S Benzyl thiouronium Hexonate m.p 159°C.

(6) NE 116

(7) Hexo- p- toluidide m.p 73°C

b.p 223°C **Heptanoic acid** (n-Heptylic acid) $CH_3(CH_2)_5COOH$

(1) Colourless oily liquid, sparingly soluble in water, soluble in alcohol and ether.

(2) Heptamide m.p 94°C
(3) Heptanilide m.p 70°C.
(4) NE 130

b.p 237°C **n-Caprylic acid** (Octanoic acid) $CH_3(CH_2)_6COOH$

(1) Colourless oily liquid, insoluble in water, easily soluble in alcohol and ether.
(2) n-Caprylamide m.p106°C.
(3) n-Caprylanilide m.p 57°C.
(4) n-Capryl-p-toluidide m.p 70°C.
(5) S Benzyl thiouronium caprylate m.p 157°C.
(6) NE 144

b.p 254°C **Pelargonic acid** (Nonanoic acid) $CH_3(CH_2)_7COOH$

(1) Oily liquid, insoluble in water, soluble in alcohol and ether.
(2) Pelargonamide m.p 99°C
(3) Pelargonanilide m.p 57°C
(4) Pelargon-p-toluidide m.p 84°C
(5) NE 158

b.p 286°C (100mm) **Oleic acid** (cis 9-octadecenoic acid) $CH_3(CH_2)_7CH{=}CH(CH_2)_7COOH$

(1) Colourless oily liquid, insoluble in water, miscible with alcohol and ether.
(2) Slowly add 50gm of acid to a well stirred solution of 200 ml of peracetic acid (add 250gm of Ac_2O with stirring to 50gm of 30% H_2O_2. Continue stirring for 4hrs but keep temp below 40°C and keep ovenight) at 20°C. Continue stirring for 3 hrs, pour in 500ml water and then keep at room temperature for few hours. Filter the precipitated 9: 10 epoxy steraic acid, recrystallise from acetone, m.p 56°C.
(3) Oleamide 76°C (3) Oleanilide m.p 41°C (4) Oleic-p-toluidide m.p 42°C. (5) NE 282

b.p 122°C (15mmHg), m.p 17°C **dl Lactic acid** (2-hydroxy propanoic acid) $CH_3CHOHCOOH$

(1) Viscous, highly hygroscopic liquid, miscible with water, alcohol and ether.

(2) To 1ml of compound add 5ml aqueous resorcinol solution, allow it to flow slowly onto 5ml conc H_2SO_4 and then leave it for 2 minutes. A red colour develops at the interface.

(3) To 1ml of compound add few drops of conc H_2SO_4 and warm until the solution turns brown. Cool and add NaOH solution until the solution is alkaline and heat. Disagreable odour of acetaldehyde resin is observed.

(4) Reflux a mixture of 5.3 gm of lactic acid, 10ml propan-2-ol, 30ml benzene and 2gm zerolit 225 (acid resin) for 5 hrs. Filter off the resin, wash the filtrate with calcium carbonate and distill off the benzene under reduced pressure first and then collect isopropyl lactate at 76°C/24mmHg.

(5) Warm 2ml of compound with iodine reagent and then add NaOH to make it alkaline. Cool and collect crystals of iodoform, m.p 119°C.

(6) Heat 25ml of lactic acid at 140-150°C. Cool the distillate in the receiver which soon solidifies, collect crystals of lactide, m.p 125°C.

(7) dl lactamide m.p 74°C.

(8) dl lactanilide m.p58°C.

(9) dl lacto-p-toluidide m.p 107°C.

(10) S Benzyl thiouronium dl lactate m.p 153°C.

(11) NE 90

(12) Benzoate m.p 112°C

Aliphatic Solid

m.p 31°C, b.p 270°C **n-Capric acid** (decanoic acid) $CH_3(CH_2)_8COOH$

(1) Crystalline compound insoluble in water, soluble in alcohol and ether.

(2) n-Capramide m.p 108°C (3) n-capranilide m.p 70°C. (4) n-capri-p-toluidide m.p 78°C. (5) NE 172.

m.p 33-35°C, b.p 246°C **Levulinic acid** (4-oxopentanoic acid) $CH_3COCH_2CH_2COOH$

(1) Deliquescent solid, soluble in water, alcohol and ether.
(2) Levulinic acid 2, 4 dinitrophenylhydrazone m.p 206°C.
(3) Levulinic acid p-nitrophenylhydrazone m.p175°C.
(4) Levulinic acid semicarbazone m.p 184°C.
(5) Levulinic acid phenylhydrazone m.p 108°C.
(6) Levulinic acid oxime m.p 95°C.
(7) Levulinamide m.p 108°C.
(8) Levulinanilide m.p 102°C.
(9) Levulin-p-toluidide m.p 109°C.
(10) NE 116.

m.p 35°C, b.p 163°C **Trimethylacetic acid** (Pivalic acid) $(CH_3)_3$ C. COOH

(1) Crystalline solid, insoluble in water, soluble in alcohol and ether.
(2) Trimethylacetamide m.p 155°C
(3) Trimethylacetanilide m.p 128°C
(4) Trimethylaceto-p-toluidide m.p 120°C
(5) NE 102

m.p 44°C **Lauric acid** (n-dodecanoic acid) $CH_3(CH_2)_{10}COOH$

(1) Colourless compound, insoluble in water, soluble in alcohol and ether.
(2) To 12gm lauric acid add 10gm thionyl chloride and heat. Cool and recover lauryl chloride by distillation at b.p 145°C. To 13ml ether add with stirring 2gm lauryl chloride. Cool and add dropwise with stirring 1ml triethyl amine. Continue stirring for an hour, cool and extract with 25ml aqueous $H_2SO_4$to remove amine. Distill off the ether and then distill the residual oily layer under reduced pressure and collect laurone which soon solidifies, m.p 68-69°C.
(3) Lauramide m.p 99°C.
(4) Lauranilide m.p 77°C.
(5) Lauro-p-toluidide m.p 87°C.
(6) S Benzyl thiouronium laurate m.p 141°C
(7) NE 200

m.p 45°C, b.p 185°C **Angelic acid** (α-methyl isocrotonic acid) $CH_3CH=C(CH_3)COOH$

(1) Crystalline solid with spicy odour, sparingly soluble in water, soluble in alcohol and ether.
(2) Dissolve 500mg of compound in CS_2 and add 3ml of bromine solution (in CS_2). Shake thoroughly and leave it at room temperature for 4 hrs. Evaporate off the solvent and rub the oily layer with glass rod, angelic acid dibromide separates out, crystallise from pet ether, m.p 86°C.
(3) Angelamide m.p 127°C (4) Angelanilide m.p 126°C.

m.p 58°C **Myristic acid** (n-Tetradecanoic acid) $CH_3(CH_2)_{12}COOH$

(1) Crystalline compound, insoluble in water, easily soluble in alcohol and chloroform.
(2) n-Myristamide 103°C
(3) n-Myristanilide m.p 84°C
(4) n-Myristo-p-toluidide m.p 93°C
(5) NE 228
(6) S Benzyl thiouronium myristate m.p 139°C.

m.p 64°C **Palmitic acid** (n-Hexadecanoic acid) $CH_3(CH_2)_{14}COOH$

(1) Colourless greasy crystalline solid, insoluble in water, soluble in hot alcohol and ether.
(2) To 2gm palmitic acid and 6ml CCl_4 add dropwise 2ml liquid sulphur trioxide. Heat the mixture at 50-60°C for 1hr. Cool and place in refrigerator overnight. Filter the solid and collect α sulphopalmitic acid, m.p 90-91°C.
(3) Add dropwise 2ml of bromine to a mixture of 10gm of compound and red phosphorous. Heat the mixture on water bath for 6hrs and then keep overnight. Pour in water and heat to boiling. Cool and collect α bromopalmitic acid, m.p 51-52°C.
(4) Palmitamide m.p 106°C. (5) Palmitanilide m.p 90°C. (6) Palmito-p-toluidide m.p 98°C. (7) NE 256.
(8) S Benzyl thiouronium palmitate m.p141°C.

m.p 69°C Stearic acid (n-Octadecanoic acid) $CH_3(CH_2)_{16}COOH$

(1) Greasy, colourless, tasteless, odourless solid, insoluble in water, soluble in ether, benzene and chloroform.

(2) Heat a mixture of 4.4 gm steraic acid and 2.0gm of magnesium oxide at 335-340°C. After 1 hr add in portion 24gm of stearic acid and continue refluxing for 7hrs. Cool and pour in 100ml 4N H_2SO_4. Heat for 2hrs. Remove the lower layer and boil the upper layer with 10ml water for 1hr. Remove the aqueous layer and boil the organic layer with 100ml 10% NaOH solution for 1hr with stirring. Remove the ketonic layer and boil with hot water. Cool and collect steranone, m.p 89°C.

(3) Stearamide m.p 108°C.

(4) Stearanilide m.p 94°C.

(5) Stearo-p-toluidide m.p 102°C.

(6) NE 284

(7) S Benzyl thiouronium sterate m.p 143°C.

m.p 72°C α-Crotonic acid (trans butene–2-oic acid) $CH_3CH{=}CHCOOH$

(1) Colourless crystalline, steam volatile, soluble in water.

(2) Reduces alkaline $KMnO_4$ solution.

(3) Reduces Tollens reagent.

(4) To 400mg of compound dissolved in carbon disulphide add 2ml of bromine solution (1ml Br_2 in 1ml CS_2). Keep the mixture in direct sunlight for 1-2 days. Evaporate off the solvent and collect α Crotonic acid dibromide, m.p 87°C.

(5) Reflux 5gm of acid with 7.5 gm sec-butyl alcohol, .5ml of conc H_2SO_4 and 8ml of benzene for 12hrs. Cool and neutralise with sodium bicarbonate. Extract with ether, distill off ether under reduced pressure and collect sec-butyl crotonate at 74°C (30mmHg)

(6) Heat 4.3 gm of compound with 10ml aniline at 180-190°C for 4hrs. Cool and pour the mixture in ice containing conc HCl. Filter the precipitated hydrochloride derivative, dissolve in water and add sodium bicarbonate to it. Cool and collect β-phenylaminobutyranilide, m.p 93°C.

(7) α-Crotonamide m.p 158°C

(8) α-Crotonanilide m.p 116°C.

(9) α-Crotono-p-toluidide m.p 132°C.
(10) S-Benzyl thiouronium α-Crotonate m.p 172°C
(11) NE 86

m.p (α) 63°C (β) 79°C **Glycollic acid** (Ethanol acid, hydroxy acetic acid) $CH_2OH.COOH$

(1) Colourless deliquescent crystalline solid, soluble in water, alcohol, ether and acetone.
(2) To a mixture of 13.5gm of KOH and 17.5gm of indole add 18gm of glycollic acid solution and heat it at 250°C for 18hrs. Cool to 50°C, add 25ml water and again heat at 100°C for 30min. Extract the aqueous layer with ether. Acidify the residual aqueous mixture with 12N HCl, cool to 0°C and collect indole 3-acetic acid, m.p 164-166°C (d).
(3) To 1gm of compound add 5ml acetyl chloride and warm the solution, cool and collect acetoxy acetic acid, m.p 66°C.
(4) Glycolamide m.p 120°C.
(5) Glycolic anilide m.p 96°C.
(6) Glycolic-p-toluidide m.p 143°C.
(7) S Benzyl thiouronium glycolate m.p 142°C. (8) NE 76

m.p 91°C **Citraconic acid** (Methylmaleic acid, cis methyl butene diolc acid) $COOHCH = C(CH_3)COOH$

(1) Hygroscopic crystalline solid, soluble in water and ether. Cis isomer
(2) To 500mg of compound add 5ml dil HNO_3 and evaporate it to dryness. Cool and collect mesaconic acid, m.p 204°C.
(3) Citraconic diamide m.p 185°C
(4) Citraconic dianilide m.p 175°C.
(5) Citraconic -p-toluidide (mono) m.p 170°C.
(6) NE 65

m.p 98°C **Glutaric acid** (Propane 1, 3-dicarboxylic acid) $COOH(CH_2)_3COOH$

(1) White odourless solid, soluble in water, alcohol and ether, sparingly soluble in benzene.

(2) To 50mg of compound add 100mg of resorcinol and 2ml conc H_2SO_4. Heat the mixture until it is red brown. Cool, pour it in 50ml water made alkaline by adding NaOH. A greenish yellow fluorescence is observed.
(3) Glutaric diamide m.p 174°C.
(4) Glutaric dianilide m.p 224°C.
(5) Glutaric di-p-toluidide m.p 217°C.
(6) NE 66

m.p 100°C **(-)Malic acid** (hydroxy succinic acid, hydroxy butane dioic acid) $COOHCH_2$ CH(OH)COOH

(1) Crystalline deliquescent solid, soluble in water and alcohol.
(2) In a small porcelain dish place 100mg of compound and add 20-30 drops of freshly prepared 2-naphthol solution (200mg in conc H_2SO_4). Heat the dish on boiling water bath. A greenish yellow to intense yellow colour change is observed.
(3) Heat 25mg of acid with 50mg resorcinol and 1ml conc H_2SO_4. Cool, pour the mixture in beaker containing water made alkaline by adding NaOH. An intense blue fluorescence is observed.
(4) To 1ml aqueous solution of the acid add 1ml $FeCl_3$ solution. A yellow coloured solution is obtained.
(5) Add 2gm of malic acid to 5ml acetyl chloride and heat gently on water bath so that the exothermic reaction sets in. Then heat on water bath for 1-2hrs. Cool, pour in ice and collect malic anhydride, m.p 54°C.
(6) Add 2ml of conc $H_2SO_4$to 2gm malic acid at an interval of 30min. Add three portion of .5ml fuming H_2SO_4 so that the exothermic reaction sets in. When the evolution of gases have subsided heat the reaction mixture on water bath for 90min. Cool, pour in ice and collect coumalic acid, m.p 206°C.
(7) (-) malamide (di) m.p 157°C.
(8) (-) malanilide (di) m.p 197°C.
(9) (-) malic di-p-toluidide m.p 207°C.
(10) Di- (S Benzyl thiouronium) malate m.p 124°C.
(11) NE 67

m.p 100°C, anhydrous 153°C **Citric acid** (2-hydroxy propane 1, 2, 3 tricarboxylic acid)

$$\begin{array}{c} CH_2COOH \\ | \\ HO-C-COOH \\ | \\ CH_2COOH \end{array}$$

(1) White crystalline solid, soluble in water and alcohol, sparingly soluble in ether.

(2) Add 1ml calcium chloride to 3ml neutral aqueous solution of the acid and boil for 2minutes. White shining crystals of calcium citrate are obtained.

(3) To 50mg of compound add 10-15 drops of 2-naphthol solution (200mg in conc H_2SO_4). Heat it on boiling water bath. A pale greenish blue colour which soon changes to blue green is obtained.

(4) Heat 50mg of compound with 100mg urea to 150°C. Cool and take small amount of the solidified melt in 2-3ml water. A blue fluorescence due to the formation of citrazininc acid is observed.

(5) To 50mg of acid add 1ml acetic anhydride and 1ml pyridine and then heat. A caramine red colouration is observed.

(6) To 2ml aqueous solution of the acid add Deniges reagent (HgO in H_2SO_4). Cool and add dropwise $KMnO_4$solution. Purple colour of $KMnO_4$ disappears and white precipitate is obtained.

(7) To 1ml aqueous solution of acid add 1ml freshly prepared $FeCl_3$ soution. A yellow coloured solution is obtained.

(8) Citric acid triamide m.p 210°C.

(9) Citric acid trianilide m.p 192°C.

(10) Citric acid tri-p-toluidide m.p 189°C.

(11) NE (anhydrous) 64, (hydrated) 70.

m.p 101°C, anhydrous m.p 189°C. **Oxalic acid** (Ethane dioic acid) $(COOH)_2.2H_2O$

(1) White solid, soluble in water, moderately soluble in alcohol. Sublimes at 150°C.

(2) To 2ml aqueous solution of acid add few drops of dil H_2SO_4, warm and add 4-5 drops of $KMnO_4$ solution. Pink colour of $KMnO_4$ disappears.

(3) To 2ml neutral solution of the acid add 2ml calcium chloride solution. White precipitate of calcium oxalate is obtained.

(4) Heat few milligram of compound with diphenylamine. Cool and

take the melt in a drop of alcohol. A blue colouration due to the formation of aniline blue takes place.

(5) Reflux 5gm of compound with 10ml absolute ethanol, 19ml benzene and 2ml conc H_2SO_4 for 24hrs. Cool and pour in water. Isolate with ether (2×25ml). Distill off the ether under reduced pressure and recover diethyl oxalate, b.p 182-183°C.

(6) Mix 9 gm of compound with 40gm of PCl_5 and cool the mixture in ice bath. After the reaction has subsided keep at room temperature until the whole mass is liquid. Distill and collect oxalyl chloride at b.p 63-64°C.

(7) Oxamide (di) m.p 419°C (d)

(8) Oxanilide (di) m.p 246°C

(9) Oxalic di-p-toluidide m.p 268°C.

(10) Di-(S Benzyl thiouronium) oxalate m.p 195°C (d)

(11) NE 63

m.p 105°C **Pimelic acid** (Pentane 1, 5-dicarboxylic acid, Heptanedioic acid) $COOH(CH_2)_5COOH$

(1) Crystalline compound, soluble in water, alcohol and ether.

(2) Reflux 500mg of compound with 5ml Ac_2O for 4-5 hrs. Distill off the acetic acid formed during the process under reduced pressure and recover pimelic α-anhydride, m.p 53-56°C from the residue.

(3) Pimelic diamide m.p 175°C.

(4) Pimelic dianilide m.p 155°C.

(5) Pimelic di-p-toluidide m.p206°C

(6) NE 80

m.p 106°C **Azelaic acid** (Heptane 1, 7-dicarboxylic acid) $COOH(CH_2)_7COOH$

(1) Crystalline solid, insoluble in water, soluble in alcohol.

(2) Reflux 2gm of compound with 7-8ml acetic anhydride for 4-6 hrs. Cool and collect the precipated azelaic α-anhydride, m.p 54°C.

(3) Azelaic diamide m.p 175°C.

(4) Azelaic dianilide m.p 186°C

(5) Azelaic di-p- toluidide m.p 201°C.

(6) Di-(S Benzyl thiouronium) azelate m.p 163°C.
(7) NE 94

m.p 133°C **Sebacic acid** (Octane-1, 8-dicarboxylic acid)
$COOH(CH_2)_8COOH$

(1) Crystalline compound, insoluble in water, easily soluble alcohol and ether.
(2) Reflux 5gm of acid with 12ml absolute ethanol and 1ml conc H_2SO_4 for 12 hrs. Cool and add 25ml water, remove the upper layer of crude ester and extract the aqueous layer with ether. Combine the ether and ester layers and distill under reduced pressure, collect diethyl sebacate b.p 155°C/6mm. [b.p 308°C]
(3) Reflux 2gm of acid with 6-7ml acetic anhydride for 5hrs. Cool and recover sebacic α/anhydride, m.p 79°C.
(4) Sebacic diamide m.p 210°C.
(5) Sebacic dianilide m.p 201°C
(6) Sebacic-p-toluidide m.p 207°C.
(7) NE 101

m.p 135°C(d) **Malonic acid** (Methane di carboxylic acid, Propan di acid) $COOH\ CH_2\ COOH$

(1) Colourless crystalline solid, easily soluble in water, alcohol and ether.
(2) In a test tube place 50mg of compound, 3-4ml acetic anhydride and boil for 3-4 minutes. Dilute the reaction mixture by adding 2-ml acetic acid. A yellow –red coloured solution with greenish-yellow fluorescence is obtained.
(3) Heat together 4ml glacial acetic acid, 3ml benzaldehyde and 3gm of malonic acid at 100°C for 10 hrs. On cooling benzalmalonic acid precipitates out, m.p 196°C.
(4) To 5ml benzaldehyde and 20ml 10% ammonia solution in alcohol add 5gm of malonic acid, heat it on water bath until a clear solution is obtained. Remove ethanol by evaporation and continue heating until evolution of CO_2 ceases. Dissolve residue in water, add dil HCl and recover cinnamic acid, m.p 132°C.
(5) Heat a mixture of 9ml crotonaldehyde, 12gm malonic acid and 12ml pyridine on steam bath for 3hrs. Cool and add a solution of

4.2ml conc H_2SO_4 in 10ml water with shaking. Filter the crude sorbic acid and recrystallise from hot water, m.p 134°C.

(6) Heat a mixture of 8.8ml furfural, 10.4gm malonic acid and 4.8ml pyridine on water bath for 2hrs. Cool, dilute the mixture with 10ml water, add concentrated ammonia solution and filter. Acidify the filtrate with dil HCl and then cool in ice bath. Collect crystals of furylacrylic acid, m.p 141°C.

(7) Malonic diamide m.p 170°C

(8) Malonic dianilide m.p 224°C.

(9) Malonic di-p-toluidide m.p 252°C.

(10) S Benzyl thiouronium malonate m.p 147°C.

(11) NE 52

m.p 135°C(d) **Acetone dicarboxylic acid** (β-keto glutaric acid) $COOHCH_2COCH_2COOH$

(1) Crystalline compound, soluble in water and alcohol, sparingly soluble in ether.

(2) To few milligrams of compound add neutral $FeCl_3$ solution. A violet coloured solution is obtained.

(3) To 500mg of compound first add 4-5 drops of conc iodine reagent and then add NaOH solution till the brown colour changes to yellow. Warm the solution on water bath for 2 minutes and cool. Yellow crystals of iodoform, m.p119°C are produced.

(4) Acetone dicarboxylic acid dianilide m.p 155°C.

(5) NE 73.

m.p 138°C **Maleic acid** (Toxilic acid, cis betenedioic acid) COOHCH=CHCOOH (Cis)

(1) Colourless crystalline solid, soluble in water and alcohol.

(2) Reflux 1gm of compound with 7ml acetic anhydride for 1hr. Distill off the excess anhydride, cool and recover maleic anhydride, m.p 56°C.

(3) Dissolve 1gm of maleic acid in 2ml water and add 2ml con HCl. Reflux the mixture for 30 minutes. Crystals of fumaric acid, m.p 286 °C crystallise out in the hot mixture.

(4) Mix 50mg of acid with 100mg of resorcinol and add 1ml conc H_2SO_4. to it. Heat the solution until red brown colour is obtained.

Cool, pour the mixture in beaker containing 100ml water made alkaline by adding NaOH. An orange red solution exhibiting fluorescence is obtained.

(5) Maleic dianilide m.p 187°C.
(6) Maleic di-p-toluidide m.p 142°C. .
(7) S Benzyl thiouronium maleate m.p 163°C
(8) NE 58

m.p 140-144°C **Suberic acid** (Octanedioic acid)

$COOH(CH_2)_6COOH$

(1) Colourless crystalline solid, insoluble in water, soluble in alcohol.
(2) Reflux 2gm of compound with 8ml acetic anhydride for 4-6 hrs. Cool and collect polymeric suberic α-anhydride, m.p 66°C.
(3) Suberic diamide m.p 216°C.
(4) Suberic dianilide m.p 186°C.
(5) Suberic di-p-toluidide m.p 218°C.
(6) NE 87

m.p 151°C **Adipic acid** (Butane 1, 4 dicarboxylic acid, Hexan di acid) $COOH(CH_2)_4COOH$

(1) White crystalline solid, soluble in water and ethanol.
(2) In a dry test tube place 50mg of compound, 100mg of resorcinol and 1ml of conc H_2SO_4. Heat for few minutes, cool and pour the contents in a beaker containing 100ml water. Add dil NaOH in excess to the resultant solution. Violet red solution which exhibits fluorescence is obtained.
(3) Heat together 3gm of adipic acid with 8ml absolute ethanol 4ml toluene containing3-4 drops of conc H_2SO_4at 120°C. Distill off the solvent and collect diethyl adipate, b.p 245°C.
(4) Heat on water bath at 50-60°C 7.3gm of compound with 11ml thionyl chloride for 4hrs and recover adipyl chloride. Add this acid chloride with stirring to ice cold mixture of 15gm $AlCl_3$ and 75ml benzene. Continue stiring for 2hr and then add 50ml benzene. Heat on water bath and then remove the benzene layer. Distill off the benzene and recover dibenzoyl butane, m.p 106-107°C.
(5) Adipic di amide m.p 220°C.
(6) Adipic dianilide m.p 240°C.

(7) Adipic di-p-toluidide m.p 241°C.
(8) S Benzyl thiouronium adipate m.p 159°C.
(9) NE 73

m.p 158°C **Tartronic acid** (Hydroxy malonic acid, 2-hydroxy propane dioic acid) COOH(CHOH)COOH

(1) Colourless crystalline compound, easily soluble in water and alcohol, sparingly soluble in ether.
(2) To 1ml concentrated solution of compound add 2ml hot mixture of 10ml acetic acid, 10ml conc H_2SO_4 and 1ml resorcinol. A dark green coloured solution is obtained.
(3) Tartrondiamide m.p 195°C.

m.p 162°C(d) **Itaconic acid** (Methylene succinic acid, methylene butane dioic acid) CH_2=C(COOH)-CH_2COOH

(1) Crystalline compound, soluble in water and alcohol, very slightly soluble in ether.
(2) To 6.4gm of compound add 23.4gm of PCl_5. Shake the mixture thoroughly, vigorous reaction takes place resulting in partial liquification of mixture. Reflux for 15min and then distill off the phosphorous oxychloride. Distill the residual portion and collect itaconyl chloride at 71-72°C/2mmHg
(3) Boil 2.5gm of compound in 15ml water with 2ml aniline for 30min. Cool and collect 1-phenylpyrrolidone-5-carboxylic acid; recrystallise from aqueous alcohol, m.p190°C
(4) Itaconic diamide m.p 191°C
(5) NE 65
(6) Itaconic dianilide m.p 190°C.

m.p 170°C **d-Tartaric acid** (2, 3-dihydroxy butanedioic acid) $COOH(CHOH)_2COOH$

(1) Colourless crystalline solid, readily soluble in water, moderatly in alcohol and sparingly in ether.
(2) Reduces Tollens reagent
(3) Warm 100mg of compound with acetic anhydride and pyridine. An emerald green colouration occurs.

(4) To 1ml neutral solution of acid add few drops of $CaCl_2$ and shake thoroughly. A white precipitate is produced which is soluble in hot glacial acetic acid.

(5) To 2ml aqueous solution of acid add 5-6 drops of freshly prepared $FeSO_4$ solution, 2-4 drops of H_2O_2 and 4-5ml NaOH solution. A deep violet to black colouration due to formation of dihydroxymaleic acid is immediately produced.

(6) To 2ml aqueous solution of acid add 2ml cobalt nitrate, 5ml NaOH solution and heat. A blue colouration appears which disappears on cooling.

(7) To 8gm of acid add with stirring a solution of .3ml of conc H_2SO_4 in 25ml acetic anhydride. Heat the mixture under reflux for 10min. Cool and collect diacetyl d-tartaric anhydride, m.p 133-134°C.

(8) Reflux together 7.5 gm of tartaric acid, 1.0gm zerolit 225 (acid resin), 13.5 ml of butanol and 15ml benzene for 10hrs. Filter to remove resin. Wash the filtrate, dry and distill under reduced pressure, collect dibutyl tartarate at 150°C/ 91.5mmHg) which on cooling solidifies, m.p 22°C.

(9) Tartaric acid diamide m.p 195°C (d)

(10) Tartaric acid dianilide m.p264°C (d)

(11) NE 75

m.p 185 °C **Succinic acid** (Ethane-1, 2-dicarboxylic acid, Butan di acid) $COOH(CH_2)_2COOH$

(1) White crystalline solid, soluble in water, insoluble in chloroform.

(2) To 2ml neutral solution of the acid add 2ml freshly prepared $CaCl_2$ solution. Heat the solution to boiling and scratch the sides of the test tube. White precipitate of calcium succinate is obtained.

(3) In a dry test tube place 500mg of compound, 1gm of resorcinol and 1ml of conc H_2SO_4. Heat for few minutes, cool, pour the contents in a beaker containing 100ml water. Add dil NaOH in excess to the resultant solution. Red coloured solution which exhibits green fluorescence is obtained.

(4) To 2gm succinic acid add 500mg of red phosphorous and 3ml bromine. Heat the mixture on water bath until the colour of bromine disappears. Pour in ice water and extract with ether. Evaporate off the ether and collect monobromosuccinic acid, m.p 160°C.

(5) Heat on steam bath a mixture of 5.9gm of succinic acid, 9.9 gm of phenol and 4.1ml phosphorous oxychloride for 90min. Add 25ml benzene and reflux for additional 60 min. Decant off the hot benzene layer and extract the syrupy residue with benzene. Combine the benzene solutions, concentrate and cool. Diphenyl succinate separates out as colourless crystals, m.p 120-121°C.

(6) To 5.9 gm succinic acid add 11ml açetyl chloride, reflux on steam bath for 90 minutes. Cool the mixture in ice bath and collect succinic anhydride, wash with ether and dry, m.p 118-119°C.

(7) Succinic acid diamide(succinamide) m.p 260°C (d)

(8) Succinic acid dianilide (succinanilide) m.p 228°C.

(9) Succinic acid di-p-toluidide m.p 255°C.

(10) S Benzyl thiouronium succinate m.p154°C.

(11) NE 59

m.p 191°C(d) **Aconitic acid** (trans propene-1, 2, 3-tricarboxylic acid)

$$\begin{array}{c} HOOCCH_2-\underset{\Vert}{C}-COOH \\ HOOC-C-H \end{array}$$

(1) Crystalline solid, soluble in water and alcohol, sparingly soluble in ether.

(2) Reduces alkaline $KMnO_4$ solution.

(3) Decolourises bromine water slowly on warming.

(4) Heat few milligrams of compound with acetic anhydride and pyridine. A violet red colouration is obtained.

(5) Aconitic amide m.p 250°C.

(6) NE 58

m.p 204°C **Mesaconic acid** (methyl fumaric acid, trans methyl betenedioic acid) $COOHC(CH_3)=CHCOOH$

(1) Crystalline solid, soluble in water and ether, sublimes readily.

(2) Heat the compound at 250°C, cool and recover citraconic anhydride, b.p 213°C.

(3) Mesaconic diamide m.p 176°C

(4) Mesaconic dianilide m.p 185°C.

(5) Mesaconic di-p-toluidide m.p 212°C.

(6) NE 65.

m.p 206-210°C (d) **Mucic acid** (tetrol-hexan-diacid)

$COOH(CHOH)_4COOH$

(1) White crystalline powder, soluble in water, insoluble in alcohol.
(2) In a test tube mix 100mg of compound with 10-15 drops of NH_4OH solution and evaporate it to dryness. Dip a pine splinter in conc HCl and hold it at the top of test tube and ignite the ammonium mucate residue. The pyrrole vapours evolved during the reaction develop bright red colour in the splinter.
(3) Dissolve 5gm of mucic acid in 15ml absolute ethanol and .5ml conc H_2SO_4. Cool and pour in ice water. Collect crystals of diethyl mucate, recrystallise from alcohol, m.p 164°C.
(4) Reflux 500mg of compound with 8ml acetic anhydride and 1gm anhydrous zinc chloride. Cool, pour in ice water with stirring and collect tetracetyl mucic acid, m.p 266°C.
(5) Mucic acid diamide m.p 220°C.
(6) Di-(S Benzyl thiouronium mucate) m.p 194°C.
(7) NE 105
(8) Tetraacetate m.p 266°C

m.p 287°C (sealed tube) **Fumaric acid** (trans ethylene 1, 2-dicarboxylic acid)

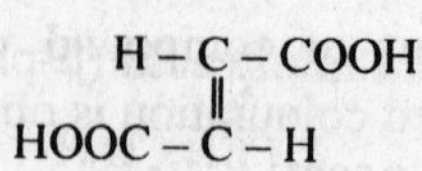

(1) Crystalline solid having tendency to sublime, sparingly soluble in alcohol and ether, insoluble in benzene.
(2) Reflux a mixture of 7gm of acid, 15ml absolute ethanol, 22ml benzene and 4-5 drops of conc H_2SO_4 for 12hrs. Cool, pour in 100ml ice water. Separate the benzene layer, wash with Na_2CO_3 and water. Remove the solvent on steam bath. Distill the residue and recover diethyl fumerate at b.p 213-215°C.
(3) Heat a well stirred solution of 2gm fumaric acid, 5ml water, and add with stirring 1ml bromine (dropwise). After the addition is complete cool the reaction mixture and collect crystals of 2, 3 dibromo succinic acid, m.p 167°C.

(4) Reflux 500mg of compound with 15ml absolute methanol saturated with HCl gas. Distill off the excess solvent and collect dimethyl fumarate, m.p 101°C.
(5) Fumaric diamide m.p 266°C
(6) Fumaric dianilide m.p 314°C
(7) S Benzyl thiouronium fumarate m.p 178°C.
(9) NE 58.

Solid Aromatic Acids

m.p 19.5°C, b.p 174°C (17mm) **o-Ethoxybenzoic acid** (Salicylic acid ethyl ether)

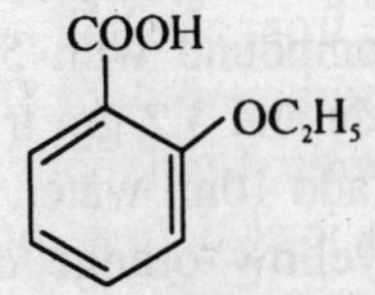

(1) Crystalline compound, sparingly soluble in water, easily soluble hot water.
(2) To 500mg of compound add 3ml conc H_2SO_4 and 3.5ml conc HNO_3. Heat on water bath at 60-70°C. Cool and collect 5-nitrosalicylic acid ethyl ether, m.p 163°C.
(3) o-Ethoxybenzamide m.p 132°C.

m.p 48°C **Hydrocinnamic acid** (β-phenyl propionic acid)

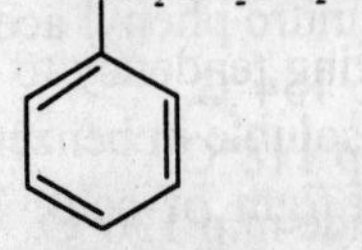

(1) Crystalline compound, steam volatile, insoluble in water, soluble in alcohol and ether.
(2) To 1gm of compound add 5ml fuming H_2SO_4. Heat at 130-140°C for 10-15 minutes. Cool and recover indanone-1, m.p 42°C.
(3) Hydrocinnamide m.p 105°C.
(4) Hydrocinnamanilide m.p 96°C
(5) Hydrocinnamo-p-toluidide m.p 136°C.
(6) NE 150.

m.p 76°C **Phenyl acetic acid** (Phenyl ethan acid)

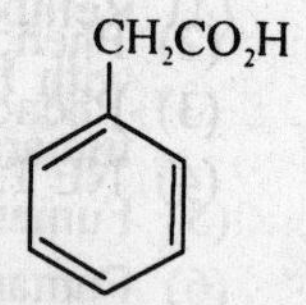

(1) White crystalline solid with a persistent and disagreeable odour, soluble in hot water, alcohol and ether. Sublimes readily.

(2) In a dry test tube place 200mg of acid and 200mg of MnO_2. Add 3ml of conc H_2SO_4 heat the mixture to boiling. Smell of bitter almond due to formation of benzaldehyde is observed.

(3) To 3ml neutral solution of acid add 5ml of alkaline $KmnO_4$ solution. Heat the mixture on water bath. Cool, acidify with dil HCl and collect benzoic acid, m.p 121°C.

(4) Reflux 6.6gm of compound with 5gm of o-nitrobenzaldehyde, 16ml acetic anhydride and 3.2 gm triethylamine for 15min. Cool to 90°C and slowly add 10ml water. Filter the solution while hot and cool. Collect yellow orange crystals of trans o-nitro α-phenylcinnamic acid, crystallise from toluene, m.p 198°C.

(5) Reflux 11gm of compound with 9ml of benzaldehyde, 16ml acetic anhydride and 8ml triethylamine for 5hrs. Steam distill and collect about 10ml distillate. Cool the residue and collect the solid, dissolve in 100ml hot alcohol and 100ml water. Heat the mixture to boiling and filter hot. Concentrate the filtrate, acidify with dil HCl, cool and collect α-phenylcinnamic acid, recrystallise from ethanol, m.p 172°C.

(6) To cooled 5gm of compound add dropwise 25ml of fuming nitric acid. After the addition is complete reflux for 1hr. Cool, pour in ice and collect 2, 4-dinitro phenyl acetic acid, m.p 183°C.

(7) Phenylacetamide m.p 154°C.

(8) Phenylacetanilide m.p 118°C.

(9) Phenylaceto-p-toluidide m.p 136°C.

(10) S Benzyl thiouronium salt m.p 165°C.

(11) NE 136.

m.p 98°C **Phenoxy acetic acid** (Glycolic acid phenyl ether)

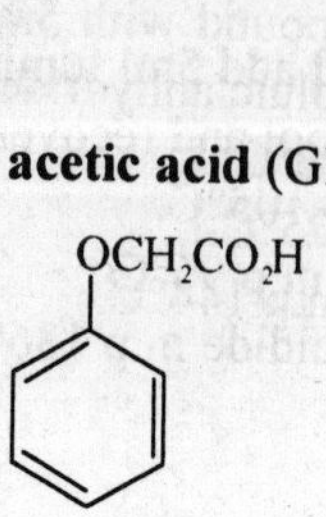

(1) Crystalline compound, insoluble in water, soluble in alcohol and ether.
(2) Phenoxyacetamide m.p 101°C
(3) Phenoxyacetanilide m.p 99°C
(4) NE 152

m.p 100°C **o-Anisic acid** (o-methoxy benzoic acid, salicylic acid methyl ether)

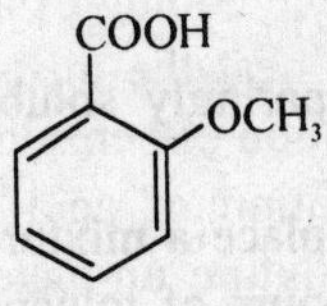

(1) Crystalline compound, soluble in hot water, alcohol and ether.
(2) o-Anisamide m.p 128°C.
(3) o-Anisanilide m.p 131°C.
(4) NE 152

m.p 104°C **o-Toluic acid** (o-Methyl benzoic acid)

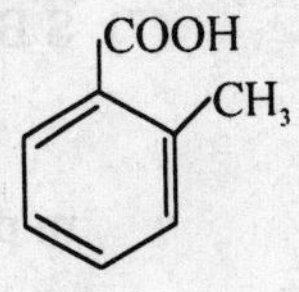

(1) Crystalline compound, sparingly soluble in water, easily soluble in alcohol.
(2) Heat 500mg of compound with 5% alkaline $KMnO_4$ at 60°C until the colour of $KMnO_4$ has discharged. Cool, add dil HCl and collect the precipitated phthalic acid, m.p 206-208°C (d).
(3) Dissolve by warming 500mg of compound in 5ml conc H_2SO_4. Add dropwise 3ml fuming HNO_3 and then heat the mixture at 100-110°C. Keep the mixture overnight and then pour in ice with stirring, collect 3, 5-dinitro-2 methyl benzoic acid, m.p 205-205°C.
(4) Reflux 1gm of compound with 5ml acetic anhydride for 2-3 hrs. Cool and collect o-toluic anhydride, m.p 39°C.
(5) o-Toluamide m.p 142°C
(6) o-Toluanilide m.p 125°C.
(7) o-Tolu-p-toluidide m.p 144°C.

(8) S Benzyl thiouronium o-toluate m.p 145°C.
(9) NE 136.

m.p 110-112°C **m-Toluic acid** (m-methyl benzoic acid)

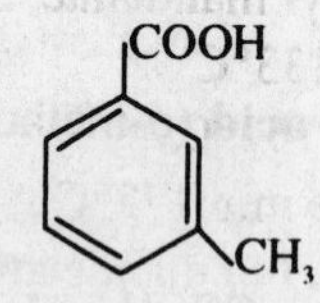

(1) Crystalline solid, sparingly soluble in water, easily soluble in alcohol and ether.
(2) In an ignition tube place a mixture of acid and soda lime in 1:1 ratio and heat. Vapours of toluene evolved which burns with a smoky flame on ignition.
(3) Heat 500mg of compound with 8ml alkaline $KMnO_4$ solution. Cool and acidify with dil HCl, collect isophthalic acid, m.p 347°C.
(4) m-Toluamide m.p 94°C.
(5) m-Toluanilide m.p 126°C.
(6) m-Tolu-p-toluidide m.p 118°C.
(7) S Benzyl thiouronium m-toluate m.p 164°C (8) NE 136.

m.p 118°C ± **Mandelic acid** (α-hydroxy phenyl acetic acid)

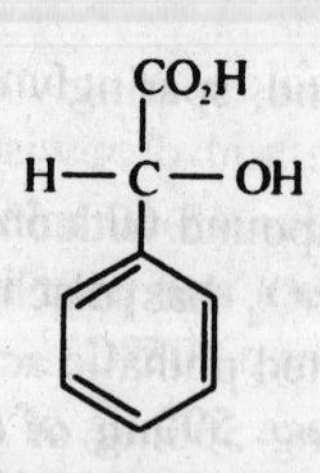

(1) White crystalline solid, soluble in water, alcohol and ether.
(2) Heat 50mg of compound with 1ml aqueous $KMnO_4$ and 1ml conc H_2SO_4. Smell of bitter almond due to formation of benzaldehyde is observed.
(3) To 500mg of compound add 10ml acetyl chloride, heat is evolved. After the reaction has subsided distill off the excess acetyl chloride, keep the residual oily portion for 1-2 days and cool, acetyl mandelic acid, m.p 80°C separate out

(4) To 1ml aqueous solution of acid add 1ml freshly prepared $FeCl_3$ soution. A yellow coloured solution is obtained.
(5) Reflux 15.2 gm of compound with 30ml of absolute ethanol (saturated with HCl gas) for 5hrs. Cool, pour in ice. Add Na_2CO_3 until alkaline. Isolate with ether. Evaporate off the ether, cool and collect crystals of ethyl mandelate, m.p 32°C.
(6) ± Mandelamide m.p 133°C
(7) ± Mandelanilide m.p 151°C
(8) ± Mandelo-p-toludide m.p 172°C
(9) S Benzyl thiouronium (±) mandelate m.p 166°C. (10) NE 152.

m.p 122°C **Benzoic acid** (Benzene mono carboxylic acid)

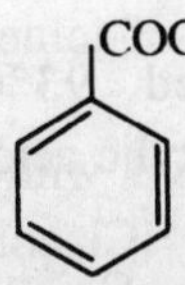

(1) White shinning crystalline solid, sparingly soluble in water, completely soluble in alcohol, ether and chloroform.
(2) To 1ml neutral solution of acid add $FeCl_3$ solution. A buff coloured precipitate of ferric benzoate is obtained.
(3) Heat together 50mg of acid with 4ml of absolute ethanol and 2-4 drops of conc H_2SO_4. A pleasant smell of ethyl benzoate is observed.
(4) Dissolve 1gm of benzoic acid in 5ml conc H_2SO_4. Cool the mixture in ice bath and then add dropwise 2ml of fuming HNO_3. First heat on water bath for 45-60 min and then on oil bath at 130-140°C for 30min. Cool and pour in ice with stirring, collect 3, 5 dinitrobenzoic acid, m.p 205°C.
(5) In a dry test tube place 500mg of acid and add in portion 3ml of nitrating mixture (1.5ml conc HNO_3 + 1.5ml conc H_2SO_4). Heat the test tube on water bath for 5-10 min, cool, pour in ice and collect 3-nitro benzoic acid, m.p 140°C.
(6) Reflux 3gm of benzoic acid with 10ml absolute methanol and .4ml conc H_2SO_4 for 4hrs. Cool, distill off the excess solvent and pour the oily residue in water. Extract the ester with chlorofrom. Distill off the solvent and collect methyl benzoate at b.p 198-200°C.

(7) To 4gm benzoic acid add 10ml fuming H_2SO_4, heat on oil bath at 240-250°C for 5hrs. Keep overnight and then pour in ice. Neutralise with $BaCO_3$ and filter. Evaporate the filtrate and add the barium salt to a melt of 12gm KOH and 12gm NaOH. Heat at 280-310°C for 90min. Cool and pour in water, acidify with dil HCl, extract with ether. Evaporate off the ether and collect 3, 5 dihydroxy benzoic acid, m.p229°C.
(8) Benzamide m.p128°C
(9) Benzanilide m.p 164°C
(10) Benz-p-toluidide m.p 158°C.
(11) S Benzyl thiouronium benzoate m.p 167°C.
(12) NE 122.

m.p 127°C, hydrated 93°C **o-Benzoyl benzoic acid** (benzophenone-o-carboxylic acid)

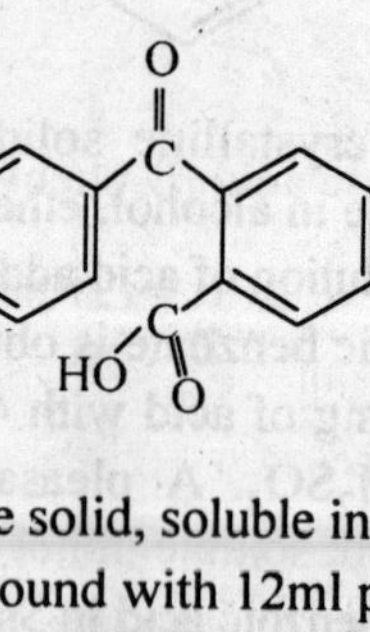

(1) Colourless crystalline solid, soluble in hot water, alcohol and ether.
(2) Reflux 1gm of compound with 12ml polyphosphoric acid for 2hrs. Cool, add 20ml water. Filter the residue, boil with 5ml ammonia solution and filter. Recrystallise from glacial acetic acid and collect anthraquinone, m.p 286°C.
(3) To 1gm of compound add 5ml of thionyl chloride, reflux for 30min. Cool, distill off the excess thionyl chloride and collect o-benzoylbenzoyl chloride, m.p 60°C.
(4) Add 5gm of compound to 35ml ether containing .5ml conc H_2SO_4 and 50ml isobutylene. Keep the mixture overnight, cool and pour in cold NaOH solution. Remove the ethereal layer. Distill off the solvent and collect t-butyl o-benzoylbenzoate, m.p 68-70°C
(5) o-Benzoylbenzamide m.p 165°C.
(6) o-Benzoylbenzanilide m.p 195°C.
(7) NE 226

m.p 133°C **Furoic acid** (Furan –2-carboxylic acid)

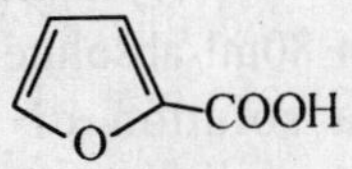

(1) Crystalline compound, sparingly soluble in water, easily soluble in hot water, alcohol and ether. Sublimes.
(2) To few milligrams of compound in test tube add ammonia solution and heat. Bring a pine splinter soaked in conc HCl near the mouth of test tube. The splinter turns red.
(3) To few milligrams of compound add neutral $FeCl_3$ solution. Red yellow coloured precipitate is obtained.
(4) Furoamide m.p 142°C.
(5) Furoanilide m.p 123°C
(6) Furo-p-toluidide m.p 107°C.
(7) NE 112.
(8) S Benzyl thiouronium furoate m.p 211°C.

m.p 133°C **Cinnamic acid** (β-phenyl acrylic acid)

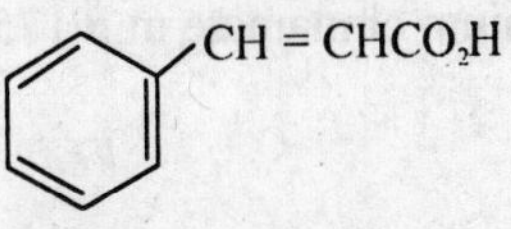

(1) Cream coloured solid, soluble in hot water, alcohol and ether.
(2) To 50mg of acid add 1ml aqueous $KMnO_4$ and 1ml conc H_2SO_4. On heating the purple colour disappears and smell of bitter almond is observed due to formation of benzaldehyde.
(3) Dissolve 2gm of cinnamic acid in 300ml 5% caustic soda solution and 200ml 2% $KMnO_4$ solution. Cool to –5°C and stir. Filter, neutralise with 20% HCl and concentrate until the dissolved salt begins to separate. Neutralise with conc HCl and extract with ether. Distill off the ether and recrystallise the recovered phenyl dihydroxypropionic acid, m.p 141°C.
(4) Dissolve 7.4gm cinnamic acid in 30ml hot $CHCl_3$. Cool and add in portion 2.6ml bromine in 5ml $CHCl_3$. Cool the mixture in ice bath and collect precipitated 2, 3-dibromo 3-phenyl propionic acid, m.p 204°C (d).
(5) Reflux 5.9gm of acid with 16ml absolute methanol and .6ml conc H_2SO_4 for 5hrs. Distill off the excess solvent and pour the residue

in water. Extract the ester with ether. Evaporate off the ether and collect methyl cinnamate. Recrystallise from methanol, m.p 36°C.

(6) Dissolve 2.3gm Na in 80ml absolute ethanol (heat if necessary). Add with stirring a solution of 6.9gm of hydroxylamine hydrochloride in 5ml water and cool immediately. Filter the precipitated NaCl. To the filtrate add 7.9gm of cinnamic acid and reflux on steam bath for 9 hrs. Cool and leave overnight. Filter the precipitated β amino phenylpropionic acid and recrystallise from hot water, m.p 221°C (d).

(7) Reflux 5gm of acid with 25ml acetic anhydride for 8hrs. Remove the excess anhydride by distillation. Cool and extract with ether. Evaporate off the ether and collect cinnamic anhydride, m.p 136°C.

(8) Stir a solution of 1gm of cinnamic acid in 10ml water and add excess of dilute alkali to it. Continue stirring and add 20gm of 2% Sodium amalgam. Decant the solution from mercury and acidify with dil HCl. Cool and collect β-phenylpropionic acid, recrystallise from hot water m.p 47°C.

(9) Cinnamamide m.p 147°C.

(10) Cinnamanilide m.p 151°C.

(11) Cinnamo-p-toluidide m.p 168°C.

(12) S Benzyl thiouronium cinnamate m.p 175°C.

(13) NE 148.

m.p 135°C **Acetyl Salicylic acid** (Aspirin) 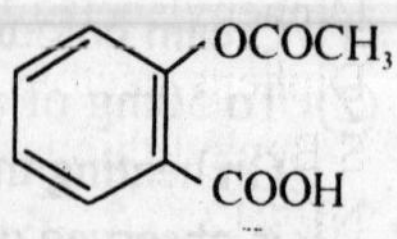

(1) White crystalline solid, soluble in hot water, alcohol and ether.

(2) In a test tube place 500mg of compound and 10ml of dil NaOH. Boil the mixture for 5min, cool, acidify with dil HCl. White crystals of Salicylic acid are obtained, m.p 156°C.

(3) Acetylsalicylamide m.p 138°C.

(4) Acetylsalicylanilide m.p 136°C.

(5) S Benzyl thiouronium acetyl salicylate m.p 144°C.

(6) NE 180.

m.p 137°C **Phenyl Propiolic acid** 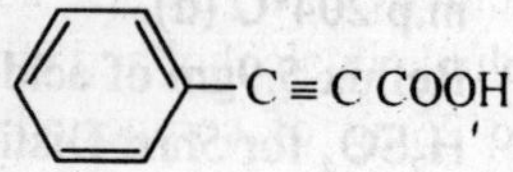

(1) Hair like crystalline compound, insoluble in water, soluble in alcohol and ether. Sublimes.
(2) Reflux 1gm of compound with 1gm zinc and 15ml NaOH solution for 16hrs. Cool, acidify with dil HCl and collect cinnamic acid, m.p 132°C.
(3) Phenylpropiolamide m.p100°C
(4) Phenylpropiolanilide m.p126°C
(5) Phenylpropiol-p-toluidide m.p 142°C
(6) NE 146

m.p 148°C **Diphenylacetic acid** (Diphenyl ethan acid)

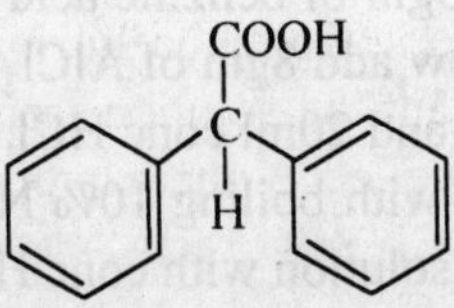

(1) Colourless crystalline compound, insoluble in water, soluble in alcohol, ether and chloroform. Sublimes
(2) Reflux 1gm of compound with 10ml acetic anhydride for 2hrs. Distill off the excess solvent, cool and recover diphenyl acetic anhydride, crystallise from ether, m.p 98°C.
(3) Diphenylacetamide m.p 168°C.
(4) Diphenylacetanilide m.p180°C.
(5) Diphenylaceto-p-toluidide m.p 173°C.
(6) S Benzyl thiouronium diphenylacetate m.p 145°C.
(7) NE 212.

m.p 150°C **Benzilic acid** (Diphenyl glycolic acid)

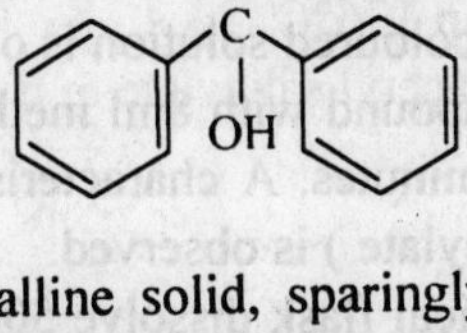

(1) Colourless crystalline solid, sparingly soluble in water, readily soluble in alcohol and ether.
(2) To 25mg of compound add 1ml conc H_2SO_4. Immediately an

orange-red colouration is produced which soon becomes red – violet.

(3) To 1ml aqueous solution of acid add 1ml freshly prepared $FeCl_3$ soution. A yellow coloured solution is obtained.

(4) Reflux 4gm of benzilic acid, 6ml of 60% HI and 2gm of red phosphorous for 2hrs. Filter while still hot and pour the filtrate in ice water. Collect the precipated diphenylacetic acid and recrystallise from ethanol, m.p 146°C.

(5) Reflux 5.9gm of acid with 16ml absolute methanol and .6ml conc H_2SO_4 for 5hrs. Distill off the excess solvent and pour the residue in water. Extract the ester with ether. Evaporate off the ether and collect methyl benzilate. Rrecrystallise from methanol, m.p 75°C.

(6) Stir a mixture of 4.5gm of benzilic acid in 70ml benzene and then cool in ice bath. Now add 8gm of $AlCl_3$ and reflux for 3hrs. Cool, add 90ml ice water and 20ml conc HCl. Distill off the solvent and extract the reisidue with boiling 10% Na_2CO_3 solution. Cool and acidify the alkaline solution with conc HCl. Collect the prcipitated 9-fluorene carboxylic acid and recrystallise from benzene, m.p 219-222°C.

(7) Benzilamide m.p 154°C.

(8) Benzilic anilide m.p 175°C.

(9) Benzilic-p-toluidide m.p 189°C.

(10) NE 228

(11) Acetyl benzilic acid m.p 98°C.

m.p 157-159°C **Salicylic acid** (1-hydroxy 2-carboxy benzene)

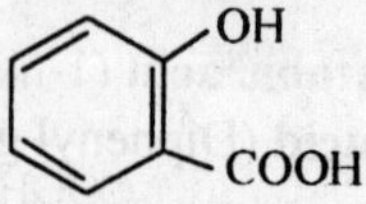

(1) Colourless crystalline solid, soluble in hot water, alcohol and ether.

(2) To 1ml aqueous solution of acid add 1ml freshly prepared $FeCl_3$ soution. A yellow coloured solution is obtained.

(3) Heat 20mg of compound with 8ml methyl alcohol and .8ml conc H_2SO_4 for 10-15 minutes. A characteristic smell of oil of winter green (methyl salicylate) is observed.

(4) In a stoppered conical flask dissolve 50mg of acid in alcohol and add few drops of conc H_2SO_4 to it. Dip the stopper in 3% KOH solution and again place it on the flask. Heat the flask at 130°C

for 5 minutes. Violet fluorescence on the stopper will be observed.

(5) To 2.5gm of salicylic acid in 23ml glacial acetic acid add with stirring 16ml of iodine chloride solution (6.2gm iodine chloride in 16ml acetic acid) and then add 50ml water. Heat the mixture to 80°C for 30-40 minutes with stirring. Cool and recover diiodosalicylic acid, m.p 235-236°C.

(6) To 1gm of compound add 5ml acetic anhydride and 1-2 drops of conc H_2SO_4. Warm the mixture on water bath at 50-60°C for 10-15 minutes. Cool, pour in ice with stirring and collect acetyl salicylic acid, m.p 135°C.

(7) Dissolve 500mg of compound in 5ml boiling water and add dropwise 5ml conc HNO_3 to it. Boil the mixture for 5-10 minutes. Cool, pour in ice with stirring and collect 5-Nitro Salicylic acid m.p 220°C.

(8) To 500mg of compound in 5ml ether add 5ml of benzoyl chloride and 1ml pyridine. Heat the mixture on water bath, cool and collect benzoyl salicylic acid, m.p 132°C.

(9) Heat a mixture of 6.9gm salicylic acid, 6.4gm p-chlorophenol and 2.9gm phosphorous oxychloride at 70-80°C for 4hrs. Pour the mixture with stirring into a solution of 6gm Na_2CO_3 in 40ml water. Collect the precipitated p-chloro phenylsalicylate and recrystallise from ethanol, m.p 65-66°C.

(10) Salicylamide m.p 139°C.

(11) Salicylanilide m.p 134°C.

(12) Salicyl-p-toluidide mp 156°C.

(13) S Benzyl thiouronium salicylate m.p 146°C. (14) NE 138.

m.p 162°C **1-Naphthoic acid** (1-naphthalene carboxylic acid)

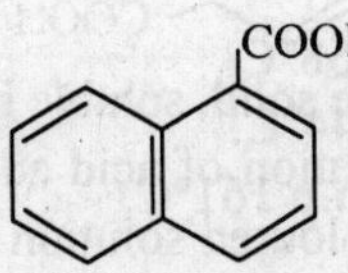

(1) Crystalline compound, insoluble in water, soluble in hot alcohol and ether.

(2) 1-Naphthoamide m.p 205°C.

(3) 1-Naphthoanilide m.p 160°C.

(4) S Benzyl thiouronium naphthoate m.p 151°C

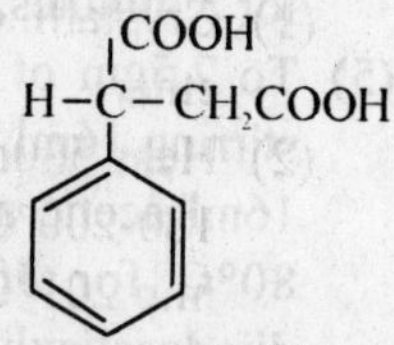

m.p 167-169°C **± Phenyl succinic acid**

(1) White crystalline solid, insoluble in benzene and pet ether, sparingly soluble in water, soluble in alcohol, ether and acetone.
(2) Reflux 500mg of compound with 10ml absolute methanol saturated with HCl gas. Cool and recover dimethyl phenyl succinate, m.p 58°C.
(3) ± Phenyl succin diamide m.p 211°C (4) ± Phenyl succin dianilide m.p 22°C (5) NE 97

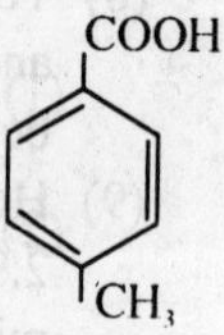

m.p 180°C **p-Toluic acid** (p-methyl benzoic acid)

(1) Crystalline compound sublimes readily, volatile with steam, insoluble in water but soluble in alcohol, ether and methanol.
(2) Reflux 1gm of compound with 15ml acetic anhydride for 6hrs. Distill off the excess acetic anhydride, cool and recover p-toluic anhydride, m.p 95°C.
(3) Dissolve 1gm of compound in 5ml conc H_2SO_4(heat if necessary) and then add dropwise 3ml fuming HNO_3, heat the mixture at 100-110°C. Cool, keep the mixture overnight and then pour in ice water with stirring and recover 3, 5 dinitro-4-methyl benzoic acid, m.p 159°C.
(4) p-Toluamide m.p 158°C
(5) p-toluanilide m.p 144°C.
(6) p-tolu-p-toluidide m.p 161°C.
(7) NE 136.
(8) S Benzyl thiouronium p toluate m.p 190°C.

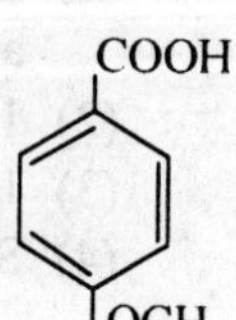

m.p 184°C **p-Anisic acid** (p-methoxy benzoic acid)

(1) Crystalline compound, insoluble in water, soluble in alcohol and ether
(2) Heat 500mg of compound with 2gm of aniline hydrochloride at 180-200°C for 45-60min. Cool and recover p-hydroxy benzanilide, m.p 201°C.
(3) Anisamide m.p 162°C.
(4) Anisanilide m.p168°C.
(5) Anis-p-toluidide m.p 186°C.
(6) S Benzyl thiouronium anisate m.p 185°C.
(7) NE 152

m.p 185°C **2-Naphthoic acid**

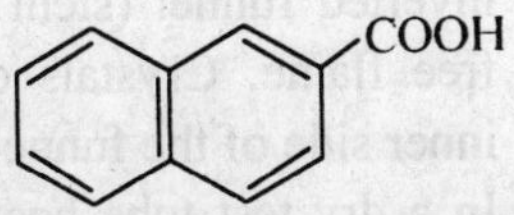

(1) Crystalline compound, insoluble in water, soluble in alcohol, ether and chloroform.
(2) Dissolve 500mg of compound in 5ml absolute methanol saturated with HCl gas. Heat the mixture on water bath, cool and recover methyl 2-naphthoate, m.p 77°C.
(3) Reflux with stirring at 160°C a mixture of 2-naphthoic acid, hydroxylamine hydrochloride and polyphosphoric acid until evolution of CO_2 ceases. Cool and recover 2 naphthylamine, m.p 112°C.
(4) Heat 5.8gm of compound with 7 gm of PCl_5 on steam bath. As soon as the reaction sets in remove the heat source and allow the reaction to take place. After the reaction has subsided heat on water bath for 30min. Distill off the phosphorous oxychloride under reduced pressure, cool and collect 2-naphthyl chloride, m.p 52°C
(5) 2-Naphthoamide m.p 192°C.
(6) 2-Naphthoanilide m.p 170°C.
(7) 2-Naphtho-p-toluidide m.p 192°C.
(8) NE 172.

m.p 206-208°C (d) **Phthalic acid** (1, 2-dicarboxy benzene)

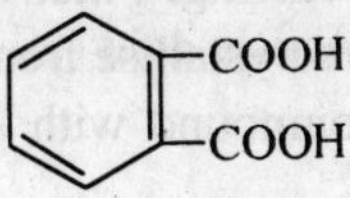

(1) White crystalline solid, insoluble in water and chloroform, soluble in hot water, alcohol and ether.
(2) In a dry test tube place 50mg of acid, 50mg of resorcinol and 1ml of conc H_2SO_4. Heat at 160°C for 3-4 minutes. Cool, pour in 50ml water and then add dil NaOH in excess. Characteristic green fluorescence is observed.
(3) To 50mg of acid add 50mg of phenol and few drops of conc H_2SO_4. Heat the mixture gently, cool, dilute by adding water and then add excess of dil NaOH solution. A red coloured solution is obtained.
(4) In a porcelain dish place 1gm of compound, cover it with an inverted funnel (stem closed with cotton). Heat the dish over a free flame. Crystals of phthalic anhydride will deposit on the inner side of the funnel, m.p 131°C.
(5) In a dry test tube heat 500mg of compound until it melts. Cool and add 1gm of urea to it and again heat for 15-20 minutes. Cool and collect phthamimide, m.p 220°C.
(6) Heat over a low flame for 15min 500mg of compound with 3ml aniline on wire gauze. Add 20ml 50% alcohol and boil again. Cool and pour in ice. Collect the precipitated phthalanil, recrystallise from alcohol, m.p 207°C.
(7) Phthalic diamide m.p 220°C.
(8) Phthalic dianilide m.p 250°C.
(9) Phthalic di-p-toluidide m.p 201°C.
(10) Di (S Benzyl thiouronium) phthalate m.p 157°C.
(11) NE 83.

m.p 197-199°C **p-Ethoxy benzoic acid**

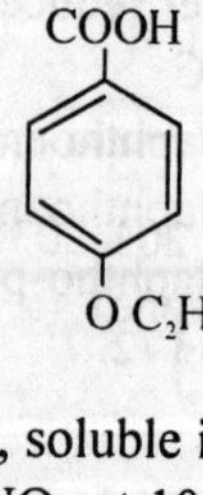

(1) Crystalline compound insoluble in water, soluble in hot water.
(2) Reflux 1gm of compound with 10ml HNO_3 at 100°C. Cool, pour in ice water with stirring. Filter the precipitated 3-nitro-4 ethoxy benzoic acid and recrystallise from alcohol, m.p 200-201°C.
(3) Reflux 1gm of compound with 10ml acetic anhydride for 2hrs.

Distill off the excess anhydride and pour the oily residue with stirring in ice water, collect p-ethoxy benzoic anhydride, crystallise from pet ether, m.p 108°C.

(4) p-Ethoxybenzamide m.p 202°C.
(5) p-Ethoxybenzanilide m.p 170°C.
(6) NE 166

m.p 200°C(d) **Protocatechuic acid** (3, 4-dihydroxy benzoic acid)

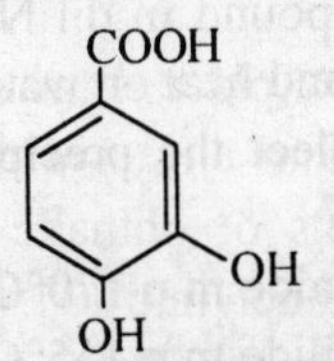

(1) Crystalline compound, sparingly soluble in water, soluble in alcohol and ether.
(2) Reduces Tollens reagent.
(3) To few milligrams of compound in water add neutral $FeCl_3$ solution. An intense green coloured solution is obtained which changes to red on additon of NH_4OH.
(4) To 1gm of compound add 15ml of 10% NaOH solution, cool and then add 5ml of dimethyl sulfate (exothermic reaction). After the reaction has subsided reflux at 100°C for 2hrs. Cool, acidify with dil HCl and collect 3, 4 dimethoxybenzoic acid, m.p181°C.
(5) Protocatechuic amide m.p 212°C
(6) Protocatechuic anilide m.p 166°C
(7) 3, 4 diacetoxy benzoic acid m.p 162°C
(8) Benzoyl (3, 4 dibenzoyloxy) benzoate m.p 198°C
(9) NE 154

m.p 201-203°C **m-Hydroxy benzoic acid** (1-hydroxy 3-carboxy benzene)

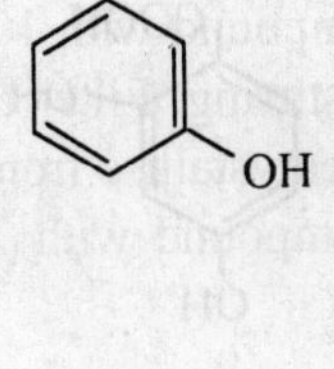

(1) Crystalline solid, sparingly soluble in water, completely soluble in hot water, ether and acetone. Sublimes readily.
(2) Heat 50mg of compound with 5ml conc H_2SO_4. An orange red coloured solution is obtained.
(3) Dissolve 5gm of compound in 18ml nitrobenzene. Cool the solution to 30°C and add dropwise with stirring 4ml fuming HNO_3 solution (2ml in 2ml nitrobenzene) Continue stirring for 2-3hrs. Cool, filter the precipitated 4-nitro 3-hydroxy benzoic acid, recrystallise from ethanol, m.p 228°C.
(4) To 500mg of compound in dil NaOH add 5ml acetic anhydride. Shake thoroughly and heat on water bath at 50-60°C. Cool, acidify with dil HCl. Collect the precipitated m-acetoxy benzoic acid, m.p 131°C.
(5) m-Hydroxy benzamide m.p 170°C.
(6) m-hydroxy benzanilide m.p 155°C.
(7) m-hydroxy benzo-p-toluidide m.p 163°C.
(8) NE 138

m.p 207°C (d) **o-Coumaric acid** (trans o-hydroxy cinnamic acid)

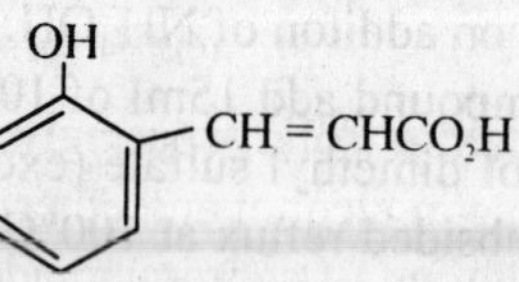

(1) Crystalline solid, sparingly soluble in water and ether, easily soluble in alcohol. Sublimes.
(2) To few milligrams of compound add dil NaOH solution. A characteristic green fluorescence is observed.
(3) Heat 1gm of compound with 7-8 ml acetic anhydride. Cool, pour in ice with stirring and recover acetyl coumaric acid, m.p 149°C.
(4) o-Coumaramide m.p 209°C (d)
(5) NE 164

m.p 213°C (d) **β-Resorcylic acid** (2, 4-dihydroxybenzoic acid)

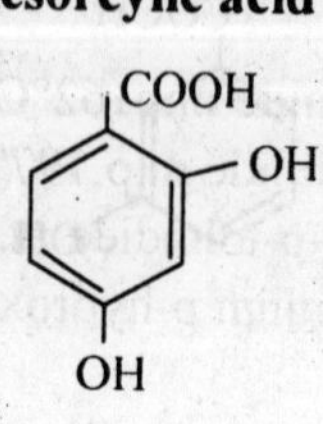

(1) Crystalline compound, insoluble in water, soluble in alcohol and ether.
(2) To few milligrams of compound add neutral $FeCl_3$ solution. A red coloured solution is obtained.
(3) Reflux 500mg of compound with 10ml methanol and 1ml conc H_2SO_4 for 10 hrs. Distill off the excess methanol and add 5ml cold water to the residue. Collect methyl 2, 4-dihydroxy benzoate and recrystallise from methanol, m.p 119°C.
(4) β-Resorcylamide m.p 222°C
(5) β-Resorcylanilide m.p 126°C.
(6) NE 154

m.p 213°C **p-Hydroxy benzoic acid** (1-hydroxy 4-carboxy benzene)

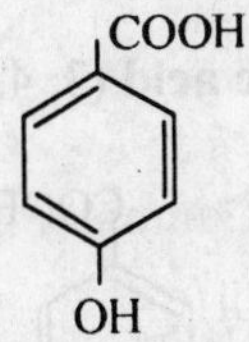

(1) White crystalline solid, sparingly soluble in water and benzene, easily soluble in alcohol and ether.
(2) To few milligram of compound add neutral $FeCl_3$. A red coloured solution is obtained.
(3) To 500mg of compound in dil NaOH add 5ml acetic anhydride. Shake thoroughly and heat on water bath at 50-60°C. Cool, acidify with dil HCl. Collect the precipitated p-acetoxy benzoic acid, m.p 191°C.
(4) To a well cooled solution of 2.5gm of compound in 30ml N NaOH solution add in portion 2gm of methylchloro carbonate. Acidify with dil HCl. Filter the precipitated solid. Dissolve in hot acetone and gradually add water till turbidity results. Cool, collect fine crystals of methylcarbonate of p-hydroxybenzoic acid, m.p 179°C.
(5) p-hydroxybenzamide m.p162°C.
(6) p-hydroxy benzanilide m.p 197°C.
(7) p-hydroxy benzo-p-toluidide m.p 204°C.
(8) S Benzyl thiouronium p-hydroxy benzoate m.p 143°C

(9) NE 138

m.p 228°C **Diphenic acid** (Biphenyl-2, 2'-dicarboxylic acid)

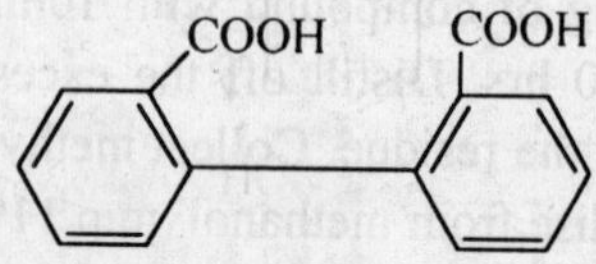

(1) Crystalline compound which sublimes, sparingly soluble in water, soluble in alcohol and ether.
(2) Heat 500mg of compound with 2gm PCl_5. Cool and recover diphenic dichloride, m.p 97°C.
(3) Diphenic diamide m.p 212°C.
(4) Diphenanilide m.p 229°C. (5) NE 121

m.p 235°C (d) **Gallic acid** (3, 4, 5-trihydroxy benzoic acid)

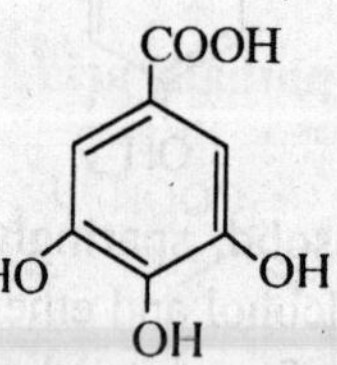

(1) White crystalline solid, insoluble in water, benzene and chloroform, soluble in hot water and ether.
(2) To few milligrams of compound add neutral $FeCl_3$ solution. A blue –black coloured precipitate is obtained.
(3) Reduces Tollens reagent.
(4) Reduces Fehling's solution.
(5) Heat 500mg of compound with 8ml acetic anhydride and 200mg of anhydrous zinc chloride. Cool and recover 3, 4, 5 triacetoxy benzoic acid, m.p 172°C.
(6) Dissolve 4gm of NaOH in 25ml water, add 2.5gm of gallic acid to it. Add in portion 7-8ml dimethyl sulfate (exothermic reaction, temp below 35°C). Reflux for 2hrs. Add 1gm of NaOH in 2ml water and heat again for 2hrs. Cool, acidify with dil HCl and collect 3, 4, 5 trimethoxy benzoic acid, m.p 167°C.
(7) Gallamide m.p 244°C (7) Gallanilide m.p 207°C (8) NE 170

m.p → Sublimes without metting above 300°C **Terephthalic acid** (Benzene-1, 4-dicarboxylic acid)

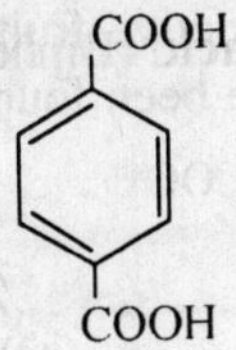

(1) Crystalline compound, insoluble in water, alcohol and chloroform, sparingly soluble in hot alcohol.
(2) Reflux 1gm of acid with 13.2 ml BF_3 etherate and 10ml anhydrous methanol for 6hrs. Cool, pour in 50ml $NaHCO_3$. Collect the precipitated dimethyl terephthalate, recrystallise from methanol, m.p 140°C.
(3) Terephthalic dianilide m.p 337°C.
(4) Di (S Benzyl thiouronium) terephthalate m.p 202°C.
(5) NE 83.

m.p 345-348°C **Isophthalic acid** (Benzene 1, 3-dicarboxylic acid)

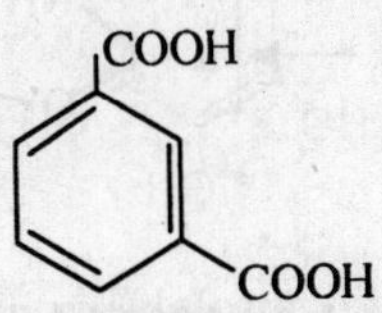

(1) Crystalline compound which sublimes below its melting point, insoluble in water, soluble in alcohol.
(2) Fuse 500mg of compound with 1.5gm PCl_5 over a low flame, cool and dissolve it in 10ml methanol, shake thoroughly. Pour the mixture in ice, collect the precipated dimethyl isophthalate, recrystallise from methanol, m.p 65°C.
(3) Isophthalic diamide m.p 280°C.
(4) Isophthalic dianilide m.p 250°C.
(5) S Benzyl thiouronium hydrogen isophthalate m.p 215°C.
(6) NE 83

(vii) Esters

Esters are derivatives of carboxylic acids in which the hydroxyl group is replaced by an alkoxy group (OR). They can be hydrolysed under acidic or

basic conditions but are not soluble in either dilute acid or base at room temperature. This solubility behaviour is helpful in distinguishing them from carboxylic acids.

Esters have a strong and distinctive fruity odour. The pleasant flavour and odour of fruits or flowers have been found to be due to the presence of esters.

Identification tests

(1) To few milligrams of compound add 1-2 ml dil NaOH solution and 1 drop of phenolphthalein indicator. The solution turns pink. Heat the solution to boiling. Disappearnce of pink colour indicates the presence of esters.

(2) Hydroxamic test :- To 100mg of solid or 1-2ml of liquid sample add 2ml of hydroxylamine hydrochloride in ethanol followed by 5ml of 10% NaOH or until the solution is alkaline. Reflux the solution for few minutes, cool and acidify with dil HCl. Add 2-3 drops of $FeCl_3$ to the acidified solution. A bright pink or magenta or reddish blue colour indicates the presence of esters.

$$RCOOR' + NH_2-OH \longrightarrow R-\overset{O}{\overset{\|}{C}}-NHOH + R'-OH$$

Derivatives

(1) **Hydrolysis of esters:** In a 100ml hydrolysis flask place 1gm of ester and add 10-15ml of 20% NaOH solution to it. Reflux the reaction mixture for an hour and then cool. Add dil HCl to the aqueous solution until the resulting solution is distinctly red to litmus. Carboxylic acid is precipitated, filter it on a buchner funnel, wash with cold water and recrystallise from ethanol. Identify the acid from its melting point.

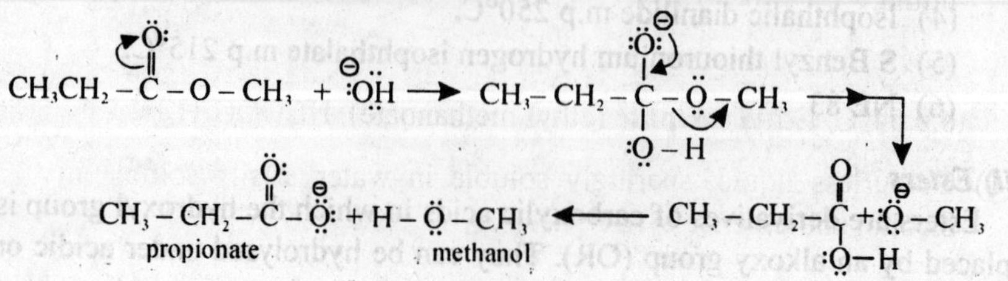

(2) **3, 5 dinitrobenzoate :** In a 50ml round bottom flask place 2-3ml of ester (1gm in case of solid), 1.5 gm of 3, 5 dinitrobenzoic acid and 3-4 drops of conc H_2SO_4. Reflux on steam bath if the b.p/m.p of ester is below 100°C and on oil bath if it is above 100°C for 45-60 min. Cool the flask and add 25ml ether to it. Remove the ethereal layer, wash first with Na_2CO_3 and then with water. Distil off the ether, recrystallise the solid from ethanol and determine its melting point.

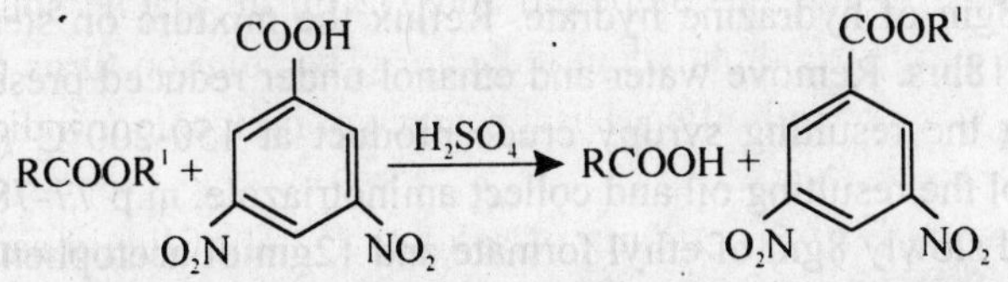

(3) **Amides:** In a 50ml conical flask place 3ml of ester (1gm in case of solid) and 5ml of ammonia solution. Stopper the flask and shake thoroughly for about 30 minutes. Filter the precipitated solid, wash with water, recrystallise and determine its melting point.

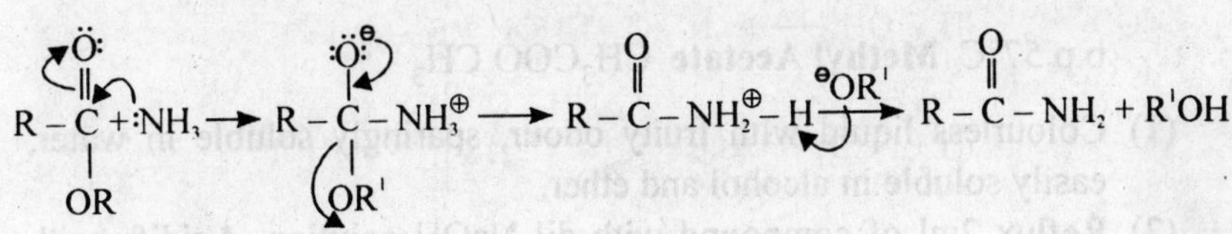

Aliphatic Liquids

b.p 32°C **Methyl formate** (methyl methanoate) $HCOOCH_3$

(1) Colourless liquid, soluble in water and ether.
(2) Reflux 5ml of compound with dil NaOH solution. Cool, acidify with dil HCl, methyl alcohol and formic acid are produced. Distil off the alcohol and perform the Tollens test with the residual liquid.

b.p 54°C **Ethyl formate** (ethyl methanoate) $HCOOC_2H_5$

(1) Colourless liquid, sparingly soluble in water, easily soluble in alcohol and ether.

(2) Reflux 2ml of compound with dil NaOH for 15-20min. Acidify with dil HCl, ethyl alcohol and formic acid will be produced. Perform the iodoform test in the above solution.
(3) To 13gm of ethyl formate add with stirring 20gm of cyclohexylamine (cool in ice bath). After the exothermic reaction has subsided reflux the solution for 2hrs, distill and collect N-cyclohexyl formamide, b.p 137-138°C.
(4) To 14.8gm ethyl formate and 15ml ethanol add with stirring 12.8gm of hydrazine hydrate. Reflux the mixture on steam bath for 18hrs. Remove water and ethanol under reduced pressure and heat the resulting syrupy crude product at 150-200°C for 3hrs. Cool the resulting oil and collect aminotriazole, m.p 77-78°C.
(5) Add slowly 8gm of ethyl formate and 12gm of acetophenone to a suspension of 7gm of sodium ethoxide in 20ml ether. Cool in ice bath. Filter the sodium salt of hydroxy ketone, dissolve in ice water, acidify with acetic acid and then add excess of cupric acetate. Collect copper salt of hydroxy ketone as olive green precipitate.

b.p 57°C **Methyl Acetate** $CH_3COO\ CH_3$

(1) Colourless liquid with fruity odour,. sparingly soluble in water, easily soluble in alcohol and ether.
(2) Reflux 2ml of compound with dil NaOH solution. Acidify with dil HCl, methyl alcohol and acetic acid will be produced. Add a small piece of copper wire to the solution and heat. A pungent odour will be produced.

b.p 71°C **Isopropyl formate** (isopropyl methanoate) $HCOOCH(CH_3)_2$

(1) Sparingly soluble in water, miscible with alcohol and ether.
(2) Reflux 5ml of ester with dil NaOH solution, cool, acidify with dil acid. Isopropyl alcohol and formic acid are obtained. Distil off the alcohol and react the residual portion with ammonical silver nitrate solution.

b.p 77°C **Ethyl acetate** (Ethyl ethanoate) $CH_3COOC_2H_5$

(1) Colourless liquid having fruity odour, sparingly soluble in water, easily soluble in alcohol and ether.

(2) Reflux 10ml of compound with dil NaOH solution. Cool, acidify with dil HCl, ethyl alcohol and acetic acid will be produced. Distil off the alcohol and perform the iodoform test.

(3) To 2.4gm powdered sodium add 30ml ethyl acetate and 1ml absolute ethanol. Cool the solution to 0°C and add 7.2ml benzaldehyde and stir. When all sodium has dissolved add 6.5ml acetic acid followed by equal volume of water. Separate the ethereal layer and extract the aqueous layer with ethyl acetate. Combine the organic layers, distil off the solvent and collect ethyl cinnamate at 126-131°C.

(4) To 50ml liquid ammonia add a small piece of lithium metal, few crystals of ferric nitrate hydrate followed by 0.8gm of lithium. After all lithium has been converted into amide add 3ml of ethyl acetate and stir for 30min. Add 3.6gm of benzophenone dissolved in 10ml ether and stir again for 30min. Neutralise with NH_4Cl. Add 20ml ether and let ammonia to evaporate overnight. Add 20ml water and extract with ether. Distil off the ether and collect β hydroxy β, β-diphenyl propionate, m.p 46-47°C.

b.p 79°C **Methyl propionate** $C_2H_5COOCH_3$

(1) Colourless liquid sparingly soluble in water, miscible with alcohol and ether.

(2) Reflux 5ml of compound with dil NaOH solution. Cool, acidify with dil HCl and collect methyl alcohol and propionic acid.

b.p 81°C **n-Propyl formate** $HCOOCH_2C_2H_5$

(1) Colourless liquid, sparingly soluble in water, miscible with alcohol and ether.

(2) Reflux 5ml of ester with dil NaOH solution, cool, acidify with dil acid. n-Propyl alcohol and formic acid are obtained. Distil off the alcohol and react the residual portion with ammonical silver nitrate solution.

b.p 88°C **Isopropyl acetate** $CH_3COOCH(CH_3)_2$

(1) Colourless liquid, sparingly soluble in water, miscible with alcohol and ether.

(2) Reflux 5ml of ester with dil NaOH solution, cool, acidify with dil acid. Iso propyl alcohol and acetic acid are obtained. Distil off the alcohol and react the residual portion with neutral $FeCl_3$ solution.

b.p 91°C **Dimethyl carbonate** $CH_3OCOOCH_3$

(1) Colourless liquid, insoluble in water, miscible with alcohol and ether.

(2) Reflux 5ml of compound with dil NaOH solution. Cool, acidify with dil HCl and collect methyl alcohol and carbonic acid.

b.p 97°C **t-butyl acetate** (trimethyl carbinyl acetate) $CH_3COOC(CH_3)_3$

(1) Colourless liquid, insoluble in water, soluble in alcohol and ether.

(2) Reflux 10ml of ester with dil NaOH solution. Cool and on acidification t-butyl alcohol and acetic acid are obtained.

b.p 98°C **Isobutyl formate** $HCOOCH_2CH(CH_3)_2$

(1) Colourless liquid, sparingly soluble in water, miscible with alcohol and ether.

(2) Reflux 5ml of ester with dil NaOH solution, cool, acidify with dil acid. Isobutyl alcohol and formic acid are obtained. Distil off the alcohol and react the residual portion with ammonical silver nitrate solution.

b.p 99°C **Ethyl propionate** $C_2H_5COOC_2H_5$

(1) Colourless liquid, sparingly soluble in water, miscible with alcohol and ether.

(2) Reflux 5ml of compound with dil NaOH solution. Cool, acidify with dil HCl and collect ethyl alcohol and propionic acid.

(3) Add a solution of 6ml of ester in 6ml ether to an ethereal solution of methyl magnesium iodide (from 15gm methyl iodide). Cool and stir the mixture during the addition and then keep overnight.

Add crushed ice and sulphuric acid. Remove the ethereal layer and extract the aqueous layer with ether. Distil off the combined ethereal layer and collect dimethylethyl carbinol at b.p 100-102°C.

b.p 101°C **n-Propyl acetate** $CH_3COOCH_2CH_2CH_3$

(1) Colourless liquid, sparingly soluble in water, miscible with alcohol and ether.
(2) Reflux 4ml of compound with dil NaOH solution. Acidify with dil HCl, n-Propyl alcohol and acetic acid will be produced.

b.p 105°C **Methyl orthoformate** (trimethoxy methane) $HC(OCH_3)_3$

(1) Reflux 10ml of ester with dil NaOH solution, cool, acidify with dil acid. Methyl alcohol and formic acid are obtained. Distil off the alcohol and react the residual portion with ammonical silver nitrate solution.

b.p 106°C **n-butyl formate** $HCOO(CH_2)_3CH_3$

(1) Colourless liquid, sparingly soluble in water, miscible with alcohol and ether.
(2) Reflux 15ml of ester with dil NaOH solution, cool, acidify with dil acid. n-butyl alcohol and formic acid are obtained. Distil off the alcohol and react the residual portion with ammonical silver nitrate solution.

b.p 111°C **Ethyl isobutyrate** (ethyl 2-methyl propanoate) $(CH_3)_2CHCOOC_2H_5$

(1) Colourless liquid, slightly soluble in water, miscible with alcohol and ether.
(2) Reflux 5ml of ester with 2gm of hydrazine hydrate in absolute alcohol for 72hrs. Cool and collect isobutyrohydrazide, recrystallise from ether, m.p 104°C.
(3) Reflux 15ml of ester with dil NaOH solution, cool, acidify with dil acid. Ethyl alcohol and isobutyric acid are obtained. Distil off the alcohol and perform the iodoform test.

b.p 112°C **sec-Butyl acetate** (α-methyl propyl ethanoate) $CH_3COOCH(CH_3)C_2H_5$

(1) Colourless liquid, insoluble in water, soluble in alcohol and ether.
(2) Reflux 14ml of compound with dil NaOH solution. Acidify with dil HCl, s-butyl alcohol and acetic acid will be produced.

b.p 116°C **Isobutyl acetate** (β-methyl propyl ethanoate) $CH_3COOCH_2CH(CH_3)_2$

(1) Colourless liquid, insoluble in water, miscible with alcohol and ether.
(2) Reflux 14ml of compound with dil NaOH solution. Acidify with dil HCl, isobutyl alcohol and acetic acid will be produced.

b.p 121°C **Ethyl Butyrate** (ethyl butanoate) $C_3H_7COOC_2H_5$

(1) Colourless liquid, sparingly soluble in water, soluble in alcohol and ether.
(2) Reflux 5ml of ester with dil NaOH solution, cool, acidify with dil acid. Ethyl alcohol and n-butyric acid are obtained. Distil off the alcohol and perform the iodoform test.
(3) To 2.5gm Mg turning add 5ml dry ether and 7ml of CH_3I in 15ml dry ether. Warm the mixture on water bath and when the reaction starts remove the heat source. When all Mg has reacted add 7ml of ethyl butyrate in 5ml ether. Reflux the mixture on water bath for 1hr and then pour in ice cooled 10ml 2M H_2SO_4. Remove the ethereal layer and extract the aqueous layer with ether. Distil off the ether and collect 2-methyl pentan-2 ol, b.p 117-120°C.

b.p 124°C **Isoamyl formate** (υ-methyl butyl methanoate) $HCOO(CH_2)_2CH(CH_3)_2$

(1) Colourless liquid, very sparingly soluble in water, soluble in alcohol, miscible with ether.
(2) Reflux 5ml of ester with dil NaOH solution, cool, acidify with dil acid. Isoamyl alcohol and formic acid are obtained. Distil off the alcohol and react the residual portion with ammonical silver nitrate solution.

b.p 126°C **n-Butyl acetate** (butyl ethanoate) $CH_3COO(CH_2)_3CH_3$

(1) Colourless liquid, insoluble in water, miscible with alcohol and ether.
(2) Reflux 4ml of compound with dil NaOH solution. Acidify with dil HCl, n-butyl alcohol and acetic acid will be produced.

b.p 126°C **Diethyl Carbonate** $C_2H_5OCOOC_2H_5$

(1) Colourless, highly inflammable liquid, insoluble in water, miscible in alcohol and ether.
(2) Reflux 5ml of ester with dil NaOH solution. Acidify with dil HCl. Ethyl alcohol and carbonic acid are produced.
(3) Stir 12.5 gm of ester with sodium ethoxide (from 500mg Na) and 3gm neopentyl ketone. Heat first at 50°C and finally at 110°C for 2½ hr. Distil off the solvent, cool and acidify the residue with acetic acid. Remove the upper layer and distil. Collect 4:4 dimethylhex-3-one-anoic ester at b.p 104-105°C/15mmHg
(4) Stir 30ml of ester with 5gm methyl isopropyl ketone and then cool to 2°C. Add a suspension of 3gm of sodium ethoxide in diethyl carbonate. After 2 hrs warm the mixture and distil off the alcohol. Cool the residue and acidify with acetic acid. Separate the organic layer and distil. Collect 4 methylpent-3-one-anoic ester at b.p 75-81°C/10mmHg.

b.p 128°C **Isopropyl n-butyrate** $C_3H_7COOCH(CH_3)_2$

(1) Colourless liquid
(2) Reflux 15ml of ester with dil NaOH solution, cool, acidify with dil acid. Isopropyl alcohol and n-butyric acid are obtained.

b.p 134°C **Ethyl isovalerate** (ethyl 3-methyl butyrate)
$(CH_3)_2CHCH_2COOC_2H_5$

(1) Insoluble in water, miscible with alcohol and ether.
(2) Reflux 5ml of ester with dil NaOH solution, cool, acidify with dil acid. Ethyl alcohol and isovaleric acid are obtained. Distil off the alcohol and perform the iodoform test.

b.p 137°C **Methyl pyruvate** (methyl 2-oxopropanoate) $CH_3COCOOCH_3$

(1) Colourless liquid, slightly soluble in water, miscible with alcohol and ether.
(2) Reflux 5ml of compound with dil NaOH solution. Cool, acidify with dil HCl and collect methyl alcohol and pyruvic acid.
(3) Methyl pyruvate 2, 4 dinitrophenylhydrazone, m.p 187°C.
(4) Methyl pyruvate semicarbazone, m.p 208°C (5) Methyl pyruvate oxime, m.p 68°C

b.p 142°C **Isoamyl acetate** (3-methyl butyl acetate) $CH_3COOCH_2CH_2CH(CH_3)_2$

(1) Colourless liquid with banana like odour, insoluble in water, miscible with alcohol and ether.
(2) Reflux 15ml of compound with dil NaOH solution. Cool, acidify with dil HCl. Isoamyl alcohol and acetic acid will be produced.

b.p 143°C **n-Propyl-n butyrate** $C_3H_7COOC_3H_7$

(1) Colourless liquid, sparingly soluble in water, miscible with alcohol and ether.
(2) Reflux 15ml of ester with dil NaOH solution, cool, acidify with dil acid. n-propyl alcohol and n-butyric acid are obtained.

b.p 145°C **Ethyl orthoformate** (triethoxy methane) $HC(OC_2H_5)_3$

(1) Colourless liquid with pungent odour, decomposes in water, soluble in alcohol and ether.
(2) Reflux 15ml of ester with dil NaOH solution, cool, acidify with dil acid. . Ethyl alcohol and formic acid are obtained.

b.p 148°C **n-Amyl acetate** $CH_3COOCH_2CH_2CH_2CH_2(CH_3)_2$

(1) Colourless liquid, very slightly soluble in water, miscible with alcohol and ether.
(2) Reflux 5ml of compound with dil NaOH solution. Cool, acidify with dil HCl. Amyl alcohol and acetic acid will be produced

b.p 151°C **Methyl n-caproate** (methyl hexanoate)

$CH_3CH_2CH_2CH_2CH_2COOCH_3$

(1) Colourless liquid, insoluble in water, easily soluble in alcohol and ether.

(2) Reflux 15ml of ester with dil NaOH solution, cool, acidify with dil acid. Methyl alcohol and n-caproic acid are obtained.

b.p 144°C **Ethyl Pyruvate** (ethyl 2-oxopropionate)

$CH_3COCOOC_2H_5$

(1) Colourless liquid, soluble in alcohol and ether.

(2) Reflux 5ml of compound with dil NaOH soution. Cool, acidify with dil HCl, ethyl alcohol and pyruvic acid will be produced. Distil off the alcohol and perform the iodoform test.

(3) Ethyl pyruvate 2, 4 dinitrophenylhydrazone, m.p 154°C.

(4) Ethyl pyruvate semicarbazone, m.p 206°C (d).

(5) Ethyl pyruvate p-nitrophenylhydrazone, m.p 185°C.

(6) Ethyl pyruvate phenylhydrazone, m.p 118°C

b.p 168°C **Ethyl n-caproate** (ethyl hexanoate)

$CH_3CH_2CH_2CH_2CH_2COOC_2H_5$

(1) Colourless to yellow coloured liquid, insoluble in water, soluble in alcohol and ether.

(2) Reflux 15ml of ester with dil NaOH solution, cool, acidify with dil acid. Ethyl alcohol and n-caproic acid are obtained.

b.p 170°C **Methyl aceto acetate** $CH_3COCH_2COOCH_3$

(1) Colourless liquid, capable of exhibiting keto enol tautomerism, sparingly soluble in water, soluble in alcohol.

(2) To 1ml of ester add few drops of neutral $FeCl_3$ solution. A dark red coloured solution is obtained.

(3) Reflux 5ml of compound with dil NaOH solution. Cool, acidify with dil acid. Methyl alcohol and acetone are obtained.

(4) Methyl acetoacetate semicarbazone, m.p 152°C.

b.p 175°C **Cyclohexyl acetate** $CH_3COOC_6H_{11}$

(1) Reflux 15ml of ester with dil NaOH solution, cool, acidify with dil acid. Cyclohexanol and acetic acid are obtained.

b.p 181°C **Dimethyl malonate** (dimethyl propane dioate) $CH_2(COOCH_3)_2$

(1) Colourless liquid, insoluble in water, soluble in alcohol and ether.
(2) Reflux 5ml of compound with dil NaOH solution. Cool, acidify with dil acid. Methyl alcohol and malonic acid are obtained.

b.p 181°C **Ethyl aceto acetate** (aceto acetic ester) $CH_3COCH_2COOCH_2CH_3$

(1) Colourless liquid having pleasant odour, sparingly soluble in water, easily soluble in most organic solvents, capable of exhibiting keto-enol tautomerism.
(2) To 1ml of compound add neutral $FeCl_3$ solution. A violet red coloured solution is obtained.
(3) To 1ml of compound add dilute neutral copper acetate solution. A green precipitate is obtained.
(4) To 9ml conc H_2SO_4 add dropwise 6.5ml ethyl aceto acetate (temp 10-15°C) and then keep at room temperature for 5-6 days. Pour in crushed ice, collect the precipitated solid and extract the aqueous filtrate with ether. Dissolve the solid in the ether and extract the ethereal layer with sodium carbonate solution. Acidify the alkaline extract with conc HCl and collect isodehydroacetic acid, m.p 154-155°C.
(5) Heat a mixture of 10gm phenol and 14gm ethyl aceto acetate in 17ml nitrobenzene at 100°C in an oil bath. Stir and add dropwise 27gm $AlCl_3$ in 100ml nitrobenzene. Heat the mixture at 130°C for 3hrs. Cool and add 25ml of HCl and water mixture(equal volume). Steam distil the reaction mixture. Separate the nitrobenzene layer from aqueous layer. Distil off the nitrobenzene and distil the residual oily liquid under reduced pressure, collect fraction boiling at 180-195°C. On cooling 4-methyl coumarin solidifies, m.p 83-84°C.
(6) To 3.5gm of KOH in 110ml water add with stirring 7ml of compound and keep the mixture at room temperature for 24hrs.

Cool a solution of 4.8ml aniline in 20ml dil HCl and add 100ml $NaNO_2$ solution to it (temp not below 10°C). Cool the potassium acetoacetate solution to 0°C and add with stirring4.5ml conc HCl followed by cold diazonium salt solution. Add 30ml of sodium acetate solution(8.2gm in 30ml water) to the reaction mixture and heat it to 50°C for 2hrs. Cool and collect methyl glyoxal ω phenyl hydrazone, m.p 148-150°C.

(7) Cool a mixture of 10ml ester and 22ml glacial acetic acid to 0°C. Add a solution of 2.7gm of sodium nitrite in 5ml water (temp below 7°C). Remove the ice bath and stir at room temperature for 30min. Add in portion 5gm zinc dust and 7ml acetic acid. Reflux for 60min. Cool and collect 2, 4 diethoxycarbonyl 3, 5 dimethyl pyrrole, m.p 138-139°C.

b.p 185°C **Diethyl oxalate** (diethyl ethanedioate) $(COOC_2H_5)_2$

(1) Colourless sweet smelling liquid, sparingly soluble in water, soluble in alcohol and ether.

(2) To 2ml of compound add concentrated ammonia solution and shake thoroughly. A white precipitate due to the formation of oxamide, m.p 417°C, will be obtained.

(3) To 2ml of compound add 2ml aniline and heat. On cooling white precipitate of oxanilide, m.p 247 °C, is obtained.

(4) Reflux 10ml of compound with dil NaOH solution. Cool, acidify with dil HCl; ethyl alcohol and oxalic acid will be produced. Distil off the alcohol and perform the iodoform test.

(5) To 30ml absolute ethanol add in portions 2.3gm sodium and then cool in an ice bath. Now add dropwise a mixture of 3.6gm butan-2-one and 16gm diethyl oxalate. Heat the resulting mixture for 30min. Cool and add 5.5ml dil H_2SO_4. Remove the precipitated Na_2SO_4, concentrate the filtrate and cool for several hours. Collect the precipitated glyoxalate ester and recrystallise from ethyl acetate, m.p 162°C.

(6) To a stirred solution of 2.3gm Na in 30ml MeOH add 3gm urea. When urea has dissolved add 5ml diethyl oxalate. Continue stirring for an hour and then add 10ml conc HCl dropwise. Continue stirring for 5-10min and filter. Concentrate the filtrate, keep it overnight in refrigerator and collect white crystals of parabanic acid, m.p 241-243°C.

(7) Heat a solution of 2.3gm Na in 30ml absolute ethanol, cool and add a solution of 16gm diethyloxalate in 24gm acetone to it. Stir the mixture for 30min. Distil to remove ethanol. Cool and treat the residual portion with 15ml conc HCl and 20gm crushed ice. Heat this yellow suspension with 15ml conc HCl on steam bath for 20hrs. Cool and collect chelidonic acid, m.p 257°C (d).

(8) To a solution of 2.5gm Na in 80ml ethanol add 6ml diethyl oxalate and 5.4 gm of o-hydroxyacetophenone. Reflux for 30 min. Cool, pour in 150ml ether. Filter the solid, dissolve in 100ml 2N HCl and extract with ether. Distil off ether and reflux the residue with 10ml glacial acetic acid and 10ml conc HCl. Cool and collect chromone-2-carboxylic acid, m.p 250-251°C. (d)

b.p 187°C **Ethyl Methyl Acetoacetate** (ethyl 2-methyl-3-oxobutanoate) $CH_3COCH(CH_3)COOCH_2CH_3$

(1) Colourless liquid, very slightly soluble in water, soluble in alcohol and ether.
(2) To 1ml of compound add few drops of neutral $FeCl_3$ solution. A blue coloured solution will be obtained.
(3) Heat 5ml of compound with 2-3ml aniline at 150-160°C for 4-5 minutes. Cool and collect α-methyl acetoacetanilide, m.p 138°C.
(4) Ethyl methylacetoacetate 2, 4 dinitrophenylhydrazone, m.p 56°C.
(5) Ethyl methylacetoacetate semicarbazone, m.p 86°C.

b.p 195-200°C **Dimethyl Succinate** (dimethyl butanedioate) $(CH_2COOCH_3)_2$

(1) Colourless liquid, sparingly soluble in water, soluble in alcohol.
(2) Reflux 5ml of ester with dil NaOH solution for 15-20min. Cool, acidify with dil HCl. Methyl alcohol and succinic acid will be produced. Distil off the alcohol and perform the fluorescence test with succinic acid.
(3) To 5ml of compound add 20ml concentrated ammonia solution and shake thoroughly. Keep the mixture at room temperature for 24hrs and filter the precipitated succinamide, m.p 254°C (d).

b.p 198°C **Diethyl malonate** (ethyl malonate, diethyl propane dioate) $CH_2(COOC_2H_5)_2$

(1) Colourless pleasant smelling liquid, sparingly soluble in water, miscible with alcohol and ether.
(2) To 25ml sodium ethoxide solution (1.5gm Na in 25ml absolute ethanol) add 7.5ml diethyl malonate followed by 3gm urea in 25ml ethanol. Reflux at 110°C for 7hrs. Add first 45ml of hot water and then add conc HCl until acidic. Filter, leave the filtrate overnight in a refrigerator and collect barbituric acid, mp 245°C.
(3) Reflux 10ml diethyl malonate with 5.5ml salicyaldehyde, 20ml absolute ethanol, 0.5ml piperidine and 0.1ml glacial acetic acid for 3hrs. Dilute the hot solution with 10ml ethanol and 10ml water. Cool and collect 3-carbethoxy coumarin, recrystallise from ethanol, m.p 92-94°C.
(4) Cool a mixture of 18ml diethyl malonate, 4gm of formalin solution (0-5°C) and the add 0.7ml diethylamine. Keep the mixture at room temperature for 15hrs, reflux on water bath for 6hrs. Remove the organic layer, add 12ml dil HCl and reflux for 5-7hrs. Evaporate off the solvent and collect glutaric acid, recrystallise from chloroform, m.p 96-97°C.
(5) Reflux 10ml of compound with dil NaOH solution. Cool, acidify with dil HCl, ethyl alcohol and malonic acid will be produced. Distil off the alcohol and perform the iodoform test.

b.p 198°C **Ethyl ethyl acetoacetate**

$CH_3COCH(C_2H_5)COOCH_2CH_3$

(1) Sparingly soluble in water, miscible with alcohol and ether.
(2) To 1ml of compound add neutral $FeCl_3$ solution. A blue coloured solution is obtained.
(3) Reflux 10ml of compound with dil NaOH solution. Cool, acidify with dil HCl. Ethyl alcohol and methyl n-propyl ketone will be produced. Distil off the alcohol and perform the iodoform test.

b.p 207°C **Di-n-butyl carbonate** $C_4H_9OCOOC_4H_9$

(1) Colourless liquid, insoluble in water, soluble in alcohol and ether.
(2) On saponification the products are n-butyl alcohol and carbon dioxide.

b.p 215°C **Di n-Propyl oxalate** $(COOCH_2C_2H_5)_2$

(1) Colourless liquid, sparingly soluble in water, soluble in alcohol and ether.

(2) Reflux 5ml of ester with dil NaOH solution for 15-20min. Cool and acidify with dil HCl. Propyl alcohol and oxalic acid will be produced.

b.p 217°C **Diethyl Succinate** (Succinic acid diethyl ester) $(CH_2COOC_2H_5)_2$

(1) Colourless liquid, insoluble in water, soluble in alcohol and ether.

(2) Reflux 5ml of ester with dil NaOH solution for 15-20min. Cool, acidify with dil HCl; ethyl alcohol and succinic acid will be produced. Distil off the alcohol and perform the iodoform test. With succinic acid perform the fluorescence test.

(3) Add 1gm of sodium to 10ml absolute ethanol, heat the mixture under reflux for 3-4hrs. To the hot solution add 3.5gm of diethyl succinate and heat the mixture under reflux for 24hrs. Remove ethanol under reduced pressure and then add 20ml 2N H_2SO_4 and stir for 3-4hrs. Collect the precipitated pink coloured 2, 5 dicarbethoxy-1, 4-cyclohexanedione, recrystallise from ethyl acetate, m.p 121-125°C.

(4) Dissolve by heating 2.3gm Na in 35ml absolute ethanol. Distil off excess ethanol and add toluene to the pasty mixture. Add to this solution 65ml ether, 15gm of diethyl oxalate and 18gm diethyl succinate. Keep the mixture for 12hrs at room temperature. Add 50ml water and remove the ethereal layer. To the aqueous layer add dil HCl and extract with ether. Combine the ethereal layers, distil off ether and collect triethyl oxalyl succinate as yellow oil. Reflux this oil with 30ml 12N HCl and 60ml water for 4hrs. Distill off the liquid and add 20ml nitroethane to the residue. Warm and filter. Cool the filtrate and collect α-ketoglutaric acid, m.p 110°C.

(5) Add a mixture of 13gm diethysuccinate and 3gm acetophenone in 12ml benzene to 1.8ml of sodium hydride in 12ml benzene. Keep the mixture at room temperature for 30min, cool and acidify with glacial acetic acid. Add 10ml water. Extract with ether. Extract the ether layer with 5% sodium carbonate solution. Acidify the

alkaline layer. Extract with ether. Distil off ether and collect the condensation product, m.p 111-112°C.

b.p 237°C **Diethyl glutarate** (diethyl pentanedioate) $CH_2(CH_2COOC_2H_5)_2$

(1) Syrupy liquid, very slightly soluble in water, soluble in alcohol and ether.
(2) Reflux 10ml of compound with dil NaOH solution. Cool and acidify with dil HCl. Ethyl alcohol and glutaric acid will be produced. Distil off the alcohol and perform the iodoform test.

b.p 245°C **Diethyl adipate** $(CH_2CH_2COOC_2H_5)_2$

(1) Colourless liquid with characteristic odour, sparingly soluble in water, soluble in alcohol and ether.
(2) Reflux 10ml of compound with dil NaOH solution. Cool, acidify with dil HCl, ethyl alcohol and adipic acid will be produced. Distil off the alcohol and perform the iodoform test.
(3) To 1gm lithium aluminium hydride add 40ml dry ether and 6gm diethyl adipate, stir the mixture for 10min. Cool, dilute by adding 10ml water. Remove the ethereal layer, distil off the ether. The colourless viscous residue soon solidifies on cooling is hexane 1, 6 diol, m.p 42°C.

b.p 246°C **Di-n-propyl succinate** $(CH_2COOCH_2C_2H_5)_2$

(1) Reflux 5ml of ester with dil NaOH solution for 15-20min. Cool, acidify with dil HCl, propyl alcohol and succinic acid will be produced. Distil off the alcohol. With succinic acid perform the fluorescence test.

b.p 250°C **Diethyl Acetone dicarboxylate** (diethyl β-oxoglutarate) $CO(CH_2COOC_2H_5)_2$

(1) Sparingly soluble in water and alcohol.
(2) To 1ml of compound add copper acetate. Green precipitate of copper enolate is obtained.

(3) To 1ml of compound add neutral $FeCl_3$ solution. A red coloured solution is obtained.
(4) Diethyl acetonedicarboxylate semicarbazone, m.p 94°C.

b.p 269°C **Ethyl Laurate** (ethyl dodecanoate)

$CH_3(CH_2)_{10}COOC_2H_5$

(1) Oily liquid, insoluble in water, easily soluble in alcohol and ether.
(2) Reflux 10ml of compound with dil NaOH solution. Cool, acidify with dil HCl. Ethyl alcohol and lauric acid will be produced. Distil off the alcohol and perform the iodoform test.
(3) Reflux 13gm of compound with 60ml diethyl carbonate and sodium ethoxide (from 1.5gm Na). Distil off the alcohol formed during the reaction. Cool, acidify with acetic acid and distil. Collect decylmalonic ester at b.p 130-132°C/1.5mmHg

b.p 280°C **Diethyl d-tartrate** (diethyl d, 2, 3 dihydroxy biltanedioate) $(CHOHCOOC_2H_5)_2$

(1) Colourless hygroscopic liquid, soluble in water and alcohol, miscible with ether.
(2) Reflux 10ml of compound with dil NaOH solution. Cool, acidify with dil HCl, ethyl alcohol and tartaric acid will be produced. Distil off the alcohol and perform the iodoform test.
(3) Diacetate, m.p 67°C.

b.p 282°C **Diethyl Suberate** $(CH_2CH_2CH_2COOC_2H_5)_2$

(1) Colourless liquid, insoluble in water, soluble in alcohol and ether.
(2) Reflux 10ml of compound with dil NaOH solution. Cool, acidify with dil HCl, ethyl alcohol and suberic acid will be produced. Distil off the alcohol and perform the iodoform test.

Aromatic Liquids

b.p 196°C **Phenyl acetate** 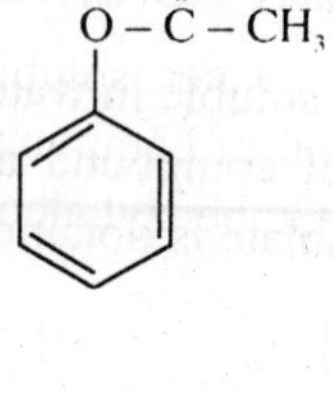

(1) Colourless sweet smelling liquid, insoluble in water, soluble in alcohol and ether.

(2) Boil 2ml of compound with 2ml water and add few drops of neutral $FeCl_3$ solution. A violet coloured solution is obtained.

(3) Reflux 10ml of compound with dil NaOH solution. Cool, acidify with dil HCl. Phenol and acetic acid will be produced. Carry out libermann test for phenol.

(4) Reflux 5.6gm $AlCl_3$, 6ml CS_2 and 4.8ml phenyl acetate on water bath for 2hrs. Distil off the solvent and reflux the residual portion on oil bath for 3hrs. Cool, add dil HCl and 10ml water. Leave the reaction mixture overnight. Extract with toluene. Distil off the solvent and distil the residue under reduced pressure, Collect o-hydroxy acetophenone at 200-210°C.

b.p 199°C **Methyl benzoate** (Niobe oil)

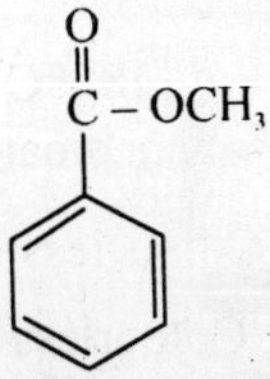

(1) Colourless liquid sparingly soluble in water, soluble in alcohol and ether.

(2) Reflux 10ml of compound with dil NaOH solution. Cool, acidify with dil HCl, methyl alcohol and white crystals of benzoic acid, m.p 121°C, will be precipitated.

(3) To ice cooled 4ml of ester add dropwise 6ml of nitrating mixture (3ml each of conc HNO_3 and conc H_2SO_4) After the addition is complete keep the flask at room temperature for 30 min and then pour in ice. Collect the precipitated methyl m-nitro benzoate, m.p 75-76°C.

b.p 203°C **Benzyl formate** (benzyl methanoate) $HCOOCH_2C_6H_5$

(1) Insoluble in water, soluble in alcohol, miscible with ether.

(2) Reflux 10ml of compound with dil NaOH solution. Cool, acidify with dil HCl, benzyl alcohol and formic acid will be produced.

b.p 211°C **Phenyl propionate** 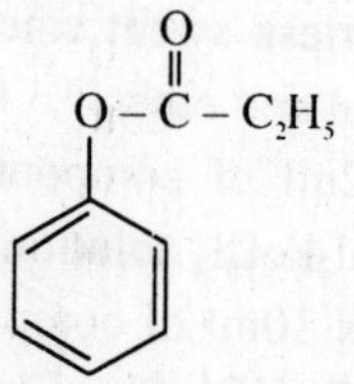

(1) Reflux 10ml of compound with dil NaOH solution. Cool, acidify with dil HCl, phenol and propionic acid will be produced. Carry out libermann test for phenol.

b.p 212°C **m-Tolyl acetate** (m-Cresyl acetate)

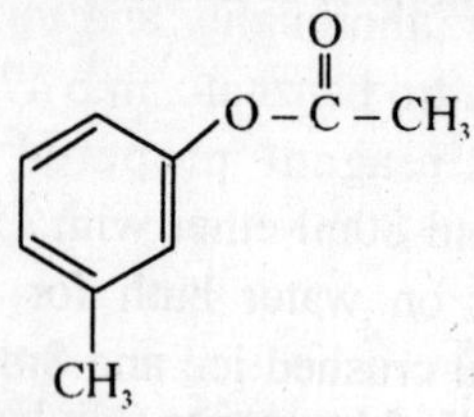

(1) Colourless oily liquid, insoluble in water, miscible with alcohol and ether.
(2) On saponification the products are m-cresol and acetic acid.

b.p 212°C **p-Tolyl acetate** (p-Cresyl acetate) 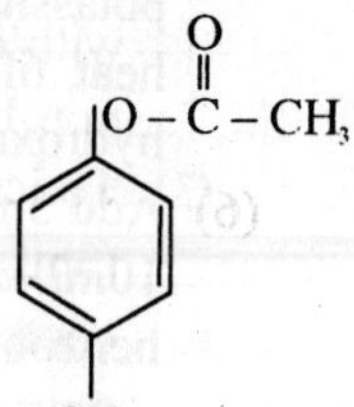

(1) Colourless liquid with floral odour, insoluble in water, sparingly soluble in alcohol.
(2) On saponification the products are p-cresol and acetic acid.

b.p 212°C **Ethyl benzoate** (ethyl benzene carboxylate)

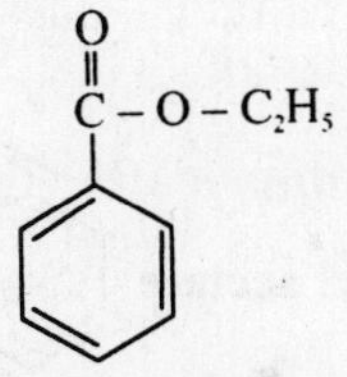

(1) Colourless sweet smelling liquid, insoluble in water easily soluble in alcohol and ether.

(2) Reflux 10ml of compound with dil NaOH solution. Cool and acidify with dil HCl, ethyl alcohol and white crystals of benzoic acid, m.p 121°C, will be precipitated.

(3) To 3ml of ester add 4ml of conc H_2SO_4 and 2ml of conc HNO_3. Shake the mixture thoroughly and then pour in ice with stirring. Collect ethyl m-nitro benzoate, m.p 47°C.

(4) Treat grignard reagent prepared from 4.2gm Mg, 20ml bromobenzene and 80ml ether with 12ml ethyl benzoate in 30ml benzene. Reflux on water bath for 1hr, cool and pour into a mixture of 250ml crushed ice and 8ml conc H_2SO_4. Separate the benzene layer and distil off the solvent. Cool and filter the crude triphenylmethanol, recrystallise from CCl_4, m.p 162°C.

(5) Heat solution of 4.67gm hydroxylamine hydrochloride in 24ml methanol to its b.p, cool and then add KOH solution to it (5.6gm in 14ml methanol). Cool, add 5ml of ethylbenzoate with stirring and filter. Keep the filtrate for 48hrs and collect crystals of potassium hydroxymate. Dissolve the salt in 16ml acetic acid and heat till a clear solution is obtained. Cool and collect benzo-hydroxamic acid, m.p 122-128°C.

(6) Add 5ml ethyl benzoate to a solution of 5ml hydrazine hydrate in 10ml ethanol. Reflux for 1hr, cool and wash the precipitated benzohydrazine with cold ethanol, m.p 112°C.

b.p 213°C **Methyl o-toluate**

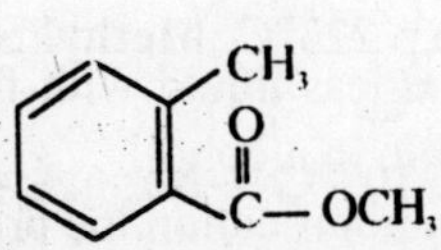

(1) Insoluble in water, miscible with alcohol and ether.

(2) Reflux 10ml of compound with dil NaOH solution. Cool, acidify with dil HCl, methyl alcohol and o-toluic acid will be produced.

b.p 215°C **Benzyl acetate** 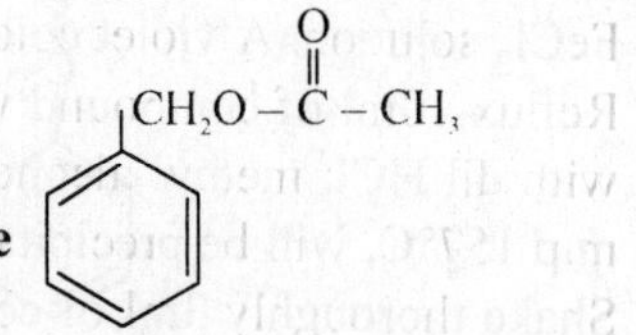

(1) Colourless liquid, insoluble in water, soluble in alcohol and ether.

(2) Reflux 10ml of compound with dil NaOH solution. Cool, acidify with dil HCl, benzyl alcohol and acetic acid will be produced.

b.p 215°C **Methyl m-toluate** 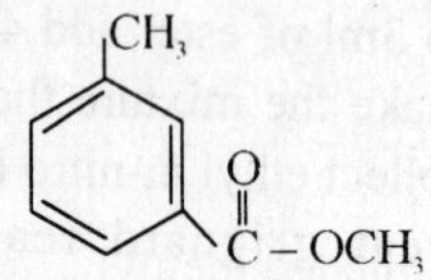

(1) Reflux 10ml of compound with dil NaOH solution. Cool, acidify with dil HCl, methyl alcohol and m-toluic acid will be produced.

b.p 215°C **Methyl Phenyl acetate** 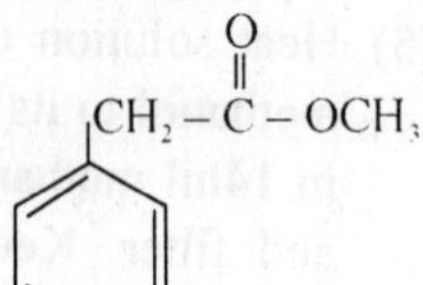

(1) Colourless liquid, insoluble in water, soluble in alcohol and ether.

(2) Reflux 10ml of compound with dil NaOH solution. Cool, acidify with dil HCl; methyl alcohol and white crystals of phenylacetic acid, m.p 76°C, will be precipitated.

b.p 223°C **Methyl Salicylate** (Methyl o-hydroxy benzoate)

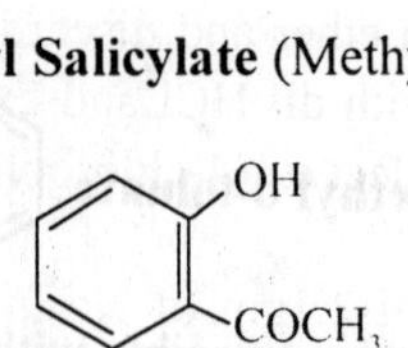

(1) Colourless liquid with characteristic smell of oil of winter green, insoluble in water, soluble in alcohol and ether.

(2) To 2ml alcoholic solution of compound add few drops of neutral $FeCl_3$ solution. A violet coloured solution is obtained.

(3) Reflux 10ml of compound with dil NaOH solution. Cool, acidify with dil HCl; methyl alcohol and white crystals of salicylic acid, m.p 157°C, will be precipitated.

(4) Shake thoroughly 3ml of compound with 3ml dil NaOH solution and 3ml acetic anhydride. On cooling crystals of methyl –o – acetoxy benzoate precipitate out, m.p 52°C.

(5) Shake thoroughly 5ml of compound with 10ml dil NaOH solution and 7-8ml benzoyl chloride until the smell of benzoyl chloride persists. Cool and collect methyl o-benzoxy benzoate, m.p 92°C.

(6) Keep the flask containing 3ml of compound in ice bath and then add dropwise 6ml of nitrating mixture(3ml each of fuming HNO_3 and conc H_2SO_4). After the addition is complete keep the flask at room temperature for 30 min and then pour in ice. Collect the precipitated methyl 3, 5-dinitrosalicylate, m.p 127°C.

b.p 226°C **Ethyl Phenyl acetate**

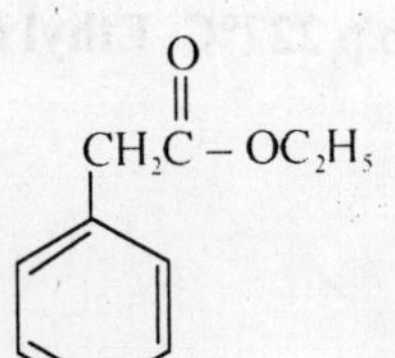

(1) Colourless liquid, insoluble in water, soluble in alcohol and ether.

(2) Reflux 10ml of compound with dil NaOH solution. Cool, acidify with dil HCl; ethyl alcohol and white crystals of phenylacetic acid, m.p 76°C, will be precipitated.

(3) To a solution of 1.15 gm of Na in 15ml ethanol add dropwise with stirring 3.78gm of 4-chlorophenyl acetonitrile and 5ml of ethyl phenyl acetate. Reflux the mixture for 3hrs. Cool pour in ice. Extract with ether and discard the ethereal layer. Acidify the aqueous layer with dil HCl and extract with ether. Evaporate off the ether and collect α–(4-chlorophenyl)-υ-phenylacetonitrile, m.p 131°C.

(4) To suspension of 2.1gm of Na in 6ml toluene add a solution of 2.5ml ethylphenylacetate in 30ml ethanol. Heat the solution on steam bath until all sodium dissolves. Distil off the ethanol and toluene and dilute the residue with water. Extract the aqueous

layer with ether, distil off the ether and recover phenyl ethanol, b.p 203°C.

b.p 227°C **Ethyl o-toluate** 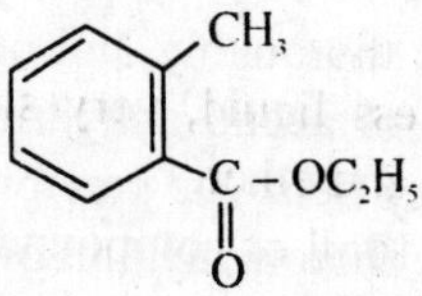

(1) Colourless liquid, insoluble in water, miscible with alcohol and ether.
(2) Reflux 10ml of compound with dil NaOH solution. Cool, acidify with dil HCl; ethyl alcohol and o-toluic acid are obtained.

b.p 227°C **Ethyl m-toluate** 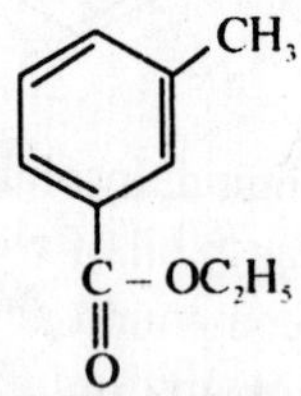

(1) Colourless liquid, insoluble in water, miscible with alcohol and ether.
(2) Reflux 10ml of compound with dil NaOH solution. Cool, acidify with dil HCl; ethyl alcohol and m-toluic acid are obtained.

b.p 228°C **Ethyl p-toluate** 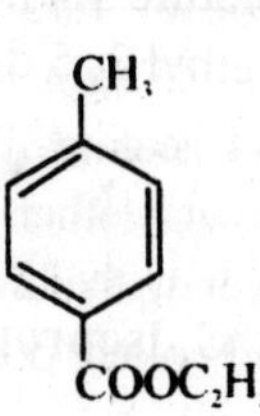

(1) Colourless liquid, insoluble in water, miscible with alcohol and ether.
(2) Reflux 10ml of compound with dil NaOH solution. Cool, acidify with dil HCl; ethyl alcohol and p-toluic acid are obtained.

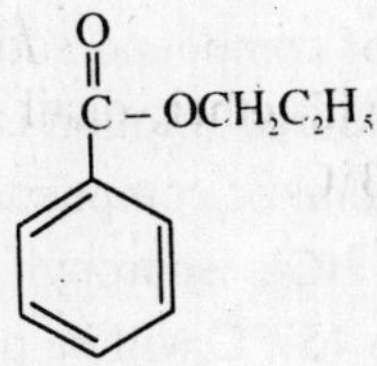

b.p 231°C **n-Propyl benzoate**

(1) Colourless liquid, very slightly soluble in water, miscible with alcohol and ether.
(2) Reflux 10ml of compound with dil NaOH solution. Cool, acidify with dil HCl; propyl alcohol and white crystals of benzoic acid, m.p 121°C, will be precipitated.

b.p 234°C **Ethyl Salicylate** (Ethyl o-hydroxy benzoate)

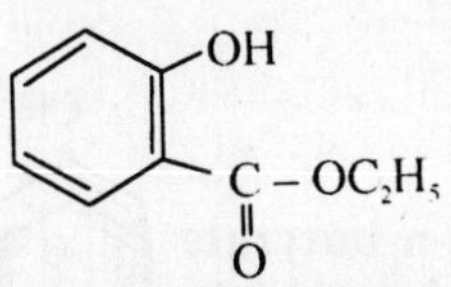

(1) Colourless liquid, insoluble in water, soluble in alcohol and ether, odour similar to oil of wintergreen.
(2) To 1ml of compound add few drops of neutral $FeCl_3$ solution. A red-violet coloured solution is obtained.
(3) Reflux 10ml of compound with dil NaOH solution. Cool, acidify with dil HCl; ethyl alcohol and white crystals of salicylic acid, m.p 157°C, will be precipitated.
(4) Keep the flask containing 3ml of compound in ice bath and then add dropwise 6ml of nitrating mixture (3ml each of fuming HNO_3 and conc H_2SO_4). After the addition is complete keep the flask at room temperature for 30 min and then pour in ice. Collect the precipitated ethyl 3, 5 dinitrosalicylate, m.p 92°C.

b.p 240-242°C **Isopropyl Salicylate**

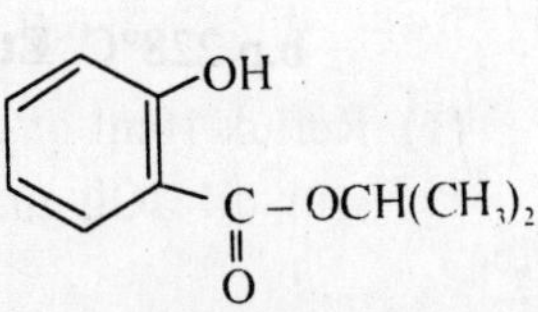

(1) Oily liquid, insoluble in water, miscible with alcohol and ether, odour similar to oil of winter green.

(2) To 1ml of compound add few drops of neutral $FeCl_3$ solution. A violet coloured solution is obtained.
(3) Reflux 10ml of compound with dil NaOH solution. Cool, acidify with dil HCl, isopropyl alcohol and white crystals of salicylic acid, m.p 157°C will be precipitated.
(4) Keep the flask containing 3ml of compound in ice bath and then add dropwise 6ml of nitrating mixture (3ml each of fuming HNO_3 and conc H_2SO_4). After the addition is complete keep the flask at room temperature for 30 min and then pour in ice. Collect the precipitated Isopropyl 3, 5 dinitrosalicylate, m.p 101-102°C.

b.p 240°C **Benzyl-n-butyrate**

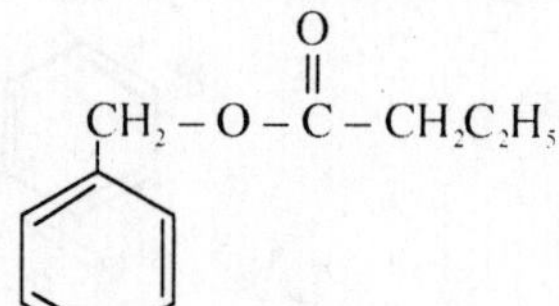

(1) Insoluble in water, soluble in alcohol and ether.
(2) Reflux 10ml of compound with dil NaOH solution. Cool, acidify with dil HCl; benzyl alcohol and n-butyric acid will be produced.

b.p 240°C **Guaiacol acetate** (o-methoxy phenyl acetate)

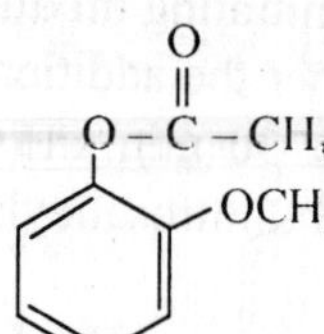

(1) Reflux 10ml of compound with dil NaOH solution. Cool, acidify with dil HCl; guaiacol and acetic acid will be produced.

b.p 237°C **Isobutyl benzoate** (eglantine, β-methylpropyl benzene carboxylate)

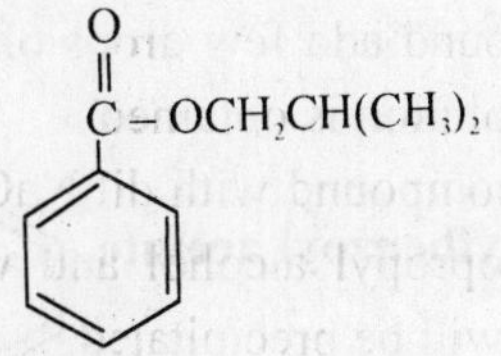

(1) Colourless liquid with characteristic odour, insoluble in water, miscible with alcohol and ether.

(2) Reflux 10ml of compound with dil NaOH solution. Cool, acidify with dil HCl; isobutyl alcohol and white crystals of benzoic acid, m.p 121°C, will be precipitated.

b.p 250°C **Butyl benzoate** (butyl benzene carboxylate)

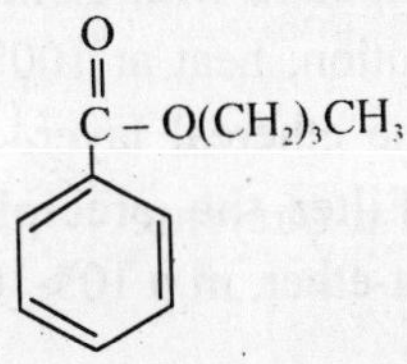

(1) Reflux 10ml of compound with dil NaOH solution. Cool, acidify with dil HCl; n-butyl alcohol and white crystals of benzoic acid, m.p 121°C, will be precipitated.

b.p 262°C **Isoamyl benzoate** (3-methyl-1-butanol benzoate)

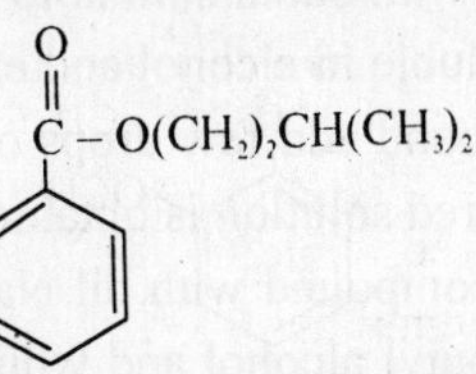

(1) Colourless liquid, insoluble in water, soluble in alcohol, miscible with ether.

(2) Reflux 10ml of compound with dil NaOH solution. Cool, acidify with dil HCl, isoamyl alcohol and white crystals of benzoic acid, m.p 121°C, will be precipitated.

b.p 265°C (d) **Ethylbenzoyl acetate** 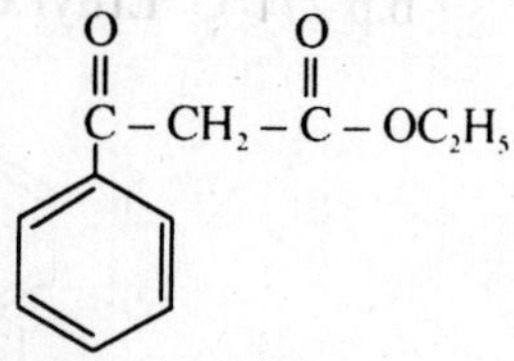

(1) Liquid, insoluble in water, soluble in alcohol and ether.
(2) To 1ml alcoholic solution of compound add add few drops of $FeCl_3$ solution. A red violet coloured solution is obtained.
(3) To a solution of 4.5ml of compound in 10ml ethanol add a solution of 1.75gm of hydroxylamine hydrochloride and 2.5gm sodium acetate in 6ml water. Reflux for 2hrs. Cool and collect 3-Phenyl-5-isoxazolone, recrystallise from ethanol, m.p 152°C.
(4) Reflux 4.5ml of compound with 2.5ml aniline for 1hr. Cool, add 25ml dil NaOH solution, heat at 100°C for 5 min. Extract with ether and discard the ethereal layer. Acidify the aqueous layer with acetic acid. Filter the precipitated benzoylacetanilide, recrystallise from pet-ether, m.p 105-106°C.

b.p 268°C **n-Butyl Salicylate** 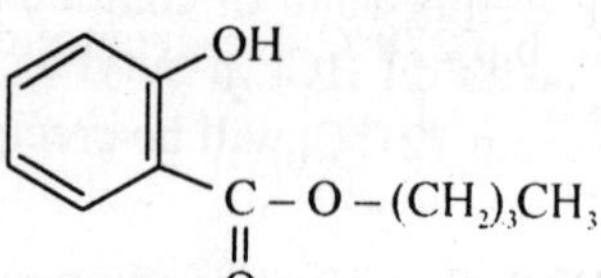

(1) Colourless liquid with odour similar to oil of winter green, difficult soluble water, soluble in alcohol and ether.
(2) To 1ml of compound add few drops of neutral $FeCl_3$ solution. A faint violet coloured solution is obtained.
(3) Reflux 10ml of compound with dil NaOH solution. Cool, acidify with dil HCl; n-butyl alcohol and white crystals of salicylic acid, m.p 157°C, will be precipitated.
(4) Keep the flask containing 5ml of compound in ice bath and then add dropwise 8ml of nitrating mixture (4ml each of fuming HNO_3 and conc H_2SO_4). After the addition is complete keep the flask at room temperature for 30 min and then pour in ice. Collect the precipitated n-butyl 3, 5 dinitrosalicylate, m.p 60-61°C.

b.p 271°C **Ethyl Cinnamate** (ethyl trans-3-phenyl propenoate)

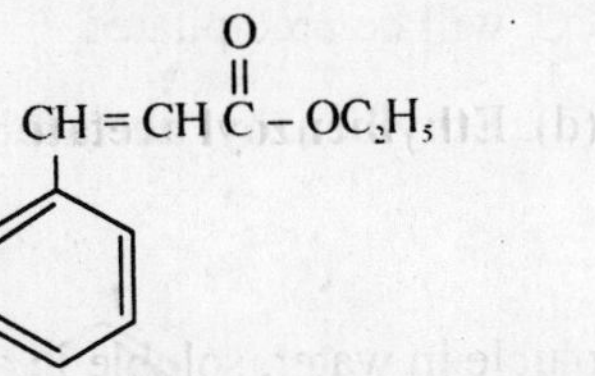

(1) Colourless liquid, insoluble in water, soluble in alcohol and ether.
(2) Reflux 10ml of compound with dil NaOH solution. Cool, acidify with dil HCl; ethyl alcohol and cream coloured crystals of cinnamic acid, m.p 132°C, will be precipitated.
(3) Cool 5ml of ethyl cinnamate in 5ml CCl_4 in an ice bath and add dropwise with stirring 2.5ml of bromine. After keeping at room temperature for an hour pour it in ice. Collect the precipitated ethyl α, β dibromo β phenyl propionate, m.p 65-71°C

b.p 278°C **Resorcinol diacetate**

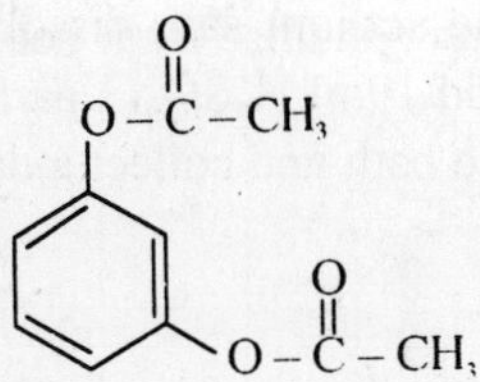

(1) Reflux 10ml of compound with dil NaOH solution. Cool, acidify with dil HCl; acetic acid and pinkish coloured crystals of resorcinol, m.p 110°C, will be precipitated.

b.p 282°C **Dimethyl Phthalate** (dimethyl 1, 2 benzene dicarboxylate)

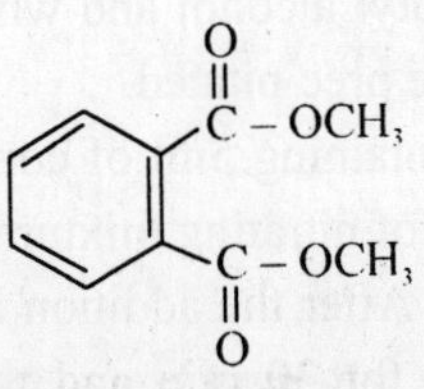

(1) Colourless liquid, insoluble in water.

(2) Reflux 10ml of compound with dil NaOH solution. Cool, acidify with dil HCl; methyl alcohol and white crystals of Phthalic acid, m.p 206-208°C, will be precipitated

b.p 296°C **Diethyl Phthalate**

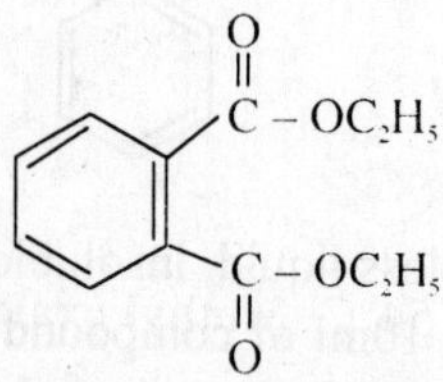

(1) Colourless liquid, insoluble in water, soluble in alcohol and ether.
(2) Reflux 10ml of compound with dil NaOH solution. Cool, acidify with dil HCl; ethyl alcohol and white crystals of Phthalic acid, m.p 206-208°C, will be precipitated.
(3) Heat 12.5 ml diethyl phthalate and 2.5gm sodium on steam bath, add with stirring 13.6ml ethyl acetate and 2-4 drops of absolute ethanol. Continue heating for 6hrs. Cool and add 5ml ether. Filter the sodium salt, dissolve in 100ml hot water, cool to 70°C and add 10ml H_2SO_4(3 part acid and 1part water) Cool the mixture in ice bath and collect indane-1, 3 dione, m.p 130°C.

b.p 309°C **Ethyl –1-naphthoate**

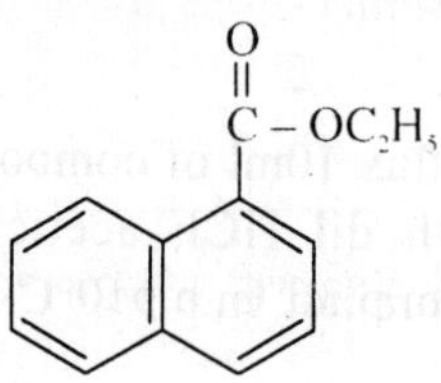

(1) Reflux 10ml of compound with dil NaOH solution. Cool, acidify with dil HCl, ethyl alcohol and naphthoic acid, m.p 162°C, will be precipitated.

b.p 323°C, m.p 21°C **Benzyl benzoate**

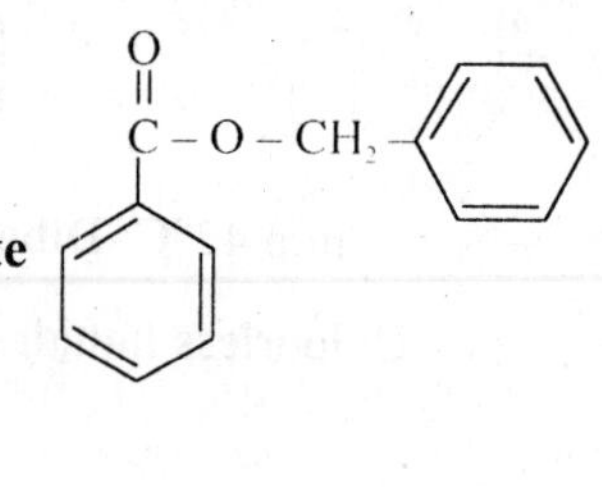

(1) Colourless oily liquid, insoluble in water, soluble in alcohol and ether.
(2) Reflux 10ml of compound with dil NaOH solution. Cool, acidify with dil HCl; benzyl alcohol and white crystals of benzoic acid, m.p 121°C, will be precipitated.

Aliphatic Solid

m.p 54°C **Methyl oxalate** (dimethyl ethanedioate) 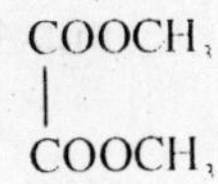

(1) White solid, soluble in water and alcohol.
(2) Reflux 500mg of compound with dil NaOH solution. Cool, acidify with dil HCl. Oxalic acid, m.p 101°C, separates out.
(3) Reduces Fehling's solution.
(4) Dissolve 500mg of compound in minimum amount of water, add concentrated ammonia solution dropwise and shake thoroughly. There will be an immediate precipitation of oxamide, m.p 417°C.
(5) Dissolve 1gm of compound in alcohol and then cool. Add concentrated ammonia solution and keep at 0°C overnight. Heat the solution to boiling and filter hot (to remove oxamide). Cool the filtrate and collect methyl oxamate, m.p 120-123°C.

Aromatic Solid

m.p 36°C, b.p 261°C **Methyl Cinnamate** 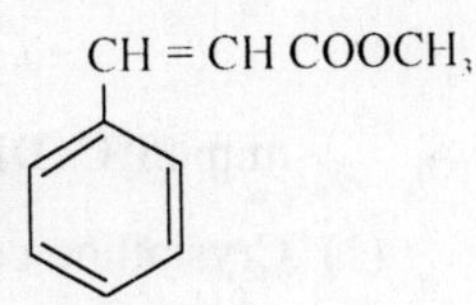

(1) Colourless solid, insoluble in water, soluble in alcohol and ether
(2) Reflux 5 gm of compound with dil NaOH solution. Cool, acidify with dil HCl. Cinnamic acid, m.p 132°C, will separate out.

m.p 42°C **Dibenzyl Phthalate** 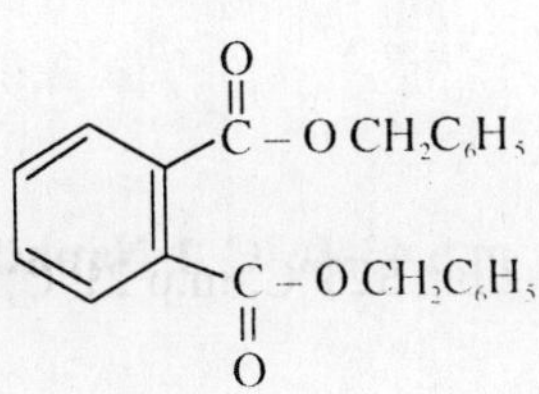

(1) Insoluble in water, soluble in alcohol and ether.
(2) Reflux 5gm of compound with dil NaOH solution. Cool, acidify with dil HCl. Benzyl alcohol and white crystals of Phthalic acid, m.p 206-208°C (d), will be precipitated.

m.p 43°C **Phenyl Salicylate (Salol)**

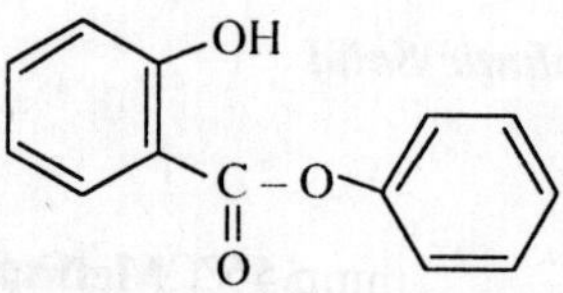

(1) Colourless solid, insoluble in water, soluble in alcohol and ether.
(2) Reflux 5gm of compound with dil NaOH solution. Cool, acidify with dil HCl. White crystals of salicylic acid, m.p 157°C, will be precipitated.
(3) To 1ml alcoholic solution of compound add few drops of neutral $FeCl_3$ solution. A violet-red coloured solution will be obtained.
(4) Shake thoroughly 1gm of compound with 10ml dil NaOH solution and 7-8ml benzoyl chloride until the smell of benzoyl chloride persists. Cool and collect Phenyl o-benzoxy benzoate, m.p 81°C.
(5) Heat to boiling a mixture of 4.2gm phenyl salicylate, 2.67gm o-toluidine and 6.0gm 1, 2, 4 trichloro benzene so that the phenol formed slowly distils. Slowly raise the temperature to 200°C. Continue distilling and collect salicyl-o-toluidide, cool and crystallise from hexane, m.p 143-144°C.

m.p 45°C **Dibenzyl Succinate** $(CH_2COOCH_2C_6H_5)_2$

(1) Crystalline compound, insoluble in water, soluble in alcohol and ether.
(2) Reflux 10 gm of compound with dil NaOH solution. Cool, acidify with dil HCl. Benzyl alcohol and succinic acid, m.p 185°C, will be precipitated.

m.p 43-46°C **1-Naphthyl acetate**

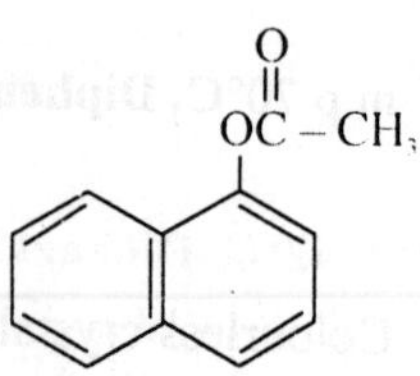

(1) Crystalline compound, soluble in alcohol and ether.
(2) Reflux 5gm of compound with dil NaOH solution. Cool, acidify with dil HCl. 1-naphthol and acetic acid will be obtained.

m.p 56°C **1-Naphthyl Benzoate**

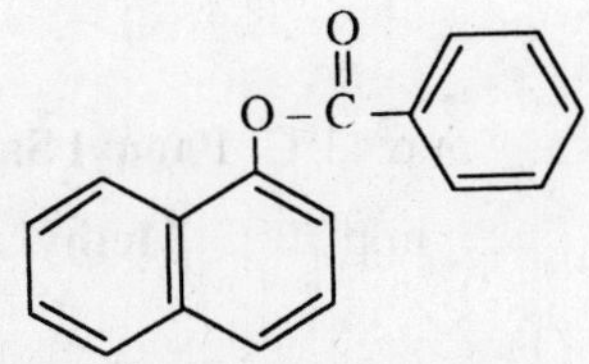

(1) Reflux 10mg of compound with dil NaOH solution. Cool, acidify with dil HCl. 1-Naphthol m.p 94°C and benzoic acid, m.p 121°C, will be precipitated. Separate and identify them

m.p 70°C **Phenyl benzoate**

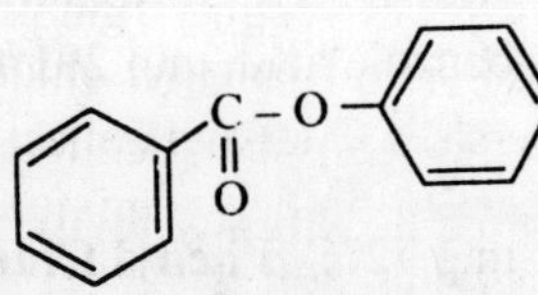

(1) White crystalline solid, insoluble in water, soluble in hot alcohol and ether.
(2) Reflux 5gm of compound with dil NaOH solution. Cool, acidify with dil HCl. White crystals of benzoic acid, m.p 121°C, will be precipitated.
(3) Heat 2gm of compound with 1.5 gm of $AlCl_3$ at 145°C for 15 minutes. Cool and collect p-hydroxy benzophenone, recrystallise from methanol, m.p 135°C.

m.p 70°C **Diphenyl Phthalate**

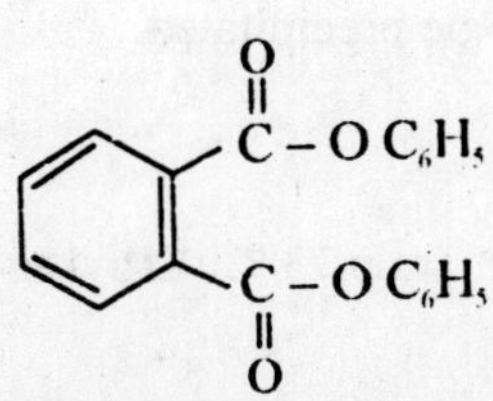

(1) Colourless crystalline solid, insoluble in water, slightly soluble in alcohol and ether.

(2) Reflux 5gm of compound with dil NaOH solution. Cool, acidify with dil HCl. Phenol and white crystals of Phthalic acid, m.p 206-208°C (d), will be precipitated.

m.p 70°C **Methyl m-hydroxy benzoate**

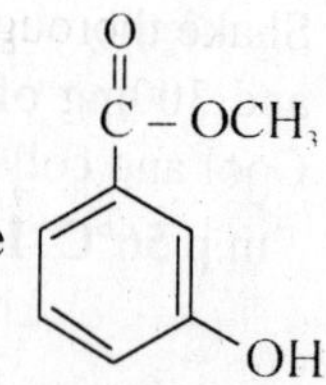

(1) Colourless crystalline solid.
(2) To 1ml aqueous solution of compound add few drops of neutral $FeCl_3$ solution. A violet coloured solution is obtained.
(3) Reflux 5gm of compound with dil NaOH solution. Cool, acidify with dil HCl. Methyl alcohol and white crystals of m-hydroxy benzoic acid, m.p 201°C, will be precipitated.

m.p 72°C **Phenyl Cinnamate**

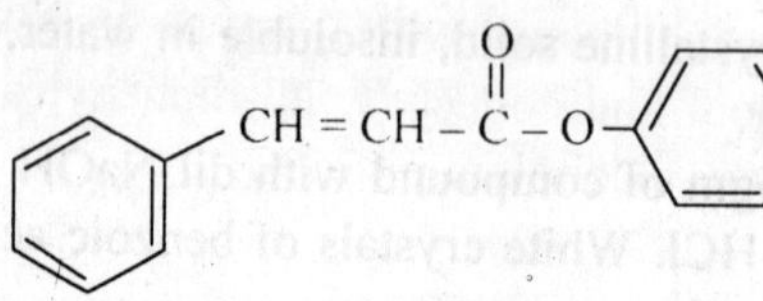

(1) White crystalline solid, insoluble in water.
(2) Reflux 5gm of compound with dil NaOH solution. Cool, acidify with dil HCl. Cinnamic acid, m.p 132°C, separates out.

m.p 73°C **Ethyl m-Hydroxy benzoate**

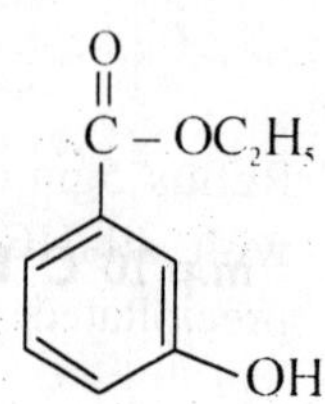

(1) Colourless solid, sparingly soluble in water, easily soluble in alcohol and ether.

(2) Reflux 5gm of compound with dil NaOH solution. Cool, acidify with dil HCl. Ethyl alcohol and white crystals of m-hydroxy benzoic acid, m.p 201°C, will be precipitated.

(3) To 1ml aqueous solution of compound add few drops of neutral $FeCl_3$ solution. A violet coloured solution is obtained.

(4) Shake thoroughly 1gm of compound with 7-8ml benzoyl chloride and 300mg of $AlCl_3$ until the smell of benzoyl chloride persists. Cool and collect ethyl m-benzoxy benzoate, m.p 58°C.

m.p 78°C **Diphenyl Carbonate**

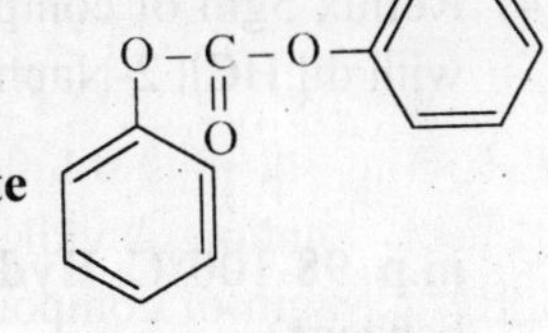

(1) Crystalline compound, insoluble in water, soluble in alcohol and ether.

(2) Reflux 10ml of compound with dil NaOH solution. Cool, acidify with dil HCl; phenol and carbon dioxide will be produced. Carry out libermann test for phenol.

(3) Heat 1gm of compound with 7-8ml phenylhydrazine at 160-170°C for 1hr. Cool and collect N, N'-diphenylcarbazide, recrystallise from aqueous alcohol, m.p 175°C.

m.p 83°C **Benzoin acetate**

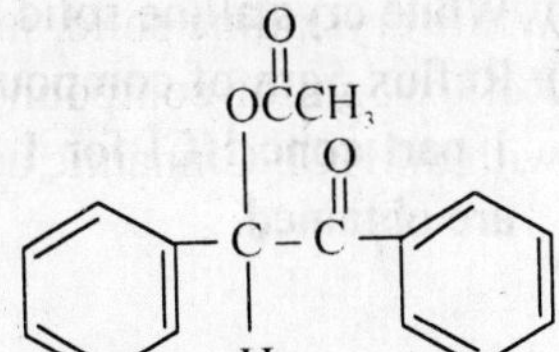

(1) Reflux 5gm of compound with dil NaOH solution. Cool, acidify with dil HCl. Acetic acid and benzoin, m.p 137°C, will be precipitated.

m.p 95°C **2-Naphthyl Salicylate** (Betol, Naphthol Salol)

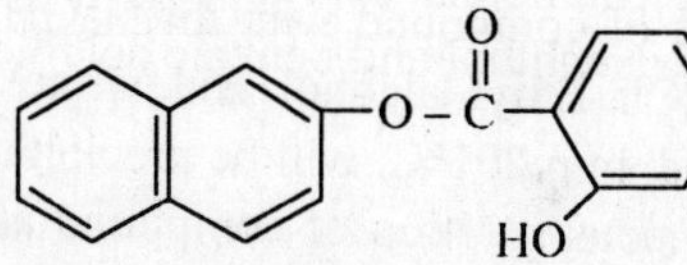

(1) Crystalline compound, insoluble in water, soluble in alcohol.
(2) To few milligrams of compound add few drops of neutral $FeCl_3$ solution. A violet coloured solution is obtained.
(3) Reflux 500mg of compound with 7-8ml acetic anhydride and 400mg of fused sodium acetate for 3-4 hrs. Cool, pour in ice with stirring and collect 2-Naphthyl o-acetoxy benzoate, m.p 136°C.
(4) Reflux 5gm of compound with dil NaOH solution. Cool, acidify with dil HCl. 2-Naphthol and salicylic acid will be precipitated.

m.p 98-100°C **Hydroquinone triacetate** (1, 2, 4 triacetoxy benzene)

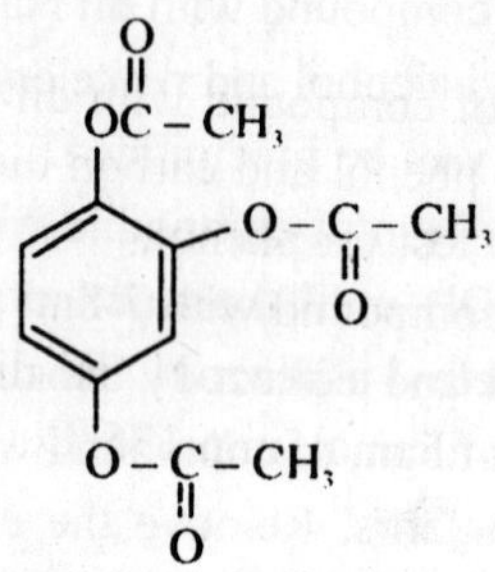

(1) White crystalline solid.
(2) Reflux 5gm of compound with a mixture of 2 parts methanol and 1 part conc HCl for 1 hr. Hydroxyhydroquinone and acetic acid are obtained.

m.p 107°C **2-Naphthyl benzoate**

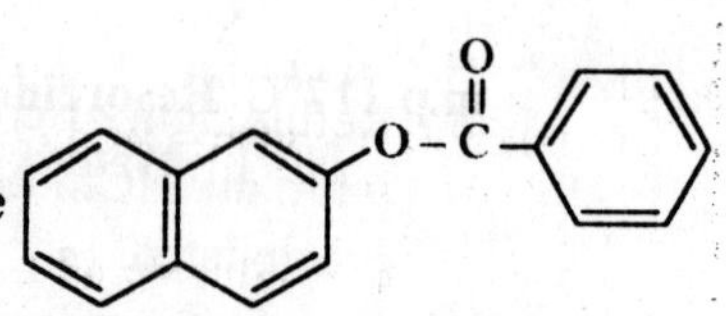

(1) Crystalline compound, insoluble in water, slightly soluble in ether, soluble in hot alcohol.

(2) Reflux 5gm of compound with dil NaOH solution. Cool, acidify with dil HCl. 2-Naphthol and benzoic acid will be precipitated.

m.p 116°C Ethyl p-hydroxy benzoate

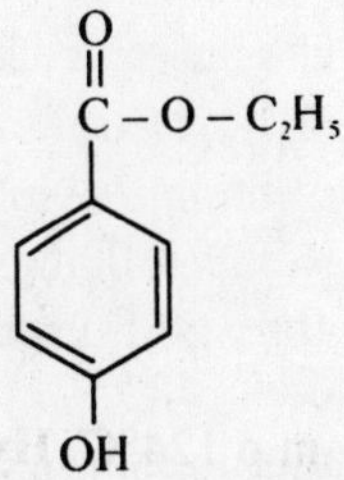

(1) Crystalline solid, sparingly soluble in water, easily soluble in alcohol and ether.
(2) To few milligrams of compound add few drops of neutral $FeCl_3$ solution. A violet coloured solution is obtained
(3) Reflux 500mg of compound with dil NaOH solution. Cool, acidify with dil HCl. Ethyl alcohol and white crystals of p-hydroxy benzoic acid, m.p 214°C, will be precipitated.
(4) Shake thoroughly 1gm of compound with 7-8ml benzoyl chloride and 10ml dil NaOH until the smell of benzoyl chloride persists. Cool and collect ethyl p-benzoxy benzoate, m.p 94°C.
(5) Warm a mixture of 5gm of compound with 6.6 ml sulfuryl chloride on steam bath for 2hrs. Remove the excess sulfuryl chloride on steam bath. Recrystallise the residue from ethanol and recover 3, 5 dichloro-4 hydroxy benzoate hydrate, m.p 110-114°C.

m.p 117°C Resorcinol dibenzoate

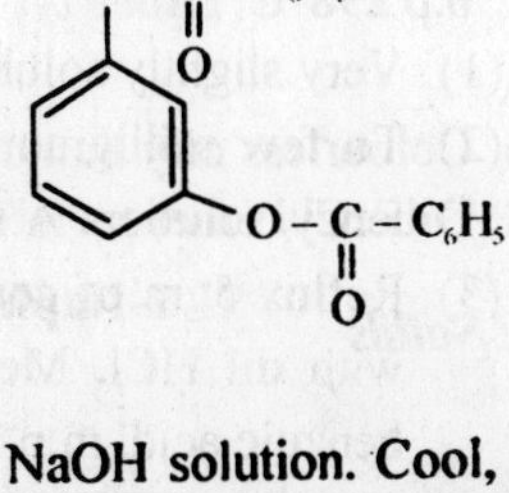

(1) Reflux 5gm of compound with dil NaOH solution. Cool, acidify with dil HCl. Resorcinol and benzoic acid will be precipitated.

m.p 121°C **Diphenyl Succinate** $(CH_2COOC_6H_5)_2$

(1) Reflux 5gm of compound with dil NaOH solution. Cool, acidify with dil HCl. Phenol and succinic acid will be precipitated.

m.p 124°C **Hydroquinone diacetate** 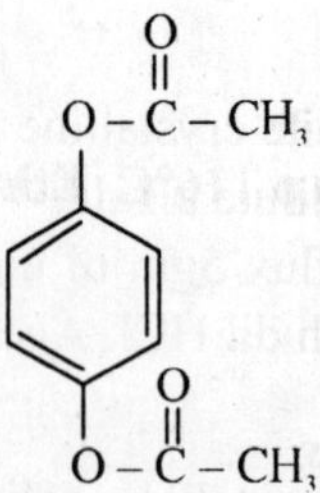

(1) Crystalline compound, slightly soluble in hot water and alcohol, soluble in ether.
(2) Reflux 5gm of compound with dil NaOH solution and cool. Hydroquinone and acetic acid will be obtained after acidification.

m.p 126-128°C **Methyl p-hydroxy benzoate** 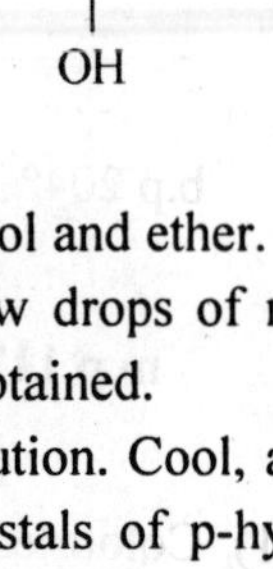

(1) Very slightly soluble in water, soluble in alcohol and ether.
(2) To few milligrams of compound add add few drops of neutral $FeCl_3$ solution. A violet coloured solution is obtained.
(3) Reflux 5gm of compound with dil NaOH solution. Cool, acidify with dil HCl. Methyl alcohol and white crystals of p-hydroxy benzoic acid, m.p 214°C will be precipitated.
(4) Heat 1gm of compound with 7ml acetic anhydride on water bath. Cool and collect Methyl p-acetoxy benzoate, m.p 85°C.

m.p 165°C **Pyrogallol triacetate**

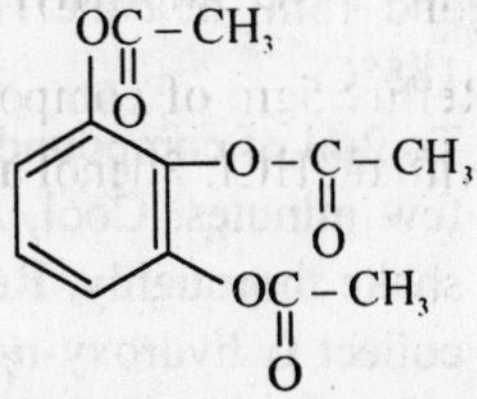

(1) White crystalline powder, very slightly soluble in alcohol, soluble in dilute alkali.

(2) Reflux 5gm of compound with dil NaOH solution. Cool, acidify with dil HCl. Acetic acid and pyrogallol will be obtained.

(viii) Lactones

Cyclic ethers are also known as lactones. They have properties which are quite similar to other esters, but the smaller lactones (strained) are more reactive than the larger lactones or their open chain counterpart.

Identification test

(*i*) To few milligrams of compound add 1-2 ml dil NaOH solution and 1 drop of phenolphthalein indicator. The solution turns pink. Heat the solution to boiling. Disappearnce of pink colour indicates the presence of lactone or esters.

Distinction between esters and lactones

Boil the given compound with alkali solution, lactones dissolves and on acidification the orignal (lactone) compound is obtained. However esters under similar reaction conditions give carboxylic acid and not the orignal compound.

Liquids

b.p 204°C υ **Butyrolactone** (4-hydroxy butanoic acid lactone)

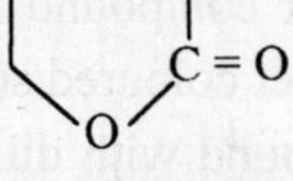

(1) Colourless oily liquid with characteristic odour, miscible with water, soluble in alcohol and ether.

(2) Reduces ammonical silver nitrate solution.

(3) Reflux 5ml of compound with 7-8gm of potassium dichromate

and 15ml of conc H_2SO_4. Cool and recover succinic acid, m.p 185°C.

(4) To 2ml of compound add 5ml phenylhydrazine, heat at 100°C for few minutes. Cool, add 10ml ether to the reaction mixture and shake thoroughly. Remove the organic layer, distill off ether and collect υ hydroxy-n-butyric phenylhydrazide, recrtystallise from chloroform, m.p 94°C.

(5) Heat 4.3gm of compound with 3.6gm of KCN in an oil bath at 190-200°C for 2hrs. Cool and dissolve the potassium salt of cyano acid in 10ml hot water. Acidify with conc HCl, cool and collect glutaric acid monoamide, m.p 174°C.

(6) Heat 5.2 gm of compound with 50ml of benzene and 30gm of $AlCl_3$ for 16hrs on steam bath. Cool, pour in crushed ice containing 25ml conc HCl. Remove the organic layer and extract the aqueous layer with toluene. Combine the two organic layer and distill. Collect the α tetralone as red-purple viscous oil.

b.p 207°C **δ Valerolactone (4-hydroxy pentanoic acid lactone)**

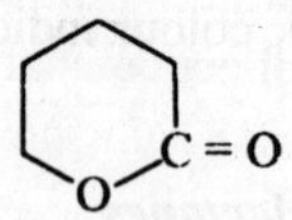

(1) Colourless liquid miscible with water and alcohol.

(2) Reflux 2ml of compound with 5ml phenylhydrazine at 100°C for 8hrs. Cool, collect υ hydroxy-n-valeric phenylhydrazide, recrtystallise from chloroform, m.p 79°C.

(3) Heat 2ml of compound with 3-5ml dil HNO_3 until the brown fumes subsides. Cool and collect succinic acid, m.p 185°C.

Solid

m.p 59°C Chromone (1, 4 benzopyrone)

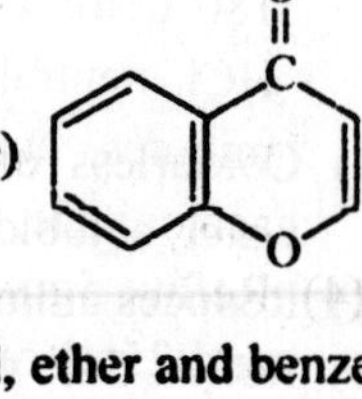

(1) Crystalline compound, soluble in alcohol, ether and benzene.

(2) To few milligrams of compound add conc H_2SO_4. A yellow coloured solution is obtained which exhibits blue-violet fluorescence.

(3) Chromonehydrazone m.p 96°C.

m.p 67°C **Coumarin** (1, 2 benzopyrone) 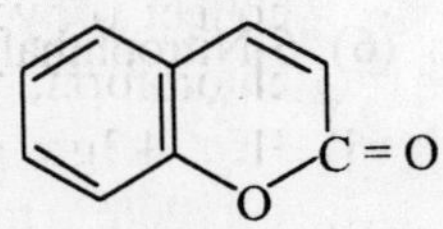

(1) Crystalline compound with fragrant odour like that of sweet grass, insoluble in water, soluble in alcohol, ether and chloroform. Sublimes.

(2) Reduces $KMnO_4$ solution.

(3) Dissolve few milligrams of compound in hot water, cool to room temperature, add few drops of I_2+KI solution and shake thoroughly. A clear brown supernatant liquid is obtained after few seconds.

(4) To 150mg of compound in 1ml $CHCl_3$ add a solution of .5ml Br_2 in 1ml $CHCl_3$ and shake thoroughly for few minutes. Wash the $CHCl_3$ layer with water. Remove the $CHCl_3$ layer, dry in a porcelain dish and collect coumarin dibromide, m.p 105°C.

(5) 6-Nitrocoumarin m.p 183°C (fuming HNO_3)

m.p 73°C **Phthalide** 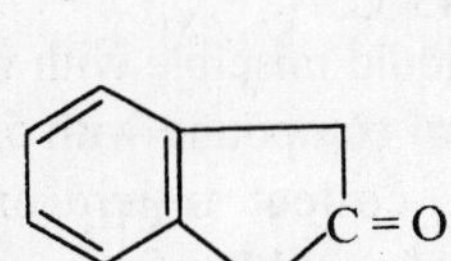

(1) White solid, insoluble in water, soluble in alcohol.

(2) Reflux 1gm of compound with 10ml dil NaOH solution. Cool, acidify with dil HCl and collect o-hydroxy methyl benzoic acid, m.p 120°C

(3) To 1gm of compound add 1gm KCN and heat the mixture at 180°C for 4-5hrs. Cool, add 5ml water and then add .5ml of conc HCl, neutralise with sodium carbonate. Stir the mixture for few minutes, acidify with conc HCl and collect o-carboxy phenyl acetonitrile, m.p 115°C.

(4) Reflux 5gm of compound with 7.5gm of potassium phthalimide and 25ml of DMF for 5hrs. Cool, add a solution of 15ml glacial

acetic acid in 25ml water. Stir for 30min and filter. Dissolve the solid product in 20ml ethanol and boil for 10min, filter hot. On cooling α-phthalimido-o-toluic acid is precipitated, m.p 265°C.

(5) Heat 6.7gm of compound with 2.8ml of bromine on oil bath (temp 130-140°C)in CO_2 gas atmosphere for 10hrs. Cool and collect 2-Bromophthalide, m.p 73°C.

(6) 6-Nitrophthalide m.p 143°C.

m.p 86-87°C **Glycollide** (2, 5-p-dioxanedione) 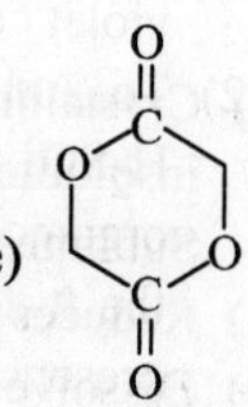

(1) Crystalline compound, insoluble in water, soluble in acetone and hot alcohol.

(2) Heat 2.9 gm of compound with 5ml aniline for 4hrs at 100°C. Cool, collect the precipitated glycolanilide, recrystallise from water, m.p 97°C.

(3) Heat 3gm of compound with 3gm of p-toluidine for 2-3 hrs at 100°C. Cool and collect glycolic-p-toluidide, recrystallise from water, m.p 143°C.

m.p 124°C **Lactide** 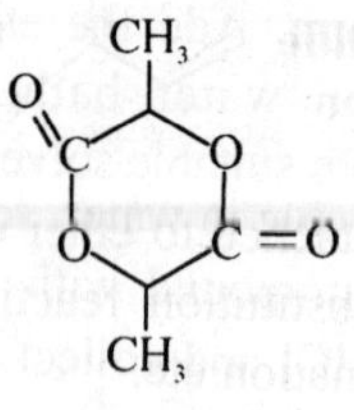

(1) Crystalline compound, insoluble in water, soluble in acetone and benzene.

(2) Lactamide m.p 74°C (3) Lactanilide m.p 58°C.

(ix) Ethers

Ethers are regarded as derivatives of water in which the oxygen atom is attached to two groups. Either or both the groups may be aliphatic or aromatic.

R-O-R

Ethers are colourless, volatile inflammable liquid with a pleasant odour. They are generally distinguished from hydrocarbon due to their solubility in concentrated sulphuric acid.

Identification Tests

(1) Dissolve 1ml of compound in 5ml benzene and add 5ml dilute solution of iodine in benzene. Ethers form charge transfer complexes with iodine to give a brown solution. Hydrocarbon and its halogen derivatives however react with iodine to give violet colouration.

(2) To 2ml of compound add 6ml glacial acetic acid and 2ml conc H_2SO_4. Reflux the mixture for 5-10minutes. Cool and extract the organic compound with ether. Distill off the ether and perform the test for esters with the residue. A positive test indicates the presence of ethers.

(3) To 1ml of compound in a test tube add 5ml of glacial acetic acid and 5ml of 57% hydroiodic acid. Cover the mouth of test tube with cotton and then place a piece of filter paper moistened with mercuric nitrate over the cotton plug. Heat the reaction mixture. The volatile alkyl iodide reacts with mercuric nitrate to produce a light orange or vermilion colour due to the formation of mercuric iodide.

$$R—O—R' + 2HI \rightarrow R'I + RI + H_2O$$

Derivatives

(1) Picrate:-Dissolve picric acid (500mg) in minimum amount of hot alcohol/chloroform. Add the ether to the hot solution. Heat for few minutes on water bath, cool and collect the picrate. Recrystallise from suitable solvent and determine its melting point.

(2) Benzene ring attached to ether oxygen is highly reactive towards electrophilic substitution reaction and respond to reaction like nitration, bromination etc.

(3) **3, 5-dinitrobenzoate:** Reflux 5ml of ether with 500mg of anhydrous zinc chloride and 2.5gm of 3, 5 dinitrobenzoyl chloride for 1 hr. Treat the reaction product with 20ml sodium carbonate solution, heat on water bath for 2-3 min. Cool and collect the precipitated ester, recrystallise from chloroform or carbon tetrachloride.

(4) Ethers containing alcoholic groups respond to all the reaction of alcohols and derivatives like benzoate, acetate can also be prepared.

Aliphatic Liquids

b.p 34°C **Diethyl ether** (ethoxy ethane) $CH_3CH_2OCH_2CH_3$

(1) Colourless liquid with characteristic odour, insoluble in water, miscible with alcohol and chloroform.
(2) Ethyl 3, 5-dinitrobenzoate m.p 93°C

b.p 39°C **Divinyl ether** (ethynyloxy ethane) $CH_2{=}CHOCH{=}CH_2$

(1) Colourless liquid, insoluble in water, miscible with alcohol and ether.
(2) To 1ml of compound add conc H_2SO_4. A black tarry resinous product is obtained.
(3) To 1ml of compound add conc HCl. A yellow coloured solution is obtained.
(4) Reduces aqueous $KMnO_4$ solution.
(5) Iodoform m.p 119°C.

b.p 56°C **t-Butyl Methyl ether** $(CH_3)_3COCH_3$

(1) Liquid with camphor like odour, sparingly soluble in water, soluble in alcohol and ether.

b.p 65°C **Tetrahydrofuran** (tetramethylene oxide) 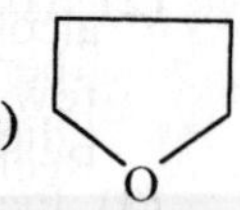

(1) Colourless liquid with characteristic odour, soluble in water and alcohol.
(2) Heat a mixture of 10ml THF, 1.75gm of red phosphorous and 3ml water. Add 6.5ml of bromine and then heat the mixture at 100-150°C for 60min. Cool, dilute with water, add 50ml ether and filter. Separate the ethereal layer, ditill off ether and collect 1, 4-dibromobutane at 83-84°C/12mmHg. (b.p 198°C)
(3) Cool 8ml of THF and add with stirring 3ml of phosphorous oxychloride. Cool and add 1ml of conc H_2SO_4. Heat to 80°C for 15 min, dilute with water and reflux for another 30min. Extract

with ether, distill off ether and collect 4, 4' dichlorodibutyl ether at 116-118°C/10mmHg.

(4) Reflux 3.6gm of THF with 33gm KI, 14ml of 85% orthophosphoric acid and 6.5 gm phosphoric anhydride for 3hrs. Cool, add 15ml water and 25ml ether. Remove the ether layer, distill and collect 1, 4 diiodobutane at 108-110°C/10mmHg.

b.p 68-69°C **t-Butyl ethyl ether** (2-ethoxy-2-methyl propane) $(CH_3)_3COC_2H_5$

(1) Insoluble in water, soluble in alcohol and ether.

b.p 90°C **Di n-propyl ether** (1-propoxy propane) $C_3H_7-O-C_3H_7$

(1) Colourless liquid, slightly soluble in water, miscible with alcohol and ether.

(2) n-propyl 3, 5-dinitrobenzoate m.p 74°C.

b.p 102°C **p-Dioxan** (diethylene dioxide) 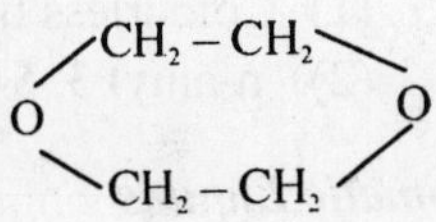

(1) Colourless liquid, miscible with water and most organic solvents.

(2) To 1ml of compound in benzene add 1ml of iodine solution in benzene and shake thoroughly. Cool, collect red violet coloured dioxane-iodine complex, m.p 84-85°C

(3) Picrate (pale yellow) m.p 66°C

b.p 124°C **Ethylene glycol monomethyl ether** (2-methoxy ethanol) $CH_3OCH_2CH_2OH$

(1) Colourless liquid, miscible with water, ether and benzene

(2) 2-methoxy ethyl p-nitrobenzoate m.p 50°C.

b.p 135°C **Ethylene glycol monoethyl ether** (2-ethoxy ethanol) $CH_3CH_2OCH_2CH_2OH$

(1) Colourless liquid miscible with water, alcohol and ether
(2) 2-ethoxyethyl 3, 5-dinitrobenzoate m.p 75°C

b.p 142°C **Di-n-butyl ether** (1-butoxy butane) $C_4H_9OC_4H_9$

(1) Colourless liquid, slightly soluble in water, soluble in organic solvents.
(2) n-butyl 3, 5-dinitrobenzoate m.p 63°C

b.p 172°C **Di isoamyl ether** [3-methyl-1-(υ-methyl butoxy) butane] $(CH_3)_2CHCH_2CH_2O\ CH_2CH_2CH(CH_3)_2$

(1) Colourless liquid, insoluble in water, soluble in alcohol and ether.
(2) Isoamyl 3, 5-dinitrobenzoate m.p 61°C

b.p 190°C **Di-n-amyl ether** (pentoxy pentane)
$CH_3(CH_2)_4\text{-O-}(CH_2)_4CH_3$

(1) Colourless liquid, insoluble in water, soluble in alcohol and ether.
(2) n-amyl 3, 5-dinitrobenzoate m.p 43°C.

Aromatic Liquids

b.p 31°C **Furan** 

(1) Colourless liquid, insoluble in water, soluble in alcohol and ether.
(2) To a solution of 5gm of lead tetraacetate in 10ml furan add dropwise a solution of 700mg of 1-aminobenzotriazole in 20ml furan. Stir for 5 min, filter and evaporate the filtrate to dryness. Dissolve the residue in ether and filter. Evaporate the filtrate to dryness. Extract with pet ether. Cool and collect 1, 4 dihydro-endo-1, 4-oxanaphthalene m.p 54°C.

b.p 154°C **Anisole** (methyl phenyl ether, methoxy benzene)

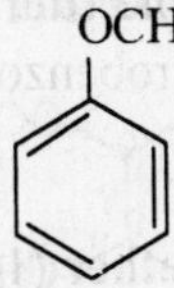

(1) Pleasant smelling liquid, insoluble in water, soluble in alcohol and ether.

(2) To 2ml of compound add 1ml conc H_2SO_4. Add dropwise 2ml of nitrating mixture (1ml fuming HNO_3 and 1ml conc H_2SO_4). After addition is complete keep at room temperature for sometime with occasional shaking and then pour in ice water. Collect the precipitated 2, 4 dinitroanisole, recrystallise from hot water, m.p 87°C.

(3) To 1ml of compound in 1ml $CHCl_3$ add saturated picric acid solution in $CHCl_3$. Heat on water bath, cool and collect bright yellow crystals of anisole picrate, m.p 81°C.

(4) To 5gm of anisole add 6gm of chloroacetyl chloride, 20ml CS_2 and 15gm $AlCl_3$. Reflux the mixture on steam bath. Distill off the solvent and then heat the residue on water bath for 4hrs. Cool, add 20gm crushed ice and conc HCl. Collect the precipitated ω chloro-4-hydroxyacetophenone, m.p 148°C.

(5) Reflux 2ml of compound with 1.6gm p-nitrobenzoyl chloride, 2ml CS_2 and 200mg of $AlCl_3$ for 30min. Cool, add 5ml water to the reaction mixture. Extract with ether. Evaporate off the ether and collect 4'-Nitro-4-methoxybenzophenone, recrystallise from alcohol, m.p 120°C.

(6) 2, 4-dibromoanisole m.p 61°C.

b.p 170°C **Benzyl methyl ether** (α-methoxy toluene)

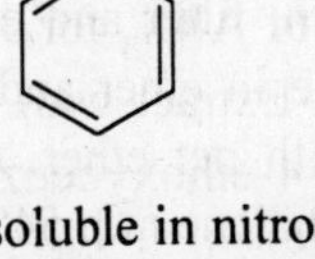

(1) Insoluble in water, soluble in nitromethane, alcohol and ether.

(2) To 1ml of compound in 1ml $CHCl_3$ add saturated picric acid solution in $CHCl_3$. Heat on water bath, cool and collect cream coloured crystals of benzyl methyl ether picrate, m.p 116°C.

b.p 172°C **Methyl-o-tolyl ether** (o-cresyl methyl ether)

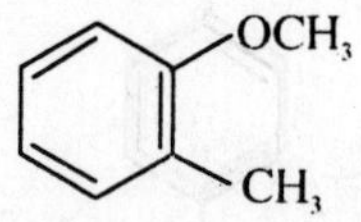

(1) Colourless liquid with odour similar to oil of winter green, insoluble in water, soluble in alcohol and ether.
(2) To 3ml of compound in 10ml alcohol add dropwise 1.5 ml bromine. After addition is complete continue stirring for 10 minutes. Cool and collect 5-bromo-2-methoxy toluene, recrystallise from alcohol, m.p 63°C
(3) To 1ml of compound in 1ml $CHCl_3$ add saturated picric acid solution in $CHCl_3$. Heat on water bath, cool and collect bright yellow crystals of methyl-o-tolyl ether picrate, m.p 113°C.
(4) Cool 2ml of compound (5-10°C) and add dropwise 4ml of fuming nitric acid. After addition is complete stir at room temperature and then pour in ice. Collect 3, 5-dinitro-2-methoxy toluene as yellow crystals, m.p 69°C.

b.p 172°C **Phenetole** (Ethyl Phenyl ether)

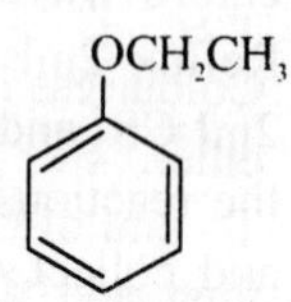

(1) Colourless liquid with aromatic odour, insoluble in water, soluble in alcohol and ether.
(2) To 1ml of compound in 1ml $CHCl_3$ add saturated picric acid solution in $CHCl_3$. Heat on water bath, cool and collect light yellow crystals of Phenetole picrate, m.p 92°C.
(3) Reflux 2ml of compound with 1.6gm p-nitrobenzoyl chloride, 2ml CS_2 and 200mg of $AlCl_3$ for 30min. Cool, add 5ml water to the reaction mixture. Extract with ether. Evaporate off the ether and collect 4'-Nitro-4-ethoxybenzophenone, recrystallise from alcohol, m.p 110°C.

b.p 174°C **Methyl-p-tolyl ether** (4-methoxy toluene)

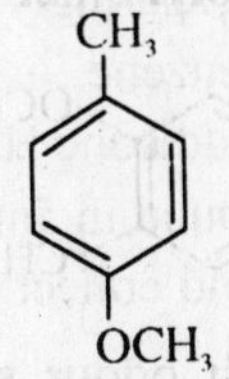

(1) Colourless liquid, soluble in nitromethane.

(2) To 2ml of compound add 1.75ml conc H_2SO_4, 20ml water and 1.75gm potassium dichromate. Reflux for 4hrs, cool and add 20ml water to it. Isolate three times with 20ml ether. Combine the ether extract, wash first with Na_2CO_3 and then with dil HCl. Distill off the ether, cool and collect p-anisic acid, m.p 184°C.

(3) Methyl p-tolyl ether picrate (yellow coloured) m.p 89°C.

b.p 206°C, m.p 22°C **Veratrole** (1, 2-dimethoxy benzene, pyrocatechol dimethyl ether)

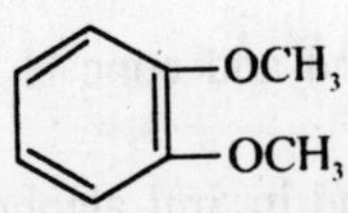

(1) Colourless liquid, slightly soluble in water, soluble in alcohol and ether.

(2) To 2ml of compound add 2ml glacial acetic acid, cool to 0°C and then add dropwise with stirring 1.5ml conc HNO_3. Continue stirring at 0°C for 2hrs and then at room temperature for 30min. Cool, collect 4-nitro-1, 2-dimethoxy benzene, m.p 95°C.

(3) Dissolve 2ml compound in 5ml alcohol and then add dropwise 5ml of bromine solution (1.5ml in 3.5ml alcohol). After addition is complete keep at room temperature for 2hrs, cool and collect 4, 5 dibromo veratrole, recrystallise from alcohol, m.p 92°C.

(4) Veratrole picrate (red coloured) m.p 57°C.

b.p 217°C **Resorcinol dimethyl ether** (1, 3-dimethoxy benzene)

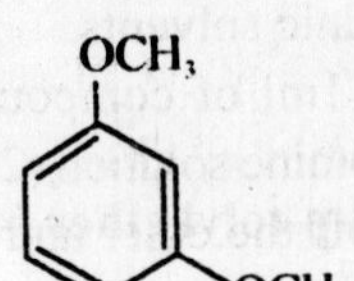

(1) Colourless oily liquid, sparingly soluble in water, easily soluble in alcohol, ether and benzene.
(2) To 1ml of compound add conc H_2SO_4. The solution turns yellow.
(3) Dissolve 2ml of compound in 5ml alcohol and then add dropwise 1.5ml bromine. Cool and collect 4, 6 dibromoresorcinol dimethyl ether, m.p140°C.
(4) 2, 4, 6 trinitro 1, 3 dimethoxy benzene m.p 124°C.
(5) Resorcinol dimethyl ether picrate (yellow coloured) m.p 58°C.

b.p 232-234°C **Safrole** (4-allyl-1, 2 methylene-dioxybenzene)

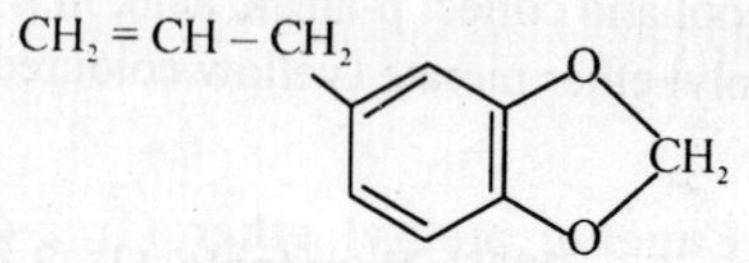

(1) Oily liquid with pleasant odour, insoluble in water, soluble in alcohol and ether.
(2) To 1ml of compound add conc H_2SO_4. An intense red coloured solution is obtained.
(3) To 2ml of compound in 5ml alcohol add dropwise 4ml bromine. After addition is complete heat on water bath for 15-20min. Cool, collect pentabromosafrole, recrystallise from benzene, m.p 169°C.
(4) Safrole picrate (red coloured) m.p 104°C.

b.p 235°C, m.p 22°C **Anethole** (p-propenylanisole)

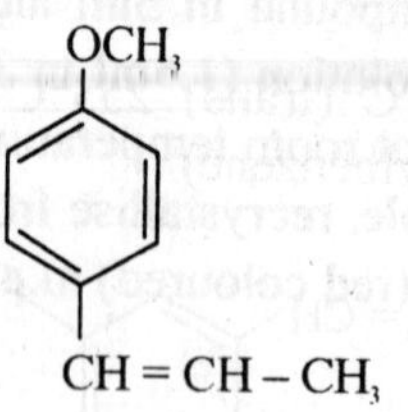

(1) Liquid with odour like oil of anise, insoluble in water, miscible with almost all organic solvents.
(2) Cool a solution of 1ml of compound in 2ml ether and then add dropwise 5ml of bromine solution (2ml in 3ml ether). After addition is complete distill off the ether and treat the residue with alcohol,

recrystallise from pet ether and collect 2 bromo anethole dibromide, m.p 108°C

(3) Anethole picrate (red coloured0 m.p 70°C

b.p 245°C **Ethylene glycol monophenyl ether**

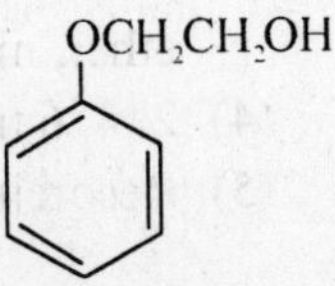

(1) Colourless liquid, sparingly soluble in water, easily soluble in alcohol and ether

(2) 2-Phenoxyethyl benzoate m.p 64°C

b.p 244°C **Eugenol methyl ether** (1, 2-dimethoxy –4-allyl benzene)

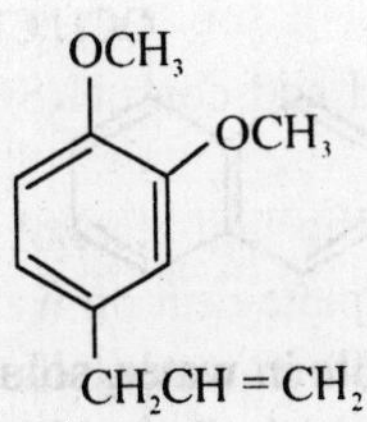

(1) Colourless to pale yellow Oily liquid, with odour similar to eugenol, insoluble in water, soluble in nitromethane.

(2) Eugenol methyl ether picrate (red brown coloured) m.p 114°C.

b.p (cis) 242-243°C (trans) 253°C **Isosafrole** (1, 2 methyl-enedioxy-4-propenylbenzene)

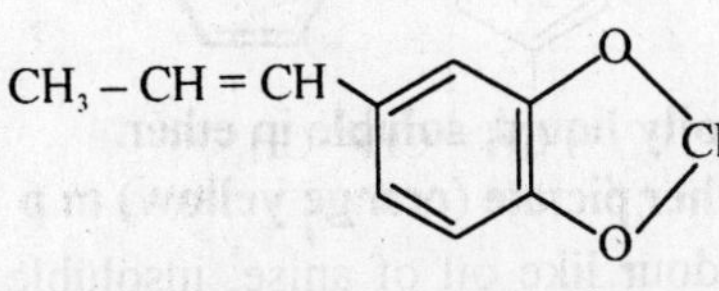

(1) Oily liquid, miscible with alcohol, ether and benzene.

(2) Dissolve 800mg of compound in 4ml carbon disulphide and then add dropwise 1.5ml bromine. Leave the mixture overnight, cool and collect Bromo isosafrole dibromide, m.p 109°C.

(3) Picrate (dark red) m.p 75°C.

b.p 269°C **1-Naphthyl methyl ether** (1-methoxy naphthalene)

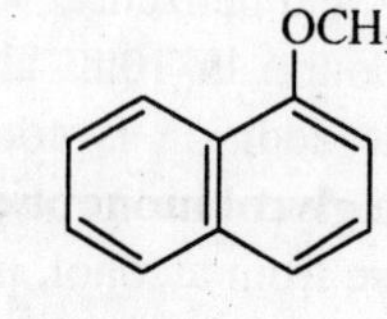

(1) Colourless oily liquid, insoluble in water, soluble in alcohol, ether, benzene and chloroform.
(2) Methyl 1-naphthyl ether picrate (yellow orange) m.p 130°C.
(3) 2, 4 dibromo-1-methoxynaphthalene m.p 54°C

b.p 280°C **Ethyl 1-naphthyl ether** (1-ethoxy naphthalene)

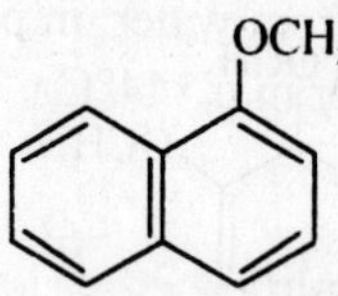

(1) Oily liquid, insoluble in water, soluble in alcohol and ether.
(2) To 2ml of compound in 5ml chloroform, cool and add dropwise 1ml of bromine. Keep the mixture overnight and collect 4 bromo-1-ethoxy naphthalene, m.p 48°C
(3) Ethyl 1-naphthyl ether picrate m.p 118°C.

b.p 298°C **Dibenzyl ether** 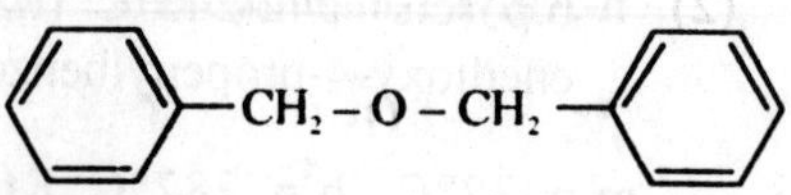

(1) Colourless oily liquid, soluble in ether.
(2) Dibenzyl ether picrate (orange yellow) m.p 78°C.

Aromatic Soilds

m.p 27°C, b.p 259°C **Diphenyl ether** (phenoxy benzene)

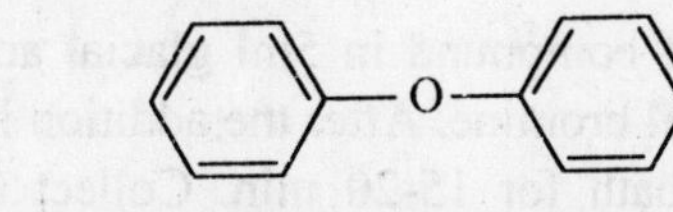

(1) Crystalline compound with geranium like odour, insoluble in water, soluble in alcohol, ether and benzene.

(2) Dissolve 2gm of compound in 10ml alcohol and add dropwise 2ml bromine. After the addition is complete keep the reaction mixture overnight. Collect the precipitated 4, 4'dibromodiphenylether, recrystallise from alcohol, m.p 55°C.

(3) Heat a mixture of 8.5gm of diphenylether, 1.3gm of sulphur and 2.6 gm of aluminium chloride on water bath at 100°C for 4hrs. Pour the molten mixture into ice cold solution of 5ml conc HCl in 50ml water. Extract with ether. Distill off the ether, cool the residual oil and collect Phenoxathin, m.p 56°C.

(4) Heat 13gm of diphenylether and add dropwise 4ml of β anilinocrotonate. Reflux the mixture for 30min, cool and add pet ether. Stir until a white solid is formed, collect 2-Methyl-4-quinolone, recrystallise from water, m.p 233-234°C.

(5) 4, 4' dinitrodiphenyl ether m.p 144°C.

m.p 30°C **n-Amyl 1-naphthyl ether**

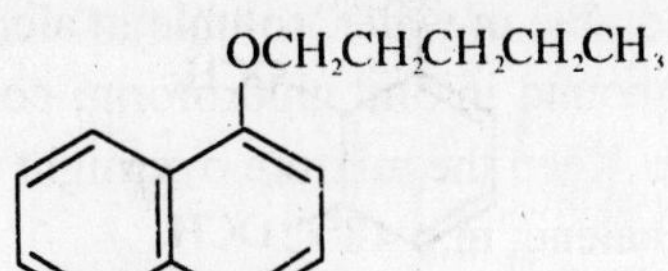

(1) Crystalline compound, insoluble in water.

(2) n-Amyl 1-naphthyl ether picrate m.p 75°C

m.p 37°C, b.p 282°C **Ethyl 2-naphthyl ether** (2-ethoxy naphthalene)

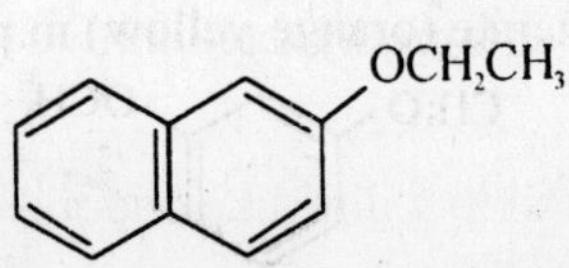

(1) Crystalline solid, insoluble in water, soluble in alcohol, ether, pet ether and toluene.

(2) Dissolve 2gm of compound in 5ml glacial acetic acid and then add dropwise 1ml bromine. After the addition is complete cool by keeping in ice bath for 15-20 min. Collect the precipitated 1-bromo-2-ethoxy naphthalene, recrystallise from pet ether, m.p 66°C.

(3) 1, 6, 8-trinitro 2-ethoxy naphthalene m.p 186°C.

(4) Ethyl 2-naphthyl ether picrate (orange-yellow)m.p 101°C

m.p 43°C **Catechol diethyl ether** (1, 2-diethoxy benzene)

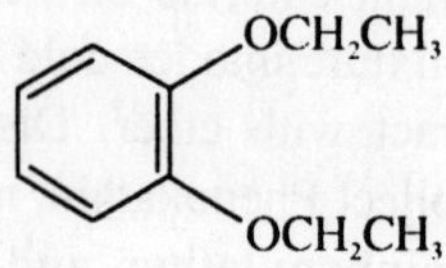

(1) Crystalline compound

(2) Catechol diethyl ether picrate (red brown) m.p 69°C

m.p 47°C, b.p 241°C **Pyrogallol trimethyl ether** (1, 2, 3-trimethoxy benzene)

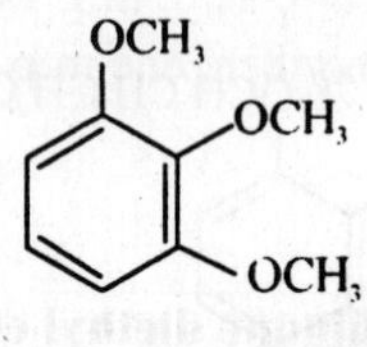

(1) Crystalline compound, easily soluble alcohol, ether and benzene.

(2) 4, 5, 6-tribromo-1, 2, 3-trimethoxy benzene m.p 74°C.

(3) Pyrogallol trimethyl ether picrate (yellow) m.p 78°C.

m.p 52°C, b.p 255°C **Phloroglucinol trimethyl ether** (1, 3, 5-trimethoxy benzene)

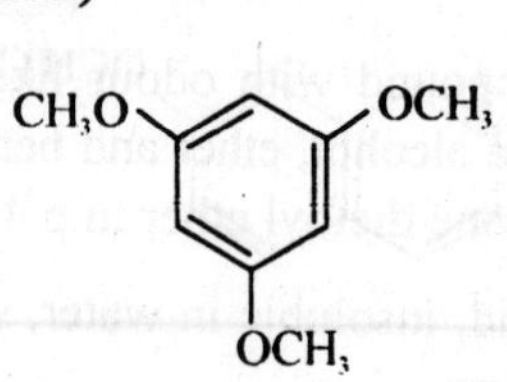

(1) Crystalline compound, insoluble in water, soluble in alcohol, ether and benzene. Sublimes.
(2) 2, 4, 6-tribromo –1, 3, 5-trimethoxy benzene m.p 145°C.

m.p 56°C, b.p 212°C **Hydroquinone dimethyl ether** (1, 4-dimethoxy benzene)

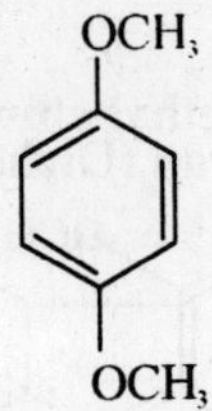

(1) Colourless crystalline solid, insoluble in water, soluble in alcohol and ether.
(2) To few milligram of compound add conc H_2SO_4. A yellow coloured solution is obtained.
(3) Dissolve 1.75 gm of compound in 5ml glacial acetic acid and add dropwise 6.5ml of bromine solution. (1.5ml in 5ml acetic acid). Cool, collect the precipitated dibromohydroquinone dimethylether, recrystallise from acetic acid, m.p 142°C.
(4) 2-nitro 1, 4-dimethoxy benzene m.p 72°C.
(5) Picrate m.p 47°C

m.p 72°C **Hydroquinone diethyl ether** (1, 4-diethoxy benzene)

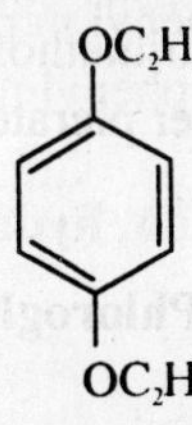

(1) Crystalline compound with odour like oil of anise, insoluble in water, soluble in alcohol, ether and benzene.
(2) Nitrohydroquinone diethyl ether m.p 49°C.

m.p 72°C **2-Naphthyl methyl ether** (2-methoxy naphthalene)

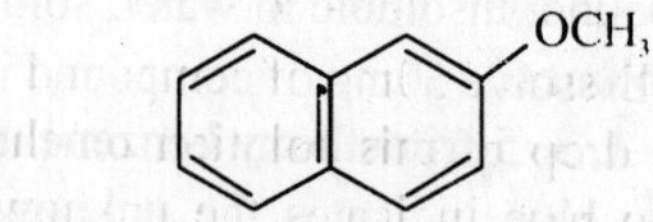

(1) Crystalline compound with odour of orange blossoms, soluble in ether, chloroform and benzene, sparingly soluble in alcohol.
(2) 1-nitro-2-methoxy naphthalene m.p 128°C.
(3) 2-Naphthyl methyl ether picrate (deep yellow) m.p 117°C.

m.p 87°C **Dibenzofuran** (Diphenylene oxide)

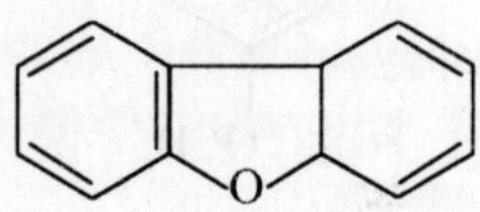

(1) White crystalline solid, insoluble in water, slightly soluble in alcohol, soluble in ether and benzene.
(2) To 500mg of compound in 5ml glacial acetic acid add 2ml of fuming nitric acid. Heat the mixture at 60-70°C. Cool and collect 3-nitrodibenzofuran, m.p 182°C.
(3) Picrate m.p 94°C

m.p 90°C **4-methoxy biphenyl** (p-phenyl anisole)

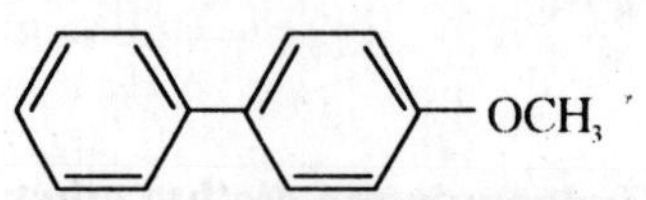

(1) Crystalline compound, soluble in hot alcohol.
(2) 3-nitro-4-methoxy biphenyl m.p 91°C.
(3) 4'-bromo-4-methoxy biphenyl m.p 144°C.

3. Compounds containing carbon, hydrogen and nitrogen (with or without oxygen)

(i) Amines

Amines are considered to be organic derivatives of ammonia in which one or more alkyl/aryl group are attached to nitrogen. They are classified according to the degree of substitution on nitrogen. If there is one alkyl/aryl group attached to nitrogen the amine is primary. If there are two groups the amine is secondary. When three alkyl or aryl groups are bonded to nitrogen the amine is tertiary.

Identification tests

(*i*) **Litmus Paper test :** Dissolve 50mg of compound in minimum amount of water or alcohol. Put a drop of this solution on the red litmus paper. Conversion of red litmus to blue indicates the unknown compound to be basic.

Primary Amines

(*i*) **Rimini's test (for primary alkyl amines) :** In a test tube place 2ml dilute solution of amine (.5ml in 2ml water) and add .5ml of acetone to it. Add 1-2drops of 1% freshly prepared sodium nitroprusside solution to it. Development of violet red colour indicate the presence of primary alkyl amine.

(*ii*) **Carbylamine or Isocyanide test :** Aliphatic as well as aromatic primary amines react with chloroform and alcoholic potash to form nauseatingly odoured isocyanides.

Dissolve 150mg of compound in 2ml of alcoholic potassium hydroxide solution. Add few drops of chloroform and heat the mixture to boiling. A foul characterisitc odour of isocyanide indicates primary amine.

$$RNH_2 + CHCl_3 + 3KOH \rightarrow RNC + 3KCl + 3H_2O$$

Mechanism

$$CHCl_3 + \overset{\ominus}{OH} \longrightarrow \overset{\ominus}{CCl_3} \xrightarrow{-\overset{\ominus}{Cl}} :CCl_2 \text{ (dichlorocarbene)}$$

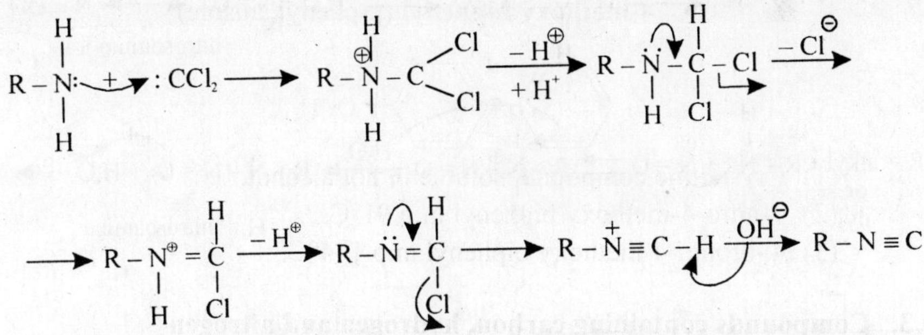

(*iii*) **Nitrous acid test :** Aliphatic and aromatic primary amines react with nitrous acid to give corresponding diazonium ions. The aliphatic diazonium compounds are quite unstable and they decompose even in cold conditions evolving nitrogen gas whereas the aromatic diazonium salts are quite stable and they react with phenols to form azo dyes.

Aliphatic primary amines : Dissolve 100mg of compound in 2-3ml 2N HCl. Cool in ice bath and add dropwise with stirring 1ml of ice cold 10% sodium nitrite solution. Warm the solution on water bath. Brisk effervescence due to evolution of nitrogen indicates the presence of aliphatic primary amine.

$$RNH_2 + HONO + HCl \rightarrow RN_2^+Cl^- + H_2O$$
$$\downarrow H_2O + heat$$
$$ROH + N_2 + HCl$$

Aromatic primary amines :-In a test tube dissolve 1gm of compound in 3-4ml conc HCl, dilute it with 6ml water. Cool in an ice bath. In another test tube dissolve 1.1 gm of sodium nitrite in 6ml water. Add this nitrite solution to the cold amine hydrochloride solution taking care that the temperature does not rise above 5°C. Now add this cold diazonium solution to the cold 2-naphthol solution (800mg in 8ml of 5% NaOH solution). A coloured dye (orange red) indicate the presence of aromatic amine.

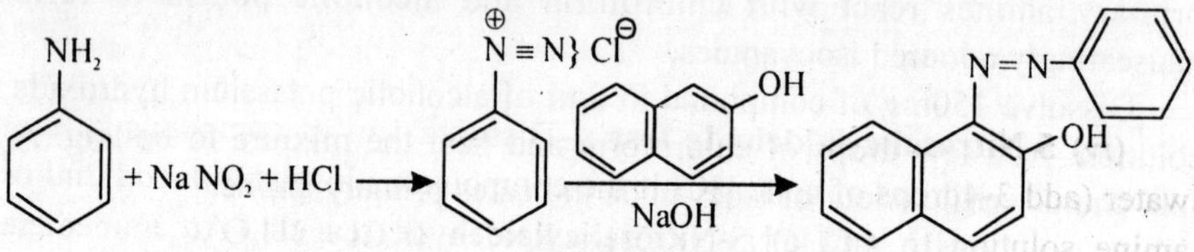

Mechanism

$$NaNO_2 + HCl \longrightarrow H-O-N=O + NaCl$$

$$H-O-N=O + H^{\oplus} \rightleftharpoons H-\overset{\oplus}{O}(H)-N=O \rightleftharpoons H_2O + [\overset{\oplus}{N}=\ddot{O}: \longleftrightarrow \overset{\oplus}{N}\equiv O:]$$

nitrosonium ion

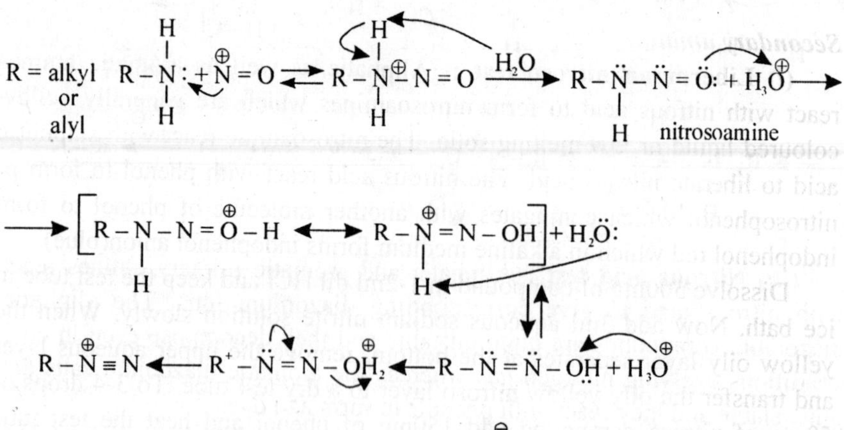

$$R = Alkyl \quad R-\overset{\oplus}{N}\equiv N \xrightarrow{heat} R^+ + N_2 \xrightarrow{OH^{\ominus}} R-OH$$

R = aryl

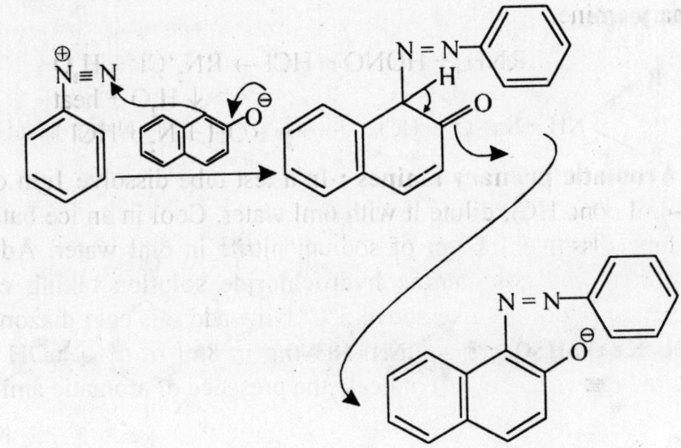

(*iv*) **5-Nitrosalicylaldehyde test :** -Dissolve 1ml of compound in 6ml water (add 3-4drops of conc HCl if the compound is insoluble). Add 2ml of amine solution to 4ml of 5-Nitrosalicylaldehyde reagent. An immediate precipitation of nickel derivative of schiff's base indicates the presence of primary amine (aromatic amines take 2-3minutes)

5 Nitrosalicylaldehyde reagent :-Add 1gm of 5-Nitrosalicylaldehyde to 30ml of triethanolamine and 50ml water. Shake thoroughly until the solid completely dissolves. Now add a solution of 1gm of nickel chloride dissolved in minimum amount of water. Make the whole solution to 200ml.

Secondary amine

(*i*) **Libermann nitroso test :-** Aliphatic as well as aromatic amines react with nitrous acid to form nitrosoamines which are generally yellow coloured liquid or low melting solid. The nitrosoamine react with sulphuric acid to liberate nitrous acid. The nitrous acid react with phenol to form p-nitrosophenol which conjugates with another molecule of phenol to form indophenol red which in alkaline medium forms indophenol anion(blue)

Dissolve 500mg of compound in 1-2ml dil HCl and keep the test tube in ice bath. Now add 3ml aqueous sodium nitrite solution slowly. When the yellow oily layer separates at the bottom, remove the upper aqueous layer and transfer the oily yellow nitroso layer to a dry test tube. To 3-4 drops or 50mg of nitroso compound add 150mg of phenol and heat the test tube gently on water bath for few seconds. Cool, add 2ml of conc H_2SO_4. An intense green or bluish green coloured solution results which on dilution with water changes to red. Add excess of NaOH solution, blue or green colour results.

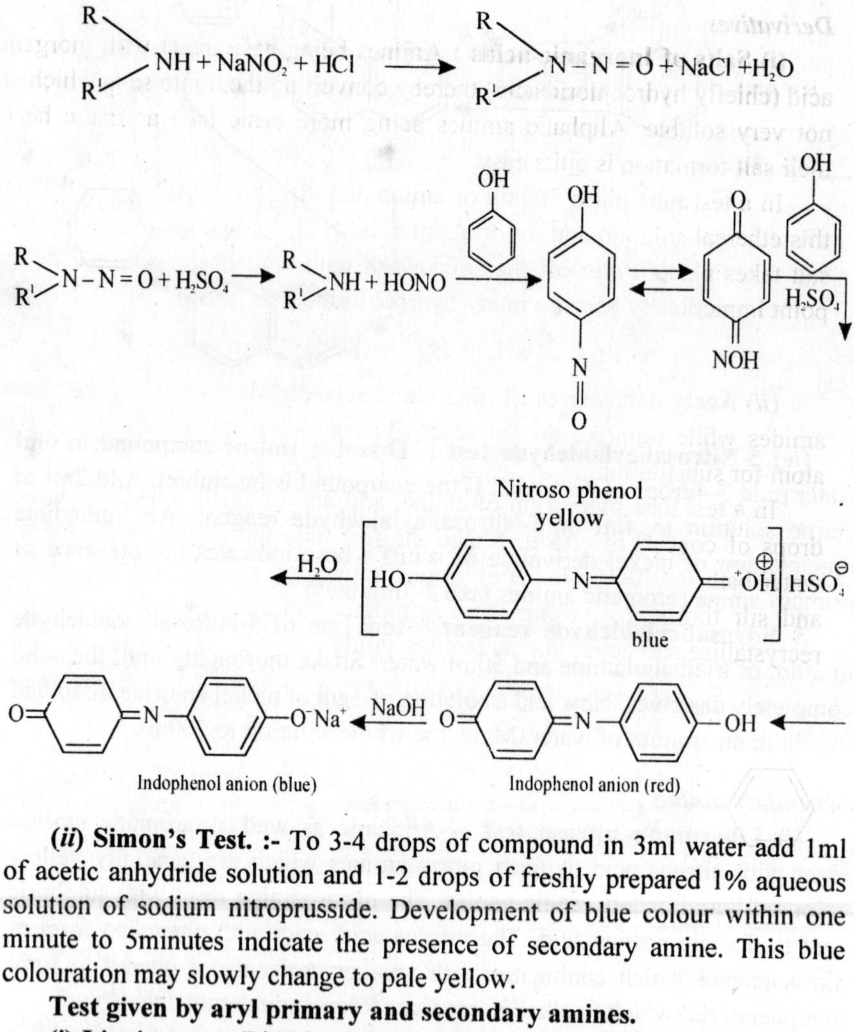

(ii) Simon's Test. :- To 3-4 drops of compound in 3ml water add 1ml of acetic anhydride solution and 1-2 drops of freshly prepared 1% aqueous solution of sodium nitroprusside. Development of blue colour within one minute to 5minutes indicate the presence of secondary amine. This blue colouration may slowly change to pale yellow.

Test given by aryl primary and secondary amines.

(i) Lignin test :- Dissolve 50mg of compound in 1-2ml ethanol. Take a small piece of newspaper and moisten a small part of it with this solution. Put 2-3 drops of 6N HCl on the moistened spot. Development of yellow or orange colour is a positive test for aromatic primary and secondary amines.

Tertiary amines

In a test tube place 1ml of t-amine or its alcoholic solution and add to it 1ml of citric acid-acetic anhydride solution (20gm citric acid dissolved in

100ml acetic anhydride). Heat the mixture on water bath. Development of red to purple colour within 1-2minutes indicates t-amine.

Derivatives

(i) Salts of Inorganic acids : Amines being basic react with inorganic acid (chiefly hydrochloric acid) thereby converting them into salts which are not very soluble. Aliphatic amines being more basic then aromatic hence their salt formation is quite easy.

In a test tube place 500mg of amine and dissolve it in 2ml of ether. To this ethereal solution add 3-4drops of conc HCl. Immediate precipitation of salt takes place. Filter off the solid wash with ether, determine its melting point immediately because many hydrochloride salts are hygroscopic.

$$RNH_2 + HCl \rightarrow RNH_3^+Cl^-$$

(*ii*) Acety derivatives : Primary and secondary amines yield substituted amides while t-amines do not react since they do not possess a hydrogen atom for substitution.

In a test tube place 1gm of amine, add 3ml of acetic anhydride and 1-2 drops of conc H_2SO_4 to it. Shake the test tube thoroughly and heat it on water bath for 10-15min. Cool, pour the reaction mixture in ice cold water and stir thoroughly. Filter the precipitated solid, wash with cold water, recrystallise and determine its melting point.

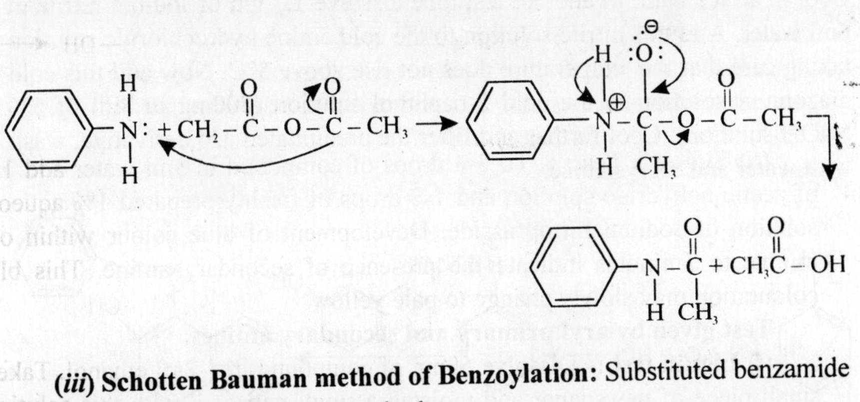

(*iii*) Schotten Bauman method of Benzoylation: Substituted benzamide derivatives are formed by this method.

In a 50ml conical flask place 1gm of amine and add 15-20ml of 10% NaOH solution to it. . Now add 3-4 ml benzoyl chloride. Stopper the flask and shake thoroughly until the smell of benzoyl chloride ceases. Filter the precipitated solid, wash with cold water, recrystallise with suitable solvent and determine its melting point.

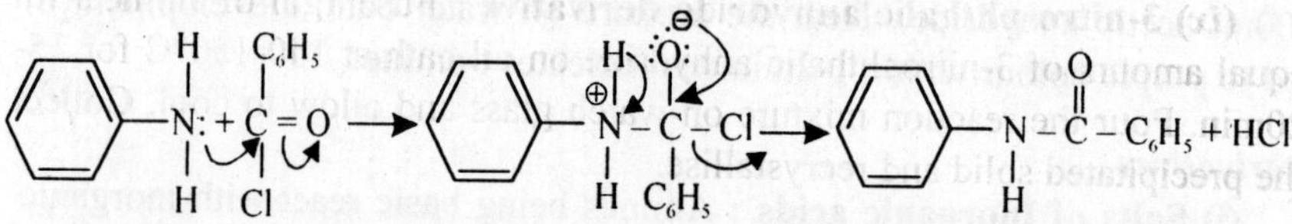

(iv) Picrates : Dissolve 1gm of amine in 7-8 ml of ethanol and add 10ml saturated picric acid solution to it. Heat the reaction mixture on water bath for 5 minutes. Cool, filter the precipitated solid, recrystallise from suitable solvent and determine its melting point.

(v) N-Substituted phthalimides (for primary amines) : Dissolve 1gm of primary amine and 1gm of phthalic anhydride in 10ml glacial acetic acid. Reflux the mixture for 25-30min. Cool, filter the precipitated solid, recrystallise from suitable solvent and determine its melting point.

(vi) Benzylidene derivatives or Schiff's bases (for primary aromatic amines) : Heat 1gm of amine with 4-5 ml freshly distilled benzaldehyde at 100°C for 10-15 min. Cool, filter the precipitated solid, recrystallise from suitable solvent and determine its melting point.

$$RNH_2 + O=\overset{\displaystyle H}{\overset{|}{C}}-C_6H_5 \longrightarrow R-N=\overset{\displaystyle H}{\overset{|}{C}}-C_6H_5 + H_2O$$

(vii) Dye formation (for primary aromatic amines) : In a test tube dissolve 1gm of compound in 3-4ml conc HCl, dilute it with 6ml water. Cool in an ice bath. In another test tube dissolve 1.1 gm of sodium nitrite in 6ml water. Add this nitrite solution to the cold amine hydrochloride solution taking care that the temperature does not rise above 5°C. Now add this cold diazonium solution to the cold 2-naphthol solution (800mg in 8ml of 5% NaOH solution). Cool further and filter the precipitated azo derivative, wash with water and recrystallise.

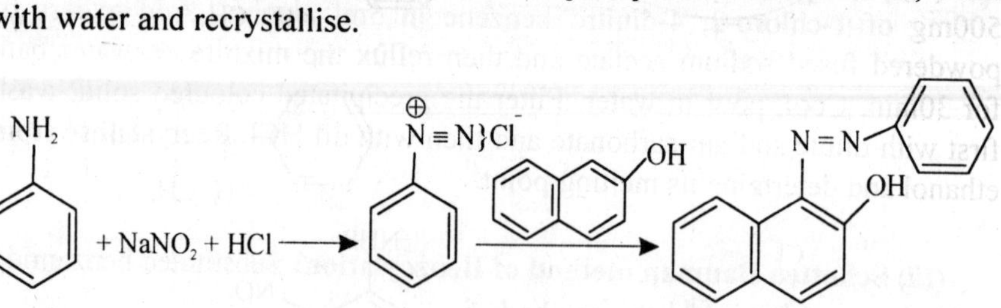

(viii) Formyl derivative : Reflux 1gm of amine with 10ml of 90% formic acid for 10-15 minutes. Add 5-10ml ice cold water and cool further. Filter the precipitated solid, wash with water and recrystallise from suitable solvent.

$$C_6H_5NHR + HCOOH \rightarrow C_6H_5N(CHO)R + H_2O$$

(ix) 3-nitro phthalic anhydride derivative : Fuse 1gm of amine with equal amount of 3-nitrophthalic anhydride on oil bath at 140-150°C for 15-20min. Pour the reaction mixture on watch glass and allow to cool. Collect the precipitated solid and recrystallise.

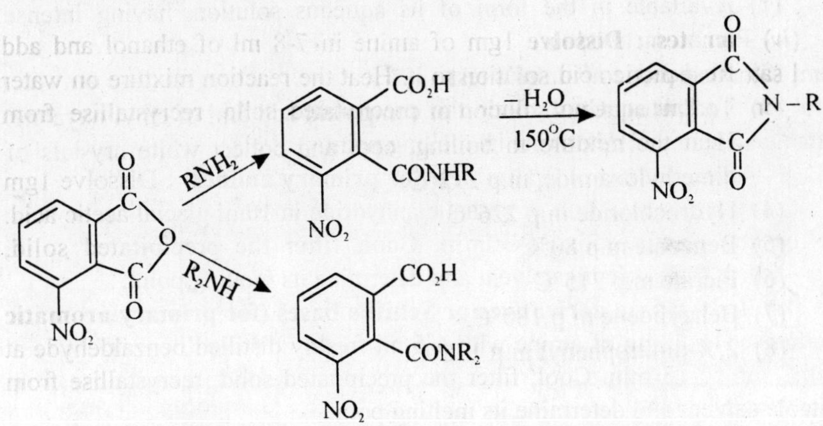

(x) Methiodides (for tertiary amines) : Add 1.5-2ml of methyl iodide to 1gm of compound. Shake slowly to ensure thorough mixing and then keep at room temperature for 5 minutes. If no precipitation occurs then warm the mixture under reflux on water bath for 5minutes, cool in ice bath and filter the precipitated solid, recrystallise from a suitable solvent and determine its melting point

$$RR^1R^2N + \quad CH_3I \rightarrow [RR^1R^2NCH_3]^+I^-$$

(xi) 2, 4-Dinitrophenyl derivative : Dissolve 500mg of amine and 500mg of 1-chloro-2, 4-dinitro benzene in 5ml alcohol. Add excess of powdered fused sodium acetate and then reflux the mixture on water bath for 30min. Cool, pour in water. Filter the precipitated coloured solid, wash first with dilute sodium carbonate and then with dil HCl. Recrystallise from ethanol and determine its melting point.

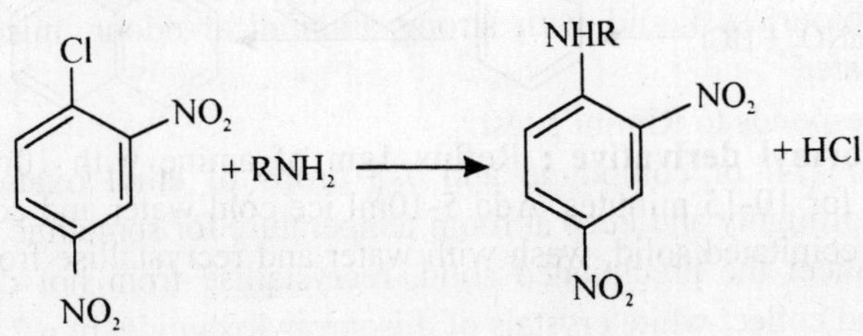

Aliphatic Liquids

b.p –6°C (Gas) **Methyl amine** (amino methane) CH_3NH_2

(1) Available in the form of its aqueous solution, having intense ammonical odour.
(2) Responds to Rimini's test.
(3) To 8ml aqueous solution of compound add 5ml of ethyl oxyalate. Heat the mixture to boiling, cool and collect white crystals of dimethyloxamide, m.p 212°C.
(4) Hydrochloride m.p 226°C
(5) Benzoate m.p 80°C
(6) Picrate m.p 215°C
(7) Benzylidene m.p 180°C.
(8) 2, 4-dinitrophenyl m.p 178°C

b.p 17°C **Ethylamine** (amino ethane) $CH_3CH_2NH_2$

(1) Colourless liquid with intense ammonical odour, miscible with water.
(2) Responds to Rimini's test.
(3) To 1ml of compound add 7-8 drops of ethyl oxalate. Heat is evolved and precipitation takes place. Filter, wash with water and collect diethyl oxamide, m.p 179°C
(4) Hydrochloride m.p 108°C
(5) Benzoate m.p 69°C.
(6) Picrate (yellow) m.p 165°C
(7) 2, 4-dinitrophenyl m.p 114°C

b.p 32°C **Isopropyl amine** $(CH_3)_2CHNH_2$

(1) Colourless liquid with strong ammonical odour, miscible with water.
(2) Responds to Rimini's test.
(3) To 1ml of compound add 7-8 drops of ethyl oxalate. Shake thoroughly and keep at room temperature for sometime. Cool and collect the precipitated solid, recrystallise from hot dil alcohol and collect white crystals of diisopropyloxamide, m.p 213-214°C.
(4) Benzoate 100°C.

(5) Picrate 150°C.
(6) 2, 4-dinitrophenyl m.p 94°C

b.p 44°C **t-Butylamine** (trimethyl carbinyl amine) $(CH_3)_3CNH_2$

(1) Colourless liquid, soluble in water, alcohol and ether.
(2) Add 10gm of compound to a stirred solution of 65gm of $KMnO_4$ in 300ml water. Stir for 8hrs (temp 50- 60°C). Steam distill and collect t-Nitrobutane at b.p 127-128°C which on cooling soon solidifies, m.p 25-26°C.
(3) Hydrochloride m.p 270°C.

b.p 49°C **Propylamine** $CH_3CH_2CH_2NH_2$

(1) Colourless liquid with characteristic ammonical odour, soluble in water.
(2) Responds to Remini's test.
(3) Add slight excess of alcoholic solution of propylamine to alcoholic solution of 2, 4-dinitrobromobenzene Heat the mixture on water bath. Cool and add cold water. Collect yellow crystals of 2, 4 dinitropropyl aniline, m.p 95°C
(4) Hydrochloride m.p 157°C.
(5) Benzoate 85°C
(6) Picrate m.p 135°C
(7) Benzylidene b.p209°C.
(8) 2, 4-dinitrophenyl m.p 96°C

b.p 53°C **Allyl amine** (2-propenyl amine) $CH_2{=}CHCH_2NH_2$

(1) Colourless liquid with irritating odour, miscible with water.
(2) Responds to Remini's test.
(3) To 1ml of compound add 1ml ethyl oxalate. Shake thoroughly and keep at room temperature for sometime. Cool, filter and dissolve the precipitae in dil alcohol. Heat on water bath and cool. Collect white crystals of diallyloxamide, m.p 154°C.
(4) Picrate (lemon –yellow) m.p 140°C.
(5) 2, 4-dinitrophenyl m.p 76°C

b.p 56°C **Vinyl amine** (ethenyl amine) $CH_2{=}CHNH_2$

(1) Unsaturated liquid with ammonical odour, miscible with water
(2) Picrate m.p 142°C.

b.p 63°C **s-Butyl amine** (2-amino butane) $CH_3CH(NH_2)CH_2CH_3$

(1) Colourless liquid, soluble in water, alcohol and ether.
(2) Responds to Remini's test.
(3) Benzoate m.p 76°C. (4) Picrate m.p 140°C.

b.p 68°C **Isobutylamine** (1-amino-2-methyl propane) $(CH_3)_2CHCH_2NH_2$

(1) Colourless liquid with characteristic ammonical odour, miscible with water.
(2) Respond's to Remini's test.
(3) To 1ml of compound add 1.2ml ethyl oxalate. Shake thoroughly and keep at room temperature for sometime. Cool, filter and dissolve the precipitate in dil alcohol. Heat on water bath and cool. Collect crystals of diisobutyloxamide, m.p 169°C.
(4) Hydrochloride m.p 160°C
(5) Benzoate m.p 57°C.
(6) Picrate m.p 151°C.
(7) 2, 4-dinitrophenyl m.p 80°C.

b.p 78°C **n-Butyl amine** (1-amino butane) $CH_3(CH_2)_3NH_2$

(1) Liquid with unpleasant odour, soluble in water.
(2) Respond to Remini's test.
(3) Hydrochloride 195°C. (4) Picrate m.p 151°C. (5) 2, 4-dinitrophenyl m.p 58°C (6) Benzoate m.p 42°C

b.p 95°C **Isoamyl amine** (Isopentyl amine, 1-amino-3-methyl butane) $(CH_3)_2CHCH_2CH_2NH_2$

(1) Liquid with ammonical odour, slightly similar to amyl alcohol. Soluble in water.
(2) Responds to Remini's test.

(3) To 1ml of compound add 1.2ml ethyl oxalate. Shake thoroughly and keep at room temperature for sometime. Cool, filter and dissolve the precipitate in dil alcohol. Heat on water bath and cool. Collect crystals of diisoamyloxamide, m.p 140°C.

(4) Picrate m.p 138°C. (5) 2, 4-dinitrophenyl m.p 91°C.

b.p 104°C **n-Amyl amine** (Pentyl amine) $CH_3(CH_2)_3CH_2NH_2$

(1) Colourless liquid, soluble in water, alcohol and ether.

(2) Responds to Remini's test.

(3) Picrate m.p139°C (4) 2, 4-dinitrophenyl m.p 81°C.

b.p 106-108°C **Amino cyclopentane** 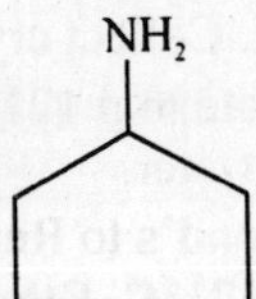

(1) Colourless liquid with ammonical odour, miscible with water.

b.p 117°C **Ethylenediamine** (1, 2-diamino ethane) $NH_2CH_2CH_2NH_2$

(1) Colourless liquid with ammonical odour, soluble in water, insoluble in ethanol and ether.

(2) Mix 8ml ethylenediamine with 14.5ml glacial acetic acid. Distill at 110°C to remove water and excess acid (about 5ml). Cool the residual liquid and stir, 1, 2-diacetamidoethane soon solidifies, recrystallise from ethanol, m.p 172-174°C.

(3) Acetate (di) m.p 172°C

(4) Benzoate (di)m.p 249°C

(5) Picrate m.p 233°C

(6) 2, 4-dinitrophenyl (di) m.p 302°C

b.p 134°C **Aminocyclohexane** 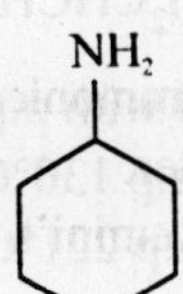

(1) Colourless liquid with coniine like odour, slightly soluble in water, soluble in alcohol and ether.
(2) Hydrochloride m.p 206°C (3) Acetate m.p 104°C (4) benzoate m.p 148°C. (5) Picrate m.p 158°C.
(6) 2, 4-dinitrophenyl m.p 156°C.

b.p 156°C **Heptylamine** $CH_3(CH_2)_5CH_2NH_2$

(1) Colourless liquid with unpleasant odour, insoluble in water.
(2) Respond to Remini's test.
(3) To 1ml of compound add 1ml ethyl oxalate. Shake thoroughly and keep at room temperature for sometime. Cool, filter and dissolve the precipitate in dil alcohol. Heat on water bath and cool. Collect crystals of diheptyloxamide, m.p 132°C.
(4) Picrate m.p 121°C

b.p 171°C **Ethanolamine** (2-amino ethanol) $NH_2CH_2CH_2OH$

(1) Colourless highly viscous liquid with faint odour, miscible with water and alcohol.
(2) Mix 6ml ethanolamine with 5.7ml glacial acetic acid. Heat the mixture so that water formed during the reaction distill off (temp 165°C, 3-4mlwater). Distill the residual liquid under reduced pressure and collect N-acetylethanolamine at b.p 144-149°C/ 1.5mmHg.
(3) Dissolve 5ml ethanolamine in 17ml benzene and add 8.4ml cyclohexanone to it. Reflux the mixture for 30min and distill off water formed during the reaction (2-4ml). Evaporate off the benzene and distill the residual oily portion under reduced pressure and collect cyclohexnespiro-2'(1', 3'oxazolidine b.p 90-91°C/ 16mmHg.
(4) Picrate m.p 160°C.

b.p 178-180°C **Pentamethylenediamine** (Cadaverine) $NH_2CH_2(CH_2)_3CH_2NH_2$

(1) Syrupy liquid with piperidine like odour, soluble in water.
(2) Dibenzoate m.p 130°C
(3) Picrate (di) m.p 221°C

b.p 207°C (–) **Menthylamine**

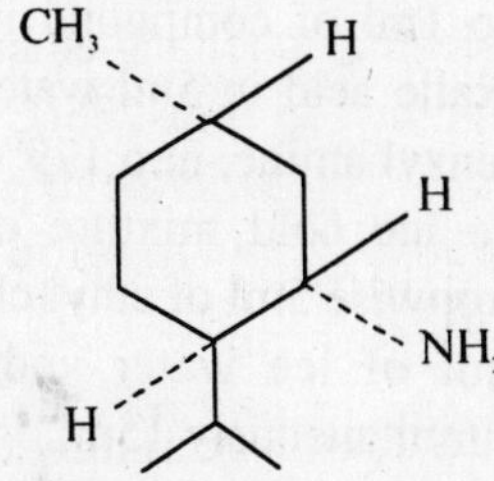

(1) Colourless liquid with strong characteristic coniine like odour, burning bitter taste, difficult soluble in water.
(2) Acetate m.p 145°C.
(3) Benzoate m.p 156°C
(4) Picrate m.p 215°C.

Aliphatic Solid

m.p 27°C, b.p 158°C **Tetramethylenediamine** (Putrescine)
$NH_2(CH_2)_4NH_2$

(1) Colourless solid with piperidine like odour, soluble in water, difficult soluble ether.
(2) Acetate (di) m.p 137°C (3) Benzoate (di) m.p 177°C (4) Picrate (greenish –yellow) m.p 249°C.

Aromatic Liquids

b.p 185°C **Benzylamine** (α-amino toluene)

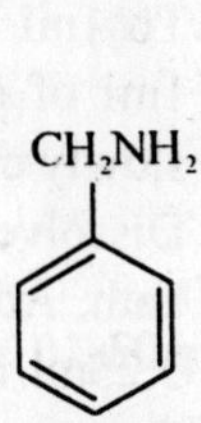

(1) Colourless liquid with ammonia like odour, sparingly soluble in water, easily soluble in alcohol.
(2) Responds to Carbylamine test.
(3) Dissolve 1ml of compound in 5ml alcohol. Add 3ml CS_2 and heat to boiling. Heat on water bath until crystals begin to separate. Cool, filter and dry the white crystals of benzylamine benzyldithiocarbamate, m.p 128°C.

(4) To 1ml of compound in 4ml water add a solution of 200mg of oxalic acid in 5ml water. Heat to boiling. On cooling oxalate of benzyl amine, m.p 179°C is obtained.
(5) To ice cold mixture of 10ml benzylamine in 5ml water add dropwise 5ml of ethyl chloro carbonate (temp below 10°C). Again 5ml of ice water and 5ml ethyl chloro carbonate is added. Simultaneously 15ml of ice cold NaOH solution is added. Stir the mixture for 30 min and collect Ethyl N-benzyl carbamate, m.p 47°C.
(6) Hydrochloride m.p 248°C.
(7) Acetate m.p 60°C
(8) Benzoate m.p106°C
(9) Picrate m.p 199°C.
(10) 2, 4-dinitrophenyl m.p 116°C

b.p 184°C **Aniline** (amino benzene)

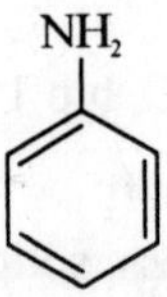

(1) Colourless oily liquid which turns brown on exposure to air or light, insoluble in water, soluble in organic solvents.
(2) Responds to Carbylamine test.
(3) Responds to dye test.
(4) To 1ml of compound add 5-6 drops of dilute $K_2Cr_2O_7$ solution, 1ml of conc H_2SO_4 and shake. A red coloured solution which slowly changes to deep blue is obtained.
(5) Dissolve 1ml aniline in 2ml glacial acetic acid and cool in ice bath. Add dropwise 5ml of bromine solution (1.75ml bromine in 3.25ml acetic acid). Keep the mixture at room temperature and then pour in ice. Collect the precipitated 2, 4, 6-tribromoaniline, m.p 119°C.
(6) To 5ml of cooled aniline add in portion 10ml of conc H_2SO_4. After the addition is complete reflux on oil bath at 180-190°C for 5hrs. Cool, pour in crushed ice, recrystallise the Sulphanilic acid from hot water.
(7) Cool a mixture of 5ml aniline, 7gm sodium bicarbonate and 50ml water to 10°C. Stir and add in portion 12gm of pure iodine. After addition is complete continue stirring for 30min. Filter the

precipitated p-iodo aniline, recrystallise from pet-ether, m.p 62-63°C.

(8) To 12.5gm of aniline add 100gm of conc H_2SO_4 and 300ml water. Cool in ice bath and add in portion with stirring a solution of 12.5gm of sodium dichromate in 50ml water(temp below 10°C). Cool overnight and then add in portion a solution of 50gm of sodium dichromate. Keep for 2hrs and then extract with ether. Distill off ether and collect yellow crystals of p-benzoquinone, m.p 116°C.

(9) Hydrochloride m.p 198°C.

(10) Acetate (acetanilide) m.p 114°C

(11) Benzoate (Benzanilide) m.p 163°C.

(12) Picrate (yellow) m.p 165°C.

(13) Formyl m.p 47°C.

(14) 2, 4-dinitrophenyl m.p 156°C

b.p 185°C **(±) Phenylethylamine**

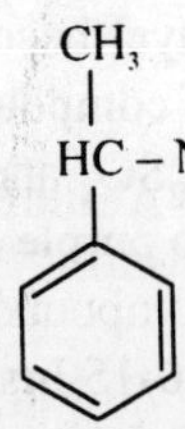

(1) Sparingly soluble in water.

(2) Hydrochloride m.p 158°C (3) Acetate m.p 57°C (4) Benzoate m.p 120°C (5) Picrate m.p 189°C (6) 2, 4-dinitrophenyl m.p 118°C.

b.p 195°C **2-Phenylethylamine** (1-amino-2-phenyl ethane)

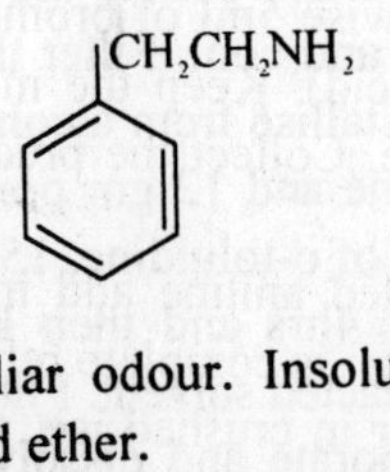

(1) Liquid with peculiar odour. Insoluble in water, very difficult soluble alcohol and ether.

(2) Hydrochloride m.p 217°C

(3) Acetate m.p 51°C

(4) Benzoate m.p 116°C

(5) Picrate m.p 167°C
(6) 2, 4-dinitrophenyl m.p 154°C

b.p 200°C **o-toluidine** (o-amino toluene) 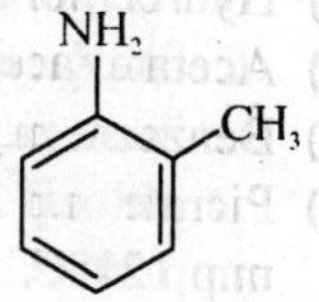

(1) Colourless oily liquid with aniline like odour, slightly darkens on exposure to air or light, insoluble in water, soluble in organic solvents.
(2) Responds to Carbylamine test.
(3) Responds to dye test.
(4) Dissolve 1ml of compound in 2ml ether, add 6-7ml of water to it. Add 1ml of dry bleaching powder and shake thoroughly. The ethereal layer becomes brown.
(5) To 1ml of compound add 1ml of dilute $K_2Cr_2O_7$ solution and 2ml of 50% H_2SO_4 and shake. A blue coloured solution which slowly changes to purple on dilution is obtained.
(6) To 5ml compound add 10ml benzoyl chloride and heat it on oil bath for 10-15 hrs at 230°C. Add 15ml alcoholic solution (7.5ml alcohol+ 7.5ml conc HCl) and boil the mixture for 14hrs. Distill off alcohol and ethyl benzoate. Add 10ml water and boil. Filter and add NaOH solution. Cool and collect 4-amino 3-methyl benzoate, m.p 112°C.
(7) Cool a solution of 6gm of o-toluidine in 90ml conc H_2SO_4 to 0°C. Add dropwise with stirring the nitrating mixture (3.5gm conc HNO_3+6.3gm conc H_2SO_4). Keep at room temperature for 2hrs. Pour in ice and collect the precipitated solid, dissolve it in NaOH solution and cool. Filter the precipitated yellow 4-nitro-o-toluidine, recrystallise from alcohol, m.p 107°C.
(8) Add 12gm iodine and 12 gm precipitated calcium carbonate to a mixture of 6gm of o-toluidine, 15ml ether and 10ml water. Reflux the mixture for 4hrs and then keep overnight. Steam distill to remove the unreacted solvent. Filter and boil the solid with alcohol and filter. Evaporate and dissolve the residue in 10ml dil HCl, filter and add 5ml conc HCl to the filtrate. Crystals of 5-iodo-o-toluidine separate out, m.p 216°C.
(9) Dilute a solution of 10.7gm of compound in 10ml glacial acetic

acid with 100ml water. Add a solution of 8gm of potassium cyanate in 20 ml water to it. Stir and collect precipitated o-tolylurea, recrystallise from alcohol, m.p 191°C.

(10) Hydrochloride m.p 215°C.
(11) Acetate (aceto o-toluidide) m.p 110°C.
(12) Benzoate m.p 144°C.
(13) Picrate m.p 200°C. (14) Formyl m.p 59°C. (15) 2, 4-dinitrophenyl m.p 120°C.

b.p 203°C **m-toluidine** (m-amino toluene)

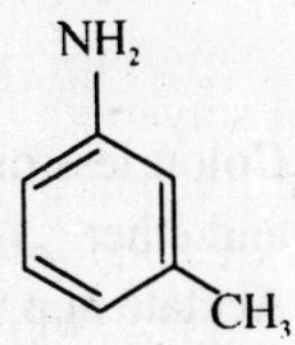

(1) Colourless oily liquid with aniline like odour, slightly darkens on exposure to air or light, insoluble in water, soluble in organic solvents.
(2) Responds to Carbylamine test.
(3) Responds to dye test.
(4) Dissolve 1ml of compound in 2ml ether and add 6-7ml of water to it. Add 1ml of dry bleaching powder and shake thoroughly. The ethereal layer becomes red, while the aqueous layer becomes yellow or brown.
(5) To 1ml of test sample add 10ml dil H_2SO_4. Now add 1ml of conc HNO_3. After 5 min a red colour develops. Heat gently for few seconds. The colour of solution will now be intense red.
(6) Hydrochloride m.p 228°C
(7) Acetate (aceto m-toluidide) m.p 65°C.
(8) Benzoate m.p 125°C.
(9) Picrate m.p 200°C. (10) 2, 4-dinitrophenyl m.p 160°C

b.p 210°C **2-Phenylpropylamine**

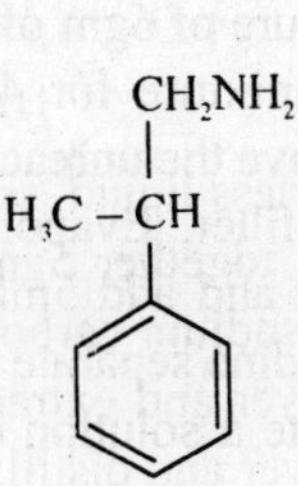

(1) Oily liquid with fish like odour, very difficult soluble in water, soluble in alcohol and ether.

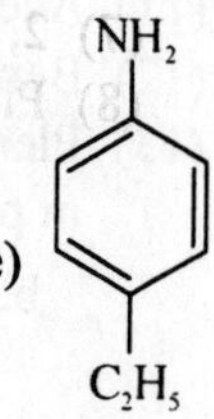

b.p 216°C **p-ethyl aniline** (1-ethyl-4 amino benzene)

(1) Colourless oily liquid, slightly soluble in water, soluble in alcohol and ether.
(2) Acetate m.p 94°C
(3) Benzoate m.p 151°C

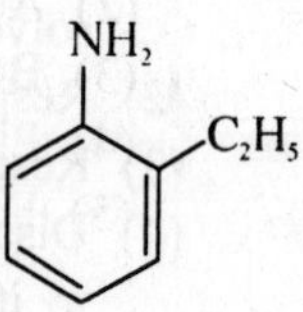

b.p 216 °C **o-Ethylaniline** (o-amino ethyl benzene)

(1) Slightly soluble in water, soluble in alcohol and ether.
(2) Acetate m.p 112°C (3) Benzoate m.p 146°C. (4) Picrate m.p 194°C

b.p 218°C **p-Xylidine** (2, 5 dimethyl aniline,1, 4-dimethyl –2-aminobenzene)

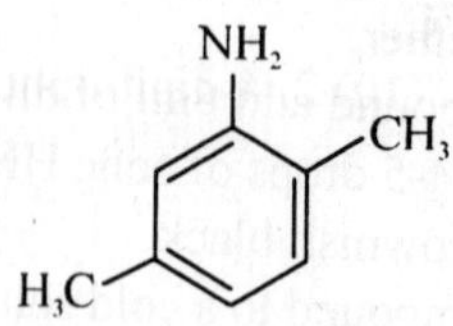

(1) Colourless liquid, sparingly soluble in water
(2) Reflux together 5gm of compound with 15gm of methyl iodide, 13gm sodium carbonate and 125 ml water for 6hrs. Separate the oily layer and extract the aqueous layer with ether. Combine the two layer and distill. Collect dimethyl p-xylidine at b.p 205°C.

(3) To 1ml of compound add conc HNO_3. An intense red-brown colouration is produced.
(4) Hydrochloride m.p 228°C.
(5) Acetate (aceto –p-xylide) m.p 139°C.
(6) Benzoate m.p 140°C.
(7) 2, 4-dinitrophenyl m.p 150°C
(8) Picrate m.p 171°C

b.p 220°C **sym-m-Xylidine** (3, 5-dimethyl aniline)

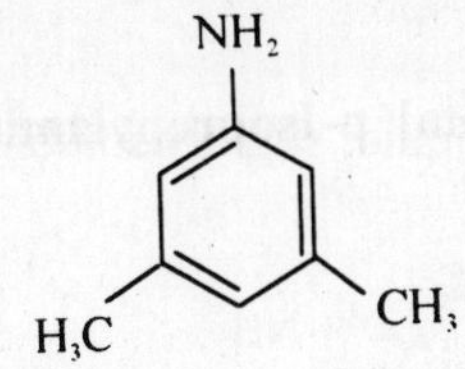

(1) Liquid, slightly soluble in water.
(2) Acetate m.p 144°C
(3) Benzoate m.p 136°C
(4) Picrate m.p 209°C
(5) Formyl m.p 77°C.

b.p 225°C **o-Anisidine** (o-methoxy aniline)

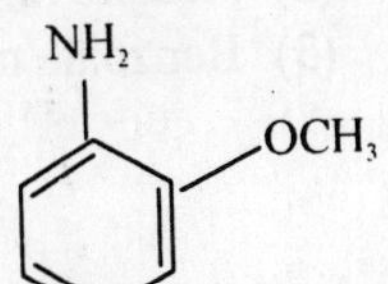

(1) Colourless liquid with characteristic mild aromatic odour, soluble in alcohol and ether.
(2) To 1ml of compound add 6ml of dil H_2SO_4(4ml conc H_2SO_4 and 2ml water) and 4-5 drops of conc HNO_3. The solution immediately turns opaque brownish black.
(3) Add 5gm of compound to a cold solution of 5gm of nitrosobenzene in 19ml acetic acid. Keep the mixture at room temperature for 40hrs and then pour in ice water. Steam distill and collect o-methoxyazo benzene at b.p 195-197°C which on cooling soon solidifies, m.p 40-41°C.

(4) Add well cooled diazonium solution prepared from 11gm of o-anisidine, 13ml conc H_2SO_4, 50ml water and a solution of 7gm of $NaNO_2$ in 20ml water to a mixture of 70gm of sodium sulphate, 50ml conc H_2SO_4 and 50ml water. Heat at 135-140°C and distill. Saturate the distillate with ammonium chloride. Extract with ether. Evaporate off the ether and cool the oily residue, guaiacol precipitates out, m.p 33°C.

(5) Acetate (aceto-o-anisidide) m.p 85°C (6) Benzoate m.p 60°C (7) Picrate m.p 200°C. (8) formyl m.p 83°C. (9) 2, 4-dinitrophenyl m.p 151°C

b.p 225°C [745mm] **p-Isopropylaminobenzene** (cumidine, p-isopropyl aniline)

(1) Colourless liquid, insoluble in water, soluble in alcohol and ether.
(2) Acetate m.p 102°C.
(3) Benzoate m.p 162°C.

b.p 229°C **o-Phenetidine** (o-ethoxy aniline)

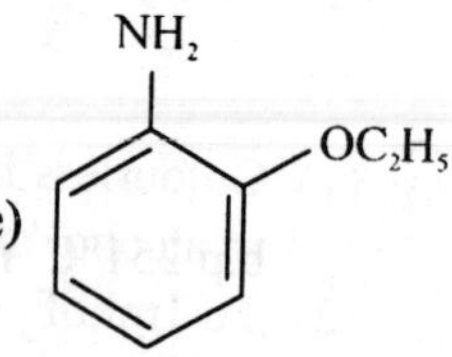

(1) Colourless liquid, insoluble in water, soluble in alcohol and ether.
(2) Hydrochloride m.p 214°C.
(3) Acetate m.p 79°C
(4) Benzoate m.p104°C.
(5) Formyl m.p 62°C.
(6) 2, 4-dinitrophenyl m.p 164°C

b.p 233°C **2, 4, 6-trimethylaniline** (mesidine)

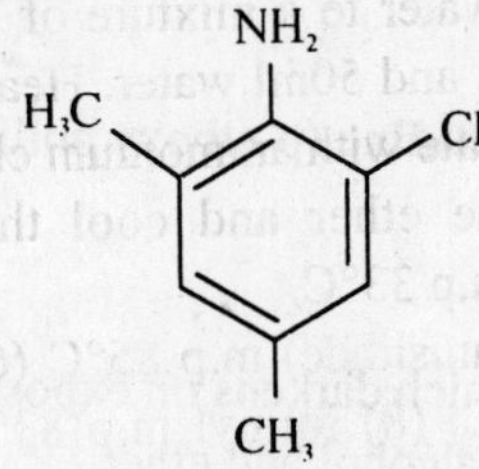

(1) Acetate m.p 216°C (2) Benzoate m.p 204°C (3) Picrate m.p 193°C

b.p 248°C **m-Phenetidine** (m-ethoxy aniline)

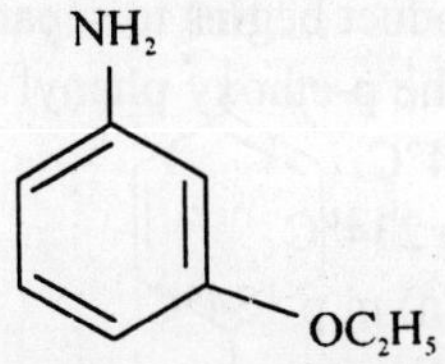

(1) Colourless liquid, insoluble in water.
(2) Acetate m.p 97°C
(3) Benzoate m.p 103°C
(4) Picrate m.p 158°C
(5) Formyl m.p 52°C

b.p 251°C **m-anisidine** (m-methoxy aniline)

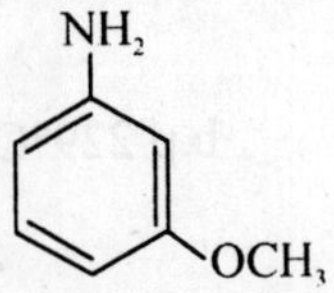

(1) Colourless liquid, soluble in alcohol and ether.
(2) Acetate m.p 81°C.
(3) Benzoate m.p 110°C
(4) Picrate m.p 169°C
(5) Formyl m.p 57°C.
(6) 2, 4-dinitrophenyl m.p 138°C

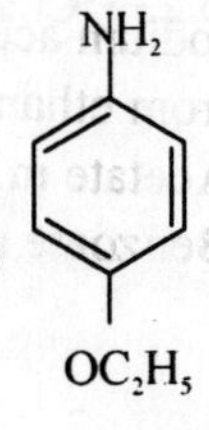

b.p 254°C **p-Phenetidine** (p-ethoxy aniline)

(1) Colourless liquid which darkens on exposure, very slightly soluble in water, soluble in alcohol and ether.
(2) To 4-5 drops of compound add 2-3 drops of conc HNO_3. A blue coloured solution is obtained.
(3) To a mixture of 8.7gm of phenetidine hydrochloride and 12gm of urea add 20ml water, 1ml conc HCl and 1ml glacial acetic acid. Heat the mixture thoroughly and heat to boiling for 70-90 min. After the crude product begins to separate remove the heat source and cool. Collect the p-ethoxy phenyl urea, recrystallise from hot water, m.p 173-174°C.
(4) Hydrochloride m.p 234°C.
(5) Acetate (Phenacetin) m.p 134°C.
(6) Benzoate m.p173°C.
(7) Picrate m.p 69°C
(8) Formyl m.p 76°C.
(9) 2, 4-dinitrophenyl m.p118°C

Aromatic Solid

m.p 20°C, b.p 251°C **o-Aminoacetophenone** (o-acetyl aniline)

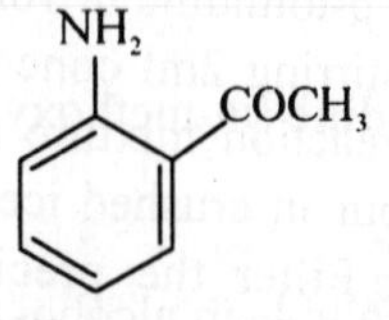

(1) Crystalline solid or yellow oil, insoluble in water, soluble in ether.
(2) Cool a solution of 2.0gm of o-aminoacetophenone in 175ml conc HCl to 0°C. Add with stirring a solution of 1.14gm of sodium nitrite in 10ml water. Heat on water bath at 70-75°C for 3hrs. Evaporate to dryness. Dissolve the residue in 10ml water and add

sodium acetate solution. Cool and collect 4-cinnolone, recrystallise from ethanol, m.p 233-234°C.

(3) Acetate m.p 76°C

(4) Benzoate m.p 98°C.

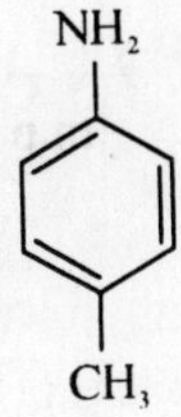

m.p 43°C, b.p 200°C **p-toluidine** (p-amino toluene)

(1) Light coloured solid, odour similar to aniline, soluble in alcohol and ether.

(2) Responds to Carbylamine test.

(3) Responds to dye test.

(4) To 100mg of compoud add 10ml water and 4-5 drops of conc HCl. Add 3-4 drops of 10% $FeCl_3$ solution and heat so that it boil. Cool, the colour of solution becomes red orange.

(5) To 100mg of compound add 6ml conc H_2SO_4 and 3ml water. Now add 2-3drops of conc HNO_3 without shaking. A blue ring appears immediately.

(6) Dissolve 5gm of compound in 25ml 10% HNO_3 and diazotise it with 30% of 3gm of sodium nitrite solution (temp below 8°C). Keep the mixture at 0°C for 3hrs and then reflux on oil bath. Steam distill the solution, cool the distillate and collect nitro cresol (3-nitro 4-hydroxy toluene) m.p 37°C.

(7) Dissolve 5gm of p-toluidine in 10ml conc H_2SO_4 and then cool to 0°C. Add with stirring 2ml conc HNO_3 and 5ml conc H_2SO_4 at 0°C. Keep the reaction mixture at room temperature for 30-40 min and then pour in crushed ice. Add sodium carbonate to this acidic solution. Filter the precipitated m-nitro toluidine and recrystallise from alcohol, m.p 78°C

(8) To a solution of 2.5gm of compound in 10ml glacial acetic acid add dropwise with stirring a solution of 1.5ml of bromine in 15ml acetic acid. Continue stirring for 5 minutes and then add water. Filter the solid and disslove it in water. Add sodium acetate to the solution and cool. Collect 3, 5-dibromo p-toluidine, m.p 79°C.

(9) Reflux 2gm of compound with 2ml n-butyric acid for 2hrs. Cool and pour in ice water, acidify with dil HCl. Filter the precipitated n-butyro-p-toluidide, recrystallise from alcohol, m.p 73-74°C.

(10) Hydrochloride m.p 240°C (11) Acetate m.p 148°C (12) Benzoate m.p 158°C (13) Picrate m.p 181°C. (14) Formyl m.p 53°C. (15) 2, 4-dinitrophenyl m.p 136°C.

m.p 49°C **4-o-Xylidine** (4-amino-1, 2 xylene, 3, 4-dimethy aniline)

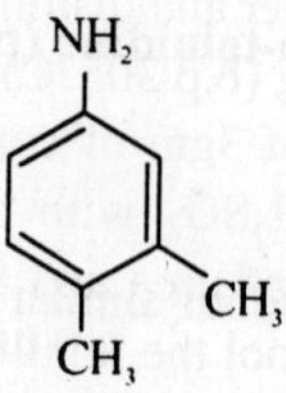

(1) Crystalline compound, difficult soluble water.
(2) To few milligrams of compound add 3-4 drops of conc HNO_3. A strong green colour which slowly changes to light greenish brown is obtained.
(3) Hydrochloride m.p 256°C.
(4) Acetate m.p 99°C.
(5) Benzoate m.p 118°C.
(6) 2, 4-dinitrophenyl m.p 141°C.

m.p 50°C **1-Naphthylamine**

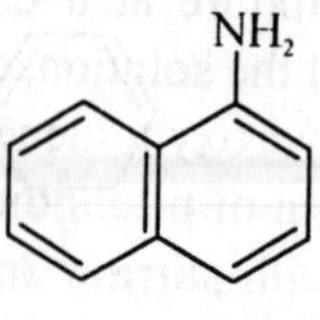

(1) Light pink coloured solid which turns dark brown black on standing, insoluble in water, soluble in alcohol and ether, pungent odour. Sublimes readily.
(2) Responds to Carbylamine test.
(3) Responds to dye test.
(4) To few milligrams of compound in 5ml alcohol add a drop of KNO_2 and one drop of conc HCl. A red coloured solution is obtained within few seconds.
(5) Heat few milligram of compound (15-20mg) with 10ml water,

1-2drops of conc HCl and 3-4 drops of $FeCl_3$. A blue coloured solution is obtained.

(6) Heat 5gm of compound with 15ml conc H_2SO_4 in an oil bath for 4-5hrs. Cool and pour the reaction mixture with stirring in crushed ice and collect the solid naphthionic acid (1-naphthylamine 4-sulphonic acid)

(7) Add a solution of 8.5gm of ethyl bromide in 8ml alcohol to a boiling solution of 10gm of 1-naphthylamine in 24ml alcohol. Slowly precipitation of ethyl 1-naphthylamine hydrochloride takes place. Treat the pecipitated solid with excess of potash solution. Collect the organic layer and distill. Collect ethyl 1-naphthylamine at b.p 191°C/16mmHg (b.p 305°C)

(8) Diazotise a solution of 3gm of compound in 18ml glacial acetic acid and 8ml conc H_2SO_4 with aqueous solution of 1.5gm of $NaNO_2$. Add 500mg of urea and a solution of 1.5gm of sodium azide in 5ml water. Cool the mixture to 0°C and then keep at this temperature for several hours. Extract with ether. Distill off ether, cool the residual oily portion, triazonaphthalene, m.p 13°C separate out.

(9) Hydrochloride m.p 286°C. (10) Acetate m.p 160°C.

(11) Benzoate m.p 161°C

(12) Picrate m.p 161°C,

(13) Formyl m.p 139°C.

(14) 2, 4-dinitrophenyl m.p 190°C.

m.p 53°C **2-Aminobiphenyl** (o-Phenylaniline)

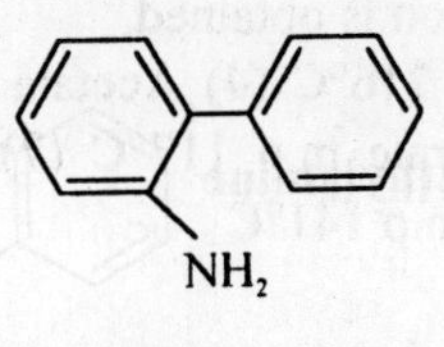

(1) Crystalline solid, insoluble in water.

(2) Dissolve 1gm of compound in 9ml warm conc HCl and 25ml water. Cool and add a solution of 700mg of sodium nitrite in 5ml water. Stir the mixture at 0°C for 5min. Pour the diazonium solution into a solution of 30gm sodium acetate and 1.25gm sodium azide in 50ml water. Stir for 20min. Extract with ether. Evaporate off the solvent, cool the residue, 2-azidobiphenyl separates out, m.p 49-50°C.

(3) Reflux 1gm of compound with 8ml formic acid for 1hr. Cool and pour with stirring into 20ml ice cold water. Collect o-formamidobiphenyl, recrystallise from ethanol, m.p 75°C.

(4) Acetate m.p 119°C (5) Benzoate m.p 102°C

m.p 54°C **4-Aminobiphenyl** (p-Phenylaniline, Xenylamine)

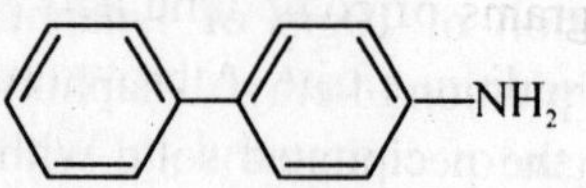

(1) Crystalline compound, insoluble in water, soluble in alcohol, ether and chloroform.

(2) Acetate m.p 172°C (3) Benzoate m.p 230°C.

m.p 60°C **p-Anisidine** (p-methoxy aniline)

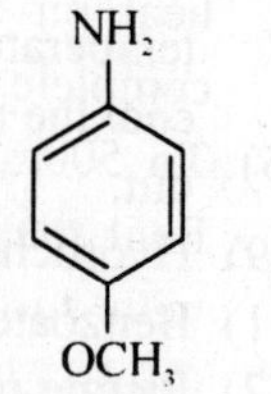

(1) Crystalline solid which darkens on exposure, insoluble in water, soluble in alcohol and ether.

(2) Heat few milligram of compound (15-20mg) with 10ml water, 1-2 drops of conc HCl and 3-4 drops of $FeCl_3$. Cool, a violet coloured solution is obtained.

(3) Hydrochloride 216°C (4) Acetate m.p 127°C (5) Benzoate m.p 154°C (6) Picrate m.p 117°C (7) Formyl m.p 81°C. (8) 2, 4-dinitrophenyl m.p 141°C

m.p 63°C **m-Phenylenediamine** (m-diamino benzene)

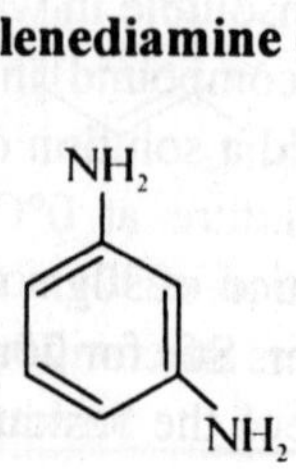

(1) Crystalline compound which darkens on exposure to light. Soluble in hot water.
(2) To few milligrams of compound add 2-3 drops of conc HCl and 2ml of water. To this aqueous solution add 10% acetaldehyde solution in 50% alcohol and warm. A yellow coloured solution which exhibits green fluorescence is obtained.
(3) To few milligrams of compound add 5-6 drops of conc HCl and few drops of sodium nitrite. A brown precipitate is obtained.
(4) Dissolve 1.2gm of compound in 12ml conc HCl, cool and add with stirring benzene diazonium chloride solution (prepared from 2ml aniline). Then add 14ml sodium acetate solution (4gm in 14ml water) with stirring until the precipitation of dye stuff is complete, continue stirring for further 45-60min. Heat the mixture to boiling and filter hot. Add 10gm of NaCl to the filtrate and heat on steam bath until the precipitation of chrysoidine is complete.
(5) To 500mg of compound add 4ml Ac_2O and heat it to boiling. Pour the mixture in 6ml boiling water. Cool and add dropwise with stirring bromine water until the solution remains yellow. Cool, filter the precipitated diacetoxy dibromo derivative and recrystallise from alcohol, m.p 257°C.
(6) Acetate (di) m.p 191°C. (7) Benzoate (di) m.p 240°C. (8) Formyl m.p 155°C (9) Picrate m.p 184°C. (10) 2, 4-dinitrophenyl (di) m.p 172°C

m.p 64°C **2, 5 Toluylenediamine** (2, 5-diaminotoluene)

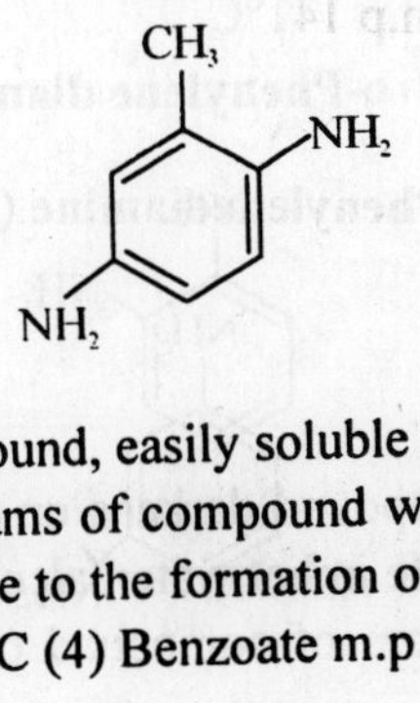

(1) Crystalline compound, easily soluble in water, alcohol and ether.
(2) Heat few milligrams of compound with dil H_2SO_4 and MnO_2. A pungent odour due to the formation of toluquinone is observed.
(3) Acetate m.p 220°C (4) Benzoate m.p 307°C

m.p 66°C **s-Pseudocumidine** (2, 4, 5-trimethyl aniline)

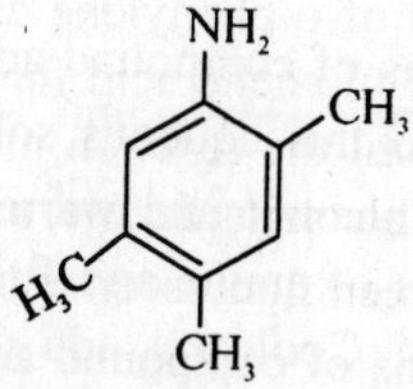

(1) Colourless crystalline solid with odour similar to aniline, soluble in alcohol and ether.
(2) Acetate (acetyl pseudocumidide) m.p 162°C.

m.p 99°C **2, 4 Toluylenediamine** (2, 4-diamino toluene)

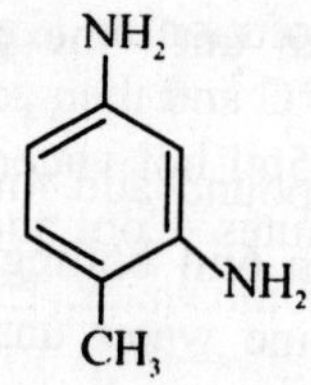

(1) Colourless crystalline odourless solid, soluble in water, alcohol and ether. Saturated solution has a bitter and burning taste.
(2) Hydrochloride m.p 305°C
(3) Acetate(di) m.p 220°C
(4) Benzoate (di) m.p 224°C
(5) Formyl m.p 177°C.
(6) 2, 4-dinitrophenyl (di) m.p 184°C.

m.p 103-104°C **o-Phenylene diamine** (1, 2-diamino benzene)

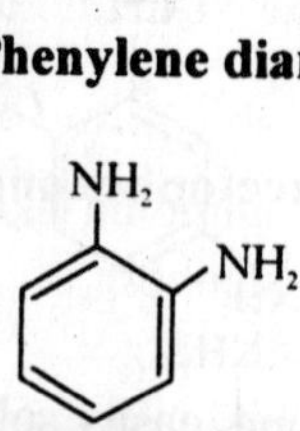

(1) Crystalline compound darkens on exposure to light, insoluble in water but soluble in hot water, alcohol, ether and chloroform.
(2) To few milligram of compound (15-20mg) in 1-2ml water add

2-3drops of conc HCl and 3-4 drops of $FeCl_3$. Red coloured precipitate is obtained within few seconds.

(3) Reflux together 1gm of o-phenylene diamine hydrochloride with 4ml of water and 1.1ml acetic acid for 45 minutes. Cool and add ammonia solution until the reaction mixture is basic. Collect the precipitated 2-methyl benzimidazole, m.p 176°C.

(4) Heat together 2.7 gm of amine and 1.6ml of 90% formic acid on water bath for 2hrs. Cool and add 10% NaOH solution with stirring until the solution is alkaline. Filter the crude benzimidazole, recrystallise from hot water, m.p 171-172°C.

(5) Dissolve 2.1gm of amine in a mixture of 2.5ml acetic acid and 6ml water(heat if necessary). Cool to 10°C and add with stirring a solution of 1.5gm of sodium nitrite in 3ml water. The temperature may rise to 80°C. Continue stirring for 10-15 minutes, cool in ice bath for 20-30min. Collect the precipitated benzotriazole, recrystallise from hot water, m.p 99-100°C.

(6) Add 9ml glyoxal to a solution of 11gm sodium metabisulphite in 50ml water. Heat to 70°C and then add to a solution of 6gm of o-phenylenediamine in 75ml hot water. Keep the mixture at room temperature for 15 minutes. Cool and add a solution of 22gm of sodium carbonate in 50ml water. Extract with ether. Distill off ether and collect quinoxaline, m.p 30°C.

(7) Dissolve 5gm compound in 100ml toluene and then add 15ml thionly chloride to it. Reflux for 10-12 hrs. Distill off the solvent, cool the residual oil which soon solidifies, collect benzo-2, 1, 3-thiadiazole, m.p 42°C.

(8) Mix alcoholic solution of o-phenylenediamine with equivalent amount to phenanthraquinone in hot glacial acetic acid. Cool and collect phenanthrazine, m.p 217°C.

(9) Acetate (di) m.p 185°C

(10) Benzoate (di) m.p 301°C

(11) Picrate m.p 208°C.

m.p 106°C **p-Aminoacetophenone** (p-amino phenyl methyl ketone)

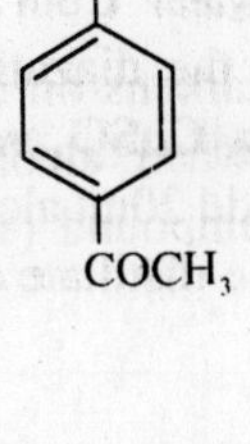

(1) Yellow coloured crystalline compound, very slightly soluble in water, soluble in alcohol, ether and benzene.
(2) Acetate m.p 167°C
(3) Benzoate m.p 205°C.

m.p 113°C **2-Naphthylamine**

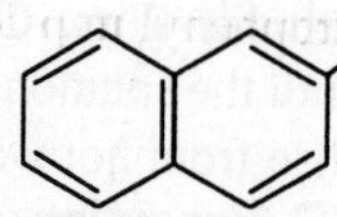

(1) Pink coloured solid, insoluble in water, soluble in alcohol and ether.
(2) Reduces Tollens reagent.
(3) To 3.5ml conc HCl and 4ml water add 1gm of compound. Cool the mixture to 5°C, add .5gm of sodium nitrite and filter. A solution of 2gm of $HgCl_2$ in 3ml conc HCl mixed with ice is added to the stirred solution. Diazonium salt separates, continue stirring for 30min and filter. To the solid addition compound add 7ml acetone and .4gm Cu powder. Cool, stir for 45min and then keet for overnight. The insoluble 2-naphthyl mercuric chloride m.p 266°C is then collected.
(4) Heat 2gm of 2-naphthylamine and 100mg of iodine at 230°C for 4hrs. Cool and collect 2, 2' dinaphthylamine, recrystallise from benzene, m.p 170°C.
(5) Reflux 10gm of compound with 16gm of ethyl iodide and a solution of 20gm of sodium carbonate in 200ml water for 3hrs. Cool and extract with ether. Distill under reduced pressure and collect ethyl 2-naphthylamine at b.p 191°C/25mmHg. (b.p 315-316°C)
(6) Add about 8gm of potassium cyanate to a solution of 10gm of compound in 10ml glacial acetic acid. Warm the resulting mixture until the cyanate has completely dissolved and then pour in 50ml ice water. Collect the precipitated 2-naphthylurea, m.p 213-214°C.
(7) Add 3.5gm of compound to boiling solution of 5.5ml of conc HNO_3 and 100ml of water. Cool and diazotise with 6gm of $NaNO_2$ in 10ml water. Add the diazotised solution to a suspension of cuprous oxide (25gm $CuSO_4$ reduced by glucose and aqueous alkali). After 20min add 20ml alcohol. Keep the mixture overnight. Steam distill. Cool the distillate and then heat the solid in an dish

on water bath for few minutes. Cool and collect 2-nitronaphthalene, m.p 79°C.

(8) Hydrochloride m.p 260°C. (9) Acetate m.p 134°C

(10) Benzoate m.p 162°C

(11) Picrate m.p 195°C.

(12) Formyl m.p 129°C.

(13) 2, 4-dinitrophenyl m.p 179°C.

m.p 128°C **Benzidine** (4:4' diamino diphenyl)

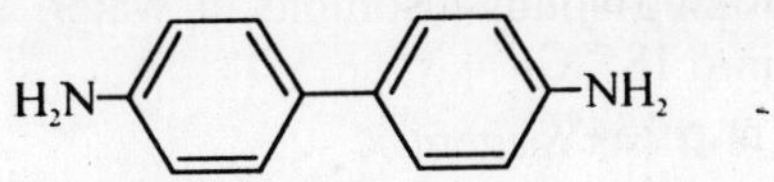

(1) Buff coloured solid, insoluble in water, soluble in alcohol.

(2) Responds to dye test. (Violet coloured dye)

(3) Dissolve few milligrams of compound in 2-3ml CS_2 and first add 2ml of water and then few drops of bromine solution (bromine in CS_2). The aqueous layer will first turn blue, then green and finally colourless while the CS_2 layer will be red.

(4) To 100mg of compound add 2ml conc H_2SO_4, few crystals of $K_2Cr_2O_7$ and heat. A pungent smell due to the formation of p-benzoquinone is observed.

(5) Reflux 500mg of compound in 5ml alcohol with 5ml benzaldehyde. Cool, filter the precipitated solid, recrystallise and collect dibenzylidenebenzidene, m.p 232°C.

(6) Reflux a mixture of 4gm of benzidine with 25ml ethanol and 1.6gm of raney nickel for 15hrs. Add 25ml ethanol boil again and filter rapidly. Concentrate the filtrate, cool and keep overnight in a refrigerator. Collect N, N-diethyl benzidene, m.p 110-115°C.

(7) Dissolve 5gm of benzidene in 90ml of 2% HCl and 500ml of 5% H_2SO_4. Diazotise with 4gm of 20% sodium nitrite solution. Pass steam for 30min or until the sample gives no precipitate with alkaline solution of phenol. Filter hot, cool the filtrate and collect 4, 4'dihydroxy diphenyl, recrystallise from hot water, m.p 275°C.

(8) Acetate m.p 317°C (9) Benzoate m.p 352°C.

m.p 129°C **o-Tolidine** (4, 4' diamino-3, 3' dimethylbiphenyl)

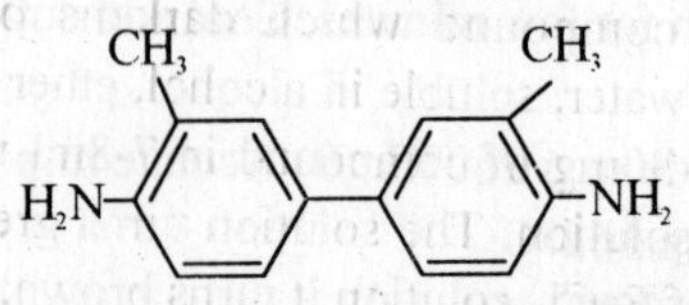

(1) Colourless crystalline compound which darkens on exposure to air, insoluble in water, soluble in alcohol and ether.
(2) Reflux 500mg of compound in 5ml alcohol with 5ml benzaldehyde. Cool, filter the precipitated solid, recrystallise and collect dibenzylidenetolidine, m.p 150°C.
(3) Benzoate (di) m.p 265°C
(4) Picrate m.p 185°C
(5) Formyl m.p 254°C

m.p 137°C **o-Dianisidine** (4, 4'-diamino-3, 3'-dimethoxybiphenyl)

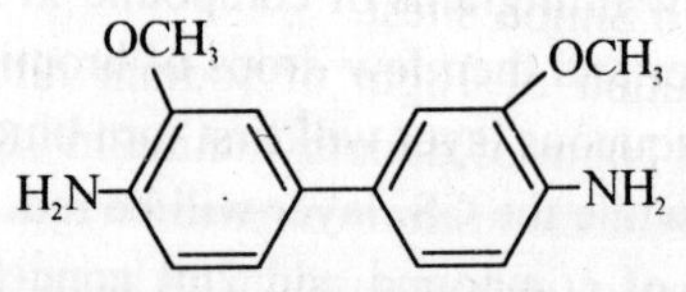

(1) Colourless solid which turns violet on exposure to air, insoluble in water, soluble in alcohol, ether and benzene.
(2) Dissolve 10gm of compound in 75ml boiling water containing conc HCl. Cool to 15°C and add 9ml conc HCl. Cool further to 5^0C and diazotise with 6gm of sodium nitrite in 12ml water. Pour the diazotised solution in 80ml ice cold 30% hypophosphorous acid. Keep the reaction mixture at 5°C for 8-10hrs. Extract with ether. Distill off ether and cool the residue, 3, 3' dimethoxydiphenyl, m.p 42-43°C separate out.
(3) Acetate (di) m.p 243°C. (3) Benzoate(di) m.p 236°C

m.p 145-147°C **p-Phenylene diamine** (1, 4-diamino benzene)

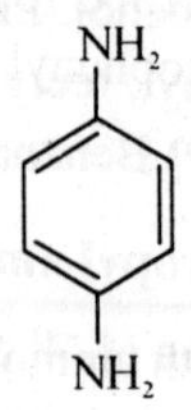

(1) Crystalline compound which darkens on exposure to light, insoluble in water, soluble in alcohol, ether and chloroform.
(2) Dissolve 25-30mg of compound in 7-8ml water, add 1-2 drop of 10% $FeCl_3$ solution. The solution turns green and on adding one more drop of $FeCl_3$ solution it turns brown.
(3) Acetate (di) m.p 303 °C
(4) Picrate m.p 202°C
(5) Formyl m.p 206°C.
(6) 2, 4-dinitrophenyl (di) m.p 177°C.

SECONDARY AMINES

Aliphatic Liquid

b.p 7°C **Dimethyl amine** $(CH_3)_2NH$

(1) Liquid with ammonical odour, soluble in water.
(2) Responds to Simon's test.
(3) Add a solution of 18gm of sodium nitrite in 20ml water to a solution of 20gm of dimethylamine in 10ml water acidified with H_2SO_4. Distill the mixture almost to dryness. Acidify the distillate and distill again. Add potassium carbonate to the distillate. Remove the yellow oily layer, distill to recover nitrosoamine at b.p 148°C.
(4) Benzoate m.p 42°C (5) Picrate (yellow) m.p 158°C. (6) 2, 4-dinitrophenyl m.p 87°C.

b.p 55°C **Diethylamine** $(C_2H_5)_2NH$

(1) Colourless liquid with ammonical odour, sparingly soluble in water, soluble in alcohol and ether.
(2) Respond's to Simon's test.
(3) Dissolve 1ml of compound in 3ml saturated alcoholic oxalic acid solution. Shake thoroughly, filter the precipitated dioxalate derivative, m.p 206°C.
(4) Hydrochloride m.p 215°C (5) Benzoate m.p 42°C (6) Picrate m.p 155°C. (7) 2, 4 dinitrophenyl m.p 80°C

b.p 62°C **Methyl propyl amine** $CH_3NHCH_2CH_2CH_3$

(1) Colourless liquid with ammonical fishy odour, soluble in water.

(2) Respond's to Simon's test.

b.p 84°C **Diisopropylamine** $[(CH_3)_2CH]_2NH$

(1) Colourless liquid with ammonical odour, slightly soluble in water.
(2) Picrate m.p 140°C.

b.p 119°C **Di-n-propyl amine** $(C_2H_5CH_2)_2NH$

(1) Colourless liquid with ammonical odour, soluble in water.
(2) Respond to Simon's test.
(3) Dissolve 1ml of compound in 3ml pet ether. Add 1ml of CS_2. After 2-3 minutes white crystals of dipropylamine dipropyldithiocarbamate are precipitated, recrystallise from pet ether, m.p 118°C.
(4) Picrate m.p 75°C. (5) 2, 4-dinitrophenyl m.p 40°C

b.p 139°C **Di-isobutylamine** [bis (β-methyl propyl) amine] $[(CH_3)_2CHCH_2]_2NH$

(1) Liquid with unpleasant ammonical odour, slightly soluble in water.
(2) Responds to Simon's test.
(3) Add 1ml of compound to 3ml saturated aqueous oxalic acid solution. Heat to dissolve the precipitate. Cool and collect the dioxalate derivative, m.p 250°C.
(4) Acetate m.p 86°C (5) Picrate m.p 119°C. (6) 2, 4-dinitrophenyl m.p 112°C.

b.p 159°C **Di –n-butyl amine** $[CH_3(CH_2)_3]_2NH$

(1) Colourless liquid, soluble in water, alcohol and ether.
(2) Responds to Simon's test.
(3) Picrate m.p 60°C

b.p 190°C **Diisoamyl amine** [bis (υ-methyl butyl) amine] $[(CH_3)_2CHCH_2CH_2]_2NH$

(1) Colourless liquid with unpleasant odour similar to amyl alcohol. Very difficult soluble in water, soluble in alcohol and ether.

(2) Add 1ml of compound to 3ml saturated aqueous oxalic acid solution. Heat to dissolve the precipitate. Cool and collect the dioxalate derivative, m.p 253°C.

Aromatic Liquids

b.p 196°C **N-Methyl aniline**

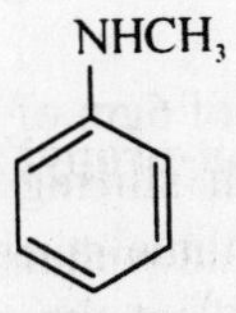

(1) Colourless liquid which turns yellowish after keeping for longer duration, odour aniline like, insoluble in water, soluble in alcohol.
(2) Responds to Libermann test.
(3) To 2-3drops of compound add .5ml dil HCl and few drops of $FeCl_3$ solution. A greenish blue coloured solution is obtained.
(4) Dissolve 1ml of compound in 5ml water and 2ml conc HCl. Add 2ml of saturated sodium nitrite solution. A light yellow oily layer of nitrosomethyl phenylamine will separate out.
(5) To 1ml of compound add 6ml conc H_2SO_4 and cool. Add with stirring 2.5 ml conc HNO_3. Continue stirring for 2-3 minutes and then pour dropwise on crushed ice and stir vigorously. Tetranitromethyl aniline precipitates out, m.p 128°C.
(6) Heat 5.3gm of compound with 9gm of methyl ester of p-toluenesulphonic acid and 10ml acetone. Cool and collect by filteration trimethylphenylammonium p-toluenesulfonate, m.p 159°C. Evaporate off the filtrate and add sodium hydroxide solution to it. Distill and collect dimethylaniline at b.p 193°C.
(7) Hydrochloride m.p 120°C
(8) Acetate m.p 102°C
(9) Benzoate m.p 63°C
(10) Picrate m.p 147°C.
(11) 2, 4-dinitrophenyl m.p 167°C.

b.p 205°C **N-Ethyl aniline** (N-ethyl phenyl amine)

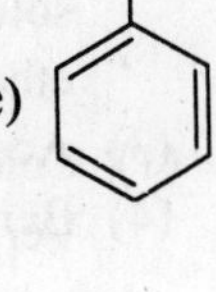

(1) Colourless liquid with aniline like odour, darkens on exposure, sparingly soluble in water, soluble in organic solvents.
(2) Cool a mixture of 1ml of conc HNO_3 and 3ml of conc H_2SO_4 in an ice bath and add dropwise to it a mixture of 1ml of compound in 5ml conc H_2SO_4. After the addition is complete, pour it on crushed ice with stirring. Tetranitroethylaniline seperates out, m.p 96°C.
(3) To a solution of 6gm of compound in 25ml glacial acetic acid add dropwise with stirring 7ml of bromine. After the addition is complete continue stirring for 5min, add 10ml water and leave overnight. Collect the precipitated 2, 4, 6-tribromoethylaniline, recrystallise from alcohol, m.p 45°C.
(4) Hydrochloride m.p 176°C
(5) Acetate m.p 54°C
(6) Benzoate m.p 60°C
(7) Picrate m.p 132°C
(8) 2, 4-dinitrophenyl m.p 95°C.

b.p 206°C **N Methyl-m-toluidine**

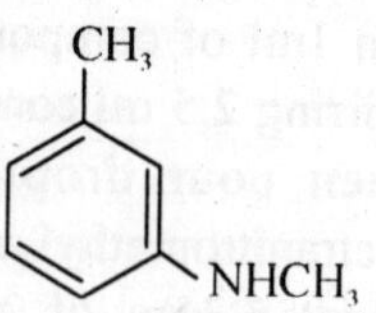

(1) Insoluble in water, soluble in alcohol and ether.
(2) Acetate m.p 66°C.

b.p 207°C **N Methyl-o-toluidine**

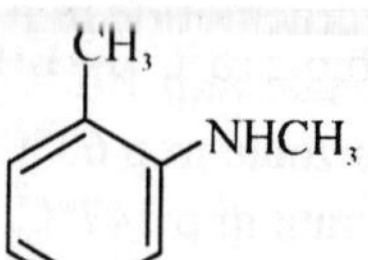

(1) Insoluble in water, soluble in alcohol and ether.
(2) Cool a solution of 12gm of compound in conc HCl to 0°C. Add a solution of 8gm of sodium nitrite in 35ml water. Extract with ether. Distill off the ether and collect o-Tolylmethylnitrosoamine.
(3) Acetate m.p 55°C
(4) Benzoate m.p 66°C

(5) Picrate m.p 90°C
(6) 2, 4-dinitrophenyl m.p 155°C.

b.p 208°C **N Methyl-p-toluidine** 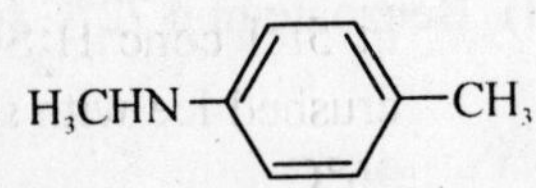

(1) Insoluble in water, soluble in alcohol and ether.
(2) Cool a solution of 12gm of compound in conc HCl to 0°C. Add a solution of 8gm of sodium nitrite in 35ml water. Extract with ether. Distill off the ether and collect p-Tolylmethylnitrosoamine, m.p 53°C.
(3) Hydrochloride m.p 119°C
(4) Acetate m.p 83°C
(5) Benzoate m.p 53°C
(6) Picrate m.p 131°C.

b.p 217°C **N Ethyl-p-toluidine** (N-ethyl-4-methyl aniline)

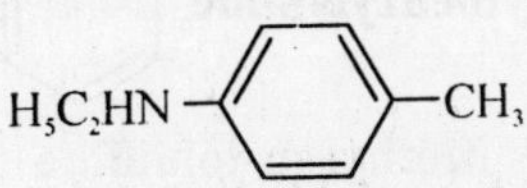

(1) Dissolve 1ml of compound in 2ml conc H_2SO_4. Cool and add dropwise 1ml conc HNO_3. Collect the precipitated nitro derivative, m.p 48°C.
(2) 2, 4-dinitrophenyl m.p 120°C.

b.p 218°C **N Ethyl-o-toluidine** (N-ethyl-2-methyl aniline)

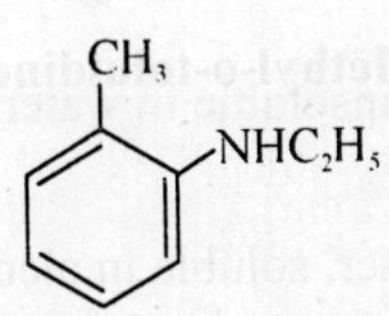

(1) Colourless or yellow coloured liquid, insoluble in water, soluble in alcohol and ether.
(2) Cool a solution of 5gm of compound in 20ml conc HCl and add with stirring a concentrated aqueous solution of of 2.5gm of sodium

nitrite. Add about 10ml conc HCl and keep at room temperature for 20min. Dilute with water and then neutralise with aqueous ammonia. Cool and collect 5-Nitroso-ethyl-o-toluidine, recrystallise from benzene, m.p 140°C.

(3) Benzoate m.p 72°C (4) 2, 4 dinitrophenyl m.p 114°C.

b.p 222°C **N-Propylaniline**

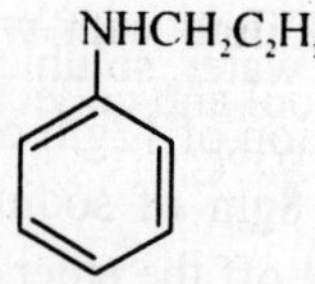

(1) Colourless liquid, insoluble in water, soluble in alcohol and ether.
(2) Hydrochloride m.p 125°C
(3) Acetate m.p 48°C

b.p 228°C **Butylbenzylamine**

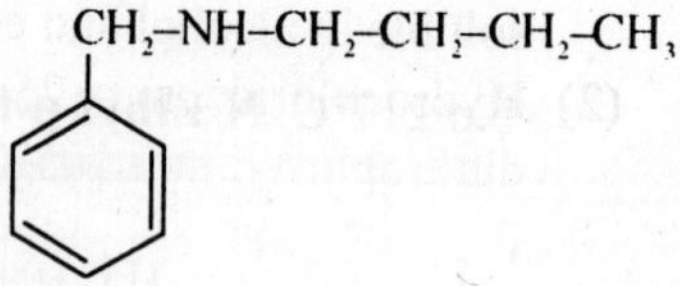

(1) Colourless liquid, insoluble in water.
(2) Hydrochloride m.p 241°C

b.p 242°C **N-Butylaniline**

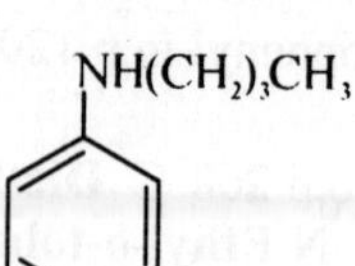

(1) Colourless liquid, insoluble in water, soluble in alcohol and ether.
(2) Benzoate m.p 56°C.

b.p 293°C **Methyl 1-naphthylamine**

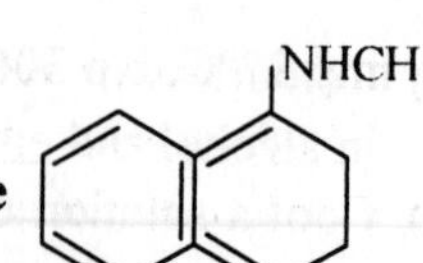

(1) Reddish oily liquid, insoluble in water, soluble in alcohol and ether.

(2) To 1ml of compound add few drops of alcoholic $FeCl_3$. A dark violet coloured solution is obtained.

(3) Cool a solution of 5gm of compound in 20ml conc HCl and add with stirring a concentrated aqueous solution of of 2.5gm of sodium nitrite. Add about 10ml conc HCl and keep at room temperature for 20min. Dilute with water and then neutralise with aqueous ammonia. Cool and collect the nitroso derivative, m.p 157°C.

(4) Acetate m.p 94°C.

b.p 300°C (d) **Dibenzylamine**

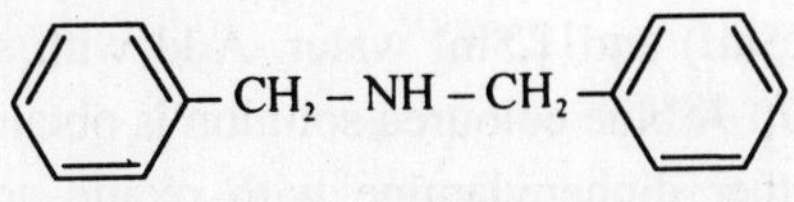

(1) Colourless liquid with bromotoluene like odour, insoluble in water, soluble in alcohol and ether.

(2) Hydrochloride m.p 256°C (3) Benzoate m.p 112°C. (4) 2, 4-dinitrophenyl m.p 105°C

Aliphatic Solid

m.p 28°C **Iminoethyl alcohol** $NH(CHCH_2OH)_2$

(1) Deliquescent solid, soluble in water, alcohol and ether.

m.p 28°C, b.p 268°C **Diethanolamine** (2, 2'-imino di-ethanol) $NH(CH_2CH_2OH)_2$

(1) Liquid or crystalline solid, soluble in alcohol and ether.

(2) Picrate m.p 110°C

Aromatic Solid

m.p 37°C, b.p 306°C **N-Benzylaniline**

(1) Colourless solid, soluble in alcohol.

(2) Acetate m.p 58°C (3) Benzoate m.p107°C (4) 2, 4-dinitrophenyl m.p 168°C. (5) Hydrochloride m.p 197°C.

m.p 53°C **Diphenylamine** (N-Phenyl aniline)

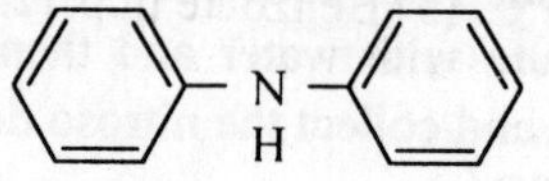

(1) Lustrous colourless solid, insoluble in water, soluble in alcohol and ether.

(2) Dissolve few milligram of compound in 5ml mixture of conc H_2SO_4 (3.5ml) and 1.5ml water. Add with stirring one drop of conc HNO_3. A blue coloured solution is obtained.

(3) Heat together diphenylamine with oxalic acid. The product is intense blue called diphenylamine blue.

(4) Dissolve 200mg of compound in alcohol and then add 20-25 drops conc HCl (heat if necessary). Cool and add 10-15 drops of saturated aqueous sodium nitrite solution. The solution first becomes green and then pale yellow precipitate is obtained. Dilute with 1-2ml water, cool in ice and filter. Diphenyl nitrosoamine is obtained as yellow crystals, m.p 66°C.

(5) Mix together 2.2gm of diphenylamine, 900mg of sulphur and 400mg of anhydrous $AlCl_3$ by heating at 140-150°C. The melt when cool is extracted first with water and then with dil alcohol. The residue consists of pure thiodiphenylamine, m.p 190°C.

(6) Acetate m.p 101°C.

(7) Benzoate m.p 180°C.

(8) Picrate m.p 182°C.

m.p 79°C **Di-p-tolylamine**

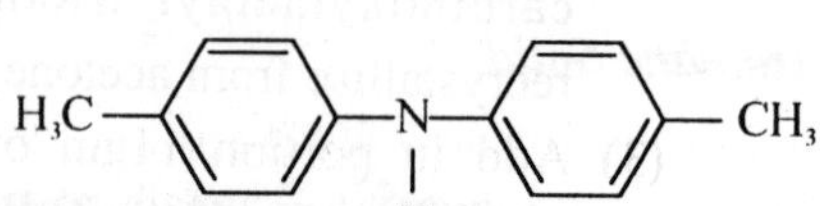

(1) Crystalline solid, very slightly soluble in water.

(2) Dissolve few milligrams of compound in conc H_2SO_4 and then

add 2-3 drops of conc HNO_3. A yellow coloured solution is obtained.

(3) Dissolve 200mg of compound in alcohol and then add 20-25 drops conc HCl (heat if necessary). Cool and add 10-15 drops of saturated aqueous sodium nitrite solution. Cool, yellow crystals of di-p-tolylnitrosoamine are obtained, m.p 101°C.

(4) Acetate m.p 85°C. (5) Benzoate m.p 125°C.

TERTIARY AMINES

Aliphatic Liquid

b.p 3°C **Trimethylamine** $(CH_3)_3N$

(1) Colourless gas with intense ammonical and fishy odour, soluble in water.

(2) Mix 10ml aqueous solution of compound (33%) with 60ml of 3% H_2O_2 solution. Keep the mixture at room temperature for 24hrs (add more H_2O_2 if smell of amine persist). Evaporate off the solution under reduced pressure, crystallise the residue from ether-alcohol and collect the hydrate of trimethyl amine oxide, m.p 96°C.

(3) Hydrochloride m.p 278°C (4) Picrate m.p 216°C. (5) Methiodide m.p 230°C (d)

b.p 89°C **Triethylamine** $(C_2H_5)_3N$

(1) Colourless liquid with ammonical odour, soluble in alcohol and ether.

(2) Mix thoroughly 2gm of compound with 4gm of iodoacetic ester and then keep at room temperature for sometime. Precipitation of carbthoxymethyl triethyl ammonium iodide takes place, recrystallise from acetone, m.p 139°C.

(3) Add in portion 11gm of compound to a solution of 8gm of propionyl chloride in 50ml of ether such that the mixture refluxes. Keep overnight. Remove the ethereal layer and distill. Collect methyl ketene dimer at b.p 57-58°C/12mmHg.

(4) Hydrochloride m.p 253°C (4) Picrate m.p 173°C.

(5) Methiodide m.p 280°C

b.p 155-158°C **Tri-n-propyl amine** $(C_2H_5CH_2)_3N$

(1) Colourless liquid with ammonical odour, very slightly soluble in water, soluble in alcohol and ether.
(2) Picrate m.p 116°C. (3) Methiodide m.p 208°C

Aromatic Liquid

b.p 185°C **N, N-Dimethyl-o-toluidine**

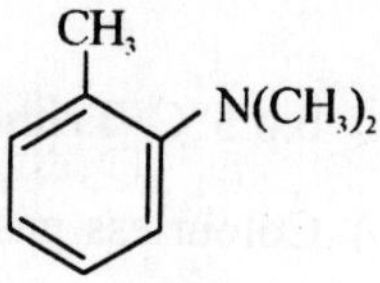

(1) Colourless liquid, soluble in alcohol.
(2) Hydrochloride mp 156°C. (3) Picrate m.p 122°C (4) Methiodide m.p 210°C

b.p 194°C **N, N-Dimethyl aniline**

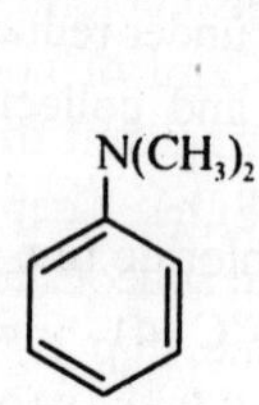

(1) Colourless liquid with strong persistent unpleasant odour, insoluble in water soluble in organic solvents.
(2) To 1ml of compound add 4ml of benzaldehyde and 50mg of $ZnCl_2$. Heat the mixture over a low flame for 15-30 seconds. Cool and add 9ml alcohol and 1ml conc HCl. Solution when viewed by direct light will be intense green.
(3) Dissolve 2ml of compound in 7ml conc HCl and stir. Cool to 0°C and then add 2ml of 10% sodium nitrite solution(1.3gm in 3ml water). Keep at room temperature for sometime and then filter. Collect the solid p-nitrosodimethylaniline hydrochloride (m.p 177°C) dissolve in 15ml water and then add 1ml saturated K_2CO_3 solution and shake. Add 10ml ether, shake and remove the organic layer. Repeat the extraction process twice. Combine the ethereal

layer, distill off ether and collect p-nitrosodimethyl aniline, m.p 85°C.

(4) Mix thoroughly 4.8gm of compound with 7gm of allyl iodide and then keep at room temperature for several days. Collect the precipitated phenyl dimethyl allylammonium iodide, m.p 88°C.

(5) Stir 10gm of compound with 250ml of 3% H_2O_2 at 70-80°C for 5hrs. Filter and concentrate the filtrate. To this filtrate add aqueous solution of picric acid, warm on water bath and cool. Collect the picrate of dimethylaniline oxide, m.p 137-138°C.

(6) Hydrochloride m.p 90°C (7) Picrate m.p 162°C (8) Methiodide m.p 220°C (d)

b.p 210°C **N, N-Dimethyl-p-toluidine**

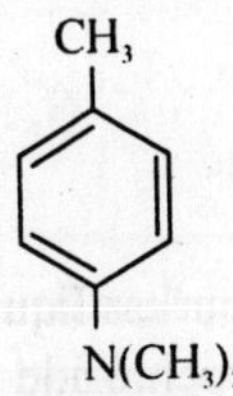

(1) Liquid, insoluble in water, soluble in alcohol and ether.

(2) Dissolve 2ml of compound in 5ml conc H_2SO_4. Add dropwise with stirring 5ml of nitrating mixture (2.5ml conc HNO_3+2.5ml conc H_2SO_4). Keep the mixture overnight, pour in ice and then add dilute aqueous ammonia solution to it. Collect 2-nitrodimethyl-p-toluidine, m.p 27°C.

(3) Add a solution of 10gm of compound in 10ml alcohol to a solution of 4.6gm of boric acid in 30ml alcohol. Cool the mixture in ice and add a solution of 15gm of tetranitromethane in 15ml alcohol. Keep the mixture in ice for 1hr and filter. Dissolve the precipitated solid in 30ml sodium carbonate solution. Extract with ether. Distill off ether and collect 3-Nitrodimethyl-p-toluidine at b.p 110-113°C/ .5mmHg.

(4) Picrate m.p 129°C

(5) Methiodide m.p 220°C.

b.p 212°C **N, N-Dimethyl-m-toluidine**

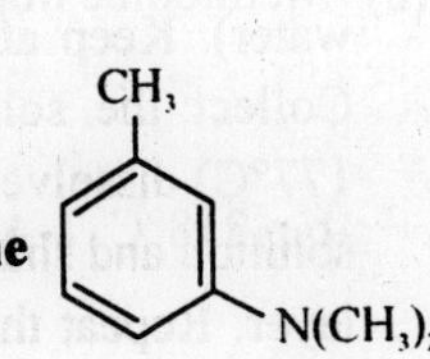

(1) Very slightly soluble in water, soluble in alcohol and ether.
(2) Dissolve 13gm of compound in 25ml conc HCl and 25ml water. Cool and add 7gm of aqueous solution of sodium nitrite. Keep at room temperature and collect the precipitated p-nitrosodimethyl – m-toluidine hydrochloride. Dissolve it in water and add dilute NaOH solution. Extract with ether. Distill off ether and collect p-nitrosodimethyl –m-toluidine, m.p 92°C.
(3) Picrate m.p 131°C. (4) Methiodide m.p 177°C.

b.p 216°C **N, N-Diethyl aniline** (N-phenyl diethyl amine)

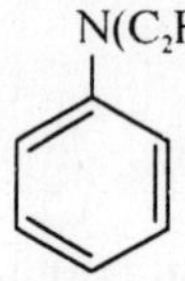

(1) Colourless liquid with aniline like odour, insoluble in water, soluble in alcohol and ether.
(2) Heat together 1ml of compound with equal amount of $ZnCl_2$ on oil bath at 200°C for 3-4 minutes. . The fused mass will give blue or violet colour. Dissolve it in 3-5ml alcohol, the colour changes to bright pink
(3) Dissolve 1ml of compound in 5ml glacial acetic acid. Add 1.5 ml conc HNO_3 drop by drop and heat the mixture over a lowflame until it is light yellow. Remove the heat source and when the colour of mixture is light brown, cool and pour in crushed ice. Collect the precipitated p-nitrodiethylaniline, m.p 78°C.
(4) To a cold solution of 2ml absolute ethanol add 5ml of diethylaniline in 10ml ether, with stirring add a solution of 1.5 ml of phosphorous trichloride in 5ml pet ether. Reflux for 30 min. Cool and collect the precipitated diethylaniline hydrochloride while the liquid contains triethylphosphite.
(5) Picrate m.p 142°C
(6) Methiodide m.p 104°C.

b.p 218°C **2-Amino dimethylaniline**

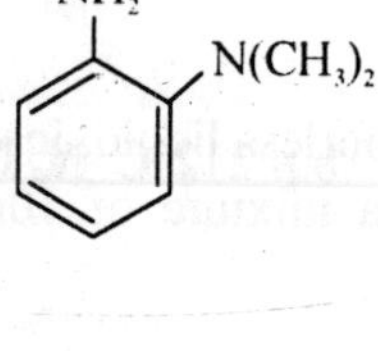

(1) Colourless liquid with menthol like odour
(2) To 1ml of compound add few drops of $FeCl_3$ solution. A brown coloured solution is obtained which on boiling turns red.
(3) Picrate m.p 138°C.

b.p 227°C **Diethyl-p-toluidine** (1-diethyl amino-4-methyl benzene)

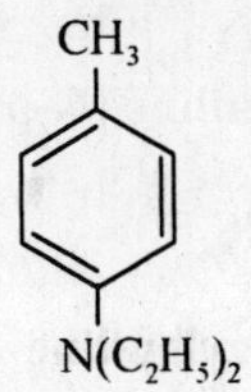

(1) Colourless liquid, very slightly soluble in water, miscible with alcohol and ether.
(2) Hydrochloride m.p. 157°C (3) Picrate m.p110°C. (4) Methiodide m.p 184°C

b.p 231°C **Diethyl-m-toluidine**

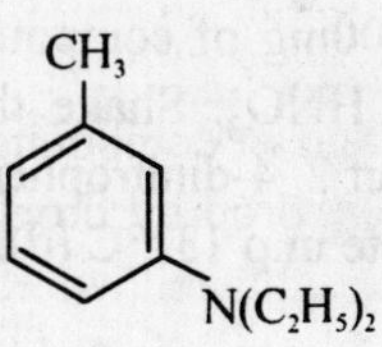

(1) Light amber coloured oily liquid.
(2) Picrate m.p 97°C.

b.p 267-275°C **N, N-Di-n-butylaniline** (N-Phenyl dibutyl amine)

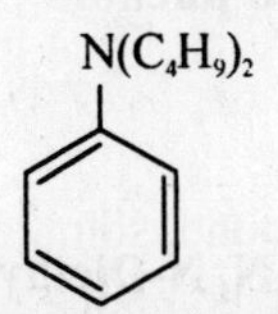

(1) Colourless liquid, insoluble in water, soluble in alcohol and ether.
(2) To a mixture of 10ml of fuming nitric acid and 10ml glacial

acetic acid add with stirring 5ml of compound. Warm the mixture until the dark colour of the mixture changes to pale yellow or red. Cool and pour in ice. Filter the precipitated solid, dissolve it in 5ml of cold fuming nitric acid and then leave for 2 days. Collect the pecipitated 2, 4, 6-trinitrophenyl-n-butylamine, recrystallise from glacial acetic acid, m.p 100°C.

(3) Picrate m.p 125°C.

Aromatic Solid

m.p 70°C **N, N-Dibenzylaniline** 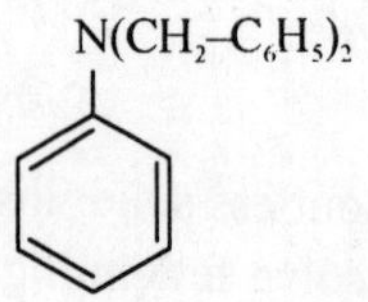

(1) Crystalline solid, difficult soluble alcohol, easily soluble ether and benzene.
(2) Dissolve 200mg of compound in 5ml conc H_2SO_4 and add $K_2Cr_2O_7$ to it. Heat the mixture for 10-15 min. Smell of bitter almond due to formation of benzaldehyde is observed.
(3) To 500mg of compound in 5ml glacial acetic acid add 1.5ml of conc HNO_3. Shake thoroughly and heat at 30-40°C. Cool and collect 2, 4-dinitrophenyl dibenzylamine, m.p 106°C.
(4) Picrate m.p 131°C (d). (5) Methiodide m.p 135°C

m.p 91°C **Tribenzylamine** $(C_6H_5–CH_2)_3$ N

(1) Crystalline compound, insoluble in water, sparingly soluble in alcohol, easily soluble in ether.
(2) Hydrochloride m.p 227°C. (3) Picrate m.p 190°C. (4) Methiodide m.p 184°C

m.p 72°C, b.p 209°C **N, N-Diethyl-o-toluidine** 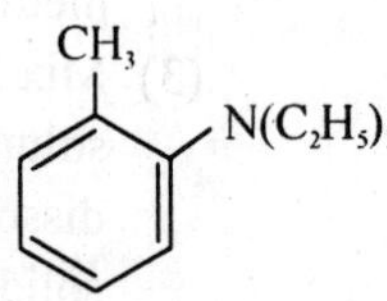

(1) Crystalline solid, very slightly soluble in water, soluble in alcohol and ether.

(2) Add a solution of iodine in KI to a solution of compound in dil HCl. Shake thoroughly, keep at room temperature and collect the solid complex derivative m.p 100°C.
(3) Picrate m.p 180°C.
(4) Methiodide m.p 224°C

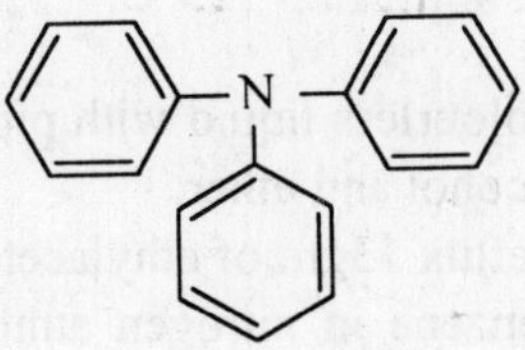

m.p 127°C **Triphenylamine**

(1) Colourless solid, insoluble in water, soluble in benzene and ether.
(2) Dissolve few milligrams of compound in glacial acetic acid. The solution turns violet. On adding conc. H_2SO_4 the solution turns blue.
(3) Dissolve few milligrams of compound in glacial acetic acid. The solution turns violet. On adding conc HNO_3 the solution turns green. .
(4) Trinitro m.p 280°C.

Aliphatic Solid

m.p 280°C **Hexamethylene tetramine** (urotropin) $(CH_2)_6N_4$

(1) Crystalline solid, sparingly soluble in water, insoluble in ether. Sublimes at 263°C.
(2) Dissolve 200mg of compound in 7ml water and then add 7-8 drops of conc HCl. Add 2ml of sodium nitrite solution (4gm in 10ml water). Keep the mixture at room temperature for sometime with occassional shaking. Filter the precipitated dinitrosopenta-methylenetetramine, m.p 209°C.
(3) Mix 2ml of 1% aqueous solution of compound with 10ml saturated solution of $HgCl_2$. Add 5ml water and boil until the precipitate dissolves. Filter hot and cool the filtrate. Collect the precipitated solid, m.p 232°C.
(4) Picrate m.p 179°C.

(ii) Heterocyclic Bases

Liquids

b.p 89°C **Pyrrolidine** (tetrahydro pyrrole) 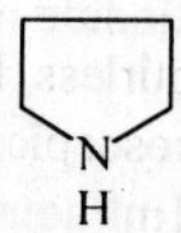

(1) Colourless liquid with piperidine like odour, miscible with water, alcohol and ether.
(2) Reflux 13gm of ethylacetoacetate, 7.1gm of pyrrolidine and 40ml benzene in nitrogen atmosphere for 45minutes. Cool, distill off the excess benzene and collect ethyl β pyrrolidinocrotonate
(3) To 12gm isobutyraldehyde add dropwise with stirring 13gm of pyrrolidine. Reflux the mixture under nitrogen for 3.5 hrs. Distill the resultant mixture under reduced pressure and recover 1-(2-methylpropenyl) pyrrolidine b.p 94-107°C /114-118mm.
(4) Picrate m.p 112°C

b.p 106°C **Piperidine** (hexa hydro pyridine) 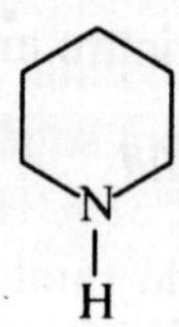

(1) Colourless liquid with characteristic unpleasant ammonical odour, miscible with water and alcohol.
(2) Responds to Simon's test.
(3) To 4ml of compound add carefully 3ml of CS_2, a vigorous reaction sets in. After the reaction has subsided dissolve it in 10ml alcohol. Concentrate the resultant mixture, cool and collect piperyldithiocarbamate, m.p 172°C.
(4) To 1gm of compound add 4ml of 25% H_2SO_4, cool to 0°C and then diazotise with sodium nitrite solution. Continue stirring for 5min and then isolate with ether. Distill off ether and collect nitrosopiperidine as yellow oil at b.p 218°C.
(5) Hydrochloride m.p 237°C (6) Benzoate m.p 48°C (7) Picrate m.p 152°C (8) 2, 4-dinitrophenyl m.p 92°C.

b.p 115°C **Pyridine** 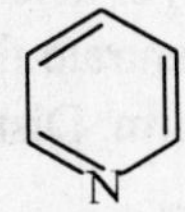

(1) Colourless liquid with strong characteristic unpleasant odour, hygroscopic, miscible with water, alcohol, ether and chloroform.

(2) To 1ml aqueous solution of compound add alcoholic solution of 2, 4-dinitrochlorobenzene. Shake thoroughly and add NaOH solution. A red-violet colouration appears.

(3) To a solution of 2.3 gm of hydroxylamine-o-sulphonic acid in 13ml water add 5ml pyridine. Heat the mixture at 90°C for 20 min, cool and add 2.7gm of K_2CO_3. Distill off excess water and pyridine, add 20ml ethanol and filter. Add 3ml of HI to the filtrate and keep the solution at –20°C for 1hr. Collect the precipitated aminopyridinium iodide and recrystallise from ethanol, m.p 160-162°C.

(4) Cool 25ml of pyridine to 10°C and add in portion 3.4 gm of CrO_3. After the addition is complete stir the resultant mixture. After one hour deep red coloured crystals appear. Decant the supernatant liquid and collect crystals of dipyridine chromium (*vi*) oxide.

(5) To a solution of 20gm of pyridine in 150ml acetic acid add 25ml of 33% H_2O_2. Heat the mixture on water bath for 3hrs and then add 15ml of H_2O_2, continue heating for another 9hrs. . Concentrate the solution under reduced pressure, dilute with 10ml water and again concentrate. Make the concentrate alkaline by adding sodium carbonate. Add 100ml chloroform to the alkaline solution. Remove the chloroform layer and distill. Collect pyridine –1-oxide at b.p 138-140°C/15mmHg. (b.p 270°C)

(6) Picrate m.p 167°C. (7) Methiodide m.p 117°C

b.p 129°C **2-methyl pyridine** (α-picoline) 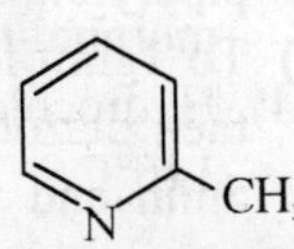

(1) Colourless liquid with strong unpleasant pyridine like odour, miscible with water, alcohol and ether.

(2) Heat a solution of 11ml of compound in 100ml water to 70°C on water bath. Add in portions 45gm of $KMnO_4$. Continue heating

until the colour of $KMnO_4$ is discharged and raise the temperature to 95°C. Filter hot. Concentrate the filtrate and acidify with conc HCl. Extract with chloroform. Distill off the chloroform and collect picolinic acid, m.p 138°C.

(3) To a 8gm suspension of sodamide in 30ml xylene add 8ml of compound. Stir the mixture and add 10ml of butylchloride. Reflux for 2-3hrs, cool and add 20ml water. Discard the aqueous layer and extract the xylene solution with 1:1 hydrochloric acid: water. Steam distill the acidic layer, cool and make it alkaline. Steam distill again. Extract with ether. Distill off ether and collect 2-pentyl pyridine at 122-124°C /50mmHg.

(4) Picrate m.p 167°C.

(5) Methiodide m.p 227°C

b.p 129°C **Morpholine** (tetrahydro-1, 4-oxazine)

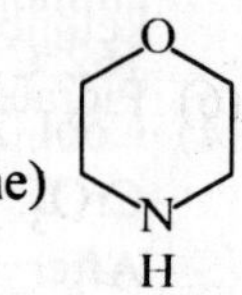

(1) Colourless liquid with piperidine like odour, very volatile and hygroscopic, soluble in all solvents.

(2) Heat a mixture of 7gm of morpholine and 5.2gm of acetone cyanohydrin nitrate to 60°C. Vigorous reaction occurs and temperature reaches 110°C, cool to 80°C and maintain this temperature for an hour. Pour in 40ml 10% HCl and extract with methylene chloride. Distill off the solvent, pour the residual oil in 15ml ethanol and cool to 0°C. Recover white crystals of N-nitromorpholine, m.p 52-54°C.

(3) Heat to boiling a mixture of 14.7gm of cyclohexanone, 15.7 gm of morpholine and 0.3gm of p-toluene sulphonic acid in 30ml toluene for 4hrs. Distill off the excess toluene and recover 1-morpholine 1-cyclohexene at 118-120°C/10mmHg.

(4) Hydrochloride m.p 175°C. (5) Benzoate m.p 75°C (6) Picrate m.p 146°C.

b.p 130°C **Pyrrole** (azole)

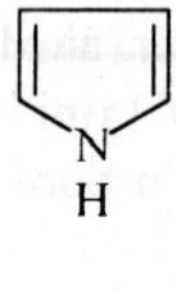

(1) Colourless liquid with chloroform like odour, turns yellow on exposure to air, insoluble in water, soluble in alcohol and ether.
(2) In a test tube heat 1ml of compound with 2ml water. Bring a pine splinter dipped in HCl near the mouth of test tube. The pine splinter turns red.
(3) To 1ml of compound add 1ml of HCl, a vigorous reaction occurs forming a red brown resin which slowly becomes hard on cooling.
(4) To 1ml of compound add 1ml aqueous isatin and 1ml dil H_2SO_4. Stir, an indigo-blue precipitate is obtained.
(5) To 22.5 gm of trichloroacetyl chloride in 20ml ether add with stirring and in portion7.7 gm of pyrrole in 64ml ether. Heat of reaction causes the mixture to reflux during each addition. Continue stirring for 1hr and add 10gm K_2CO_3 in 30ml water. Remove the organic layer and distill off the ether. Dissolve the residue in minimum hexane, cool to 0°C and collect 2-pyrrole trichloromethyl ketone, m.p 73-75°C.
(6) Picrate m.p 69°C (d).

b.p 143°C **3-Methyl pyridine** (β-picoline) 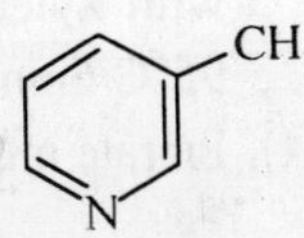

(1) Colourless liquid, miscible with water, alcohol and ether.
(2) Heat a solution of 11ml of compound in 100ml water to 70°C on water bath. Add in portions 45gm of $KMnO_4$. Continue heating until the colour of $KMnO_4$ is discharged and raise the temperature to 95°C. Filter hot. Concentrate the filtrate, acidify with conc HCl (pH 3.4) and leave overnight. Collect precipitated nicotinic acid, recrystallise from hot water, m.p 235°C.
(3) To 60ml ammonia solution add with stirring 50mg of Iron (*III*) nitrate. Add in small pieces 1.2gm Na metal. Introduce 5ml of compound and stir for 20-25 min. Add dropwise 5.3ml of butyl chloride such that the reaction does not become too vigorous. Keep overnight, allow ammonia to evaporate. Extract with xylene. Distill off xylene and recover 3-pentyl pyridine at 100°C/9mmHg.
(4) Picrate m.p 150°C. (5) Methiodide m.p 92°C

b.p 143°C **4-Methylpyridine** (υ picoline) 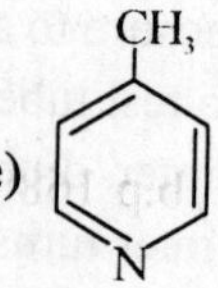

(1) Colourless liquid, miscible with water, alcohol and ether.
(2) Heat a solution of 11ml of compound in 100ml water to 70°C on water bath. Add in portions 45gm of $KMnO_4$. Continue heating until the colour of $KMnO_4$ is discharged and raise the temperature to 95°C. Filter hot. Concentrate the filtrate, acidify with conc HCl (pH 3.6). Collect precipitated isonicotinic acid, recrystallise from hot water, m.p 308°C.
(3) To 60ml ammonia solution add with stirring 50mg of Iron (*III*) nitrate. Add in small pieces 1.2gm Na metal. Introduce 5ml of compound and stir for 20-25 min. Add dropwise 5.3ml of butyl chloride such that the reaction does not become too vigorous. Keep overnight, allow ammonia to evaporate. Extract with xylene. Distill off xylene and recover 4-pentyl pyridine at 95°C/6mmHg.
(4) Picrate m.p 167°C (5) Methiodide m.p 152°C.

b.p 149°C **2-Ethyl pyridine** (α-ethyl pyridine) 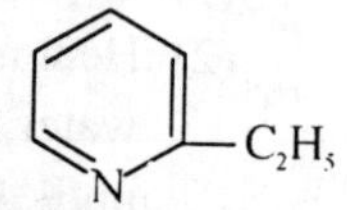

(1) Slightly soluble in water, easily soluble alcohol and ether.
(2) Hydrochloride m.p 103°C.

b.p 165°C **3-Ethyl pyridine** (β-ethyl pyridine) 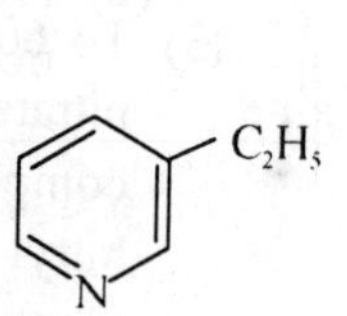

(1) Colourless liquid with unpleasant odour, insoluble in water, soluble in alcohol and ether.
(2) On oxidation with CrO_3 mixture, nicotinic acid is obtained.
(3) Picrate m.p 128°C.

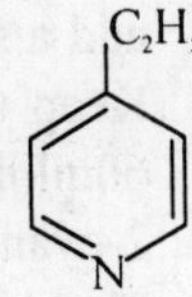

b.p 168°C **4-Ethyl pyridine** (υ-ethyl pyridine)

(1) Colourless liquid, soluble in dilute acid.
(2) On oxidation with $KMnO_4$, isonicotinic acid is obtained, m.p 308°C.
(3) Picrate m.p 168°C.

b.p 170.5°C **2, 4, 6 Trimethylpyridine** (υ Collidine)

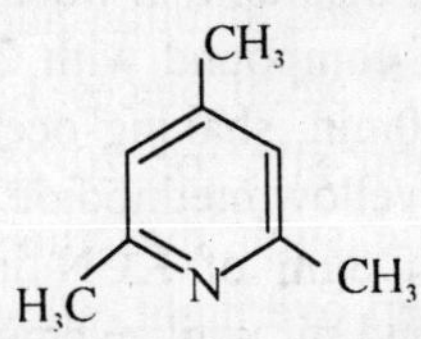

(1) Colourless liquid, insoluble in water, soluble in alcohol and ether.
(2) To 1ml of compound add $AgNO_3$ solution. Crystalline precipitate is obtained.
(3) Picrate m.p 156°C.

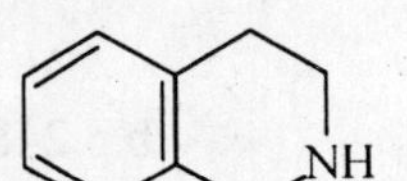

b.p 233°C **1, 2, 3, 4-Tetrahydroisoquinoline**

(1) Reduces ammonical silver nitrate solution
(2) Hydrochloride m.p 195°C (3) Acetate m.p 46°C (4) Benzoate m.p 129°C (5) Picrate m.p 195°C.

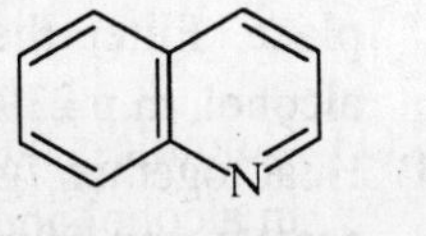

b.p 237°C **Quinoline** (1-benzazine)

(1) Colourless liquid with strong basic unpleasant characteristic odour, sparingly soluble in water, soluble in alcohol, ether and chloroform.

(2) To a cold solution of 7gm of quinoline in 2ml hydrogen cyanide add a solution of 5gm of o-nitrobenzoyl chloride in 5ml benzene. Keep the resultant mixture at room temperature for 12hrs. Add 60ml benzene and concentrate. Cool and collect l-o-nitrobenzoyl-2-cyano-1:2 dihydroquinoline, m.p 173°C.

(3) Picrate m.p 203°C (4) Methiodide m.p 133°C.

b.p 243°C, m.p 26°C **Isoquinoline** (2-benzazine, leucoline)

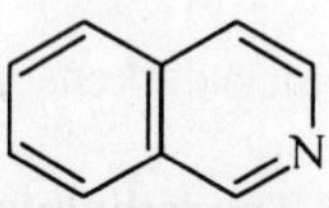

(1) Colourless liquid, insoluble in water.

(2) Warm 100mg of compound with 2-3ml methyl iodide and then keep aside for 10min, shaking occassionally during this period. Cool and collect yellow methiodide, m.p 159°C.

(3) To a solution of 8gm of KCN in 50ml water add 5.2gm of compound, cool and stir until an emulsion is formed. Add dropwise with stirring 10.5gm of benzoyl chloride. After addtion is complete continue stirring for 1hr. Cool and collect 1-cyano-2-benzoyl-1, 2 dihydroisoquinoline, m.p 125-127°C.

(4) Picrate m.p 222°C.

b.p 248°C **Quinaldine** (2-Methylquinoline)

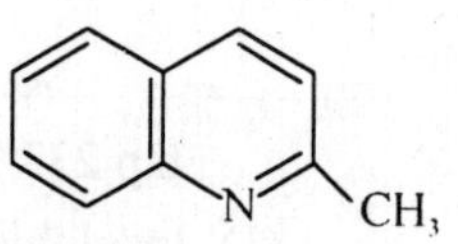

(1) Colourless liquid, insoluble in water, soluble in alcohol, ether and chloroform.

(2) Heat 6ml of compound with 1.2 gm of phthalic anhydride in an oil bath at 170°C for 5-10 min. Dissolve in 10ml of warm conc H_2SO_4. Cool and add water until no further precipitation takes place. Filter the yellow quinophthalon, recrystallise from hot alcohol, m.p 239°C.

(3) Heat together 7gm of compound with 25gm of anhydrous sodium acetate and 50gm of glacial acetic acid on water bath to 70°C. Add dropwise 8ml of bromine in 20ml acetic acid. Remove the water bath and heat the solution to boiling for few minutes and

again heat on water bath for 30min. Cool, pour in ice and collect ω tribromoquinaldine, recrystallise from ethanol, m.p 128°C.

(4) Picrate m.p 191°C (5) Methiodide m.p 195°C.

b.p 249°C **1, 2, 3, 4 Tetrahydroquinoline**

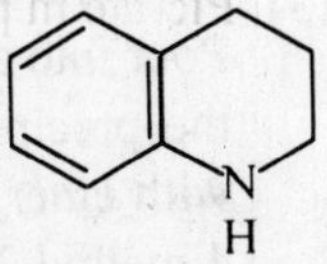

(1) Colourless liquid which darkens on exposure to air, insoluble in water.

(2) Dissolve 2ml of compound in 8ml of dil HCl. Add 2ml of saturated aqueous sodium nitrite solution. Keep at room temperature for sometime and separate the brownish oily layer. Shake the oily layer with 6ml of dil HNO_3 and cool. Collect the precipitated yellow crystals of nitrosotetrahydroquinoline, m.p 143-145°C.

(3) Hydrochloride m.p 181°C (4) Benzoate m.p 75°C (5) Picrate m.p 141°C.

b.p 252°C **8-Methylquinaldine** (2, 8-dimethyl quinoline)

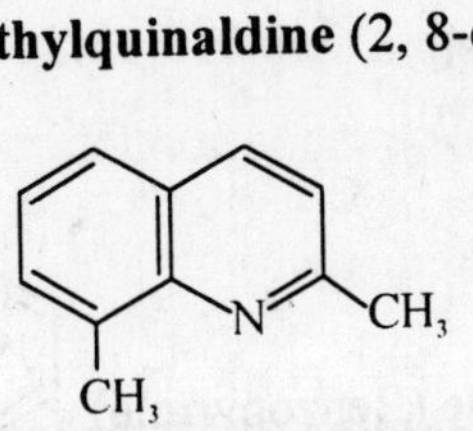

(1) Insoluble in water, soluble in alcohol and ether.

(2) Heat 6ml of compound with 1.2 gm of phthalic anhydride in an oil bath at 170°C for 5-10 min. Dissolve in 10ml of warm conc HCl. Cool and collect 8-methyl quinophthalone, recrystallise from hot alcohol, m.p 271°C.

(3) Picrate m.p 180°C.

b.p 263°C **Lepidine** (4-Methylquinoline)

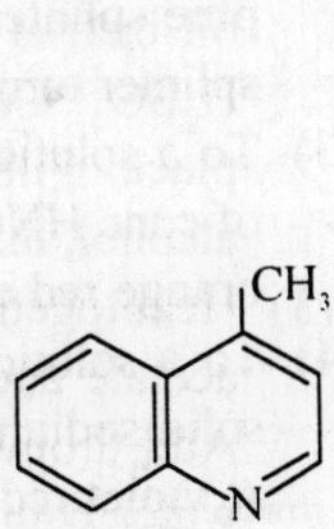

(1) Colourless liquid with unpleasant characteristic quinoline like odour, insoluble in water, soluble in alcohol, ether and benzene.

(2) Stir 10.5ml bromobenzene in 30ml ether with 770mg of lithium metal and then reflux for 2hrs to get a suspension of phenyllithum. Add 6.5ml lepidine to this suspension and reflux for 30min. Pour into a solution of 30ml 6N HCl and 100ml ice water. Collect the precipitated solid, dissolve in 15ml NaOH solution. Extract with ether, distill off the ether, cool the pale yellow coloured oil, 4-methyl 2-phenylquinoline soon solidifies, recrystallise from pet ether, m.p 64-65°C.

(3) Picrate m.p 211°C (4) Methiodide m.p 173°C.

Solid

m.p 20-22°C, b.p 124°C **Pyrimidine** (1, 3-diazine) 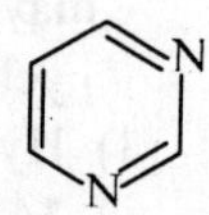

(1) Crystalline mass with penetrating narcotic odour, soluble in water, and alcohol.

(2) Picrate m.p 156°C.

m.p 52°C **Indole** (Benzopyrrole) 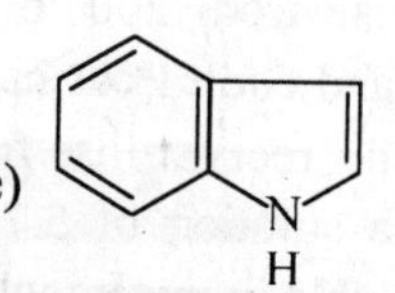

(1) Colourless solid with 1-naphthylamine like odour, insoluble in water, soluble in hot water, alcohol, ether and chloroform.

(2) In a test tube heat 100mg of compound with 2ml water. Bring a pine splinter dipped in HCl near the mouth of test tube. The pine splinter turns red.

(3) To a solution of 10mg of compound in 5ml water add 4-5 drops of conc HNO_3 and 2-3 drops of 6% potassium nitrite solution. An orange red coloured solution is obtained.

(4) To a solution of 10mg of compound in 2ml water add 10mg of solid sodium nitroprusside and 9-10 drops of dil NaOH solution. A violet red colouration which changes to violet blue on addition of 9-10 drops of conc HCl.

(5) To cooled 6ml DMF add dropwise with stirring 1.8ml of phosphorous oxychloride. Introduce dropwise a solution of 2gm of indole in 3ml DMF(temp below 10°C). Remove the cooling source, stir for 1hr and then add 10gm of crushed ice. Add 20ml NaOH solution (7.5gm) and heat the mixture to boiling. Cool and leave overnight. Collect indole 3-aldehyde, recrystallise from ethanol, m.p 196-197°C.

(6) Stir 40ml dimethyl sulfoxide with 5.2gm KOH and then add 2.3gm of indole. Stir the mixture for 45 min and then add 7gm of benzyl bromide. Continue stirring for another 45min and dilute with10ml water. Extract with ether. Distill off the ether and collect 1-Benzyl indole, m.p 43°C.

(7) Acetate m.p 157°C (8) Benzoate m.p 68°C (9) Picrate m.p 187°C.

m.p 58°C, b.p 210°C **2-Aminopyridine** (α-pyridyl amine)

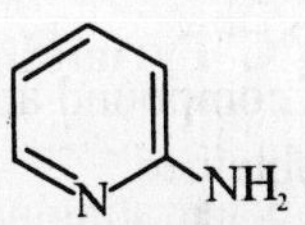

(1) Crystalline compound, insoluble in water, soluble in alcohol

(2) Boil together 9.4gm of compound with 15gm benzyl alcohol and 1gm of KOH for 30min. Some water may distill off and temperture may rise from 200 to 250°C. Maintain at this temperature for 5min and cool. Pour in ice cold water and collect 2-benzylamino pyridine, recrystallise from isopropyl alcohol, m.p 97-98°C.

(3) Cool a solution of 5.7gm of compound in 10ml acetic acid to 20°C. Add dropwise with stirring 3ml of bromine in 7ml of acetic acid. After half bromine has been added remove the cooling source. Continue stirring for 1hr after the addition is complete and then add 15ml water followed by 25ml 40% NaOH solution. Collect the precipitated 2-amino 5-bromopyridine, m.p 132-135°C.

(4) To 2.9gm of 2-aminopyridine in 30ml ethanol add 5gm of phenacyl bromide and 4.2gm of sodium bicarbonate. Stir the mixture for 2hrs and then reflux at 100°C for 2hrs. Cool, pour in ice water and collect 2-Phenylimidazo [1, 2-a]pyridine, recrystallise from aqueous ethanol, m.p 134°C.

(5) Acetate m.p 71°C (6) Benzoate m.p 169°C (7) Picrate m.p 216°C.

m.p 69°C **2, 2' Bipyridyl**

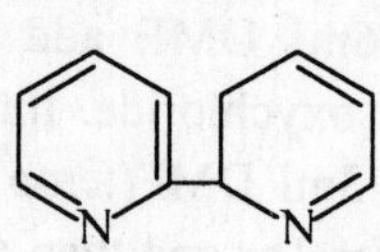

(1) Colourless solid, sparingly soluble in water, soluble in alcohol and ether.

(2) Picrate m.p 158°C

m.p 74°C **8 Hydroxyquinaldine** (8-quinolinol)

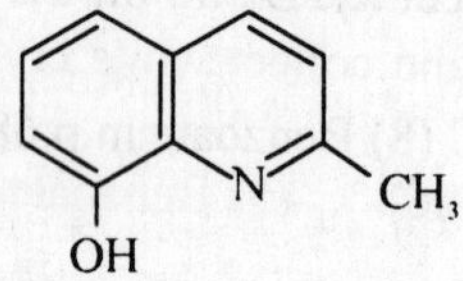

(1) Crystalline solid, soluble in hot alcohol, ether and benzene. Sublimes at 100°C.

(2) To few milligrams of compound add $FeCl_3$ solution. A faint blue-green colouration is obtained

(3) Picrate m.p 219°C.

m.p 95°C **3-Methyl Indole** (skatole)

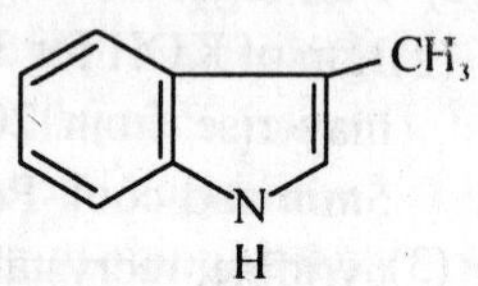

(1) Colourless crystalline solid with disagreeable odour, soluble in alcohol, ether, chloroform and benzene.

(2) In a test tube heat 100mg of compound with 2ml water. Bring a pine splinter dipped in HCl near the mouth of test tube. The pine splinter turns red.

(3) To 10mg of compound in 3ml water add 10mg of sodium nitroprusside followed by 9-10 drops of 10% NaOH solution. Dilute the orange yellow solution with 10ml water and add 5ml acetic acid. Boil for 1-2 min and keep at room temperature for 5 min. A red to violet-red colouration gradually develops.

(4) Hydrochloride m.p 167°C. (5) Acetate m.p 68°C (6) Picrate m.p 170°C.

m.p 104°C **Piperazine** (hexahydro pyrazine) 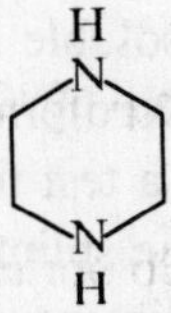

(1) Crystalline odourless compound with bitter taste, easily soluble in water and alcohol.
(2) Responds to Simon's test.
(3) Warm together a solution of 500mg of compound in 5ml water with 4ml saturated aqueous solution of sodium nitrite and 2ml conc HCl. Cool and collect white crystals of dinitrosopiperazine, m.p 157°C.
(4) Acetate (di) m.p 134°C (5) Benzoate (di) m.p 191°C. (6) Picrate m.p 280°C.

m.p 108-110°C **Acridine** 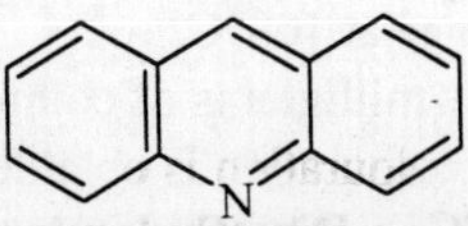

(1) Yellow coloured solid which causes skin irritation, insoluble in water, soluble in alcohol and ether. Sublimes at 110°C.
(2) Dissolve 10mg of compound in 2ml HNO_3 and dilute with 20ml water. Pale yellow coloured solution shows blue green fluorescence with black background.
(3) Methiodide m.p 224°C. (4) Picrate m.p 208°C.

m.p 120°C **2, 6-diaminopyridine** 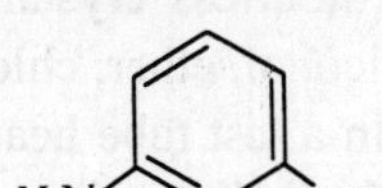

(1) Soluble in water and alcohol.
(2) Acetate (di) m.p 203°C (3) Picrate m.p 240°C (4) Benzoate(di) m.p 176°C

m.p 246°C **Carbazole** (Dibenzopyrrole) 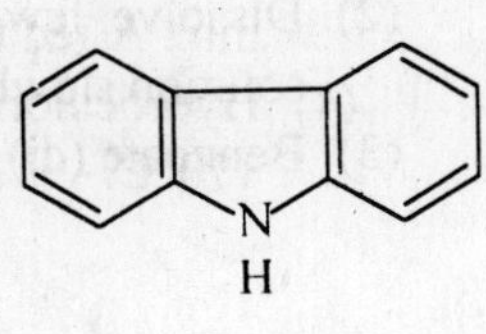

(1) White shinning crystalline solid with faint indole like odour, insoluble in water, sparingly soluble in alcohol, benzene and chloroform, soluble in conc H_2SO_4.
(2) In a test tube heat 100mg of compound with 2ml water. Bring a pine splinter dipped in HCl near the mouth of test tube. The pine splinter turns deep red.
(3) Fuse 50mg of compound with 50mg oxalic acid. A blue mass commonly known as carbazole blue will be obtained.
(4) Dissolve few milligrams of compound in conc H_2SO_4. A yellow coloured solution is obtained. Add 1-2 drops of conc HNO_3. The colour changes to blue green.
(5) Acetate m.p 69°C (6) Benzoate m.p 98°C (7) Picrate m.p 185°C.

(iii) Amino Phenols

Aromatic Solid

m.p 45°C **o-Dimethylamino phenol**

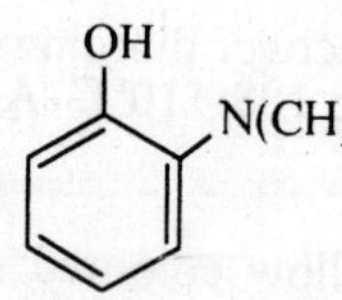

(1) Colourless crystalline solid with penetrating tar like odour, insoluble in water, soluble in alcohol and ether.
(2) To few milligrams of compound add $FeCl_3$ solution. A violet red coloured solution is obtained.

m.p 75°C **p-Dimethylamino phenol**

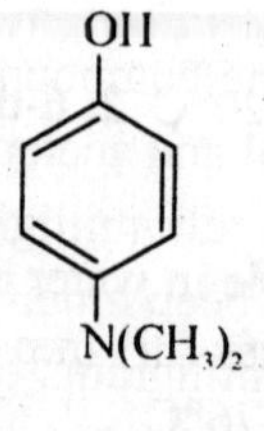

(1) Colourless crystalline compound, soluble in water and ether.
(2) Dissolve few milligrams of compound in NaOH solution. The solution slowly darkens and smell of methyl isocyanide is emitted.
(3) Benzoate (di) m.p 158°C (4) Acetate m.p 78°C.

m.p 76°C **8-Hydroxyquinoline** (oxine)

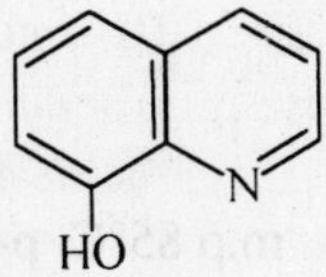

(1) Crystalline compound with phenolic odour, insoluble in water, easily soluble alcohol and benzene.
(2) Dissolve few milligrams of compound in NaOH solution. The solution turns yellow.
(3) To 5ml saturated aqueous solution of compound add 1drops of 10% $FeCl_3$ solution. An intense green colour (approximately blue-green) develops at once.
(4) Dissolve 100mg of compound in 10ml water and add 5-6 drops of conc HCl (heat if necessary). Add about 20ml bromine water and shake thoroughly. Yellow precipitate is obtained. Add sulphorous acid until the solution is colourless. Decant the liquid and recover dibromoxyquinoline, m.p 204°C.
(5) Benzoate m.p 118°C (6) Picrate m.p 204°C. (7) Methiodide m.p 143°C

m.p 79°C (d) **2, 4-Diaminophenol** (Amidol)

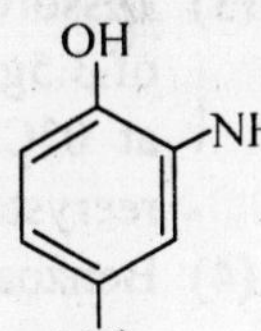

(1) Crystalline compound which darkens on exposure to air, soluble in alcohol and acetone.
(2) Dissolve few milligrams of compound in NH_4OH. The solution becomes beep blue.
(3) To few milligrams of compound add $FeCl_3$ solution. An intesnse red coloured solution is obtained.
(4) Benzoate (di) m.p 253°C (5) Benzoate (tri)m.p 231°C (6) Acetate (tri) m.p 180°C.

m.p 85°C **p-Methylamino phenol.**

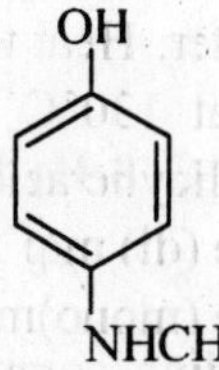

(1) Colourless solid, sparingly soluble in water, soluble in ether.
(2) To few milligrams of compound add $FeCl_3$ solution. A purple coloured solution is obtained.
(3) Acetate m.p 240°C (4) Benzoate (di) m.p 175°C

m.p 96°C **o-Methylamino phenol**

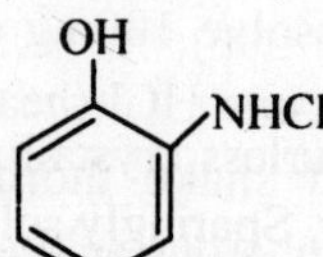

(1) Colourless solid, sparingly soluble in water, soluble in hot water, alcohol and benzene.
(2) Dissolve few milligrams of compound in dil HCl and add $FeCl_3$ solution to it. A deep red brown coloured solution is obtained.
(3) Dissolve 6gm of compound in 21ml HCl. Cool and add a solution of 3.5gm of $NaNO_2$ in 6ml water, cool and then keep the mixture at 0°C for 1hr. Filter the precipitated nitroso derivative and recrystallise from alcohol, m.p 130°C.
(4) Benzoate (di) m.p 150°C (5) Acetate (mono) m.p 150°C.

m.p 122°C **m-Amino phenol** (m hydroxy aniline)

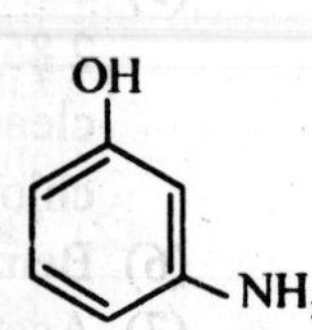

(1) Crystalline compound which darkens on exposure to air. Sparingly soluble in water, easily soluble alcohol and ether.
(2) Dissolve 100mg of compound in 2-3ml alcohol. Add 5ml of alcoholic picryl chloride solution and shake thoroughly. If no precipitate occurs heat on water bath, cool and recover picramide as deep red crystals, m.p 203-204°C.

(3) To 6gm of compound add 30gm of potassium carbonate and 60ml water. Heat the mixture gently and pass CO_2 gas. Continue heating at 130°C for 6hrs. Cool, acidify with HCl, collect p-amino salicylic acid, m.p 151°C.
(4) Benzoate (di) m.p 153°C
(5) Benzoate (mono)m.p 174°C.
(6) Acetate(di) m.p 101°C
(7) Acetate (mono) m.p 148°C.
(8) Hydrochloride m.p 229°C.

m.p 174°C(d) **o-Amino phenol** (o-hydroxy aniline) 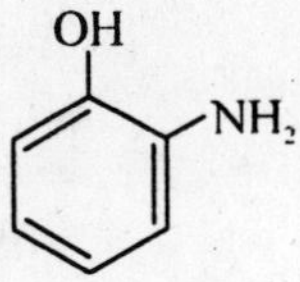

(1) Colourless crystalline compound which turns brown on exposure to air. Sparingly soluble in water, easily soluble in ether.
(2) To few milligrams of compound add $FeCl_3$ solution. A dark brown precipitate is obtained.
(3) Heat 11.2gm of compound with 17ml conc H_2SO_4 and 29gm glycerol at 80°C. Add this fluid mixture in portions to a mixture of 7.2gm o-nitro phenol and 1gm $FeSO_4$. Reflux the resulting mixture at 130-140°C for 4hrs. Neutralise with cold NaOH solution (temp below 40°C) and collect 8-hydroxy quinoline, m.p 76°C.
(4) Heat 5.5gm of compound with 15ml acetic anhydride until a clear solution results. Distill and collect 2-methyl benzoxazole as colourless liquid at b.p 200-208°C.
(5) To a suspension of 2.75gm of compound in 50ml benzene add 2.85gm of chloracetyl chloride. Heat the mixture at 100°C until a clear solution results. Filter hot and cool the filtrate. o-(N-chloroacetamido)phenol separates out, m.p 137-138°C.
(6) Benzoate (di) m.p 184°C
(7) Acetate (mono) m.p 201°C

m.p 175°C **4-Amino 2-methyl phenol** 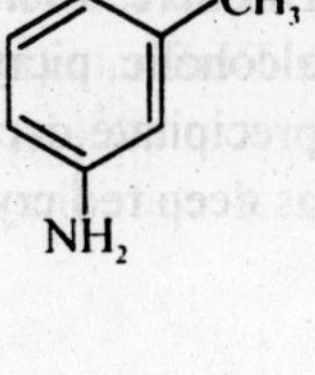

(1) Crystalline compound, sparingly soluble in water, easily soluble in ether.
(2) To few milligrams of compound add $FeCl_3$ solution. A purple coloured solution is obtained.
(3) Benzoate(di) m.p 194°C
(4) Acetate (mono) m.p 179°C
(5) Acetate (di) m.p 103°C.

m.p 190°C **p-Amino phenol** (p-hydroxy aniline, rodinal)

(1) Crystalline compound, slightly soluble in ether.
(2) To few milligrams of compound add $FeCl_3$ solution. A purple coloured solution is obtained.
(3) To 2ml aqueous solution of compound add 5-6 drops of NH_4OH. A red violet coloured solution is obtained after few seconds.
(4) To 5.5 gm of compound in 15ml water add 6ml acetic anhydride. Stir vigorously and heat on water bath for 10min. Cool, filter the precipitated p-hydroxy acetanilide, recrystallise from hot water, m.p 169°C.
(5) Cool a mixture of 5gm of compound in 4ml conc H_2SO_4. Add with stirring a solution of 3.4gm of $NaNO_2$ in 8ml water. Continue stirring for 25min. Add 1ml conc H_2SO_4 and add the diazonium solution to ice cold 10ml KI solution(10gm in 10ml water). Add 100mg of copper –bronze and warm to 80°C until evolution of N_2 is complete. Extract with $CHCl_3$. Distill off $CHCl_3$ and collect p-iodo phenol, m.p 94°C.
(6) Dissolve 500mg of compound in 10ml glacial acetic acid and add 3-4 ml benzaldehyde to it. Shake until the solution is complete. Continue shaking for 5-10min. Cool and recover benzylidene p-amino phenol, m.p 182°C.
(7) Benzoate (di) m.p 234°C
(8) Benzoate (mono) m.p 216°C.
(9) Acetate (di) m.p 150°C
(10) Acetate (mono) m.p 168°C.

m.p 200°C **2-Hydroxyquinoline** (Carbostyril, 2-quinolinol)

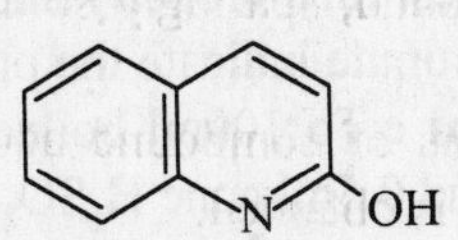

(1) Crystalline compound, insoluble in water, soluble in hot water and ether. Sublimes

(2) 6 Nitro carbostyril, m.p 280°C

m.p 207°C **8-Aminonaphthol-2** (7-hydroxy-1-naphthylamine)

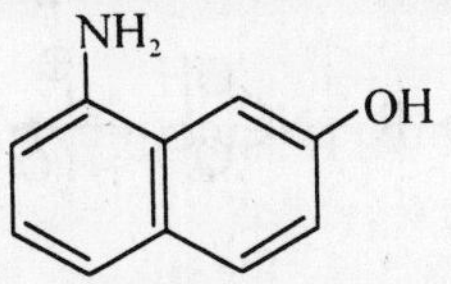

(1) Crystalline compound, soluble in water, alcohol and ether.

(2) To few milligrams of compound add $FeCl_3$ solution. A green-blue coloured solution is obtained.

(iv) Nitro Compounds

Nitro compounds are those in which the "NO_2" is attached to an alkyl or aryl group. These compounds are characterised by their odour and colour. Most of them are generally yellow in colour while the intensity of colour increases with the increase in the number of nitro groups.

Identification tests

***(i)* Diphenylamine test (aliphatic nitro compounds) :** To 10ml of diphenylamine reagent add few milligrams of powdered sample compound. Shake thoroughly. Development of blue colour indicates the presence of aliphatic nitro compound.

Diphenylamine reagent : To a mixture of 80ml conc H_2SO_4 and 20ml water add 20mg of diphenylamine

***(ii)* Ferrous hydroxide test :** This test is based on the principle that nitro compounds are good oxidising agents and they oxidise ferrous hydroxide to ferric hydroxide.

$$\underset{}{ArNO_2} + 4H_2O + \underset{\text{Blue-green}}{6Fe(OH)_2} \rightarrow ArNH_2 + \underset{\text{Red-brown}}{6Fe(OH)_3}$$

In a test tube mix 40mg of test sample with 3ml of freshly prepared ferrous sulphate reagent. To this solution add 3ml of alcoholic potassium hydroxide solution. Cork the test tube properly and shake vigorously. A red brown precipitate within one minute indicate the presence of nitro group.

Ferrous Sulphate reagent : To 100ml boiled distilled water add 5gm ferrous ammonium sulphate and 0.5ml conc H_2SO_4

Potassium Hydroxide Solution : Dissolve 6gm KOH in 6ml water and add this solution to 40ml 95% ethanol.

***(iii)* Reaction with Zinc and Ammonium chloride :** In this test nitro group on treatment with NH_4Cl and Zn is partially reduced to hydroxylamine which then acts as a reducing agent to give a positive Tollens test.

$$RNO_2 + 4NH_4Cl + 2Zn \rightarrow RNHOH + 2ZnCl_2 + 4NH_3 + H_2O$$

$$RNHOH + 2Ag[(NH_3)_2]OH \rightarrow RNO + 2Ag\downarrow + 4NH_3$$

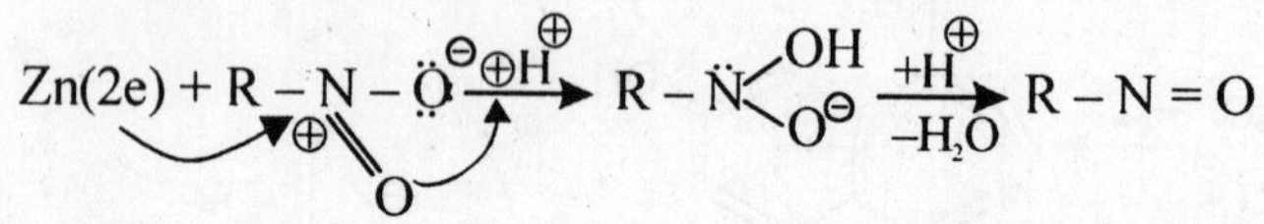

Dissolve 400mg of test sample in 10ml ethanol (heat if necessary). To this solution add 400mg of solid ammonium chloride and 400mg of Zn dust. Boil the resultant reaction mixture on water bath for 3-5 minutes. Cool and filter directly in a test tube containing 5ml of ammonical silver nitrate solution (Tollens reagent). A grey or black precipitate or silver mirror indicates the presence of nitro group.

Distinction between primary and secondary nitro compounds

To few milligram of compound in 10% caustic soda solution add sodium nitrite solution. Any precipitate which may be formed is redissolved by dropwise addition of water. To this solution carefully add dropwise with cooling conc H_2SO_4.

A blood red colouration in weakly alkaline medium which disappears on addition of acid indicates the presence of primary nitro compound.

A blood red colouration on acidification forms intensely blue to green colouration indicates the presence of secondary nitro compound

Distinction between mono, di or tri nitro compounds

Dissolve 200mg of compound in 20ml acetone and add 4ml of dil NaOH solution to it. Mononitro compounds however do not produce any change in colour, dinitro normally produce purplish colour whereas trinitro compounds produce blood red colouration.

Derivatives

***(i)* Reduction by metal in acidic medium :** In a boiling test tube place 500mg of nitro compound, 5ml of conc HCl, 1.5gm of tin granules or 3gm of solid stannous chloride. Boil the mixture under reflux for 30min or until the smell of nitro compound disappears. Decant the hot solution, cool and make it alkaline by adding sodium hydroxide solution. Add 10ml ether to the mixture and shake thoroughly. Remove the organic layer and repeat the process again using 10ml ether. Combine the two organic layer, distill off the ether and collect the amino compound. Recrystallise and determine its melting point.

$$RNO_2 + 6H[Sn+HCl] \rightarrow RNH_2 + 2H_2O$$

Mechanism : The nitro group is first reduced to nitroso compound. Because of higher carbonyl activity the nitroso compound is hydrogenated more rapidly to hydroxylamine. The hydroxylamine is finally reduced to amine.

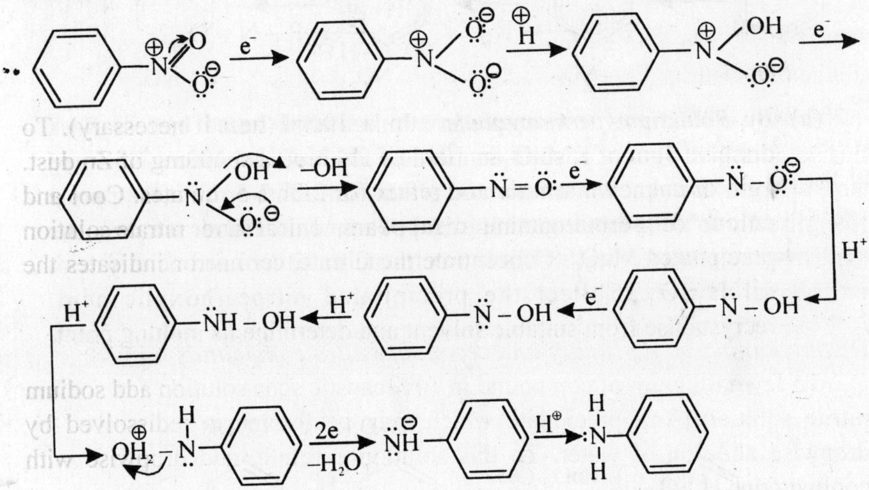

(*ii*) Partial reduction of di or trinitro compounds : Dissolve 1gm of di or tri nitro compound in 6ml methanol. Add with swirling motion methanolic solution of Sodium hydrogen sulphide and boil the mixture for 20 minutes. Cool, distill off the methanol on water bath and pour the liquid concentrate with stirring on crushed ice. Filter the precipitated amino nitro derivative, recrystallise and determine its melting point.

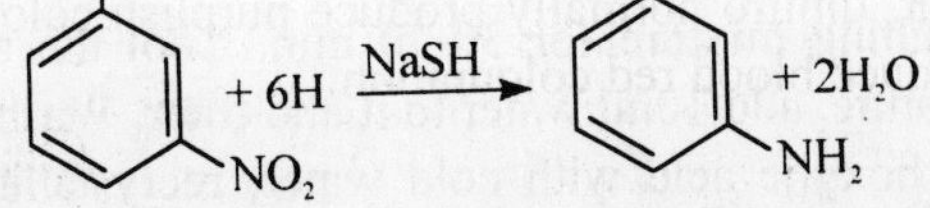

Sodium Hydrogen Sulphide reagent : To 2gm of sodium sulphide in 5ml water add in portion 600mg of sodium bicarbonate. After the addition of bicarbonate is complete add 6ml of methanol. Cool the resulting solution to 10°C. Filter the precipitated sodium carbonate and retain the filtrate as it contains the reducing agent Sodium Hydroden Sulphide (NaSH).

$$Na_2S + NaHCO_3 \rightarrow NaSH + Na_2CO_3$$

***(iii)* Oxidation of side chain:** The side chain alkyl group attached to aromatic ring for eg nitro toluene is oxidised to nitrobenzaldehyde and further to nitro benzoic acid. The mechanism by which this oxidation occur is however complex. Various intermediates like the chromate and permanganate esters are involved but it also appears to involve an intermediate benzyl cation.

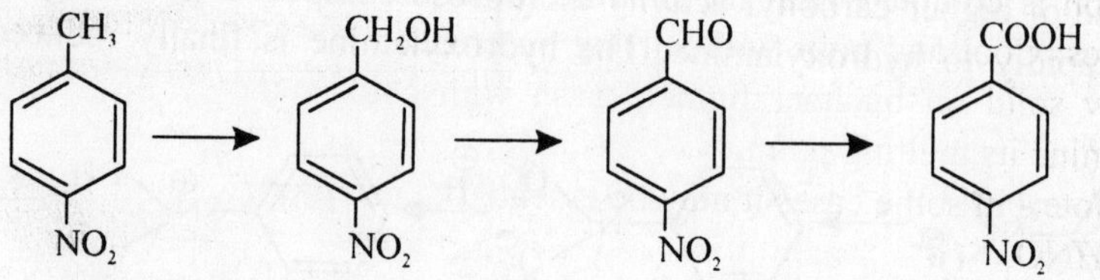

(*a*) *By Potassium permanganate* : In a 100ml round bottom flask dissolve 3gm of $KMnO_4$ in 10ml of 5% NaOH solution. Now add 1gm of unknown sample and reflux for 2-3hrs or until the purple colour of permanganate disappears. Filter and remove the precipitated MnO_2. Concentrate the filtrate, cool and acidify with dil H_2SO_4. Collect the precipitated nitrocarboxylic acid, recrystallise from suitable solvent and determine its melting point.

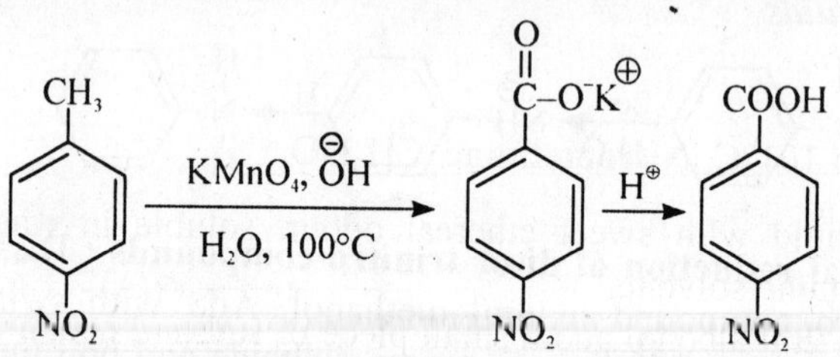

(*b*) *By Potassium dichromate* : In a 50ml round bottom flask place 1gm of nitro compound and 2.75 gm of potassium or sodium dichromate in 6ml water. Add with stirring 4ml conc H_2SO_4 to it. The reaction usually starts at once, if not then warm on water bath to initiate the reaction. After the reaction has subsided boil the resulting mixture for 30-40 min. Cool the mixture to room temperature, add 10ml water to it and filter. Wash the precipitated nitrocarboxylic acid with cold water, recrystallise from suitable solvent and determine its melting point.

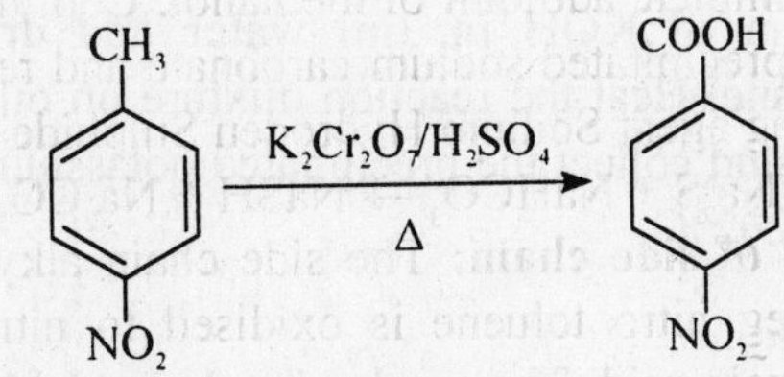

***(iv)* Nitration of the Nitro Compounds :** In a 50ml round bottom flask place 1gm of nitro compound and add 4ml of conc H_2SO_4 to it. Now add dropwise with stirring 4ml of conc HNO_3 to the acidic solution. After the addition is complete heat the mixture on water bath at 50-60°C for 15-20 minutes. Cool and pour with stirring on crushed ice. Filter the precipitated yellow solid on buchner funnel, wash with cold water, recrystalllise and determine its melting point.

Note:- In some cases it may be essential to use fuming HNO_3 in place of conc HNO_3

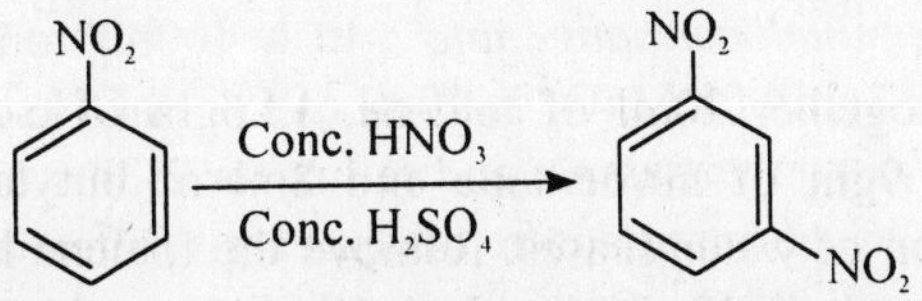

Nitrohydrocarbons

Aliphatic Liquids

b.p 101°C **Nitromethane** CH_3NO_2

(1) Liquid with sweet ethereal odour, soluble in water and most organic solvents

(2) To 2ml alcohol add a small piece of sodium, cool and add 1ml of nitro compound. Shake thoroughly, filter the precipitated solid. Dissolve the solid in 2ml water and add 3-4 drops of saturated sodium nitrite solution. Acidify with dil H_2SO_4 and then make it alkaline with NaOH solution. The alkaline solution shows orange-yellow colouration due to formation of sodium methyl nitrolate.

(3) Cool a mixture of 5.4ml nitromethane, 10.5ml benzaldehyde and 20ml methanol to −10°C. Now add dropwise with stirring cold NaOH solution (4.2gm in 10ml cold water) to it (temp 10-15°C). Add this cold mixture into 50ml 4M HCl. Collect the precipitated ω nitrostyrene, recrystallise from hot ethanol, m.p 57-58°C.

(4) To 4.48gm of KOH in 3ml water add dropwise 1.2gm of nitromethane. Heat the reaction mixture on oil bath at 160°C for 1hr. Cool and collect the precipitated potassium salt of nitroacetic acid, m.p 262°C.

b.p 114°C **Nitroethane** $CH_3CH_2NO_2$

(1) Liquid with sweet chloroform like odour, soluble in most organic solvents.
(2) To 2ml alcohol add a small piece of sodium, cool and add 1ml of nitro compound. Shake thoroughly, filter the precipitated solid. Dissolve the solid in 2ml water and add 3-4 drops of saturated sodium nitrite solution. Acidify with dil H_2SO_4 and then make it alkaline with NaOH solution. The alkaline solution shows red to orange-yellow colouration due to formation of sodium ethyl nitrolate.
(3) Reflux together 20ml of toluene, 13.6gm of o-methoxybenzaldehyde, 9gm of nitroethane and 2ml of butylamine until the separation of water ceases. Remove the toluene layer. Distill off the toluene, distill the residual oil under reduced pressure and collect 1-(o-Methoxyphenyl)-2-nitro-1-propene at 135-138°C/1mmHg which on cooling soon solidifies, m.p 52°C.
(4) To cooled 6ml of nitroethane add 15ml aqueous potassium hydroxide solution. Add excess of sodium nitrite solution and cool to 0°C. Add dilute sulphuric acid. Extract with ether. Distill off ether and collect ethyl nitrolic acid m.p 81-82°C.

b.p 120°C **2-Nitropropane** $CH_3CH(NO_2)CH_3$

(1) Colourless liquid, insoluble in water.
(2) Warm to 70°C a mixture of 9gm of 2-nitropropane, 5ml dioxane and 1ml aqueous solution of benzyltrimethylammonium hydroxide (triton B). Add in portion 6.8 gm of methyl acrylate. Continue heating for 4hrs. Cool, acidify with dil HCl. Extract with ether. Distill off the ether and collect methyl υ methyl-υ-nitrovalerate under reduced pressure as colourless oil at 79°C/1mmHg.

b.p 126°C **Tetranitromethane** $C(NO_2)_4$

(1) Colourless liquid, insoluble in water, soluble in alcohol and ether.
(2) Reflux together 11gm of tetranitromethane with 7.3gm of tripropylamine, 6.2gm of acetic acid in 55ml alcohol for 3hrs. Dilute with water and add caustic potash solution. Extract with ether. Distill off ether and collect dipropylnitrosamine at b.p 59-61°C/1.5mmHg

b.p 132°C **1-Nitropropane** $CH_3CH_2CH_2NO_2$

(1) Colourless liquid with sweet ethereal like odour, insoluble in water, soluble in alcohol and ether.
(2) To 2ml alcohol add a small piece of sodium, cool and add 1ml of nitro compound. Shake thoroughly, filter the precipitated solid. Dissolve the solid in 2ml water and add 3-4 drops of saturated sodium nitrite solution. Acidify with dil H_2SO_4 and then make it alkaline with NaOH solution. The alkaline solution shows red to orange-yellow colouration due to formation of sodium propyl nitrolate.
(3) Stir and heat a mixture of 4ml of 1-nitropropane, 2ml of diisopropylamine and 20ml of chloroform to 60°C. Add dropwise 3gm of methyl vinyl ketone. Stir the mixture for another 16hrs at 60°C. Cool, concentrate the chloroform solution, distill the residual portion under reduced pressure and collect 5-nitroheptan-2-one at 65-70°C/.2mmHg.

Aromatic Liquids

b.p 210°C **Nitrobenzene** 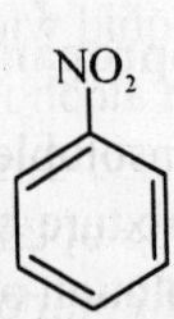

(1) Pale yellow coloured liquid with bitter almond like odour, insoluble in water, soluble in alcohol, ether and benzene.
(2) To 1ml of compound add 3ml glacial acetic acid and a pinch of zinc dust. Boil, cool and add 2-3ml of water and then make it alkaline with NaOH. Now add 2-3 drops of sodium nitroprusside. A violet coloured solution is obtained.
(3) To 1ml of compound add 10ml conc HCl. Boil for 3-5 minutes

and add 20ml of 10% NaOH solution, cool. Distill this solution. To 5 drops of this distllate add 5ml water, 5-6 drops of saturated bleaching powder and shake thoroughly. Filter, on dilution the filtrate will show violet red colouration.

(4) Boil gently on water bath for 30-40 min 5ml of nitro compound with 4.5gm of tin and 10ml conc HCl. Cool and add 10ml of 10% NaOH solution and distill off the aniline, b.p 183°C.

(5) To a mixture of 3.2 ml conc H_2SO_4 and 2.5ml fuming HNO_3 add in portion with stirring 2.1ml nitrobenzene. Reflux on water bath for 30 min. Cool, pour with stirring on crushed ice and collect m-dinitro benzene, m.p 90°C.

(6) Heat 5ml of nitrobenzene on oil bath at 130°C and add 200mg Fe powder. Now add dropwise 1.3ml Br_2. Heat for 1hr and again add 200mg Fe powder and 1.3ml Br_2. Repeat this process two more times. Cool and pour the mixture in crushed ice and collect m-bromo nitro benzene, m.p 55°C.

(7) Reflux on water bath a mixture of 3ml nitrobenzene, 35ml methanol, 4gm zinc powder and 10ml NaOH solution (4.2gm) for 10hrs. Filter the mixture while hot. To the filtrate add conc HCl and distill off the methanol. Cool and collect the precipitated azobenzene, m.p 68°C.

b.p 222°C **o-Nitro toluene**

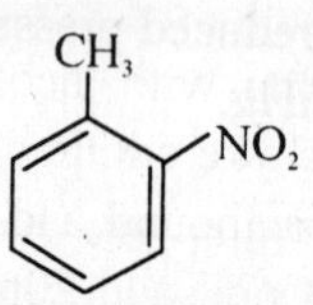

(1) Pale yellow coloured liquid with bitter almond like odour. Insoluble in water, miscible with alcohol, ether and benzene.

(2) To 1ml of compound add 2ml NaOH solution and warm. A red coloured solution will be obtained.

(3) Warm together 1ml of compound, 5ml alcohol, a small piece of sodium and 500mg of zinc dust for 1-2 min. Cool and add 2ml conc HCl. Heat for 30-40 sec, keep at room temperature for 2-3 min and filter. The filtrate shows permanent red colouration.

(4) Cool a mixture of 8ml Ac_2O and 1ml compound to 0°C, slowly add with stirring 2ml conc H_2SO_4. Now add a solution of 2gm CrO_3 in 9ml Ac_2O and continue stirring for 3hrs (temp 5-10°C). Pour the reaction mixture on crushed ice and collect o-

nitrobenzilidene diacetate, m.p 84°C. To the crude diacetate add 3.5ml conc HCl, 6ml water and 1ml ethanol and then boil under reflux for 30min. Cool to 0°C and collect o-nitro benzaldehyde, m.p 45°C.

(5) To 1ml compound add 1.5 ml conc H_2SO_4 and 1.5ml fuming HNO_3. Boil for 3-5 min. Cool, pour on crushed ice with stirring and collect 2, 4-dinitro toluene, m.p 70°C.

(6) To a solution of 3gm of $KMnO_4$ in 25ml water add 3ml of compound and reflux gently. When the colour of $KMnO_4$ has disappeared add 1.5gm of $KMnO_4$ and reflux until the colour of $KMnO_4$ fades. Repeat this process once more. Cool, add conc HCl and collect the precipitated o-nitro benzoic acid, m.p 147°C.

(7) Reflux 10.2 gm of compound with 12 gm N-Bromosuccinimide, 100mg of benzoyl peroxide and 45ml carbon tetrachloride for 6-7 hrs. Cool and filter the precipitated o-nitro benzylpyridinium, m.p 206-208°C.

b.p 232°C **m-Nitrotoluene**

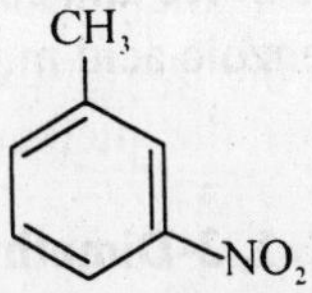

(1) Pale yellow coloured liquid with sweet burning taste, insoluble in water, miscible with alcohol, benzene and ether. It has a faint bitter almond like odour.

(2) To 5ml of compound add 7gm tin and 20ml conc HCl. Heat for 20-30 min. Cool, add 20ml NaOH solution and distill to collect m-toluidine b.p 203°C. To this liquid add 5ml Ac_2O and heat to boiling. Cool, pour in ice and collect m-acetotoluidide, m.p 65°C.

(3) To 2ml of compound add 2gm of $KMnO_4$, few drops of NaOH and 15ml water. Boil the mixture until the colour of $KMnO_4$ disappears. Cool, acidify with HCl and collect white precipitate of m-nitro benzoic acid, m.p 141°C.

b.p 240°C **2-Nitro, 1, 4-xylene** (2-nitro-p-xylene)

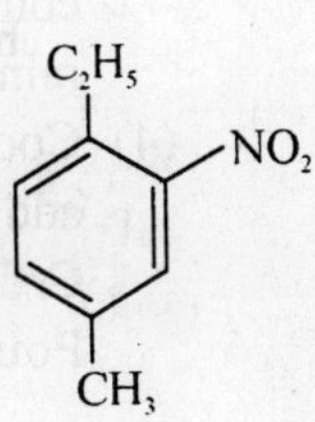

(1) Pale yellow coloured liquid, insoluble in water, soluble in alcohol and ether.
(2) 1, 4-dimethyl 2, 3, 5-trinitrobenzene m.p 139°C (heat at 100°C with conc $H_2SO_4$and HNO_3 mixture for 5-7 min).

b.p 243°C **p-nitroethyl benzene** (1-ethyl-4-nitrobenzene)

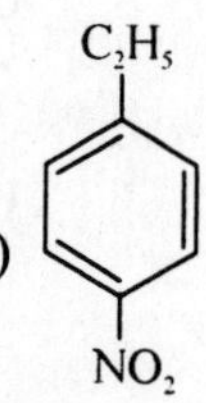

(1) Pale yellow coloured liquid, insoluble in water, soluble in organic solvents
(2) To 5ml of compound add 7gm tin and 20ml conc HCl. Heat for 20-30 min. Cool, add 20ml NaOH solution, distill to collect p-ethyl aniline b.p 214°C. To this liquid add 5ml Ac_2O, heat to boiling. Cool, pour in ice and collect the acetyl derivative, m.p 94°C.
(3) p-nitro benzoic acid m.p 239°C.

b.p 244°C **1, 3-Dimethyl 4-nitrobenzene** (p-Nitro m-xylene)

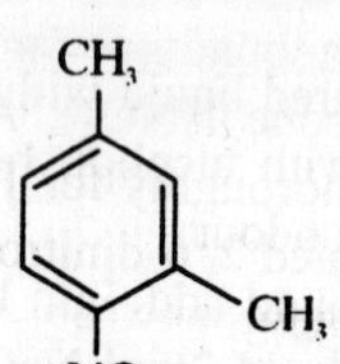

(1) Insoluble in water, soluble in alcohol and ether.
(2) 4, 6-dinitro m-xylene m.p 93°C (fuming HNO_3 at 0°C)
(3) 2, 4, 6-trinitro m-xylene m.p 182°C (conc H_2SO_4+conc HNO_3, 100°C for 5-10min)

b.p 245°C **1, 2-dimethyl 3-nitrobenzene** (3-nitro o-xylene)

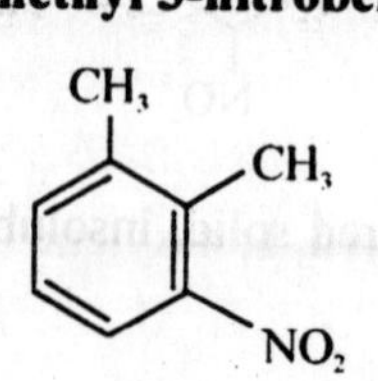

(1) Pale yellow coloured liquid, insoluble in water, soluble in alcohol.
(2) 1, 2-dimethyl 3, 4-dinitrobenzene m.p 82°C (conc H_2SO_4+fuming HNO_3, 100°C for 5-10min)

b.p 264°C **2-nitro p-cymene** (4-isopropyl-1-methyl-2-nitro benzene)

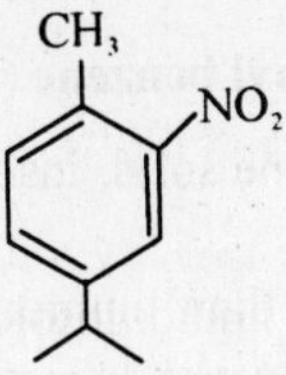

(1) Yellow coloured liquid, soluble in organic solvent.
(2) To 16.7ml of compound add 30ml absolute ethanol and 600-800mg raney nickel. Pass H_2 gas for 30 min and maintain the reaction temperature from 100-120°C (exothermic reaction). Cool, filter off the catalyst, distill off the alcohol and water from the reaction mixture and collect 2-amino p-cymene distilling at 239-240°C. To this liquid add 10ml Ac_2O and heat to boiling. Cool, pour in ice and collect the acetyl derivative, m.p 115°C.
(3) To 5ml ice cold compound add 5ml nitrating mixture (2.5ml conc H_2SO_4 and 2.5ml conc HNO_3). After the addition is complete shake the mixture thoroughly for 10-15 min and then pour in ice. Collect the precipitated 2, 6-dinitro p-cymene, m.p 54°C.

Aromatic Solid

m.p 30°C, b.p 258°C **p-Nitro o-Xylene** (4-Nitro 1, 2 xylene)

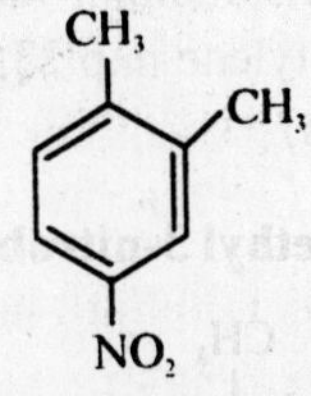

(1) Pale yellow coloured solid, insoluble in water, soluble in alcohol and ether.

(2) 1, 2-dimethyl 3, 4-dinitrobenzene m.p 82°C (conc H_2SO_4+fuming HNO_3, 100°C for 5-10min)
(3) 1, 2-dimethy aniline m.p 50°C (Sn+HCl)

m.p 37°C, b.p 320°C **o-Nitrobiphenyl** 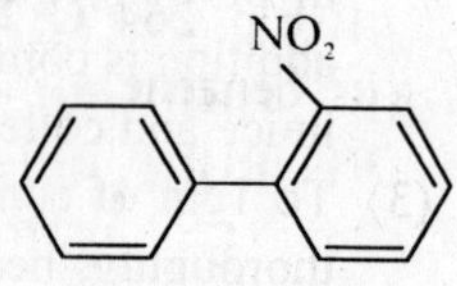

(1) Colourless crystalline solid, insoluble in water, soluble in alcohol and ether.
(2) Heat a mixture of 6gm ammonium chloride, 6gm reduced iron powder and 75ml water, stir and add in portion a solution of 4.1gm of o-nitrobiphenyl in 20ml ethanol. Reflux for 4hrs. Steam distill and cool the distillate. Filter the solid, dissolve in ether, dry over $MgSO_4$, evaporate off ether and collect 2-aminobiphenyl m.p 44°C.
(3) o-amino biphenyl m.p 56°C (Sn+HCl)
(4) 2, 4-dinitrobiphenyl m.p 93°C (conc H_2SO_4+fuming HNO_3, 100°C for 5-10min).

m.p 44°C, b.p 255°C **Nitromesitylene** (1, 3, 5-trimethyl-2-nitro benzene)

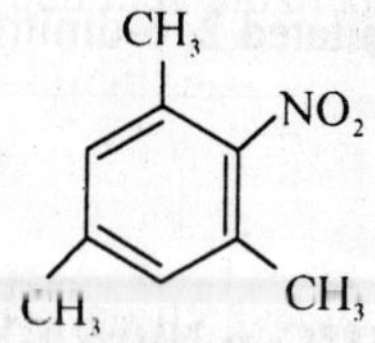

(1) Crystalline solid, insoluble in water, soluble in hot alcohol.
(2) 2, 4, 6 trinitromesitylene m.p 235°C (conc H_2SO_4+fuming HNO_3, 100°C for 5-10min)

m.p 53-54°C **p-Nitro toluene** 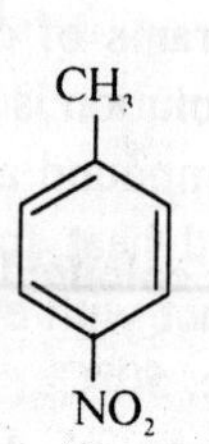

(1) Crystalline solid with strong aromatic odour, insoluble in water, soluble in alcohol and ether.

(2) To a mixture of 1.2 ml conc H_2SO_4 and 0.8ml fuming HNO_3 add in portion 1gm of p-nitro toluene(temp below 50°C). After the addition is complete heat on water bath for 10-15 min. Cool, pour in ice and collect 2, 4-dinitro toluene, m.p 70°C.

(3) To 1gm of compound add 1.75 gm tin and 4ml conc HCl, shake thoroughly, heat for 20-30min. Cool, add 10ml NaOH solution. Extract with ether thrice. Distill off ether and collect p-toluidine, m.p 45°C.

(4) To 2gm p-nitrotoluene add 1.5ml bromine in portion and heat it on oil bath at 145-150°C for 90min. Cool, pour the reaction mixture in ice and collect p-nitro benzyl bromide, m.p 97°C.

(5) Cool a mixture of 7ml Ac_2O and 1gm p-nitrotoluene to 0°C, add with stirring 1.5ml conc H_2SO_4. Now add with stirring a solution of 2gm CrO_3 in 9ml Ac_2O (temp 5-10°C) and continue stirring for 90 min. Pour the mixture on crushed ice and collect p-nitro benzilidene diacetate. Reflux the diacetate with 3ml ethanol, 3ml water and 0.3ml conc H_2SO_4. for 20-30 min. Cool and collect p-nitro benzaldehyde, m.p 106°C.

(6) To a mixture of 2.3gm of compound, 6.8gm of potassium dichromate and 15ml water add dropwise 9ml of conc H_2SO_4. Reflux for 30min, cool and dilute with 10ml water, filter the precipitatd p-Nitro benzoic acid and recrystallise, m.p 237°C.

m.p 60°C **1-Nitro naphthalene**

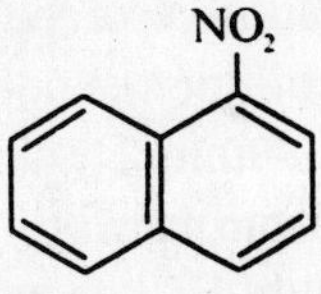

(1) Pale yellow coloured solid, insoluble in water, soluble in alcohol and ether.

(2) To few milligrams of compound add 1ml conc H_2SO_4. A deep red coloured solution is obtained.

(3) To 1gm of compound add 1.75 gm tin and 4ml conc HCl, shake thoroughly and heat for 20-30min. Cool and add 10ml NaOH solution. Extract with ether thrice. Distill off ether and collect 1-naphthylamine, 50°C.

(4) Heat a mixture of 4gm compound, 10gm hydroxylamine

hydrochloride and 150ml ethyl alcohol at 50-60°C. Add slowly a filtered solution of 20gm of KOH in 100ml methanol. Continue stirring with heating for 1hr. Cool, pour on crushed ice and collect 4-nitro 1-naphthylamine m.p 191°C.

(5) Picrate m.p 71°C.

(6) 1, 3, 8 trinitro naphthalene m.p 218°C (conc H_2SO_4 + conc HNO_3)

m.p 66°C **2, 6-dinitro toluene** (2-methyl-1, 3-dinitro benzene)

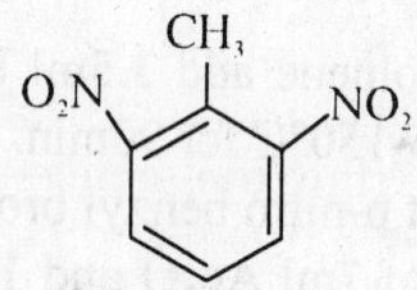

(1) Crystalline compound, soluble in alcohol

(2) To few milligrams of compound add dil NaOH solution. A purple coloured solution is obtained.

(3) 2, 6-diamino toluene (Sn+HCl) m.p 105°C.

(4) 6-nitro-o-toluidine (NH_4SH) m.p 92°C.

m.p 70°C **2, 4-dinitro toluene** (1-methyl-2, 4 dinitro benzene)

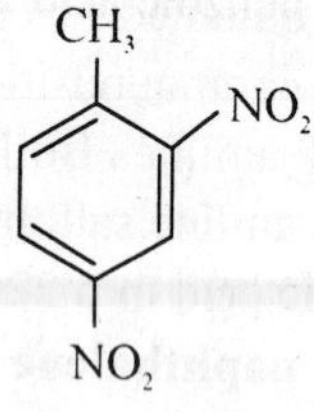

(1) Pale yellow crystalline odourless solid, insoluble in water, sparingly soluble alcohol, easliy soluble ether.

(2) To few milligram of compound in acetone add 1ml KOH solution. A blue coloured solution is obtained.

(3) To 500mg of compound add 5ml conc HCl, 4gm of granulated tin and heat. When the compound has liquified continue heating until all the oily drops disappear. Cool, add 20ml KOH solution and shake thoroughly. Cool and extract with 15ml ether twice. Evaporate off the ether and dissolve the residue in 10-15ml hot

ethanol and filter. Cool the filtrate and collect 2, 4 diamino toluene, m.p 99°C

(4) Reflux 5gm of compound with 5gm p-nitrosodimethyl aniline, 9gm Na_2CO_3 and 30ml alcohol for 6hrs. The precipitated anil is filtered and then mechanically shaken with 20ml dil HNO_3 and 30ml benzene. The organic layer is removed and benzene distilled off on water bath. The residue is again shaken with HNO_3 and benzene, the residue obtained is 2, 4-dinitro benzaldehyde, m.p 72°C.

(5) 2, 4 dinitro benzoic acid ($KMnO_4$) m.p 183°C.

m.p 82°C **2, 4, 6-trinitrotoluene** (sym trinitro toluene, TNT)

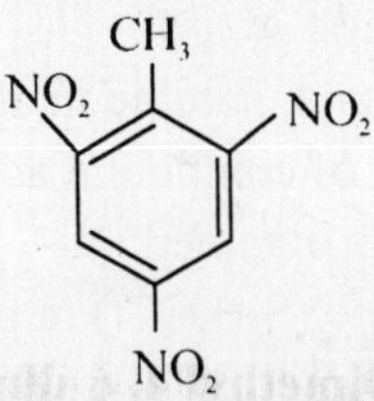

(1) Crystalline very pale yellow coloured solid, insoluble in water sparingly soluble in alcohol, soluble in ether.

(2) To few milligrams of compound add dil NaOH. A red orange coloured solution is obtained.

(3) Dissolve 100mg of compound in 20ml boiling alcohol and then add 10-15 drops of aniline. Boil the mixture gently, filter hot. Cool and collect the aniline salt, m.p 84°C.

(4) 2, 4, 6 trinitrobenzoic acid m.p 220°C (CrO_3+ H_2SO_4)

m.p 90°C **m-Dinitro benzene**

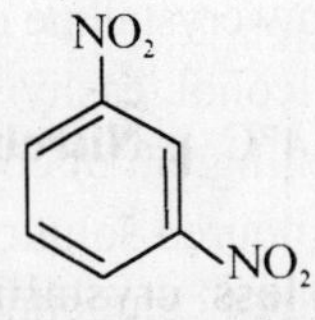

(1) Light yellow odourless crystalline solid, insoluble in water, soluble in alcohol, ether and benzene.

(2) To few milligrams of compound add dil NaOH. A light brown coloured solution is obtained.

(3) To few milligrams of compound in acetone add a drop of NaOH solution. A violet coloured solution is obtained which changes to violet-red on adding acetic acid.
(4) To 500mg of compound in 20ml ethanol add 5ml NH_4OH and pass H_2S gas. Heat the mixture to boiling and pass H_2S gas again. Repeat this process twice. Concentrate the mixture, acidify with dil HCl and filter. To the filtrate add NH_4OH and collect m-nitro aniline, m.p 114°C.
(5) Reflux a mixture of 2.5gm of m-dinitrobenzene, 10gm granulated tin and 20ml conc HCl. Dilute with 10ml water and heat to boiling, pass H_2S gas. Repeat the process twice and filter the precipitated tin sulphide. Add NaOH solution to the filtrate until alkaline. Extract with ether. Distill off ether and collect m-phenylenediamine, m.p 63°C.

m.p 93°C **1, 3-Dimethyl 4, 6-dinitrobenzene**

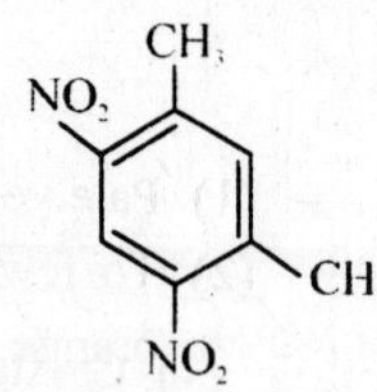

(1) Crystalline compound, insoluble in water, sparingly soluble in alcohol.
(2) To few milligrams of compound add NaOH solution. A violet colouration is obtained.
(3) 1, 3-dimethyl 2, 4, 6-trinitrobenzene m.p 125°C (conc H_2SO_4 + fuming HNO_3, warm).
(4) 1, 3-dimethyl 4, 6-diaminobenzene m.p 105°C (Sn+HCl)

m.p 114°C **p-Nitrobiphenyl**

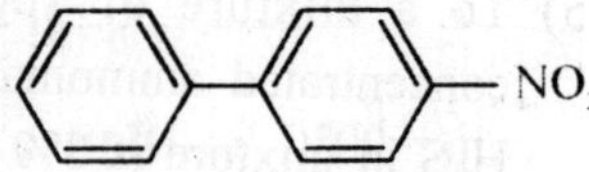

(1) Colourless crystalline solid, insoluble in water, soluble in chloroform and ether.
(2) 4, 4' dinitro biphenyl m.p 233°C (conc H_2SO_4 + fuming HNO_3, warm).
(3) p-amino biphenyl m.p 53°C (Sn+HCl)

m.p 118°C **o-Dinitrobenzene**

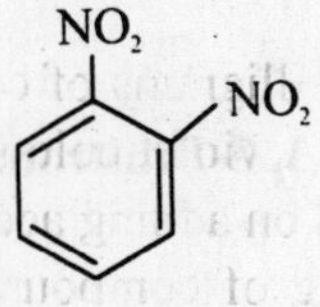

(1) Light yellow coloured solid, insoluble in water, soluble in chloroform and ether.
(2) o-Phenylene diamine m.p 102°C (Sn+HCl)
(3) o-nitroaniline m.p 71°C (NH_4SH)

m.p 122°C **1, 3, 5-trinitrobenzene** (sym-trinitro benzene)

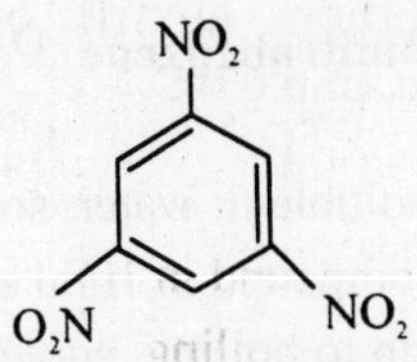

(1) Pale yellow coloured solid, insoluble in water, soluble in alcohol.
(2) To few milligram of compound add dil NaOH solution. A red-orange coloured solution is produced.
(3) Dissolve 500mg of compound in 20ml hot alcohol and then add 10-15 drops of aniline to it. Heat the mixture to boiling, filter hot and allow the salt to crystalllise. Collect the precipitated addition product, m.p 124°C.
(4) To 2.5gm of compound add 10gm tin and 10ml conc HCl. Heat for 20-30 min. Cool, add 20ml NaOH solution. Extract with ether, distill off the ether and collect triaminobenzene. To this residue add 10ml Ac_2O, heat to boiling. Cool, pour in ice and collect triacetyl derivative, m.p 208°C.
(5) To a mixture of 10gm of absolute alcohol and 5.4gm of concentrated ammonia pass H_2S gas until the concentration of H_2S in mixture is 8% by weight. Add this alcoholic ammonium sulphide solution to a solution of 1.5gm of trinitrobenzene in 30ml boiling alcohol. Boil the resultant mixture for 90min. Cool and filter. Distill off the alcohol and extract the residue with hot water, cool and collect 3, 5-dinitroaniline, m.p 154-157°C.

m.p 170°C **1, 8-Dinitronaphthalene**

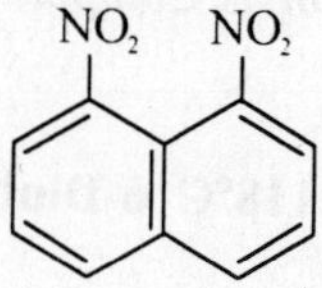

(1) Crystalline solid, insoluble in water, sparingly soluble in benzene and chloroform.
(2) 1, 3, 8 trinitronaphthalene m.p 218°C (conc H_2SO_4 + fuming HNO_3, 160-170°C for 5-10min)
(3) 1, 8 diaminonaphthalene m.p 66°C (Sn+HCl)

m.p 173-174°C **p-Dinitrobenzene**

(1) Crystalline solid, insoluble in water, soluble in benzene.
(2) Dissolve 500mg of compound in 10ml alcohol and add naphthalene to it. Heat the solution to boiling, cool and collect the naphthalene adduct, m.p 118°C.
(3) p-Phenylene diamine m.p 140°C (Sn+HCl)
(4) p-nitroaniline m.p 147°C (NH_4SH)

m.p 216°C **1, 5-Dinitronaphthalene**

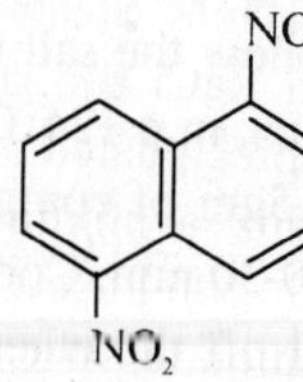

(1) Crystalline solid, insoluble in water, difficult soluble benzene and other organic solvents. Soluble in ether.
(2) To few milligrams of compound in alcohol add KCN. A blue green coloured solution is obtained.
(3) 1, 5 diamino naphthalene m.p 190°C. (Sn+HCl)
(4) 1-amino, 5-nitro naphthalene m.p 118°C (NH_4SH)
(5) 1, 4, 5 trinitronaphthalene m.p 154°C(conc H_2SO_4 + fuming HNO_3, 200-210°C for 5-10min).

(v) Nitrocarbonyls Compounds

Solid

m.p 44°C **o-Nitrobenzaldehyde** 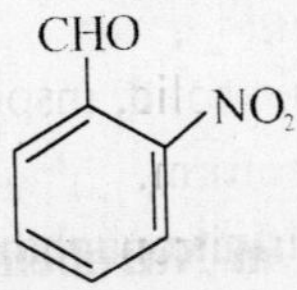

(1) Yellow crystalline solid with pungent odour, slighltly soluble in water, easily soluble in ether.

(2) Dissolve 50mg of compound in 10ml hot water, cool and add 1.5ml acetone followed by 3-4 drops of 10% NaOH solution. The solution first becomes yellow and then green. Heat over a low flame for 1-2 min, the colour becomes blue.

(3) Dissolve 100mg of compound in 2ml dil NaOH solution. An orange coloured solution is obtained.

(4) Heat a mixture of 7gm aniline and 11gm of nitro compound on water bath for 1hr. Cool and dissolve in 20ml ether. Evaporate off the ether and collect o-nitrobenzalaniline, recrystallise from ethanol, m.p 64-66°C.

(5) Heat a mixture of 3gm of compound, 50ml water, 50gm $FeSO_4$ containing 2-3 drops of conc HCl on steam bath When the temperature reaches 90°C add 12ml NH_4OH and continue heating for 10min. Steam distill the reaction mixture. Saturate the distillate with NaCl and stir at 5°C until precipitation takes place. Collect the precipitated o-amino benzaldehyde, m.p 39°C.

(6) Reflux a mixture of 3gm of compound, 4gm of phenylacetic acid, 10ml Ac_2O and 2gm triethylamine for 15 min. Cool the solution to 90°C and add 10ml water, filter hot. Cool the filtrate and collect trans o-nitro α-phenylcinnamic acid, recrystallise from toluene, m.p 197-198°C.

(7) Dissolve 2gm of compound in a mixture of 10ml acetone and 10ml water. Add dropwise 10ml of 1N NaOH solution. Keep the mixture at room temperature for 15min, cool and collect indigotin, m.p >350°C.

(8) o-Nitrobenzoic acid m.p 144°C ($KMnO_4$). (9) o-Nitrobenzaldehyde 2, 4-dinitrophenylhydrazone m.p 265°C.

(10) o-Nitrobenzaldehyde semicarbazone m.p 256°C

(11) o-Nitrobenzaldoxime m.p 103°C.
(12) o-Nitrobenzaldehyde p-nitrophenylhydrazone m.p 263°C (13) o-Nitrobenzaldehyde phenylhydrazone m.p 156°C.

m.p 58°C **m-Nitrobenzaldehyde**

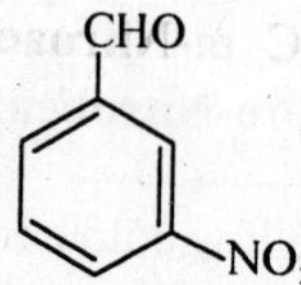

(1) Pale yellow coloured solid with faint bitter almond like odour, provoking sneezing, insoluble in water, soluble in alcohol and ether.
(2) Gives crystalline bisulphite addition compound with sodium metabisulphite.
(3) Dissolve 100mg of compound in 2ml dil NaOH solution. A yellow coloured solution is obtained.
(4) Heat under reflux 5gm of compound with 7.5gm malonic acid, 15ml pyridine and 0.2ml piperidine for 1hr. Cool, pour in ice water containing HCl. Filter the precipitated m-nitrocinnamic acid, m.p 197°C.
(5) Heat a mixture of 3.5gm of compound, 2.5gm of aniline and 10ml of rectified spirit on water bath for 20min. Add water until cloudiness persists. Cool and scratch the oily layer with glass rod to induce crystallisation. Collect m-nitrobenzylidene aniline, recrystallise from methanol, m.p 65-66°C.
(6) Cool a suspension of 3gm of compound in 4ml water to 0°C. Add with stirring 3ml of benzoyl chloride followed by a solution of 1.2gm of sodium cyanide in 2.5ml water (exothermic reaction). After the reaction has subsided cool and collect m-nitro-o-benzoyl mandelonitrile, m.p 100-101°C.
(7) Add 2.5gm of compound to a solution of 1.25gm of KOH in 8ml water. Keep the mixture overnight. Saturate the mixture with CO_2 gas. Extract with ether. Evaporate off the ether, cool the residue and collect m-nitrobenzyl alcohol, m.p 27°C.
(8) m-Nitrobenzoic acid m.p 140°C ($KMnO_4$).
(9) m-Nitrobenzaldehyde 2, 4-dinitrophenylhydrazone m.p 292°C.
(10) m-Nitrobenzaldehyde semicarbazone m.p 246°C
(11) m-Nitrobenzaldoxime m.p 120°C.
(12) m-Nitrobenzaldehyde p-nitrophenylhydrazone m.p 247°C

(13) m-Nitrobenzaldehyde phenylhydrazone m.p 120°C.

m.p 80°C **m-Nitroacetophenone** 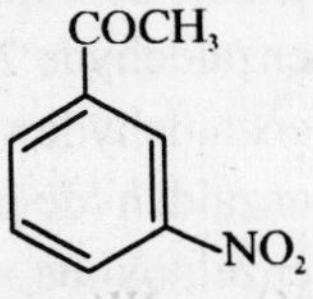

(1) Light yellow coloured solid, insoluble in water, soluble in alcohol and ether.
(2) m-Nitroacetophenone 2, 4-dinitrophenylhydrazone m.p 233°C.
(3) m-Nitroacetophenone semicarbazone m.p 261°C
(4) m-Nitroacetophenone oxime m.p 132°C
(5) m-Nitroacetophenone phenylhydrazone m.p 135°C.

m.p 105°C **o-Nitrobenzophenone** (o-nitrophenyl phenyl ketone)

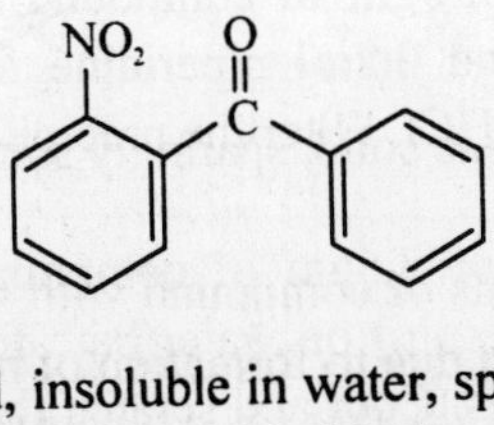

(1) Crystalline solid, insoluble in water, sparingly soluble in alcohol.
(2) Heat 3gm of compound with 24ml acetic acid, 4gm iron fillings and 10ml water at 90-95 C for 2hrs. Cool and dilute with water. Extract with ether. Wash the organic layer with sodium carbonate. Distill off ether and collect o-amino benzophenone, m.p 103-105°C.

m.p 106°C **p-Nitrobenzaldehyde** 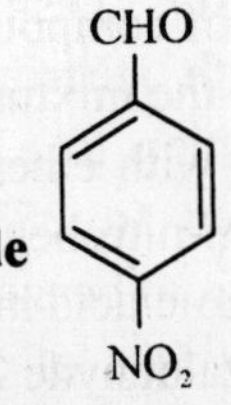

(1) Crystalline compound with faint bitter almond like odour, insoluble in water, soluble in alcohol. Slightly soluble in ether.

(2) Reduces Tollens reagent.
(3) Dissolve 100mg of compound in 2ml dil NaOH solution. A deep red brown coloured solution is obtained.
(4) p-Nitrobenzoic acid m.p 241°C ($K_2Cr_2O_7$+H_2SO_4)
(5) p-Nitrobenzaldehyde 2, 4-dinitrophenylhydrazone m.p 320°C
(6) p-Nitrobenzaldehyde semicarbazone m.p 220°C
(7) p-Nitrobenzaldehyde p-nitrophenylhydrazone m.p 249°C.
(8) p-Nitrobenzaldoxime m.p 130°C.
(9) p-Nitrobenzaldehyde phenylhydrazone m.p 159°C.

(vi) Nitrocarboxylic Acids

Solid

m.p 141.4°C **m-Nitrobenzoic acid**

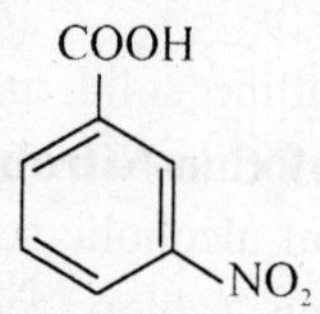

(1) Colourless crystalline solid, sparingly soluble in water and alcohol, soluble in ether.
(2) Heat few milligrams of compound with soda lime. Smell of bitter almond is observed due to formation of nitrobenzene.
(3) To 500mg of compound add 5ml conc HCl and 1gm of granulated tin. Heat until all the metal dissolves. Concentrate the mixture and pour in 10ml water. Pass H_2S gas and filter the precipitated tin sulphide. Evaporate the filtrate to dryness, dissolve the residue in 5ml water and add 2ml NH_4OH to it. Cool, acidify with dil HCl and collect m-amino benzoic acid, m.p 174°C.
(4) Heat on steam bath 10gm of m-nitrobenzoic acid and 15 thionyl chloride for 3hrs. Distill off the excess thionyl chloride and add dropwise with stirrring m-nitrobenzoyl chloride(dissolved in acetone) solution to a solution of sodium azide (4gm in 25ml water) cooled to 25°C. m-Nitro benzazide separates out, recrystallise from benzene, m.p 68-69°C.
(5) Heat together 5gm of compound with 13gm of thionyl chloride for 3hrs. Distill off the excess thionyl chloride and then distill the residue under reduced pressure. Collect m-nitrobenzoyl chloride

at b.p 153-154°C/12mmHg which on cooling soon solidifies, m.p 33°C.

(6) m-Nitrobenzamide m.p 143°C (7) m-nitrobenzanilide m.p 154°C.

(8) m-nitro benz-p-toluidide m.p 162°C. (9) s-Benzyl thiouronium salt m.p 163°C (10) NE 167

m.p 144°C (hydrated m.p 125°C) **3-NitroSalicylic acid** (2-hydroxy-3-nitro benzoic acid)

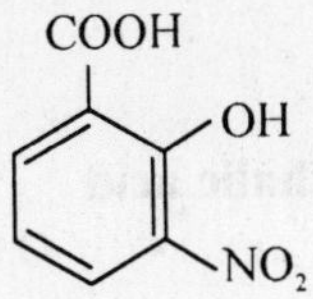

(1) Crystalline solid, insoluble in water, soluble in alcohol, ether, chloroform and benzene.

(2) To 2ml alcoholic solution of compound add few drops of $FeCl_3$ solution. A blood red coloured solution is obtained.

(3) Heat few milligrams of compound with CaO. A strong phenolic odour due to the formation of o-nitrophenol is observed.

(4) 3-amino salicylic acid m.p 235°C (Sn+AcOH)

(5) 3-NitroSalicylamide m.p 145°C.

(6) NE 183

m.p 146°C **o-Nitrobenzoic acid**

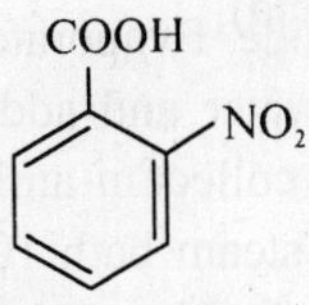

(1) Colourless solid, soluble in hot water.

(2) Heat few milligram of compound with soda lime. Smell of bitter almond is observed due to formation of nitrobenzene.

(3) To 500mg of compound add 5ml conc HCl and 1gm of granulated tin. Heat until all the metal dissolves. Concentrate the mixture and pour in 10ml water. Pass H_2S gas and filter the precipitated tin sulphide. Evaporate the filtrate to dryness, dissolve the residue

in 5ml water and add 2ml NH_4OH to it. Cool, acidify with dil HCl and collect o-amino benzoic acid(anthranilic acid), m.p 144°C.

(4) o-Nitrobenzamide m.p 176°C
(5) o-nitrobenzanilide m.p 155°C
(6) o-nitro benz-p-toluidide m.p 203°C.
(7) s-Benzyl thiouronium salt m.p 159°C
(8) NE 167

m.p 164°C **4-Nitrophthalic acid**

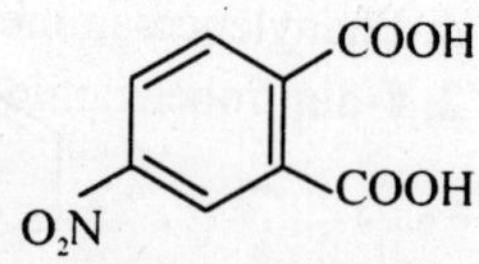

(1) Pale yellow coloured solid, soluble in alcohol and ether.
(2) Heat 3gm of compound with 7gm granulated tin and 20ml conc HCl for 20-30min. Decant and to the filtrate add NH_4OH solution, heat on steam bath for 20min. Cool, add dil HCl and recover m-amino benzoic acid, m.p 170°C.
(3) 4-Nitrophthalamide m.p 200°C (d)
(4) 4-Nitrophthalanilide m.p 192°C
(5) NE 211
(6) 4-Nitro phthal-p-toludide m.p 172°C

m.p 174°C **3, 5-Dinitro Salicylic acid** (3, 5-dinitro-2-hydroxy benzoic acid)

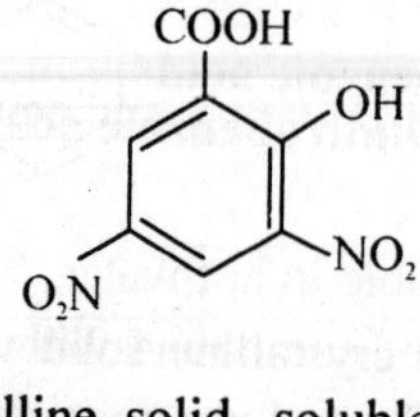

(1) Colourless crystalline solid, soluble in hot water, alcohol and ether. Sublimes
(2) To 2ml alcoholic solution of compound add few drops of $FeCl_3$ solution. A blood red coloured solution is obtained.
(3) 3, 5-dinitrosalicylamide m.p 181°C.
(4) NE 288

m.p 184°C **2, 4-dinitrobenzoic acid** 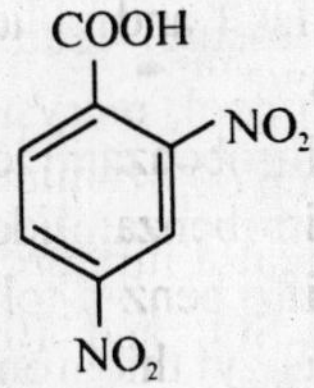

(1) Colourless crystalline solid with bitter taste, insoluble in water, soluble in alcohol.
(2) m-Phenylenediamine m.p 63°C (Sn+ HCl)
(3) 2, 4-dinitrobenzamide m.p 203°C (4) NE 212

m.p 199°C **m-nitrocinnamic acid** (trans) 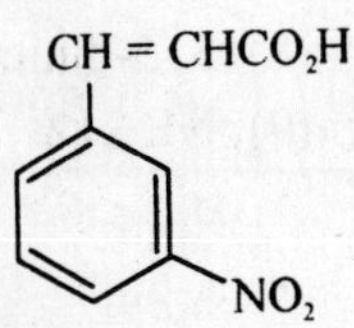

(1) Colourless, or yellow crystalline solid, very slightly soluble in water, soluble in alcohol.
(2) m-Aminocinnamic acid m.p 180°C (Sn+HCl)
(3) m-Nitrocinnamide m.p 196°C
(4) NE 193

m.p 205°C **3, 5-Dinitrobenzoic acid** 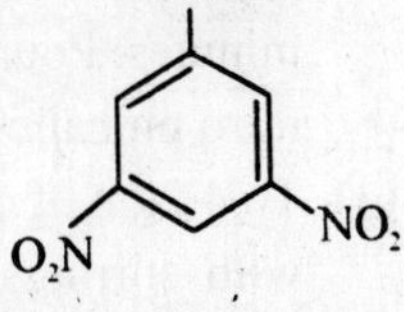

(1) Faintly yellowish crystalline solid with bitter taste, insoluble in water, soluble in alcohol and acetone.
(2) To few miligrams of compound in acetone add NaOH solution drop by drop until it is alkaline. A deep blue colouration is produced.
(3) Reflux 3gm of compound with 3.5gm of PCl_5 on oil bath for 60-70 min. Cool and distill off phosphorous oxychloride under reduced

pressure. Transfer the residual liquid on watch glass and allow to cool, recrystallise 3, 5-dinitrobenzoyl chloride from pet ether, m.p 69°C.

(4) Heat a mixture of 5gm of compound with 9ml oleum, 2ml conc H_2SO_4 and 10ml chloroform at 45°C. Slowly add in portions 1.75gm of sodium azide while keeping the temperature at 30-40°C. Heat the mixture for 3hrs at temperature at which chloroform refluxes. Cool, pour in crushed ice and keep for 1hr. Collect yellow crystals of 3, 5-dinitro aniline, m.p 162°C.

(5) 3, 5-diamino benzoic acid m.p 236°C (Sn+HCl)

(6) 3, 5-dinitrobenzamide m.p 183°C

(7) 3, 5-dinitrobenzanilide m.p 234°C

(8) 3, 5-dinitrobenz-p-toluidide m.p 280°C

(9) NE 212.

m.p 216°C(d) **3-Nitrophthalic acid**

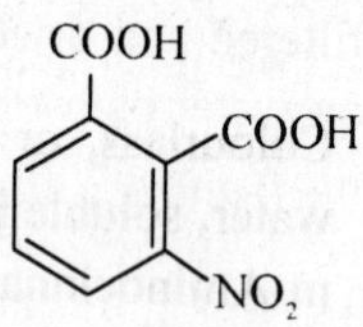

(1) Colourless or slightly yellow crystalline solid, insoluble in water and chloroform, soluble in alcohol and ether.

(2) Heat to gentle boiling 2.1gm of compound with 4ml Ac_2O until a clear solution is obtained and then heat further for another 10 minutes. Pour the mixture in a porcelain dish and cool. Collect 3-nitro phthalic anhydride, m.p 164°C.

(3) To 4.2gm of compound add 5ml water and 7-10ml NaOH solution with stirring. Now add 3gm of hydrazine sulphate to the well stirred solution and evaporate it to dryness on sand bath. Mix the residue with 3-5ml tetralin and heat at 160-170°C for 3hrs. Add 5ml benzene and filter. Collect 5-nitro 2, 3 dihydro-1, 4-phthalazinedione and recrystallise from hot water, m.p 315-316°C.

(4) 3-Nitro phthalamide m.p 201°C(d)

(5) 3-Nitrophthalanalide m.p 234°C.

(6) 3-Nitrophthal-p-toluidide m.p 224°C

(7) NE 105

m.p 220°C(d) **2, 4, 6-trinitrobenzoic acid** (sym-trinitro benzoic acid)

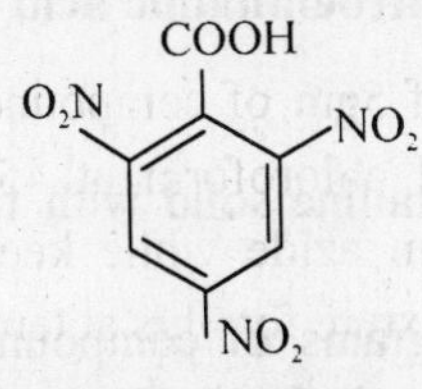

(1) Pale yellow coloured odourless solid, insoluble in water.
(2) To few milligrams of compound add NaOH solution. A red coloured solution is obtained.
(3) Dissolve 500mg of compound in 20ml hot alcohol and then add 10-15 drops of aniline to it. Heat the mixture to boiling, filter hot and allow the salt to crystalllise. Collect the precipitated addition product, m.p 124°C.
(4) To 5gm of compound add 10ml NH_4OH solution and aqueous solution of 3.5gm of silver nitrate. The precipitated silver salt is filtered and dried. Reflux the silver salt with 10ml methyl iodide on water bath for 1hr. Distill off the excess methyl iodide, extract the residue with boiling alcohol and filter. Concentrate the filtrate, cool and collect the precipitated methyl trinitrobenzoate, m.p 157°C.
(5) 2, 4, 6-trinitrobenzamide m.p 264°C.

m.p 230°C **5-Nitrosalicylic acid** (2-hydroxy-5-nitro benzoic acid)

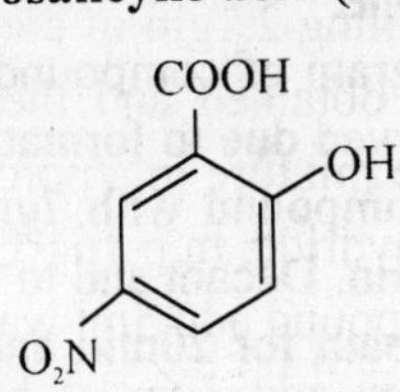

(1) Colourless crystalline solid, insoluble in water, soluble in hot water and ether.
(2) Heat few milligrams of compound with CaO. A strong phenolic odour due to the formation of p-nitrophenol is observed.
(3) To 2ml alcoholic solution of compound add few drops of $FeCl_3$ solution. A red coloured solution is obtained.
(4) 5-Amino salicylic acid m.p 283°C(Sn+HCl) (5) 5-Nitro salicylamide m.p 225°C (6) NE 183 (7) 5-Nitro salicylanilide m.p 224°C

m.p 240°C **o-Nitrocinnamic acid** 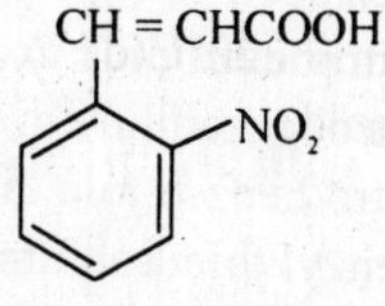

(1) Colourless crystalline solid with faint aromatic odour, insoluble in water
(2) Heat few milligrams of compound with conc H_2SO_4. A deep purple coloured solution is obtained.
(3) Heat 1gm of compound with 30ml $KMnO_4$ solution containing few drops of NaOH solution Filter hot, concentrate the filtrate. Cool, acidify with conc HCl and collect o-nitrobenzoic acid, m.p 146°C.
(4) o-Nitrocinnamide m.p 185°C. (5) NE 193 (6) o-Amino cinnamic acid m.p 158°C (d) (Sn+HCl)

m.p 242.4°C **p-Nitrobenzoic acid** 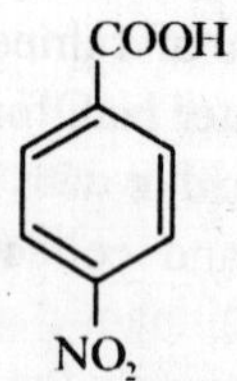

(1) Colourless odourless solid, sparingly soluble in water and alcohol, soluble in acetone.
(2) Heat few milligram of compound with soda lime. Smell of bitter almond is observed due to formation of nitrobenzene.
(3) Heat 3gm of compound with 7gm granulated tin and 20ml conc HCl for 20-30min. Decant and to the filtrate add NH_4OH solution, heat on steam bath for 20min and filter. Concentrate the filtrate. Cool, add dil HCl and recover p-amino benzoic acid, m.p 186°C.
(4) Heat together 5gm of compound with 5.5gm of p-toluene sulphonamide and 13gm PCl_5 until all contents have become liquid. Raise the temperature to 200-205°C or until no more contents distill. Cool and add with stirring 12ml pyridine followed by 100ml water. Filter the solid benzonitrile and recrystallise from acetic acid, m.p 147-148°C.
(5) Heat 5gm of compound with 6.3gm PCl_5 until a yellow homogenous liquid is formed. Distill the liquid under reduced

pressure and collect p-nitrobenzoyl chloride. Cool the yellow liquid which soon solidifies, m.p 71°C.

(6) p-Nitrobenzamide m.p 201°C
(7) p-nitrobenzanilide m.p 211°C
(8) p-nitro benz-p-toluidide m.p 204°C.
(9) s Benzyl thiouronium salt m.p 182°C (10) NE 167

m.p 286°C **Trans p-Nitrocinnamic acid**

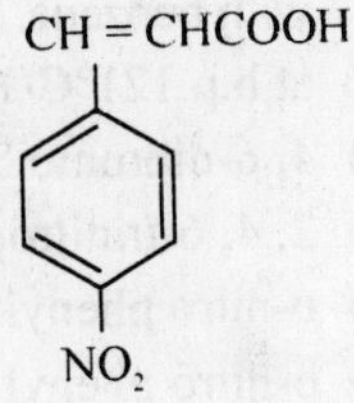

(1) Colourless or yellow crystalline solid, insoluble in water, alcohol and ether.
(2) To few milligrams of compound add conc H_2SO_4 and heat. An orange coloured solution is obtained.
(3) Heat 1gm of compound with 30ml $KMnO_4$ solution containing few drops of NaOH solution. Filter hot and concentrate the filtrate. Cool, acidify with conc HCl and collect p-nitrobenzoic acid, m.p 241°C.
(4) p-amino cinnamic acid m.p 175°C (Sn+HCl)
(5) p-nitrocinnamide m.p 204°C.
(6) NE 193

(*vii*) *Nitro Phenols*

Solid

m.p 45°C, b.p 214°C **o-Nitrophenol**

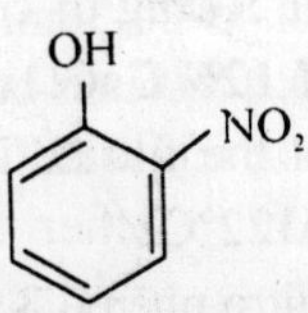

(1) Bright yellow coloured solid with strong phenolic odour, insoluble in water, soluble in hot water, alcohol and ether.
(2) To few milligrams of compound add dil NaOH solution. A yellow coloured solution is obtained.

(3) Boil 500mg of compound with 1.25gm Zn dust, 10ml water and 5ml 10% $CaCl_2$ until the solution becomes colourless. Filter hot, cool the filtrate in ice. Crystals of o-aminophenol separate out, m.p174°C.
(4) Reflux a mixture of 5.6gm of compound, 6gm of n-butyl bromide, 5.6 gm of potassium carbonate and 40ml of acetone for 48hrs. Distill off the acetone and add 15ml water to the residue. Extract with benzene. Distill off benzene and collect o-butoxy nitrobenzene at b.p 121°C/1mmHg
(5) 4, 6-dibromo 2-nitrophenol m.p 117°C
(6) 2, 4, 6-trinitrophenol m.p 122°C.
(7) o-nitro phenyl 3, 5-dinitrobenzoate m.p 155°C
(8) o-nitro phenyl p-nitrobenzoate m.p 141°C
(9) o-nitro phenyl benzoate m.p 59°C
(10) o-nitro phenyl acetate m.p 41°C.

m.p 97°C **m-Nitrophenol**

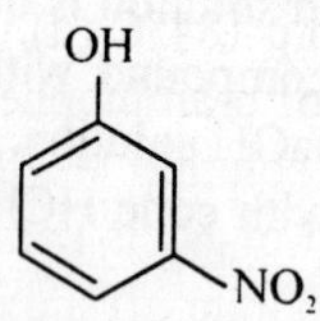

(1) Pale yellow crystalline solid with aromatic odour, insoluble in water, soluble in alcohol and ether.
(2) Dissolve 20mg of compound in 8ml 10% NaOH solution. An orange coloured solution is obtained.
(3) To 2ml saturated aqueous solution of compound add 2-3 drops of $FeCl_3$ solution. A violet red coloured solution is obtained.
(4) Boil 500mg of compound with 1.25gm Zn dust, 10ml water and 5ml 10% $CaCl_2$ until the solution becomes colourless. Filter hot, cool the filtrate in ice. Crystals of m-aminophenol separate out, m.p122°C.
(5) m-nitro phenyl 3, 5-dinitrobenzoate m.p 159°C
(6) m-nitro phenyl p-nitrobenzoate m.p 174°C
(7) m-nitro phenyl benzoate m.p 95°C
(8) m-nitro phenyl acetate m.p 56°C.

m.p 114°C **p-Nitrophenol** 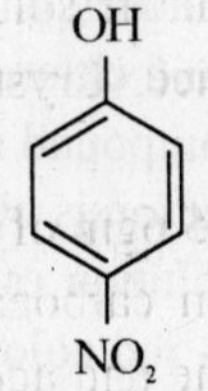

(1) Colourless solid with faint aromatic odour, insoluble in water, soluble in hot water and ether.

(2) To 2ml saturated aqueous solution of compound add 2-3 drops of $FeCl_3$ solution. A violet red coloured solution is obtained.

(3) Dissolve 20mg of compound in 8ml 10% NaOH solution. A bright yellow coloured solution is obtained.

(4) Boil together 400mg of compound with 5ml 10% $CaCl_2$ solution and 1gm of Zn dust for 5min or until the mixture is colourless. Filter hot. Cool the filtrate and collect the precipitated solid, dry and dissolve in 15ml boiling benzene. Cool and collect p-aminophenol, m.p 184°C (d).

(5) Diazotise 3gm of p-aminophenol (as synthesized above) and add the diazonium solution to 40ml boiling 25% $CuSO_4$ solution. When evolution of N_2 is complete, cool the solution and extract with ether. Evaporate off the ether and crystallise the quinol from dilute H_2SO_4, m.p 169°C.

(6) Dissolve 2.78gm of compound in 9ml glacial acetic acid and add dropwise with stirring 9.4ml of Br_2 solution (2.4ml bromine in 7ml glacial acetic acid). Heat the mixture on water bath for 1hr. Cool and collect pale yellow 2, 6-dibromo-4-nitrophenol, m.p 138-140°C.

(7) Picric acid (2, 4, 6 trinitrophenol) m.p 122°C.

(8) p-nitro phenyl 3, 5 dinitrobenzoate m.p 186°C

(9) p-nitro phenyl p-nitrobenzoate m.p 159°C

(10) p-nitro phenyl benzoate m.p 142°C

(11) p-nitro phenyl acetate m.p 83°C.

m.p 114°C **2, 4-dinitrophenol** 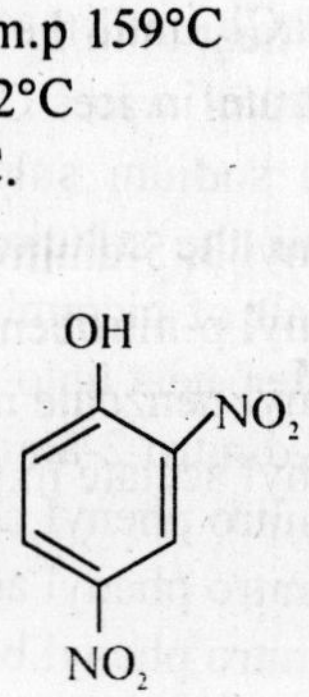

(1) Yellow crystalline odourless solid, insoluble in water, soluble in hot water, benzene and chloroform.
(2) Dissolve 20mg of compound in 8ml 10% NaOH solution. A bright yellow coloured solution is obtained.
(3) To 2ml alcoholic solution of compound add 2-3 drops of $FeCl_3$ solution. An orange- red coloured solution is obtained.
(4) To a solution of 3gm of compound in 25ml water add with stirring and heating to 85°C, 6gm of NH_4Cl and 1ml conc ammonia solution. Lower the temperature to 70°C and add in portion 7gm of fused sodium sulphide. Continue heating at 80-85°C for 15 minutes. Filter and keep the filtrate overnight. Collect the crystals of 2-amino 4-nitro phenol m.p 140-142°C.
(5) Picric acid m.p 122°C. (6) 6-bromo 2, 4-dinitrophenol m.p 118°C. (7) 2, 4-dinitro phenyl p-nitrobenzoate m.p 139°C (8) 2, 4-dinitro phenyl benzoate m.p 132°C (9) 2, 4-dinitro phenyl acetate m.p 72°C. (10) 2, 4-diaminophenol m.p 79°C (d) (Sn+HCl)

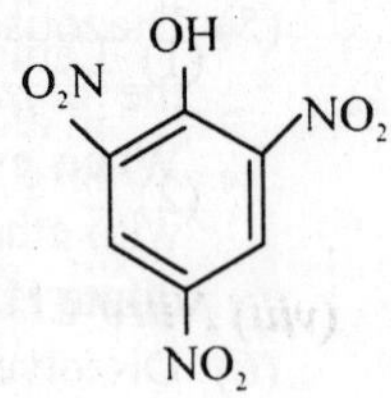

m.p 122°C **Picric acid** (2, 4, 6-trinitro phenol)

(1) Light yellow coloured solid, sparingly soluble in water, soluble in alcohol and benzene.
(2) Boil aqueous solution of compound with 5-6 drops of 1% KCN solution. A red coloured solution is obtained.
(3) Dissolve 1gm of compound and 400mg of caustic soda in 60ml water and heat upto 55°C. Add 10ml of sodium sulphide solution (4gm in 10ml water). Then add concurrently 12gm picric acid and 40ml sodium sulphide(22gm in 40ml water)(temp below 65°C). Stir the solution for 10-15min, cool and add ice to it. Sodium salt of picramic acid separates out. Dissolve the sodium salt in water, heat upto 80°C, acidify with dilute H_2SO_4, Picramic acid (4, 6-dinitro-2-aminophenol), m.p 168°C separate out.
(4) 2, 4, 6-trinitro phenyl p-nitrobenzoate m.p 143°C.
(5) 2, 4, 6-trinitro phenyl acetate m.p 76°C.
(6) 2, 4, 6-trinitro phenyl benzoate m.p 163°C.

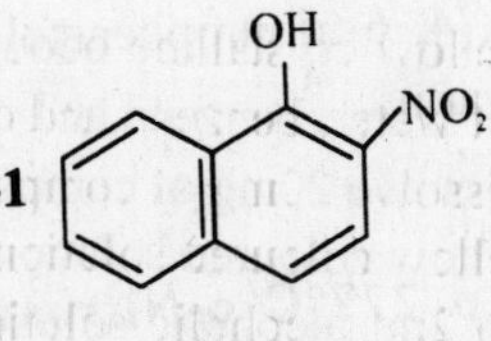

m.p 128°C **2-Nitronaphthol-1**

(1) Yellow coloured solid, insoluble in water, sparingly soluble in alcohol.
(2) Dissolve 20mg of compound in 8ml 10% NaOH solution. A red coloured solution is obtained

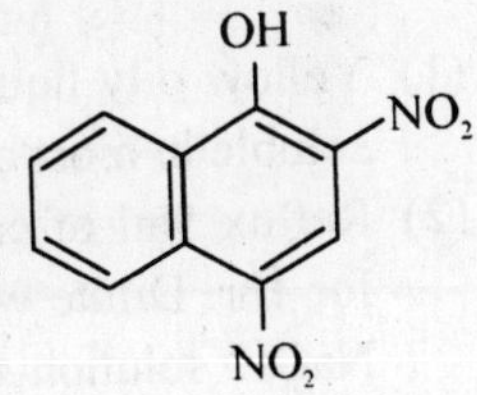

m.p 138°C **2, 4-Dinitro-1-naphthol**

(1) Lemon yellow coloured solid, insoluble in water, sparingly soluble in alcohol, ether and benzene, soluble in acetone.
(2) 2, 4 dinitro 1-naphthyl benzoate m.p 174°C

(viii) Nitro Ethers

Liquids

b.p 268°C **o-Nitrophenetole** (Ethyl-o-nitrophenyl ether)

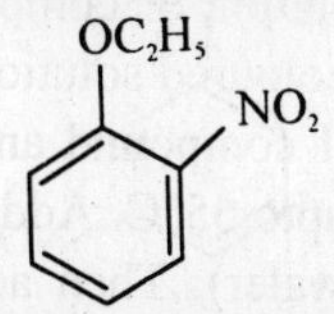

(1) Yellow oily liquid with mild aromatic odour, insoluble in water, soluble in most organic solvents.
(2) Reflux 5ml of compound with 9.2gm of tin and 20ml conc HCl for 1hr. Dilute with 15ml water and pass H_2S gas. Filter and add NaOH solution to the filtrate until alkaline. Extract with ether, distill off ether and collect o-phenetidine, b.p 229°C.
(3) 2, 4-dinitrophenetole m.p 86°C. (conc H_2SO_4 + conc HNO_3, 0°C)

(4) 2, 4, 6-trinitrophenetole m.p 78°C (conc H_2SO_4 + conc HNO_3, warm)

b.p 277°C **o-Nitroanisole** (Methyl -o-nitrophenyl ether, 1-methoxy-2-nitro benzene)

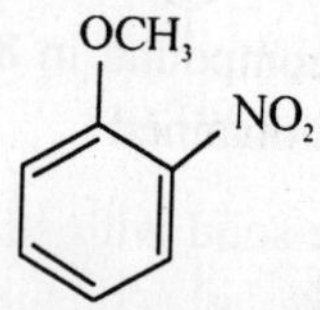

(1) Yellow oily liquid with mild aromatic odour, insoluble in water, soluble in most organic solvents.

(2) Reflux 5ml of compound with 9.2gm of tin and 20ml conc HCl for 1hr. Dilute with 15ml water and pass H_2S gas. Filter and add NaOH solution to the filtrate until alkaline. Extract with ether, distill off ether and collect o-anisidine, b.p 218°C.

(3) 2, 4-dinitroanisole m.p 88°C. (conc H_2SO_4 + conc HNO_3, 0°C)

(4) 2, 4, 6-trinitroanisole m.p 68°C (conc H_2SO_4 + conc HNO_3, warm on water bath)

Solid

m.p 38°C, b.p 258°C **m-Nitroanisole** (1-methoxy-3-nitro benzene)

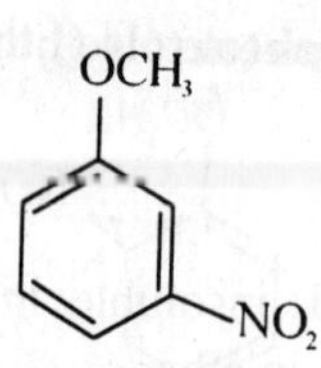

(1) Crystalline compound, insoluble in water, soluble in most organic solvents.

(2) To 3gm of compound add 9gm granulated tin and 20ml conc HCl. Heat the mixture until all the oily drops disappear. Dilute with 15ml water and pass H_2S gas. Filter and add NaOH solution to the filtrate until alkaline. Extract with ether, distill off the ether and collect m-anisidine, b.p 251°C

m.p 54°C, b.p 274°C **p-Nitroanisole** (1-methoxy-4-nitro benzene)

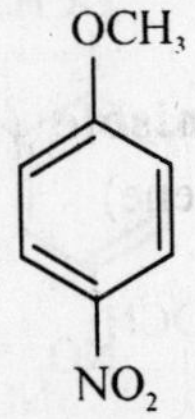

(1) Colourless crystalline solid with strong aromatic odour, insoluble in water, soluble in alcohol and ether.
(2) To few milligrams of compound add dil NaOH solution. A yellow coloured solution is obtained.
(3) To 1gm of compound add 3gm granulated tin and 7ml conc HCl. Heat the mixture until all the oily drops disappear. Then add 30ml KOH solution. Warm and filter. Collect the precipitated solid, wash with water. Dissolve in hot water, filter hot. Cool the filtrate and collect crystals of p-anisidine, m.p 58°C
(4) 2, 4-dinitroanisole m.p 88°C. (fuming HNO_3, 0°C)
(5) 2, 4, 6-trinitroanisole m.p 68°C (conc H_2SO_4 + conc HNO_3, warm)

m.p 58°C **p-Nitrophenetole** (1-ethoxy-4-nitro benzene)

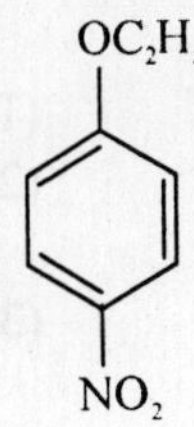

(1) Crystalline compound, insoluble in water, sparingly soluble in alcohol, easily soluble in ether.
(2) To 3gm of compound add 8gm granulated tin and 20ml conc HCl. Heat the mixture until all the oily drops disappear. Dilute with 15ml water and pass H_2S gas. Filter and add NaOH solution to the filtrate until alkaline. Extract with ether, distill off the ether and collect p-Phenetidine, b.p 254°C
(3) 2, 4-dinitrophenetole m.p 86°C (conc H_2SO_4 + conc HNO_3, 0°C)
(4) 2, 4, 6-trinitrophenetole m.p 78°C (conc H_2SO_4 + conc HNO_3, warm)

m.p 68°C **2, 4, 6-Trinitroanisole** (Methyl picrate)

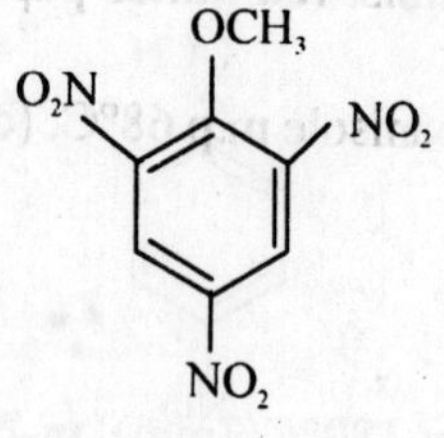

(1) Lemon–yellow or colourless solid, insoluble in water, soluble in alcohol, ether and acetone.
(2) Heat 500mg of compound with 5ml NaOH solution. Cool, acidify with dil HCl and collect picric acid, m.p 122°C.
(3) Picramide m.p 188°C

m.p 78°C **2, 4, 6-Trinitrophenetole** (Picric acid ethyl ether)

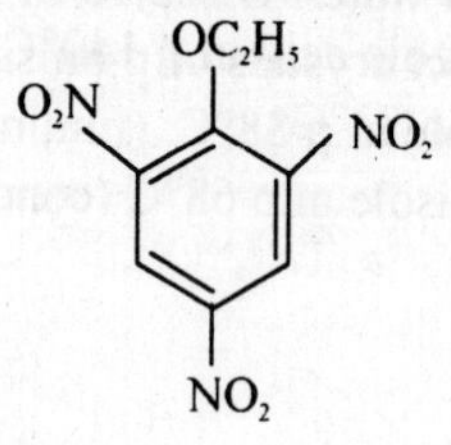

(1) Colourless crystalline solid, insoluble in water, soluble in ether.
(2) Heat 500mg of compound with 5ml NaOH solution. Cool, acidify with dil HCl and collect picric acid, m.p 122°C.
(3) Picramide m.p 188°C.

m.p 95°C **2, 4-Dinitroanisole** (1-methoxy, 2, 4-dinitrobenzene)

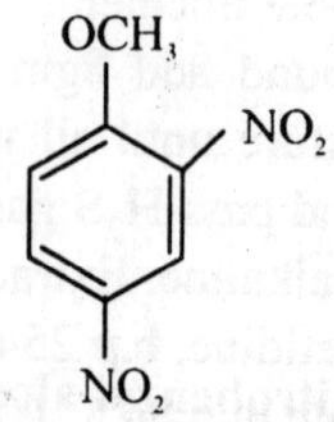

(1) Pale yellow crystalline compound, insoluble in water, soluble in ether.

(2) Boil few milligrams of compound with alcoholic KOH solution and bring a moist red litmus paper near the mouth of test tube. It turns blue.

(3) 2, 4, 6-trinitroanisole m.p 68°C. (conc H_2SO_4 + conc HNO_3, warm)

(ix) *Nitro Alcohols*

m.p 27°C, b.p 180°C(3mm) **m-Nitrobenzyl alcohol**

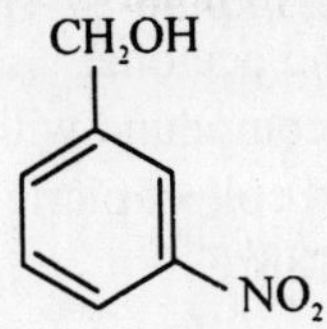

(1) Pale yellow coloured crystalline solid, slightly soluble in water, soluble in alcohol and ether.

(2) m-Nitro benzoic acid m.p 140°C

(3) Benzoate m.p 72°C

m.p 70-72°C **o-Nitrobenzyl alcohol**

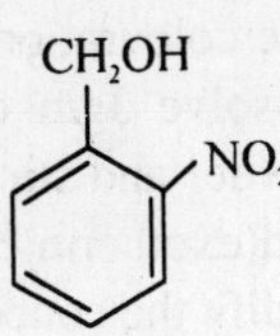

(1) Crystalline compound, insoluble in water, soluble in alcohol and ether.

(2) Acetate m.p 35°C

(3) Benzoate m.p 102°C.

m.p 94°C **p-Nitrobenzyl alcohol**

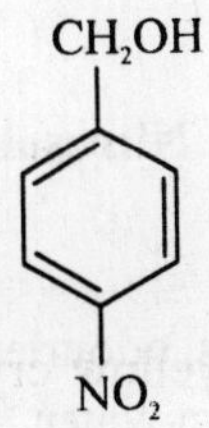

(1) Colourless crystalline solid, insoluble in water, soluble in hot water, alcohol and ether.

(2) Cool a mixture of 15ml of dichloromethane and 2.1ml diethylaminosulfur trifluoride to 10°C. Add a solution of 2.3gm of p-nitrobenzyl alcohol in 45ml dichloromethane. Keep the resultant mixture at room temperature for sometime and then pour in ice. Remove the organic layer and extract the aqueous layer with dichlormethane. Combine the organic layers, evaporate off the solvent, crystallise the residue from pentane and collect p-nitrobenzyl fluoride, m.p 36-37°C.

(3) Acetate m.p 78°C

(4) Benzoate m.p 94°C.

(x) Nitroso Compounds

m.p 67°C **N-Nitroso diphenylamine** (nitrous diphenyl amide) $(C_6H_5)_2NNO$

(1) Pale yellow coloured solid, insoluble in water, souble in hot alcohol and benzene.

(2) Dissolve few milligrams of compound in conc H_2SO_4. An intense blue colouration is obtained.

(3) Dissolve 3gm of compound in 15ml alcohol, add 7gm of zinc powder and stir vigorously. Add dropwise 4.5ml of glacial acetic acid(exothermic reaction). Filter, concentrate the filtrate. Cool and acidify the concentrate with 7ml of conc HCl. Collect blue crystals of phenylhydrazine hydrochloride. Dissolve it in dilute NaOH solution and extract with toluene. Distill off the toluene, cool the residue and collect N, N-diphenylhydrazine, m.p 35°C.

(4) Diphenyl amine m.p 54°C (Sn+HCl in alcohol)

m.p 68°C **Nitrosobenzene** 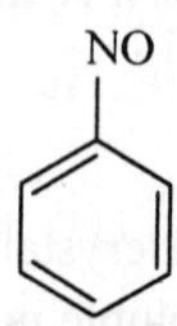

(1) Colourless, odourless solid which turns emerald green on melting, insoluble in water, soluble in nearly all organic solvent.

(2) Add a solution of 2.5gm of compound in 5ml glacial acetic acid to a solution of 2.5 gm of aniline in 4ml glacial acetic acid. Temperature rises immediately. Aftersome time when the reaction is complete, cool and collect azobenzene, m.p 68°C.

(3) To a cooled solution of 2.5gm of compound in 10ml glacial acetic acid add 2.8 gm of o-anisidine. Keep the mixture at room temperature for 40hrs and then pour in water. Collect the precipitated solid and steam distill. Collect o-methoxyazobenzene at b.p 195-197°C /15mmHg which on cooling scon solidifies, m.p 41°C.

m.p 84°C **p-Nitrosodiethylaniline**

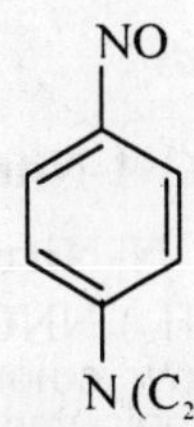

(1) Dark green coloured crystalline solid with bluish reflection, insoluble in water, soluble in alcohol and ether.

(2) To 500mg of compound add 10ml conc HNO_3 and 10ml water. Stir for 30 second and then pour with stirring in 15ml cold water. Collect the precipitated dinitrodiethylaniline, m.p 80°C.

(3) p-Aminodiethylaniline b.p 262°C. (Sn+HCl)

m.p 85°C **p-Nitroso N, N-dimethylaniline**

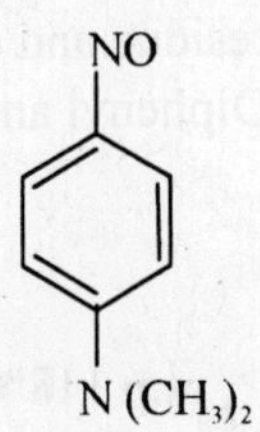

(1) Dark green coloured crystalline solid with yellowish reflection, insoluble in water, soluble in alcohol and ether.

(2) Dissolve 200mg of compound in 3ml alcohol and then add 10ml conc HNO_3 and 5ml water to it. Stir the mixture vigorously, cool

and then pour in ice water. Collect the yellow dinitrodimethylaniline, m.p 89°C.

(3) Heat together 6.6gm of compound with 2gm of sodium carbonate and 30ml alcohol for 30min and filter. Heat the filtrate with 6gm 2, 4-dinitrotcluene for 5hrs. Cool and collect 2, 4-dinitrobenzylidene-p-amino dimethylaniline. Dissolve it in 5ml conc HCl and 5ml water, heat at 105°C for 15min. Cool filter the precipitated solid, recrystallise from pet ether and collect 2, 4-dinitrobenzaldehyde, m.p 71°C.

(4) p Aminodimethylaniline m.p 41°C. (Sn+HCl)

m.p 109°C **1-Nitroso-2-Naphthol**

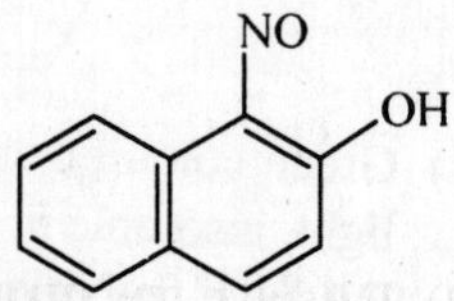

(1) Dull yellow coloured solid, insoluble in water, soluble in hot alcohol, ether and benzene.

(2) To few milligrams of compound in acetic acid add $FeCl_3$ solution. A black precipitate is obtained. Dissolve this precipitate in aniline. A deep blue colouration is obtained.

(3) Dissolve 200mg of compound in 3ml alcohol and then add 10ml conc HNO_3 and 5ml water to it. Stir the mixture vigorously, cool and then pour in ice water. Collect the yellow 1- nitronaphthol-2, m.p 103°C.

(4) 1, 6-dinitro-2-naphthol m.p 195°C(d) (fuming HNO_3, 45-55°C).

m.p 118°C **p-Nitroso N-Methyl aniline**

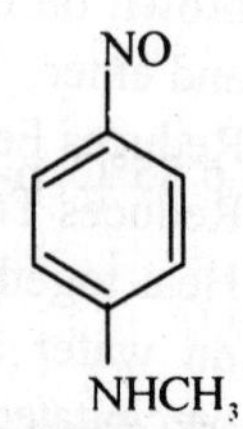

(1) Green coloured crystalline solid with bluish reflection, insoluble in water.

(2) To 8gm of compound add 15gm of granulated tin and in portion

30ml of conc HCl. Heat the mixture on water bath for 45min. Add a solution of 26gm of NaOH in 35ml water and steam distill. Collect the distillate, saturate with NaCl. Separate the organic layer and extract the aqueous layer with ether. Combine the organic layers, distill and collect methyl aniline at b.p 193-194°C.

(3) Dissolve 2.5gm of compound in 5ml ether and add 10ml of alcohol saturated with HCl gas. Keep at room tempeature for some time and collect hydrochloride of p-nitroso N-methylaniline.

m.p 145°C **p-Nitrosodiphenylamine**

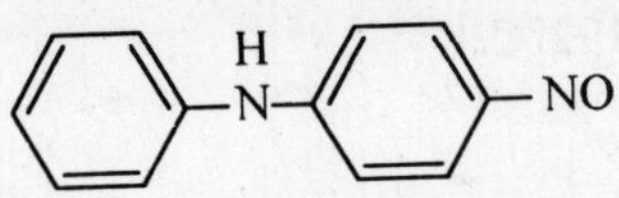

(1) Green coloured crystalline solid with dark steel blue by reflected light, insoluble in water, soluble in ether.

(2) Dissolve few milligrams of compound in conc H_2SO_4. An orange or red-orange coloured solution is obtained.

(3) p-Aminodiphenylamine m.p 66°C (Zn+ acetic acid).

(xi) *Azo, Azoxy and Hydrazine Compound*

m.p 19°C, b.p 243°C **Phenylhydrazine**

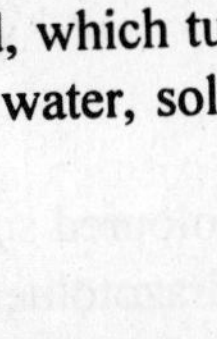

(1) Pale yellow coloured solid or oily liquid, which turns yellow and brown on exposure to air, insoluble in water, soluble in alcohol and ether.

(2) Reduces Fehling's solution.

(3) Reduces Tollens reagent.

(4) Heat together 4gm of compound with 4.9ml of ethylacetoacetate on water bath for 2hrs. Cool and add 10ml ether. Filter the precipitated 3-methyl-1-phenylpyrazol-5-one, recrystallise from hot water, m.p 127°C.

(5) Acetate m.p 128°C

(6) Benzoate m.p 168°C.

m.p 34°C **N, N-Diphenylhydrazine** $(C_6H_5)_2$ N.NH_2

(1) Yellow crystalline solid or liquid, insoluble in water, soluble in alcohol and ether.
(2) Reduces Tollens reagent.
(3) Dissolve few milligrams of compound in conc H_2SO_4. An intense blue coluration is obtained.
(4) Dissolve 20mg of compound in 5ml alcohol and then add 7-8 drops of benzladehyde. Heat on water bath, cool and collect benzaldehydediphenylhydrazone, m.p 123°C.
(5) Diphenylamine m.p 54°C (Sn+HCl)
(6) Acetate m.p 184°C
(7) Benzoate m.p 192°C.

m.p 36°C **Azoxybenzene** 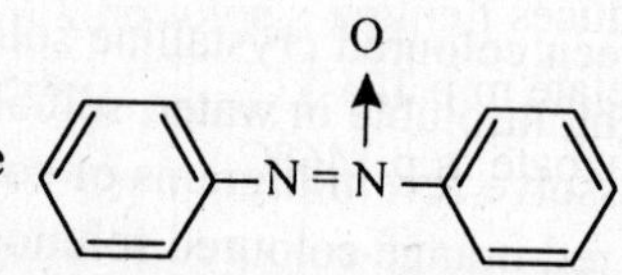

(1) Yellow coloured solid, insoluble in water, soluble in ether.
(2) Mix 200mg of compound with 400mg of PCl_5 and heat over a small flame until the fusion is complete. Cool, remove the excess PCl_5, wash the residue with water, recrystallise from hot alcohol and collect azobenzene, m.p 68°C.

m.p 55°C **o-Azotoluene** (2, 2' dimethylazobenzene)

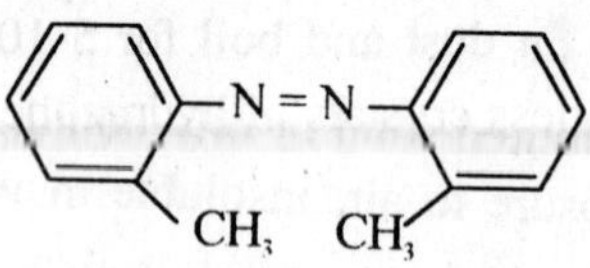

(1) Red coloured solid, insoluble in water, soluble in benzene.
(2) o-Hydrazotoluene m.p 165°C (Zn+ ethanolic NaOH).

m.p 59°C **o-Azoxytoluene** 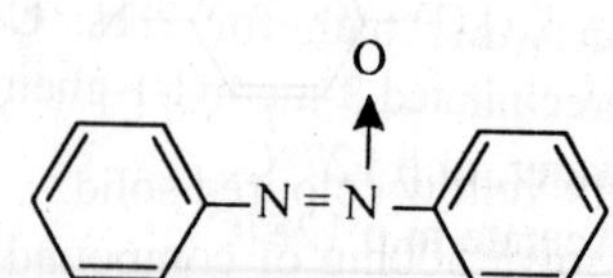

(1) Light yellow coloured solid.
(2) Heat 500mg of compound with conc H_2SO_4 at 100-125°C. Cool, pour in cold water and collect o-azotoluene m.p 55°C.
(3) o-Toluidine b.p 200°C (Sn+HCl)

m.p 61°C, b.p 240°C **p-Tolylhydrazine**

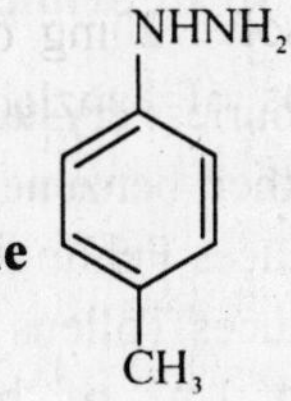

(1) Colourless crystalline solid, soluble in alcohol, ether and benzene.
(2) Reduces Fehling's solution.
(3) Acetate m.p 121°C
(4) Benzoate m.p 146°C.

m.p 68°C **Azobenzene** (diphenyl diimide)

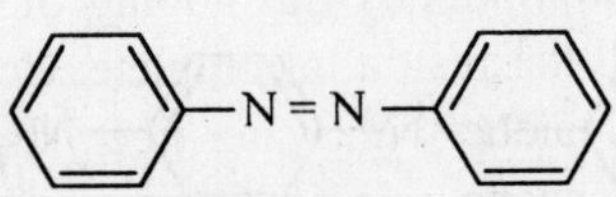

(1) Reddish-Orange coloured solid, insoluble in water, soluble in ether.
(2) Reduces Tollens reagent.
(3) Dissolve 500mg of compound in 15ml 70% alcohol with 500mg KOH. Add 5gm Zn dust and boil for 5-10 min. Filter hot. Cool and collect colourless crystals of hydrazobenzene, m.p 130°C.

m.p 75°C **p-Azoxy toluene**

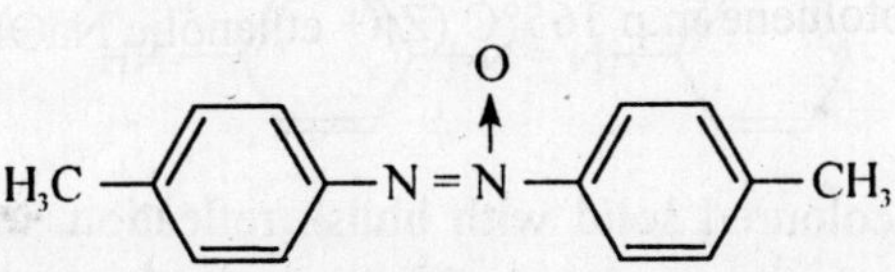

(1) Light yellow coloured solid.
(2) Dissolve 500mg of compound in 15ml 70% alcohol with 500mg

KOH. Add 5gm Zn dust and boil for 5-10 min. Filter hot. Cool and collect crystals of p-azotoluene, m.p 144°C.

(3) p-Toluidine m.p 45°C (Sn+HCl)

m.p 96°C **Semicarbazide** (amino urea) $NH_2\overset{O}{\overset{\|}{C}}NHNH_2$

(1) Colourless crystalline solid, soluble in water and alcohol, insoluble in ether, benzene and chloroform.

(2) Reduces Fehling's solution.

(3) Reduces Tollens reagent.

(4) Add 1ml of benzaldehyde to 100mg of semicarbazide hydrochloride in 5ml water. Heat the mixture on water bath. Cool and shake vigorously. Collect the precipitated benzalsemicarbazone, m.p 213°C.

(5) Hydrochloride m.p 173°C(d).

m.p 117°C **p-Dimethylaminoazobenzene** (N, N-dimethyl-p-phenyl azoaniline)

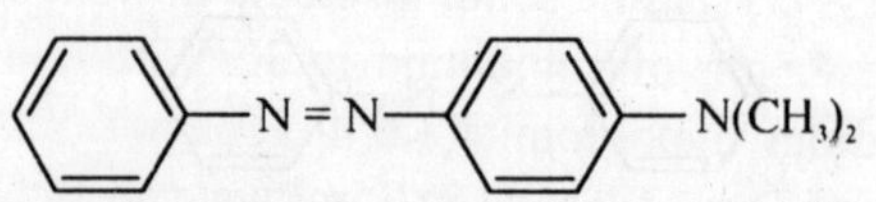

(1) Yellow coloured solid, soluble in nearly all organic solvents.

(2) Dissolve few milligrams of compound in conc H_2SO_4. A red coloured solution is obtained.

(3) Methoidide m.p 174°C

m.p 126°C **p-Aminoazobenzene** (p-phenyl azoaniline)

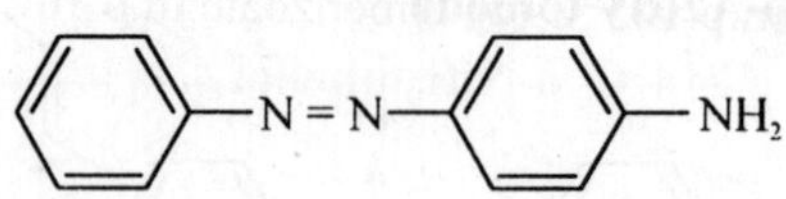

(1) Yellow coloured solid with bluish reflection, insoluble in water soluble in hot alcohol and ether.

(2) Acetate m.p 146°C.

(3) Benzoate m.p 211°C.

m.p 128°C **Acetylphenylhydrazine** (hydracitin, pyrodin)

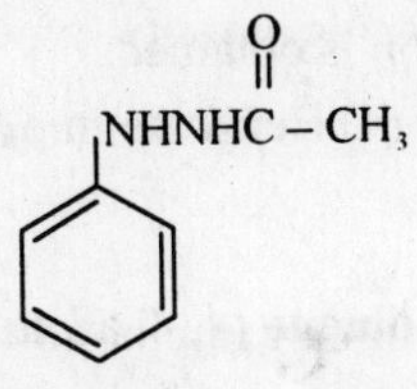

(1) Crystalline solid, soluble in hot water, alcohol, benzene and chloroform.
(2) Reduces Tollens reagent.
(3) Hydrolyse with NaOH, phenylhydrazine and acetic acid are obtained.
(4) Dibromo m.p 146°C (Br_2+HCl)

m.p 131°C **Hydrazobenzene** (sym-diphenyl hydrazine)

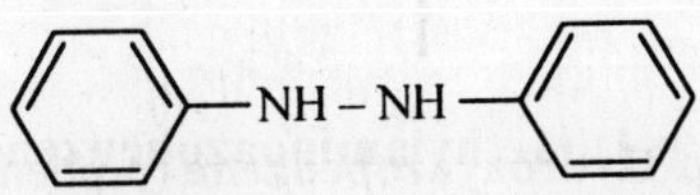

(1) Colourless crystalline solid, insoluble in water, soluble in ether.
(2) Dissolve 4.6gm of compound in 30ml ether and add in portion cold sodium hypobromite(1.6ml bromine added dropwise to a solution of 3gm of NaOH in 35ml water) solution to it. Shake the mixture for 10min, separate the ether layer. Distill off ether and dissolve the residue in 10ml alcohol. Heat to boiling and then add water dropwise until azobenzene commences to separate. Cool and collect the orange crystals, m.p 68°C.
(3) Dissolve by heating 200mg of compound in 3ml conc HCl and 5ml water. Add 4-5 drops of conc H_2SO_4. Cool and collect crystals of benzine sulphate.
(4) Diacetate m.p 105°C (5) Dibenzoate m.p 162°C.

m.p 132°C **Benzeneazo-2-naphthol**

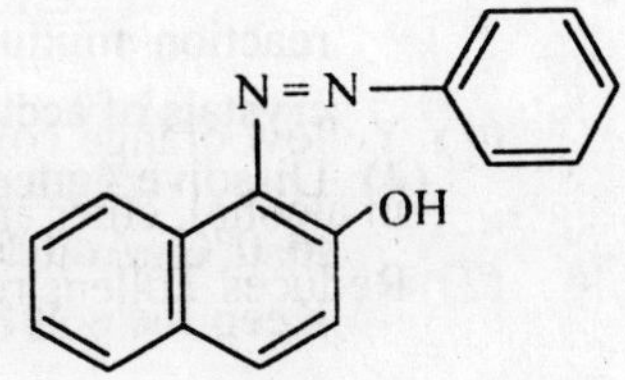

(1) Red -gold coloured crystalline solid, soluble in ether and benzene.

(2) Dissolve few milligrams of compound in conc H_2SO_4. A fuchsine red coloured solution is obtained.

(3) Acetate m.p 117°C (4) Benzoate m.p 125°C.

m.p 144°C **p-Azotoluene** (4, 4' azotoluene)

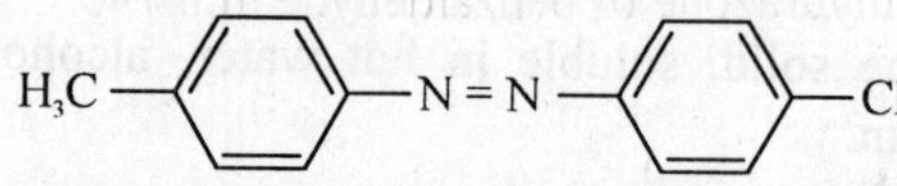

(1) Orange-yellow coloured solid, insoluble in water, sparingly soluble in alcohol, soluble in ether.

(2) Dissolve 500mg of compound in 15ml 70% alcohol with 500mg NaOH. Add 5gm Zn dust and boil for 5-10 min. Filter hot. Cool and collect crystals of p-hydrazotoluene, m.p 133°C.

(3) p-Toluidine m.p 45°C (Sn+HCl)

m.p 152°C **p-Hydroxyazobenzene** (p-phenyl azo phenol)

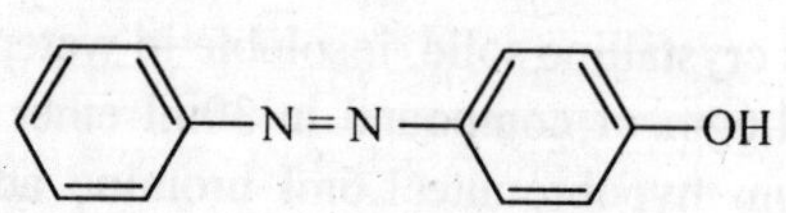

(1) Orange coloured solid, insoluble in water, soluble in alcohol and ether.

(2) Acetate m.p 84°C (3) Benzoate m.p 138°C.

m.p 157°C **p-Nitrophenylhydrazine**

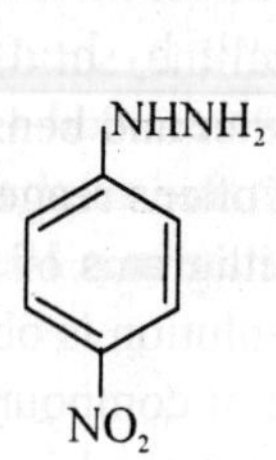

(1) Yellow orange coloured solid, sparingly soluble in water, soluble in alcohol, ether and chloroform.

(2) Reduces Tollens reagent.

(3) To few milligrams of compound add NaOH solution. A dark red brown coloured solution is obtained.
(4) p-Phenylene diamine m.p 140°C (Sn+HCl)
(5) Acetate m.p 205°C
(6) Benzoate m.p 193°C
(7) Picrate m.p 119°C.
(8) p-Nitrophenylhydrazone of benzaldehyde m.p 190°C.

m.p 168°C **Benzoylphenylhydrazine** (1-benzoyl-2-phenyl hydrazine)

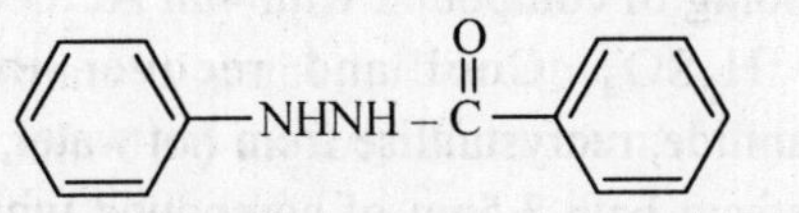

(1) Colourless crystalline solid, insoluble in water, sparingly soluble in ether, soluble in hot alcohol.
(2) Hydrolyse with conc HCl, phenylhydrazine hydrochloride and benzoic acid m.p 121°C are obtained.

m.p 198°C(d) **2, 4 dinitrophenylhydrazine**

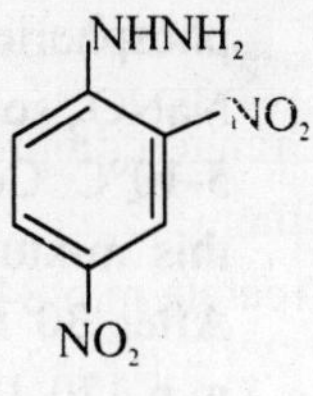

(1) Red crystalline solid, insoluble in water, sparingly soluble in alcohol, ether and benzene.
(2) Reduces Tollens reagent.
(3) To few milligrams of compound add NaOH solution. A deep red coloured solution is obtained.
(4) To 500mg of compound add 10ml NaOH solution and boil. Bring a glass rod dipped in conc HCl at the mouth of test tube. White fumes due to formation of NH_4Cl takes place.
(5) Acetate m.p 198°C.
(6) Benzoate m.p 206°C.
(7) 2, 4 dinitrophenylhydrazone of benzaldehyde m.p 237°C.

(xii) Nitro Amino Compounds

m.p 71°C **o-Nitroaniline** (1-amino-2-nitro benzene)

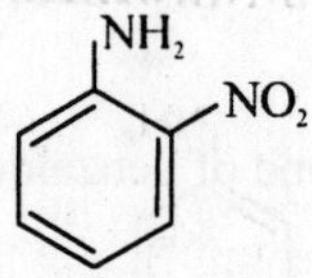

(1) Red orange coloured solid, difficult soluble water, easily soluble ether.

(2) Reflux 500mg of compound with 4ml Ac_2O containing 2-3 drops of conc H_2SO_4. Cool and recover the precipitated o-nitroacetanilide, recrystalllise from hot water, m.p 93°C.

(3) Heat on steam bath 3.5gm of compound with 4ml of NaOH and 10ml of ethyl alcohol until the solution boils. Remove the heat source and add in portion 6.5gm of Zn powder. Reflux the mixture for 1hr and filter while hot. Cool and extract with ether, distill off the ether and collect o-phenylene diamine recrystallise from hot water, mp 99-101°C.

(4) Dissolve 1.3gm of compound in 7.5ml conc H_2SO_4, 10ml phosphoric acid and 5ml water. Cool the mixture and add dropwise $NaNO_2$ solution (800mg in 3ml water) with temperature between 5-10°C. Cool the mixture to –10°C and add 5ml liquid SO_2. Pour this mixture on 5.5gm $FeSO_4$ and 100mg Cu powder mixture. After 30 min solid 2-nitro benzenesulfinic acid is precipitated, m.p 120-125°C(d)

(5) Heat together 1.5gm of compound with 500mg potassium carbonate, 6ml bromobenzene and a trace of cuprous iodide for 12hrs. Distill off the excess bromobenzene. Add HCl and warm, filter and collect o-nitrodiphenylamine, recrystallise from alcohol, m.p 75°C.

(6) Dissolve 3gm of compound in 15ml conc HCl and then pour in ice. Add a solution of 1.5 gm of sodium nitrite in 5ml water. Pour on 1.5gm copper bronze moistened with conc HCl(vigorous reaction occurs). After the reaction has subsided distill the mixture and collect o-nitrochlorobenzene at b.p 242°C which on cooling soon solidifies, m.p 32°C.

(7) Benzoate m.p 98°C.
(8) Picrate m.p 73°C.

m.p 78°C **4-Methyl-3-Nitroaniline** (3 nitro-p-toluidine) [NH_2 = 1]

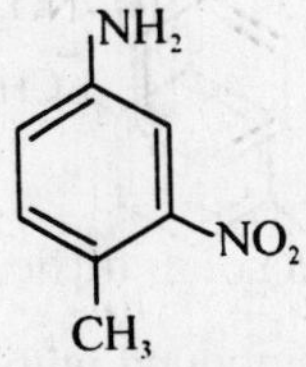

(1) Yellow coloured solid, insoluble in water, soluble in ether.
(2) m-Toluylenediamine m.p 99°C(Sn+HCl).
(3) Acetate m.p 148°C (4) Benzoate m.p 172°C.

m.p 91°C **2-Methyl-3-nitroaniline** (3 nitro-o-toluidine) [NH_2 = 1]

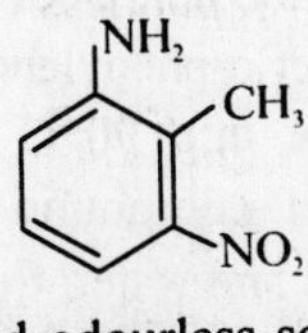

(1) Pale yellow coloured odourless solid, insoluble in water, soluble in alcohol, ether and benzene.
(2) Toluylene 2, 6-diamine m.p 104°C (Sn+HCl)
(3) Acetate m.p158°C
(4) Benzoate m.p 168°C

m.p 96°C **2-methy-6-nitroaniline** (6 nitro-o toluidine) [NH_2 = 1]

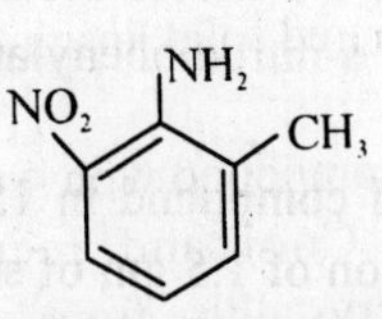

(1) Orange-yellow coloured solid, insoluble in water, soluble in alcohol, ether, chloroform and benzene.
(2) Toluylene 2, 3 diamine m.p 61°C (Sn+HCl)
(3) Acetate m.p 158°C

(4) Benzoate m.p 167°C.

m.p 107°C **2-methyl-5-nitroaniline** (5-nitro-o-toluidine) [NH_2 = 1]

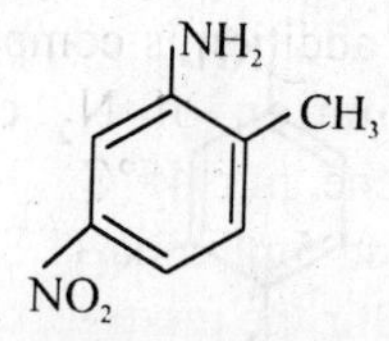

(1) Yellow crystalline compound with sweet taste, soluble in alcohol and ether.
(2) Dilute a solution of 3gm of compound in 4ml conc H_2SO_4 with 100ml water, cool and diazotise with 1.4gm sodium nitrite. Warm on water bath and then boil. Filter hot. Concentrate the filtrate and cool. Dissolve the precipitated solid in hot water containing 2ml of dil NaOH solution, boil and filter hot. Saturate the filtrate with CO_2. Collect the colourless crystals of 4-nitroindazole, m.p 181°C.
(3) m-Toluylenediamine m.p 90°C (4) Acetate m.p 151°C (5) Benzoate m.p 183°C.

m.p 114°C **m-Nitroaniline** (1-amino-3-nitro benzene)

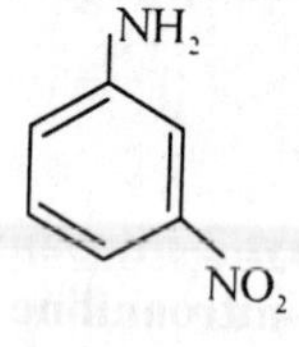

(1) Orange-yellow coloured solid, insoluble in water, soluble in alcohol and ether.
(2) Reflux 500mg of compound with 4ml Ac_2O containing 2-3 drops of conc H_2SO_4. Cool and recover the precipitated m-nitroacetanilide, recrystalllise from hot water, m.p 155°C.
(3) Heat under reflux 2.6gm of benzaldehyde with 3.5gm of compound in 15ml recitified spirit for 20 min. Cool and stir the solution to induce crystallisation and collect the schiff's base, m.p 72°C. Dissolve the schiff's base in 15ml methanol. Warm the solution

to 40°C and add in portion with stirring 800mg of $NaBH_4$. Then heat under reflux for 15min. Cool and collect N-benzyl m-nitroaniline, m.p 106-107°C.

(4) To 5gm of Copper (I)chloride (synthesized from $CuSO_4$) add diazonium salt solution prepared from 5gm of m-nitroaniline(temp 20-25°C). After addition is complete warm the solution on water bath until evolution of N_2 ceases. Cool and collect m-chloronitrobenzene, m.p 45°C.

(5) To 7ml water add 5ml conc H_2SO_4. To this solution add 3.5gm m-nitroaniline. Stir the mixture, cool by adding crushed ice. Now add a cold solution of 1.6gm of $NaNO_2$ in 4ml water and continue stirring for 5-10min. Keep the mixture at room tempeature for 5min and then decant the supernatant liquid. In a separate flask containing boiling H_2SO_4(15ml in 14ml water) add the supernatant liquid at such a rate that the mixture boils for 30min. Continue boiling for another 5min, cool and pour in crushed ice. Collect the precipitated m-nitrophenol, m.p 96°C.

(6) To a cooled solution of 1.65gm of compound in 7ml glacial acetic acid add dropwise with stirring a solution of 2ml of bromine in 15ml glacial acetic acid. After the addition is complete keep at room temperature for sometime, add 10ml of cold water and collect 2, 4, 6-tribromo 3-nitroaniline, recrystallise from alcohol, m.p 102-103°C.

(7) m-phenylene diamine m.p 63°C (Sn+HCl)

(8) Benzoate m.p 155°C.

(9) Picrate m.p 143°C.

m.p 117°C **4-Methyl-2-nitroaniline** (2-Nitro-p-toluidine)

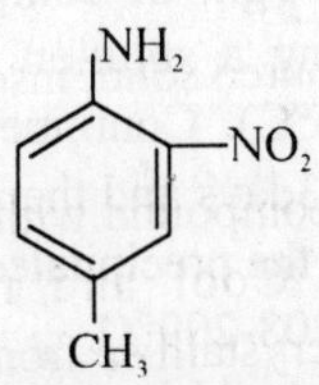

(1) Red coloured solid, insoluble in water, soluble in alcohol.

(2) Cool a solution of 3.4gm of compound in 10ml alcohol containing 3ml conc H_2SO_4 to 10°C. Diazotise with 1.7gm of conc aqueous solution of 1.7gm of sodium nitrite. Heat the mixture under reflux.

Distill to remove alcohol and acetaldehyde. Distill the residue under reduced pressure and collect m-nitrotoluene at b.p 112-114°C/15mmHg (b.p 231°C)

(3) Acetate m.p 96°C

(4) Benzoate m.p 148°C.

m.p 123°C 2, 4 dimethyl-5-nitroaniline

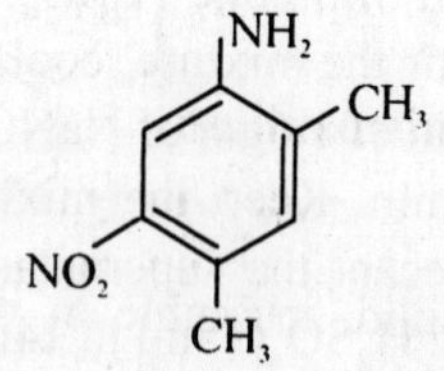

(1) Orange red coloured solid, insoluble in water, soluble in hot alcohol.

(2) Acetate m.p 160°C. (3) Benzoate m.p 200°C.

m.p 130°C 2-Methyl 4-nitroaniline (4-nitro-o-toluidine)

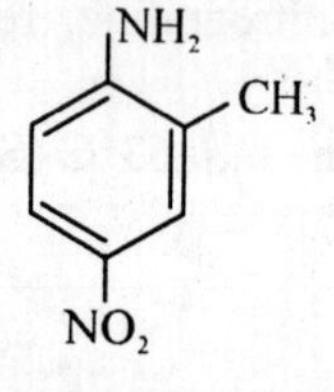

(1) Lemon-yellow coloured solid, insoluble in water, sparingly soluble in hot water, soluble in ether and alcohol.

(2) To a solution of 2.75gm of compound in 125ml glacial acetic acid add with stirring a solution of 1.25gm of $NaNO_2$ in 3ml water(temp below 5°C). Continue stirring for 20min and filter. Keep the filtrate for 3days and then concentrate. Dilute with 10ml water, cool and filter the precipitated 5-nitro indazole, recrystallise, from methanol m.p 208-209°C.

(3) To 5gm of compound add ice cold solution of 20ml conc H_2SO_4 in 5ml water. Now add well stirred Caro's acid solution(17ml conc H_2SO_4 added to 30gm potassium persulfate followed by 90gm crushed ice and 30ml water) to it. Stir the mixture and heat to 40°C. Remove the heat source and continue stirring for 2hrs.

Dilute with 50ml water and filter. Distill the filtrate and collect 2-nitroso-5-nitro toluene. Cool the distillate, the product crystallises out, m.p 135-136°C.

(4) p-toluylenediamine m.p 64°C (Sn+HCl).

(5) Acetate m.p 202°C.

(6) Benzoate m.p 178°C.

m.p 138°C **2, 6-dinitroaniline**

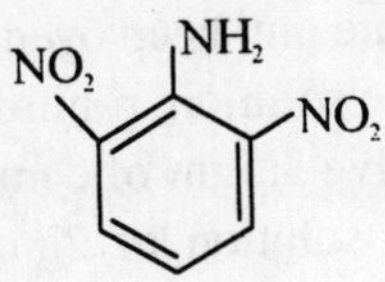

(1) Yellow coloured solid, insoluble in water, sparingly soluble in alcohol, soluble in ether.

(2) Dissolve by heating 1.6gm $NaNO_2$ in 16ml conc H_2SO_4. Cool to 25°C and add a solution of 3.6gm of compound in 40ml glacial acetic acid(temp below 40°C). Stir the solution at 40°C for 20min and then add this diazonium salt to ice cold cuprous chloride solution (4.4gm Cu_2Cl_2 in 40ml conc HCl)with stirring. Dilute with 10ml water, cool and collect yellow crystals of 1-chloro-2, 6-dinitrobenzene, m.p 87°C.

(3) Acetate m.p 197°C

m.p 147°C **p-Nitroaniline** (1-amino-4-nitro benzene)

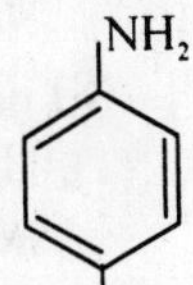

(1) Yellow coloured, tasteless, odourless solid, insoluble in water, sparingly soluble in alcohol and ether.

(2) Reflux 500mg of compound with 4ml Ac_2O containing 2-3 drops of conc H_2SO_4. Cool and recover the precipitated p-nitroacetanilide, recrystallise from hot water, m.p 207°C.

(3) Dissolve 1.4gm of compound in 5ml boiling glacial acetic acid. Remove the burner and add dropwise 3.25gm of iodinemonochloride in 2ml glacial acetic acid. Heat the mixture

on water bath for 2hrs. Cool and recover diiodo p-nitroaniline, m.p 243-245°C.

(4) Dissolve 1gm compound in 2.1ml conc HCl and 2.5ml water, cool to 0°C. Diazotise by adding 0.5gm of $NaNO_2$ solution in 2ml water. Stir, filter and add it to ice cold sodium sulphite solution (4.1gm in 10ml water containing 0.4gm NaOH). Keep the solution for 5min, acidify with 7ml HCl and heat on steam bath for 25-30min when crystals of p-nitrophenylhydrazine hydrochloride separate out, keep ovenight. Treat the precipitate with conc sodium acetate solution, p-nitrophenylhydrazine separate out, m.p 158°C.

(5) Dissolve 3.5gm of compound in 31ml conc HCl at 50°C and then add a solution of 2gm of potassium chlorate in 45ml water at room temperature. Dilute with water and collect the precipitated 2, 6-dichloro 4-nitroaniline, recrystallise from glacial acetic acid, m.p 189°C.

(6) Add with stirring cold solution of 2.2gm of cuprous chloride in 10ml HCl to diazonium solution prepared from 3gm of p-nitroaniline, 3ml conc H_2SO_4, 6ml water and 1.6gm $NaNO_2$. After the evolution of N_2 has ceased the liquid is distilled on steam bath until no more p-chloronitrobenzene passes over. Cool and recrystallise from benzene, m.p 83°C.

(7) Stir 2gm of compound with 1.5ml conc H_2SO_4 and 6ml water for 1hr. Cool to 0°C and diazotise with a solution of 1gm $NaNO_2$ in 3ml water. Filter the cold solution and add the filtrate to a solution of 4gm KI in 12ml water. Collect the precipitated p-iodonitrobenzene, m.p 171°C.

(8) Dissolve 1gm of compound in hot mixture of 1.3ml conc HCl and 1.3ml water. Cool to 0°C, diazotise the precipitated hydrochloride with a cold solution of 0.34gm of $NaNO_2$ in 2ml water and stir(temp below 5°C). Add diazotised solution slowly to cold 1-naphthol solution(.5gm in 1.2gm NaOH in 5ml water) and stir. Acidify with conc HCl. Filter and collect Magneson *II* dye.

(9) Cool a solution of 5.6gm of compound in 20ml conc HNO_3 to 0°C and then diazotise with aqueous solution of sodium nitrite(15gm). Pour this diazotised solution over freshly prepared copper oxide solution (from 20gm $CuSO_4$). Keep the mixture for 2 days and distill. Collect p-dinitrobenzene which on cooling soon solidifies, m.p 172-173°C.

(10) p-Phenylene diamine m.p 140°C (Sn+HCl)

(11) Acetate m.p 215°C

(12) Benzoate m.p 199°C.
(13) Picrate m.p 100°C

m.p 169°C **Picramic acid** (4, 6-dinitro 2-amino phenol)

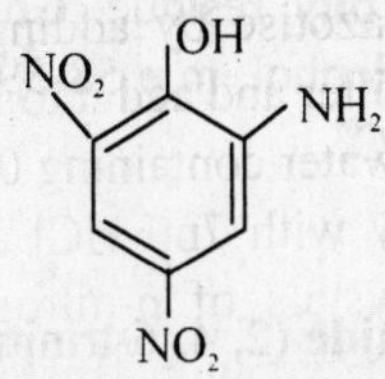

(1) Dark red coloured solid, sparingly soluble in water, easily soluble in benzene, chloroform and ether.
(2) To few milligrams of compound add NaOH solution. An orange coloured solution is obtained.
(3) N-Acetate m.p 201°C
(4) O-Acetate m.p 193°C.
(5) N-Benzoate m.p 230°C

m.p 188°C **2, 4-dinitroaniline**

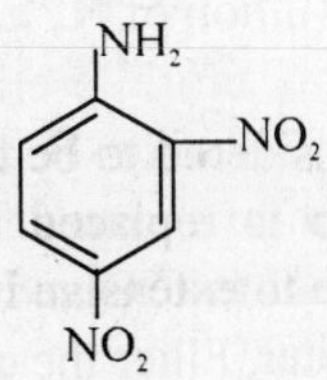

(1) Yellow coloured solid with bluish reflection, insoluble in water, slightly soluble in alcohol.
(2) Heat few milligrams of compound with dil KOH solution. A red coloured solution is obtained.
(3) To 500mg of compound add 10ml KOH solution and boil. Bring a glass rod dipped in conc HCl at the mouth of test tube. White fumes due to formation of NH_4Cl takes place.
(4) Reflux 500mg of compound with 5ml Ac_2O. Cool, pour in crushed ice and collect 2, 4-dinitro acetanilide, m.p 120°C.
(5) Heat a mixture of 2.38gm of compound with 25ml ethanol and 12ml NH_4OH at 45°C and then pass H_2S gas. Repeat this process 3-4 times or until all the nitroamino compound is dissolved. Cool, keep the reaction mixture for 16-18hrs and collect 1, 2-diamino 4-nitrobenzene, m.p 197-198°C.
(6) To a cooled suspension of 2.7ml of 90% H_2O_2 in 25ml chloroform

add 17ml of trifluoroacetic acid. Add in portion 4.6gm of compound to this freshly prepared peroxy-trifluorocetic acid and then heat under reflux. Distill off the excess solvent and add 25ml cold water to the oily residue. Collect 1, 2, 4-trinitrobenzene, recrystallise from alcohol, m.p 58-59°C.

(7) Benzoate m.p 202°C.

m.p 188°C **Picramide** (2, 4, 6-trinitroaniline)

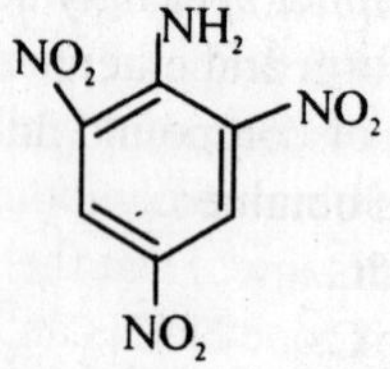

(1) Dark yellow coloured solid, soluble in acetone.
(2) Acetate m.p 230°C
(3) Benzoate m.p 196°C.

(xiii) Amides

Amides are considered to be to derivatives of carboxylic acids in which the hydroxyl group is replaced by an amino group. They are generally soluble in water due to extensive hydrogen bonding.

Identification Tests

(*i*) **Hydrolysis with alkali :** Place 200mg of compound in a test tube and add 6ml of 10% NaOH solution to it. Heat the mixture to boiling. Evolution of ammonia takes place which turns red litmus blue or gives white fumes with HCl indicating the presence of amide group.

$RCONH_2 + NaOH \rightarrow RCOO^-Na^+ + NH_3$

$NH_3 + HCl \rightarrow NH_4Cl$

(*ii*) **Hydroxamate test :** Amides which are not N-substituted generally respond to this test

Add 100mg of compound to 3-4ml of 1N hydroxylamine hydrochloride in propylene glycol/ethanol. Boil the mixture for 2-3 minutes. Cool and add 4-5ml of 5% freshly prepared $FeCl_3$ solution. A red to violet colour due to the formation of ferric hydroxymate is indicative of the positive test.

$$RCONH_2 + H_2NOH.\,HCl \rightarrow RCONHOH + NH_4Cl$$

$$3RCONHOH + FeCl_3 \rightarrow \left[R-C \begin{matrix} \diagup NH-O \\ \diagdown\!\!\diagdown O \rightarrow \end{matrix} \right]_3 Fe + 3HCl$$

(*iii*) Reaction with phenylhydrazine : Urea and N-substituted urea react with phenylhydrazine forming diphenylcarbazide, ammonia or an amine. Thiourea also reacts with phenylhydrazine forming diphenycarbazide which reacts with nickel ion to form violet coloured complex.

$$H_2NCONHR + 2H_2NNHC_6H_5 \rightarrow (C_6H_5NHNH)_2CO + NH_3 + RNH_2$$

(*a*) *Urea and N-Substituted urea :* In a dry test tube place 100mg of compound and add 4-5 drops of phenylhydrazine, heat the mixture to 200°C. Ammonia is liberated which can be identified.

(*b*) *Thioureas* : Place 50mg of compound in a test tube and add 4-5 drops of phenylhydrazine. Heat the test tube in oil bath for 5min. Cool and add 2ml of conc NaOH and 10% nickel sulphate solution. Shake the test tube thoroughly and keep it at room temperature for 3-5 minutes. Extract the mixture with chloroform. A violet to red violet colouration of chloroform layer is a positive test.

(*iv*) Biuret test for Urea and monosubstituted Urea : In a dry test tube place 100mg of urea and heat it gently until the solid melts. Continue the heating process further for 1-2 minutes or until the liquid appears to solidify with the formation of biuret. Cool and add 2-4 drops of $CuSO_4$ solution and shake. A pink violet colouration due to formation of complex between cupric ion and biuret takes place.

$$2\,H_2NCONH_2 \rightarrow H_2NCONHCONH_2 + NH_3$$

Mechanism

$$H_2NCONH_2 \rightarrow NH_3 + HO-C=N \leftrightarrow O=C=NH$$

$$H_2NCONH_2 + O=C=NH \rightarrow H_2NCONHCONH_2$$

Distinction between Aliphatic and Aromatic Amides

This test is based on the conversion of aliphatic or aromatic amides to hydroxamic acid. Aromatic amides are converted to hydroxamic acid by treating with hydrogen peroxide whereas aliphatic amides can only be converted by treatment with hydroxyamine.

Aliphatic amide : Add 100mg of compound to 3-4ml of 1N hydroxylamine hydrochloride in propylene glycol/ethanol. Boil the mixture for 2-3 minutes. Cool and add 4-5ml of 5% freshly prepared $FeCl_3$ solution. A red to violet colour due to the formation of ferric hydroxymate is indicative of the positive test.

$$RCONH_2 + H_2NOH.\,HCl \rightarrow RCONHOH + NH_4Cl$$

Aromatic amide : Add 100mg of substance to 5-6ml water in a test tube and shake it thoroughly for few seconds. Add 10-12 drops of 6% H_2O_2 and heat the mixture to boiling. Cool and add 4-5 drops of freshly prepared $FeCl_3$ solution. Development of bluish red colour is a positive test.

$$ArCONH_2 + H_2O_2 \rightarrow ArCONHOH + H_2O$$

Derivatives

(1) **Hydrolysis of Amides :** Place in a 50ml round bottom flask fitted with a reflux condensor 2gm of substance and 25ml of 10% NaOH solution. Add a piece of porcelain chip, heat the mixture for 30 min. Cool and acidify the reaction mixture with dilute HCl. Filter off the precipitated solid on buchner funnel, wash with cold water, recrystallise from a suitable solvent and determine its melting point.

$$RCONH_2 + NaOH \rightarrow RCOO^-Na^+ + NH_3$$

$$RCOO^-Na^+\ HCl \rightarrow RCOOH + NaCl$$

Mechanism

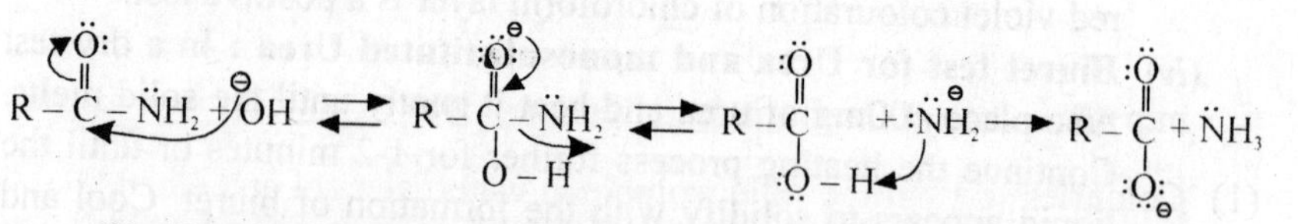

(2) **Xanthylamides :** Dissolve 75mg of xanthydrol in 9ml glacial acetic acid (filter if necessary) and add 75mg of amide to it. Shake the mixture thoroughly and keep aside for 5-10min. If a crystalline compound does not precipitate out then warm the mixture on water bath for 30min. Cool and recover the precipitated xanthylamide derivative, recrystallise from 50% acetic acid and determine its melting point.

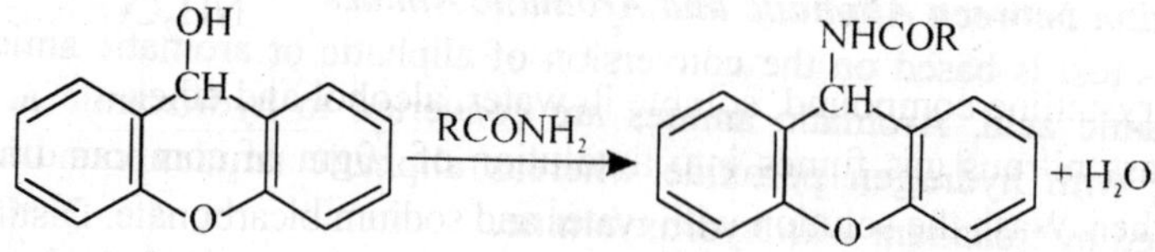

Aliphatic Liquid

b.p 210°C(d) **Formamide** (Methanamide) $HCONH_2$

(1) Colourless liquid, miscible with water and alcohol, sparingly soluble in ether.

(2) Reflux together 7.5ml formamide with 1ml water, 500mg of ammonium chloride in oil bath at 180-190°C. Add dropwise 4gm of 4, 4-dimethoxy –2-butanone. After addition is complete continue heating for 1hr. Cool and pour in 10ml 1N NaOH solution. Extract with chloroform. Distill off chloroform and collect 4-methylpyrimidine, b.p 140-142°C.

(3) Heat 3ml of compound with aniline. Ammonia gas is evolved, cool the mixture and add 10ml water, collect the precipitated formanilide, m.p 46°C.

(4) Heat 8ml formamide with 5gm o-nitrobenzaldehyde on water. Pass HCl gas to the hot solution for 20 minutes. Keep the mixture overnight and then add with stirring 15ml ethanol to it, Cool and collect α, α-Diformadido-o-nitrotoluene, m.p 172°C (d)

(5) Xanthylamide m.p 184°C.

Aliphatic Solid

m.p 49°C **Ethyl Carbamate** (Urethane) $NH_2COOC_2H_5$

(1) Colourless crystalline solid, soluble in water, alcohol and ether.

(2) Dissolve 500mg of compound in 5ml of dil NaOH solution followed by I_2-KI reagent. Warm on water bath, cool and collect iododorm, m.p 119°C.

(3) Carbanilide m.p 238°C.

(4) Xanthylamide m.p 169°C

m.p 52°C **Methyl Carbamate** (Urethylan, methyl urethane) NH_2COOCH_3

(1) Crystalline compound, soluble in water, alcohol and ether.

(2) Pass nitrous gas fumes into a solution of 10gm of compound in ether. Wash the solution with water and sodium bicarbonate. Distill and collect nitrosomethyl urethane as yellowish red liquid at b.p 70°C/27mmHg

(3) Carbanilide m.p 238°C.

(4) Xanthylamide m.p 193°C

m.p 60°C **n-Propyl Carbamate** (n-Propyl urethane) $NH_2COOC_3H_7$

(1) Crystalline compound soluble in water, alcohol and ether.
(2) Carbanilide m.p 238°C.

m.p 64°C **Isoamyl carbamate** (Isoamyl urethane) $NH_2COOC_5H_{11}$

(1) Crystalline compound with aromatic odour, soluble in hot water, alcohol and ether.
(2) Carbanilide m.p 238°C.

m.p 76°C **asy Dipropylurea** $NH_2CON(CH_2CH_2CH_3)_2$

(1) Colourless crystalline solid, soluble in water.
(2) Picrate m.p 135°C

m.p 79°C **Propionamide** $CH_3CH_2CONH_2$

(1) Colourless solid, soluble in water, alcohol and ether.
(2) Heat 500 mg of compound with 3-4ml of aniline. Ammonia gas is evolved and on cooling the mixture add 5-7ml water and collect the precipitated propionanilide, m.p 103°C.
(3) Xanthylamide m.p 214°C

m.p 82°C **Acetamide** (Ethanamide) CH_3CONH_2

(1) Colourless solid with smell similar to mice urine. Soluble in water and alcohol. Sparingly soluble in ether.
(2) Heat 50mg of compound with conc H_2SO_4. Smell of vinegar due to formation of acetic acid is observed.
(3) Dissolve 6gm of acetamide in 3ml Br_2 and then add dropwise with stirring16ml NaOH solution (4gm in 16ml water). Heat the reaction mixture on water bath for 5-10min. Cool and collect crystals of acetyl methyl urea, m.p 169-170°C.
(4) Dissolve 4gm of acetamide in 1ml Br_2 and then cool the solution to 0°C. Add 5ml of 50% aqueous KOH solution with stirring. Keep the solution at 0°C for 2-3hrs and then add 2gm NaCl and

10ml $CHCl_3$. Heat the mixture on water bath for 3-4 minutes and then decant off the chloroform layer. Repeat the extraction process twice. Combine the organic layers, concentrate and cool. Crystals of N-Bromoacetamide precipitates out, m.p 102-105°C.

(6) Heat 500 mg of compound with aniline. Ammonia gas is evolved and on cooling the mixture add 10ml cold water and collect the precipitated acetanilide, m.p 114°C.

(7) To a solution of 5.9gm of acetamide in 50ml water add 20gm of zinc oxide and then cool to –5°C. Pass chlorine gas into the reaction mixture until there is an increase in weight by 70gm. Acetochloramide separates out, recrystallise from pet ether, m.p 110°C.

(8) To 10ml benzene add a mixture of 15gm acetamide and 10gm phosphorous pentasulphide. Now add in 5ml portion a solution of 20ml chloroacetone in 7ml benzene. Heat the mixture, as soon as the reaction starts (exothermic reaction) remove the heat source. Again add 5ml solution and repeat the same process. After the addition is complete heat the mixture for 30min, cool and add 30ml water to it. . Remove the benzene layer and basify the aqueous layer with NaOH solution. Extract with ether. Combine the organic extracts and distill. Collect 2, 4-dimethylthiazole at b.p 141-144°C.

(9) Picrate m.p 107°C

(10) Xanthylamide m.p 245°C.

m.p 106°C **s-Dimethyl urea** (N, N'-Dimethyl urea) $CH_3NHCONHCH_3$

(1) Colourless solid, soluble in water and alcohol, insoluble in ether.

(2) Heat 50mg of compound with conc HCl. Cool, hydrochloride will be precipitated, m.p 124°C

(3) To saturated aqueous solution of compound add conc HNO_3. Heat the mixture on water bath and cool. White crystals of s-Dimethylurea nitrate, m.p 65°C are obtained.

m.p 109°C **Stearamide** (Octadecanamide) $CH_3(CH_2)_{16}CONH_2$

(1) Colourless solid insoluble in water, soluble in hot water, slightly soluble in ether.

(2) Xanthylamide m.p 140°C

m.p 116°C **n-Butyramide** (Butanamide) $CH_3(CH_2)_2CONH_2$

(1) Colourless solid, soluble in water and alcohol sparingly soluble in ether.
(2) n-Butyranilide m.p 90°C.
(3) Xanthylamide m.p 186°C

m.p 125°C **Succinimide** (Butanimide) 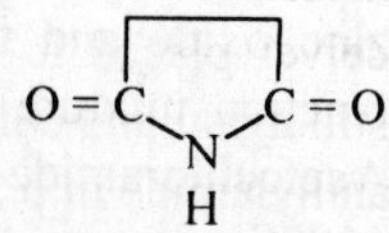

(1) Colourless crystalline odourless solid, soluble in water, sparingly soluble in ether and alcohol.
(2) Mix 50mg of compound with 1gm of Zn dust and heat the mixture in an ignition tube. Bring a pine splinter dipped in conc HCl near the mouth of ignition tube. The pine splinter turns bright red.
(3) To 50mg of compound add 100mg of resorcinol and 1-2ml conc H_2SO_4. Heat the mixture for 5minutes. Cool and pour in NaOH solution. An orange green fluorescence is produced.
(4) To a solution of 1.5gm of compound in 60ml absolute ethanol add in one portion a solution of 2.5gm of silver nitrate in 10ml dimethyl sulfoxide. Stir and add dropwise 35ml of .4M sodium ethoxide in ethanol. After the addition is complete continue stirring for 45min and then keep overnight in a refrigerator. Collect the silver salt of succinimide.
(5) Heat 500 mg of compound with 5 ml of aniline for 15 minutes. Ammonia gas is evolved, cool the mixture and add 10ml cold water, succinanil is precipitated, m.p 156°C.
(6) Reflux 500mg of compound with 10ml alcoholic KOH solution. Ammonia gas is evolved. Cool and acidify with dil HCl, succinic acid is precipitated, m.p 185°C.
(7) Dissolve 16gm of compound in an ice cooled NaOH solution (6.4gm in 40ml water). Add at once with stirring 8.5ml of Br_2. Cool and collect the precipitated N-Bromosuccinimide, m.p 173°C(d).
(8) Xanthylamide m.p 246°C.

m.p 129°C **Isobutyramide** (2-Methyl propionamide) $(CH_3)_2CHCONH_2$

(1) Colourless solid, soluble in water, alcohol and ether.

(2) Heat 1 gm of compound with 9-10ml of aniline for 15 minutes. Ammonia gas is evolved and on cooling the mixture add 15ml cold water, isobutyranilide is precipitated, m.p 105°C.

(3) Heat 8.5gm of compound with 16gm of P_2O_5 in an oil bath for 8-10 hrs. Set the apparatus for downward distillation, raise the temperature to 200-220°C and distill. Collect the distillate in a receiver placed in ice bath. Add 2gm P_2O_5 to the distillate and distill again and collect isobutyronitrile b.p 101-103°C.

(4) Xanthylamide m.p 211°C.

m.p 132°C **Urea** (Carbamide) H_2NCONH_2

(1) White crystalline solid, soluble in water and alcohol. Insoluble in chloroform.

(2) Heat 500mg of compound with 2-3 ml NaOH solution. Bring a glass rod dipped in conc HCl near the mouth of test tube. White fumes due to the formation of NH_4Cl are produced.

(3) Heat 500mg of compound in dry test tube over a low flame. The solid will melt and boil giving white sublimate and smell of ammonia. Cool the test tube, a white residue of biuret, m.p 192°C is obtained. Dissolve this residue in 3-5ml water and then add 10-15 drops of NaOH solution and 2-3 drops of $CuSO_4$ solution. A violet coloured solution is obtained.

(4) To saturated aqueous solution of compound add conc HNO_3. Heat the mixture on water bath and cool. White crystals of Urea nitrate, m.p 163°C are obtained.

(5) To saturated aqueous solution of compound add saturated oxalic acid solution and shake thoroughly. Crystals of Urea oxalate, m.p 171°C are obtained.

(6) Heat 500 mg of compound with 5-6ml of aniline for 15 minutes. Ammonia gas is evolved and on cooling the mixture add 10ml cold water, carbanilide is precipitated, m.p 238°C.

(7) To 30ml absolute methanol add 2.3gm of sodium in pieces. Now add 3gm of urea followed by 7gm of diethyl oxalate with stirring. Continue stirring for 1hr and then add dropwise 10ml of conc HCl. Continue stirring for few minutes and then filter. Distill off the alcohol and dissolve the residue in minimum amount of hot water and cool. White crystals of parabanic acid, m.p 241-243°C are obtained.

(8) Xanthylamide m.p 274°C

m.p 170°C **Malonamide** (Propanediamide) $CH_2(CONH_2)_2$

(1) Colourless solid, soluble in water, insoluble in alcohol and ether.
(2) To 100mg of compound add 5ml NaOH solution and 2-3 drops of $CuSO_4$ solution. A red coloured solution is obtained.
(3) Reflux 500mg of compound with NaOH solution. Ammonia gas is evolved. Cool and acidify with dil HCl, malonic acid is precipitated, m.p 133°C.
(4) Xanthylamide m.p 270°C. (5) malonanilide m.p 224°C

m.p 182°C **asym Dimethyl urea** $(CH_3)_2NCONH_2$

(1) Crystalline compound with a sweet taste, soluble in water, sparingly soluble in alcohol, insoluble in ether.
(2) Picrate m.p 130°C(d) (3) Xanthylamide m.p 250°C.

m.p 192°C **Biuret** (allophanamide, carbamyl urea) $H_2NCONHCONH_2$

(1) White odourless solid, insoluble in water and ether, soluble in alcohol.
(2) To 100mg of compound add 5ml NaOH solution and 2-3 drops of $CuSO_4$ solution. A red coloured solution is obtained.
(3) Heat 500mg of compound with 2-3 ml NaOH solution. Bring a glass rod dipped in conc HCl near the mouth of test tube. White fumes due to the formation of NH_4Cl are produced.
(4) Heat 500 mg of compound with 5-6ml of aniline for 15 minutes. Ammonia gas is evolved and on cooling the reaction mixture add 10ml cold water, diphenylbiuret is precipitated, m.p 260°C.
(5) Xanthylamide m.p 260°C.

m.p 243°C **Succinamide** (butanediamide) $H_2NCOCH_2CH_2CONH_2$

(1) Colourless crystalline solid, insoluble in alcohol and ether, soluble in hot water.
(2) Reflux 500mg of compound with 10ml NaOH solution. Ammonia

gas is evolved. Cool and acidify with dil HCl, succinic acid is precipitated, m.p 185°C.

(3) Heat 1gm of compound at 260-270°C. Ammonia gas is evolved, cool and allow the liquid to solidify, succinimide will be precipitated, m.p 125°C

(4) Succinanil m.p 156°C.

(5) Xanthylamide m.p 275°C.

m.p 245°C **Barbituric acid** (Malonyl urea)

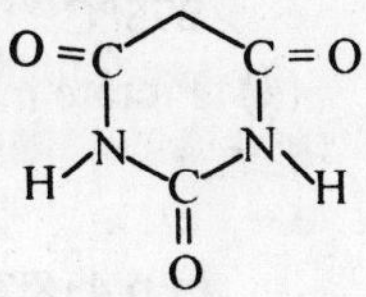

(1) Crystalline solid, soluble in hot water, slightly soluble in alcohol, easily soluble in ether.

(2) To few milligrams of compound in water add $AgNO_3$ solution. Yellow coloured precipitate is obtained.

(3) Dissolve 2.5gm of barbituric acid in 25ml water by heating on water bath. Then add 2.2ml benzaldehyde and continue the heating and stirring process for 1hr. Cool and collect benzal barbituric acid, m.p 254-256°C.

(4) To 7ml of cooled fuming HNO_3 add with stirring and in portion 5gm of barbituric acid. Continue the stirring process for another one hour. Cool and add 20ml of ice cold water. Collect the crude nitro barbituric acid and recrystallise from hot water, m.p 181-183°C.

m.p 260°C(d) **Creatinine** (1-methyl glyco cyamidine)

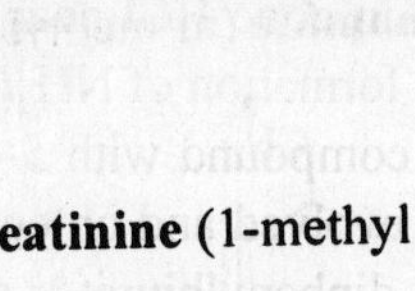

(1) Crystalline compound, soluble in water, sparingly soluble in alcohol.

(2) Dissolve about 15mg of compound in 1.5ml water and add 2drops

of NaOH solution followed by 10-15 drops of saturated picric acid solution. The solution soon becomes orange and slowly changes to orange –red in colour.

(3) Heat together 5.7gm of compound with 12gm of vanillin in an oil bath at 170°C. The mixture melts, continue heating for 15minutes, during this period the mixture solidifies. Heat further for 3min and then cool, add about 25ml ethanol to the mixture and heat on oil bath and filter. Collect the crude 5-(3-methoxy-4-hydroxybenzal) creatinine, recrystallise from acetic acid, m.p 273°C.

(4) Picrate m.p 220°C.

m.p 418°C(d) **Oxamide** (ethanediamide) $NH_2COCONH_2$

(1) Colourless crystalline, tasteless, odourless solid, insoluble in water, slightly soluble in alcohol and ether.

(2) To 100mg of compound add 5ml NaOH solution and 2-3 drops of $CuSO_4$ solution. A red coloured solution is obtained.

(3) Reflux 1gm of compound with 20ml NaOH solution. Ammonia gas is evolved. Cool and acidify with dil HCl, oxalic acid is precipitated, m.p 101°C.

(4) Oxanilide m.p 254°C.

Aromatic Solid

m.p 97°C **m-Toluamide** (m-methyl benzamide)

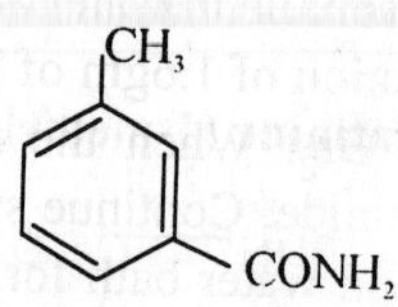

(1) White crystalline solid, insoluble in water, soluble in alcohol, slightly soluble in ether.

(2) Reflux 1gm of compound with 20ml NaOH solution. Ammonia gas is evolved. Cool and acidify with dil HCl, m-Toluic acid is precipitated, m.p 112°C.

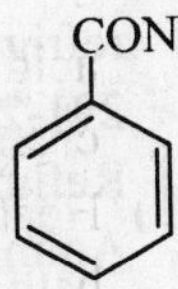

m.p 128°C **Benzamide** (benzene carbonamide)

(1) White crystalline tasteless, odourless solid, sparingly soluble in water, soluble in alcohol and ether.

(2) Heat 1gm of compound with 1.7gm of P_2O_5 over a free flame. Smell of bitter almonds is produced due to the formation of benzonitrile.

(3) Reflux 1gm of compound with 20ml NaOH solution for 20-30min. Ammonia gas is evolved. Cool, acidify with dil HCl, white crystals of benzoic acid are precipitated, m.p 121°C.

(4) Heat 500 mg of compound with 5ml of aniline. Ammonia gas is evolved, cool the reaction mixture and add 10ml cold water to it, white precipitate of benzanilide is obtained, m.p 162°C.

(5) Add 1.2gm of benzamide to a solution of 2ml of bromine in 16ml chloroform. Heat the mixture to boiling, cool to 40°C and then add a solution of 500mg of KOH in 3ml water. On cooling benzobromamide separates out.

(6) Xanthylamide m.p 224°C.

m.p 129-131°C **Nicotinamide** (3-pyridine carboxamide)

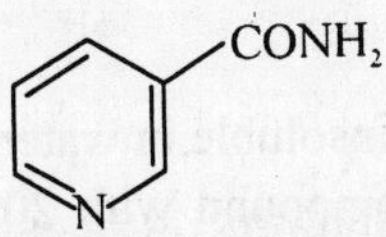

(1) Crystalline solid, soluble in water and alcohol.

(2) To ice cooled solution of 1.6gm of NaOH in 16ml water add with stirring 0.7ml of Br_2. When the temperature reaches 0°C add 1.2gm of nicotinamide. Continue stirring for 15minutes and then heat the solution on water bath for 70-80 minutes. Cool, saturate with NaCl and extract with ether. Concentrate the ether extract, cool and collect crystals of 3-amino pyridine, m.p 62-64°C.

(3) Dissolve 5gm of compound in 50ml acetic acid (heat if necessary) and then add 8ml H_2O_2. to it. Heat the mixture on steam bath for 4hrs. Distill under reduced pressure almost to dryness and dissolve the residue in minimum hot water and leave overnight in a

refrigerator. Collect white crystals of nicotinamide 1-oxide, m.p 291-293°C.

(4) Reflux 1gm of compound with 20ml NaOH solution for 20-30min. Ammonia gas is evolved. Cool and acidify with dil HCl, nicotinic acid is precipitated, m.p 232°C.

m.p 139°C **Salicylamide** (o-hydroxy benzamide)

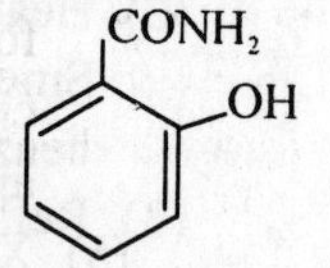

(1) Crystalline solid, insoluble in water, soluble in alcohol.
(2) To few milligrams of compound add $FeCl_3$ solution. A violet coloured solution is obtained.
(3) Reflux 1gm of compound with 20ml NaOH solution for 20-30min. Ammonia gas is evolved. Cool, acidify with dil HCl, white crystals of salicylic acid are precipitated, m.p 155°C.
(4) Acetate m.p 135°C (5) Benzoate m.p 143°C.

m.p 147°C **o-Toluamide** (o-methyl benzamide)

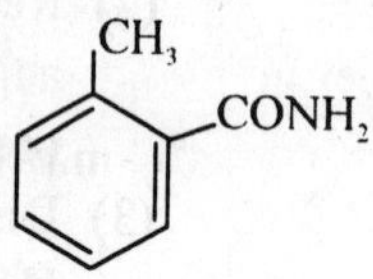

(1) Crystalline solid, insoluble in water, soluble in alcohol and ether.
(2) Reflux 1gm of compound with 20ml NaOH solution. Ammonia gas is evolved. Cool, acidify with dil HCl, o-Toluic acid is precipitated, m.p 108°C.
(3) Xanthylamide m.p 200°C (4) Benzoate m.p 158°C.

m.p 147°C **Phenyl urea** (N-Phenylcarbamide)

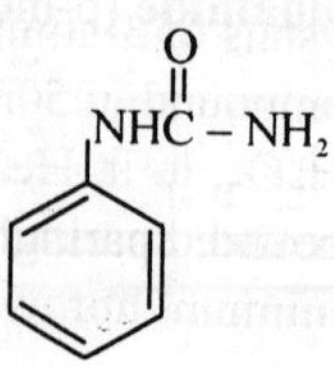

(1) Colourless crystalline tasteless, odourless solid, insoluble in water, soluble in alcohol and ether.
(2) Respond to carbylamine test.
(3) Reflux 500 mg of compound with aniline. Ammonia gas is evolved. Cool the mixture, add dil HCl, shake thoroughly and collect white crystalline carbanalide, m.p 238°C.
(4) Heat 3.4 gm of phenylurea with 10ml of hydrazine hydrate solution for 12hrs. Filter and concentrate the filtrate, cool and collect crystals of 4-Phenyl semicarbazide, recrystallise from hot water and HCl, m.p 120°C.
(5) Xanthylamide m.p 225°C.

m.p 155°C **2-Phenylacetamide**

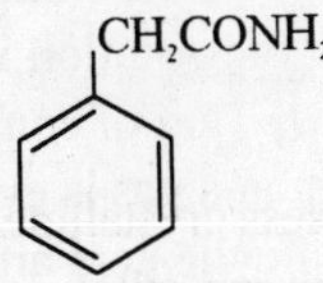

(1) Colourless crystalline solid, sparingly soluble in water and ether, souble in hot water and alcohol.
(2) Reflux 1gm of compound with 20ml NaOH solution. Cool and acidify with dil HCl, white crystals of phenylacetic acid are precipitated, m.p 76°C.
(3) Dissolve 200mg of compound in 2-3ml water and then add 200mg HgO. Shake thoroughly and collect the crystalline compound, m.p 208°C.
(4) Phenylacetanilide m.p 117°C.
(5) Xanthylamide m.p 196°C.

m.p 160°C **p-Toluamide** (p-methyl benzamide)

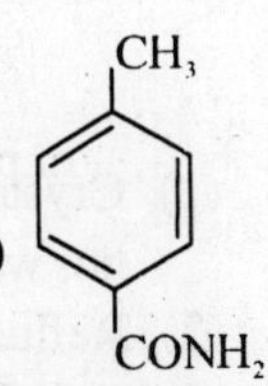

(1) Crystalline compound, sparingly soluble in water, easily soluble in alcohol.

(2) Reflux 1gm of compound with 20ml NaOH solution for 20-30min. Ammonia gas is evolved. Cool, acidify with dil HCl, p-Toluic acid is precipitated, m.p 180°C.
(3) N-Acetate m.p 147°C.
(4) Xanthylamide m.p 224°C.
(5) N-Benzoate m.p 119°C

m.p 174°C **p-Ethoxyphenyl urea** (p-Phenetylurea, Dulcin, Sucrol)

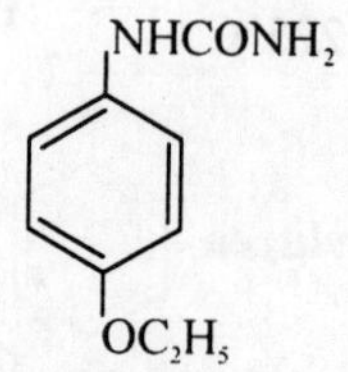

(1) Colourless crystalline compound with sweet taste, sparingly soluble in water and ether.
(2) Heat 50mg of compound in 5ml water on water bath and add 2-4 drops of mercuric nitrate to it. A violet blue coloured solution is obtained.
(3) Reflux 1gm of compound with 20ml NaOH solution. Ammonia gas is evolved. Cool and acidify with dil HCl, p-phenetidine b.p 254°C is obtained.

m.p 187°C **p-Tolylurea**

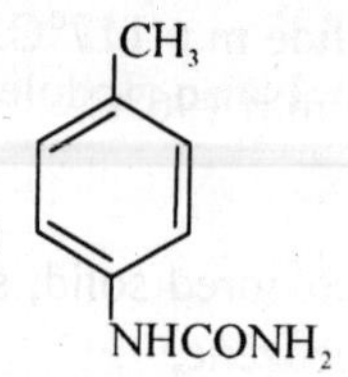

(1) Crystalline compound, insoluble in water and ether, soluble in hot water and alcohol.
(2) Reflux 1gm of compound with 12ml conc HCl for 30 min. Cool and dilute with water, p-toluidine hydrochloride, m.p 240°C is obtained.
(3) N-Acetate m.p 200°C

(4) N-Benzoate m.p 223°C

m.p 189°C **asy-Diphenylurea** $(C_6H_5)_2NCONH_2$

(1) Crystalline compound, very slightly soluble in water, soluble in alcohol and ether.
(2) Heat few milligrams of compound with conc H_2SO_4, add 1-2 drop of conc HNO_3. A blue coloured solution is obtained.
(3) Xanthylamide m.p 180°C.

m.p 191°C **o-Tolylurea** 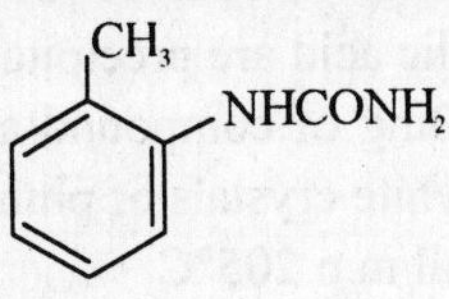

(1) Crystalline compound, insoluble in water, sparingly soluble in acetone, soluble in ether.
(2) Reflux 1gm of compound with 12ml conc HCl for 30 min. Cool and dilute with water, o-toluidine hydrochloride, m.p 215°C is obtained.
(3) N-Acetate m.p 168°C (4) N-Benzoate m.p 210°C (5) Xanthylamide m.p 228°C.

m.p 202°C(d) **Isatin** (Indole –2, 3 dione) 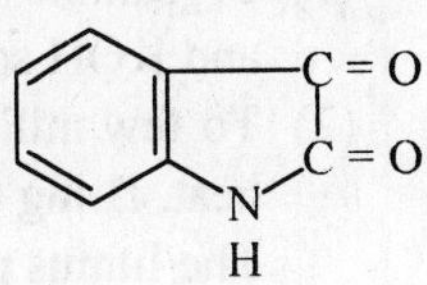

(1) Red-orange coloured solid, sparingly soluble in water, soluble in alcohol and benzene.
(2) Fuse few milligrams of compound with KOH. A distinct odour of anline is observed.
(3) Reflux 3gm of isatin with 7ml of acetic anhydride for 4hrs. Cool the mixture to 0°C. Collect the precipitated N-acetyl isatin, m.p 142°C.
(4) Isatin semicarbazone m.p 266°C(d)
(5) Isatinoxime m.p 225°C (d)

m.p 220°C **Phthalamide** (Phthalic diamide) 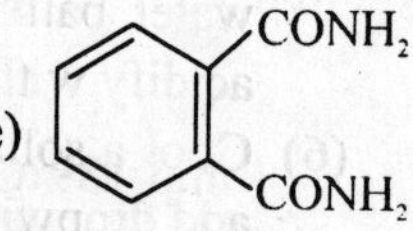

(1) Colourless crystalline solid, insoluble in water, alcohol and ether.
(2) To 250mg of compound add 250mg of resorcinol and 0.5ml conc H_2SO_4. Heat the mixture for 5 minutes or until the mixture is red brown. Cool and pour the contents in dil 50ml NaOH solution. An orange green fluorescence is produced.
(3) Reflux 1gm of compound with 20ml NaOH solution for 20-30min. Ammonia gas is evolved. Cool, acidify with dil HCl, white crystals of phthalic acid are precipitated, m.p 206-208°C.
(4) Heat 500mg of compound at 220°C. Allow the melt to cool and collect white crystals of phthalimide, m.p 238°C.
(5) Phthalanil m.p 205°C.

m.p238°C **Phthalimide** (1, 3-isoindoledione) 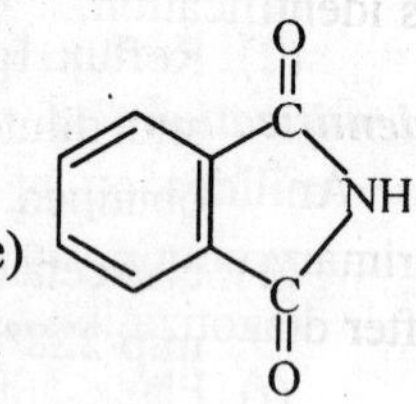

(1) Crystalline solid, insoluble in water and benzene, soluble in alcohol and KOH solution.
(2) To few milligrams of compound add alcoholic KOH solution and heat. Bring the moist red litmus paper near the mouth of test tube. The litmus paper turns blue.
(3) Heat 50mg of compound with 500mg of resorcinol and 10ml conc H_2SO_4. A blue green coloured solution is obtained.
(4) To 250mg of compound add 250mg of resorcinol and 0.5ml conc H_2SO_4. Heat the mixture for 5 minutes or until the mixture is red brown. Cool and pour the contents in dil 50ml NaOH solution. An orange green fluorescence is produced.
(5) Stir 9gm of Zn with 50mg of $CuSO_4$ in 2ml water and then add 16ml of 20% NaOH solution. Cool the flask and then add in porton with stirring 7gm of phthalimide (temp below 5°C). Continue stirring for 30min and then add 20ml water. Heat on

water bath for 3hrs, concentrate under reduced pressure. Cool, acidify with dil HCl and collect phthalide, m.p 72-73°C

(6) Cool a solution of 3gm of NaOH in 12ml water to 0°C and then add dropwise 1ml of Br_2. Add with stirring 2.4 gm of phthalimide to the hypobromite solution. Stir the mixture at room temperature and then add 8ml NaOH solution (2.2gm NaOH) and heat on water bath for 2-5min. Cool, add dilute HCl and collect anthranilic acid, m.p 145°C.

(7) Reflux 1gm of compound with 20ml NaOH solution for 20-30min. Ammonia gas is evolved. Cool, acidify with dil HCl, white crystals of phthalic acid are precipitated, m.p 206-208°C.

(8) Xanthylamide m.p 177°C

(*xiv*) *Anilides*

Anilides are considered to be N substituted derivatives of amides. There are not any specific colour reaction for their identification. However if the additional element nitrogen is present and the test for amines, nitro and amides are negative then the hydrolysis of anilide should be carried out for its identification.

Identification test

Anilides on hydrolysis yield a primary amine and carboxylic acid. Primary amine can very easily be identified by the formation of a red dye after diazotization and coupling with alkaline 2-naphthol.

(*i*) Place 500mg of compound and 5ml conc H_2SO_4 in a dry test tube. Heat the test tube for 15 minutes. Cool the test tube and perform the dye test or carbylamine test with the resultant product.

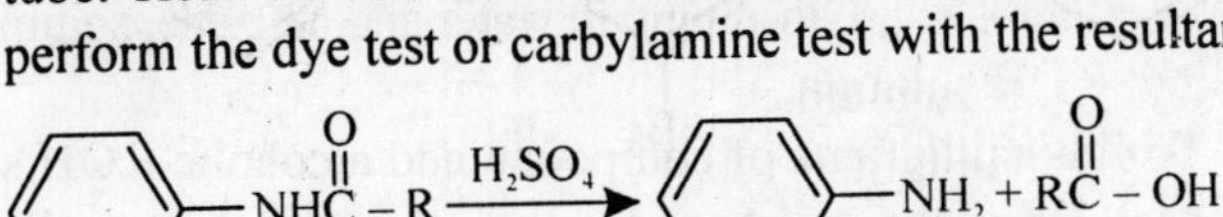

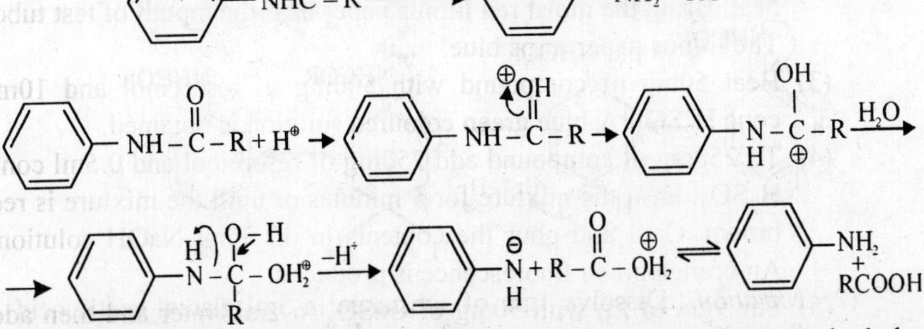

(*ii*) Dissolve 50mg of compound in 2ml conc $H_2SO_4$and shake thoroughly. Add 25mg of potassium dichromate to the acid solution. A bluish pink colour indicates the presence of anilides.

(*iii*) In a dry test tube mix 200mg of compound with 2 pellets of NaOH and cover the mouth of test tube with a filter paper moistened with 3-4 drops of glacial acetic acid. Heat the test tube on a free flame and fuse the mixture. Aniline and its substituted derivatives generated during the reaction vapourise and react with acetic acid on the filter paper to form aniline acetate.
In a separate test tube place 100mg of sugar (glucose/sucrose) and cover the mouth of test tube with moistened filter paper of the fusion mixture tube. Heat the test tube until the sugar chars. Furfural produced during the reaction reacts with aniline acetate to give pink colour.

Derivatives

(1) *Bromination* : Dissolve 1gm of substance in 5ml glacial acetic acid (heat if necessary) in a 100ml conical flask. Cool the flask by keeping it in an ice bath. Add drop by drop 6ml of bromine solution (1ml bromine in 5ml glacial acetic acid). After the addition is complete keep the flask at room temperature for 10-15 min and shake it occassionally. Pour the reaction mixture in 50gm of crushed ice. Filter the precipitated solid, recrystallise from ethanol and determine its melting point.

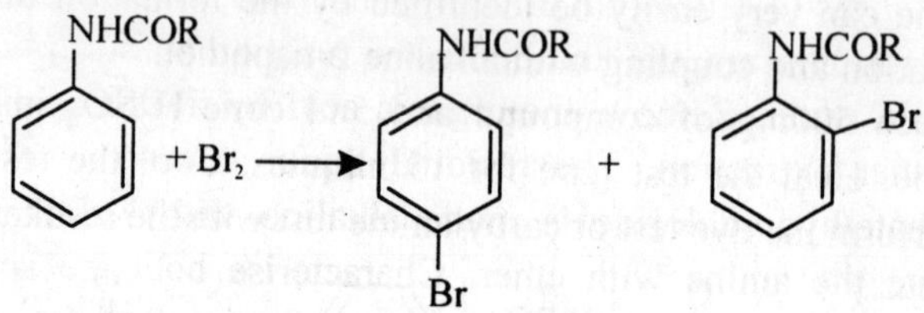

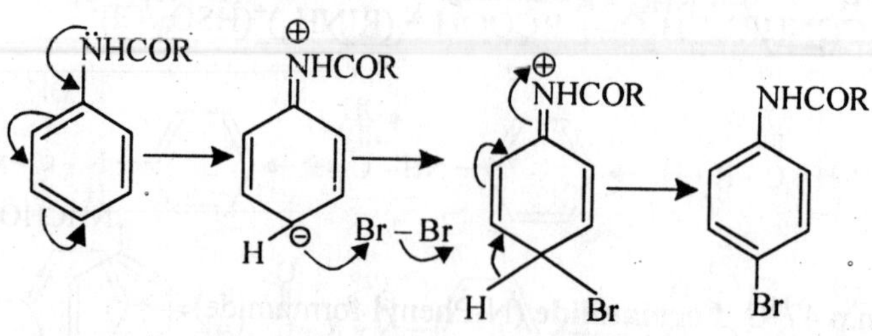

(2) *Nitration* : Dissolve 1gm of substance in 5ml glacial acetic acid (heat if necessary) in a 100ml conical flask. Cool the flask by keeping it in an ice bath. Add drop by drop 5ml of nitrating mixture (2.5ml conc HNO_3 +2.5ml conc H_2SO_4). After the addition is complete keep the flask at room temperature for 30 min Pour

the reaction mixture in 50gm of crushed ice. Filter the precipitated solid, wash with water, recrystallise from ethanol and determine its melting point.

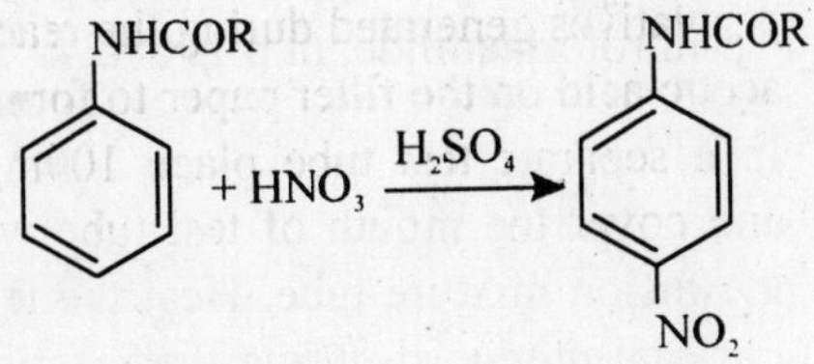

$$HNO_3 + H^{\oplus} \longrightarrow H_2O - NO_2^{\oplus} \longrightarrow H_2O + NO_2^{\oplus}$$

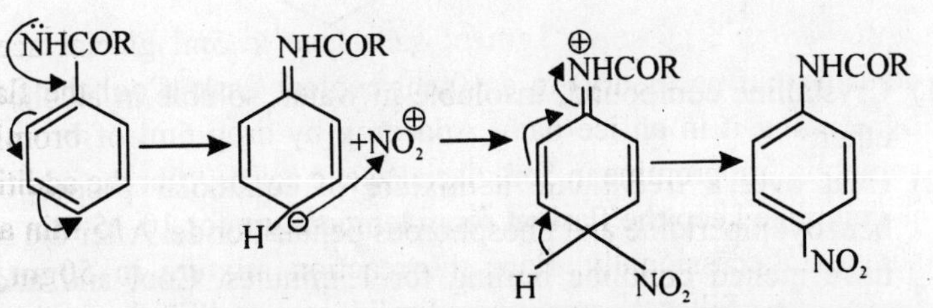

(3) *Hydrolysis* : Reflux 1gm of compound with 12ml of conc H_2SO_4/HCl for 30min. Cool and dilute the reaction mixture with 6-7ml of cold water. Remove the organic acid by filtration, distillation or ether extraction. The residual liquor containing the amine sulphate/hydrochloride is made alkaline with NaOH solution. Isolate the amine with ether. Characterise both the amine and organic acid.

$RCONHR' + H_2O \rightarrow RCOOH + (R'NH_3)^+(HSO_4/Cl)^-$

Solid

m.p 47°C **Formanilide** (N-Phenyl formamide)

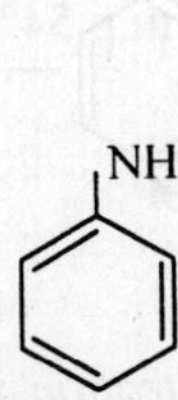

(1) Colourless crystalline solid with characteristic odour, burning taste, soluble in water and alcohol, insoluble in pet ether.

(2) Responds to carbylamine test.

(3) Hydrolyse with HCl, formic acid and aniline are obtained, perform the dye test.
(4) Dissolve 500mg of compound in 2ml conc H_2SO_4. Add dropwise with stirring 5ml of nitrating mixture (2ml conc HNO_3+3ml conc H_2SO_4) (temp 10-15°C). Cool, pour in ice with stirring and collect p-nitroformanilide, m.p 196°C.

m.p 48°C **N-Benzoyl piperidine**

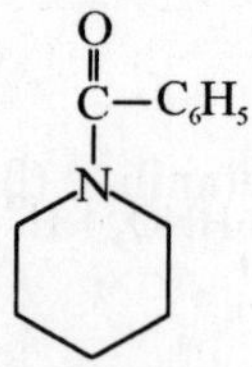

(1) Crystalline compound, insoluble in water, soluble in alcohol and ether.
(2) Heat over a free flame a mixture of equimolar proportion of benzoyl piperidine and phosphorous pentachloride. After the solids have melted continue boiling for 15minutes. Cool and add ice water. Distill to remove benzonitrile and dichloropentane. Cool the non volatile oily layer which soon solidifies. Triturate with pet-ether and collect Benzoyl –5-chloro amylamine, m.p 66°C.
(3) Heat over a free flame a mixture of equimolar proportion of benzoyl piperidine and phosphorous pentachloride. After the solids have melted continue boiling for 15minutes. Distill and collect the fraction boiling at 100-110°C and 170-195°C. Pour the two distillate in ice water. Extract with ether and distill. Collect 1, 5 dichloropentane at b.p 177°C and benzonitrile at b.p 190°C.

m.p 54°C **N-Ethylacetanilide** (Ethyl phenyl acetamide)

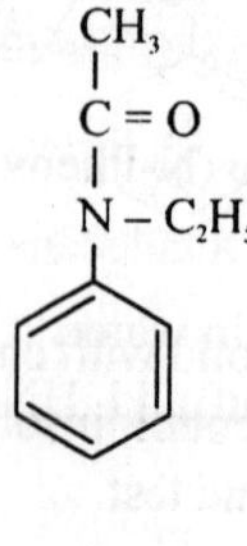

(1) White crystalline solid, insoluble in water, soluble in ether.
(2) Hydrolyse with HCl, acetic acid and aniline are obtained, perform the dye test.
(3) p-Nitro N-ethyl acetanilide m.p 118°C. (conc H_2SO_4 + conc HNO_3, at 40°C)

m.p 85°C **Acetoacetanilide** (β-ketobutyranilide)

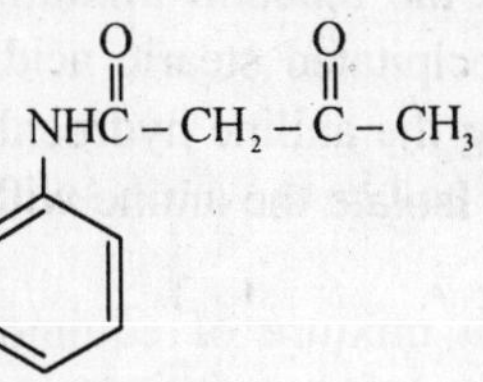

(1) Crystalline compound, insoluble in water, soluble in alcohol and ether, easily soluble acid or alkali.
(2) Reduces ammonical silver nitrate solution.
(3) To few milligrams of compound add aqueous $FeCl_3$ solution. A dark violet coloured solution is obtained.
(4) Heat 5ml of conc H_2SO_4 to 75°C and add in portion 4gm of compound to it (temp below 75°C). After the addition is complete heat the mixture at 95°C for 30min. Cool, pour in ice and recover 4-methyl carbostyril, m.p 221°C.
(5) Carbanilide m.p 238°C.

m.p 87°C **Phenyl-p-toluidine**

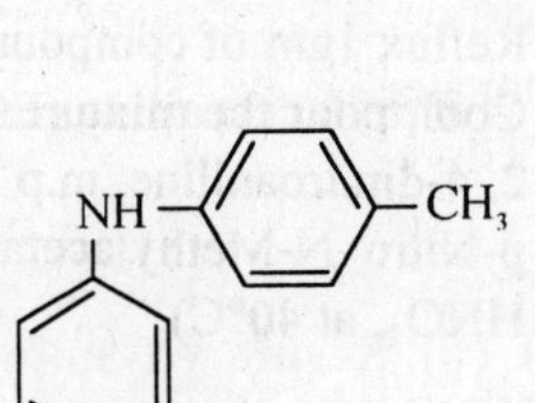

(1) Crystalline compound, insoluble in water.
(2) To few milligrams of compound add HNO_3. A blue coloured solution is obtained.

m.p 94°C **Steranilide** (N-phenyl steramide)

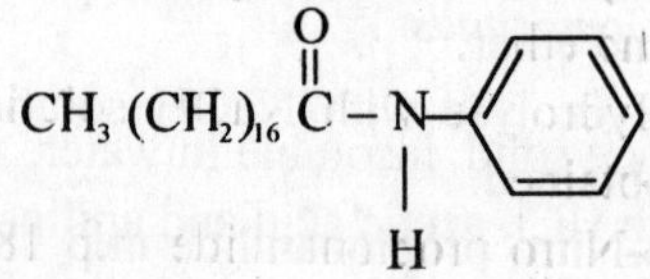

(1) Crystalline compound, insoluble in water, soluble in alcohol, ethe and benzene.
(2) Hydrolyse with HCl, stearic acid and methyl aniline hydrochlorid are obtained.
(3) Reflux 4gm of compound with 30ml of conc HCl for 30min Cool and dilute the reaction mixture with 15ml of cold water Remove the precipitated stearic acid by filtration. The residua liquor containing the aniline hydrochloride is made alkaline with NaOH solution. Isolate the amine with ether and perform the dye test.

m.p 102°C **N-Methylacetanilide** (exalgin)

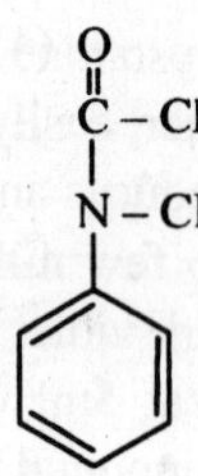

(1) Colourless solid, insoluble in water, soluble in alcohol.
(2) Hydrolyse with HCl, acetic acid and methyl aniline hydrochlorid are obtained.
(3) Reflux 1gm of compound with 10ml dil HNO_3 for 2hrs at 100°C Cool, pour the mixture in ice and collect yellow coloured Methy 2, 4-dinitroaniline, m.p 178°C
(4) p-Nitro N-Methylacetanilide m.p 153°C (conc H_2SO_4 + con HNO_3, at 40°C)

m.p 104°C **Propionanilide** (N-phenyl propionamide)

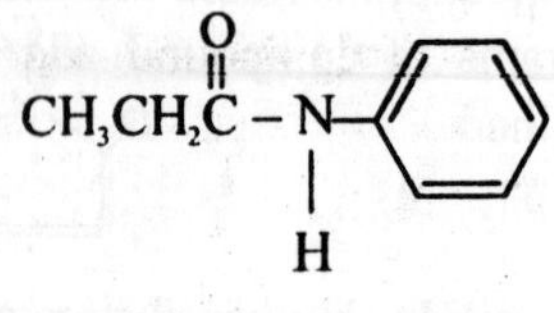

(1) Crystalline solid, sparingly soluble in water, soluble in alcohol and ether.
(2) Hydrolyse with NaOH solution, propionic acid and aniline are obtained.
(3) p-Nitro propionanilide m.p 182°C. (conc H_2SO_4 + conc HNO_3, at 0°C)

m.p 110°C **Aceto-o-toluidide** (o-methyl acetanilide)

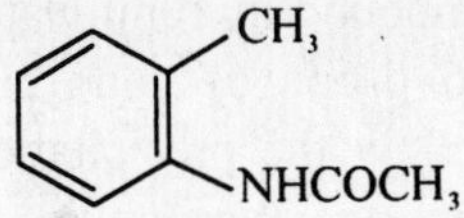

(1) Crystalline compound, insoluble in water, soluble in ether.
(2) Responds to Carbylamine test.
(3) Hydrolyse with HCl, acetic acid and o-toluidine hydrochloride are obtained.
(4) Dissolve 500mg of compound in 5ml glacial acetic acid, add dropwise with stirring 1ml Br_2. Shake thoroughly and collect m-bromo acetotoluidide, m.p 157°C.
(5) Heat a mixture of 3.2gm sodium amide, 5gm of compound and 3ml ether at 240-260°C for 15 min. Cool, add 5ml ethanol and 25ml water. Heat the mixture for 2-3 minutes. Cool and extract with ether. Distill off the ether and collect 2-methyl indole, m.p 57°C.
(6) Add 4gm of compound in portion to 15ml of conc HNO_3 (temp below 20°C). Heat the mixture at 40-50°C for 1hr. Cool, pour in ice water and collect the precipitated mixture. Dissolve it in 10ml conc HCl and steam distill. Cool the distillate and collect 3-nitro-o-toluidine, m.p 97°C. Add excess of sodium carbonate to the residue and collect 5-nitro-o-toluidine, recrystallise from alcohol, m.p 133°C.

m.p 114°C **Antipyrine** (2, 3-dimethyl-1-phenylpyrazol-5-one, analgesin)

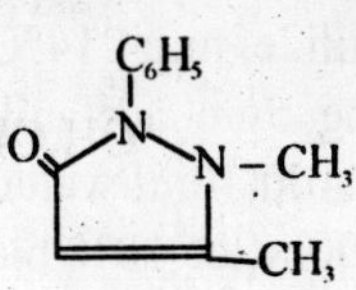

(1) Colourless, odourless crystalline compound with faint bitter taste, soluble in water and alcohol, sparingly soluble in ether.
(2) To 4ml of 1% aqueous solution of compound add 2-3 drops of $FeCl_3$ solution. An orange red coloured solution is obtained.
(3) To 4ml of 1% aqueous solution of compound add 3-4 drops of fuming HNO_3. A green coloured solution is obtained which changes to reddish brown on further addition of acid.
(4) To 500mg of compound in 10ml water add 3ml of 10% $NaNO_2$ solution and 5-6 drops of conc H_2SO_4. Shake the mixture thoroughly and allow the precipitation to take place. Cool and collect nitrosoantipyrine, recrystallise from acetone, m.p 200°C.
(5) p-Nitroantipyrine m.p 273°C.
(6) Picrate m.p 188°C.

m.p 114°C **Acetanilide** (N-phenyl acetamide, antifebrin)

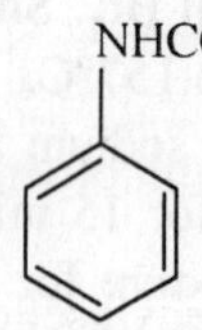

(1) Colourless crystalline odourless solid, insoluble in water, soluble in alcohol, chloroform and ether.
(2) Responds to carbylamine test.
(3) Heat few milligrams of compound with a pinch of potassium dichromate and few drops of conc H_2SO_4. A red coloured solution which slowly changes to green is obtained.
(4) To a mixture of 11ml CS_2, 1.02gm acetanilide and 3gm chloroacetyl chloride add in portion and with stirring 6gm of anhydrous $AlCl_3$. After the addition is complete reflux the mixture for 30min and then keep at room temperature for 3hrs. Remove the CS_2 layer and then pour the red brown layer in crushed ice to which 2ml conc HCl is added. Collect the precipitated p-chloroacetyl acetanilide, m.p 214°C.
(5) Dissolve by heating 5gm acetanilide in 15ml alcohol and 7ml glacial acetic acid. Add 10ml water, heat the solution to 50°C and add with stirring 100ml of 10% bleaching powder solution. Cool and collect white precipitate of p-chloroacetanilide, recrystalllise from alcohol, m.p 179°C.

(6) Hydrolyse with HCl, acetic acid and aniline hydrochloride are obtained

(7) Stir at 18°C a solution of 2.5gm of compound in 17ml glacial acetic acid and 7.5ml Ac_2O containing 2.5gm fused potassium acetate and 250mg phosphoric oxide. Add 25% solution of nitrosyl chloride (1.5gm) in Ac_2O. Stir for 15min and then pour in ice water. Yellow crystals of nitrosoacetanilide separate out, m.p 50-51°C.

(8) Reflux 5gm of compound with 2.5gm potassium carbonate, 60gm bromobenzene and some cuprous iodide in 20ml nitrobenzene for 15hrs. Distill off the excess solvent. Extract with ether. Evaporate off the solvent and collect acetyldiphenylamine, recrystallise from alcohol, m.p 102°C.

(9) p-Nitroacetanilide m.p 210°C (10) p-bromoacetanilide m.p 167°C.

m.p 117°C **Phenylacetanilide** (α-phenyl acetanilide)

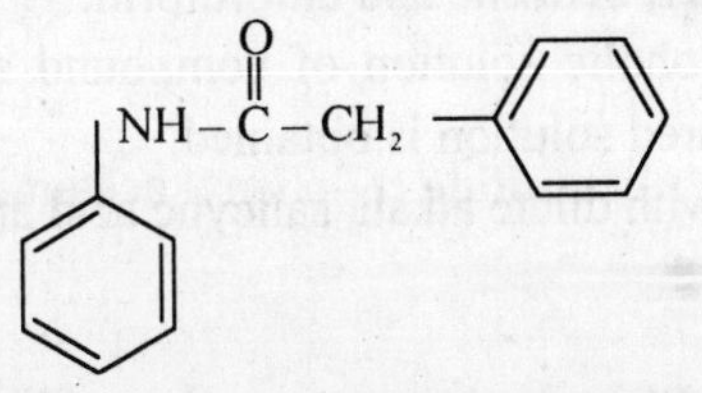

(1) Crystalline compound, insoluble in water, soluble in alcohol and ether.

(2) Hydrolyse with HCl, phenylacetic acid, m.p 76°C and aniline hydrochloride are obtained.

(3) 2, 4-dinitrophenylacetic acid m.p 189°C.

m.p 129°C **Piperine** (Piperonylpiperidine)

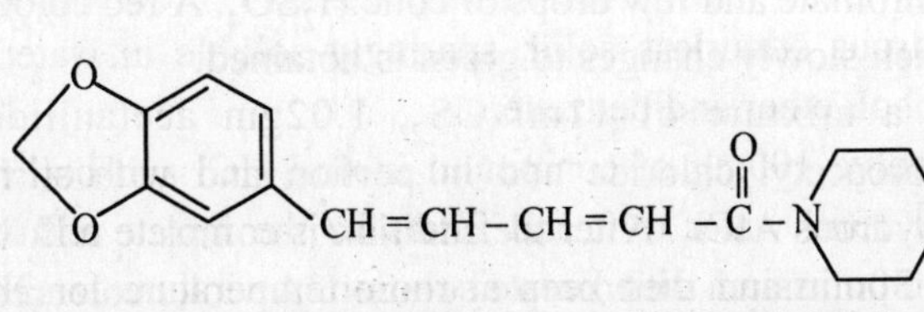

(1) Colourless crystalline, odourless solid, insoluble in water and pet ether, sparingly soluble in ether, soluble in chloroform.

(2) To few milligrams of compound add conc H_2SO_4, yellow colouration which soon changes to reddish brown is observed.
(3) Hydrolyse with NaOH solution, piperidine and piperic acid, m.p 216°C are obtained.

m.p 135°C **Salicylanilide** (N-phenyl salicylamide)

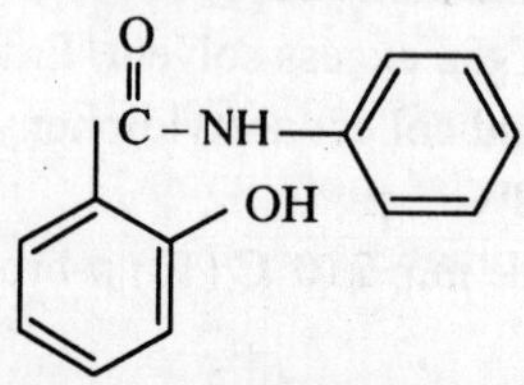

(1) Colourless crystalline compound, insoluble in water, soluble in alcohol, ether, benzene and chloroform.
(2) To 2ml alcoholic solution of compound add $FeCl_3$ solution. A violet coloured solution is obtained.
(3) Hydrolyse with dilute alkali, salicylic acid and aniline are obtained.

m.p 135°C **Phenacetin** (aceto-p-phenetidine, p-acetamino ethoxy benzene)

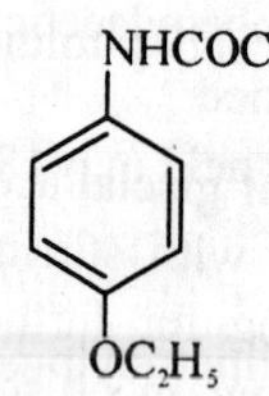

(1) Lustrous odourless solid, sparingly soluble in water, soluble in alcohol, ether and benzene.
(2) Dissolve 100mg of compound in 4ml HCl and boil for 1-2 min. Add 15ml water to it and filter. To the filtrate add 1-2 drops of N/10 potassium dichromate solution. A violet coloured solution is obtained which slowly changes to red.
(3) Hydrolyse with HCl, acetic acid and phenetidine hydrochloride are obtained.
(4) 3-Nitrophenacetin m.p 103°C.

m.p 146°C **Benzo-o-toluidide**

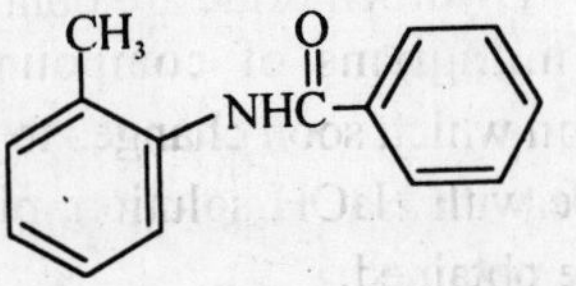

(1) Colourless odourless solid, insoluble in water, soluble in alcohol and ether.
(2) Responds to carbylamine test.
(3) Hydrolyse with sulphuric acid, benzoic acid, m.p 121°C and o-toluidine are obtained.
(4) To solution of 2gm of compound in 5ml acetic acid pass nitrous acid fumes. Keep the mixture at room temperature for 5min and then cool. Collect the precipitated N-Nitrosobenzoyl-o-toluidine, m.p 63°C.

m.p 153°C **Aceto-p-toluidide**

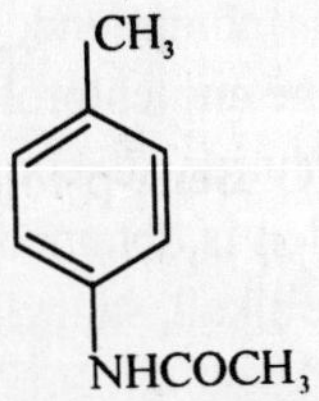

(1) Colourless solid, sparingly soluble in water, soluble in alcohol and ether.
(2) Responds to carbylamine test.
(3) Hydrolyse with HCl, acetic acid and p-toluidine hydrochloride are obtained.
(4) Dissolve 1gm of compound in 5ml glacial acetic acid by heating to 40-45°C. Then add dropwise with stirring 1ml of bromine (temp 50-55°C). Continue stirring for further 30min after the addition is complete and then pour in crushed ice. Collect the precipitated 4-acetamido 3-bromotoluene, m.p 117°C.
(5) 4-acetamido 3-nitrotoluene m.p 144°C.

m.p 153°C **Cinnamanilide**

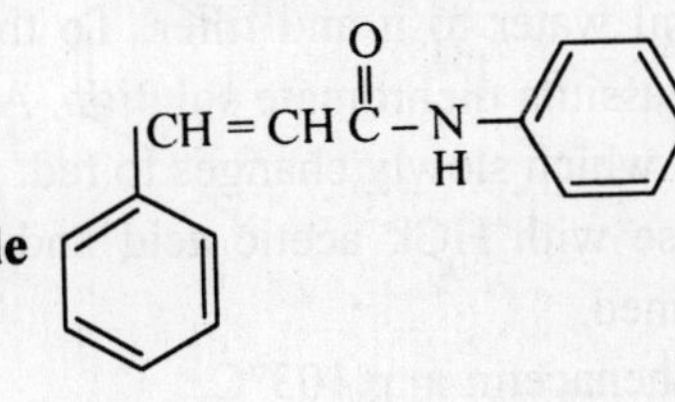

(1) Crystalline solid, insoluble in water
(2) Hydrolyse with HCl, cinnamic acid m.p 132°C and aniline hydrochloride are obtained.

m.p 156°C **Succinanil** (N-phenyl succinimide)

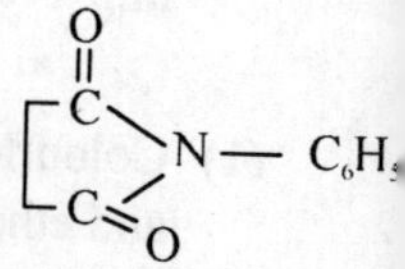

(1) Crystalline compound, insoluble in water, soluble in hot alcohol and ether.
(2) Hydrolyse with HCl, succinic acid, m.p 185°C and aniline hydrochloride are obtained.

m.p 158°C **Benz-p-toluidide**

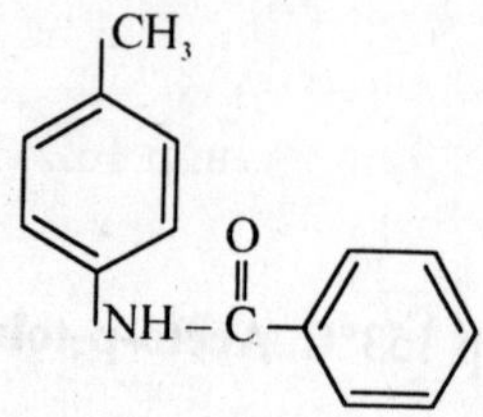

(1) Crystalline, tasteless, odourless compound, insoluble in water, soluble in ether.
(2) Respond to carbylamine test.
(3) Hydrolyse with HCl, benzoic acid, m.p 121°C and p-toluidine hydrochloride are obtained.
(4) Dinitro m.p 186°C (fuming HNO_3 at 0°C)

m.p 163°C **Benzanilide** (N-phenyl benzamide)

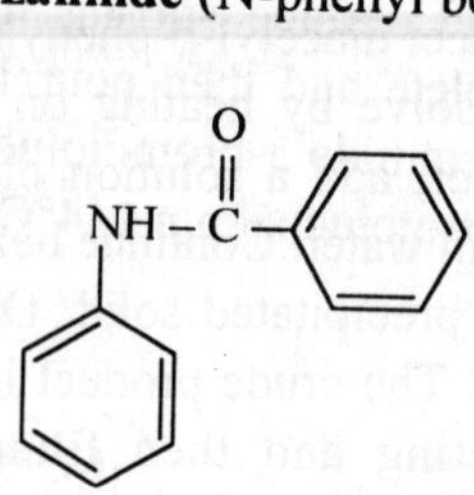

(1) Colourless, odourless solid, insoluble in water, sparingly soluble in ether, soluble in warm alcohol.
(2) Responds to carbylamine test.
(3) Reflux 10gm of compound with 100ml of 70% H_2SO_4 for 30min. Add 20ml hot water to the mixture. Cool and collect the precipitated benzoic acid, m.p 121°C. To the filtrate add dil NaOH solution. Cool, extract with ether. Distill and collect aniline at b.p 184°C.
(4) Heat at 110°C 2gm of benzanilide with PCl_5 for 30min, then at 160°C for 90min. Now add 1gm pyridine and 1.2gm aniline to this reaction mixture and heat at 160°C for 20min. Cool and add 5ml water to the mixture. Collect crude diphenylbenzamidine hydrochloride, treat with 5ml NH_4OH and warm gently for an hour. Cool and collect diphenybenzamidine, recrystallise from ethanol, m.p 144-145°C.
(5) p-Bromobenzanilide m.p 204°C (6) p-nitrobenzanilide m.p 199°C.

m.p 162°C **p-Aminoacetanilide** (N-acetyl-p-phenylenediamine)

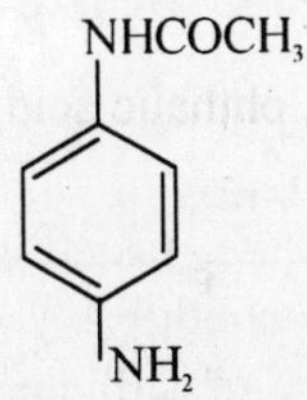

(1) Crystalline compound, sparingly soluble in water, soluble in alcohol and ether.
(2) Hydrolyse with HCl, acetic acid and p-phenylenediamine hydrochloride are obtained.
(3) Heat 1gm of compound with 5ml of acetic anhydride. Cool and collect diacetyl-p-phenylenediamine, m.p 304°C.
(4) Dissolve by heating on water bath 3gm of compound in 40ml water, add a solution of 4.5gm of trithiocarbodiglycolic acid in 50ml water. Continue heating for 5hrs. Cool in ice bath and collect the precipitated solid. Dissolve in water, heat to 80°C and filter hot. The crude product is then dissolved in glacial acetic acid by heating and then filter, cool and collect crystals of N-(p-Acetylaminophenyl) Rhodanine which decomposes above 240°C.

m.p 167°C **N, N'-Dibenzylurea**

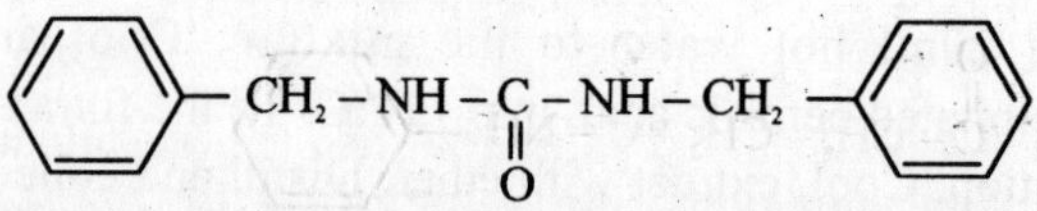

(1) Crystalline compound, insoluble, in water, soluble in alcohol.
(2) Hydrolyse with HCl, benzylamine hydrochloride is obtained.

m.p 207°C **Phthalanil** (N-Phenyl phthalimide)

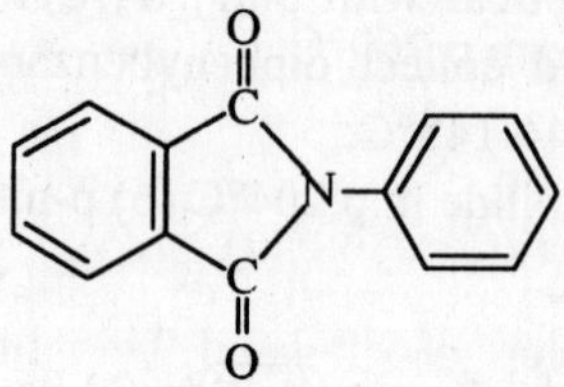

(1) Crystalline compound, insoluble in water, soluble in chloroform and benzene.
(2) Hydrolyse with HCl, phthalic acid, m.p 206-208°C is obtained

m.p 215°C **p-Nitroacetanilide**

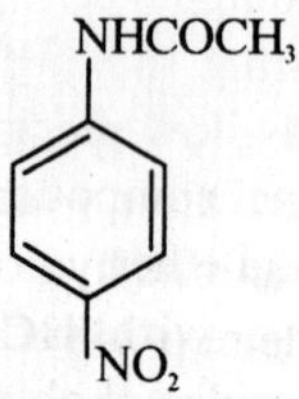

(1) Yellow coloured compound, insoluble in water, soluble in chlorofrom, alcohol and ether.
(2) To few milligrams of compound add 10% NaOH solution. A yellow coloured solution is obtained.
(3) To few milligram of compound add KOH solution. An intense red-brown coloured solution is obtained.
(4) Warm 1gm of compound with 15ml KOH solution and then keep the solution for 12hrs. Cool and collect p-nitroaniline, m.p 147°C.
(5) Add 3gm of KOH to a warm solution of 2.5gm of compound in 50ml acetone. Reflux and add a solution of 3gm of methyl iodide in 7ml acetone. Cool, distill off the solvent and then add cold water to it. p-Nitro –N-methylacetanilide separates out, m.p 154°C.

m.p 230°C **Succinanilide**

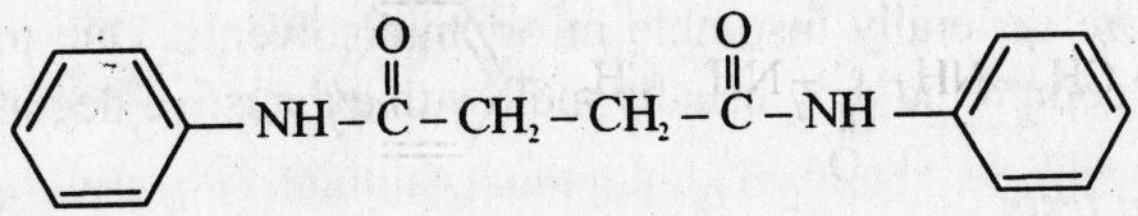

(1) Colourless crystalline compound, insoluble in water, soluble in ether.
(2) Hydrolyse with HCl, succinic acid m.p 185°C and aniline hydrochloride are obtained.
(3) Heat 500mg of compound with 2ml fuming HNO_3. Cool and collect the dinitroderivative, m.p 260°C.

m.p 238°C **s-Diphenylurea** (Carbanilide)

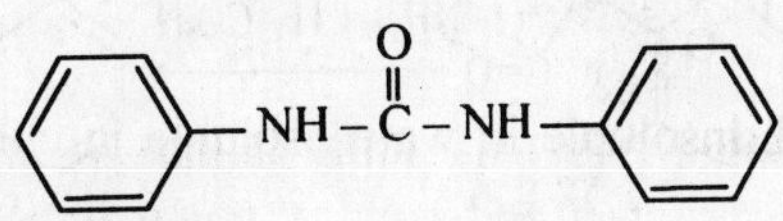

(1) Silky white crystalline solid, sparingly soluble in water, soluble in alcohol and ether.
(2) Responds to Carbylamine test
(3) Dissolve 500mg of compound in 10ml boiling alcohol. Cool and add dropwise with stirring 1ml of Br_2. After the addtion is complete shake thoroughly and allow precipitation to take place. Filter the precipitated bromo derivative which sublimes above 290°C. .
(4) Hydrolyse with HCl, aniline hydrochloride is obtained.
(5) m-dinitrocarbananilide m.p 248°C.

m.p 250°C **Oxanilide** (N, N-diphenyl oxamide)

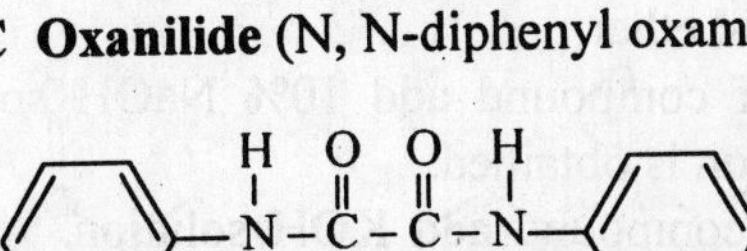

(1) Crystalline compound, insoluble in water, sparingly soluble in alcohol.
(2) Responds to Carbylamine test
(3) Hydrolyse with HCl, oxalic acid, m.p 101°C and aniline hydrochloride are obtained.
(4) di-p-nitro oxanilide m.p 260°C.

(xv) Amino Acids

Amino acids commonly referred as α amino acids are compounds that contain both an amino group and an acid group. Nearly all amino acids are chiral and they have been found to posses (S) configuration at the α carbon atom. They are generally insoluble in organic solvents. Due to the dipolar nature of the compound they usually melt with exhaustive decomposition.

Identification tests

(i) **Ninhydrin Test :** Dissolve 10mg of compound in 1-2ml water and add 7-8 drops of 1% Ninhydrin solution. Heat the mixture to boiling for 1-2 minutes. Development of violet, blue or purple colour indicates the presence of α amino acids

Note:- Proline, o, m, p-amino benzoic acids however give yellow colour.

Nihydrin Reagent : Dissolve .5gm of triketohydrindene hydrate in 100ml of 90% ethanol

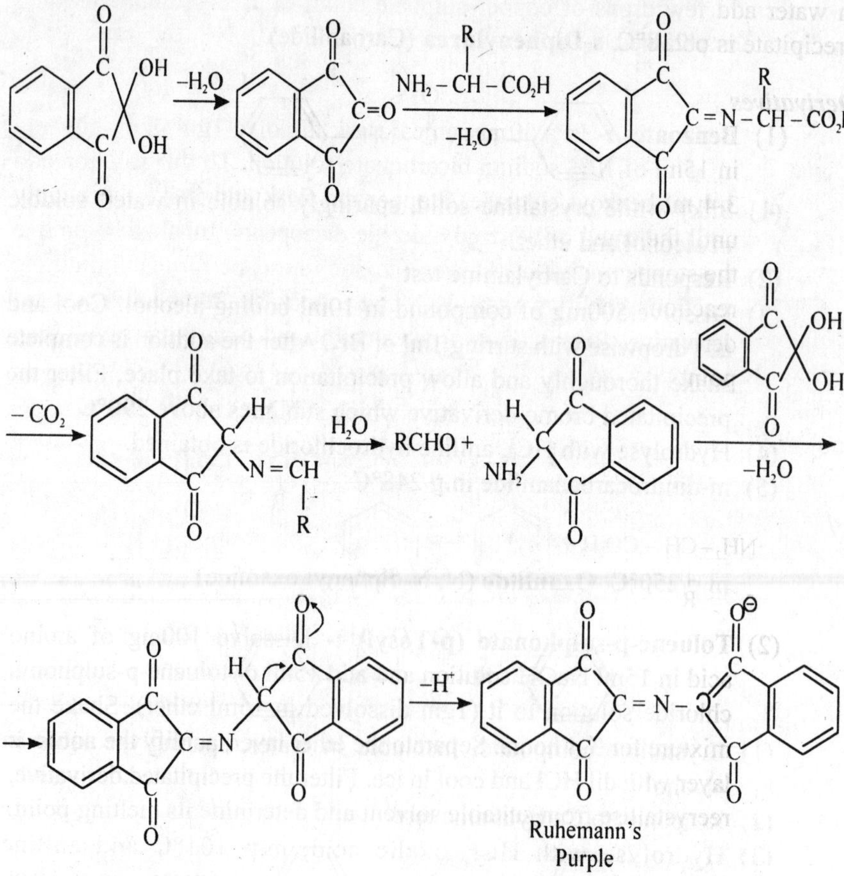

(*ii*) Ferric Chloride test : To aqueous solution of compound add $FeCl_3$ solution. A red coloured solution is obtained

(*iii*) Carbylamine test : Dissolve 100mg of compound in 2ml alcoholic KOH solution. Add few drops of chloroform and heat the mixture to boiling. A foul characteristic odour due to the formation of isonitrile is observed.

$$\underset{\displaystyle\overset{|}{R}}{HOOC-CH}-NH_2 + CHCl_3 + KOH \rightarrow \underset{\displaystyle\overset{|}{R}}{HOOC-CH}-N\equiv C + 3KCl + 3H_2O$$

(*iv*) Copper complex formation test: To few milligrams of compound in water add few drops of copper sulphate solution. A deep blue coloured precipitate is obtained.

Derivatives

(1) **Benzoates :-** In a 50ml conical flask dissolve 1gm of amino acid in 15ml of 10% sodium bicarbonate solution. To this solution add 3-4 ml benzoyl chloride. Stopper the flask and shake thoroughly until the smell of benzoyl chloride disappears. Inbetween remove the stopper and allow carbon dioxide to escape formed during the reaction. Acidify with dil HCl and collect the precipitated benzoyl derivative. Recrystallise from alcohol and determine its melting point.

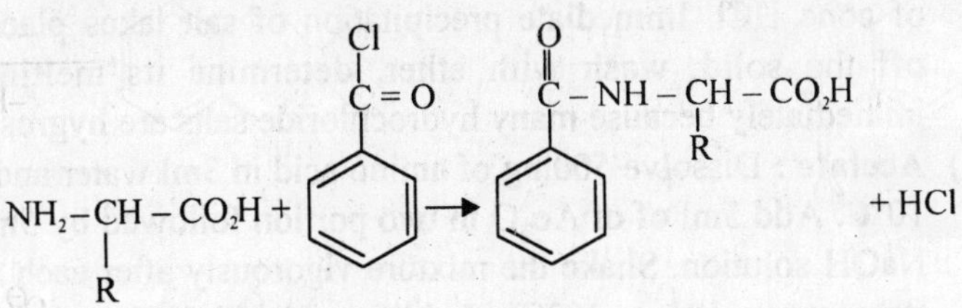

(2) **Toluene-p-sulphonate (p-Tosyl) :-** Dissolve 100mg of amino acid in 15ml NaOH solution and add 15ml of toluene-p-sulphonyl chloride solution to it (1gm dissolved in 15ml ether). Shake the mixture for 3-4 hours. Separate the ether layer, acidify the aqueous layer with dil HCl and cool in ice. Filter the precipitated derivative, recrystallise from suitable solvent and determine its melting point.

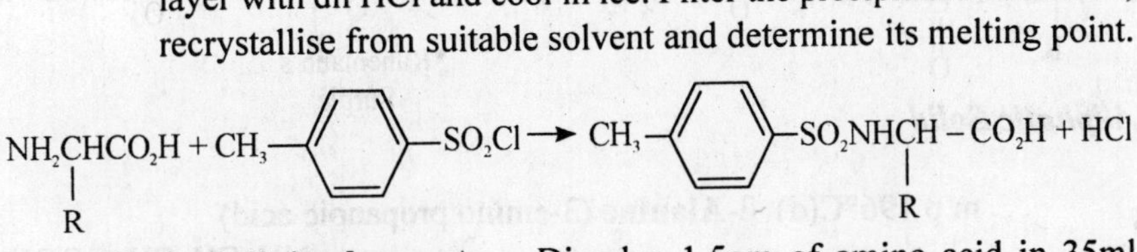

(3) **3, 5 dinitrobenzoate :-** Dissolve 1.5gm of amino acid in 35ml NaOH solution and add 4.5gm of 3, 5 dinitrobenzoyl chloride to it. Shake the mixture vigorously for 5-10 minutes. Acidify with

dil HCl and cool. Filter the precipitated derivative, recrystallise from suitable solvent and determine its melting point.

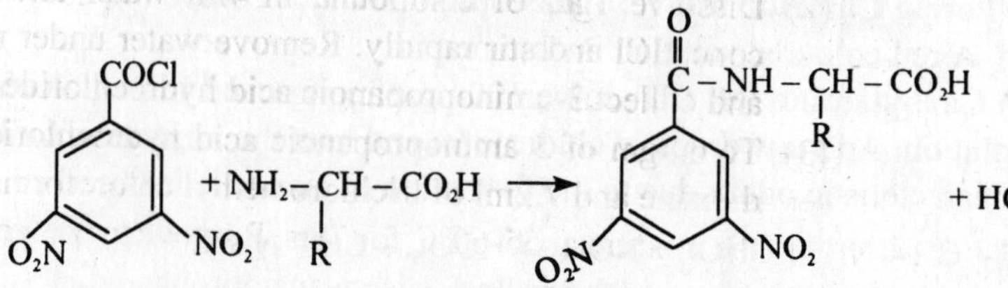

(4) **Phthaloyl :-** Heat together 1gm of amino acid with 2gm of phthalic anhydride in an oil bath at 175-185°C for 15-20 min (Shake occassionally during this period). Remove the mixture from oil bath and cool the molten liquid which soon solidifies. Recrystallise from ethanol and determine its melting point.

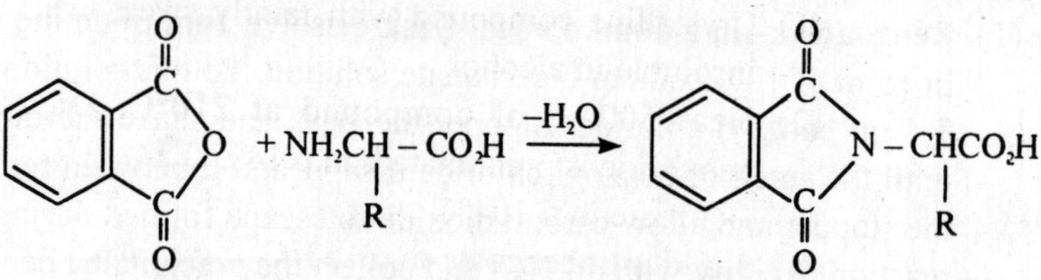

(5) **Salt Formation :** In a test tube place 500mg of amino acid and dissolve it in 2ml of ether. To this ethereal solution add 3-4 drops of conc HCl. Immediate precipitation of salt takes place. Filter off the solid, wash with ether, determine its melting point immediately because many hydrochloride salts are hygroscopic.

(6) **Acetate :** Dissolve 500mg of amino acid in 3ml water and cool to 10°C. Add 5ml of of Ac_2O in two portion followed by 5ml of 6N NaOH solution. Shake the mixture vigorously after each addition (tempeature below 10°C). Acidify with dil HCl and then keep overnight in a refrigerator. Filter the precipitated solid, recrystallise from hot water or ethanol and determine its melting pooint.

$$\underset{\substack{|\\R}}{NH_2CHCO_2H} + CH_3-\overset{\substack{O\\\|}}{C}-O-\overset{\substack{O\\\|}}{C}-CH_3 \longrightarrow CH_3-\overset{\substack{O\\\|}}{C}-NH-\underset{\substack{|\\R}}{CH}-CO_2H + CH_3CO_2H$$

Aliphatic Solid

m.p 196°C(d) β-**Alanine** (3-amino propanoic acid)

$NH_2CH_2CH_2COOH$

(1) Colourless solid with faintly sweet taste, soluble in water, insoluble in alcohol and ether.
(2) Dissolve 1gm of compound in 4ml water and then add 1.5ml conc HCl and stir rapidly. Remove water under reduced pressure and collect 3-aminopropanoic acid hydrochloride as white solid.
(3) To 6.3gm of 3 aminopropanoic acid hydrochloride add 125ml of dioxane and 7.2ml of trichloromethyl chloroformate. Stir and heat the mixture at 55-60°C for 7hrs. Remove the solvent under reduced pressure and collect 3-isocyanatopropanoyl chloride as colourless liquid b.p 92-94°C (25mmHg).
(4) Benzoate m.p 120°C

m.p 210°C (d) **Sarcosine** (N-Methyl glycine) CH_3NHCH_2COOH

(1) Crystalline compound with faintly sweet taste, soluble in water, insoluble in alcohol.
(2) Heat 500mg of compound at 210°C. On cooling sarcosine anhydride is precipitated, m.p 149°C
(3) Benzoate m.p 104°C
(4) 3, 5-dinitrobenzoate m.p 153°C
(5) Toluene-p-sulphonate m.p 102°C.
(6) Acetate m.p 135°C
(7) Hydrochloride m.p 169°C
(8) Picrate m.p 149°C.

m.p 223°C **L(+) Arginine**
$NH_2C{=}(NH)NH(CH_2)_3CH(NH_2)COOH$

(1) Colourless, crystalline odourless solid with faintly bitter taste, soluble in hot water. [DL-Arginine m.p 218°C]
(2) Benzoate (di) m.p 235°C (3) Picrate (di) m.p 190°C.

m.p 232°C **Glycine** (Aminoacetic acid) NH_2CH_2COOH

(1) White crystalline odourless solid, soluble in water, insoluble in ethanol.
(2) To aqueous solution of compound add a drop of phenol followed by sodium hypochlorite. A blue coloured solution is obtained.
(3) Heat 100mg of compound in 2ml water and CuO for 1 minute.

Filter the blue coloured solution while hot and cool the filtrate. Silky sky blue crystals are obtained.

(4) Reflux 1gm of compound with 6ml pyridine and 15ml Ac_2O for 6hrs. Distill off the excess pyridine, Ac_2O and AcOH under reduced pressure and recover acetamido acetone as pale yellow oil, b.p 120-125°C.

(5) Reflux 5gm of glycine with 25ml ethylene glycol at 175°C for 1hr. Leave the mixture overnight in refrigerator and then centrifuge. Collect and recrystallise piperazine 2, 5-dione from boiling water, m.p 310-312°C (d)

(6) Heat a suspension of 3.75gm of glycine in 15ml ethanol on water bath. Pass HCl gas into the reaction mixture for 30-40min. Cool and collect white crystals of ethylaminoacetate hydrochloride, m.p 145°C.

(7) Benzoate (Hippuric acid) m.p 187°C

(8) 3, 5-dinitrobenzoate m.p 179°C

(9) Phthalyl m.p 192°C

(10) Acetate m.p 206°C

(11) Picrate m.p 198°C

m.p 236°C(d) **L-Asparagine** $NH_2COCH_2CH(NH_2)COOH.H_2O$

(1) Tasteless crystalline compound, insoluble in alcohol and ether, soluble in water.

(2) Benzoate m.p 189°C

(3) 3, 5-dinitrobenzoate m.p196°C

(4) Toluene-p-sulphonate m.p 175°C.

(5) Picrate m.p 180°C.

m.p 240°C (d) **(±) Serine** (α-amino β-hydroxy propionic acid) $HOCH_2CH(NH_2)COOH$

(1) Crystalline solid with sweet taste, soluble in hot water, insoluble in alcohol and ether. [L-Serine m.p 222°C (d)]

(2) To aqueous solution of compound add $FeCl_3$ solution. A red coloured solution is obtained.

(3) Benzoate m.p 149°C (d)

(4) Toluene-p-sulphonate m.p 213°C

(5) Picrate m.p 169°C.

(6) 3, 5-dinitrobenzoate m.p 183°C

m.p 247°C (d) **L(+) Glutamic acid** (α-amino glutaric acid) $COOHCH_2CH_2CH(NH_2)COOH$

(1) Colourless crystalline solid, sparingly soluble in water, insoluble in alcohol. Sublimes at 200°C. [DL-Glutamic acid m.p 225°C]
(2) Heat 500mg of compound in an ignition tube and bring a pine splinter dipped in conc HCl near the mouth of ignition tube. The splinter turns red.
(3) To few milligrams of compound in water add copper sulphate solution. An intense blue coloured solution is obtained.
(4) Benzoate m.p 138°C
(5) Acetate m.p 199°C
(6) 3, 5-dinitrobenzoate m.p 217°C
(7) Toluene-p-sulphonate m.p 131°C
(8) Phthalyl m.p 159°C
(9) Hydrochloride 203°C

m.p 271°C **L-Aspartic acid** $HOOCCH_2CH(NH_2)COOH$

(1) Tasteless, odourless, crystalline compound, soluble in hot water, insoluble in alcohol. [DL-Aspartic acid m.p 280°C (d)]
(2) Benzoate m.p 185°C
(3) Toluene –p-sulphonate m.p 140°C
(4) Phthalyl m.p 193°C.

m.p 280°C **(Sublimes) α-Aminoisobutyric acid** $(CH_3)_2C(NH_2)COOH$

(1) Colourless crystalline solid, soluble in water, insoluble in alcohol.
(2) Benzoate m.p 198°C.

m.p 292°C (d) **(±)Isoleucine** (2-amino-3-methyl pentanoic acid) $CH_3CH_2CH(CH_3)CH(NH_2)COOH$

(1) Crystalline solid with odour like freshly crushed green beans, sparingly soluble in water, soluble in hot alcohol.
(2) Heat 100mg of compound in 2ml water with CuO for 1 minute.

Filter the blue coloured solution white hot and cool the filtrate. Silky sky blue crystals are obtained.

(3) Benzoate m.p 118°C

(4) Toluene-p-sulphonate m.p 140°C

m.p 295°C (d) **L(-) Leucine** (2-amino-4-methyl pentanoic acid) $(CH_3)_2CHCH_2CH(NH_2)COOH$

(1) Crystalline greasy solid with faintly bitter taste, moderately soluble in water and alcohol, insoluble in ether.

(2) Heat 500mg of compound with zinc dust in an ignition tube and bring a pine splinter dipped in conc HCl near the mouth of ignition tube. The splinter turns red.

(3) Heat 100mg of compound in 2ml water with CuO for 1 minute. Filter the blue coloured solution white hot and cool the filtrate. Pale blue crystals are obtained.

(4) Benzoate m.p 107°C (5) 3, 5-dinitrobenzoate m.p 187°C (6) Toluene-p-sulphonate m.p 124°C (7) Acetate m.p 189°C (8) Phthalyl m.p 116°C.

m.p 295°C **(±) Alanine** (2-amino propanoic acid) $CH_3CH(NH_2)COOH$

(1) Crystalline solid, slightly soluble in alcohol, insoluble in ether.

(2) Heat 500mg of compound with zinc dust in an ignition tube and bring a pine splinter dipped in conc HCl near the mouth of ignition tube. The splinter turns red.

(3) To few milligrams of compound in water add copper sulphate solution. A deep blue coloured solution is obtained.

(4) Dissolve 500mg of compound in 3ml water and then add N-Bromosuccinimide to it. Heat the mixture on water bath for 30min. An apple like odour is observed due to the formation of acetaldehyde.

(5) Heat 1.75gm of compound with 8ml pyridine and 11ml Ac_2O on steam bath for 6hrs. Distill off the excess pyridine, Ac_2O and AcOH under reduced pressure and recover 2-acetamido 2-butanone at 102-106°C/2mmHg.

(6) Benzoate m.p 165°C

(7) 3, 5-dinitrobenzoate m.p 177°C

(8) Toluene-p-sulphonate m.p 139°C
(9) Phthalyl m.p 161°C.

m.p 298°C (d) **(±)Valine** (2-amino-3-methyl butanoic acid) $(CH_3)_2CHCH(NH_2)COOH$

(1) Crystalline solid with sweet taste, soluble in water, insoluble in alcohol and ether. [D-Valine m.p 293°C (d)]
(2) Heat 500mg of compound at 300°C. Cool, allow the molten liquid to solidify and collect valine anhydride, m.p > 300°C.
(3) Benzoate m.p 132°C
(4) Toluene-p-sulphonate m.p 110°C.
(5) Phthalyl m.p 102°C
(6) Acetate m.p 148°C
(7) Hydrochloride m.p 189°C.

m.p 304°C (±) α **Amino-n-butyric acid** [dl-2-amino butanoic acid] $CH_3CH_2CH(NH_2)COOH$

(1) Colourless crystalline solid, soluble in water, insoluble in alcohol. Sublimes at 300°C.
(2) To few millgrams of compound add $FeCl_3$ solution. A dark brown-red coloured solution is obtained.
(3) To 2ml aqueous solution of compound add copper acetate solution. A violet blue coloured precipitate is obtained.
(4) Benzoate m.p 147°C

m.p 315°C (d) **L(+) Valine** (α-amino isovaleric acid) $(CH_3)_2CHCH(NH_2)COOH$

(1) Crystalline silvery solid with taste feebly sweet and bitter, soluble in water, sparingly soluble in alcohol.
(2) Heat 100mg of compound in 2ml water with CuO for 1 minute. Filter the blue coloured solution white hot and cool the filtrate. Blue crystals are obtained.
(3) Benzoate m.p 127°C
(4) 3, 5-dinitrobenzoate m.p 181°C
(5) Toluene-p-sulphonate m.p 150°C
(6) Phthalyl m.p 115°C

(7) Acetate m.p 164°C.

m.p 332°C (d) (±) **leucine** (dl-α-amino isocaproic acid)
$(CH_3)_2CHCH_2CH(NH_2)COOH$

(1) Crystalline solid with faintly sweet taste, sparingly soluble in water.
(2) Heat 500mg of compound with zinc dust in an ignition tube and bring a pine splinter dipped in conc HCl near the mouth of ignition tube. The splinter turns red.
(3) Benzoate m.p 137°C
(4) Acetate m.p 157°C
(5) Phthalyl m.p 141°C
(6) Picrate m.p 134°C.

Aromatics Solid

m.p 92°C **Ethyl p-aminobenzoate** (Anaesthesine, Benzocaine)

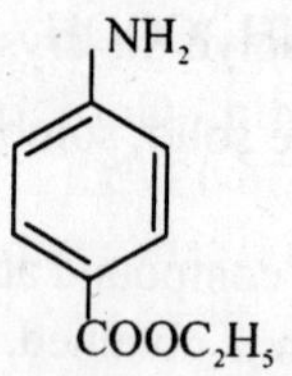

(1) Crystalline odourless solid, insoluble in water, soluble in alcohol, benzene and chloroform.
(2) On a watch glass mix 2-3mg of compound with 4-5 drops of conc H_2SO_4. Touch the acidic solution with a glass rod moistened with conc HNO_3. The solution turns green to green yellow. Diltue with water and add 1ml dil NaOH solution. The solution turns orange.
(3) Heat 3.3gm of compound with 6ml water and 5ml conc HCl, cool to 0°C and then diazotise with 1.5gm aqueous solution of $NaNO_2$(temp below 7°C). Add a solution of fluoboric acid (1.3gm of boric acid in 2.6gm of 60% hydrofluoric acid) while keeping the temp below 10°C. Cool and filter the precipitated fluoborate. Dissolve it in a solution of 1.2gm KOH in 2ml alcohol and 5ml water. Heat over a low flame, cool and acidify with dil HCl, p-fluoro benzoic acid separate out, m.p 196°C.
(4) Benzoate m.p 148°C (4) Acetate m.p 110°C (5) Picrate m.p 131°C.

m.p 127°C **Phenylglycine** (Anilinoacetic acid)

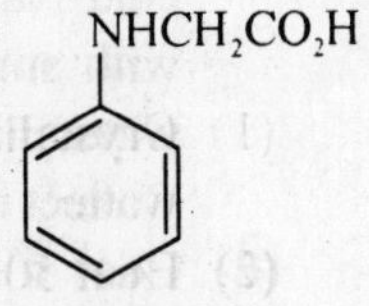

(1) Crystalline solid, moderately soluble in water, sparingly soluble in ether.
(2) Fuse few milligrams of compound with moist KOH at low temperature. Dissolve the fused mass in water and then leave it for few minutes. A blue colouration or precipitate of indigo will appear.
(3) Dissolve 5gm of compound in 60ml water, cool to 0°C. Add in portion a solution of 2.5gm of $NaNO_2$ in 15ml water(temp below 0°C). Filter the red clear solution. Cool and add with stirring 5ml of conc HCl to the filtrate and collect fluffy crystals of N-nitroso N-Phenylglycine. Dissolve the crystals in 25ml Ac_2O and heat on water bath for 90min. Cool, pour in ice and collect 3-phenylsydnone, m.p 136-137°C.
(4) Benzoate mp 63°C (5) Acetate m.p 194°C.

m.p 145°C **Anthranilic acid** (o-amino benzoic acid)

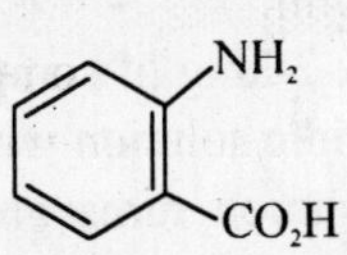

(1) Crystalline odourless solid, soluble in hot water and alcohol.
(2) Dissolve few milligrams of compound in water. A bluish fluorescence is observed.
(3) To saturated solution of compound add 1-2 drops of saturated bleaching powder solution. A violet red coloured solution is obtained.
(4) Cool a solution of 2gm compound dissolved in 15ml water containing 2ml conc H_2SO_4 to 5°C and then diazotise with 4ml cold $NaNO_2$ solution (1gm). Add with stirring a solution of 4gm of KI in 7ml conc H_2SO_4, heat to boiling for 10min. Cool and collect o-iodobenzoic acid, m.p 162°C.
(5) Cool a solution of 2.5gm compound dissolved in 7.5ml water containing 4.6ml conc HCl to 5°C and then diazotise with 17ml

cold $NaNO_2$ solution (1.3gm). Add this diazotised cold solution with stirring to cupro-ammonia reagent (6.3gm $CuSO_4$ in 25ml water +10ml NH_4OH). Heat to boiling, acidify with conc HCl. Collect diphenic acid, m.p 228°C.

(6) Cool a solution of 1.4gm compound dissolved in 10ml water containing 2ml conc HCl to 5°C and then diazotise with 2.4ml cold $NaNO_2$ solution (.7gm). Add this diazotised cold solution with stirring to copper (I) chloride solution (2.6gm $CuSO_4$ + 1.2gm NaCl in 5ml water. Heat and add 8ml concHCl+ 1.4gm Cu turnings). Keep the mixture at room temperature for 3hrs. Cool and collect o-chlorobenzoic acid, m.p 139°C.

(7) Reflux a mixture of 2gm compound, 1.4gm chloroacetic acid, 3.2gm Na_2CO_3 and 30ml water for 3hrs. Cool, acidify with conc HCl and keep overnight in a refrigerator. Collect crystals of phenylglycine-o-carboxylic acid, recrystallise from hot water, m.p 220°C.

(8) Benzoate m.p 182°C

(9) 3, 5-dinitrobenzoate m.p 278°C

(10) Toluene-p-sulphonate m.p 217°C.

(11) Acetate m.p 185°C

(12) Hydrochloride m.p 193°C

(13) Picrate m.p 104°C

(14) Amide m.p 108°C.

(15) Anilide m.p 126°C.

m.p 150°C **Oxanilic acid** (phenyl oxanic acid)

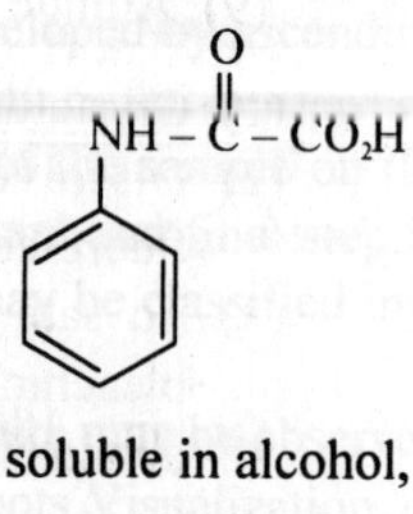

(1) Colourless crystalline solid, insoluble in water, soluble in alcohol, chloroform and ether.

(2) Dissolve few milligrams of compound in conc H_2SO_4 and add few crystals of $K_2Cr_2O_7$. A blue-violet coloured solution is obtained.

(3) Responds to carbylamine test.

(4) Oxanilide m.p 254°C

m.p 174°C **m-Aminobenzoic acid**

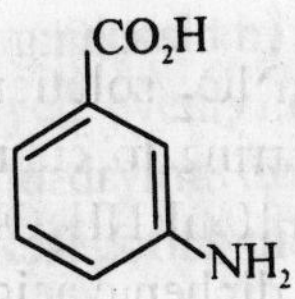

(1) Colourless crystalline odourless solid, soluble in hot water and alcohol.
(2) Responds to carbylamine test.
(3) Dissolve 100mg of compound in 1ml water and add 2-3 drops of saturated bleaching powder solution. An orange yellow coloured solution is obtained.
(4) To an ice cold mixture of 2.7gm compound, 16.5ml conc HCl and 60ml water add with stirring 4.5ml Br_2. Cool and recover tribromoderivative, m.p 170-172°C. Cool a mixture of 100ml conc H_2SO_4 and 50ml water to –5°C and then first add 3.7gm of $NaNO_2$ in small portions followed by 20ml cold 50% hypophosphorous acid. Add a solution of tribromo derivative dissolved in 100ml glacial acetic acid to the diazotised solution (temp below –10°C)Continue stirring for 30 min and then keep in refrigerator for 36hrs. Distill off acetic acid and cool the residual liquid, acidify with conc HCl and collect 2, 4, 6-tribromobenzoic acid, m.p 192°C.
(5) Benzoate m.p 248°C
(6) 3, 5-dinitrobenzoate m.p 270°C
(7) Acetate m.p 250°C
(8) Amide m.p 75°C.
(9) Anilide m.p 129°C.

m.p 186°C **p-Aminobenzoic acid**

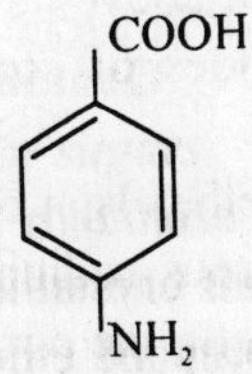

(1) Colourless crystalline odourless solid, soluble in hot water, alcohol and ether.
(2) Responds to carbylamine test.
(3) Dissolve 100mg of compound in 1ml water and add 2-3 drops of

saturated bleaching powder solution. A violet red coloured solution is obtained.

(4) Saturate 8ml absolute alcohol with HCl gas and add 1.2gm of compound to it. Reflux for 2hrs. Cool and collect the hydrochloride of p-aminobenzoate. Dissolve it in water and add Na_2CO_3 until neutal to litmus. Collect the precipitated ethyl p-aminobenzoate, m.p 91°C.

(5) Dissolve 1gm of compound in 45ml dil HCl and then add a solution of 4.8gm of iodine monochloride in 4ml of 45% HCl and then stir the mixture for 2-3 minutes. Dilute with 50ml water and then heat on water bath at 90°C for 15 minutes. Cool and collect the precipitated 4-amino 3, 5-diiodobenzoic acid, m.p >350°C.

(6) Benzoate m.p 278°C

(7) 3, 5-dinitrobenzoate m.p 290°C

(8) Toluene-p-sulphonate m.p 223°C.

(9) Acetate m.p 252°C

(10) Amide m.p 183°C.

m.p 187°C **Hippuric acid** (Benzoylglycine)

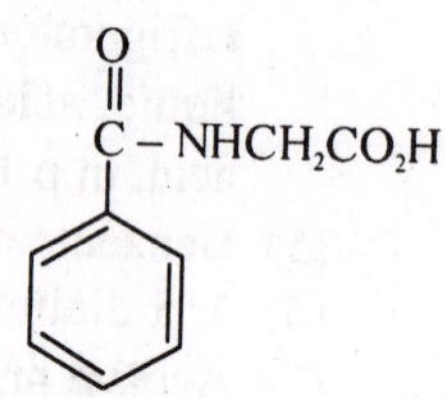

(1) Crystalline solid, insoluble in water, benzene and pet ether, sparingly soluble in ether and alcohol.

(2) Heat 1gm of compound in a round bottom flask fitted with a condensor above its melting point. Benzoic acid will be deposited on the sides of condensor and smell of bitter almond will be emitted.

(3) Heat together 1gm of compound with 5ml benzaldehyde and 4gm of anhydrous sodium acetate on water bath for 30min. Cool and collect lactimide of benzoylamino cinnamic acid, m.p 165°C.

(4) Heat 5.37 gm of compound with 16ml Ac_2O at 80°C for 40 min. Cool and distill off the excess Ac_2O. Dissolve the residue in 10ml t-butanol and keep overnight in a rerigerator. Collect crystals of 2-phenyl 5-oxazolone, m.p 89-92°C.

(5) Stir 4.5gm of compound with 7.5gm piperonal, 8ml Ac_2O and

2.5gm potassium hydrogen carbonate. The temperature slowly rises to 100°C. Keep the mixture overnight, extract with hot water and discard. Cool the residue and collect azlactone of a benzoylamino-3, 4-methylenedioxycinnamic acid, m.p 195-197°C.

(6) Amide m.p 183°C
(7) Anilide m.p 208°C.

m.p 200°C (d) **p-Aminophenylacetic acid**

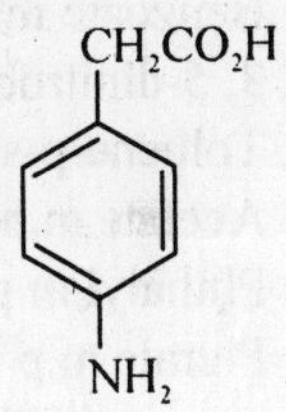

(1) Soluble in hot water, alcohol and alkali.
(2) N-Acetate m.p 168-170°C
(3) N-Benzoate m.p 205°C
(4) Hydrochloride m.p 220-240°C(d)

m.p 248°C(d) N-(p-Hydroxyphenyl) **glycine**

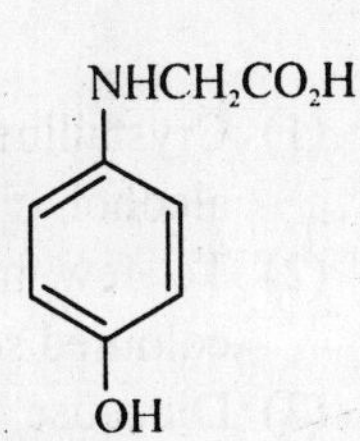

(1) Crystalline compound, sparingly soluble in water and alcohol, insoluble in ether.
(2) To few milligram of compound add $FeCl_3$ solution. A blue coloured solution is obtained.
(3) Benzoate m.p 117°C
(4) Acetate (di) m.p 175°C.

m.p 273°C (±)**Phenylalanine**

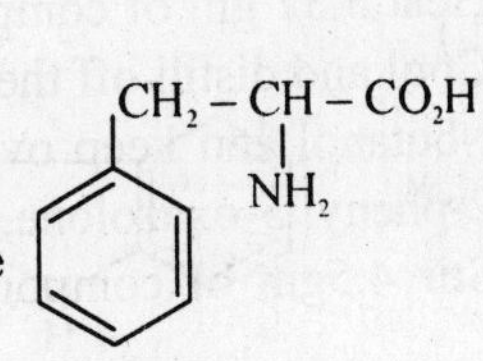

(1) Crystalline solid with sweet taste, sparingly soluble in water and alcohol, insoluble in ether.
(2) Heat 500mg of compound with zinc dust in an ignition tube and bring a pine splinter dipped in conc HCl near the mouth of ignition tube. The splinter turns red.
(3) Dissolve few milligrams of compound in conc H_2SO_4, add few crystals of $K_2Cr_2O_7$. On warming characteristic odour of phenylacetic aldehyde is emitted.
(4) Benzoate m.p 187°C
(5) 3, 5-dinitrobenzoate m.p 93°C
(6) Toluene-p-sulphonate m.p 134°C.
(7) Acetate m.p 152°C
(8) Phthalyl m.p 175°C
(9) Picrate m.p 173°C.

m.p 283°C **5-AminoSalicylic acid** (5-amino-2-hydroxy benzoic acid)

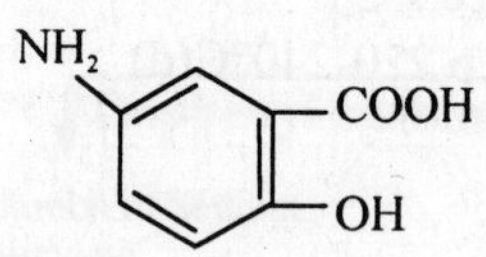

(1) Crystalline compound, sparingly soluble in water, insoluble in alcohol.
(2) To few milligrams of compound add $FeCl_3$ solution. A red coloured solution is obtained which slowly changes to brown.
(3) Diazotise 1gm of compound with cold $NaNO_2$ solution. Reduce the diazotised solution and recover 5-hydrazino salicylic acid, m.p 148°C.
(4) Benzoate m.p 252°C
(5) Acetate (di)m.p 184°C
(6) Acetate (mono) m.p 218°C.

m.p 289°C (d) **L(-) Tryptophan** (β-Indole-α-amino propionic acid)

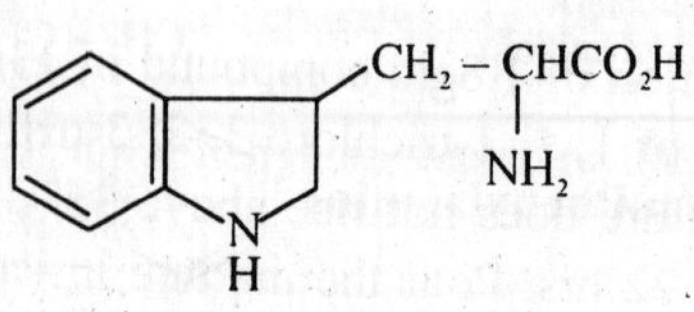

(1) Silky crystalline odourless solid with faintly bitter taste. Sparingly soluble in water and alcohol, easily soluble in hot water.
(2) Heat 100mg of compound in an ignition tube. Bring a pine splinter dipped in conc HCl near the mouth of ignition tube. The splinter turns red.
(3) Dissolve 10mg of compound in 2ml warm water. Add 8-10 drops of freshly saturated bromine water solution. Shake thoroughly with 20ml ethyl acetate. A deep violet red solution showing greenish fluorescence is obtained.
(4) Benzoate m.p 104°C
(5) 3, 5-dinitrobenzoate m.p 233°C
(6) Toluene-p-sulphonate m.p 176°C.
(7) Acetate m.p 189°C
(8) Hydrochloride m.p 257°C(d)
(9) Picrate m.p 195°C

m.p 295°C (d) **L(-) Tyrosine** (p-Hydroxyphenyl-α-amino propionic acid)

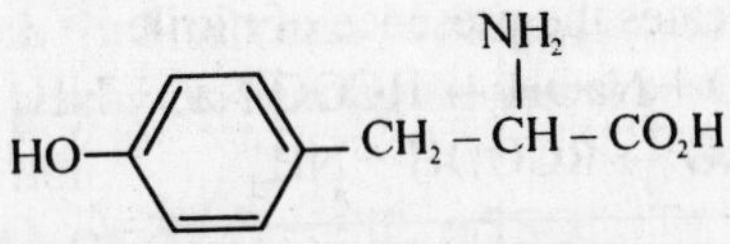

(1) Colourless silky needles type compound, odourless, tasteless. Gives smell of burnt horn on ignition. Sparingly soluble in water and alcohol, soluble in alkali and mineral acid.
(2) Heat 100mg of compound with zinc dust in an ignition tube, bring a pine splinter dipped in conc HCl near the mouth of ignition tube. The splinter turns red.
(3) Heat 100mg of compound with conc H_2SO_4 on water bath for 30min. Cool and dilute with water. Neutralise the acidic solution with $BaCO_3$, filter and concentrate the filtrate. On adding few drops of $FeCl_3$ the solution turns violet.
(4) Add few milligrams of compound to 2-3ml formaldehyde solution(1ml formaldehyde in 50ml conc H_2SO_4). The solution turns wine red. Add 2ml glacial acetic acid and boil. The solution turns green.
(5) Cool a solution of 1.8gm compound in 13ml dimethyl sulfoxide. Add 6.4gm of 1, 1, 1-trichloro 3, 3, 3-trifluoro acetone such that the temperature does not rise above 35°C. Remove the ice bath and stir for 22 hrs. Pour the mixture in ice, isolate with butanol.

Concentrate the organic extract, cool and collect N-trifluoroacetyl L-tyrosine, m.p 192-193°C.

(6) N-Benzoate m.p 165°C
(7) O, N-Dibenzoate m.p 211°C
(8) N-Toluene-p-sulphonate m.p 187°C.
(9) N-Acetate m.p 153°C
(10) O, N-Diacetate m.p 172°C.

(xvi) Nitriles

Nitriles are compound that contain cyano group C≡N. They are considered to be derivatives of hydrogen cyanide (HCN) in which the hydrogen is replaced by alkyl or aryl group. Nitriles are either liquid or low melting solid and have a characteristic pleasant odour resembling mostly with that of bitter almond.

Identification test

(*i*) In a dry test tube to 100mg of compound add 5ml dil NaOH or dil H_2SO_4. Boil the mixture and bring a glass rod dipped in HCl near the mouth of test tube. White fumes due to the formation of NH_4Cl indicates the presence of nitrile

$$RCN + H_2O + NaOH \rightarrow RCOO^-Na^+ + NH_3$$

$$RCN + 2H_2O \rightarrow RCOOH + NH_3$$

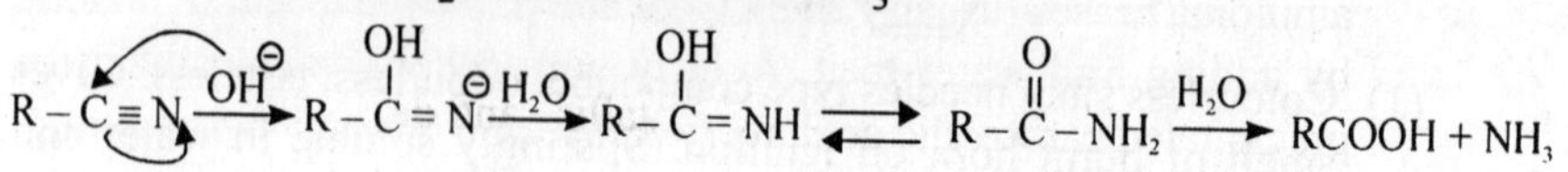

$$R-C\equiv N + H_2SO_4 \longrightarrow \left[R-\overset{\oplus}{C}=NH\right]HSO_4^- \xrightarrow{H_2O} R-\overset{O}{\overset{\|}{C}}-NH_2 \xrightarrow{H_2O} RCOOH + NH_3$$

Note : This test is to be considered positive if amide group is absent

(*ii*) Feigl test :- To 100mg (1ml in case of liquid) of compound add 1.5ml of 5% methanolic hydroxylamine hydrochloride solution. Now add saturated KOH solution until the mixture is alkaline. Boil the resultant mixture for 2 minutes and add few drops of $FeCl_3$ solution. A wine red colour indicates the presence of nitrile

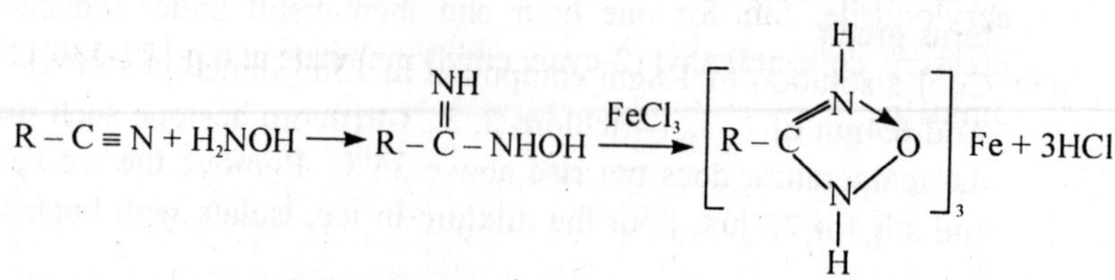

Note :-This test is to be considered positive if ester, amide groups are absent.

Derivatives

(1) **Hydrolysis with acid to organic acid :** Heat 5ml of 75% H_2SO_4/HCl and 200mg NaCl to 150-160°C in oil bath. Add in portion 1gm of compound to it. After the addition is complete heat the resultant mixture at 190°C for 1hr. Cool and pour the mixture on crushed ice. Collect the precipitated solid and dissolve it in dil NaOH solution. Filter to remove any amide that may be present. Acidify the filtrate with dil HCl and collect the carboxylic acid, recrystallise and determine its melting point.

$$RC \equiv N \xrightarrow{H_2O} \left[R-C\begin{matrix} \diagup OH \\ \diagdown\!\!\diagdown NH \end{matrix} \right] \longrightarrow RCONH_2 \xrightarrow{H_2O} RCOOH + NH_3$$

(2) **Hydrolysis with acid to amide :** Heat a mixture of 2gm of compound with 8ml conc H_2SO_4/HCl on water bath at 80-90°C for 5minutes. Cool and pour the mixture in ice water. Filter the precipitated solid and stir it with 20ml dil NaOH solution. Filter to remove the salt of acid which is soluble. Recrystallise the amide and determine its melting point.

(3) **Hydrolysis with base to carboxylic acid :** Reflux 1gm of compound with 15ml dil NaOH solution until the evolution of ammonia ceases (usually 2hrs). Cool and dilute the reaction mixture by adding 5ml water to it. Acidify with dil HCl and collect the precipitated carboxylic acid, recrystallise and determine its melting point.

Aliphatic Liquids

b.p 78°C **Acrylonitrile** (Vinyl cyanide) CH_2=CHCN

(1) Colourless liquid, soluble in water alcohol and ether. Vapours highly toxic and lachrymatory. Forms azeotropic mixture with water, b.p 70°C

(2) To 10ml diethyl malonate add with stirring 2ml sodium ethoxide solution (70mg Na in 2ml absolute ethanol) and then add 2ml acrylonitrile. Stir for one hour and then distill under reduced pressure, collect diethyl (2-cyanoethyl) malonate at b.p 127-130°C/3mmHg.

(3) Heat a mixture of 6ml diethylamine and 3.3ml acrylonitrile at 50°C for 10hrs, keep the mixture at room temperature for 2days. Distill under reduced pressure and collect 3-(diethylamino) propionitrile at 75-77°C/11mmHg.
(4) Shake together 3.3ml of compound with 2.5ml dil NaOH and 3.3 ml ethanol for 2hrs. Two layers result, remove the upper layer and neutralise with 5% acetic acid. Distill and collect 3-ethoxy propionitrile, b.p 172-174°C.
(5) Reflux 10ml of compound with 12gm 2-naphthol and 200mg of sodium for 3hrs. Cool and pour with stirring in 100ml water containing 10ml conc HCl. Extract with benzene. Distill off the benzene, cool the residue which soon solidifies, recrystallise 2-naphthyl 2-cyanoethyl ether from ethanol, m.p 109°C.

b.p 82°C **Acetonitrile** (Methyl cyanide) CH_3CN

(1) Colourless liquid with ethereal odour, miscible with water, alcohol and ether.
(2) Boil 1ml of compound with 5ml NaOH. Bring a glass rod dipped in conc HCl near the mouth of test tube. White fumes due to formation of NH_4Cl are obtained.
(3) Pass HCl gas into a cooled solution of 3ml acetonitrile, 2.5gm phloroglucinol and 10ml ether for 2hrs. Cool further, decant the ether and reflux the solid with 5ml water for 2hrs. Cool and collect phloroacetophenone, m.p 219°C.
(4) On acid or alkaline hydrolysis the products are acetic acid and ammonia.
(5) To a mixture of 3.5ml acetonitrile, 25ml ether, 5gm resorcinol and 2gm $ZnCl_2$ pass H_2S gas and then leave the mixture overnight. Filter the chlorozincate of imine, dissolve in 50ml dil HCl and boil for 30 min. Cool and collect 2, 4-dihydroxyacetophenone, m.p 145-146°C.
(6) To a saturated aqueous solution of 4.1gm of acetonitrile and 7gm of hydroxylamine hydrochloride add alcohol to get a clear solution. Cool and add a solution of sodium methoxide (2.3gm Sodium). Keep the resultant mixture at 30-40°C for 70-80hrs. Filter and concentrate the filtrate. Add HCl and again concentrate. Dilute with warm alcohol and filter. Cool the filtrate and collect acetamide oxime hydrochloride, m.p 140°C.

b.p 97°C **Propionitrile** (Ethyl cyanide) CH_3CH_2CN

(1) Colourless liquid with ethereal odour, moderatly soluble in water, soluble in alcohol and ether.
(2) On acid or alkaline hydrolysis the products are propionic acid and ammonia.

b.p 107°C **Isobutyronitrile** (Isopropyl cyanide) $(CH_3)_2CHCN$

(1) Colourless liquid, slightly soluble in water, soluble in alcohol and ether.
(2) Boil 1ml of compound with 5ml NaOH. Bring a glass rod dipped in conc HCl near the mouth of test tube. White fumes due to formation of NH_4Cl are obtained.
(3) On acid or alkaline hydrolysis the products are isobutyric acid and ammonia.

b.p 118°C **n-Butyronitrile** (Propyl cyanide) $CH_3CH_2CH_2CN$

(1) Colourless liquid, slightly soluble in water, miscible with alcohol and ether.
(2) Boil 1ml of compound with 5ml NaOH. Bring a glass rod dipped in conc HCl near the mouth of test tube. White fumes due to formation of NH_4Cl are obtained.
(3) On acid or alkaline hydrolysis the products are butyric acid and ammonia.

b.p 119°C β-**Butenonitrile** (allyl cyanide) $CH_2{=}CHCH_2CN$

(1) Liquid with highly alliaceous odour, slightly soluble in water, soluble in alcohol and ether.
(2) Boil 1ml of compound with 5ml NaOH. Bring a glass rod dipped in conc HCl near the mouth of test tube. White fumes due to formation of NH_4Cl are obtained.
(3) Warm a mixture of 3ml of compound with 4ml conc HCl for 5-10 min, remove the heat source as the mixture continues to reflux. Cool and add 5ml water, remove the upper layer and extract the aqueous layer with ether. Distill off the ether and collect vinyl acetic acid, b.p 169°C.

(4) On alkaline hydrolysis the products are crotonic acid and ammonia.

b.p 120°C(d) **α-Hydroxyisobutyronitrile** (Acetone cyanohydrin) $(CH_3)_2C(OH)CN$

(1) Colourless odourless liquid, soluble in water, alcohol and ether.
(2) Heat 5ml of compound with 7.5gm of ammonium carbonate on water bath at 70-80°C for 3hrs. Raise the temperature to 90°C and heat at this temperature for 30 min. Cool and collect colourless solid. Recrystallise from hot water and collect crystals of dimethylhydantoin, m.p 178°C.
(3) To 4.5ml warm conc H_2SO_4 containing 50mg of sulphur add with stirring 4.7ml of compound (temp 70-80°C). Heat the resultant mixture on oil bath at 150°C for 20-25min. Cool in ice bath, dilute with 30ml water and filter. Add 20gm of sodium carbonate to the cooled filtrate. Collect the precipitated methacrylamide, recrystallise from benzene, m.p 105°C.

b.p 130°C **Isovaleronitrile** (Isobutyl cyanide, 3-methyl butane nitrile) $(CH_3)_2CHCH_2CN$

(1) Colourless liquid, slightly soluble in water, soluble in alcohol and ether.
(2) On acid or alkaline hydrolysis the products are isovaleric acid, b.p 176°C and ammonia.

b.p 141°C **n-Valeronitrile** (Butyl cyanide, pentane nitrile) $CH_3CH_2CH_2CH_2CN$

(1) Liquid with nauseating odour, insoluble in water, soluble in alcohol and ether.
(2) Boil 1ml of compound with 5ml NaOH. Bring a glass rod dipped in conc HCl near the mouth of test tube. White fumes due to formation of NH_4Cl are obtained.
(3) To cooled 5ml ethyl alcohol first add 3ml conc H_2SO_4 and then 3ml butyl cyanide. Reflux the mixture for 10hrs. Cool and pour in ice water. Remove the upper ester layer. Distill and collect ethyl valerate at b.p 143-146°C.
(4) Reflux together 6.2ml valeronitrile with 13ml NaOH solution

(9.2gm in 13ml water) for 5hrs. Add 10ml water, cool and acidify with dil H_2SO_4. Separate the upper acid layer. Distill and collect valeric acid at b.p 184°C.

b.p 163°C **n-Capronitrile** (hexanenitrile)

$CH_3CH_2CH_2CH_2CH_2CN$

(1) Colourless liquid, Insoluble in water, soluble in alcohol and ether.
(2) On acid or alkaline hydrolysis the products are n-caproic acid, b.p 205°C and ammonia.
(3) Add 6ml of compound to a suspension of sodium ethoxide (from 2gm Na) in 30ml diethyl carbonate. Reflux the mixture for 5hrs and distill the alcohol formed during the reaction. Cool and acidify with acetic acid. Remove the upper layer and distill. Collect α cyanocaproic ester at b.p 128-129°C/23mmHg

b.p 167°C **Acetlactonitrile** $CH_3CH(O.\ OCCH_3)\ CN$

(1) Soluble in alcohol and ether.
(2) On heating with alkali at 140°C, the products are acetaldehyde, acetic acid and hydrogen cyanide.

m.p 183°C(d) **Glycolonitrile** (Formaldehyde cyanohydrin)

$HOCH_2CN$

(1) Benzoate m.p 195°C

b.p 207°C **Ethyl cyanoacetate** $CH_2(CN)COOC_2H_5$

(1) Colourless liquid, insoluble in water, miscible with alcohol and ether.
(2) Add 30.40gm of compound to a solution of 1.40gm of sodium in 40ml absolute ethanol followed by 9.70gm of o, α-dichlorotoluene. Shake thoroughly, hydrolyse the precipitated solid with 20ml of 10% NaOH solution and collect 2-(o-chlorobenzyl)cyanoacetic acid, m.p 129-132°C.
(3) Stir togethr 2ml of compound with 1.4gm KCN, 2ml ethanol and 2ml butanone for 2hrs. Cool and keep at room temperature for 7days. Add 20ml conc HCl and reflux for 8hrs. Extract with

ether. Distill off ether, dissolve the residue in 4ml dil HCl and heat at 100°C. Cool and collect α-ethyl α-methyl succinic acid, m.p 91°C.

Aliphatic Solid

m.p 54°C **Succinonitrile** (Ethylene cyanide, butane dinitrile) $CNCH_2CH_2CN$

(1) Colourless transparent amorphours mass, easily soluble in water, alcohol and ether.
(2) Reflux 1gm of compound with 10ml dil NaOH solution. Bring a glass rod dipped in con HCl near the mouth of test tube, white fumes are produced. Cool and acidify the mixture with dil HCl and collect succinic acid, m.p 185°C.

m.p 58°C **Dipropionyl Dicyanide** $(CH_3CH_2COCN)_2$

(1) Colourless crystalline solid, sparingly soluble in water, easily soluble alcohol, ether and benzene.
(2) On alkaline hydrolysis the products are propionic acid, hydrogen cyanide and ammonia.

m.p 67°C **Cyanoacetic acid** $CNCH_2COOH$

(1) Colourless solid, soluble in water, alcohol and ether.
(2) Reflux together 1.6gm of compound with 2.1gm furfural, 300mg ammonium acetate, 5ml toluene and 3ml piperidine for 48hrs. Cool, remove the solvent under reduced pressure and collect 3-(2 furyl) acrylonitrile, b.p 97°C.
(3) To 3.4gm of cyanoacetic acid in 20ml ether add in portion with stirring 8.64 gm of PCl_5. After addition is complete continue stirring for 30min and then distill off the solvent, cool in ice and collect the red oily residue of acid chloride. Add this acid chloride dropwise to a solution of 3gm of butyl alcohol 5gm of dimethylaniline in 6ml ether. Reflux for 2hrs and then stir for 15hrs. Dilute with 10ml water. Remove the ethereal layer, extract the aqueous layer with ether. Combine the organic layers, distill off the solvent and collect t-butyl cyanoacetate at b.p 54-56°C/ .3mmHg.

(4) On acid hydrolysis the product is malonic acid, m.p 133°C(d).

(5) Heat 2gm of compound with 5ml butaldehyde and 200mg of piperidine for 4hrs. Cool and extract with sodium carbonate. On acidifying the alkali solution α-cyano –n-hexenoic acid is precipitated, m.p 102°C.

Aromatic Liquids

b.p 190°C **Benzonitrile** (Phenyl cyanide) 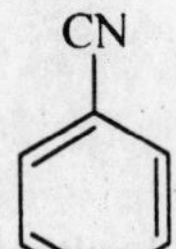

(1) Colourless oily liquid with strong bitter almond like odour, insoluble in water, soluble in alcohol and ether.

(2) Dissolve 500mg of KOH in 10ml methyl cellosolve(or any high boiling alcohol) and add 5.5gm of dicyandiamide and 5gm benzonitrile. Stir and heat the mixture(exothermic reaction). After the reaction has subsided reflux for 5hrs, cool and collect benzoguanamine, m.p 227°C.

(3) Boil 2.5ml of compound with 40ml of 10% NaOH solution for 45 min. Cool, acidify with dil HCl and recover benzoic acid, m.p 121°C.

(4) Heat together 3ml of aniline with 3.2ml benzonitrile and 2.4 gm $AlCl_3$ at 200°C for 30 min. Pour the molten mass onto a mixture of 1ml conc HCl and 160ml water. Filter, and to the filtrate add dil NaOH solution until alkaline, cool and collect N-Phenyl benzamidine, m.p 115°C.

(5) To 1ml of compound add 5ml nitrating mixture (2.5ml conc H_2SO_4+2.5ml conc HNO_3), heat on water bath. Cool and collect the m-nitro derivative, m.p 117°C.

(6) Dissolve 1.3gm of compound in 6ml alcohol and 2ml dil NaOH. Stir and add dropwise 2.5 ml H_2O_2 (temp below 50°C). Then heat the mixture at 50°C for 3hrs. Add dil H_2SO_4 until neutral, cool and recover benzamide, m.p 128°C

(7) Pass H_2S gas into a solution of 5gm benzonitrile in 5ml pyridine and 4gm triethylamine for 2-4hrs. Cool, pour in ice water, filter the precipitated thiobenzamide, m.p 116°C.

(8) Cool a mixture of 5ml of benzonitrile and 5ml absolute ethanol

saturated with HCl gas to 0°C. Keep the mixture for several days, evaporate the solvent under reduced pressure. Cool the residue, benimino ethyl ether hydrochloride crystallise out, m.p 119-120°C.

b.p 204°C **o-Tolunitrile** (2-methyl benzene carbonitrile)

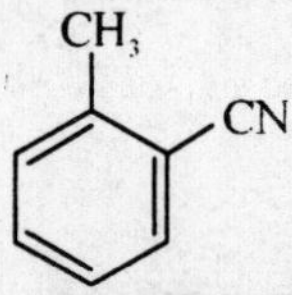

(1) Liquid with nitrobenzene like odour, insoluble in water, soluble in alcohol.

(2) Cool a mixture of 33.5gm of tetramethylene sulfone and 7.31gm of nitronium tetrafluoroborate to 10°C, add dropwise with stirring 6gm of o-tolunitrile. After the addition is complete continue stirring for another 15min and then heat at 100-115°C for 1hr. Cool, pour in ice and collect 3, 5-dinitro-o-tolunitrile, recrystallise from methanol, m.p 83°C.

(3) Reflux together 1gm of compound with 15ml dil NaOH and 3ml alcohol for 90 min. Cool, acidify with dil HCl and collect o-Toluic acid, m.p 103°C.

(4) Dissolve 1.3gm of compound in 6ml alcohol and 2ml dil NaOH. Stir and add dropwise 2.5 ml H_2O_2 (temp below 50°C). Then heat the mixture at 50°C for 3hrs. Add dil H_2SO_4 until neutral, cool and recover o-toluamide, m.p 141°C

(5) To 1ml of compound add 5ml nitrating mixture (2.5ml conc H_2SO_4+2.5ml conc HNO_3) heat on water bath. Cool and collect the nitro derivative, m.p 105°C.

b.p 214°C **m-Tolunitrilc** (m-methyl benzonitrile)

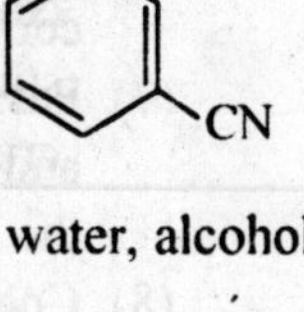

(1) Liquid, insoluble in water, sparingly soluble in hot water, alcohol and ether.

(2) Dissolve 1.3gm of compound in 6ml alcohol and 2ml dil NaOH. Stir and add dropwise 2.5 ml H_2O_2 (temp below 50°C). Continue the stirring process for 3hrs. Add dil H_2SO_4 until neutral, cool and recover m-toluamide, m.p 97°C

(3) Reflux together 1gm of compound with 15ml dil NaOH and 3ml alcohol for 90 min. Cool, acidify with dil HCl and collect m-Toluic acid, m.p 111°C.

b.p 227°C **p-Propylbenzonitrile**

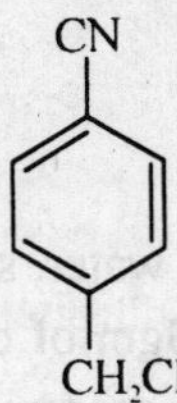

(1) On acid or alkaline hydrolysis the products are propylbenzoic acid and ammonia.

b.p 234°C **Phenylacetonitrile** (Benzyl cyanide)

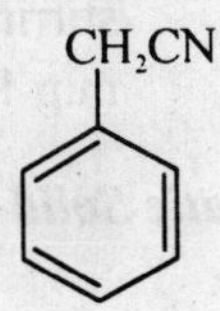

(1) Colourless pungent smelling liquid, insoluble in water, soluble in alcohol and ether.

(2) Reflux together 5ml benzyl cyanide with 5ml water, 5ml conc H_2SO_4 and 5ml glacial acetic acid for 60min. Cool and pour with stirring in 15ml cold water. Collect the preipitated phenylacetic acid, m.p 77°C.

(3) Heat 5ml of benzyl cyanide with 20ml conc HCl on water bath at 40°C for 60 min. Cool and add 15ml ice water Collect phenylacetamide, m.p 155°C.

(4) To a cooled 5.5ml nitrating mixture(2.75ml conc H_2SO_4+2.75ml conc HNO_3) add with stirring 1ml benzyl cyanide. Continue stirring at room temperature for 60min. Cool and pour in 20gm crushed ice and collect p-nitro benzylcyanide, m.p 117°C.

(5) Reflux together 7.4ml benzylcyanide with 12ml absolute ethanol and 6.8ml conc HCl for 8 hrs. Cool and pour ice water. Separate the upper layer and dissolve it in ether. Distill off ether and collect ethyl phenyl acetate at 228°C.

b.p 255°C, m.p 20°C **Trans Cinnamonitrile** (styryl cyanide)

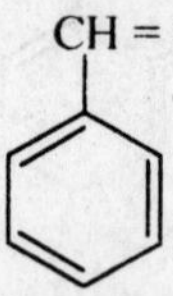

(1) Insoluble in water, soluble in alcohol.
(2) Dissolve 1.3gm of compound in 6ml alcohol and 2ml dil NaOH. Stir and add dropwise 2.5 ml H_2O_2 (temp below 50°C). Continue the stirring process for 3hrs. Add dil H_2SO_4 until neutral, cool and recover cinnamamide, m.p 147°C.
(3) Reflux together 5ml of compound with 5ml water, 5ml conc H_2SO_4 and 5ml glacial acetic acid for 60min. Cool and pour with stirring in 15ml cold water. Collect the preipitated cinnamic acid, m.p 132°C.

Aromatic Solid

m.p 29°C, b.p 217°C **p-Tolunitrile**

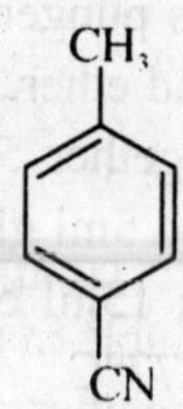

(1) Colourless crystalline compound with strong bitter almond like odour, insoluble in water, soluble in ether and alcohol.
(2) Reflux together 1gm of compound with 15ml dil NaOH and 3ml alcohol for 90 min. Cool, acidify with dil HCl and collect p-Toluic acid, m.p 180°C.
(3) Dissolve 1.3gm of compound in 6ml alcohol and 2ml dil NaOH. Stir and add dropwise 2.5 ml H_2O_2 (temp below 50°C). Stir the

resultant mixture. Add dil H_2SO_4 until neutral, cool and recover p-toluamide, m.p 158°C.

m.p 34°C, b.p 296°C **1-Naphthonitrile** (1-naphthalene carbonitrile)

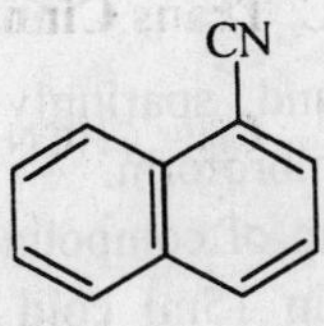

(1) Crystalline compound with aromatic odour similar to bromobenzene, insoluble in water, easily soluble in alcohol and ether.

(2) Reflux together 500mg of compound with 1.5gm of KOH and 15ml alcohol for 30 min. Evaporate to dryness and treat the residue with 10ml water, filter. Recrystallise from aqueous alcohol and collect 1-naphthionamide, m.p 202°C.

(3) Heat to 115°C 1gm of naphthonitrile with 2ml of glacial acetic acid, 2ml water and 2ml conc H_2SO_4 for 90 minutes. Cool and pour in 15ml ice water. Collect 1-naphthoic acid, m.p 161°C.

m.p 66°C **2-Naphthonitrile** (2-naphthalene carbonitrile)

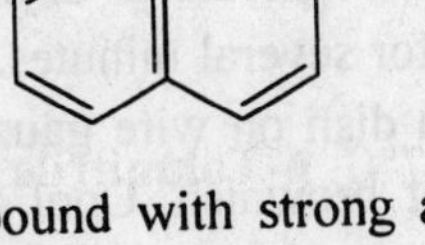

(1) Crystalline compound with strong aromatic odour, insoluble in water, soluble in alcohol and ether.

(2) Reflux together 500mg of compound with 1.5gm of KOH and 15ml alcohol for 60min. Evaporate to dryness, dissolve the residue in 15ml water, warm on water bath and filter. Acidify the filtrate with dil HCl, cool and collect 2-Naphthoic acid, m.p 184°C.

(3) Saturate 7.6gm stannous chloride and 40ml ether with HCl gas and then add a solution of 3.1gm of compound in 20ml ether to it. Stir for 60 min and then leave the mixture overnight. Decant the ether and collect aldimine-stannichloride. Heat the solid on oil

bath and collect 2-naphthaldehyde at b.p 156-158°C/15mmHg which on cooling soon solidifies, m.p 58°C.

m.p 141°C **Phthalonitrile** 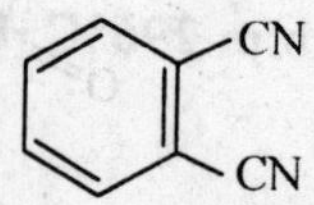

(1) Crystalline compound, sparingly soluble in water, soluble in alcohol, ether and chloroform.

(2) Reflux together 1gm of compound with 15ml of HCl. Cool and pour with stirring in 15ml cold water Collect the preipitated phthalic acid, m.p 206-208°C.

(xvii) Purines

Purines are a group of cyclic diureides consisting of structures in which the pyrimidine ring is fused to an imidazole ring.

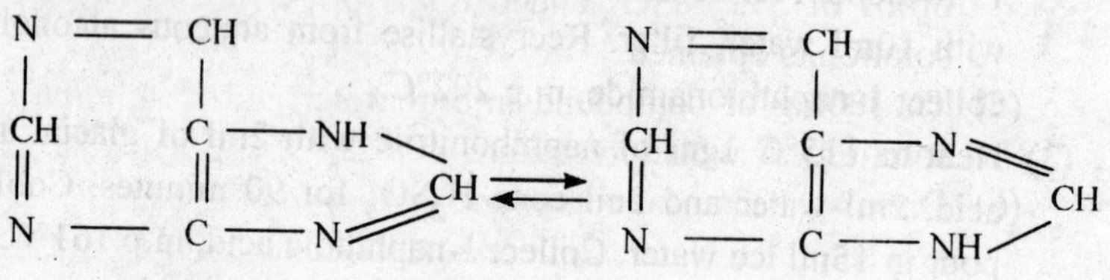

Identification test

(i) **Murexide Test :-** (Method I)To few milligrams of compound in a porcelain dish add few crystals of potassium chlorate and 2-3 drops of dil HCl. Evaporate to dryness on water bath and then continue heating for several minutes. Remove the water bath and heat the porcelain dish on wire gauze until the residue becomes slightly pinkish or brownish. Cool and add 2-3 drops of dilute NH_4OH solution. A purple red coloured solution is obtained. (Method II) :- To 100mg of compound in a porcelain dish add 2.5ml bromine water solution and evaporate it to dryness on water bath. Add 2-3 drops of NH_4OH solution. A purple coloured solution is obtained.

m.p 235°C **Caffeine** (1, 3, 7-trimethyl xanthine, 1, 3, 7-trimethyl 2, 6-dioxy purine)

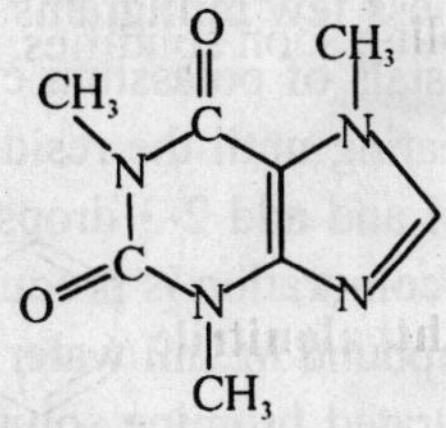

(1) White odourless solid with slight bitter taste, slightly soluble in water, alcohol and ether. Sublimes at 180°C.

(2) Respond to murexide test

(3) To 100mg of compound add 1ml potassium ferricyanide solution and conc HNO_3. A prussian blue colour is obtained on warming.

(4) To 200mg of compound in 2ml water add 2ml conc HCl and boil. Add 15ml of saturated bromine solution and boil again until the solution is colourless. Dilute the concentrated solution with water to make upto the orignal volume. To 4ml of this solution add 2 drops of 5% $FeSO_4$ and 6 drops of NH_4OH. A blue coloured solution is obtained.

(5) To 100mg of compound in 5ml water add I_2 solution (I_2 in KI solution). A white precipitate, m.p 215°C is obtained.

(6) Dissolve 100mg of compound in 10ml warm water. Cool and add 4ml saturated $HgCl_2$ solution and shake thoroughly. White crystals of mercuric salt are precipitated, m.p 243°C.

(7) Dissolve 300mg of caffeine, 225mg salicylic acid in 10ml benzene. Heat the mixture on water bath until benzene begins to boil. Add few drops of pet-ether so as to create cloudiness. Cool and collect crystals of caffeine salicylate, m.p 137°C.

m.p 270-272°C **Theophylline** (1, 3-dimethyl xanthine, 1, 3-dimethyl 2, 6-trioxy purine, Theocin)

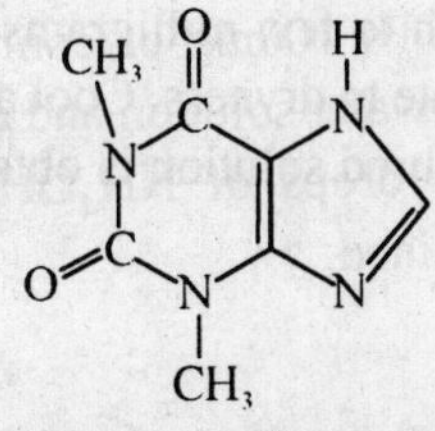

(1) White crystalline solid, sparingly soluble in water, alcohol and ether. Soluble in hot alcohol.

(2) Evaporate to dryness few milligrams of compound in a porcelain dish with 2-3 crystals of potassium chlorate and 2-3 drops of dil HCl. Continue heating until the residue becomes slightly pinkish or brownish. Cool and add 2-3 drops of dilute NH_4OH solution. An intense violet colouration is produced.

(3) To 200mg of compound in 2ml water add 2ml conc HCl and boil. Add 15ml of saturated bromine solution and boil again until the solution is colourless. Dilute the concentrated solution with water to make upto the orignal volume. To 4ml of this solution add 2 drops of 5% $FeSO_4$ and 6 drops of NH_4OH. A blue coloured solution is obtained.

(4) Acetate m.p 158°C (5) Benzoate m.p 202°C.

m.p 337°C **Theobromine** (3, 7-dimethyl xanthine)

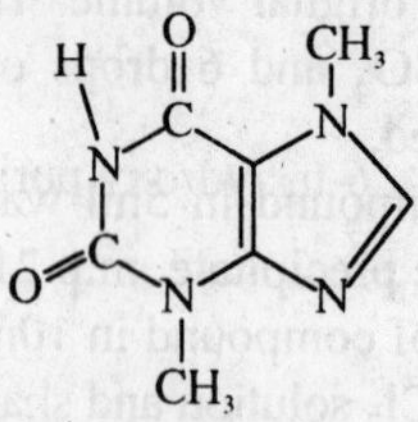

(1) White crystalline solid with slightly bitter taste, insoluble in water, alcohol and ether, soluble in mineral acid or alkali. Sublimes at 290°C

(2) To 200mg of compound in 2ml water add 2ml conc HCl and boil. Add 15ml of saturated bromine solution and boil again until the solution is colourless. Dilute the concentrated solution with water to make upto the orignal volume. To 4ml of this solution add 2 drops of 5% $FeSO_4$ and 6 drops of NH_4OH. A blue coloured solution is obtained.

(3) In a porcelain dish to few milligrams of compound add 3-5 drops of HNO_3, evaporate to dryness. Cool and add 3-4 drops of NH_4OH. A purple red coloured solution is obtained.

m.p 360-365°C **Adenine** (6-amino purine)

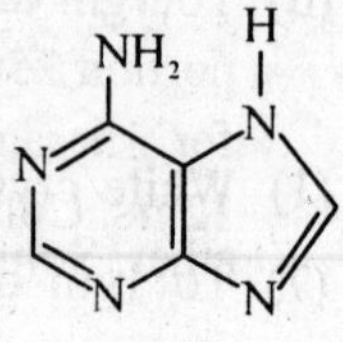

(1) Crystalline compound, insoluble in water, ether and chloroform. Soluble in hot alcohol, alkali and mineral acid. Sublimes at 260°C.
(2) Does not give murexide test.
(3) Heat 100mg of compound with 1ml conc H_2SO_4 and Zn on water bath for 30 min. A purple colour develops and soon disappears. Add 2ml water and filter. To the filtrate add excess NaOH solution. A ruby red coloured solution is obtained.
(4) To few milligrams of compound add $FeCl_3$ solution. A red coloured solution is obtained.
(5) Benzoate (mono) m.p 234°c (6) Acetate (di) m.p 195°C. (7) Picrate m.p 280°C(d).

(d) **Uric acid** (2, 4, 6-trihydroxy purine) 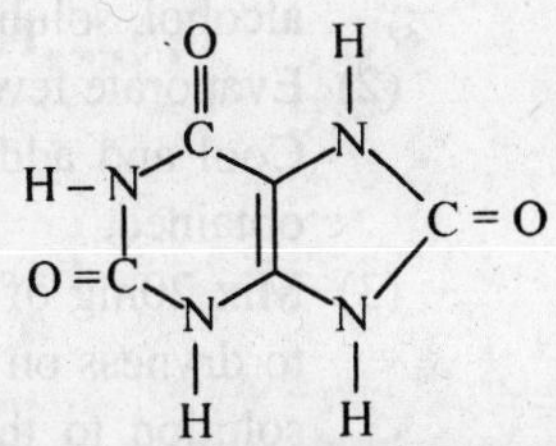

(1) White crystalline powder, sparingly soluble in water, insoluble in alcohol and ether, soluble in NaOH.
(2) Reduces Fehling's solution on heating.
(3) Evaporate uric acid with dil HNO_3 to dryness. Cool and add 2-3 drops of NH_4OH. A blue colouration is obtained.
(4) Dissolve 50mg of compound in 3ml Na_2CO_3. Add a drop of this solution to the filter paper that is moistened with $AgNO_3$ solution. An immediate black or grey stain is produced.
(5) To 1gm of uric acid add water to form paste and then add NaOH solution (200mg dissolved in 1ml water). Cool in ice bath and add in portion 50ml $KMnO_4$ solution(800mg). After the addtion is complete pass SO_2 gas. Concentrate the mixture to half its volume and leave for 2days. Cool and collect allantoin, m.p 229°C.
(6) To 1gm uric acid add 1.8ml conc HCl, heat to 35°C and add in portion 250mg of potassium chlorate. Add about 4ml water, keep for 1hr and filter. Saturate the filtrate with H_2S and leave for 12hrs. Collect colourless crystals of alloxantin, m.p 234°C.
(7) To 1gm of alloxantin add 1ml conc HNO_3 and 1.2ml fuming

HNO_3. Leave for 2 days. Cool and collect the precipitated alloxan, m.p 240-245°C (d).

m.p unchanged at 360°C **Xanthine** (2, 6-dioxypurine)

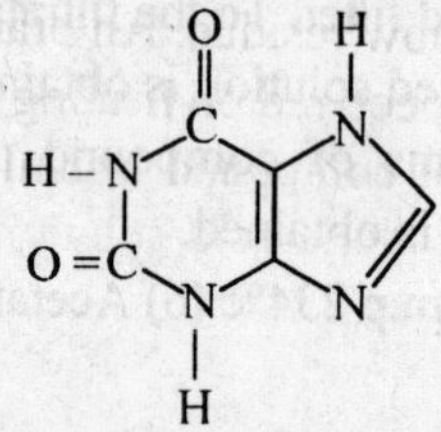

(1) Crystalline, tasteless, odourless compound, insoluble in water and alcohol, soluble in alkali.
(2) Evaporate few milligrams of compound with dil HNO_3 to dryness. Cool and add 2-3 drops of NH_4OH. A violet red colouration is obtained.
(3) Mix 20mg of compound with 4-5 drops of HNO_3 and evaporate to dryness on water bath. Cool and add 2-3 drops of 10% NaOH solution to the yellow residue. A yellow orange colouration is produced which changes to violet red on heating on steam bath.
(4) Dissolve 500mg of compound in 5ml of NH_4OH and 5ml water. To this alkaline solution add $AgNO_3$ solution. White precipitate is obtained.
(5) Dissolve 100mg of compound in 5ml hot water and then add 2-3ml of 5% $HgCl_2$ solution. Cool and recover crystals of mercury salt.

m.p 360°C (d) **Guanine** (2-amino-6-oxy purine)

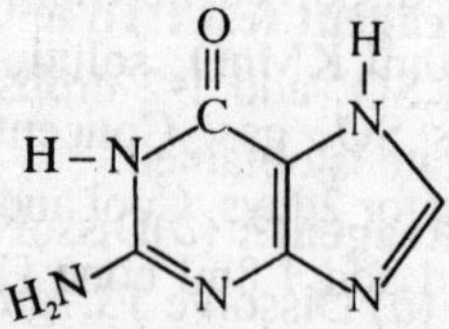

(1) Amorphous powder, odourless, insoluble in water, alcohol and ether, soluble in alkali and mineral acid.
(2) Evaporate few milligrams of compound with dil HNO_3 to dryness.

Cool and add 2-3 drops of NH_4OH. A violet red colouration is obtained.

(3) Mix 20mg of compound with 4-5 drops of HNO_3 and evaporate to dryness on water bath. Cool and add 2-3 drops of 10% NaOH solution to the yellow residue. An orange colouration is produced which changes to deep red on heating on steam bath.

(4) Dissolve 100mg of compound in 2ml conc HCl and 15ml water, add 8ml saturated picric acid solution. Heat on water bath. Cool and collect picrate, m.p 170°C.

(*xviii*) *Alkaloids*

Alkaloids are a class of organic compounds of plant origin which are basic in nature. They posses one or more nitrogen heterocyclic rings as their structural unit. Except for few exception alkaloids are usually colourless, crystalline compounds insoluble in water but soluble in organic solvents. Most of them are optically active, bitter in taste and form insoluble precipitates with picric acid, potassium mercuric iodide, phosphomolybdic acid reagent etc.

Identification tests

(*i*) **Mayer's Reagent test** : Dissolve few milligrams of alkaloid in 1-2ml dil HCl and 2-3 drops of mayer's reagent. White or yellow crystalline precipitate is produced.

Mayer's Reagent: Dissolve 2.7gm of mercuric chloride and 9.96 gm of potassium iodide in 200 ml water.

(*ii*) **Wagner's Reagent test**: Dissolve few milligrams of compound in conc H_2SO_4 and add 2-3 drops of reagent. A brown coloured precipitate is obtained.

Wagner's Reagent: Dissolve 2 parts iodine and 5 parts of potassium iodide in 100ml water.

(*iii*) **Dragendorff's reagent test** : To few milligrams of alkaloid in 1-2ml conc HCl/H_2SO_4 add 1-2 drops of reagent when an orange coloured precipitate is obtained.

Dragendorff's Reagent : (*a*)Dissolve 4gm of bismuth nitrate in 10ml 5N HNO_3. (*b*) Dissolve 13.7gm of KI in 20ml water. Mix **a** and **b** solutions and leave for a day. Collect crystals of KNO_3 and make the volume to 50ml with water.

(*iv*) **Marme's reagent test** : To few milligrams of compound in 1-2ml conc H_2SO_4 add 1-2 drops of reagent. White precipitate is obtained.

Marme's Reagent: To concentrated boiling solution of potassium iodide add cadium iodide solution, followed by equal volume of cold potassium iodide solution.

(*v*) **Picric acid test:** To few milligrams of alkaloid in 1-2ml dil HCl add saturated picric acid solution. Yellow crystalline precipitate is obtained.

Picric Acid : Dissolve 2gm picric acid in 50ml water.

In addition to the above tests alkaloids give some colour tests with different reagents like :-

(*i*) Conc sulphuric acid. (2-3 drops)

(*ii*) Conc Nitric acid (2-3drops)

(*iii*) 10% Sodium hydroxide solution.

(*iv*) Marquis reagent :- add 4 drops of 40% formadehyde solution to 6ml conc sulphuric acid.

(*v*) Erdmann's reagent:- Add 2 drops of conc HNO_3 to 50ml water. Add 10drops of this mixture to 11ml conc H_2SO_4

(*vi*) Frohde's reagent:-Dissolve 100mg of sodium molybdate in 10ml conc H_2SO_4 by gentle heating.

(*vii*) Buckingham's reagent:- Dissolve 100mg of ammonium molybdate in 100ml conc H_2SO_4.

Liquids

b.p 167°C **d-Coniine** (d-2-propyl piperidine)

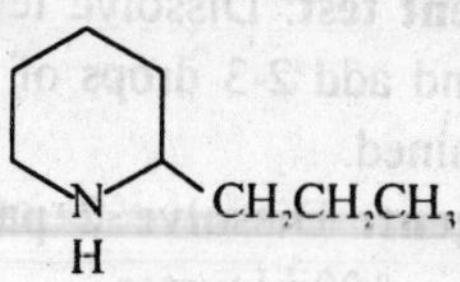

(1) Colourless oily liquid with unpleasant mouse like odour, turns brown on exposure to air, soluble in ether and ethanol.

(2) To 1ml of compound add conc HNO_3. An orange red colouration is obtained.

(3) To 1ml of compound add 1ml 50% aqueous ethanol and 2-3 drops of phenolphthalein. The solution turns pink.

(4) To 1ml of compound add 1ml conc HCl. Shake thoroughly. Add 2ml ether. Decant off the ether and recover the hydrochloride, m.p 216°C.

(5) To 1ml of compound add 5ml alcohol and 200mg of phthalic

anhydride. Heat until a clear solution is obtained. Evaporate on water bath. Dissolve the sticky residue in 7ml hot 25% alcohol solution containing 7-8 drops of conc HCl. Shake thoroughly until precipitation occurs. Cool and collect conylenephthalamic acid, m.p 154°C.

b.p 247°C **Nicotine** (1-Methyl-2-β pyridyl pyrrolidine)

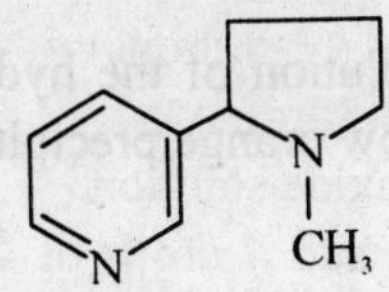

(1) Colourless oily liquid which turns brown on exposure to air with strong unpleasant odour, soluble in water and organic solvents.
(2) To 1ml of compound add conc HNO_3. Pink coluration is produced.
(3) To 1ml of compound add 1ml 50% aqueous ethanol and 2-3 drops of phenolphthalein. The solution turns pink.
(4) To 1ml of compound add 2ml conc HCl. Light violet or brown colouration on warming is produced. Add conc HNO_3, the colour changes to orange.
(5) Dissolve 1ml of compound in 5ml alcohol and add 5ml saturated aqueous picric acid solution. Heat the mixture on water bath, cool and collect picrate, m.p 219°C.

Solid

m.p 98°C **l-Cocaine** (methyl ester of benzoylecgonine)

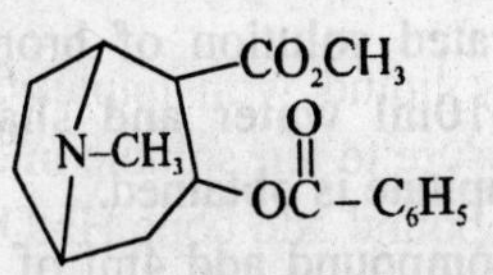

(1) Colourless crystalline odourless solid, slightly soluble in water, soluble in alcohol, ether and chloroform.
(2) To few milligrams of compound in conc H_2SO_4 add few crystals of $K_2Cr_2O_7$. A brownish –green colouration is obtained.
(3) Heat 50mg of compound with 2ml conc H_2SO_4 and 10ml alcohol. Sweet aromatic odour of ethyl benzoate will be produced.

(4) Heat 500mg of compound with 7-8ml conc H_2SO_4 on water bath for 5min. Cool and add 10ml ice water. Filter and collect the precipitated benzoic acid, m.p 121°C.
(5) Evaporate to dryness 500mg of compound with 10ml dil HCl. Cool and collect cocaine hydrochloride, m.p 197°C.
(6) To 2ml aqueous solution of the cocaine hydrochloride add 1ml of 5% CrO_3 solution. Precipitation occurs which disappears on shaking. Add 2ml HCl. Yellow precipitate of cocaine chromate appears
(7) To 2ml aqueous solution of the hydrochloride add 2-3 drops of I_2-KI reagent. Yellow orange precipitate is obtained.

m.p 115°C **dl-Atropine** (Daturine) 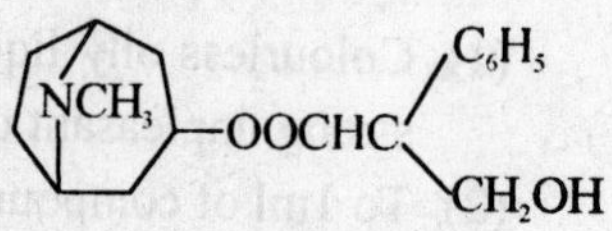

(1) Silky needles, odourless with disagreeable bitter taste, sparingly soluble in water, soluble in alcohol, ether and benzene.
(2) Dissolve 500mg of compound in 1ml conc HNO_3. Evaporate to dryness on water bath. To the residue add 3-5 drops of alcoholic potash solution (1gm KOH in 5ml alcohol). A violet red colour slowly changing to red is produced.
(3) Heat 100mg of compound until white fumes appear. Add 7ml conc H_2SO_4 and heat until the mixture begins to turn brown. Add immediately 10ml water. An intense sweet smell of honey and orange blossom will be noticed. Add a small crystal of $KMnO_4$. The odour changes to that of bitter almond.
(4) Add 5ml saturated solution of bromine in HBr to 100mg of compound in 10ml water and shake thoroughly. A yellow crystalline precipitate is obtained.
(5) To 100mg of compound add 4ml of 1% $HgCl_2$ solution in 50% alcohol. Heat the solution gently. On cooling yellow precipitate of mercuric oxide is obtained.
(6) To 100mg of compound in 5ml water add 5ml saturated aqueous picric acid solution and then add 3-5 drops of conc HCl, shake thoroughly. Yellow crystals of picrate, m.p 175°C are obtained.
(7) Atropine hydrochloride m.p 165°C.

m.p 129-130°C **Piperine** (Piperonylpiperidine)

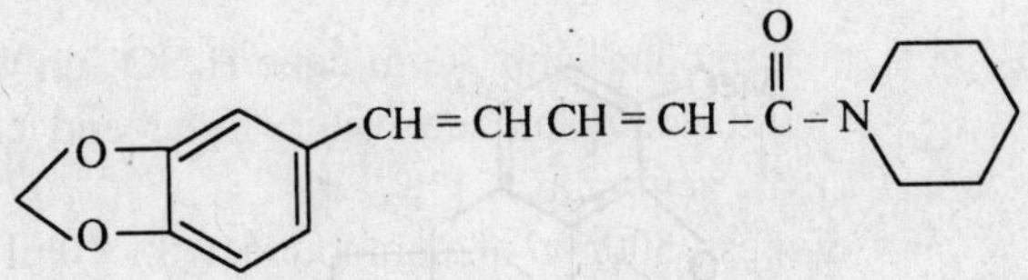

(1) Crystalline odourless compound, insoluble in water, soluble in hot alcohol, ether and chloroform.

(2) To few milligrams of compound add conc H_2SO_4. An orange coloured solution is produced which disappears on dilution.

(3) To few milligrams of compound add conc HNO_3. A greenish yellow colour is produced which slowly changes to orange.

(4) To 10mg of compound add 5-7 drops of formalin and conc H_2SO_4 mixture (1 volume formalin + 2 volume conc H_2SO_4). A greenish brown colour is produced.

m.p 147°C **Papaverine** (tetramethoxy benzylisoquinoline)

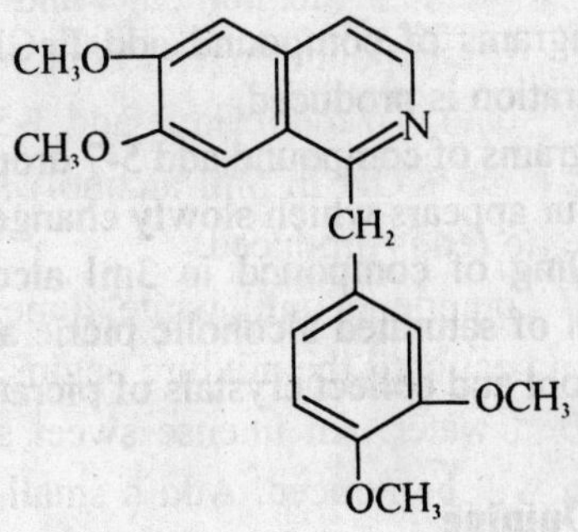

(1) Crystalline tasteless odourless solid, insoluble in water, soluble in alcohol and chloroform.

(2) To few milligrams of compound add 5-7 drops of Marquis reagent. A faint pink coloured solution is obtained which slowly changes to brown.

(3) Dissolve 100mg of compound in 3ml alcohol and add 5ml of saturated alcoholic picric acid solution. Heat on water bath, cool and collect crystals of picrate, m.p 179-181°C.

m.p155°C **Codeine** (Morphine methyl ether)

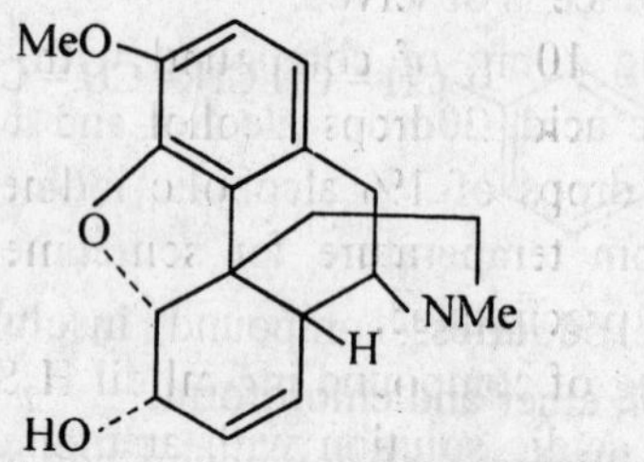

(1) **Crystalline compound with slight bitter taste, sparingly soluble in water, soluble in alcohol, ether and chloroform.**
(2) **Dissolve 1mg of compound in 5-6 drops of conc H_2SO_4. Add with stirring 10mg of sodium arsenite and heat on water bath. A red violet coloured solution is obtained which slowly changes to blue violet.**
(3) **Treat 5mg of compound with 10ml of conc H_2SO_4 and 5-6 drops of 10% aqueous sucrose solution with stirring. A violet red colour appears which changes to red on heating.**
(4) **To few milligrams of compound add conc HNO_3. A yellow colouration is produced.**
(5) **To few milligrams of compound add $FeCl_3$ solution. A yellow orange colouration is produced.**
(6) **To few milligrams of compound add 5-7 drops of Marquis reagent. A violet colour appears which slowly changes to violet red.**
(7) **Dissolve 100mg of compound in 3ml alcohol and 0.5ml conc HCl, add 5ml of saturated alcoholic picric acid solution. Heat on water bath, cool and collect crystals of picrate, m.p 195°C.**

m.p 175°C **Quinine**

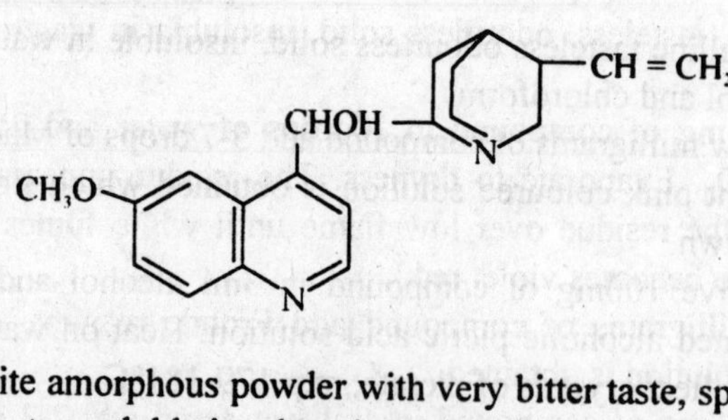

(1) **White amorphous powder with very bitter taste, sparingly soluble in water, soluble in ethanol and ether.**

(2) Dissolve 2mg of compound in 10ml dil H_2SO_4. A pronounced bluish fluorescence is observed.

(3) Heat to boiling 10mg of compound with 20drops of mixture (30drops acetic acid, 20drops alcohol and 1 drop conc H_2SO_4). Then add 2-3 drops of 1% alcoholic iodine solution. Keep the mixture at room temperature for sometime. Green crystals of herapathite are precipitated.

(4) Dissolve 100mg of compound in 1ml dil H_2SO_4 and 10ml water. Neutralise the acidic solution with ammonium hydroxide. Now add 1-2 drops of 3% H_2O_2, 1-2 drops of $CuSO_4$ and boil. Solution first becomes red, changes to blue and finally to yellowish green.

(5) Dissolve 20mg of compound and 100mg of bleaching powder in 10ml water. Add first 5-6 drops of HCl followed by 10ml of ammonium hydroxide. An emerald green coloured solution is obtained.

(6) Dissolve 100mg of compound in 5ml water containing 5-6 drops of conc HCl. Add 5ml of saturated picric acid solution. Shake thoroughly, warm on water bath, cool and collect picrate, m.p 126°C.

m.p 176°C **Narcotine (Opianine)**

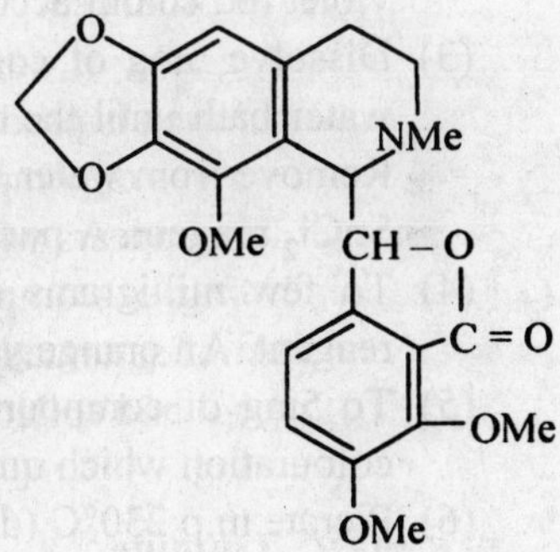

(1) Crystalline, tasteless, odourless solid, insoluble in water, soluble in alcohol.

(2) Dissolve 5mg of compound in 10drops of water and 2drops of conc H_2SO_4. Evaporate to dryness. The residue appears orange red. Heat the residue over low flame until white fumes appear. The residue becomes violet red.

(3) To few milligrams of compound add Frohde reagent. A green coloured solution is obtained.

(4) Dissolve 20mg of compound in 4ml water and add 2-3 drops of conc HCl and 2-3 drops of bromine water. A yellow coloured

solution results. Heat to boiling and add 8-10 drops of bromine water. An orange coloured solution results.

(5) Dissolve 100mg of compound in 5ml water containing 5-6 drops of conc HCl. Add 5ml of saturated picric acid solution. Shake thoroughly, warm on water bath, cool and collect picrate, m.p 141°C.

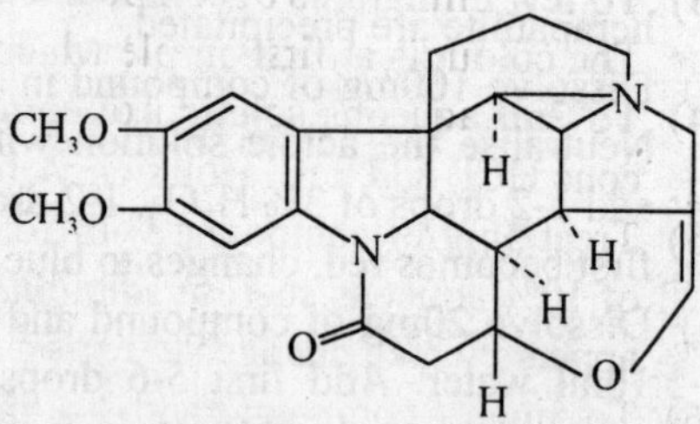

m.p 178°C **Brucine**

(1) Crystalline odourless solid with bitter taste, sparingly soluble in water and ether, soluble in alcohol and chloroform.
(2) Dissolve 5mg of compound in 10drops of conc HNO_3. An orange coloured solution is obtained. Add 20drops of saturated sodium thiosulphate solution and heat to 40-50°C. Within 3-4minutes a violet red colour accompanied by yellow precipitate develops.
(3) Dissolve 5mg of compound in 5drops of conc HNO_3. Heat on water bath until the orange red colour changes to orange yellow. Remove from water bath and add 10-15 drops of freshly prepared $SnCl_2$ reagent. A purple coloured solution is obtained.
(4) To few milligrams of compound add 4-5 drops of Erdmann's reagent. An orange yellow coloured solution is obtained.
(5) To 5mg of compound add 2-3 drops of Froehde's reagent. Red colouration which quickly changes to red brown.
(6) Picrate m.p 230°C (darkens).

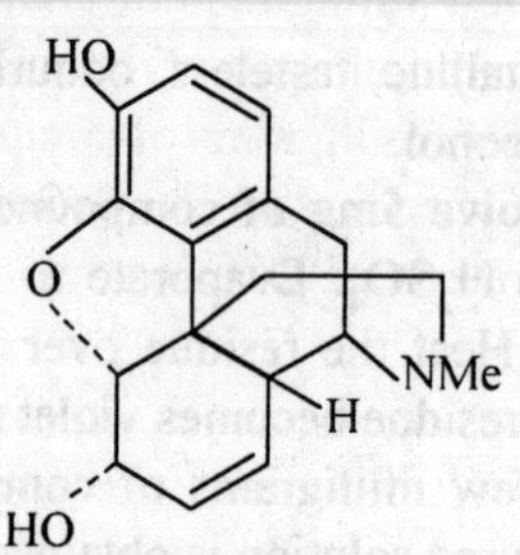

m.p 254°C (d) **Morphine**

(1) White crystalline odourless solid with bitter taste, sparingly soluble in water and ether, soluble in sodium hydroxide.
(2) Dissolve few milligrams of compound in conc HNO_3. At first the colour of solution is orange which changes to yellow on warming.
(3) To few milligrams of compound add 2-3 drops of marquis reagent. The colour is at first purple which soon changes to blue.
(4) To 2ml aqueous solution of compound (containing few drops of conc HCl) add $FeCl_3$ solution. A blue coloured solution is obtained.
(5) Treat 5mg of compound with 10ml of conc H_2SO_4 and 5-6 drops of 10% aqueous sucrose solution with stirring. A violet red colour appears which changes to red on heating.
(6) To 5mg of compound add 3-4 drops of Froehde's reagent. First a violet colouration is obtained which changes to broken green after 10minutes.
(7) Morphine benzyl ether m.p 131°C (8) Morphine ethyl ether m.p 93°C. (9) Morphine hydrochloride m.p 200°C.

m.p 264°C(d) **Cinchonine**

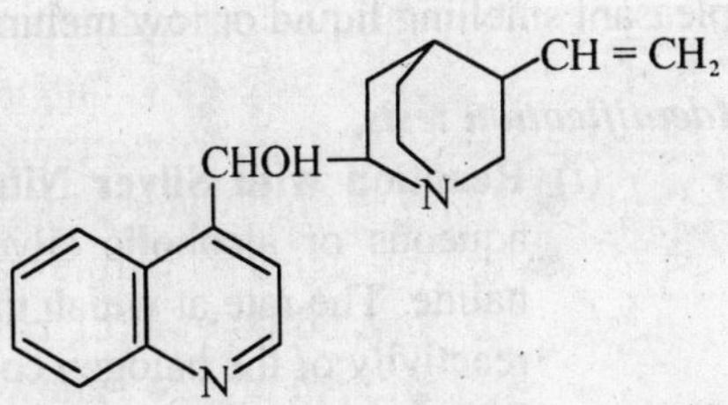

(1) White powder or crystalline solid insoluble in water and pet ether, soluble in chloroform.
(2) To few milligrams of compound add conc H_2SO_4. A pale green colouration is observed.
(3) To few milligrams of compound add conc H_2SO_4 and few crystals of $K_2Cr_2O_7$. An olive green colouration is obtained.
(4) Picrate m.p193°C

m.p 268°C **Strychnine**

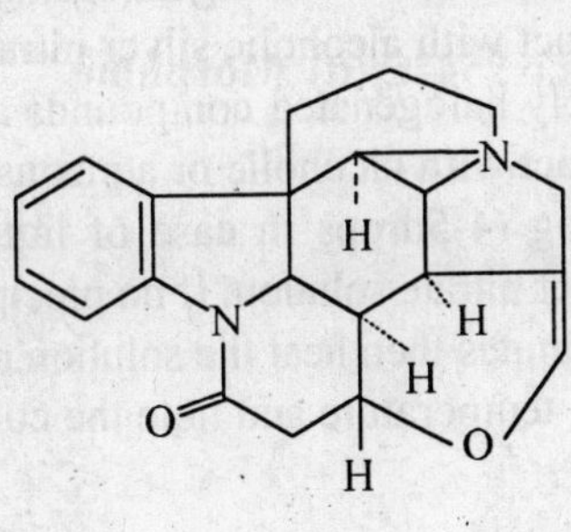

(1) Colourless crystalline odourless solid with bitter taste, insoluble in water, sparingly soluble in ethanol, soluble in chloroform.

(2) Dissolve 5mg of compound in 1-2ml conc H_2SO_4 and then add few crystals of MnO_2. First the colour of solution is violet, then violet red and finally orange.

(3) Warm 5mg of compound with 1-2ml conc HNO_3 on water bath, add with stirring 10mg of $KClO_3$. A red orange colour is produced. Add 5-6 drops of NH_4OH, the colour changes to yellow orange.

(4) Dissolve 100mg of compound in 5ml water containing 5-6 drops of conc H_2SO_4. Add 5ml of saturated picric acid solution and heat the mixture on water bath. Cool and collect the picrate, m.p 275-286°C.

(*d*) Compounds containing carbon, hydrogen and halogens (with or without the presence of nitrogen and oxygen)

Halogenated Compounds

Halogenated compounds are those in which the halogen atom (Cl, Br, I and F) is attached to the alkyl or aryl group. They are mostly colourless, pleasant smelling liquid or low melting solid

Identification tests

(*i*) **Reaction with Silver Nitrate:** Alkyl or aryl halides react with aqueous or alcoholic silver nitrate to produce insoluble silver halide. The rate at which this reaction occur is the measure of the reactivity of the halogen compound

$$RX + AgNO_3 \rightarrow AgX + RONO_2$$

(a) Water soluble halogen compound such as acyl halide of low molecular weight react immediately even with aqueous silver nitrate.

(b) Acyl halides, alkyl iodides and α halo ethers react with alcoholic silver nitrate at room temperature.

(c) Primary and secondary alkyl chlorides, dibromides and aromatic compound containing halogen in side chain or activated aryl halides react with alcoholic silver nitrate only on heating.

(d) Poly halogenated compounds and unactivated aryl halides do not react with alcoholic or aqueous silver nitrate even on heating.

To 100mg (4-5drops in case of liquid) of compound add 4ml of 2% alcoholic silver nitrate solution. If no precipitate is formed at room temperature within 2-3 minutes then heat the solution on boiling water bath for 5 minutes. Cool at room temperature and note the colour of precipitate.

Note:- Some carboxylic acid also give precipitate with silver nitrate which is soluble in nitric acid.

(*ii*) **Reaction with potassium hydroxide :** Reflux 1ml of compound with 7-8 ml of .5 M alcoholic potassium hydroxide solution for 10-15 minutes. Dilute with water and acidify with dilute nitric acid. Add few drops of silver nitrate solution. A white or yellow precipitate of silver halide indicate that the halogen present is reactive or loosely bound(eg benzyl halides)

Alcoholic Potassium Hydroxide: Dissolve 8gm KOH in 250ml ethanol.

(*iii*) **Reaction with Sodium iodide :** To 200mg of compound add 4ml of sodium iodide reagent. Shake thoroughly and then keep the mixture at room temperature for 4-5 minutes. Appearance of reddish brown colour or precipitate is a positive test. However if no change occurs heat the mixture on water bath at 50°C for 5-6 minutes. Cool at room temperature and allow precipitation to take place.

$RBr(Cl) + NaI \rightarrow RI + NaBr(Cl)$

(a) If precipitation occurs at room temperature (25°C) within 3 minutes then it indicates the presence of primary bromides.

(b) If precipitation occur only on heating to 50°C for 5-6 minutes then it may be either primary chlorides, secondary or tertiary bromide.

(c) No precipitation occur even on heating then it may be either tertiary chlorides or polychloro compounds.

(d) Iodine is liberated in addition to precipitation then it may be 1, 2 dichloro or dibromo compounds.

***Sodium Iodide*:-** Dissolve 15gm of sodium iodide in 100ml acetone. A colourless solution is obtained.

Derivatives :

(1) **S-Alkylisothiouronium picrates :** To 1gm thiourea add 1gm of alkyl halide and 10ml ethanol. Reflux the mixture (primary alkyl bromides and iodides for 15-20 minutes, secondary alkyl bromide or iodide for 2-3 hrs, alkyl chlorides for 3-5 hrs and dihalides for 30-50min). Remove the heat source and add 3ml of saturated picric acid solution. Boil the mixture for 3-5 minutes or until a clear solution is obtained. Cool in ice and recover the crystals, determine its melting point.

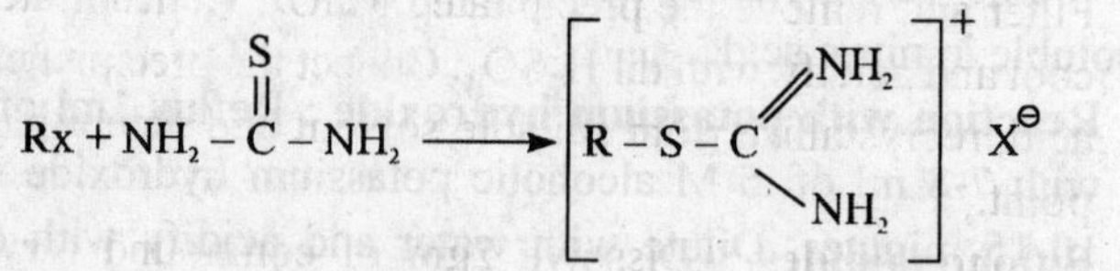

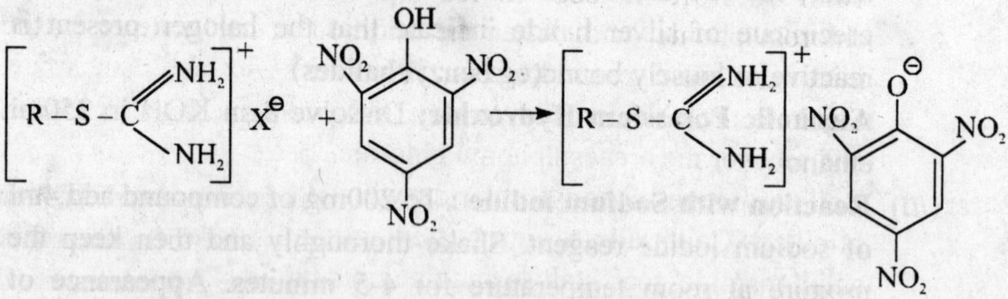

(2) **2-Naphthyl alkyl ethers and its picrates :** To 2ml of compound add 4gm of 2-naphthol, 2gm of KOH and 20ml ethanol. Reflux the mixture for 15-20 minutes. Dilute with water and cool. Collect the precipitated solid, recrystallise and determine its melting point.

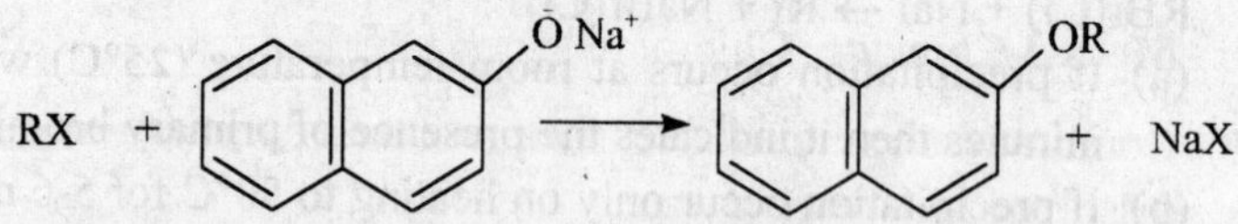

Dissolve the precipitated 2-naphthyl ether in minimum amount of alcohol and then add saturated picric acid solution. Heat the solution on water bath. Cool and collect the precipitated picrate, recrystallise from alcohol and determine its melting point.

(3) **Nitration (for aromatic compounds) :** Dissolve 500mg of compound in 2ml conc H_2SO_4 and then add dropwise 2ml of fuming HNO_3. Heat the mixture on water bath for 10-15 minutes and then pour on crushed ice. Collect the precipitated nitro derivative, wash with cold water and recrystallise.

(4) **Picrates :** To 2ml saturated solution of picric acid add 2ml solution of halogenated hydrocarbon (500mg in 2ml ethanol). Shake the mixture thoroughly and then heat on water bath for few minutes. Cool, filter the precipitated coloured solid, recrystallise and determine its melting point.

(5) **Oxidation of side chain:** By Potassium permanganate :- In a 100ml round bottom flask dissolve 3gm of $KMnO_4$ in 10ml of 5% NaOH solution. Add 1gm of unknown sample and reflux for 2-3hrs or until the purple colour of permanganate disappears.

Filter and remove the precipitated MnO_2. Concentrate the filtrate, cool and acidify with dil H_2SO_4. Collect the precipitated carboxylic acid, recrystallise from suitable solvent and determine its melting point.

(6) **Sulphonamides :** Dissolve 2gm of compound (aryl halide) in 10ml chloroform, cool in ice and add dropwise 7-10ml of chlorosulphonic acid. After the addition is complete keep the mixture at room temperature for 30 minutes and then pour in ice. Separate the chlorform layer, dry, evaporate off the chloroform and recover the arenesulphonyl chloride.

Boil the arenesulphonyl chloride with 10ml NH_4OH for 10-15 minutes. Cool, dilute with water and collect the precipitated sulphonamide, recrystallise and determine its melting point.

(i) Halogenated Hydrocarbons

Aliphatic Liquids

Methyl Chloride (Chloromethane) CH_3Cl

(1) Combustible gas, available in form of aqueous solution(1 volume dissolved in 4 volume water).

(2) S-Methyl thiouronium picrate m.p 224°C (3) Methyl 2-naphthyl ether m.p 72°C

b.p 12°C **Ethyl Chloride** (chloroethane) CH_3CH_2Cl

(1) Combustible liquid, very little soluble in water.

(2) S-Ethyl thiouronium picrate m.p 188°C.

(3) Picrate of Ethyl 2-naphthyl ether m.p 102°C

b.p 35°C **Isopropyl chloride** (2-chloropropane) $CH_3CHClCH_3$

(1) Colourless liquid, slightly soluble in water, soluble in ether.

(2) S-isopropyl thiouronium picrate m.p 196°C (3) Picrate of isopropyl 2-naphthyl ether m.p 95°C.

b.p 38°C **Ethyl Bromide** (bromoethane) CH_3CH_2Br

(1) Colourless highly refractive liquid with characteristic odour, insoluble in water, soluble in most organic solvents.

(2) Stir 25ml of chlorobenzene and 11.5gm of Na in 100ml pet-ether for 2hrs. Add dropwise with stirring 15ml ethyl bromide and continue stirring for 1 hr. Steam distill and collect ethyl benzene at 135-136°C.

(3) To 5gm Mg add first 5ml of ethyl bromide solution (17.5ml ethyl bromide in 50ml ether). Then add dropwise with stirring the remaining solution. Heat the mixture under reflux for 30min. Cool and add 3.7ml tin tetrachloride. Reflux for 1hr and then distill off the ether. Cool, add 20ml ether, 10ml cold water and 40ml ice cold 10% HCl. Stir and separate the ethereal layer. Distill off ether and collect tetraethyl tin at 63-65°C/12mmHg. (b.p 181°C)

(4) Heat on water bath 15gm thiourea with 25gm ethyl bromide and 20ml absolute ethanol at 60°C for 3hrs. Distill off excess solvent, cool the residual oily S-ethyl thiourea hydrobromide. To cold 9.2gm hydrobromide add 25ml of 2N NaOH solution. Now add 10ml of hot glycine solution(4gm in 10ml water). Remove from ice bath and after 30min add 20ml ether and leave overnight. Cool and collect crystals of guanido acetic acid, m.p 280-284°C.

(5) S-Ethyl thiouronium picrate m.p 188°C (6) Picrate of ethyl 2-naphthyl ether m.p 102°C.

b.p 40°C **Methylene Chloride** (Dichloro methane) CH_2Cl_2

(1) Colourless liquid, insoluble in water, soluble in organic solvents.

(2) Cool 30ml dichloromethane containing 2.4ml triethylamine. Add dropwise with stirring 1.8ml thiophenol. Stir the resultant mixture at 20°C for 3 hrs. Filter to remove triethylamine hydrochloride. Distill off excess dichloromethane and cool the residue. Collect Bis(phenylthio) methane as white crystals, m.p 37°C.

(3) Alkyl 2-naphthyl ether (di) m.p 133°C.

b.p 42.4°C **Methyl Iodide** (iodomethane) CH_3I

(1) Colourless liquid with sweet smell, insoluble in water, soluble in organic solvents

(2) To 2ml of compound add alcoholic $AgNO_3$ solution. Yellow coloured precipitate of AgI is obtained.

(3) To a mixture of 5.7gm of heptylamine and 18gm of tributylamine add dropwise with stirring 20gm of methyl iodide. Cool as the reaction tends to be vigorous. Keep the reaction mixture overnight

at room temperature and then add 50ml ether. Cool and collect crystals of heptyltrimethylammonium iodide, m.p 146°C.

(4) To a solution of 13.5gm of N, N dimethylbenzylamine in 20ml absolute ethanol add dropwise with stirring 19gm of methyl iodide. Reflux for 30 min on steam bath. Cool and collect benzyl trimethyl ammonium iodide, m.p 179°C.

(5) Dissolve 6gm of quinacetophenone in 60ml acetone by heating on water bath. Cool, add 5.6gm of K_2CO_3 followed by 8.2gm of methyl iodide. Heat the mixture at 60-70°C for 6 hrs. Distill off the acetone and acidify the residual liquid with 2N H_2SO_4. Steam distill the resulting mixture. Cool the distillate in ice and keep it overnight. Collect crystals of quinacetophenone mono methyl ether, m.p 50°C

(6) S-Methyl thiouronium picrate m.p 224°C (7) Picrate of Methyl 2-naphthyl ether m.p 117°C.

b.p 45°C **Allyl chloride** (3-Chloropropene) $CH_2=CHCH_2Cl$

(1) Colourless liquid with mustard like odour, insoluble in water.

(2) To 4.1gm of Mg turning add in portion a solution of 23gm of allylchloride in 120ml ether (exothermic reaction). Stir the mixture for 5 hrs. Cool in ice and add 5% HCl until evolution of heat has ceased. Separate the ethereal layer, distill off the ether. Distill the residue under reduced pressure and collect Biallyl as colourless liquid at 59-60°C.

(3) S-Allyl thiouronium picrate m.p 154°C. (4) Picrate of allyl 2-naphthyl ether m.p 99°C.

b.p 46°C **n-Propyl chloride** (1-chloro propane) $CH_3CH_2CH_2Cl$

(1) Colourless liquid insoluble in water, soluble in alcohol and ether.

(2) S-Propyl thiouronium picrate m.p 177°C.

(3) Picrate of Propyl 2-naphthyl ether m.p 81°C

b.p 51°C **t-Butyl Chloride**

(2-methyl 2-chloro propane) $(CH_3)_3CCl$

(1) Colourless liquid, very slightly soluble in water, soluble in alcohol and ether.

(2) To 5gm anhydrous $AlCl_3$ add 20ml dry benzene, cool in ice bath. Add dropwise with stirring 6ml of t-butyl chloride (temp 0-5°C). After addition is complete continue stirring for 1hr and then add 20gm crushed ice followed by 20ml water. Steam distill the resulting mixture. Remove the hydrocarbon layer and extract with ether. Distill off the ether and collect t-butyl benzene at 165-170°C.

(3) Cool 20ml of nitromethane to 0°C and then add 9gm of $AlCl_3$, 4.5gm of malonitrile in 5ml nitromethane and 15gm of t-butyl chloride in 5ml nitromethane(temp below 10°C). Stir the reaction mixture for 10hrs at 0-10°C. Add 100ml sodium bicarbonate solution. Remove the organic layer and extract the aqueous phase with dichloromethane. Combine the two layer and distill to remove the solvent. Distill the residual oil under reduced pressure and collect t-butylmalonitrile, purify by sublimation, m.p 76-79°C.

(4) To 6gm of Mg in 20ml ether add dropwise with stirring a solution of 25gm of t-butyl chloride in 110ml ether. After the addition is complete continue stirring for 20minutes. Cool to 0°C and pass CO_2 gas for 3hrs(temp below 0°C). Hydrolyse by adding dil H_2SO_4. Remove the ether layer and extract the aqueous layer with ether. Combine the two layers and wash with NaOH. Distill, cool and concentrate, acidify with dil H_2SO_4 and collect trimethylacetic acid, m.p 34-35°C.

(5) S t-Butyl thiouronium picrate m.p 151°C.

b.p 57°C **Ethylidene dichloride** (1, 1-dichloroethane) CH_3CHCl_2

(1) Colourless liquid, insoluble in water.
(2) Reduces Fehling's solution.
(3) Hydolyse with NaOH solution, the products are HCl and chloroethylene with faint aldehyde resin like odour.

b.p 61°C **Chloroform** (trichloro methane) $CHCl_3$

(1) Colourless liquid with sweet ethereal odour, insoluble in water, soluble in alcohol.
(2) Reduces Fehling's solution on boiling.
(3) Gives carbylamine odour with aniline.

(4) Boil 1ml of compound with 600mg resorcinol and 6ml water. Add 5-6 drops of 10% NaOH solution. A yellow red coloured solution with yellow green fluorescence is obtained.

(5) To a mixture of 23ml benzene and 3ml chloroform add with stirring in portion 3.5gm of anhydrous $AlCl_3$. The liquid boils on addition of $AlCl_3$. Heat on water bath for 30min, cool and pour in crushed ice containing 1ml conc HCl. Separate the benzene layer and distill. Collect the distillate at 190-215°C/10mmHg which soon solidifies. Triphenylmethane melts at 92°C.

b.p 68°C **s-Butyl chloride** (2-chlorobutane) $CH_3CH_2CHClCH_3$

(1) Clear liquid, very slightly soluble in water, soluble in ether.
(2) S-butyl thiouronium picrate m.p 166°C (3) Picrate of butyl 2-naphthyl ether m.p 85°C.

b.p 69°C **Isobutyl chloride**
(1-chloro-2-methyl propane) $(CH_3)_2CHCH_2Cl$

(1) Soluble in alcohol and ether.
(2) S-isobutyl thiouronium picrate m.p 167°C (3) picrate of isobutyl 2-naphthyl ether m.p 84°C

b.p 70°C **Allyl Bromide** (3-bromopropene) $CH_2{=}CHCH_2Br$

(1) Colourless liquid with mustard like odour, insoluble in water.
(2) Reflux 6.6gm of allyl bromide with 6.3gm guaiacol, 7 gm anhydrous K_2CO_3 and 10ml acetone for 8hrs. Cool and extract with ether. Wash organic layer with NaOH. Evaporate off the ether and distill the residual oil under reduced pressure and collect guaiacol allyl ether at 110-113°C/12mmHg. (b.p 252-254°C)
(3) To a mixture of 8gm Mg in 80ml ether add in portion with stirring 12gm allyl bromide and 25gm triphenyl tin chloride in 60ml THF. Add 50ml ether and reflux the resultant mixture at 60°C for 8-10hrs. Add 15ml saturated NH_4Cl, decant the organic layer. Extract the aqueous portion with ether Combine the organic layers and distill off the solvent. Collect allyl triphenyl tin and recrystallise from hexane, m.p 74°C.

(4) Cool a solution of 18gm of compound in 25ml CCl_4 to 0°C. Add dropwise with stirring 8ml of bromine. After addition is complete continue stirring at room temperature for 90min. Distill to remove the solvent. The residue is pure 1, 2, 3 tribromopropane b.p 100-103°C/18mmHg (b.p 220°C)

(5) S-Allyl thiouronium picrate m.p 154°C. (6) Picrate of allyl 2-naphthyl ether m.p 99°C.

b.p 71°C **n-Propyl bromide** (1-bromo propane) $CH_3CH_2CH_2Br$

(1) Soluble in water and ether.

(2) S-Propyl thiouronium picrate m.p 177°C.

(3) Picrate of Propyl 2-naphthyl ether m.p 81°C

b.p 72°C **Ethyl iodide** (iodo ethane) CH_3CH_2I

(1) Liquid with sweet smell, insoluble in water, soluble in alcohol.

(2) To 2ml of compound add cold alcoholic silver nitrate solution. Yellow coloured precipitate is obtained.

(3) Reflux 13gm of zinc copper couple with 7.8gm of ethyl iodide and 5.5gm of ethyl bromide for 90min. Remove the heat source and allow the reaction to continue. After 90 minutes distill the reaction mixture and collect diethyl zinc at 115-120°C.

(4) Mix 10gm of compound with a solution of 1.2gm of sodium in 13ml absolute ethanol and 8gm of malonic ester. The solution begins to boil and after the reaction has subsided reflux until it is neutral. Distill off the alcohol, add 10ml water and extract with ether. Distill and collect ethyl malonic ester at b.p 206-208°C.

(5) Add 13gm of ethyl acetoacetate to a solution of 2.2gm of Na in 75ml methyl alcohol and then keep at room temperture for 2hrs. Add 30gm of ethyl iodide and reflux for 5hrs. Distill off the excess solvent, add water to the residue and isolate with ether. Distill and collect the ester at b.p 79-80°C/14mmHg. Mix the ester with 200ml aqueous ammonia and shake for 24hrs. Evaporate under reduced pressure, crystallise the residue from benzene and collect α-ethyl-α acetylacetamide, m.p 96°C.

(6) Reflux 17gm of ethyl iodide with 10gm of 2-naphthylamine and a solution of 20gm of sodium carbonate in 200ml water for 3hrs.

Extract with ether. Distill off ether and collect ethyl –2-naphthylamine b.p 191°C/25mmHg (b.p 298°C)

(6) S-Ethyl thiouronium picrate m.p 188°C

(7) Picrate of ethyl 2-naphthyl ether m.p 102°C.

b.p 78°C **n-Butyl chloride** (1-chlorobutane) $CH_3CH_2CH_2CH_2Cl$

(1) Colourless liquid, insoluble in water, soluble in alcohol and ether.

(2) To 15ml dimethy sulphoxide add 3gm of NaCN. Heat the mixture to 90°C, remove the heat source and then add dropwise 5ml of butyl chloride(exothermic reaction, temp below 150°C). Pour the mixture in water, extract with ether. Distill and collect butyl cyanide at b.p 138-140°C.

(3) S-n Butyl thiouronium picrate m.p 177°C (4) Picrate of butyl 2-naphthyl ether m.p 67°C.

b.p77°C **Carbon tetrachloride** (tetrachlormethane) CCl_4

(1) Colourless liquid with chloroform like odour, insoluble in water, soluble in alcohol and ether.

(2) Gives carbylamine odour

(3) Boil 1ml of compound with few crystals of 1-naphthol and NaOH solution in cyclohexanol. On cooling a blue coloured solution is obtained.

(4) Cool 15ml of benzene and 3.5ml CCl_4 in ice bath, add in portion 3.5gm $AlCl_3$. Remove the ice bath and reflux the mixture on steam bath until evolution of HCl subsides. Cool, pour the mixture in crushed ice containing 20ml conc HCl. Separate the benzene layer and extract the aqueous layer with benzene. Combine the organic layers, distill off the solvent and collect triphenylchloromethane, m.p 112°C.

(5) Cool 9.1gm $AlCl_3$ and 3.5ml CCl_4 in ice bath. Add slowly 2ml of dry benzene. When the reaction starts add slowly a mixture of 10ml benzene and 11ml CCl_4 (temp below 10°C). Keep the mixture overnight. Cool in ice and add with stirring 50ml of water. Distill off the excess solvent and distill the residual liquid with steam. Separate the benzene layer and extract the aqueous layer with benzene. Combine the organic layer and distill. Recover benzophenone, m.p 48°C.

b.p 83.5°C **Ethylene dichloride** (1, 2-dichloro ethane)

$ClCH_2CH_2Cl$

(1) Colourless liquid with odour similar to chloroform, insoluble in water.
(2) Does not reduce Fehling's solution.
(3) Heat 8.4gm of compound with 100ml of concentrated aqueous ammonia at 115-120°C for 5hrs. Concentrate and then add 20ml alcohol, cool and collect the precipitated ethylenediamine hydrochloride. Make the paste of the salt with sodium hydroxide solution and distill. Collect ethylene diamine at b.p 118°C.
(4) S-alkyl thiouronium picrate m.p 260°C (4) Alkyl 2-naphthyl ether (di) m.p 217°C.

b.p 87°C **Trichloroethylene** $CHCl{=}CCl_2$

(1) Colourless, photoreactive liquid, with chloroform like odour, slightly soluble in water, soluble in nearly all organic solvents.
(2) Warm 2ml of compound with pyridine and NaOH solution. A red coloured solution which slowly changes to orange is obtained.

b.p 89°C **Isopropyl iodide** (2-iodo propane) $(CH_3)_2CHI$

(1) Colourless liquid, insoluble in water.
(2) Boil 20gm of compound with 23gm of acetoacetic est r and a solution of 2gm of sodium in 30ml alcohol for 24hrs under reduced pressure. Distill off the solvent, add water and isolate with ether. Distill and collect isopropylacetoacetic ester at b.p 134-136°C/ 14mmHg [b.p 205°C (d)]
(3) S-isopropyl thiouronium picrate m.p 196°C
(4) Picrate of isopropyl 2-naphthyl ether m.p 95°C

b.p 96.4°C **Propylene dichloride**

(1, 2-dichloropropane) $CH_3CHClCH_2Cl$

(1) Colourless liquid with chloroform like odour, insoluble in water, soluble in nearly all organic solvents.
(2) S-alkyl thiouronium picrate m.p 232°C (3) Picrate of alkyl 2-naphthyl ether m.p 152°C.

b.p 99°C **Isoamyl chloride** (1-chloro-3-methyl butane)
$(CH_3)_2CHCH_2CH_2Cl$

(1) Colourless or slightly yellow coloured liquid, insoluble in water, soluble in alcohol and ether.
(2) S-isoamyl thiouronium picrate m.p 173°C.

b.p 100°C (d) **t-Butyl iodide** (2-Iodo-2 methyl propane) $(CH_3)_3CI$

(1) Insoluble in water, soluble in alcohol and ether.
(2) Heat 5ml of compound with 5ml water. HI and trimethylcarbinol are obtained.

b.p 101°C **n-Butyl bromide** (1-bromo butane)
$CH_3CH_2CH_2CH_2Br$

(1) Colourless liquid, insoluble in water.
(2) Reflux 4.6ml butyl bromide with 5.6gm o-nitrophenol, 5.6gm K_2CO_3 and 40ml acetone on steam bath for 48hrs. Distill off the acetone and extract with benzene. Distill off the benzene, distill the residual liquid under reduced pressure and collect butyl o- n - butoxy nitrobenzene at 172°C/19mmHg.
(3) S-n butyl thiouronium picrate m.p 177°C. (4) Picrate of n butyl 2-naphthyl ether m.p 67°C.

b.p 102°C **n-Propyl iodide** (1-iodo propane) $CH_3CH_2CH_2I$

(1) Colourless liquid, miscible with alcohol and ether.
(2) Propyl 3, 5-dinitrobenzoate m.p 74°C
(3) S-propyl thiouronium picrate m.p 177°C
(4) Picrate of propyl 2-naphthyl ether m.p 81°C.

b.p 103°C **Allyl iodide** (3-iodopropene) $CH_2{=}CHCH_2I$

(1) Yellow coloured liquid, insoluble in water, soluble in alcohol.
(2) To 1ml of compound add saturated silver nitrate solution. A yellow coloured solution is obtained.
(3) To 5.5gm Na(in small pieces) add in portion 7.2ml of allyl iodide (exothermic reaction). After the reaction is complete keep at room

temperature for 3hrs and cool. Distill and collect biallyl at 59-60°C.
(4) S-Allyl thiouronium picrate m.p 154°C.
(5) Picrate of allyl 2-naphthyl ether m.p 99°C.

b.p 107°C **n-Amyl chloride** (1-Chloro pentane) $CH_3CH_2CH_2CH_2CH_2Cl$

(1) Insoluble in water, soluble in alcohol and ether.
(2) S-amyl thiouronium picrate m.p 154°C

b.p 119°C **s-Butyl iodide** (2-Iodo butane) $CH_3CH_2CHICH_3$

(1) Insoluble in water, soluble in alcohol and ether.
(2) sec butyl 3, 5 dinitrobenzoate m.p 74°C (3) S-s butylthiouronium picrate m.p 166°C (4) Picrate of s-butyl 2-naphthyl ether m.p 85°C.

b.p 120°C **Trimethylene dichloride** (1, 3-dichloropropane) $CH_2ClCH_2CH_2Cl$

(1) Soluble in alcohol and ether.
(2) Picrate of alkyl 2-naphthyl ether m.p 148°C

b.p 120°C **Isobutyl iodide** (1-Iodo 3-methyl propane) $(CH_3)_2CHCH_2I$

(1) Insoluble in water, soluble in alcohol and ether.
(2) Isobutyl 3, 5 dinitrobenzoate m.p 84°C
(3) S-isobutyl thiouronium picrate m.p 167°C
(4) Picrate of isobutyl 2-naphthyl ether m.p 84°C.

b.p 130°C **n-Butyl iodide** (1-iodo butane) $CH_3CH_2CH_2CH_2I$

(1) Soluble in alcohol and ether.
(2) butyl 3, 5 dinitrobenzoate m.p 64°C
(3) S-n butyl thiouronium picrate m.p 177°C.
(4) Picrate of n-butyl 2-naphthyl ether m.p 67°C.

b.p 131°C **Ethylene dibromide** (1, 2-dibromoethane)

$BrCH_2CH_2Br$

(1) Colourless liquid with sweetish odour, insoluble in water.

(2) To 200ml of anhydrous ethanol add in pieces 1.2gm of Na metal. When Na has completely dissolved add 2.1ml of ethanedithiol followed by 2.2ml of ethylene dibromide. Reflux the mixture for 4hrs. Cool and filter. Distill the filtrate to remove ethanol. When half of the distillation is over, cool the reaction mixture and collect dithiane, m.p 112-113°C.

(3) Heat to boiling a mixture of 15ml ethylene dibromide, 65ml ethanol and 22ml water. Add with stirring and in portion a solution of 6.3gm of sodium sulphite in 25ml water. Reflux the mixture for 2hrs. Distill off the excess alcohol and ethylene dibromide and evaporate the residual liquid to dryness. Extract with boiling ethanol and cool, 2-bromoethanesulfonate precipitates out.

(4) Reflux a mixture of 30gm of thiourea dissolved in 140ml ethanol. When the solution is clear remove the heat source, add 18ml of ethylene dibromide within 5 minutes. Vigorous reaction occurs. Cool and collect ethylene diisothiuronium bromide, m.p 227°C (d).

(5) Reflux 9.4gm of compound with a solution of 6.9gm of potassium carbonate in 50ml water for 24-48 hrs. Evaporate off the water, dissolve the residue in absolute alcohol and filter. Distill and collect ethylene glycol at b.p 195-197°C.

(6) Heat 90gm of ethylene dibromide with 75gm of potassium phthalimide until a brownish yellow syrup results. Add 100ml alcohol and heat the mixture under reflux till the insoluble residue is yellowish brown. Filter hot. Distill off the solvent and collect N-β-bromoethylphthalimide, recrystallise from alcohol, m.p 82-83°C.

b.p 149.5°C **Bromoform** $CHBr_3$

(1) Colourless liquid, insoluble in water, soluble in alcohol and ether.

(2) Responds to carbylamine test.

(3) Cool a mixture of 5gm of 1, 1-diphenylethylene, 20ml pentane and 5.6gm of potassium t-butoxide to 0°C. Add dropwise with stirring 13.1gm of bromoform. Continue stirring for 2hrs. Add

30ml water and filter. Collect the yellow crystals of 1, 1-dibromo-2, 2 diphenylcyclopropane, m.p 151-152°C.

(4) To 10.4gm of bromoform add a solution of sodium arsenate made by dissolving 4.6gm of arsenious oxide and 8.8gm of NaOH in 30ml water. Heat the mixture on steam bath for 4hrs. Remove the lower layer and extract the aqueous layer with ether. Combine the two organic layers and distill. Collect the yellow methylene bromide at b.p 98-100°C.

(5) S-alkylisothiouronium picrate m.p 232°C

b.p 155°C **1, 4 dichlorobutane** (tetra methylene dichloride)

$ClCH_2CH_2CH_2CH_2Cl$

(1) Colourless liquid with pleasant odour, soluble in nearly all organic solvents.

(2) To 15ml dimethy sulphoxide add 2gm of NaCN. Heat the mixture to 90°C, remove the heat source and then add dropwise 3ml of 1, 4 dichlorobutane (exothermic reaction, temp below 150°C). After the reaction has subsided maintain at 90°C for another 15min. Add 15ml chloroform and then pour into saturated sodium chloride solution. Remove the chloroform layer. Distill and collect adiponitrile at b.p 140-141°C/1.5mmHg. (b.p 295°C)

b.p 167°C **Trimethylene dibromide** (1, 3-dibromopropane)

$CH_2BrCH_2CH_2Br$

(1) Colourless liquid, insoluble in water, soluble in most organic solvents.

(2) Dissolve 6gm of sodium cyanide in 7ml water(warm for 1-2hrs). Add a solution of 10gm of compound in 20 ethanol. Reflux for 35-40hrs. Distill under reduced pressure. Extract with ethyl acetate and filter. Distill under reduced pressure and collect trimethylene cyanide at 144-147°C/13mmHg.

(3) Warm a mixture of 10gm of compound, 20ml water and 3.7gm phenol and add dropwise with stirring 5ml of NaOH solution. Reflux the mixture for 6hrs. Discard the upper aqueous layer. Distill the lower layer and collect phenoxy propyl bromide at 136-142°C/20mmHg(m.p 8°C).

(4) Alkyl 2-naphthyl ether m.p 148°C

b.p 180°C **Methylene iodide** (Diiodomethane) CH_2I_2

(1) Colourless oily liquid, insoluble in water.

(2) Quaternary salt with pyridine m.p 220°C (d)

Aliphatic Solid

m.p 16°C, b.p 219°C **Glycerol tribromohydrin** (1, 2, 3-tribromopropane) $CH_2BrCHBrCH_2Br$

(1) Colourless glistening prism, insoluble in water.

m.p 90°C **Carbon tetrabromide** (tetra bromomethane) CBr_4

(1) Crystallline solid with ethereal odour and sweet taste, insoluble in water, soluble in alcohol and ether. Sublimes slowly.

m.p 95°C(d) **Methyliodoform** (1, 1, 1-Triiodethane) CI_3CH_3

(1) Yellow crystalline solid, sparingly soluble in alcohol, soluble in benzene and ether.

m.p 119°C **Iodoform** (triiodo methane) CHI_3

(1) Lemon yellow coloured crystalline solid with characteristic penetrating saffron like odour, insoluble in water, soluble in ether and chloroform.

(2) Responds to carbylamine test by substituting iodoform for chloroform and treating it with aniline and KOH

(3) Heat to dryness a solution of iodoform in alcohol with few milligrams of phenol and KOH. A red coloured residue is obtained which is soluble in alcohol to give caramine red coloured solution.

(4) Dissolve few milligrams of compound in 1-2ml alcohol, add 1ml alcoholic $AgNO_3$ solution and boil. Yellow precipitate is obtained.

m.p 171°C (d) **Tetraiodomethane** CI_4

(1) Red crystalline compound which decomposes on exposure to heat or sunlight.

(2) Boil about 100mg gm of compound in 3ml water. Cool and collect iodoform.

m.p 187°C **Hexachloro ethane** (Carbon hexachloride) CCl_3CCl_3

(1) White crystalline solid with camphor like odour. Sublimes. Insoluble in water, soluble in alcohol and ether.
(2) Heat 500mg of compound with alcoholic KOH at 100°C. Cool and collect oxalic acid.

Aromatic Liquid

b.p 132°C **Chlorobenzene** (Phenyl chloride)

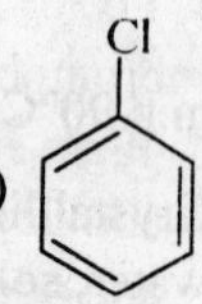

(1) Colourless pleasant smelling liquid, insoluble in water, soluble in alcohol and ether.
(2) To 13gm of Na in 9ml benzene add dropwise a mixture of 1.7gm of arsenic trichloride and 2.72gm of chlorobenzene. Heat the mixture under reflux for 12hrs. Filter hot. Extract with benzene. Distill off benzene and distill the residual oily layer at 110-112°C for 2hrs. Cool and collect triphenyl arsine, m.p 61°C.
(3) Heat a mixture of 5.1ml chlorobenzene, 20ml CS_2 and 15gm $AlCl_3$. After refluxing starts add 5ml Ac_2O and then reflux for 1hr. Disstill off the excess CS_2 and pour the mixture in ice containing 30ml conc HCl. Extract with ether and recover p-chloroacetophenone, m.p 20°C.
(4) To 3ml of compound add 15ml nitrating mixture (7.5ml conc HNO_3 and 7.5ml conc H_2SO_4). Heat the resultant mixture to 90°C for 30min. Cool, pour in ice and collect 2, 4-dinitrochlorobenzene, m.p 52°C.
(5) Sulphonamide m.p 143°C.

b.p 156°C **Bromobenzene** (Phenyl bromide)

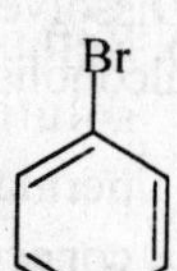

(1) Light yellow pleasant smelling liquid, insoluble in water, miscible in organic solvents.

(2) Heat a mixture of 5.2ml bromobenzene, 20ml CS_2 and 15gm anhydrous $AlCl_3$. After refluxing starts add 5ml Ac_2O and then reflux for 1hr. Distill off the excess CS_2 and pour the mixture in ice containing 30ml conc HCl. Extract with ether and recover p-bromoacetophenone, m.p 50°C.

(3) To 20ml nitrating mixture (10ml conc HNO_3 and 10ml conc H_2SO_4) add dropwise 5ml bromobenzene (temp below 60°C). After the addition is complete heat the mixture on water bath for 30min. Cool, pour in ice and collect p-bromonitrobenzene, m.p 125°C.

(4) To 6.7gm Mg and 100ml ether add with stirring 2ml bromobenzene. After the reaction has started add with stirring 40gm of bromobenzene. Reflux the mixture for 30-40min, cool, add 25gm of benzonitrile in 50ml ether. Reflux for 4-6hrs. Cool and add 50ml methanol. Filter, distill under reduced pressure and collect diphenyl ketimine at 127-128°C/3.5mmHg.

(5) To 2.5gm of Mg turnings covered with 75ml ether add 14gm of cupric chloride and 16gm of bromobenzene (exothermic reaction). After the reaction has subsided reflux for 2hrs, cool, add water and dil HCl. Remove the organic layer, extract the aqueous layer with ether. Distill off the solvent and collect diphenyl, m.p 70°C.

(6) Sulphonamide m.p 160°C.

b.p 159°C **o-Chloro toluene** (1-methyl 2-chloro benzene)

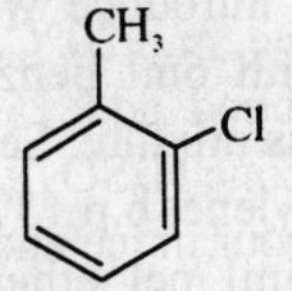

(1) Colourless liquid, insoluble in water, soluble in benzene and ether.

(2) Reflux 5gm of compound with 125ml potassium permanganate solution (14gm in 125ml water) until all the colour of permanganate has disappeared (about 4 hrs). Filter hot and concentrate the filtrate. Acidify with 7.5ml conc HCl and collect o-chlorobenzoic acid, m.p 140°C.

(3) 3, 5 dinitro, 1-methyl 2-chloro benzene (3, 5-dinitro o-chlorotoluene) m.p 63°C. (4) Sulphonamide m.p 126°C

b.p 162°C **m-Chloro toluene** (1-methyl 3-chloro benzene)

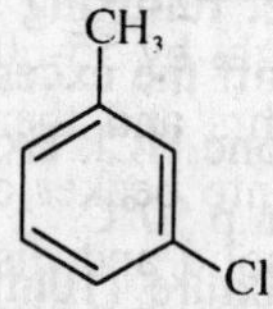

(1) Colourless liquid, insoluble in water, soluble in benzene and ether.

(2) Reflux 5gm of m-chloro toluene with 125ml potassium permanganate solution (14gm in 125ml water) until all the colour of permanganate has disappeared (about 4 hrs). Filter hot and concentrate the filtrate. Acidify with 7.5ml conc HCl and collect m-chlorobenzoic acid, m.p 158°C.

(3) 4, 6-dinitro m-chlortoluene m.p 91°C (4) Sulphonamide 185°C

b.p 162°C **p-Chloro toluene** (1-methyl 4-chloro benzene)

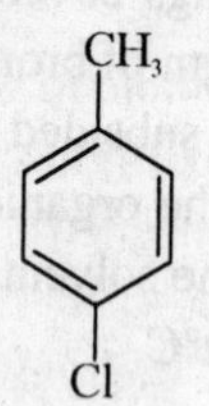

(1) Colourless oily liquid, insouble in water, soluble in ether.

(2) Stir 15gm of p-chlorotoluene with 5.2gm of Na wire in 50ml pet ether for 2hrs. Now introduce with stirring a mixture of 12ml dimethyl sulphate with 6ml benzene (temp below 30°C). Add water to the colourless mixture. Separate the organic layer. Steam distill and collect p-xylene, b.p 137-138°C.

(3) To 2.3gm Na in 25ml pet ether add with stirring 6.3gm p-chlorotoluene (temp below 30°C) and keep stirring for 3hrs. Pour the reaction mixture in 200gm cardice containing 100ml ether. Collect the precipitated p-toluic acid, m.p 176°C

(4) Reflux 5gm of p-chloro tolouene with 125ml potassium permanganate solution (15gm in 125ml water) until all the colour of permanganate has disappeared. Filter hot and concentrate the filtrate. Acidify with con HCl and collect p-chlorobenzoic acid, m.p 238°C.

(5) Heat together 12.6 gm of p-chloro toluene and .4gm of PCl_5 at 160-170°C. Now pass Cl_2 gas into the reaction mixture until there is an increase in weight by 55-60gm. Add 40ml of conc H_2SO_4, stir vigorously for 5hrs and then leave overnight. Transfer the lower viscous layer into beaker containing crushed ice. Remove the cream coloured solid and dissolve in ether. Shake the ether solution with NaOH. Distill off ether and recover p-chlorobenzaldehyde, m.p 46-47°C.

(6) Sulphonamide m.p 143°C.

b.p 172°C **m-Dichlorobenzene** (1, 3-dichloro benzene)

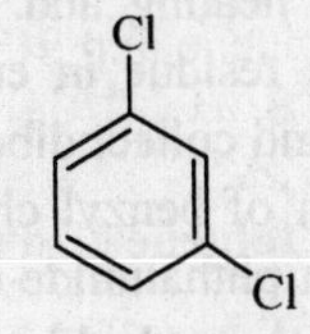

(1) Colourless liquid, insoluble in water, soluble in benzene.

(2) To a stirred solution of 14gm of KNO_3 in 50ml conc H_2SO_4 add 10gm of m-dichlorobenzene. The temperature rises to 135-140°C. Keep the temperature at 120-130°C for 1hr. Pour the mixture in crushed ice. Collect the yellow precipitated 1, 5-dichlor-2, 4-dinitro benzene, recrystallise from ethanol, m.p 104°C.

(3) Sulphonamide m.p 180°C.

b.p 179°C **Benzyl chloride** (α-chloro toluene)

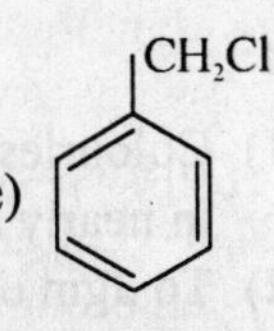

(1) Colourless liquid with irritating odour, insoluble in water, soluble in alcohol and ether. Lachrymator.

(2) Heat 1ml of compound with lead nitrate. A smell of bitter almond due to the formation of benzaldehyde is observed.

(3) Cool a solution of 7ml of benzyl chloride in 30ml benzene. Add in portion with stirring .5gm of anhydrous $AlCl_3$ (exothermic reaction). After the addition is complete add after 15min crushed

ice and 15ml water to the reaction mixture. Separate the benzene layer and distill. Collect diphenymethane at 250-275°C which soon solidifies on cooling, m.p 25°C.

(4) Reflux together 7.6gm of thiourea, 20ml warm water, 13.5ml ethanol and 12ml benzyl chloride on water bath until benzyl chloride dissolves. Cool, shake the mixture thoroughly and collect S-benzylisothiouronium chloride, m.p 174°C.

(5) Reflux 4.5ml benzyl chloride, 25ml water and 10ml conc HNO_3 for 6 hrs. Cool and shake thoroughly. Collect the precipitated benzoic acid, m.p 121°C

(6) Heat a solution of 6ml benzyl chloride in 16ml ethanol on steam bath. Add with stirring a solution of 2.9gm sodium sulphide in 8ml water. Continue heating and stirring for 72hrs. Distill off ethanol and pour the residue in crushed ice. Triturate the oily portion with ethanol and collect dibenzyl sulphide, m.p 49°C.

(7) Reflux together 15gm of benzyl chloride with 7.6gm potassium carbonate and 14.7gm phthalimide on oil bath at 190°C for 3hrs. Distill off the excess benzyl chloride, cool and collect benzyl-phthalimide, recrystallise from glacial acetic acid, m.p 116°C.

(8) S-benzylisothiouronium picrate m.p 188°C.

b.p 180°C **o-Dichlorobenzene** (1, 2-dichloro benzene)

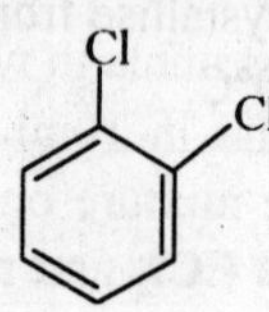

(1) Colourless liquid with pleasant odour, insoluble in water, soluble in nearly all organic solvents.

(2) To 3gm of compound in 15ml ether add grignard reagent prepared from 1.3gm of Mg and 6.9gm of 1-bromobutane. Now add 25mg of dichlor [1, 3bis(diphenylphosphino) propane] nickel (*II*) (exothermic reaction). Cool in ice. Stir the mixture at room temperature for 2hr. Reflux for 6hrs, cool and hydrolyse with 25ml 2N HCl. Remove the organic layer and extract the aqueous layer with ether. Combine the two layers, distill and collect 1-butyl 2-chlorobenzene at b.p 52-54°C /3.5mmHg

(3) Sulphonamide m.p 135°C

b.p 181°C **o-Bromo toluene** (o-tolyl bromide) 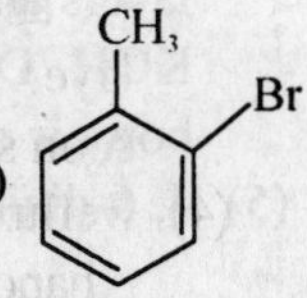

(1) Colourless liquid, insoluble in water, soluble in organic solvents.

(2) To 3ml of compound add 90ml potassium permanganate solution (7.5ml in 90ml water) and 1ml NaOH solution. Reflux the mixture for 3-4 hrs. Cool and pass SO_2 gas until the purple colour has disappeared. Filter and recover o-bromo benzoic acid, m.p 147°C

(3) To 5ml of compound in 10ml CCl_4 add dropwise with stirring 1.3ml bromine. Heat the mixture on water bath for 30min. Cool, distill off excess CCl_4 and recover o-bromobenzyl bromide, m.p 31°C.

(4) Sulphonamide m.p 168°C (5) 3, 5-dinitro o-bromo toluene m.p 83°C.

b.p 183°C **m-Bromo toluene** (m-tolyl bromide) 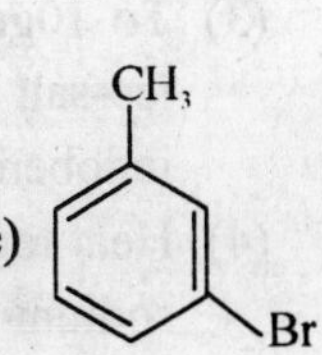

(1) Colourless liquid, insoluble in water, soluble in alcohol and ether.

(2) To 5ml of compound in 10ml CCl_4 add dropwise with stirring 1.3ml Br_2. Heat the mixture on water bath for 10-20 min. Cool, distill off the excess CCl_4 and recover m-bromo benzyl bromide, m.p 41°C.

(3) To 3ml of compound add 90ml potassium permanganate solution (7.5ml in 90ml water) and 1ml NaOH solution. Reflux the mixture for 3-4 hrs. Cool and pass SO_2 gas until the purple colour has disappeared. Filter and recover m-bromo benzoic acid, m.p 155°C

(4) Stir together 4gm of m-bromo toluene. 9gm Mg turning and 90ml ether. When nearly all the Mg has reacted decant the solution and add it slowly with stirring to cooled 8gm tributyl borate in 30ml ether(temp below –7°C). After addition is complete continue stirring for 2hrs and then keep it overnight in a refrigerator. Add the cooled solution to 24ml 10% H_2SO_4. Separate the two layers and extract the aqueous layer with ether. Combine the two organic

layers and concentrate. Add 10ml water and make it alkaline with KOH. Distill off and acidify the residual liquid, collect m-tolyl boronic acid.

(5) 4, 6-dinitro m-bromo toluene m.p 103°C.

b.p 188°C **Iodobenzene** (Phenyl iodide)

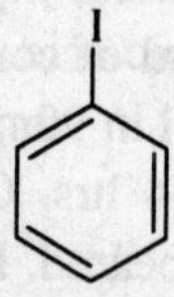

(1) Colourless liquid, insoluble in water, soluble in alcohol and ether.

(2) To 10gm of iodobenzene heated to 30°C add with stirring and dropwise 15ml of 40% peracetic acid. After the addition is complete continue stirring for another 30min. Cool in crushed ice for 1hr and collect crystals of iodosobenzene diacetate, recrystallise from acetic acid, m.p 158-159°C.

(3) To 10gm of iodobenzene add 15ml $CHCl_3$. Cool the mixture in ice-salt mixture and pass Cl_2 gas. Collect the yellow crystals of iodobenzene dichloride and wash with $CHCl_3$, m.p 80°C(d).

(4) Heat together 11gm of iodobenzene dichloride with 115ml of 5% sodium hypochlorite solution and .5ml glacial acetic acid at 75°C for 1hr. Cool and collect the precipitated iodoxy benzene.

(5) Heat 10gm of iodobenzene with 10gm of copper bronze at 230°C for 3hrs. Extract with ether. Distill off the ether, cool the residual oily portion which soon solidifies. Collect diphenyl, m.p 70°C.

(6) p-nitro iodobenzene m.p 171°C

(7) p-bromo iodobenzene m.p 92°C.

b.p 198°C(d) **Benzyl bromide** (α-bromo toluene)

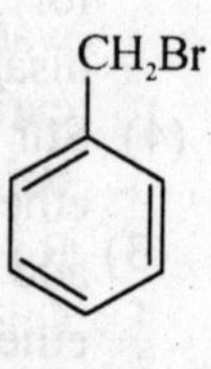

(1) Colourless strong lachrymatory liquid with irritating odour, insoluble in water, alcohol and ether.

(2) To well stirred mixture of 100ml dimethyl sulfoxide and 13gm of

KOH add 5.8gm of indole. Continue stirring for 45minutes and add 17.2gm of benzyl bromide. Continue stirring for 45 minutes and then add 100ml water. Extract with ether. Distill off the ether and cool. Recover 1-benzyl indole and recrystallise from ethanol, m.p 43°C.

(3) Reflux 4.5ml benzyl bromide, 25ml water and 10ml conc HNO_3 for 6 hrs. Cool and shake thoroughly. Collect the precipitated benzoic acid, m.p 121°C

b.p 204°C **m-Iodo toluene** (m-tolyl iodide)

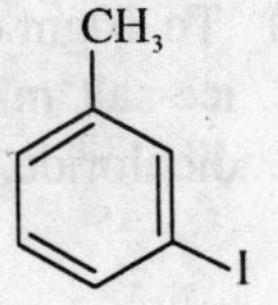

(1) Colourless liquid, insoluble in water, soluble in alcohol and ether.
(2) To 10gm of m-iodo toluene add 15ml $CHCl_3$. Cool the mixture in ice-salt mixture and pass Cl_2 gas. Collect the yellow crystals of dichloride and wash with $CHCl_3$, m.p 88°C(d).
(3) m-Iodo benzoic acid m.p 186°C.

b.p 207°C **Benzal chloride** (Benzylidene dichloride, 1, 1-dichloro toluene)

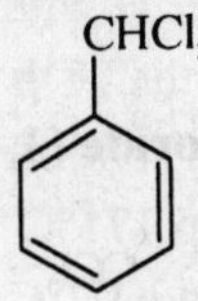

(1) Colourless liquid with pungent irritating odour, insoluble in water.
(2) Heat 15ml of compound with 250mg of Iron powder and 25ml water to 100°C. After evolution of HCl has stopped add sodium carbonate. Smell of bitter almond is observed due to formation of benzaldehyde.
(3) Reflux together 10ml of compound with 7gm of sodium acetate. Cool and recover cinnamic acid, m.p 132°C.
(4) Reflux benzal chloride with excess of phenylhydrazine in alcohol. Cool and collect benzaldehyde phenylhydrazone, m.p 156°C.

b.p 211°C **o-Iodo toluene** (o-tolyl iodide)

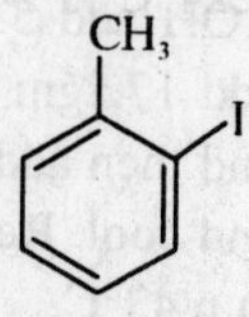

(1) Colourless liquid, insoluble in water, miscible with alcohol and ether.
(2) Heat on water bath 5ml of compound with 20ml dil HNO_3 until the evolution of oxides of nitrogen has ceased. Cool, pour in ice water. Filter and collect o-iodobenzoic acid, m.p 162°C.
(3) To 10gm of o-iodo toluene add 15ml $CHCl_3$. Cool the mixture in ice-salt mixture and pass Cl_2 gas. Filter and collect crystals of dichloride, wash with $CHCl_3$, m.p 85°C.

b.p 219°C **m-dibromobenzene** (1, 3-dibromo benzene)

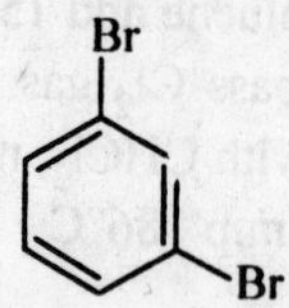

(1) Colourless liquid, insoluble in water, soluble in alcohol and ether.
(2) 4-nitro-1, 3-dibromobenzene m.p 62°C (3) Sulphonamide m.p 190°C.

b.p 220°C **Benzotrichloride** (1, 1, 1-trichloro toluene)

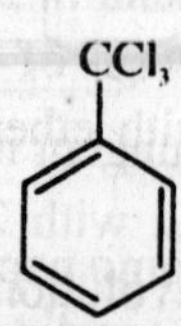

(1) Colourless oily liquid with pungent odour, insoluble in water, soluble in alcohol and ether.
(2) Warm 1ml of compound with 1ml dimethylaniline and 500mg of fused zinc chloride. An intense green coloured solution is obtained.
(3) Reflux 5ml of compound with NaOH solution. Cool, acidify with dil HCl and recover benzoic acid, m.p 121°C.

b.p 224°C **o-Dibromobenzene** (1, 2-dibromo benzene)

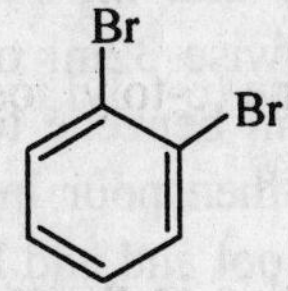

(1) Colourless liquid, insoluble in water, soluble in ether.

(2) Reflux 4.3gm of freshly prepared cuprous oxide, 6.2gm of butanethiol and 75ml of ethanol until orange red colour is changed to white. Cool and collect n-butylmercaptide and then add it immediately to a mixture of 5.9gm of o-dibromobenzene, 25ml quinoline and 8ml pyridine. Heat the mixture under reflux for 3.5hrs. Cool and pour in crushed ice containing 40ml conc HCl. Extract with ether. Distill off the ether and collect 1, 2 bis(n-butyl thio) benzene b.p 123-124°C/.3mmHg.

(3) 1, 2-dibromo-4, 5-dinitro benzene m.p 114°C

(4) Sulphonamide 175°C.

b.p 281°C **1-Bromo naphthalene**

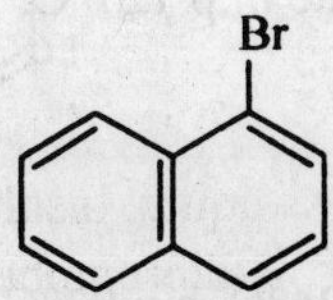

(1) Colourless liquid, miscible with alcohol, benzene and ether.

(2) Stir together a mixture of 60ml DMF, 6.1gm ethylene diamine and 60ml of 20% chromium perchlorate solution. Add a solution of 2ml of 1-bromonaphthalene in 20ml DMF and stir for 80min. Pour the mixture in a solution of 40gm ammonium sulphate in 400ml .4N HCl. Extract with ether. Distill off ether and collect naphthalene, m.p 80°C.

(3) Reflux together 6.6gm 1-bromo naphthalene, 3.5gm dry powdered cuprous cyanide and 3.5ml pyridine at 210-215°C for 15hrs. Pour the hot mixture on to a mixture of 15ml NH_4OH and 15ml water. Add 15ml benzene and shake thoroughly. Add 10ml ether and filter. Separate the benzene-ether layer, wash with NH_4OH, HCl, water and NaCl. Distill off the solvent and distill the residue under reduced pressure. Collect 1-naphthonitrile at 173-174°C/ 27mmHg (b.p 299°C, m.p 38°C)

(4) Heat on water bath 6gm Mg, 25ml ether and 2ml 1-bromonaphthalene so that the reaction starts, remove the heat source and add dropwise 32ml of 1-bromonaphthalene. Reflux for 30min. Add 135ml benzene to the grignard reagent, cool in ice salt mixture and then pour in 65gm cardice. After all solid CO_2 has evaporated cool and add H_2SO_4 until no further reaction takes place and all Mg has dissolved. Separate the upper layer and extract the aqueous layer with ether. Extract ether-benzene layer with NaOH solution. Acidify with H_2SO_4 and collect 1-naphthoic acid, recrystallise from toluene, m.p 160-161°C.

(5) Sulphonamide m.p 192°C (6) Picrate m.p 135°C

b.p 305°C **1-Iodo naphthalene**

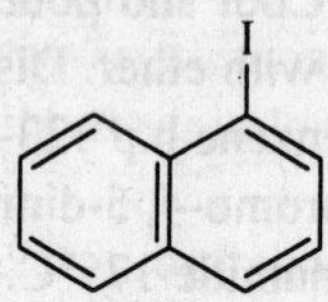

(1) Thick oily liquid, miscible with alcohol, ether and benzene
(2) Picrate m.p 127°C.

Aromatic Solid

m.p 24°C, b.p 93°C (10mm) **Benzyl iodide** (α-iodotduene)

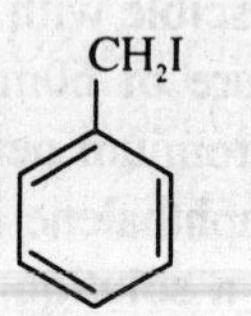

(1) Crystalline solid, insoluble in water, difficult soluble CS_2, powerful lachrymator.
(2) benzyl 3, 5-dinitrobenzoate m.p 113°C

m.p 28°C, b.p 185°C **p-Bromo toluene** (p-tolyl bromide)

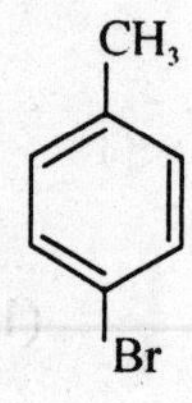

(1) Colourless compound, insoluble in water, soluble in ether

(2) Heat 10gm of compound on oil bath at 100°C. Add dropwise 7ml of Br_2. After the addition is complete raise the temperature to 150°C. To the crude p-bromo benzylidene dibromide add 20gm $CaCO_3$ and 30ml water. Reflux the mixture on wire gauze for 15hrs. Steam distill. The distillate on cooling soon solidifies, p-bromo benzaldehyde melts at 57°C.

(3) To 20ml ether add 1gm of lithium shavings and stir. Add dropwise a solution of 10gm of p-bromo toluene in 20ml ether(mixture refluxes). Stir for another 60min. Cool to –50°C and add 20ml ether Pour the mixture onto 100gm of Cardice in 100ml ether. Allow the cardice to evaporate(leave overnight). Add 100ml water followed by 5ml NaOH. Separate the ethereal layer and isolate the aqueous layer with ether. Combine the two layers and add NaOH solution. Acidify with HCl and collect p-toluic acid, m.p 175°C.

(4) Cool a mixture of 6.2gm of p-bromotoluene, 57ml glacial acetic acid and 56ml acetic anhydride(temp below 10°C). Add dropwise with stirring 8.3ml conc H_2SO_4. Cool the mixture to 5°C and add in portion 10gm of CrO_3 (temp below 10°C). Continue stirring for 15min after addition is complete and then pour in crushed ice. Filter precipitated p-bromobenzaldiacetate and recrystallise from hot alcohol, m.p 94-95°C.

(5) To 2gm of compound add 90ml potassium permanganate solution (7.5ml in 90ml water) and 1ml NaOH solution. Reflux the mixture for 3-4 hrs. Cool and pass SO_2 gas until the purple colour has disappeared. Filter and recover p-bromo benzoic acid, m.p 251°C

(6) Sulphonamide m.p 165°C.

m.p 32°C, b.p 169 (25mm) **1-Chloromethyl naphthalene** (α-naphthyl methyl chloride)

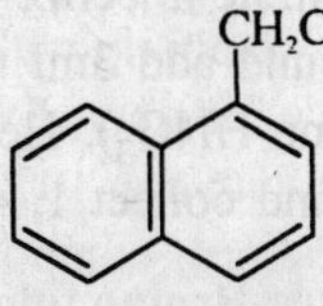

(1) Powerful lachrymator and vesicant solid, insoluble in water, soluble in organic solvents.

(2) Reflux together 10.2gm of compound with 16.8gm of hexamethylene tetramine and 25ml glacial acetic acid for 2hrs. Add 20ml conc HCl and continue refluxing for another 15min. Cool and extract with ether. Evaporate off the ether, distill the residual liquid under reduced pressure and collect 1-naphthaldehyde at 163-164°C/18mmHg. (b.p 291°C)
(3) S-Alkylisothiouronium chloride m.p 238°C.
(4) Picrate m.p 82°C.

m.p 35°C, b.p 211°C **p-Iodo toluene** (p-tolyl iodide)

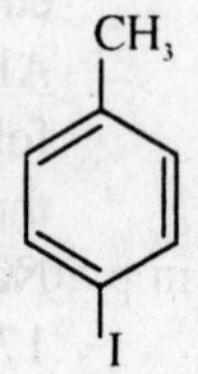

(1) Colourless crystalline solid, insoluble in water, soluble in alcohol and ether.
(2) Dissolve 5gm of compound in chloroform, cool in ice bath. Pass chlorine gas rapidly. Filter and recover the dichlorocompound, m.p 85°C(d).
(3) p-Iodo benzoic acid m.p 265°C (d)

m.p 53°C **p-Dichlorobenzene** (1, 4-dichloro benzene)

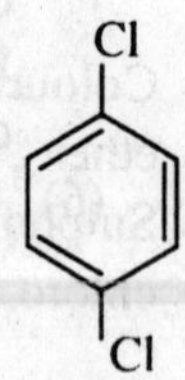

(1) Crystalline compound having anti moth and fumigant property, insoluble in water, soluble in benzene and ether.
(2) To 500mg of compound add 3ml nitrating mixture(1.5ml conc H_2SO_4 and 1.5ml conc HNO_3). Heat the mixture on water bath for 15-20min. Cool and collect 1, 4-dichlor 2-nitrobenzene, m.p 54°C.
(3) To 2gm of compound add 10gm of chlorosulphonic acid. Heat the mixture at 150°C for 1hr. Cool, pour in ice and collect 2, 5-dichlorobenzenesulphonyl chloride, m.p 39°C.

(4) Sulphonamide m.p 180°C.

m.p 55°C **2-Iodonaphthalene** 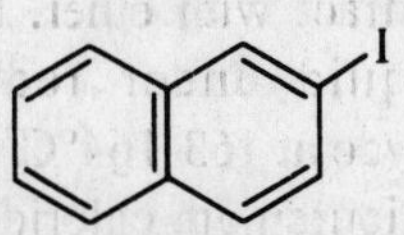

(1) Crystalline compound, soluble in alcohol, ether and glacial acetic acid.
(2) Picrate m.p 95°C.

m.p 56°C, b.p 281°C **2-Bromo naphthalene** 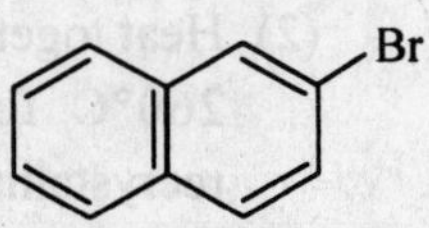

(1) **White crystalline solid, insoluble in water, soluble in ether, chloroform and benzene.**
(2) Sulphonamide m.p 208°C (3) Picrate m.p 86°C

m.p 60°C **2-Chloro naphthalene** 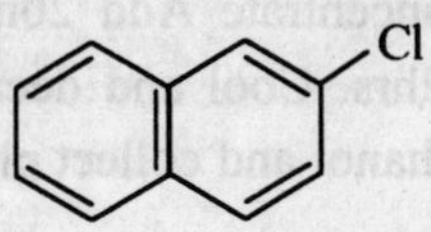

(1) Colourless compound, insoluble in water, soluble in alcohol and ether.
(2) Sulphonamide m.p 232°C (3) Picrate m.p 81°C. (4) 1, 8-dinitro-2 chloro naphthalene m.p 175°C.

m.p 61°C **p-Bromobenzyl bromide** (p-bromo-α-bromo toluene)

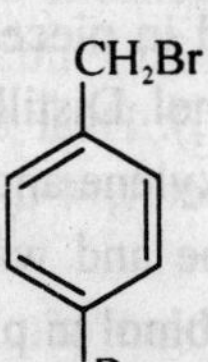

(1) Crystalline compound, insoluble in water, soluble in ether.

(2) On boiling with alcoholic KOH, p-bromobenzyl alcohol and p-bromo benzoate are obtained.

m.p 65°C **9-Bromo phenanthrene** 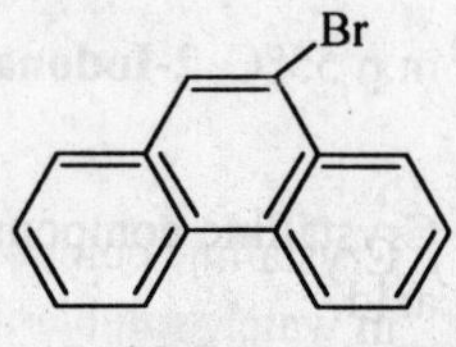

(1) Crystalline compound insoluble in water, soluble in alcohol and ether.
(2) Heat together 10gm of compound with 4gm of cuprous cyanide at 260°C for 6hrs. Cool and collect 9-cyano phenanthrene, recrystallise from ethanol, m.p 107°C.
(3) To 1gm Mg add with stirring and in portion a solution of 12gm of compound in 30ml ether. Continue stirring and add 20ml benzene. After the exothermic reaction has subsided heat the mixture with stirring for 4hrs. Cool, add 6gm ethyl orthoformate and reflux for 6hrs. Cool and add 20ml dil HCl. Remove the organic layer and concentrate Add 20ml dil H_2SO_4 to the residue and reflux for 12hrs. Cool and decant the acid. Recrystallise the residue from ethanol and collect phenanthrene 9-aldehyde, m.p 101°C.

m.p 77°C **p-Chlorobiphenyl** 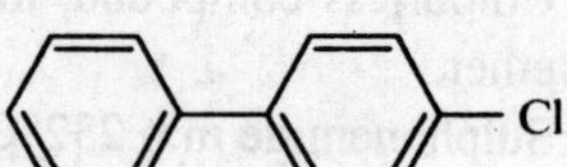

(1) Crystalline compound, insoluble in water, soluble in alcohol and ether.
(2) Heat together 10gm of compound with 2.5gm of ethyl carbonate and 75ml benzene. Add in pieces 1.5gm of Na. Reflux for 2hrs. Cool and add 10ml ethanol. Distill off the excess benzene. Dissolve the precipitate in 30ml xylene and add a pinch of animal charcoal. Distill to remove xylene and water. Filter hot, cool and collect crystals of triphenyl carbinol m.p, 207-208°C.
(3) p-Chlorobenzoic acid m.p 240°C.

m.p 87°C, b.p 219°C **p-Dibromo benzene** (1, 4-dibromo benzene)

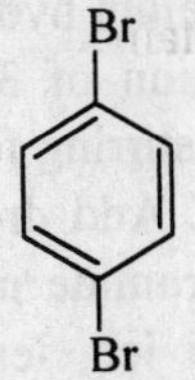

(1) Crystalline compound with characteristic aromatic odour, insoluble in water, soluble in alcohol and ether.
(2) Cool 500mg of compound and add ice cold nitrating mixture (2ml conc H_2SO_4 and 2ml conc HNO_3). After the addition is complete heat the mixture on water bath for 20-30min. Cool, pour in ice and collect 2, 5-dinitro-1, 4-dibromo benzene, recrystallise from ethanol, m.p 84°C.
(3) To 1.8gm Mg in 6ml ether add in portion 18gm of p-dibromo benzene in 100ml ether. Continue stirring for 2hrs. Cool and add a solution of 15gm chlorodiphenyl phosphine in 20ml ether(temp below 10°C). Cool further and add 30ml saturated aqueous NaCl solution. Decant the ethereal layer and acidify the aqueous layer with HCl. Extract with benzene. Combine benzene and ether layer and distill to remove the solvent. Distill the residual oil under reduced pressure and collect diphenyl p-bromophenyl phosphine as colourless oil which soon solidifies, m.p 80°C.
(4) Sulphonamide m.p 195°C.

m.p 89°C **p-Bromobiphenyl** 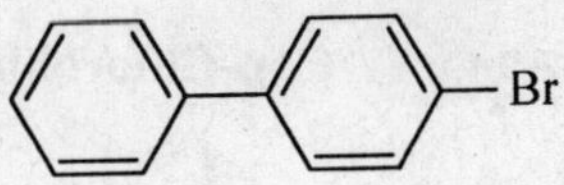

(1) Crystalline compound, soluble in alcohol, ether and benzene.
(2) p-Bromobenzoic acid m.p 251°C.

m.p 95°C (d) **o-Xylylene dibromide** (α, α'-dibromo-o-xylene)

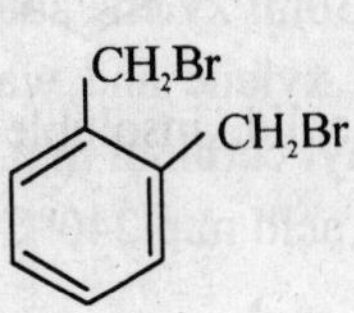

(1) Brown crystalline solid, a powerful and persistant lachrymator, insoluble in water, soluble in alcohol.

(2) To 1.9 gm of 53% Sodium hydride (in mineral oil) and 6ml DMF add dropwise a solution of 3.5gm of p-toluenesulfonamide in 10ml DMF. Continue stirring for 1hr at room temperature and for another hour at 60°C. Add dropwise with stirring a solution of 5.3gm o-xylylene dibromide in 30ml DMF(temp 60-70°C). Stir for 3hrs and pour in ice, leave overnight. Collect 2-(p-tolyl sulfonyl)dihydroisoindole, m.p 174-175°C.

m.p 107°C **1, 1, 1 Trichloro-2, 2 bis** (p-chlorophenyl) ethane (DDT)

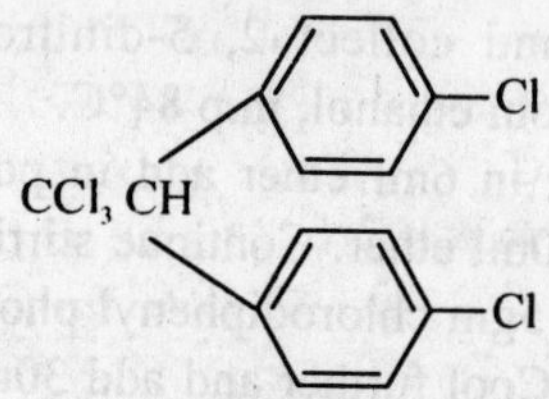

(1) Colourless crystalline solid, sparingly soluble in water. A powerful insecticide.

(2) Boil few milligrams of compound with alcoholic KOH solution for 2-3 minutes. Cool and filter, acidify the filtrate with dil HNO_3 and then add silver nitrate solution. A white precipitate is obtained.

m.p 112°C **Triphenylmethyl Chloride** (Trityl chloride)

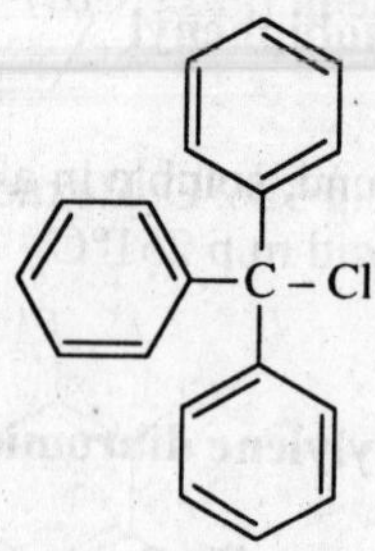

(1) White crystalline solid, insoluble in water. Soluble in benzene and ether.

(2) Add 1% sodium amalgam to a solution of compound in ether. A red coloured solution due to the formation of triphenyl methyl sodium is obtained.
(3) To 1gm of triphenylmethyl chloride add 20ml water and reflux for 10-15 minutes. Cool and collect the precipitated triphenyl carbinol, recrystallise from hot alcohol, m.p 162°C.

m.p 113°C **p-Iodobiphenyl** 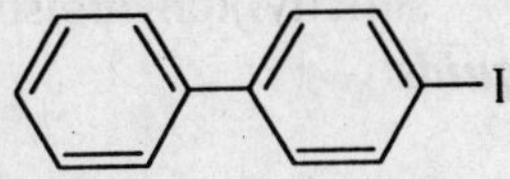

(1) Crystalline solid insoluble in water, soluble in ether and benzene.
(2) Heat 4gm of compound with 4gm of copper bronze at 270-280°C for 15 min. Cool and then heat under pressure for 15min. Cool and collect 4, 4' diphenyl diphenyl, recrystallise from nitrobenzene, m.p 320°C.
(3) Dichloro m.p 102°C.

m.p 121°C **sym-Tribromobenzene** (1, 3, 5-tribromobenzene)

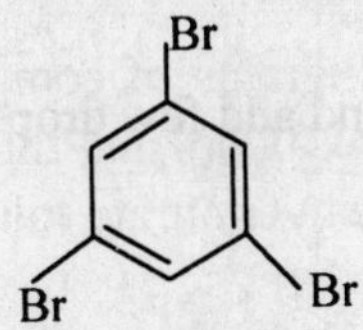

(1) Crystalline compound, insoluble in water and alcohol, difficult soluble hot alcohol, soluble in benzene, chloroform and ether.
(2) Sulphonamide m.p 222°C(d)

m.p 226°C, b.p 326°C **Hexachloro benzene** (Perchloro benzene)

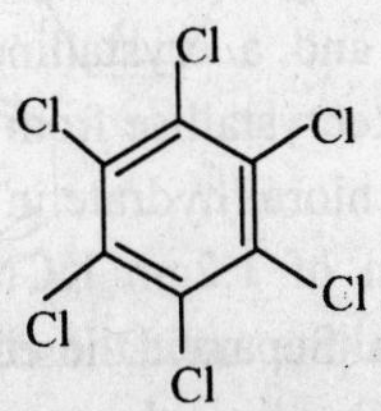

(1) Crystalline compound, insoluble in alcohol, soluble in hot benzene.

(2) Heat together 3.9gm Mg turning with 14gm hexachlorbenzene in 100ml ether. Add dropwise 19gm of ethylene dibromide in 20ml benzene. Cool and add cardice to the reaction mixture. Acidify with conc HCl. Distill off the excess ether and benzene. Filter the crude pentachloro benzoic acid and recrystallise, m.p 202-206°C.

(ii) Halogenated Carbonyls

Aliphatic Liquids

b.p 85°C **Chloroacetaldehyde** CH_2ClCHO

(1) Colourless liquid with very sharp penetrating odour, soluble in water.
(2) Reduces ammonical silver nitrate solution on warming.

b.p 90°C **Dichloroacetaldehyde** (2, 2-dichloroethanal) $CHCl_2\ CHO$

(1) Colourless liquid, soluble in water, vapours cause irritation to eyes.
(2) To 1ml of compound add few drops of conc H_2SO_4, solid polymer is formed.

b.p 98°C **Chloral** (trichloro acetaldehyde) CCl_3CHO

(1) Colourless oily liquid with characteristic pungent odour, soluble in water and alcohol, available as chloral hydrate.
(2) Reduces Fehling's solution.
(3) Reduces ammonical silver nitrate solution on warming.
(4) Add with stirring 7.4gm of chloral to 2.3 gm of cooled formamide. Heat is evolved and a crystalline mass of chloral formamide precipitates out. Recrystallise from ethanol, m.p 115°C.
(5) To a solution of chloral hydrate in 40ml ether add dropwise with stirring a solution of 1.2gm KCN in 22ml NH_4OH. Continue stirring for 20min. Separate the ether layer. Extract the aqueous layer with ether. Combine the two ethereal layer, distill off ether and collect α, α dichloro acetamide, m.p 99°C.

(6) Cool 10ml chloral hydrate solution (8gm in 10ml water). Add dropwise 10ml of $NaBH_4$ solution (.7gm in 10ml water) taking care that the temperature does not rise above 30°C. Add 5ml HCl and extract with ether. Distill off ether and collect 2, 2, 2-trichloroethanol at b.p 151-153°C.

(7) Dissolve 5ml aniline in 4.5ml conc HCl and 30ml water and to this solution add a solution of 9.1gm chloral hydrate and 135gm sodium sulphate in 135ml water. Add a solution of 11.2gm of hydroxylamine hydrochloride in 120ml water. Heat the mixture to boiling for 30min. Cool and collect isonitrosoacetanilide, m.p 178-179°C.

(8) Chloral 2, 4-dinitrophenylhydrazone m.p 131°C (9) Choral semicarbazone m.p 90°C(d).

b.p 119°C **Chloroacetone** [1-chloro–2-propanone] CH_3COCH_2Cl

(1) Colourless liquid with pungent odour, difficult soluble in water, miscible with alcohol, ether and chloroform, powerful lachrymator.

(2) Add few drops of compound to saturated K_2CO_3 solution. Caramine red coloured solution is obtained.

(3) Heat 1ml of compound with few milligrams of Zn and 1ml conc HCl. Sweet smell due to the formation of acetone is observed.

(4) Mix 7.5gm of acetamide with 5gm phosphorous pentasulphide and add 5ml benzene to it. Add dropwise a mixture of 10ml chloroacetone and 3.75ml benzene. Heat on water bath for 30min. Add 20ml water. Discard the benzene layer and add 20% NaOH solution to the aqueous layer. Extract with ether. Distill off ether and collect 2, 4-dimethyl thiazole at b.p 143-145°C.

(5) Dissolve 3.8gm thiourea in 10ml water. Add with stirring 4ml of chloroacetone. Reflux for 2hrs. Cool and add 10gm of NaOH. Remove the oily layer and extract the aqueous layer with ether. Distill off ether and collect 2-amino 4-methyl thiazole, m.p 45°C. Repeat the same process with oily layer. .

(6) Reflux 10ml chloroacetone with 10gm potassium phthalimide and 10ml xylene at 120-130°C with occassional stirring. After 60min steam distill to remove xylene and chloroacetone. Brownish oil remains which soon solidifies, recrystallise from hot water and collect acetonylphthalimide, m.p 124°C

(7) To a solution of 9.1gm of p-chlorothiophenol in 75ml NaOH solution add 5ml of chloroacetone. Stir for 45min. Extract with ether. Distill off the ether, cool the residue, (p-chlorophenylthio) acetone separates out, m.p 41°C.

(8) Chloroacetone 2, 4 dinitrophenylhydrazone m.p 125°C

b.p 120°C **unsym Dichloracetone** (1, 1-dichloroacetone)
$CH_3COCHCl_2$

(1) Colourless pleasant ethereal smelling liquid, sparingly soluble in water

(2) 1, 1 dichloroacetone semicarbazone m.p 163°C

b.p 136°C [725mm] **Bromoacetone** (1-Bromo-2-propanone)
CH_3COCH_2Br

(1) Colourless liquid, soluble in alcohol and acetone, vapours cause irritation to the eyes, provoking flow of tears

b.p 157°C **Chloroacetal** (2-chloro-1, 1-diethoxy ethane)
$CH_2ClCH(OC_2H_5)_2$

(1) Colourless liquid, slightly soluble in water, soluble in alcohol and ether.

(2) Heat 2ml of compound with dil H_2SO_4. A pungent odour due to the formation of chloroacetaldehyde is observed.

(3) Cool a solution of 3.8gm of chloroacetal in 80ml absolute methanol in dry ice –acetone bath. Add 29ml of NH_4OH and heat at 140°C for 10hrs. Cool and concentrate the coloured solution. Add 10ml of 5% KOH solution. Concentrate and add further 10ml of 5% KOH solution. Extract with ether. Distill off ether and collect amino acetal at 101-103°C/100mmHg (b.p 163°C)

(4) Chloroacetal semicarbazone m.p 134°C (d)

b.p 174°C **Bromal** (Tribromoacetaldehyde, 2, 2, 2-tribromo ethanal) CBr_3CHO

(1) Reddish yellow coloured liquid readily available as bromal hydrate, m.p 53°C. Insoluble in water, soluble in alcohol and ether.

(2) Bromaldoxime m.p 115°C.

b.p 180°C (d) **Bromoacetal** (Bromoacetaldehyde diethyl acetal, 2-bromo-1, 1-diethoxy ethane) $CH_2BrCH(OC_2H_5)_2$

(1) Strong lachrymator liquid, soluble in hot water.
(2) Heat 2ml of compound with dil H_2SO_4. A pungent odour due to the formation of bromoacetaldehyde is observed.
(3) Reflux 82ml of t-butyl alcohol with 3.9gm potassium for 8hrs. Cool and add 19.8gm of bromoacetal. Heat on oil bath and distill off butyl alcohol at 120-130°C. Raise the temperature to 150°C and heat for few minutes. Distill the residual liquid under reduced pressure and collect ketene diethyl acetal at 84-86°C/200mmHg.
(4) Bromoacetal semicarbazone m.p 130°C.

b.p 180°C [745mm] **1, 1, 3, 3-Tetrachloroacetone** $CHCl_2COCHCl_2$

(1) Liquid with pungent odour, soluble in alcohol, ether and benzene.
(2) On adding water to the compound, crystalline hydrate is precipitated m.p 48°C.

b.p 192°C **Pentachloroacetone** $CHCl_2COCCl_3$

(1) Liquid with odour similar to chloral, insoluble in water, soluble in acetone.
(2) To 2ml of compound add 2ml water, warm the solution until it becomes milky, cool to 0°C, the crystalline hydrate is precipitated m.p 17°C.

Aromatic Liquids

b.p 214°C **o-Chloro benzaldehyde** (2-chlorobenzene carbonal)

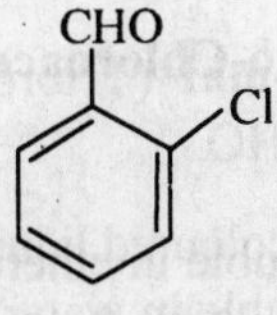

(1) Colourless liquid with pungent odour, soluble in alcohol and ether.
(2) To a solution of 5.6gm of compound in 5ml benzene add in portion a solution of 4gm of hexamethyl phosphorous triamide in 3ml ether(temp below 30°C). After addition is complete maintain the temperature at 50°C for 15min. Distill off the solvent and triturate the oily residue with 10ml water and 15ml ether. Remove the ether layer and extract the aqueous layer with ether. Combine the two organic layers, distill off ether and collect 2, 2' dichloro α, α' epoxy bibenzyl, recrystallise from methanol. (trans, m.p 76-77°C)
(3) o-Chloro benzaldehyde 2, 4-dinitrophenylhydrazone m.p 209°C.
(4) o-Chloro benzaldehyde semicarbazone m.p 226°C (5) o-Chloro benzaldehyde p-nitrophenylhydrazone m.p 249°C (6) o-Chloro benzaldehyde phenylhydrazone m.p 86°C (7) o-Chloro benzaloxime m.p 76°C(d).

b.p 214°C **m-Chlorobenzaldehyde** (3-chlorobenzene carbonal)

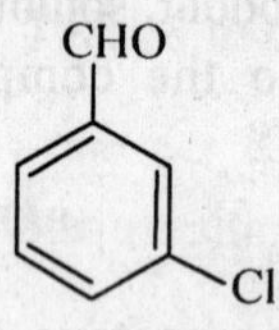

(1) Colourless liquid, soluble in alcohol and ether.
(2) m-Chloro benzaldehyde 2, 4-dinitrophenylhydrazone m.p 255°C.
(3) m-Chloro benzaldehyde semicarbazone m.p 228°C
(4) m-Chloro benzaldehyde p-nitrophenylhydrazone m.p 216°C
(5) m-Chloro benzaldehyde phenylhydrazone m.p 134°C
(6) m-Chloro benzaloxime m.p 71°C(d).
(7) m-Chlorobenzoic acid m.p 158°C.

b.p 228°C [738mm] **o-Chloroacetophenone**

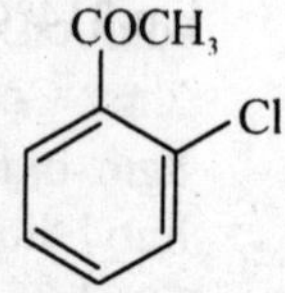

(1) Colourless liquid, soluble in ether.
(2) o-Chloroacetophenone 2, 4-dinitrophenylhydrazone m.p 206°C

(3) o-Chloroacetophenone semicarbazone m.p 160°C
(4) o-Chloroacetophenone p-nitrophenylhydrazone m.p 215°C
(5) o-Chloroacetophenone oxime m.p 113°C.

b.p 234°C **m-Bromobenzaldehyde** (3-bromo benzene carbonal)

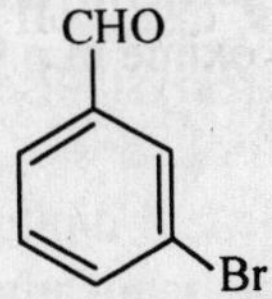

(1) Liquid which solidifies in the freezing mixture, insoluble in water.
(2) m-Bromo benzaldehyde 2, 4-dinitrophenylhydrazone m.p 257°C.
(3) m-Bromo benzaldehyde semicarbazone m.p 205°C
(4) m-Bromo benzaldehyde p-nitrophenylhydrazone m.p 220°C
(5) m-Bromo benzaldehyde phenylhydrazone m.p 141°C
(6) m-Bromo benzaloxime m.p 72°C(d).
(7) m-Bromobenzoic acid m.p 155°C.

b.p 237°C **p-Chloroacetophenone** (methyl p-chlorophenyl ketone)

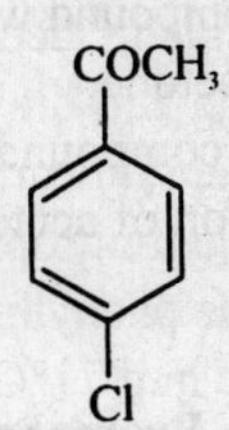

(1) Colourless liquid, insoluble in water, soluble in alcohol and ether.
(2) Reflux 10gm of compound with 30ml water, 20ml 10% NaOH solution, 5gm hydroxylamine hydrochloride and 50ml ethanol for 2hrs. Cool and recover the oxime m.p 96°C. Cool a mixture of 8gm oxime in 50ml THF. Add dropwise with stirring 70ml of butyl lithium in hexane. Then add 4.1gm of methyl anisate in 50ml THF. Continue stirring for 30min and then add 150ml 3N HCl. Heat the mixture under reflux for 1hr. Cool, neutralise with $NaHCO_3$. Separate the organic layer and extract the aqueous layer

with THF. Combine the organic extracts and concentrate, cool and collect 3-(4-chlorophenyl)-5-(4-methoxyphenyl) isoxazide, m.p 175-176°C.

(3) p-Chloroacetophenone 2, 4-dinitroacetophenone m.p 231°C
(4) p-Chloroacetophenone semicarbazone m.p 201°C.
(5) p-Chloroacetophenone p-nitrophenylhydrazone m.p 239°C
(6) p-Chloroacetophenone phenylhydrazone m.p 114°C.
(7) p-Chloroacetophenone oxime m.p 96°C.

Solid Aliphatic

m.p 45°C **sym Dichloroacetone** (1, 3-dichloro-2-propanone)

$CH_2ClCOCH_2Cl$

(1) Crystalline solid with highly irritating odour, vapours causing blisters on skin, soluble in water, alcohol and ether.
(2) Reduces Fehling's solution.
(3) sym Dichloroacetone 2, 4-dinitrophenylhydrazone m.p 133°C.
(4) sym Dichloroacetone semicarbazone m.p 120°C.

m.p 62°C **sym Diiodoacetone** (1, 3-diiodo-2-propanone)

CH_2ICOCH_2I

(1) Colourless crystalline compound with pungent odour, easily soluble in ether, benzene and acetone.
(2) Heat few milligrams of compound with Zn and acetic acid. Sweet smell due to the formation of acetone is observed.

Solid Aromatic

m.p 22°C, b.p 230°C **o-Bromobenzaldehyde** (2-bromo benzene carbonal)

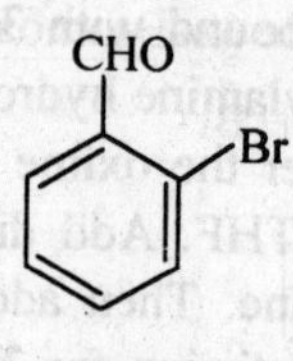

(1) Insoluble in water, soluble in alcohol and benzene.
(2) o-Bromobenzaldehyde semicarbazone m.p 214°C
(3) o-Bromobenzaldehyde p-nitophenylhydrazone m.p 240°C

(4) o-Bromobenzaloxime m.p 102°C.
(5) o-Bromobenzoic acid m.p 150°C

m.p 47°C, b.p 214 °C **p-Chlorobenzaldehyde** (4-chlorobenzene carbonal)

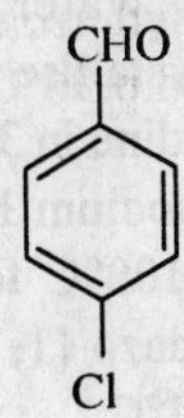

(1) Colourless compound with odour similar to benzaldehyde, soluble in hot water, alcohol and ether.
(2) Gives crystalline bisulphite addition compound with sodium metabisulphite.
(3) Reflux 5gm of compound with 120ml $KMnO_4$ solution (8gm in 15ml water) at 80-85°C. Add with stirring 10% KOH solution to make it alkaline. Filter hot. Cool, acidify with dil HCl and collect p-chlorobenzoic acid, m.p 243°C.
(4) Heat together 12gm of p-chloro benzaldehyde with 25ml methanol and 12ml formalin at 65°C. Then add in portion a solution of 17gm of KOH in 25ml water. Continue heating for another 60min. Cool, add 25ml water. Extract with ether, distill off ether and collect p-chlorobenzyl alcohol, m.p 72°C.
(5) p-Chloro benzaldehyde 2, 4-dinitrophenylhydrazone m.p 265°C.
(6) p-Chloro benzaldehyde semicarbazone m.p 231°C
(7) p-Chloro benzaldehyde p-nitrophenylhydrazone m.p 220°C
(8) p-Chloro benzaloxime m.p 106°C
(9) p-Chloro benzaldehyde phenylhydrazone m.p 127°C

m.p 50°C **Phenacyl bromide** (ω-Bromoacetophenone)

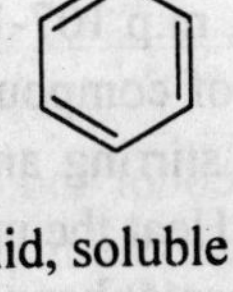

(1) White crystalline solid, soluble in alcohol and ether, vapours cause irritation to the eyes, provoking flow of tears.

(2) Dissolve 1gm of compound in 5ml alcohol, add 1ml aniline. Shake thoroughly and leave it at room temperature for sometime. Collect phenacylanilide, m.p 93°C.
(3) To a solution of 5gm phenacyl bromide and 2.6ml salicylaldehyde in 5ml ethanol add 3.45gm of anhydrous potassium carbonate. Heat the mixture on water bath at 100°C for 90 min. Pour with stirring in 75ml cold water. Collect the precipitated 2-benzoylbenzofuran, recrystallise from ethanol, m.p 89-91°C.
(4) To 2.9gm of 2-aminopyridine in 30ml ethanol add 5gm of phenacyl bromide and 4.2gm of sodium bicarbonate. Stir the mixture for 2hrs and then reflux at 100°C for 2hrs. Cool, pour in ice water and collect 2-Phenylimidazo [1, 2-a] pyridine, recrystallise from aqueous ethanol, m.p 134°C.
(5) ω-Bromoacetophenone 2, 4-dinitrophenylhydrazone m.p 220°C
(6) ω-Bromoacetophenone semicarbazone m.p 146°C.
(7) ω-Bromoacetophenone oxime m.p 90°C.

m.p 51°C **p-Bromoacetophenone** (methyl p-bromo phenyl ketone)

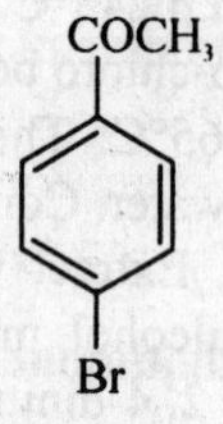

(1) Crystalline solid, insoluble in water, soluble in alcohol and ether.
(2) To a solution of 5gm of compound in 10ml glacial acetic acid add dropwise 1.3ml bromine. Shake the mixture occassionally and keep the temperature below 20°C. Cool and collect p-bromo phenacyl bromide, m.p 109°C.
(3) Reflux 5gm of compound with 1.3gm of sulphur and 5ml morpholine for 8hrs. Saponify with 25ml of 10% alcoholic NaOH solution and extract with ether. Evaporate off ether and collect p-bromo phenylacetic acid, m.p 107-108°C.
(4) Cool a solution of 5gm of compound in 15ml glacial acetic acid to 20°C. Now add with stirring and dropwise 2.6ml bromine in 10ml glacial acetic acid. Heat the mixture on water bath, cool and collect p, α, α' tribromoacetophenone, m.p 91°C.

(5) p-Bromoacetophenone 2, 4-dinitrophenylhydrazone m.p 230°C
(6) p-Bromoacetophenone semicarbazone m.p 208°C.
(7) p-Bromoacetophenone phenylhydrazone m.p 126°C
(8) p-Bromoacetophenone oxime m.p 129°C.

m.p 54°C **Phenacyl chloride** (ω-Chloroacetophenone, phenyl chloro methyl ketone)

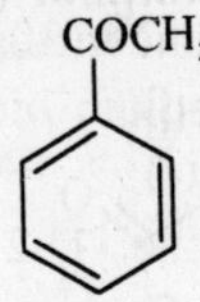

(1) Crystalline solid, insoluble in water, soluble in alcohol and ether. Powerful lachrymator
(2) To a solution of 3gm of compound in 20ml ether pass HCl gas. Then add dropwise 2.3ml n-butyl nitrite (exothermic reaction). After addition is complete pass HCl gas for another 15min. Keep the mixture overnight and then distill off ether, collect ω-Chloroisonitroso acetophenone, m.p 132°C.
(3) ω-Chloroacetophenone 2, 4 dinitrophenylhydrazone m.p 212°C
(4) ω-Chloroacetophenone oxime m.p 89°C.

m.p 55°C **Bromobenzoquinone** 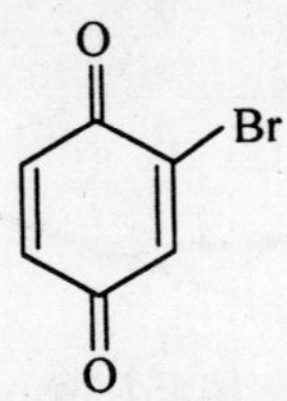

(1) Crystalline compound, easily soluble alcohol, ether, benzene and chloroform.
(2) To few milligrams of compound add NaOH solution. A green coloured solution is obtained which slowly changes to black brown.

m.p 58°C **p-Bromobenzaldehyde** (4-bromo benzene carbonal)

(1) Insoluble in water, soluble in alcohol and ether.
(2) p-Bromo benzaldehyde semicarbazone m.p 228°C
(3) p-Bromobenzaldehyde p-nitrophenylhydrazone m.p 208°C.
(4) p-Bromobenzaldehyde phenylhydrazone m.p 113°C
(5) p-Bromobenzaloxime m.p 111°C.
(6) p-Bromobenzoic acid m.p 251°C.

m.p 76°C **α-Bromo-d-Camphor** (3-bromo-d-camphor)

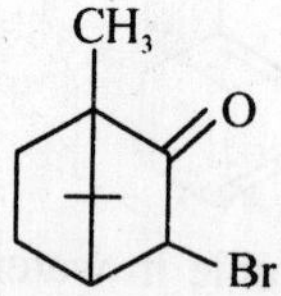

(1) Colourless solid with odour similar to camphor, insoluble in water, soluble in alcohol and ether.
(2) Camphoric acid m.p 187°C.

m.p 78°C **p-Iodobenzaldehyde** (4-iodo benzene carbonal)

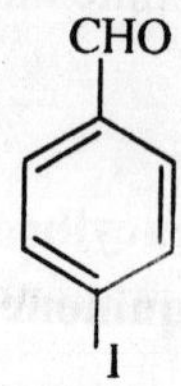

(1) p-Iodobenzaldehyde 2, 4-dinitrophenylhydrazone m.p 257°C
(2) p-Iodobenzaldehyde semicarbazone m.p 224°C.
(3) p-Iodobenzaldehyde p-nitrophenylhydrazone m.p 201°C
(4) p-Iodobenzaldehyde phenylhydrazone m.p 121°C.
(5) p-iodobenzoic acid m.p 270°C.

m.p 78°C **p-Chlorobenzophenone** (4-chloro benzophenone)

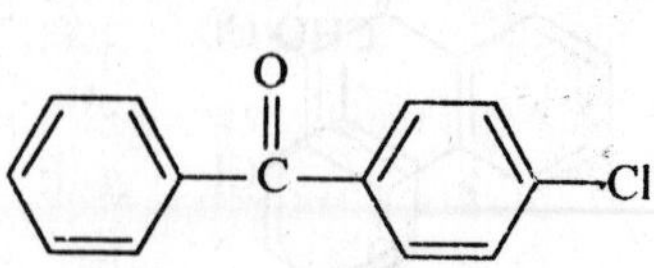

(1) Crystalline compound, insoluble in water, soluble in ether.
(2) p-Chlorobenzophenone 2, 4 dinitrophenylhydrazone m.p 185°C
(3) p-Chlorobenzophenone phenylhydrazone m.p 106°C.

m.p 87°C **Benzoquinone dibromide** (2, 3-dibromo-p-benzoquinone)

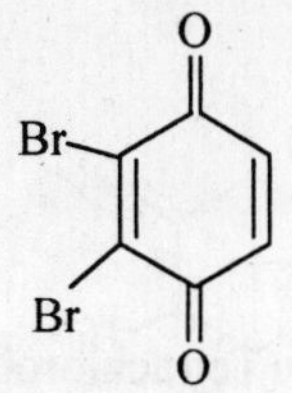

(1) Yellow crystalline solid, easily soluble in alcohol, ether and benzene
(2) Heat 200mg of compound with Zn dust and 5ml glacial acetic acid. Cool and collect hydroquinone.

m.p 108°C **p-Bromo phenacylbromide**

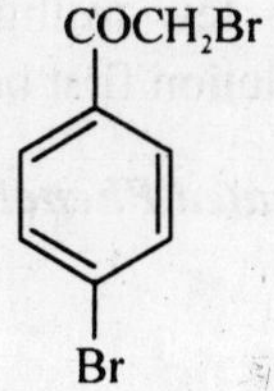

(1) Crystalline solid, insoluble in water, slightly soluble in alcohol, soluble in ether.
(2) p-Bromo phenacylbromide oxime m.p 115°C.

m.p 190-192°C **2, 3 Dichloronaphthoquinone**

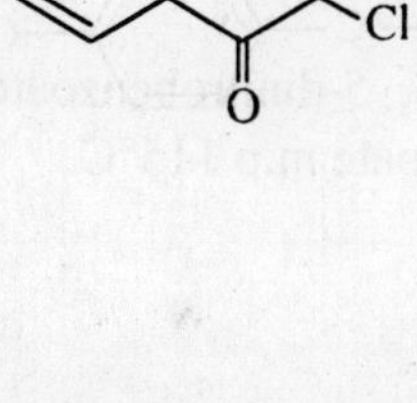

(1) Golden yellow crystalline solid, insoluble in water, soluble in hot alcohol.
(2) Boil few milligrams of compound with NaOH solution. Caramine red coloured solution is obtained.
(3) To 1gm of compound in 5ml alcohol add 2ml of aniline and warm on water bath. Cool and collect chloranilinenaphthoquinone, m.p 202°C.

m.p 290°C **Chloranil** (Tetrachlorobenzoquinone)

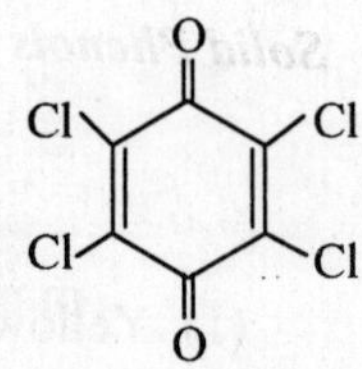

(1) Yellow crystalline solid with characteristic odour, insoluble in water, soluble in ether. Sublimes.
(2) Warm 100mg of compound with 2ml dimethylaniline. An intense deep violet blue colouration appears.
(3) To few milligrams of compound add dil KOH solution. The solution first becomes greenish black and then purple.

(iii) Halogenated Phenols

Liquid

b.p 176°C **o-Chlorophenol** (1-chloro-2-hydroxy benzene)

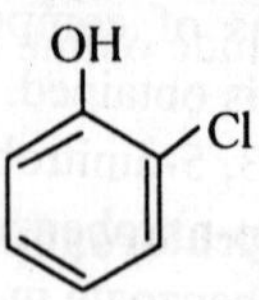

(1) Colourless liquid with unpleasant persistant odour, sparingly soluble in water, easily soluble in alcohol.
(2) To 2ml of compound add $FeCl_3$ solution. A violet coloured solution is obtained.
(3) o-Chloro phenyl 3, 5-dinitrobenzoate m.p 143°C (4) o-Chloro phenyl p-nitrobenzoate m.p 115°C.

b.p 195°C **o-Bromophenol** 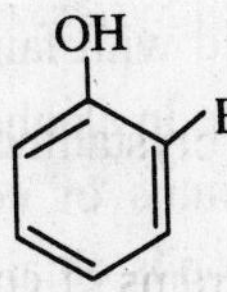

(1) Colourless liquid with unpleasant persistant odour, soluble in alcohol.
(2) To 2ml of compound add $FeCl_3$ solution. A violet coloured solution is obtained.
(3) 2, 4, 6-tribromophenol m.p 95°C

Solid Phenols

m.p 33°C, b.p 236°C **m-Bromophenol** 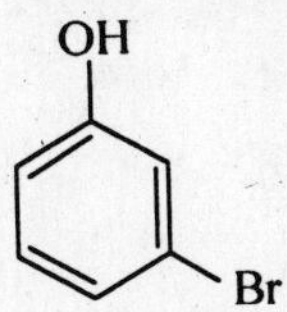

(1) Crystalline solid with unpleasant odour, soluble in alcohol
(2) m-Bromo phenyl benzoate m.p 86°C.

m.p 33°C, b.p 214°C **m-Chlorophenol** (1-chloro-3-hydroxy benzene)

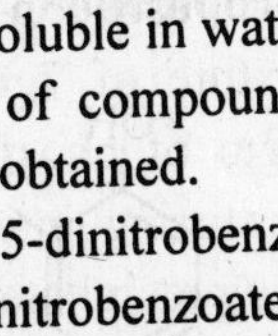

(1) Colourless solid, insoluble in water, soluble in alcohol.
(2) To few milligrams of compound add $FeCl_3$ solution. A violet coloured solution is obtained.
(3) m-Chloro phenyl 3, 5-dinitrobenzoate m.p 156°C
(4) m-Chloro phenyl p-nitrobenzoate m.p 99°C
(5) m-Chloro phenyl benzoate m.p 71°C.

m.p 36°C **2, 4-dibromophenol** 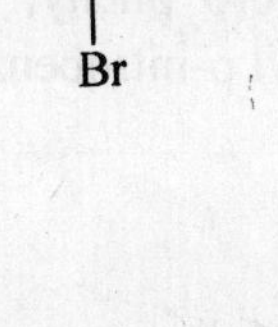

(1) Crystalline solid with faint persistant unpleasant odour, insoluble in water, soluble in alcohol, ether and benzene.
(2) To few milligrams of compound add $FeCl_3$ solution. A violet coloured solution is obtained.
(3) Reflux 5gm of compound in 15ml benzene with triphenyl phosphine at 100-150°C for 3hrs. Treat the resulting adduct with 2N NaOH solution and recover p-bromo phenol, m.p 64°C.
(4) 2, 4-dibromo phenyl p-nitrobenzoate m.p 184°C
(5) 2, 4-dibromo phenyl benzoate m.p 97°C
(6) 2, 4-dibromo phenyl acetate m.p 36°C.

m.p 44°C **p-Chlorophenol** (1-chloro-4-hydroxy benzene)

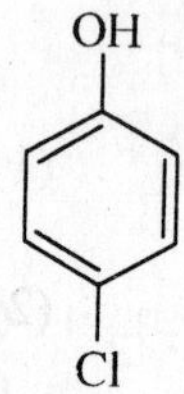

(1) Colourless solid with faint unpleasant odour, insoluble in water, soluble in alcohol and ether.
(2) To few milligrams of compound add $FeCl_3$ solution. A blue-violet coloured solution is obtained.
(3) Cool a solution of 3gm of triphenylphosphine in 10ml acetonitrile and add dropwise with stirring .6ml bromine. Distill off the excess acetonitrile and add 1gm of p-chloro phenol to it. Heat the resultant mixture on sand bath at 250-280°C for 3hrs. Cool and add 5ml water. Steam distill and recover p-bromo chlorobenzene, m.p 66°C.
(4) Heat together a mixture of 7gm of salicylic acid, 6gm of p-chlorophenol and 3gm of phosphorous oxychloride at 75-80°C for 4hrs. Pour with stirring into a solution of 6gm Na_2CO_3 in 40ml water. Filter the precipitated p-chloro phenyl salicylate and recrystallise from ethanol, m.p 70°C.
(5) Heat a mixture of 2.5gm of p-chlorophenol, 6gm phthalic anhydride, 1gm boric acid and 21ml conc H_2SO_4 in an oil bath at 200°C for 4hrs. Cool and pour with stirring in 50ml cold water and filter. Boil the precipitate with 100ml water and filter hot. Suspend the residue in 50ml boiling water, make it alkaline by

adding NaOH solution. Filter hot. Saturate the filtrate with carbon dioxide and collect quinzarin, m.p 199-200°C.

(6) p-Chloro phenyl 3, 5-dinitrobenzoate m.p 186°C.

(7) p-chloro phenyl p-nitrobenzoate m.p 168°C

(8) p-chloro phenyl benzoate m.p 87°C.

m.p 43°C **2, 4-Dichlorophenol** 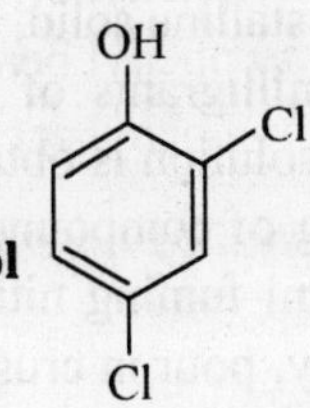

(1) Crystalline compound with unpleasant odour, insoluble in water, soluble in alcohol and ether.

(2) To few milligrams of compound add $FeCl_3$ solution. A blue-violet coloured solution is obtained.

(3) To a mixture of 4gm of 2, 4-dichlorophenol and 2.35gm of chloroacetic acid add with stirring a solution of 2.25gm of NaOH in 15ml water. Heat the mixture and evaporate off all the liquid. To the residue add 50ml water, cool and acidify with dil HCl. Extract with ether. Evaporate off the ether and collect 2, 4-dichlorophenoxy acetic acid, m.p 138°C

(4) 2, 4-dichloro phenyl benzoate m.p 97°C

(5) Bromo m.p 68°C.

m.p 43°C **o-Iodophenol** 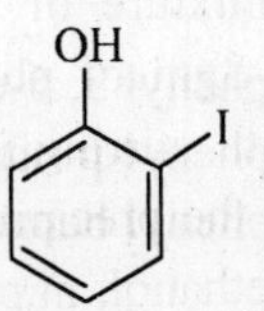

(1) Crystalline compound, insoluble in water, soluble in alcohol, ether and carbon disulphide.

(2) Boil 1gm of compound with 6-7ml conc HNO_3. Cool, pour in ice and collect picric acid.

(3) o-Iodo phenyl benzoate m.p 34°C.

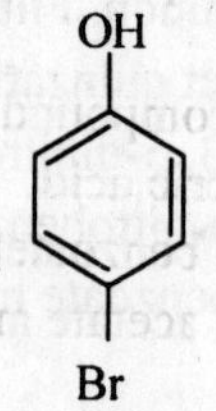

m.p 64°C **p-Bromophenol**

(1) White crystalline solid, soluble in alcohol and ether.
(2) To few milligrams of compound add $FeCl_3$ solution. A violet coloured solution is obtained.
(3) To 500mg of compound in glacial acetic acid add dropwise with stirring 2ml fuming nitric acid. After addition is complete shake thoroughly, pour in crushed ice and recover the dinitro derivative, m.p 76°C.
(4) p-Bromo phenyl 3, 5-dinitro benzoate m.p 191°C
(5) p-Bromo phenyl p-nitrobenzoate m.p 180°C.
(6) p-Bromo phenyl benzoate m.p 102°C.
(7) 2, 4, 6-tribromophenol m.p 95°C.

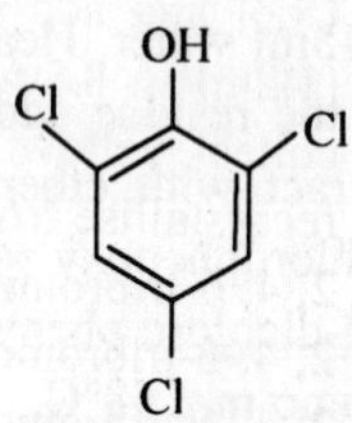

m.p 68°C **2, 4, 6-trichlorophenol**

(1) Crystalline compound, insoluble in water, soluble in alcohol and ether.
(2) 2, 4, 6-trichloro phenyl 3, 5-dinitrobenzoate m.p 136°C
(3) 2, 4, 6-trichloro phenyl-p-nitrobenzoate m.p 106°C.
(4) 2, 4, 6-trichloro phenyl benzoate m.p 74°C.

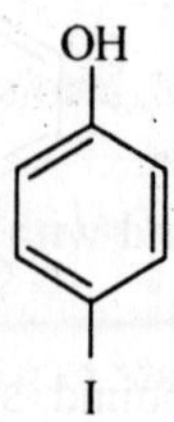

m.p 94°C **p-Iodophenol**

(1) Crystalline compound, insoluble in water, soluble in alcohol and ether.
(2) Boil 1gm of compound with 6-7ml conc HNO_3. Cool, pour in ice and collect picric acid.
(3) p-Iodo phenyl benzoate m.p 119°C.
(4) p-Iodo phenyl acetate m.p 32°C.

m.p 96°C **sym Tribromophenol** (2, 4, 6-tribromophenol)

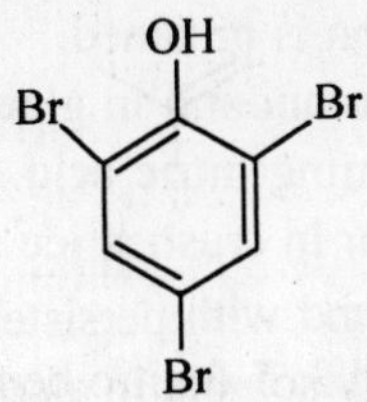

(1) Hair like crystalline compound, insoluble in water, soluble in alcohol.
(2) Heat a mixture of 6.7gm of compound, 3gm sodium acetate and 40ml glacial acetic acid at 70°C until a clear solution is obtained. Cool and add dropwise with stirring 10ml of bromine solution (1.2ml in 8.8ml acetic acid). Shake thoroughly, pour in ice and collect yellow 2, 4, 4, 6-tetrabromo 2, 5 cyclohexadien-1-one, recrystallise from chloroform, m.p 125-130°C.
(3) 2, 4, 6-tribromophenyl 3, 5-dinitrobenzoate m.p 174°C.
(4) 2, 4, 6-tribromophenyl p-nitrobenzoate m.p 153°C
(5) 2, 4, 6-tribromophenyl benzoate m.p 81°C
(6) 2, 4, 6-tribromo phenyl acetate m.p 87°C.

m.p 106°C **Chlorohydroquinone** (2-chloro-1, 4-benzenediol)

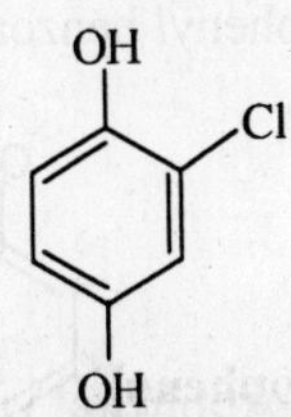

(1) Crystalline compound, soluble in water, alcohol and ether.

(2) Reduces Tollens reagent on warming.
(3) Chlorohydroquinone dibenzoate m.p 130°C
(4) Chlorohydroquinone diacetate m.p 72°C.

m.p 158°C **sym Triiodophenol** (2, 4, 6-triiodophenol)

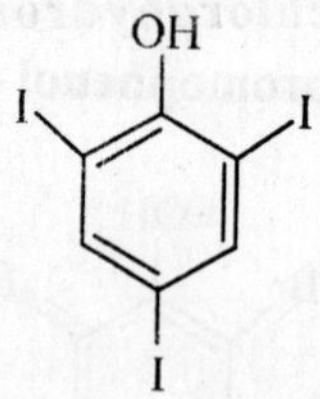

(1) Crystalline compound with persistent unpleasant odour, insoluble in water, moderately soluble in alcohol.
(2) 2, 4, 6-triiodophenyl 3, 5-dinitrobenzoate m.p 181°C.
(3) 2, 4, 6-triiodophenyl benzoate m.p 137°C
(4) 2, 4, 6-triiodo phenyl acetate m.p 156°C.

m.p 177°C **Trichloropyrogallol**

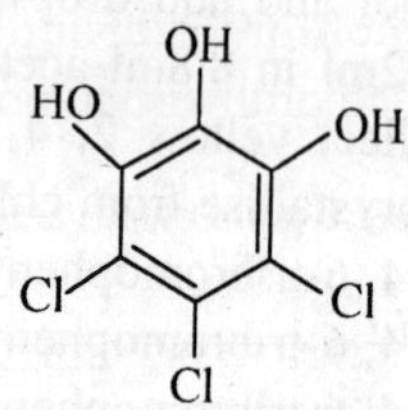

(1) Crystalline compound, soluble in hot water, alcohol and ether.
(2) Reduces ammonical silver nitrate solution.
(3) To few milligrams of compound in ether add barium hydroxide solution. Blue coloured solution is obtained.

m.p 190°C **Pentachlorophenol**

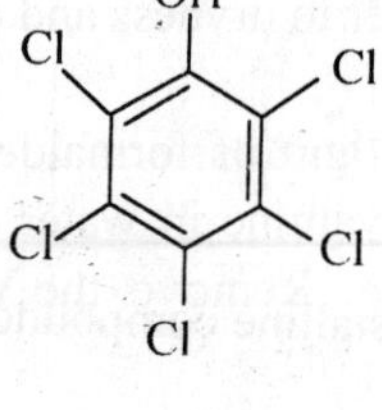

(1) Crystalline compound which turns brown on exposure to air, insoluble in water and benzene, soluble in alcohol and ether. Vapours provokes violent sneezing
(2) Pentachloro phenyl benzoate m.p 159°C
(3) Pentachloro phenyl acetate m.p 150°C.

m.p 239°C **Tetrachlorohydroquinone** (tetrachloro 1, 4-benzenediol)

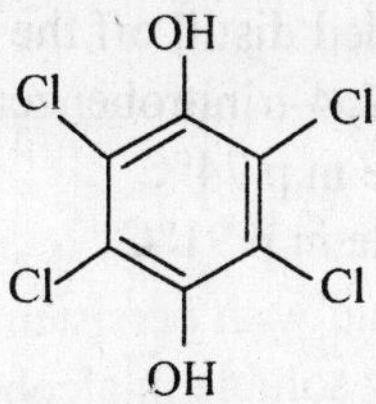

(1) Crystalline compound, insoluble in water and benzene, soluble in alcohol and ether. Sublimes
(2) Reduces ammonical silver nitrate solution.
(3) Boil 500mg of compound with 5ml conc HNO_3. Cool and recover chloranil m.p 290°C.
(4) Tetrachlorohydroquinone dibenzoate m.p 233°C (5) Tetrachlorohydroquinone diacetate m.p 245°C.

(iv) Halogenated Carboxylic Acids

Aliphatic Liquid

b.p 72°C **Trifluoroacetic acid** (trifluoro ethanoic acid)

CF_3COOH

(1) Colourless liquid, soluble in water.
(2) To a suspension of 4gm of silver oxide in 5ml water add 3.75gm of trifluoacetic acid. Filter the resulting solution and evaporate the filtrate to dryness. Dissolve the salt in ether, filter and evaporate the ether to dryness and collect silver trifluoroacetate as crystalline salt.
(2) Cool 10gm of formaldehyde and add 40% solution of 23gm of dimethyamine in water. Keep overnight and add saturated KOH solution. Remove the upper layer, distill and collect the oily

residue. Cool 5ml of trifluroacetic acid to 0°C and add the oily residue to it. Keep the temperature of mixture below –10°C and add 1 gm of 3-methyl 2-butanone. Heat at 65°C for 90min and then raise the temperature to 145°C. Cool and pour in 10ml potassium carbonate solution. Extract with dichloromethane. Distill and collect 1-(Dimethylamino)-4-methyl-3-pentanone, b.p 49°C/ 3mmHg

(3) Heat a suspension of 2.7ml of 90% H_2O_2 and 17ml trifluoroacetic acid in 25ml chloroform and add in portion 4.6gm of 2, 4-dinitroaniline to it (exothermic reaction). After the vigorous reaction has subsided distill off the solvent. Treat the oily residue with ice water, 1, 2, 4-trinitrobenzene separates out, m.p 58-59°C.
(4) Trifluoroacetamide m.p 74°C
(5) Trifluoroacetanilide m.p 91°C
(6) NE 114

b.p 186°C **2-Chloropropionic acid** (2-chloropropanoic acid)
$CH_3CHClCOOH$

(1) Liquid with sharp odour, miscible with water and alcohol.
(2) 2-Chloropropionamide m.p 80°C
(3) 2-Chloropropionanilide m.p 92°C
(4) 2-Chloro propio-p-toluidide m.p 124°C.
(5) NE 108.

b.p 194°C(d) **Dichloro acetic acid** (dichloroethanoic acid)
$CHCl_2COOH$

(1) Colourless liquid, miscible with water and alcohol.
(2) Dichloroacetamide m.p 98°C
(3) Dichloroacetanilide m.p 118°C
(4) Dichloroacet-p-toluidide m.p 153°C.
(5) S Benzylthiouronium dichloroacetate m.p 178°C
(6) NE 129.

Aliphatic Solid

m.p 25°C, b.p 203°C **2-Bromopropionic acid** (2-bromo propanoic acid) $CH_3CHBrCOOH$

(1) Colourless crystalline solid, soluble in water, alcohol and ether.
(2) To cooled 100ml NH_4OH solution add 3.85gm of 2-bromopropionic acid and leave it at room temperature for 4 days. Concentrate and filter. Concentrate further, cool and add 30ml methanol and leave overnight in a refrigerator. Collect the crude alanine, recrystallise from water-methanol mixture, m.p 295-296°C(d)
(3) 2-Bromopionamide m.p 123°C
(4) 2-Bromopropionanilide m.p 99°C
(5) 2-Bromo propio-p-toluidide m.p 125°C.
(6) NE 153.

m.p 41°C, b.p 204°C **3-Chloropropionic acid** (3-chloropropanoic acid) CH_2ClCH_2COOH

(1) Crystalline solid, soluble in water and alcohol.
(2) 3-Chloropropionamide m.p 101°C
(3) 3-Chloropropionanilide m.p 119°C
(4) 3-Chloro propio-p-toluidide m.p 121°C.
(5) NE 108.

m.p 48°C **Dibromoacetic acid** (dibromoethanoic acid) $CHBr_2COOH$

(1) Crystalline deliquescent solid, slightly soluble in water, soluble in alcohol and ether.
(2) Dibromoacetamide m.p 156°C.
(3) NE 218.

m.p 50°C **Bromoacetic acid** (bromoethanoic acid) $CH_2BrCOOH$

(1) Colourless crystalline deliquescent solid with sharp odour, causes blisters on skin. Soluble in water, alcohol and ether.
(2) To 6.12gm of compound and 10gm crushed ice add 4.5gm NaOH. To the neutral solution add 3gm of acetoxime and 4.5gm of NaOH (temp below 20°C). Extract with ether. Cool the residual aqueous solution, acidify with 5ml conc HCl (temp below 15°C). Saturate with NaCl and extract with ether. Combine the two ether layers and distill to remove ether. Collect carboxymethoxime, m.p 76°C.

(3) Bromoacetamide m.p 91°C
(4) Bromoacetanilide m.p 131°C
(5) Bromoacet-p-toluidide m.p 91°C
(6) NE 139.

m.p 56°C **3, 3-Dichloropropionic acid** $CHCl_2CH_2COOH$

(1) Crystalline compound, soluble in water, alcohol, ether, chloroform and benzene.
(2) 3, 3-dichloropropionamide m.p 140°C
(3) NE 143

m.p 57°C **Trichloroacetic acid** CCl_3COOH

(1) Crystalline deliquescent solid, soluble in water, alcohol and ether.
(2) Boil few milligrams of compound with water containing 1-2 drops of NaOH solution. Sweet smell due to the formation of chloroform is observed.
(3) To a cooled solution of 3.28gm of trichloroacetic acid in 3ml methanol add a solution of 460mg of sodium in 8ml methanol. Evaporate off the solution under reduced pressure and collect sodium trichloroacetate as white solid. Add 1ml 2, 5-dimethyl pyrrole and 10ml 1, 2-dimethoxy ethane to it. Reflux for 24 hrs, cool, filter, concentrate the filtrate and dissolve the precipitated residue in 12ml 2N HCl. Basify with NaOH solution, extract with ether. Distill off ether and collect 3-chloro-2, 6-dimethyl pyridine as colourless liquid, b.p 66-69°C/11mmHg
(4) Trichloroacetamide m.p 141°C
(5) Trichloroacetanilide m.p 94°C
(6) Trichloroacet-p- toluidide m.p 113°C
(7) S Benzylthiouronium trichloroacetate m.p 149°C.
(8) NE 163

m.p 62°C **3-Bromopropionic acid** (3-bromopropanoic acid) CH_2BrCH_2COOH

(1) Crystalline compound, soluble in water, alcohol and ether.
(2) To 6gm of compound add 12ml of CCl_4, 8ml of ethyl alcohol and 100mg of sulfosalicylic acid. Reflux the mixture for 2-3hrs. Cool,

wash the reaction mixture with dilute sodium carbonate and distill. Distill the residual liquid under reduced pressure and collect 3-bromopropionate at 62-65°C/15mmHg

(3) 3-Bromopropionamide m.p 111°C (4) NE 153

m.p 63°C **Chloroacetic acid** (chloroethanoic acid) $CH_2ClCOOH$

(1) Colourless crystalline deliquescent solid with sharp odour, causes blisters when in contact with skin, soluble in water.
(2) To cooled 100ml NH_4OH solution add 2.4gm of chloroacetic acid and leave it at room temperature for 2 days. Concentrate and filter. Concentrate further, cool and collect the crude glycine, dissolve the product in methanol and filter to remove to NH_4Cl, recrystallise from hot water, m.p 252-254°C(d).
(3) To a mixture of 4gm of 2, 4 dichlorophenol and 2.35gm of chloroacetic acid add with stirring a solution of 2.25gm of NaOH in 15ml water. Heat the mixture and evaporate off all the liquid. To the residue add 50ml water, cool and acidify with dil HCl. Extract with ether. Evaporate off the ether and collect 2, 4-dichlorophenoxy acetic acid, m.p 138°C
(4) Cool a solution of 1.89gm of compound in 2ml water and add a cold solution of 1.60gm of NaOH in 5ml water(temp below 25°C). Add aqueous solution of .6gm of methylamine(exothermic reaction, temp below 50°C). Keep at room temperature for 2hrs. Add solution of 2.57gm of $BaCl_2$ in 5ml hot water and reflux for 90min. Cool and collect the barium salt of amino acid. Dissolve it in 6ml water and add 5N H_2SO_4(1.4ml)per gm. Filter and concentrate the filtrate, cool and collect methyliminodiacetic acid, m.p 215°C.
(5) Chloroacetamide m.p 120°C
(6) Chloroacetanilide m.p 137°C
(7) Chloroacet-p-toluidide m.p162°C.
(8) S benzylthiouronium chloroacetate m.p 160°C.
(9) NE 94.

m.p 83°C **3-Iodopropionic acid** (3-iodopropanoic acid) CH_2ICH_2COOH

(1) Glassy crystalline compound, soluble in hot water, alcohol and ether.

(2) 3-Iodopropionamide m.p 101°C.
(3) NE 200.

m.p 83°C **Iodoacetic acid** (iodoethanoic acid) CH_2ICOOH

(1) Colourless leaflets which cause blisters when in contact with skin, vapours irritate the eyes. Soluble in hot water.
(2) Iodoacetamide m.p 95°C
(3) Iodoacetanilide m.p 143°C
(4) NE 186.

m.p 110°C (96°C) **Diiodoacetic acid** (diiodoethanoic acid) CHI_2COOH

(1) Yellow coloured crystalline solid, sparingly soluble in water.

m.p 113°C **Bromomalonic acid** (bromopropane dioic acid) $CHBr(COOH)_2$

(1) Crystalline deliquescent solid, soluble in alcohol and ether.
(2) NE 92

m.p 133°C **Tribromoacetic acid** (tribromoethanoic acid) CBr_3COOH

(1) Crystalline compound, soluble in water
(2) Tribromoacetamide m.p 122°C (3) NE 297.

m.p 167°C **(±)2, 3 dibromosuccinic acid** (2, 3 dibromo butan diacid) $(CHBrCOOH)_2$

(1) Colourless glistening crystals, insoluble in water, soluble in hot water, alcohol and ether.
(2) On heating at 200°C decomposition takes place with the formation of HBr and bromomaleic acid.
(3) p-Nitrobenzyl ester m.p 168°C.
(4) NE 138

Aromatic Solid

m.p 142°C **2-Chlorobenzoic acid** 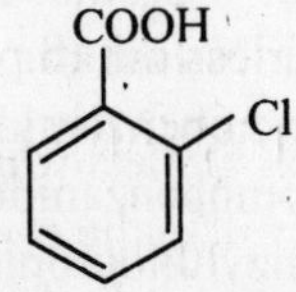

(1) White crystalline solid, soluble in hot water, alcohol and ether.
(2) Reflux together 4.1gm of compound with 15gm of aniline, 4.1gm of potassium carbonate and 100mg of copper oxide on oil bath for 2 hrs. Distill off the excess aniline, add a pinch of animal charcoal, boil for 10 min and filter. Add the filtrate to the mixture of 3ml conc HCl and 6ml water. Cool and collect the precipitated N-Phenylanthranilic acid, m.p 180°C.
(3) To a solution of 3gm of compound in 6ml conc H_2SO_4 add a mixture of 1ml conc HNO_3 and 2ml conc H_2SO_4. Keep the mixture overnight and then pour in ice water. Collect the precipitated 6-chloro-3 nitro benzoic acid, recrystallise from alcohol, m.p 165°C.
(4) 2-Chlorobenzamide m.p 142°C
(5) 2-Chlorobenzanilide m.p 118°C
(6) 2-Chlorobenz-p-toluidide m.p 131°C.
(6) NE 157.

m.p 150°C **2-Bromobenzoic acid** 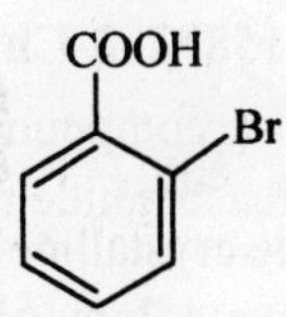

(1) White crystalline solid, soluble in hot water, alcohol, ether and chloroform.
(2) To a mixture of 15ml acetylacetone, 2.5gm o-bromobenzoic acid and 100mg of copper bromide add in portion with stirring 900mg of sodium hydride dispersed in mineral oil (exothermic reaction, temp below 55°C) Heat the mixture on water bath at 80-85°C for 5hrs. Cool, pour in 10ml water and filter after 10min. Remove the organic layer and acidify the aqueous layer with conc HCl. Collect the precipitated 2-(1-acetyl-2-oxopropyl) benzoic acid, m.p 144°C.

(3) To 15ml thionyl chloride add 4gm of compound. Heat the mixture under reflux for 3hrs. Distill off the excess thionyl chloride on water bath. Distill the residual oil under reduced pressure and collect o-bromobenzoyl chloride at b.p 132-134°C/20mmHg as colourless oil. (b.p 245°C)
(4) 2-Bromobenzamide m.p 155°C
(5) 2-Bromobenzanilide m.p 141°C
(6) S Benzylthiouronium 2-bromobenzoate m.p 171°C
(7) NE 201.

m.p 155°C **3-Bromobenzoic acid**

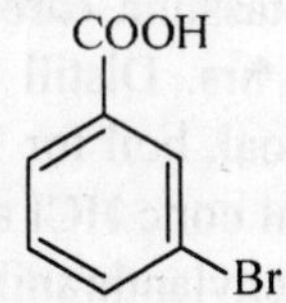

(1) Crystalline compound, soluble in hot water, alcohol and ether.
(2) 3-Bromobenzamide m.p 155°C
(3) 3-Bromobenzanilide m.p 146°C
(4) S Benzyl isothiouronium 3-bromobenzoate m.p 168°C
(5) NE 201.

m.p 158°C **3-Chlorobenzoic acid**

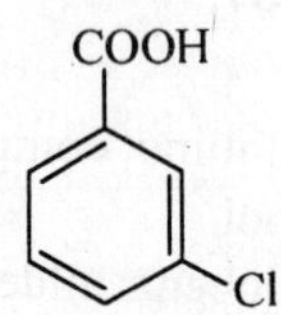

(1) White crystalline solid, soluble in hot water.
(2) Reflux 4.7gm of compound with 33ml of boron trifluoride – methanol complex on oil bath for 2hrs. Cool, pour in 125 ml saturated sodium carbonate solution. . Extract with ether. Evaporate off the ether and distill the residue under reduced pressure and collect m-chloro benzoate at 63°C/3mmHg.
(3) To 8gm of compound add 14gm (excess) of thionyl chloride and reflux. Distill and collect m-chlorobenzoyl chloride b.p 112°C/11mmHg. Add this chloride with stirring to the cooled mixture of 150mg of magnesium sulphate, 3.6gm NaOH, 36ml water 9ml 30% H_20_2 and 45ml dioxane(temp below 15°C). Continue stirring

for 15min and then add 90ml of 20% H_2SO_4. Exctract with dichloromethane. Distill off the solvent and collect m-chloroperbenzoic acid as white solid.

(4) 3-Chlorobenzamide m.p 134°C
(5) 3-Chlorobenzanilide m.p 124°C
(6) S Benzyl isothiouronium 3-chlorobenzoate m.p 155°C
(7) NE 157.

m.p 162°C **2-Iodobenzoic acid**

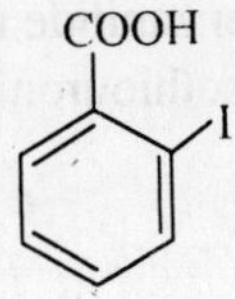

(1) Crystalline solid, insoluble in water, soluble in alcohol and ether.
(2) 2-Iodobenzamide m.p 184°C
(3) 2-Iodobenzanilide m.p 142°C
(4) NE 248.

m.p 187°C **3-Iodobenzoic acid**

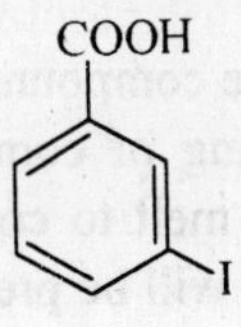

(1) Crystalline compound, sparingly soluble in water, soluble in alcohol.
(2) 3-Iodobenzamide m.p 186°C
(3) NE 248.

m.p 243°C **4-Chlorobenzoic acid**

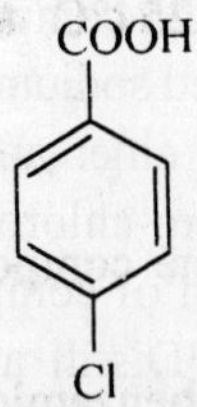

(1) Crystalline compound, soluble in alcohol and ether.

(2) Reflux 16gm of compound with 20gm of thionyl chloride until the solution is complete. Distill off the excess thionyl chloride and collect 4-chlorobenzoyl chloride at 119-120°C/22mmHg. (b.p 222°C, m.p 14°C) Warm 9gm of 4-chlorobenzoyl chloride with 25ml pyridine on steam bath for 5min. Pour in a mixture of of 50gm crushed ice and 25ml conc HCl. Collect the precipitated 4-chlorobenzoic anhydride, m.p 192°C.
(3) 4-Chlorobenzamide m.p 179°C
(4) 4-Chlorobenzanilide m.p 194°C
(5) S Benzyl isothiouronium 4-chlorobenzoate m.p 186°C
(6) NE 157.

m.p 250°C (d) **Tetrachlorophthalic acid**

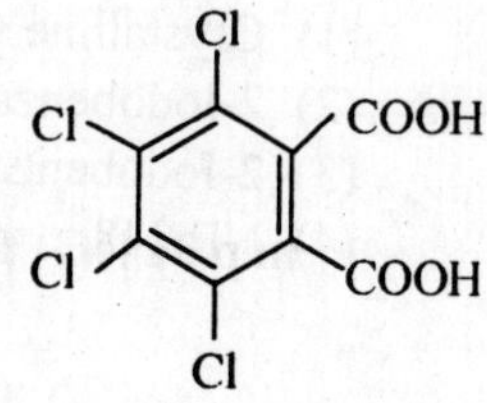

(1) Crystalline compound, soluble in alcohol.
(2) Heat 500mg of compound at 250°C for few minutes and then allow the melt to cool at room temperature. Tetrachlorophthalic anhydride will be precipitated, m.p 252°C.
(3) NE 152

m.p 251-254°C **4-Bromobenzoic acid**

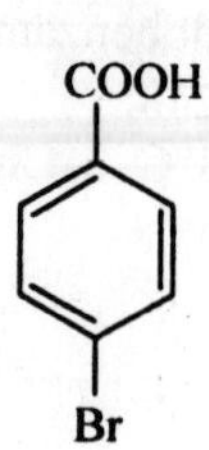

(1) Crystalline compound, insoluble in water soluble in alcohol and ether.
(2) 4-Bromobenzamide m.p 189°C
(3) 4-Bromobenzanilide m.p 197°C
(4) NE 201

m.p 270°C **4-Iodobenzoic acid**

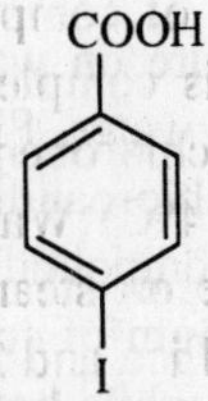

(1) Crystalline compound, soluble in alcohol. Sublimes.
(2) 4-Iodobenzamide m.p 217°C
(3) 4-Iodobenzanilide m.p 210°C
(4) NE 248

m.p 324°C **Tetraiodophthalic acid**

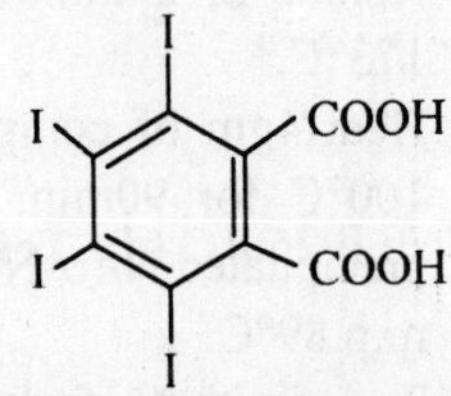

(1) Crystalline compound, difficult soluble alcohol and ether.
(2) Heat 500mg of compound above its melting point for few second. Allow the melt to cool at room temperature. Yellow crystals of tetraiodophthalic anhydride will be precipitated, m.p 325°C.
(3) NE 309

(*v*) *Halogenated Nitro Compounds*

m.p 33°C **o-Chloronitrobenzene** (1-chloro-2-nitro benzene)

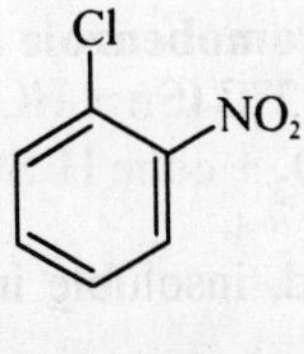

(1) Pale yellow coloured crystalline solid, slouble in alcohol and ether.
(2) Heat 48gm of iron filings in 36ml water, remove the heat source and add with stirring 4gm of compound followed by 1ml conc HCl. Again add 20gm of compound followed by 5ml conc HCl.

Heat the mixture on water bath for 1hr. Add a solution of 2gm NaOH in 4ml water. Filter hot. Remove the lower layer. Distill and collect o-chloroaniline, b.p 206-209°C.

(3) Heat 5gm of compound with 8gm of sand in an oil bath at 225°C and then add 5gm of copper bronze. Continue heating for 90min and then pour while hot on 12gm of sand. Boil the mixture with 75ml ethanol for 10 min and filter. Cool and collect 2, 2'-dinitrobiphenyl, m.p 124°C.

(4) Heat 3.8gm of sodium sulphide with 500mg of sulphur and 5ml water for 5 minutes, add it to a solution of 4.8gm of compound in 6ml ethanol at such a rate that the mixture refluxes. After the addition is complete reflux for 2 ½ hr, filter hot, wash the yellow residue 2, 2'-dinitrodiphenyldisulphide with water and dry, m.p 173°C.

(5) Heat 4gm of compound with 13ml piperidine on water bath at 100°C for 90min. Cool, pour in ice with stirring. Collect the precipitated N-(o-Nitropheny)piperidine, recrystallise from ethanol, m.p 89°C.

(6) 2, 4 dinitrochlorobenzene m.p 53°C. (conc HNO_3+conc H_2SO_4, 100°C)

m.p 42°C **o-Bromonitrobenzene** (1-bromo-2-nitrobenzene)

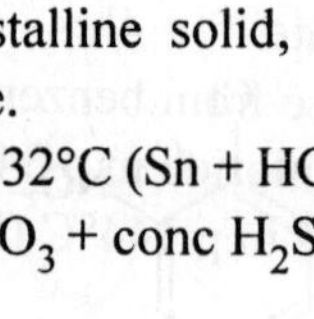

(1) Faintly yellow crystalline solid, insoluble in water, soluble in alcohol and benzene.

(2) o-Bromoaniline m.p 32°C (Sn + HCl) (3) 2, 4-dinitrobromobenzene m.p 75°C (conc HNO_3 + conc H_2SO_4, 100°C)

m.p 42°C **3, 4-dichloronitrobenzene**

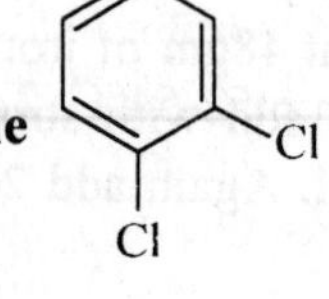

(1) Crystalline compound, insoluble in water.
(2) Heat 5gm of compound with 2gm ammonium acetate on oil bath at 210°C. Pass ammonia gas into the reaction mixture during the refluxing period. Cool, add 10ml water, heat to boiling and filter hot. Dissolve the residue in 50ml hot ethanol and add 15ml water. Heat and then leave overnight. Collect crystals of chloronitroaniline, m.p 104-105°C.

m.p 46°C **m-Chloronitrobenzene** (1-chloro-3-nitrobenzene)

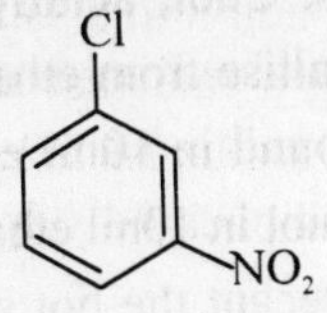

(1) Pale yellow crystalline compound, soluble in hot alcohol, ether and benzene
(2) 3, 4-dinitrochlorobenzene m.p 37°C(fuming HNO_3+conc H_2SO_4)
(3) m-chloroaniline b.p 230°C (Sn+HCl)

m.p 49°C **o-Nitrobenzyl chloride** (2-nitro benzyl chloride)

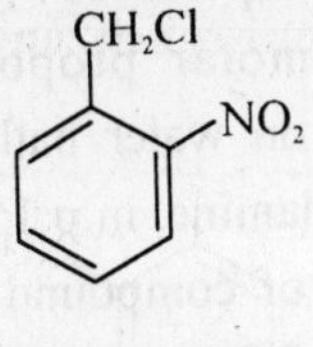

(1) Crystalline compound, insoluble in water, soluble in ether.
(2) Reflux together 5gm of compound with 75ml water and 4gm potassium carbonate for 4hrs. Cool and collect o-nitrobenzyl alcohol, recrystallise from benzene, m.p 74°C.
(3) 2, 4-dinitrobenzylchloride m.p 34°C (conc HNO_3+concH_2SO_4).
(4) o-Nitrobenzoic acid m.p 144°C ($KMnO_4$)

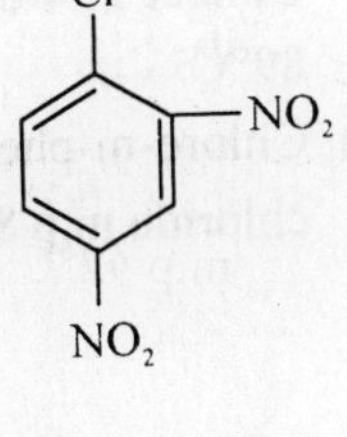

m.p 52-54°C **2, 4-Dinitrochlorobenzene**

(1) Pale yellow coloured solid, insoluble in water, sparingly soluble in alcohol, soluble in benzene and ether.

(2) Dissolve 3.5gm of hydrazine sulphate in 12ml hot water and add with stirring 12gm sodium acetate. Boil for 5min, cool to 70°C, add 8ml ethanol and filter the hydrazine solution. Add this hydrazine solution to 5gm of compound and 25ml ethanol. Reflux for 60min, cool and collect 2, 4-dinitrophenyl hydrazine, m.p 192°C(d)

(3) Reflux 5gm of compound with 6.2gm of sodium carbonate and 50ml water for 24hrs. Cool, acidify with HCl and collect 2, 4-dinitrophenol, recrystallise from ethanol, m.p 114°C.

(4) Reflux 2gm of compound in 10ml ethanol with a solution of 1ml 2-methyl-propane-1-thiol in 10ml ethanol containing 4ml dil NaOH solution for 10min. Decant the hot solution and cool. Collect the yellow crystalline isobutyl 2, 4-dinitrophenyl sulphide, m.p 76°C.

(5) Heat 5gm of compound with 2gm ammonium acetate on oil bath at 170°C for six hours. Pass ammonia gas into the reaction mixture during the refluxing period. Cool and add 10ml water, heat to boiling and filter hot. Dissolve the residue in 50ml hot ethanol and add 15ml water. Heat and then leave overnight. Collect crystals of 2, 4-dinitroaniline, m.p 175-177°C.

(6) Heat together equimolar proportion of aniline and 2, 4 dinitrochlorobenzene on water bath for few minutes. Cool and collect dinitrodiphenylamine, m.p 155°C.

(7) Reflux together 5gm of compound with 10ml methanol, 3.2gm benzyl mercaptan and 2.2gm pyridine for 16 hrs. Cool and collect 2, 4-dinitrophenyl benzyl sulfide, m.p 128°C.

(8) Reflux 4gm of compound with 20ml DMF and 15ml NaI for 15min. Pour the hot mixture on 25gm ice, add 10ml water to it and stir. Collect the solid, dissolve it in mixture of 37ml pet ether and 12ml benzene. Heat for 15min and distill off the solvent. Collect 2, 4-dinitroiodobenzene, recrystallise from pet-ether, m.p 89°C.

(9) Chloro-m-phenylenediamine m.p 91°C($SnCl_2$+HCl) (10) Picryl chloride m.p 83°C (fuming HNO_3+ concH_2SO_4 at 100°C)

m.p 56°C **m-Bromonitrobenzene** (1-bromo-3-nitrobenzene)

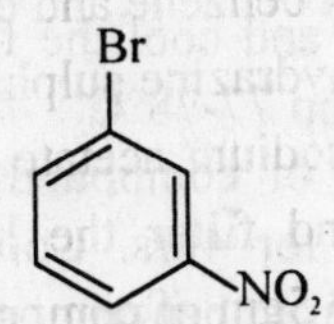

(1) Pale yellow crystalline solid, insoluble in water, soluble in ethanol.
(2) m-Bromoaniline m.p 18°C (Sn+HCl) (3) 3, 4 dinitrobromobenzene m.p 59°C (fuming HNO_3+conc H_2SO_4)

m.p 58°C **3, 4-dibromonitrobenzene**

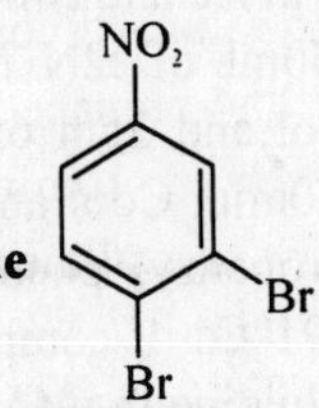

(1) Crystalline compound, insoluble in water, soluble in ethanol.
(2) Heat 5gm of compound with 2gm ammonium acetate on oil bath. Pass ammonia gas into the reaction mixture during the refluxing period. Cool, add 10ml water, heat to boiling and filter hot. Dissolve the residue in 50ml hot ethanol, add 15ml water. Heat and then leave overnight. Collect crystals of 2-bromo 4-nitroaniline.

m.p 71°C **p-Nitrobenzyl chloride** (4-nitro benzyl chloride)

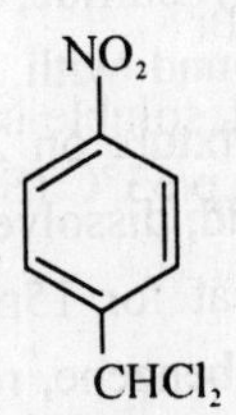

(1) Crystalline compound, soluble in alcohol.
(2) Reflux 5gm of compound with 4.5gm of fused sodium acetate and 8ml glacial acetic acid in an oil bath at 160-170°C for 8-

10hrs. Distill off the excess acetic acid and heat the residue with 10ml methanol, filter and cool the filtrate. Collect crystals of p-nitrobenzyl acetate, m.p 77-78°C.

(3) Reflux together 5gm of compound with 75ml water and 4gm potassium carbonate for 4hrs. Cool and collect p-nitrobenzyl alcohol, recrystallise from benzene, m.p 93°C.

(4) Reflux 5.7gm of compound with 5.5gm of hexamethylenetetramine and 35ml chloroform for 4hrs. Distill off the solvent, add 15ml acetone to the residue and cool. Collect the hexaminium salt and reflux with 50ml of 50% acetic acid for 60min. Add 50ml water and 12.5ml conc HCl and continue refluxing for another 10min. Cool in ice and collect crystals of p-nitrobenzaldehyde, m.p 106°C.

(5) To 50ml of .2N NaOH solution in 95% ethanol add 2gm of phenol and 2gm of compound. Heat the mixture on steam bath for 60min. Cool and add 10ml ice water. Collect the precipitated p-nitrobenzyl phenyl ether, recrystallise from aqueous ethanol, m.p 91°C.

(6) p-Toluidine m.p 45°C ($SnCl_2$+HCl +Zn dust)

(7) p-Nitrobenzoic acid m.p 241°C ($KMnO_4$).

(8) 2, 4-dinitro benzyl chloride m.p 34°C (concHNO_3+concH_2SO_4).

m.p 71°C **4-Bromo-1, 3-dinitrobenzene**

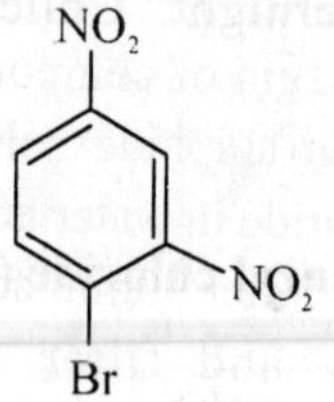

(1) Yellow crystalline solid, soluble in alcohol.

(2) m-Phenylenediamine m.p 63°C (Sn+HCl)

m.p 75°C **2, 4-Dinitrobromobenzene**

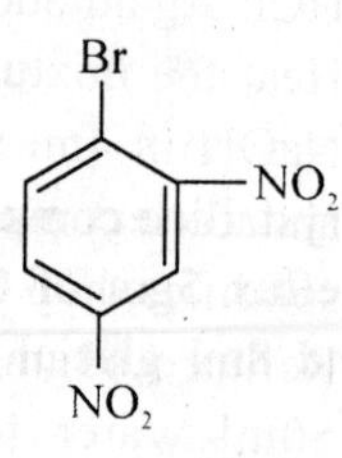

(1) Pale yellow coloured crystalline solid, soluble in alcohol.
(2) Reflux 5gm of compound with a solution of 6.5gm of sodium carbonate in 60ml water for 24hrs. Cool, acidify with dil HCl and collect 2, 4-dinitrophenol, m.p 114°C.
(3) Add in parts about 40gm of iron filings to a solution of 4gm of compound in 40ml water acidified with 1ml conc HCl. Shake the mixture gradually during the addition process. Heat the mixture on sand bath and add about 1gm of sodium carbonate. Filter the mixture hot, concentrate the filtrate and cool. Collect the precipitated bromodiamine, m.p 112°C.
(4) m-Phenylenediamine m.p 63°C (Sn+HCl).

m.p 84°C **p-Chloronitrobenzene** (1-chloro-4-nitrobenzene)

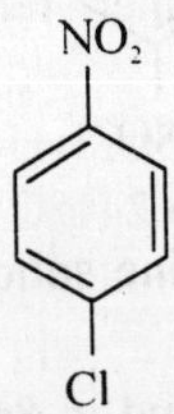

(1) Pale yellow coloured solid, insoluble in water, soluble in most organic solvents.
(2) Mix a solution of 2.2gm of compound in 4ml benzene with 22gm of sodium hydrogen sulphide solution. Now add a solution of 2gm of calcium chloride in water and shake the mixture thoroughly (temp below 30°C). After 90min add 2gm of ammonium chloride, shake thoroughly and filter. Collect the pure p-chloro phenylhydroxylamine and recrystallise from benzene, m.p 87°C.
(3) Heat 48gm of iron filings in 36ml water, remove the heat source and add with stirring 4gm of compound followed by 1ml conc HCl. Again add 20gm of compound followed by 5ml conc HCl. Heat the mixture on water bath for 1hr. Add a solution of 2gm NaOH in 4ml water. Filter hot. Remove the lower layer. Distill and collect p-chloroaniline at 129-131°C/20mmHg which soon solidifies, m.p 71°C
(4) Reflux 5.63gm of compound with 20gm of sodium sulphide and 50ml water for 20hrs. Cool and filter the precipitated p-

chloroaniline, m.p 71°C. Concentrate the filtrate and add 5ml H_2O_2 to it (temp 65-70°C). Cool and collect p-aminophenyl disulphide, recrystallise from hot ethanol, m.p 76°C.

(5) Reflux 8gm of compound with 8gm potassium xanthate and 23ml ethanol for 48hrs. Cool and collect the p-nitrophenyl sulfide, recrystallise from glacial acetic acid, m.p 160-161°C.

(6) 2, 4-dinitrochlorobenzene, m.p 53°C (concHNO_3+concH_2SO_4 at 100°C)

m.p 83°C **Picryl Chloride** (1-chloro 2, 4, 6-trinitrobenzene)

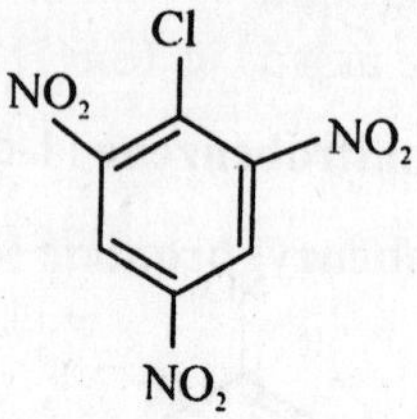

(1) Yellow coloured crystalline solid, insoluble in water, soluble in hot alcohol and benzene

(2) Dissolve 1gm of compound in 8ml $CHCl_3$ and add a solution of 1.3gm of N, N diphenylhydrazine in 2ml $CHCl_3$ to it. Shake the mixture thoroughly, warm on water bath and keep aside for 1hr. Cool, filter the amine salt, concentrate the filtrate. Treat the hot concentrate with 12ml hot ethanol. Cool and collect diphenyl picryl hydrazine, m.p 173°C.

(3) Reflux a solution of 3.7gm picryl chloride in 25ml ethanol with a solution of 1.65gm of o-amino phenol in 8ml ethanol for 2hrs. Cool and collect the precipitated 2' hydroxy-2, 4, 6-trinitrodiphenylamine, m.p 175°C.

(4) Reflux a solution of 3.7gm of picryl chloride in 25ml benzene with a hot solution of 2.85gm of aminopyridine in 25ml benzene at 100°C for 75 min. Filter the hot mixture, distill off the solvent and stir the residual oil with pet ether. Collect the 2-(2', 4', 6'trianilino) pyridine, recrystallise from benzene, m.p 135°C.

(5) To 1gm of compound add 5ml aniline and 5ml alcohol. Heat the mixture on water. Cool and collect picrylaniline, m.p 177°C.

(6) Reflux 5gm of compound with dil NaOH. Cool, acidify with HCl and collect picric acid, recrystallise from ethanol, m.p 122°C.

(7) Picramide m.p 188°C.

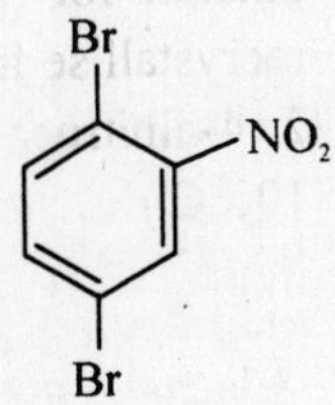

m.p 84°C 1, 4-dibromo-2-Nitrobenzene

(1) Pale yellow coloured solid.

(2) 2, 5-Dibromoaniline m.p 51°C (Sn+HCl).

m.p 100°C **p-Nitrobenzyl bromide** (4-nitro benzyl bromide)

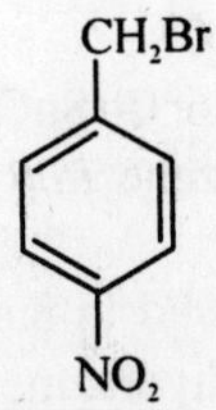

(1) Pale yellow coloured solid, soluble in alcohol.

(2) Reflux 7.2gm of compound with 5.5gm of hexamethylenetetramine and 35ml chloroform for 4hrs. Distill off the solvent and add 15ml acetone to the residue and cool. Collect the hexaminium salt and reflux with 50ml of 50% acetic acid for 60min. Add 50ml water and 12.5ml conc HCl and continue refluxing for another 10min. Cool in ice and collect crystals of p-nitrobenzaldehyde, m.p 106°C.

(3) To 50ml of .2N NaOH solution in 95% ethanol add 2gm of phenol and 2.2gm of compound. Heat the mixture on steam bath for 60min. Cool and add 10ml ice water. Collect the precipitated p-nitrobenzyl phenyl ether, recrystallise from aqueous ethanol, m.p 91°C.

(4) Reflux together 5gm of compound with 75ml water and 4gm potassium carbonate for 4hrs. Cool and collect p-nitrobenzyl alcohol, recrystallise from benzene, m.p 93°C.

(5) p-Nitro benzoic acid m.p 241°C.

m.p 127°C **p-Nitrobromobenzene** (1-Bromo-4-nitrobenzene)

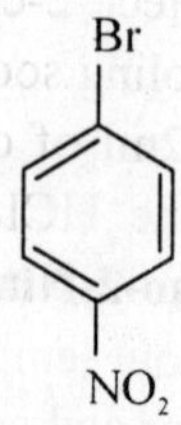

(1) Faint yellow coloured solid, sparingly soluble in ethanol.
(2) Heat 5gm of compound with 2gm ammonium acetate on oil bath at 170°C for six hours. Pass ammonia gas into the reaction mixture during the refluxing period. Cool, add 10ml water, heat to boiling and filter hot. Dissolve the residue in 50ml hot ethanol and add 15ml water. Heat and then leave overnight. Collect crystals of p-nitroaniline, m.p 147°C.
(3) p-Bromoaniline m.p 66°C ($SnCl_2$+HCl)
(4) 2, 4-dinitrobromobenzene m.p 75°C (concHNO_3+concH_2SO_4 at 100°C)

(vi) Halogenated Amines

Liquid

b.p 209°C **o-Chloroaniline** (2-chloro phenyl amine)

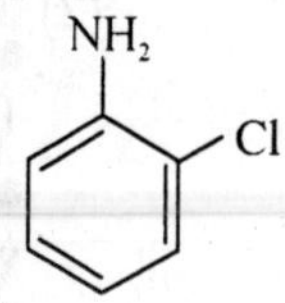

(1) Colourless liquid, soluble in nearly all organic solvents.
(2) Cool a mixture of 6.4gm of compound and 17.5ml of HBr to 0°C. Add with stirring a solution of 3.5gm of $NaNO_2$ in 7ml water(temp below 5°C). Dissolve freshly prepared copper (I) bromide (from 7.5gm $CuSO_4$) in 4ml of 40% HBr, heat to boiling and add o-chlorobenzene diazonium bromide. Pass steam until no more organic material distill. Make the distillate alkaline and distill. Collect o-bromochloro benzene at b.p 200-202°C.

(3) Diazotise 3.2gm of compound in 4ml conc HCl with conc solution of 2gm of $NaNO_2$. Cool the filtered diazonium solution and add 50ml benzene, stir and add a solution of 8gm of sodium acetate in 20ml water. Continue stirring for 48hrs. Remove the benzene layer and distill. Collect 2-chlorobiphenyl at b.p 150-155°C/ 10mmHg which on cooling soon solidifies, m.p 34°C.

(4) To a suspension of 3.2ml of compound in 30ml water add with stirring 2.75ml of conc HCl. Now add 2.5gm of ammonium thiocynate and heat the mixture on steam bath for 1hr. Cool and keep the mixture at room temperature for 1hr and then evaporate it to dryness. Add water and again evaporate it off to dryness and then heat on steam bath for 4hrs. Heat the residue with water to 70°C, cool to 40°C and filter. Collect the crude o-chlorophenyl - thiourea, recrystallise from absolute ethanol, m.p 144-146°C.

(5) Acetate m.p 88°C

(7) Benzoate m.p 99°C

(6) Picrate m.p 134°C.

b.p 230°C **m-Chloroaniline** (3-chlorophenyl amine)

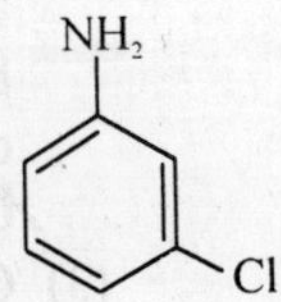

(1) Colourless liquid, soluble in nearly all organic solvents.

(2) Cool a mixture of 6.4gm of compound and 17.5ml of HBr to 0°C. Add with stirring a solution of 3.5gm of $NaNO_2$ in 7ml water(temp below 5°C). Dissolve freshly prepared copper (I) bromide (from 7.5gm $CuSO_4$) in 4ml of 40% HBr, heat to boiling and add m-chlorobenzene diazonium bromide. Pass steam until no more organic material distill. Make the distillate alkaline and distill. Collect m-bromochloro benzene at b.p 191-194°C.

(2) Heat a mixture of 12.8gm of compound with 23.3gm of ethyl ethoxymethylenemalonate on steam bath for 1hr. Distill off the alcohol formed during the reaction and heat the crude product to 250°C for 1hr. Cool and heat the crude product with 100ml dil NaOH solution for 1hr. Cool, acidify the solution with conc HCl. Collect the precipitated 7-chloro-4-hydroxy-3 quinolinecarboxylic acid, m.p 266°C.

(3) Acetate m.p 73°C
(4) Benzoate m.p 122°C
(5) Picrate m.p 177°C.

Solid

m.p 18°C, b.p 251°C **m-Bromoaniline** (1-amino-3-bromo benzene)

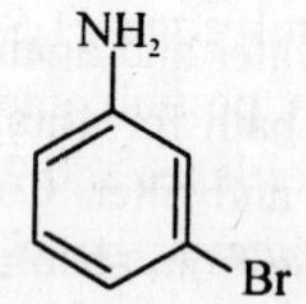

(1) Colourless solid, soluble in nearly all organic solvents.
(2) Cool a mixture of 6.6gm of compound and 17.7ml of HBr to 0°C. Add with stirring a solution of 3.6gm of $NaNO_2$ in 7ml water(temp below 5°C). Dissolve freshly prepared copper (I) bromide (from 7.8gm $CuSO_4$) in 4ml of 40% HBr, heat to boiling and add m-bromobenzene diazonium bromide. Pass steam until no more organic material distill. Make the distillate alkaline and distill. Collect m-dibromobenzene at b.p 215-217°C.
(3) Cool a solution of 20gm of compound in 200ml methyl alcohol. Add 13ml conc H_2SO_4. Cool further and add 15gm of amyl nitrite (temp below 25°C). After 1hr during which time the diazotization reaction is complete, filter and collect the diazonium sulphate. Dissolve in 200ml water, filter and keep at 40-50°C for 5hrs. Cool, extract with ether. Distill off ether and collect m-bromo phenol at b.p 135-140°C/12mmHg (m.p 33°C, b.p 236°C)
(4) Acetate m.p 88°C (5) Benzoate m.p 133°C (6) Picrate m.p 180°C.

m.p 26°C, b. 240°C, (16°C) **2-Bromo-4-methylaniline** (2-bromo-p-toluidine)

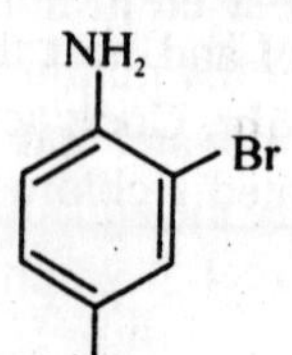

(1) Crystalline compound, insoluble in water, soluble in alcohol and ether.

(2) Cool a solution of 7.5gm of compound in 7.2ml conc H_2SO_4 and 20ml water. Add dropwise a solution of 3.3 gm of $NaNO_2$ in 8.8ml water (temp below 5°C). Stir for 5min and add 300mg of urea and 30gm of crushed ice. To a mixture of 15gm of anhydrous sodium sulphate, 11ml conc H_2SO_4 and 10ml water add the diazotised solution in portion and distill. Extract the distillate with ether. To the ether layer add 20ml of dil NaOH solution and shake thoroughly. Remove the alkaline layer and acidify with conc HCl. Extract with ether. Distill off ether and collect 3-bromo 4-hydroxy toluene at b.p 102-104°C/20mmHg.

(3) To a cold mixture of 16ml ethyl alcohol and 4ml conc H_2SO_4 add 5gm of compound. Cool to 0°C and add dropwise with stirring a solution of 2.8gmof $NaNO_2$ in 5ml water. Stir for 20 mintures and add 700mg of copper bronze(exothermic reaction). After the reaction has subsided heat on steam bath for 20 min. Add 40ml water and steam distill. Separate the oily portion and distill, collect m-bromotoluene at b.p 183°C.

(4) Acetate m.p 117°C

(5) Benzoate m.p 149°C.

m.p 26°C, b.p 237°C **3-chloro-4-methyl aniline** (3-chloro p-toluidine)

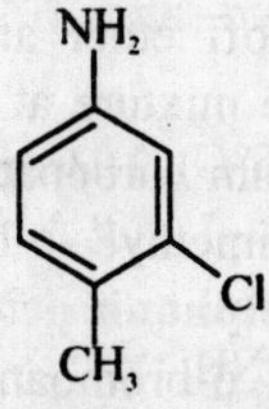

(1) Acetate m.p 105°C

(2) Benzoate m.p 122°C.

m.p 29°C, b.p 241°C **4-Chloro-2-methylaniline** (4-chloro-o-toluidine)

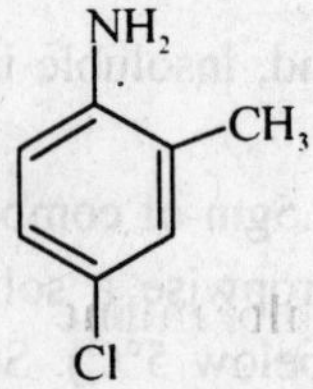

(1) Crystalline solid, insoluble in water, soluble in hot alcohol.
(2) Acetate m.p 140°C.
(3) Benzoate m.p 172°C.

m.p 32°C, b.p 229°C **o-Bromoaniline** (1-amino-2-bromobenzene)

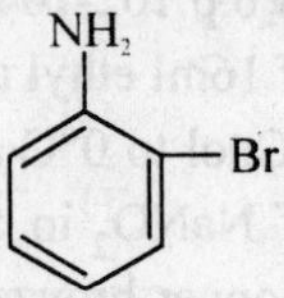

(1) Crystalline compound, slightly soluble in water, soluble in alcohol and ether.
(2) Cool a solution of 5.5gm of compound in 15ml conc HCl to 0°C, diazotize with a solution of 2.4gm of $NaNO_2$ in 10ml water. Stir for 20 min and then pour into a solution of 18gm of KI in 60ml water. Keep overnight, remove the oily layer and distill. Collect o-bromoiodobenzene at b.p 120-121°C /15mmHg. (b.p 257°C (754mmHg)
(3) Reflux a mixture of 8.5gm of compound with 18gm of sodium bicarbonate, 215ml water and 25gm of CH_3I for 20hrs. Extract with ether. Distill off ether and mix the residue with acetic anhydride. Keep the mixture at room temperature of few hours, distill and add sodium carbonate. Extract with ether. Distill off ether and collect dimethyl -o-bromoaniline at b.p 107-108°C/ 14mmHg.
(4) Treat a solution of o-bromoaniline in 20% HCl with freshly prepared solution of chlorine dissolved in 20% HCl. Cool, add water to the reaction mixture and collect 4:6-dichloro-2-bromoaniline, m.p 82°C.

(5) Acetate m.p 99°C
(6) Benzoate m.p 116°C
(7) Picrate m.p 129°C.

m.p 49-51°C **2, 5-Dichloraniline**

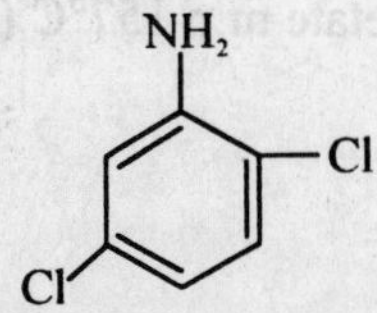

(1) Crystalline compound, slightly soluble in water, soluble in alcohol and ether.
(2) Acetate m.p 133°C
(3) Benzoate m.p 120°C
(4) Picrate m.p 86°C
(5) Hydrochloride m.p 191°C.

m.p 58°C **o-Iodoaniline**

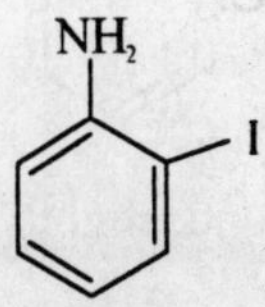

(1) Crystalline compound with unpleasant odour, sparingly soluble in water, soluble in nearly all organic solvents.
(2) Hydrochloride m.p 153°C
(3) Acetate m.p 109°C
(4) Benzoate m.p 139°C
(5) Picrate m.p 112°C.

m.p 59°C **4-Bromo-2-methylaniline** (4-bromo-o-toluidine)

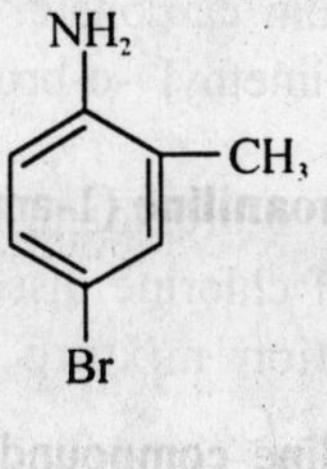

(1) Crystalline compound, insoluble in water, soluble in hot alcohol.
(2) Acetate m.p 157°C (3) Benzoate m.p 115°C.

m.p 62°C, b.p 245°C **2, 4-dichloraniline**

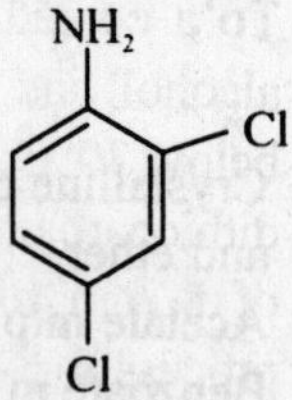

(1) Colourless crystalline solid, soluble in nearly all organic solvents.
(2) Acetate m.p 145°C
(3) Benzoate m.p 117°C
(4) Picrate m.p 106°C.

m.p 64°C **2-Chloro-p-phenylenediamine**

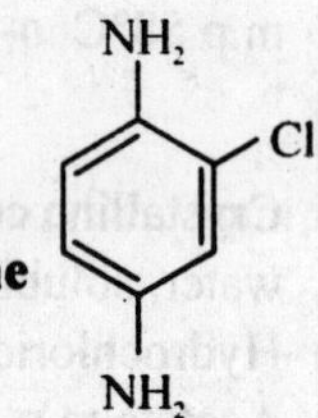

(1) Diacetate m.p 196°C
(2) Dibenzoate m.p 228°C.

m.p 66°C **p-Bromoaniline** (1-amino-4-bromo benzene)

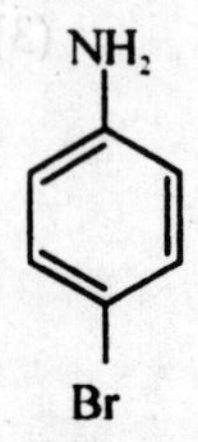

(1) Colourless crystalline compound, insoluble in water, soluble in alcohol and ether
(2) Diazotise a solution of 4.3gm of compound in 4ml conc HCl and 3ml water with a concentrated aqueous solution of 1.8gm of

$NaNO_2$. Filter, add with stirring 50ml cold benzene and a solution of 3gm of NaOH in 15ml water to the filtrate(temp below 10°C). Continue stirring for 48hrs. Separate the benzene layer, distill and collect 4-bromobiphenyl at b.p 172-175°C/8mmHg which on cooling soon solidifies, m.p 90°C.

(3) Dissolve 1.7gm of compound in warm mixture of 5ml glacial acetic acid and 10ml water. Add with stirring a solution of 600mg of sodium cyanate in 5ml water. Keep the mixture at room temperature for several hours. Filter the precipitated p-bromophenylurea, recrystallise from aqueous alcohol, m.p 228°C.

(4) To a cooled mixture of 4.5gm of compound, 3ml CS_2 and 4ml alcohol add with stirring 5ml of conc ammonia solution(temp below 30°C). Keep the mixture overnight, collect the precipitated dithiocarbamate. Dissolve in 150ml water and stir with a solution of 8.7gm of lead nitrate in 15ml water. Continue the stirring process for 20min, cool and collect p-bromophenyl isothiocynate, m.p 61°C.

(5) Acetate m.p 167°C

(6) Benzoate m.p 204°C

(7) Picrate m.p 180°C.

m.p 72°C **p-Chloroaniline** (4-chloro phenyl amino)

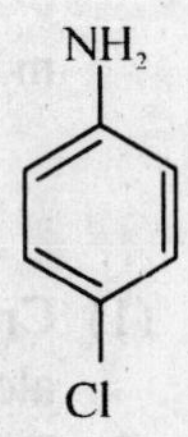

(1) Crystalline compound, sparingly soluble in water and alcohol.

(2) Heat on oil bath a mixture of 6.4gm of compound with 11.2gm of triethylorthoformate and .2ml of conc H_2SO_4 for 60min. Distill the reaction mixture under reduced pressure and collect N-Ethyl-p-chloroformanilide at b.p 123-126°C/3mmHg.

(3) Heat a mixture of 3.8gm of compound, 4ml conc aq ammonia and 3ml CS_2 at 35°C. As soon as the reaction starts, remove the heat source and maintain the temp between 30-40°C. After sometime ammonium p-chlorophenyldithiocarbamate precipitates out. Dissolve it in 25ml water, heat to 30°C and add with stirring

a solution of 2.8gm of chloroacetic acid in 3ml water (neutralised with Na_2CO_3). Continue stirring at this temperature for 1hr. Cool and add a solution of 2.1gm of $ZnCl_2$ in 7ml water (pH at 7 by adding NaOH). Stir for 1hr, collect p-chloropheny isothiocynate, m.p 44°C.

(4) To a solution of 2.7ml ethyl chloroacetate in 3ml ethanol add 3.2gm of compound and 3gm of sodium acetate. Reflux the mixture for 6hrs. Cool, pour in ice water and collect N-(p-chlorophenyl)glycine ethyl ester, recrystallise from ethanol, m.p 88-90°C

(5) Acetate m.p 179°C

(6) Benzoate m.p 193°C

(7) Picrate m.p 178°C.

m.p 76°C **4-Chlor-o-phenylenediamine**

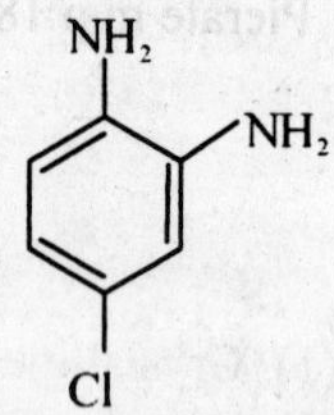

(1) Crystalline compound, sparingly soluble in water, soluble in alcohol and ether.

(2) Diacetate m.p 208°C

(3) Dibenzoate m.p 230°C.

m.p 78°C **2, 4, 6-Trichloroaniline** (sym-trichloro aniline)

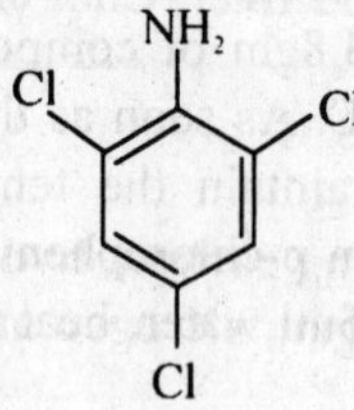

(1) White crystalline solid, soluble in alcohol and ether.

(2) Acetate m.p 206°C (3) Benzoate m.p 174°C (4) Picrate m.p 83°C.

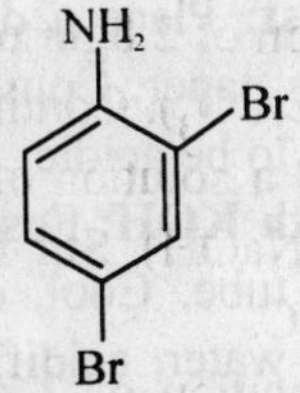

m.p 80°C **2, 4-dibromoaniline**

(1) Slightly soluble in chloroform, soluble in alcohol.
(2) Acetate 146°C
(3) Benzoate m.p 134°C
(4) Picrate m.p 124°C.

m.p 120°C **2, 4, 6-Tribromoaniline** (sym-tribromo aniline)

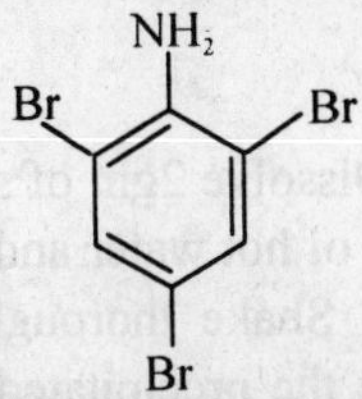

(1) Crystalline compound, soluble in alcohol and ether.
(2) Dissolve 500mg of compound in 10ml acetic anhydride and add 1drop of conc H_2SO_4 to it. Stir the mixture and after 10-15min pour it in warm water. Collect the precipitated tribromoacetanilide, recrystallise from alcohol, m.p 232°C.
(3) Dissolve by heating 2gm of compound in 12ml alcohol and 3ml benzene. Add dropwise .7ml of conc H_2SO_4 and heat on water bath until a clear solution is obtained. Add in portion 700mg of sodium nitrite. Heat on water bath, cool in an ice bath. Filter and dissolve the precipitated solid in 12ml boiling glacial acetic acid and 3ml water and filter. Cool and collect tribromobenzene, m.p 122°C.
(4) Benzoate m.p 198°C.

(e) Compounds containing carbon, hydrogen and sulphur (with or without oxygen, nitrogen and halogen)

(i) Sulphonic Acids

Sulphonic acids are derivatives of sulphuric acid in which the H group is replaced by an aryl group. Sulphonic acid group (SO_3H) is helpful in imparting water soublility and thus finds use in dyes as well as drugs Sulphonic acids are normally present in the form of their alkali salts.

Identification tests

(*i*) **Litmus test:** Place a drop of aqueous or alcoholic solution on blue litmus paper. Conversion of blue litmus to red indicates the compound to be acidic.

(*ii*) **Fusion with KOH:** Fuse 500mg of compound with 1gm KOH in a dry test tube. Cool, dissolve the fused material in minimum amount of water, acidify with dil HCl. Add few drops of $BaCl_2$ solution. A white precipitate indicates the presence of sulphonic acid group.

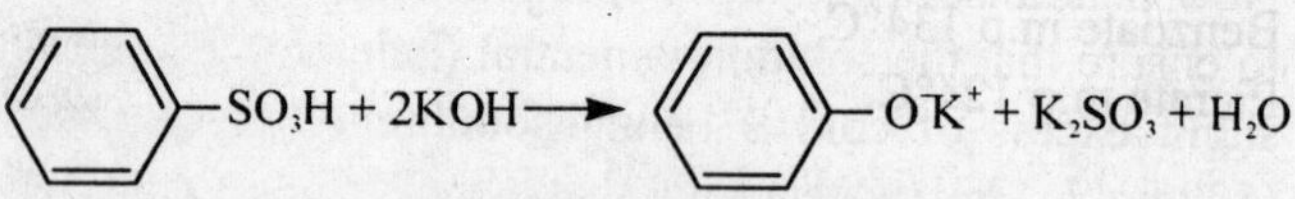

$$K_2SO_3 \xrightarrow[HCl]{BaCl_2} BaSO_3$$

Derivaties

(1) **p-Toluidine salt :** Dissolve 2gm of sodium salt of sulphonic acid in minimum amount of hot water and then add 1gm of p-toluidine and 4ml conc HCl. Shake thoroughly so that a clear solution results. Cool, collect the precipitated solid and recrystallise from hot water containing a drop of conc HCl. (In some cases if the sulphonic acid does not dissolve add more HCl and then heat on water bath).

$$RSO_3^- Na^+ + HCl + p\text{–}CH_3C_6H_4NH_2 \longrightarrow \left[p\text{–}CH_3C_6H_4NH_2 \right]^+ RSO_3^- + NaCl$$

(2) **Sulphonamide :** Fuse 750mg of compound or its salt with 2gm of PCl_5 by heating in an oil bath at 150°C for 30-35 minutes. Cool, heat the chloride derivative with 12ml benzene and filter. Add the filtrate with stirring to 10ml cold ammonium hydroxide solution. Filter the precipitated sulphonamide, recrystallise and determine its melting point.

$$RSO_3^- Na^+ + PCl_5 \longrightarrow RSO_2Cl \xrightarrow{NH_3} RSO_2NH_2$$

$$RSO_3H + PCl_5 \longrightarrow RSO_2Cl + POCl_3 + HCl$$

$$RSO_2Cl + 2NH_3 \longrightarrow RSO_2NH_2 + NH_4Cl$$

(3) **Sulphonanilides:** Fuse 750mg of compound or its salt with 2gm of PCl_5 by heating in an oil bath at 150°C for 30-35 minutes. Cool, heat the chloride derivative with 12ml benzene and filter. Heat the filtrate with 2gm aniline for 60min. Concentrate the benzene solution and cool. Collect the precipitated sulphonanilide derivative, recrystallise and determine its melting point.

$$RSO_2O^-Na^+ + PCl_5 \longrightarrow RSO_2Cl + POCl_3 + NaCl$$

$$RSO_2Cl + 2C_6H_5NH_2 \longrightarrow RSO_2NHC_6H_5 + C_6H_5NH_3^+Cl^-$$

(4) **S Benzylisothiouronium salts :** Dissolve 250mg of compound (in case of salt dissolve in water and add 2-3 drops of conc HCl) in 3-5ml water and add 2-3drops of phenolphthalein indicator. Now neutralise with NaOH solution. Add 1-2 drop of conc HCl to ensure that the solution is neutral (pale pink colour). Now add slight excess of cold S benzylisothiouronium chloride solution (500mg in 3ml water) and shake thoroughly. Cool, collect the precipitated solid and recrystallise.

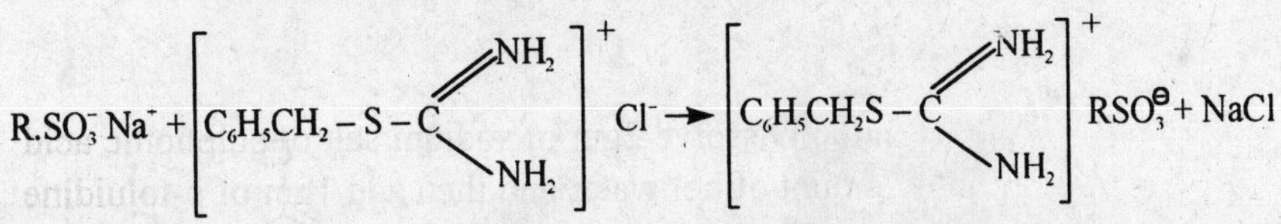

Solid

m.p 50-51°C, hydrated m.p 43°C **Benzene sulphonic acid**

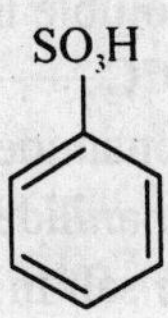

(1) Colourless deliquescent solid, soluble in water and alcohol, insoluble in ether.
(2) To 1gm of compound in 5ml water add 1gm of sodium bicarbonate followed by 2gm of NaCl, heat until it dissolves. Filter hot, cool the filtrate and collect sodium benzene sulphonate.
(3) p-Toluidine salt m.p 205°C
(4) Benzene sulphonamide m.p 153°C

(5) Benzene sulphonanilide m.p 110°C.
(6) S-Benzyl isothiouronium salt m.p 148°C.

m.p 67.5°C **o-Toluene sulphonic acid**

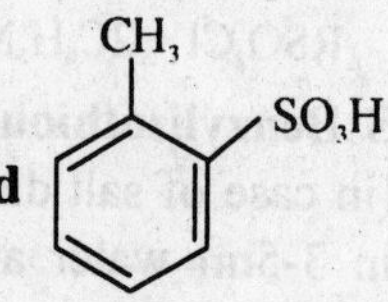

(1) Colourless crystalline hygroscopic solid, soluble in water and ether.
(2) p-Toluidine salt m.p 204°C
(3) o-Toluene sulphonamide m.p 153°C
(4) o-Toluene sulphonanilide m.p 136°C
(5) S-Benzyl isothiouronium salt m.p 170°C.

m.p 68°C **p-Chlorobenzene sulphonic acid**

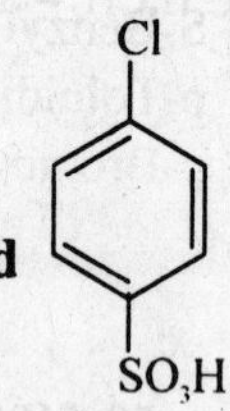

(1) Crystalline compound, soluble in water.
(2) p-Toluidine salt m.p 209°C
(3) p-Chloro benzene sulphonamide m.p 144°C
(4) p-Choro benzene sulphonanilide m.p 104°C.
(5) S-Benzyl isothiouronium salt m.p 175°C.
(6) p-Chlorobenzene sulphonyl chloride m.p 53°C

m.p 90°C **Naphthalene 1-sulphonic acid**

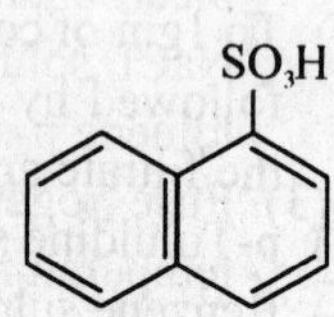

(1) Crystalline deliquescent solid, soluble in water and alcohol, sparingly soluble in ether.
(2) p-Toluidine salt m.p 181°C

(3) Naphthalene 1-sulphonamide m.p 50°C
(4) Naphthalene 1-sulphonanilide m.p 112°C.
(5) S-Benzyl isothiouronium salt m.p 137°C.
(6) Naphthalene 1-sulphonyl chloride m.p 68°C.

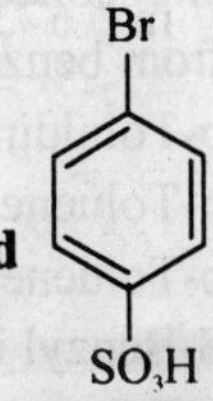

m.p 103°C **p-Bromo benzene sulphonic acid**

(1) Crystalline compound, soluble in hot water and hot alcohol.
(2) p-Bromo benzene sulphonamide m.p 166°C
(3) p-Bromo benzene sulphonanilide m.p 119°C.
(4) S-Benzyl isothiouronium salt m.p 170°C.
(5) p-Toluidine salt m.p 216°C
(6) p-Bromo benzene sulphonyl chloride m.p 75°C.

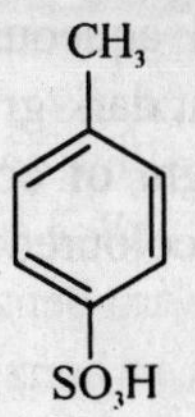

m.p 107°C **p-Toluene sulphonic acid**

(1) Colourless crystalline compound, easily soluble in water.
(2) Dissolve 1gm of compound in 5ml water and add 1.5gm sodium bicarbonate to it. Heat the solution to boiling, add 5ml saturated NaCl solution. Filter hot, cool the filtrate and collect sodium toluene p-sulphonate.
(3) Fuse together 10gm of NaOH and 4gm of KOH. When the temperature is 230°C add 1.5 gm of p-toluene sulphonate. Raise the temperature and slowly add 4.5 gm of p-toluene sulphonate. Finally raise the temperature to 330°C. Cool, dissolve the solid in 50ml water, acidify with H_2SO_4. Distill, saturate the distillate with NaCl, remove the organic layer and extract the aqueous

layer with ether. Combine the two organic layer and distill. Collect p-cresol at 95-96°C/15mmHg and cool which soon solidifies, m.p 31°C.

(4) Add 2.13gm of P_2O_5 and 140mg of kiesilguhr to a mixture of 1.90gm of compound and 70mg of asbestos. Keep the mixture at room temperature for 1hr and then heat on oil bath at 125°C for 9hrs. Add 5ml ethylene chloride and again reflux for 5 minutes. Remove the ethylene chloride layer. Repeat the extraction process twice. Combine the organic layers, distill off the solvent. Cool and collect the dark p- toluenesulfonic anhydride, recrystallise from benzene, m.p 110-127°C.

(5) p-Toluidine salt m.p 198°C

(6) p-Toluene sulphonamide m.p 137°C

(7) p-Toluene sulphonanilide m.p 103°C

(8) S-Benzyl isothiouronium salt m.p 182°C.

(9) p-Toluene sulphonyl chloride m.p 69°C.

m.p 107°C **1-Naphthol 8-Sulphonic acid**

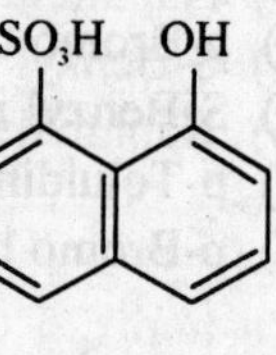

(1) Crystalline compound, soluble in water.

(2) To 1ml aqueous solution of compound add $FeCl_3$ solution. A transient dark green coloured solution is obtained.

(3) Fuse 1gm of compound with excess of PCl_5. Cool and collect yellow coloured chlornaphthosulton, m.p 174°C.

m.p 120°C **5-Sulpho Salicylic acid**

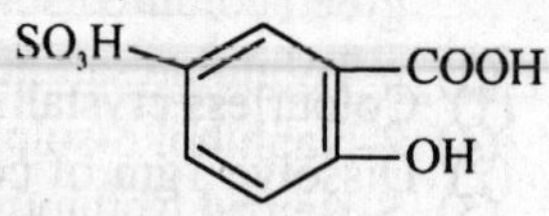

(1) Crystalline compound, soluble in water and ethanol.

(2) To 1ml aqueous solution of compound add $FeCl_3$ solution. An intense red-violet coloured solution is obtained.

(3) Boil 20mg of compound with 6ml conc HCl and 2ml conc HNO_3 for 10-15 min. Cool and collect the precipitated yellow coloured

chloranil, recrystallise from alcohol, m.p 290°C.

(4) S-Benzyl isothiouronium salt m.p 204°C.

m.p 124°C, hydrated m.p 91°C **Naphthalene 2-sulphonic acid**

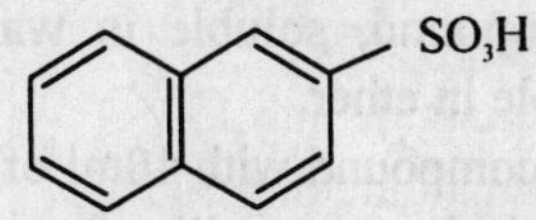

(1) White solid, soluble in water, alcohol and ether.
(2) p-Toluidine salt m.p 221°C
(3) Naphthalene 2-sulphonamide m.p 217°C
(4) Naphthalene 2-sulphonanilide m.p 132°C.
(5) S-Benzyl isothiouronium salt m.p 191°C.
(6) Naphthalene 2-sulphonyl chloride m.p 79°C

m.p 125°C **2-Naphthol 6-Sulphonic acid**

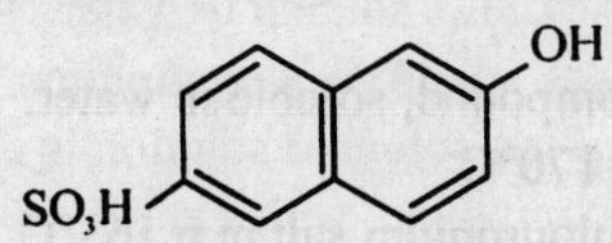

(1) Crystalline compound, soluble in water and alcohol.
(2) To 1ml aqueous solution of compound add $FeCl_3$ solution. A pale green coloured solution is obtained.
(3) p-Toluidine salt m.p 248°C
(4) 2-Naphthol 6-sulphonamide m.p 238°C.
(5) S-Benzyl isothiouronium salt m.p 217°C.

m.p 130°C **1-Chloronaphthalene 4-Sulphonic acid**

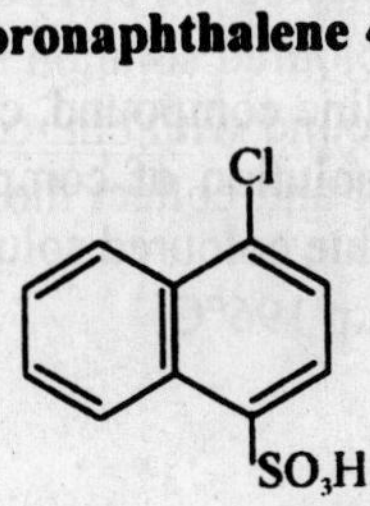

(1) Crystalline compound, sparingly soluble in water.

(2) 1-Chloronaphthalene 4-sulphonamide m.p 187°C
(3) 1-Chloronaphthalene 4-Sulphonyl chloride m.p 95°C.

m.p 134°C, hydrated 68-69°C **o-Sulphobenzoic acid**

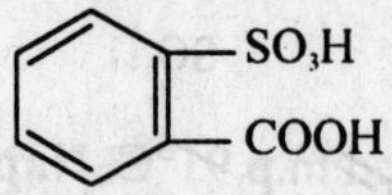

(1) Crystalline compound, soluble in water, sparingly soluble in ethanol, insoluble in ether.
(2) Reflux 1gm of compound with 10ml of acetyl chloride. Cool and collect the anhydride, recrystallise from benzene, m.p 119°C.
(3) p-Toluidine salt m.p 200°C
(4) S-Benzyl isothiouronium salt m.p 206°C.
(5) Dianilide m.p 194°C.

m.p 141°C, hydrated 98°C **m-Sulphobenzoic acid**

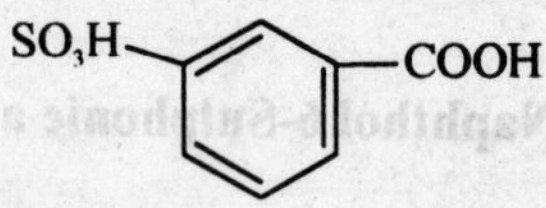

(1) Crystalline compound, soluble in water.
(2) Diamide m.p 170°C
(3) S-Benzylisothiouronium salt m.p 163°C.

m.p 170°C (d) **1-Naphthol 4-Sulphonic acid**

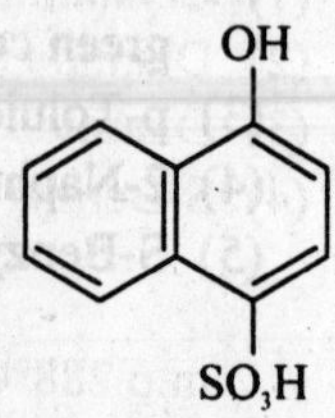

(1) Colourless crystalline compound, easily soluble in water.
(2) To 1ml aqueous solution of compound add $FeCl_3$ solution. An intense transient blue coloured solution is obtained.
(3) p-Toluidine salt m.p 196°C

(4) 1-Naphthol 4-sulphonanilide m.p 200°C
(5) S-Benzyl isothiouronium salt m.p 103°C.

(d) **Naphthalene 1, 5-Disulphonic acid** (Armstrong's acid)

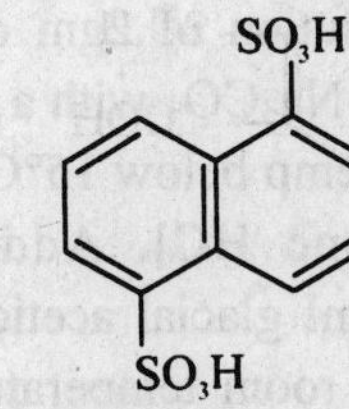

(1) Crystalline solid, soluble in water.
(2) p-Toluidine salt m.p 332°C
(3) Naphthalene 1, 5-disulphonamide m.p 310°C
(4) Naphthalene 1, 5-disulphonanilide m.p 249°C
(5) S-Benzylisothiouronium salt m.p 257°C
(6) Naphthalene 1, 5-disulphonyl chloride m.p 183°C.

m.p 260°C, hydrated m.p 94°C **p-Sulphobenzoic acid**

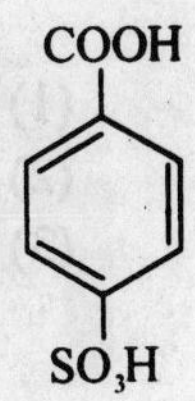

(1) Crystalline compound, soluble in water.
(2) Diamide m.p 236°C
(3) Dianilide m.p 253°C
(4) S-Benzyl isothiouronium salt m.p 213°C.

m.p 288°C (d) **Sulphanilic acid** (p-aminobenzene sulphonic acid)

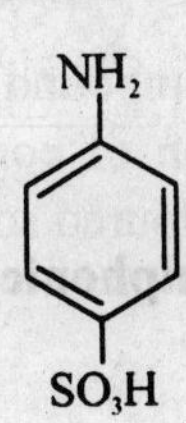

(1) Grayish white crystalline solid, sparingly soluble in water, very sparingly soluble in alcohol and ether.

(2) Diazotise a clear solution of 2gm of compound in 20ml water containing 500mg of Na_2CO_3 with a solution of 750mg of sodium nitrite in 2ml water(temp below 15°C). Pour on 20gm crushed ice containing 2ml conc HCl. Add a solution of 1.2gm of dimethylaniline in 1ml glacial acetic acid to the cold diazotized suspension. Keep at room temperature for 10min and then add 7ml of 20% NaOH solution, collect sodium salt of methyl orange.

(3) p-Amino benzene sulphonamide m.p 165°C

(4) p-Aminobenzene sulphonanilide m.p 200°C

(5) S-Benzyl isothiouronium salt m.p 182°C.

(d) **Metanilic acid** (3-aminobenzene sulphonic acid)

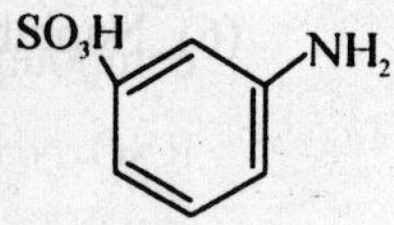

(1) Crystalline compound, sparingly soluble in water.

(2) 3-Aminobenzene sulphonamide m.p 142°C

(3) S-Benzyl isothiouronium salt m.p 148°C.

(d) **Orthanilic acid** (2-aminobenzene sulphonic acid)

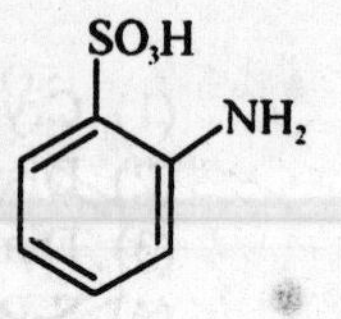

(1) Crystalline compound sparingly soluble in water.

(2) 2-Aminobenzene sulphonamide m.p 153°C.

(3) S Benzyl isothiouronium salt m.p 132°C

Naphthalene 2, 6-disulphonic acid

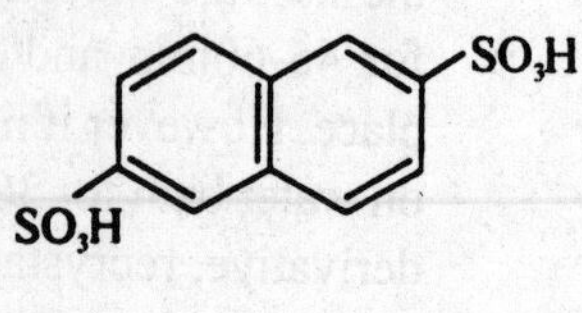

(1) Hygroscopic compound

(2) S-Benzylisothiouronium salt m.p 256°C.

(3) Naphthalene 2, 6-disulphonamide m.p 305°C

Naphthalene 2, 7-disulphonic acid

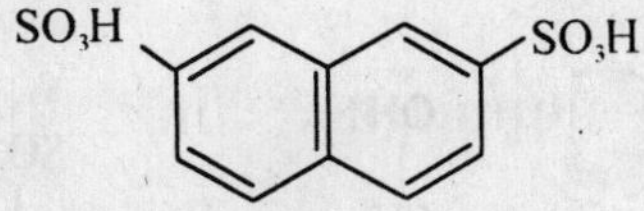

(1) White crystalline hygroscopic solid, soluble in water.
(2) S-Benzylisothiouronium salt m.p 212°C.
(3) Naphthalene 2, 7-disulphonamide m.p 243°C

(ii) Sulphonamides

Sulphonamides are the amide derivatives of sulphonic acids and are commonly known as sulpha drugs. They are important chemotherapeutic agents which have been employed for the treatment of bacterial infections.

Identification tests

(1) To 100mg of compound add 1ml conc H_2SO_4 and then heat for 1-2 minute, cool, add 1ml water to it. Add NaOH solution to the acidic mixture until alkaline. Bring a glass rod dipped in HCl near the mouth of test. If white fumes are produced then the compound is sulphonamide.

$$RSO_2NH_2 + H_2O + H_2SO_4 \rightarrow RSO_3H + (NH_4)HSO_4$$

$$(NH_4)HSO_4 + 2NaOH \rightarrow Na_2SO_4 + 2H_2O + NH_3$$

$$NH_3 + HCl \rightarrow NH_4Cl$$

(2) Heat 1gm of compound with 5ml of 80% H_2SO_4 at 160-170°C. Cool, recover the precipitated sulphonic acid and amine and identify them.

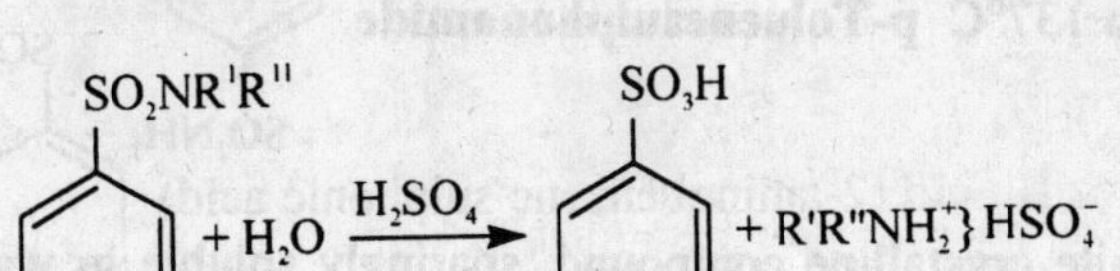

Derivatives

(1) **Xanthylsulphonamides:** Dissolve 1gm of xanthydrol in 10ml glacial acetic acid and then add 1gm of sulphonamide to it. Shake the mixture thoroughly for 5 minutes, keep at room temperature for 45-60min and then cool so that complete precipitation take place. However if no precipitaation take place then heat the mixture on water bath for 30min and cool. Collect the xanthylsulphonamide derivative, recrystallise and determine its melting point.

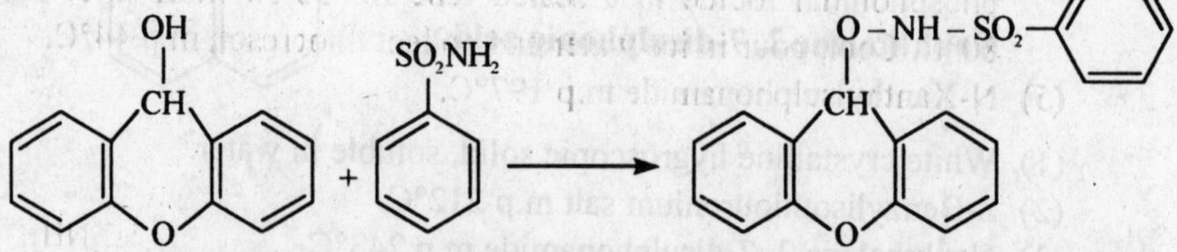

(2) **Sulphonacetamide:** Reflux 2gm of sulphonamide with 5ml acetyl chloride for 30 minutes. Distill off the excess acetyl chloride, cool and pour the residual oily portion in cold water. Collect the precipitated solid, dissolve in warm sodium bicarbonate solution and then acidify with glacial acetic acid. Collect the precipitated sulphonacetamide, recrystallise and determine its melting point.

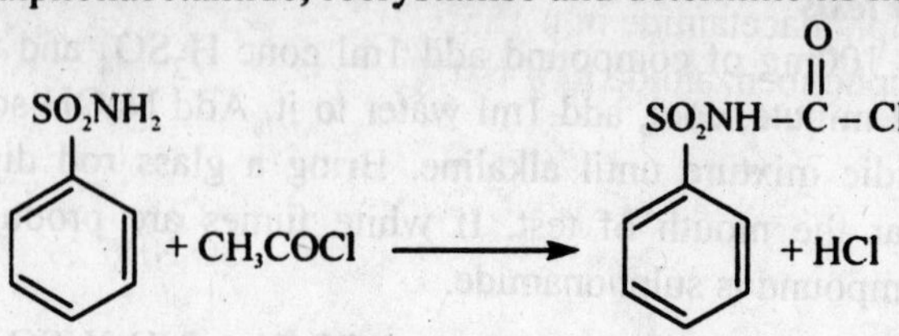

Solid

m.p 137°C p-Toluenesulphonamide

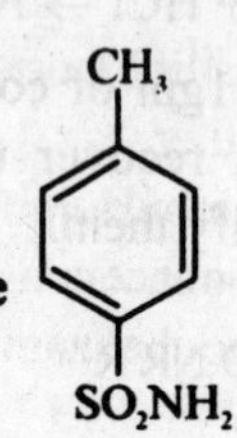

(1) **White crystalline compound, sparingly soluble in water, soluble in alcohol, ether and alkali.**
(2) **To 150ml saturated solution of bleaching powder in water add with stirring 6gm of compound. Cool and add acetic acid until precipitation is complete. Collect Dichloramine-T, m.p 83°C.**
(3) **Stir together 3.5gm of compound in 10ml DMF with 2gm of sodium hydride in 3ml DMF at 60°C for 1hr. Now add a solution of 5.2gm of o-xylene dibromide in 30ml DMF such that the temperture remains below 70°C. Continue stirring for 3hrs. Cool, pour in ice water. Collect p-tolylsulphonyl dihydroisoindole, m.p 175°C(d).**

(4) Heat 2.5gm of compound with 25ml conc HI acid and 3.5gm phosphonium iodide in a sealed tube for 30-35 minutes at 70-80°C. Cool, pour in ice water and collect thiocresol, m.p 44°C.
(5) N-Xanthylsulphonamide m.p 197°C.

m.p 150°C **Naphthalene 1-Sulphonamide**

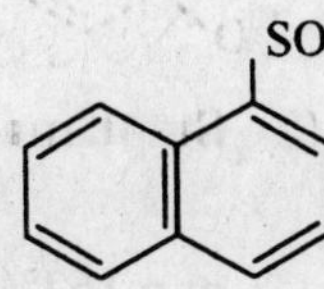

(1) **Crystalline compound, easily soluble in alcohol, ether and dilute alkali.**
(2) **Sulphonacetamide m.p 185°C**
(3) **Sulphonbenzamide m.p 194°C.**

m.p 152°C **Benzenesulphonamide**

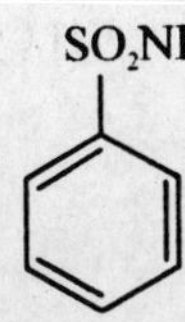

(1) **Crystalline compound, insoluble in water, soluble in hot alcohol and ether.**
(2) **N-Xanthylsulphonamide m.p 206°C**
(3) **Sulphonacetamide m.p 125°C**
(4) **Sulphonbenzamide m.p 147°C.**

m.p 155°C **o-Toluenesulphonamide**

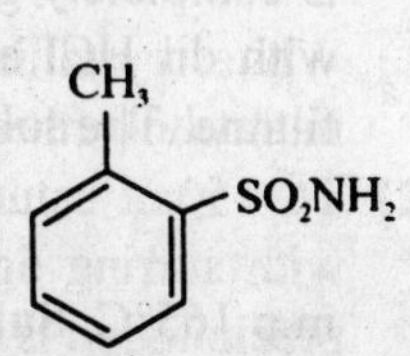

(1) **Colourless crystalline compound, insoluble in water, soluble in alcohol, ether and alkali.**
(2) **Dissolve 2.5gm of compound in 40ml NaOH solution (800mg in 40ml water) by heating to 40°C. Add in portion with stirring 4gm of powdered $KMnO_4$. After addition is complete continue stirring**

for 3hrs. Cool and collect saccharin, m.p 228°C.

(3) Boil 5gm of compound with 10ml H_2SO_4. Cool, pour in ice water. Extract with ether. Distill off ether and collect toluene, b.p 110°C

(4) N-Xanthylsulphonamide m.p183°C.

m.p 162°C **m-Nitrobenzenesulphonamide**

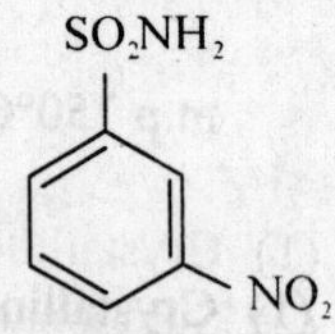

(1) Crystalline compound, sparingly soluble in alcohol, easily soluble in ether.

(2) m-Nitrobenzene sulphonacetamide m.p 189°C.

m.p 164°C **4-Toluidine –2-sulphonamide**

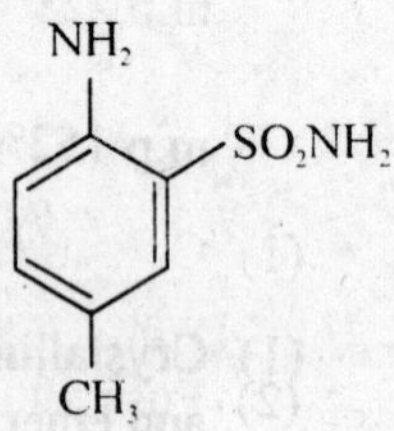

(1) Silky crystalline compound, soluble in hot water, alcohol and ether.

(2) Fuse 500mg of compound with 2gm of NaOH until ammonia gas is completely given off. Dissolve the residue in water, neutralise with dil HCl and filter. Add 1-2 drop of $FeCl_3$ solution to the filtrate. The solution turns red.

m.p 165°C **Sulphanilamide** (p-aminobenzene sulphonamide)

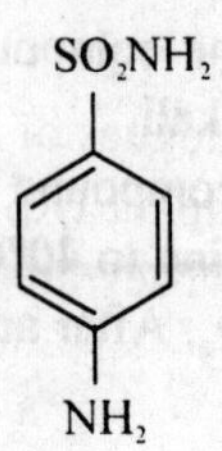

(1) Crystalline compound, soluble in alcohol and ether.
(2) N-Xanthylsulphonamide m.p 208°C.
(3) Sulphanildiacetamide m.p 217°C.

m.p 217 °C **Naphthalene 2-Sulphonamide**

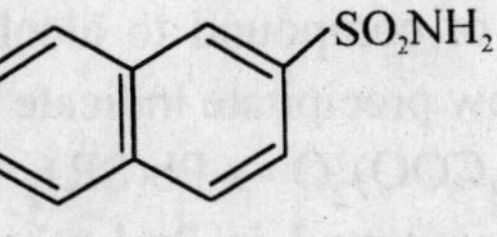

(1) Crystalline compound, soluble in alcohol, ether and alkali.
(2) Sulphonacetamide m.p 146°C.

m.p 228°C (d) **o-Sulphobenzimide** (Saccharin)

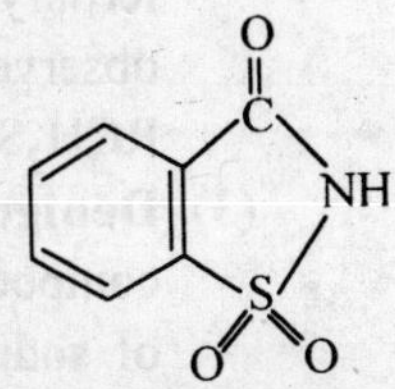

(1) Colourless crystalline compound with a sweet taste, insoluble in water, soluble in alkali.
(2) Fuse 500mg of compound with 2gm of NaOH until ammonia gas is completely given off. Dissolve the residue in water, neutralise with dil HCl and filter. Add 1-2 drop of $FeCl_3$ solution to the filtrate. The solution turns violet.
(3) Heat a mixture of 5gm of compound with 10gm of PCl_5 in an oil bath at 185-190°C for 90min. Cool, pour on crushed ice. Extract with chloroform. Distill off the chloroform, crystallise the residue from benzene and collect crystals of pseudosaccharin chloride, m.p 145°C(d).
(4) N-Xanthylsulphonamide m.p 198°C.

(iii) Thiols

Thiols are the sulphur analogues of alcohol/phenol in which the oxygen atom has been replaced by sulphur (RSH). These compounds are therefore often known as thio alcohols. They are less soluble in water in comparison to corresponding alcohol/phenol, but they are stronger acid than alcohols. Thiols are generally liquid with a penetrating and quite disagreeable odour.

ROH	RSH
Alcohol	Thiol

Identification Tests

(1) Add few drops of compound to alcoholic saturated lead acetate solution. A yellow precipitate indicate the presence of thiol group.

$$2RSH + Pb(CH_3COO)_2O \rightarrow Pb(SR)_2 + 2CH_3COOH$$

(2) To 500mg of compound in 2ml ether add 1ml conc HCl, 2ml sodium nitrite solution and shake thoroughly. A red precipitate or colouration due to the formation of thionitroso derivative indicates the presence of primary or secondary thiol. However in case of tertiary, pale green or yellow colour in the ethereal layer is observed.

$$RCH_2SH + HNO_2 \rightarrow RCH_2SN = O + H_2O$$

(3) **Deniges Nitroprusside test :** Dissolve few milligrams of compound in sodium hydroxide solution and then add 2-3 drops of sodium nitroprusside solution. Appearance of red or purple colour indicates the presence of –SH group in the compound.

(4) **Claesson's Ferric chloride test :** Dissolve few milligrams of compound in water/alcohol to which few drops of NH_4OH has been added. Now add few drops of ferric chloride solution. Appearance of blood red or red brown colouration indicate the presence of thiol group.

Derivatives

(1) **2, 4 dinitrophenyl sulphides :** Dissolve 250mg of compound in 8ml alcohol (warm if necessary) and then add 1ml of 10% NaOH solution. Mix the resultant solution with 2.5ml of 1-chloro, 2, 4-dinitrobenzene solution (.5gm in 2.5ml alcohol). Reflux the mixture for 10min on water bath, filter hot and cool the filtrate. Collect the precipitated 2, 4-dinitrophenyl sulphide, recrystallise from suitable solvent (ethanol, benzene) and determine its melting point.

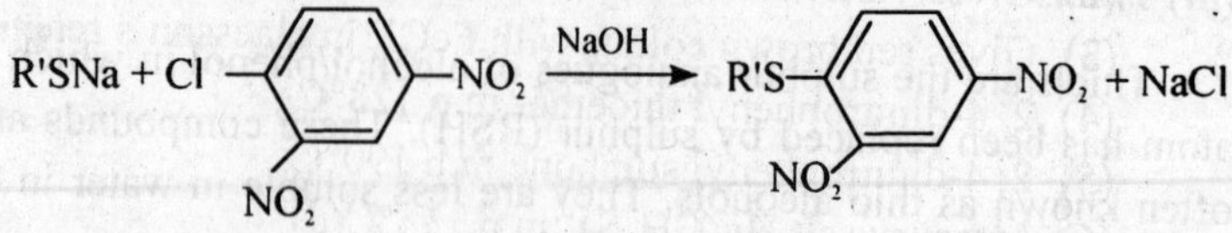

(2) **Sulphone :** Dissolve the 2, 4 dinitrophenyl sulphide in minimum amount of warm glacial acetic acid and then add 3% aqueous $KMnO_4$ solution until the colour of $KMnO_4$ persists. Decolourise the solution by adding sodium metabisulphite. Cool by adding crushed ice and collect the precipitated sulphone derivative. Recrystallise and determine its melting point.

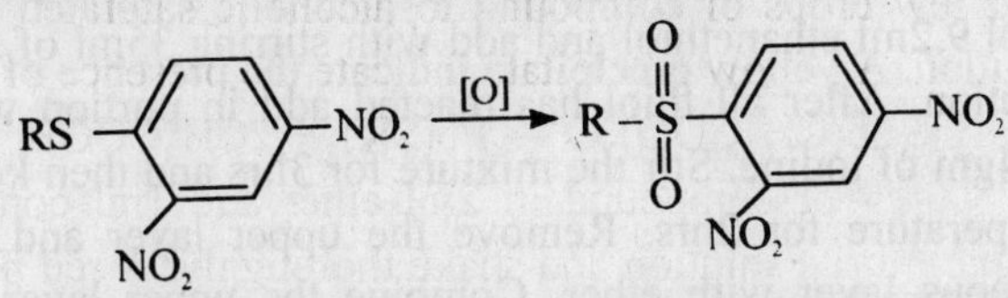

(3) **3, 5 dinitrothio benzoate :** Heat together a mixture of 200mg of compound (8-10 drops in case of liquid), 400mg of 3, 5-dinitrobenzoyl chloride and 4-5 drops of pyridine for 15 minutes. Add 1ml of water, 2-3 drops of pyridine and then stir with a glass rod. Collect the precipitated 3, 5-dinitrothiobenzoate derivative, recrystallise from suitable solvent and determine its melting point.

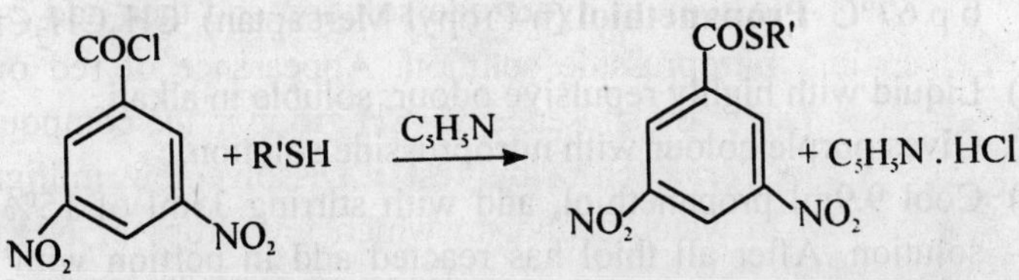

(4) **Mercury salts :** To a solution of 2ml of compound in 5ml alcohol add excess of 10% mercury cyanide solution. Cool and collect the precipitated mercaptide, recrystallise from alcohol and determine its melting point.

Aliphatic Liquid

b.p 6°C . **Methanethiol** (Methyl Mercaptan) CH_3SH

(1) Liquid with highly repulsive odour, insoluble in water, soluble in alcohol, ether and alkali.
(2) Gives violet colour with nitroprusside solution.
(3) Gives red brown colour with $FeCl_3$ in claesson's reaction.
(4) 2, 4-dinitrophenyl thioether m.p 128°C
(5) 2, 4-dinitrophenyl sulphone m.p 190°C.
(6) Mercury salt $Hg(CH_3S)_2$ m.p 175°C(d)

b.p 35°C **Ethanethiol** (Ethyl Mercaptan) CH_3CH_2SH

(1) Colourless liquid with unpleasant garlic like odour, insoluble in water.
(2) Gives dark red brown colour with $FeCl_3$ in claesson's reaction.
(3) To 1ml of compound add lead acetate solution. Shake thoroughly and collect yellow precipitate of $Pb(CH_3CH_2S)_2$, m.p 150°C
(4) Cool 9.2ml ethanethiol and add with stirring 35ml of 15% NaOH solution. After all thiol has reacted add in portion with stirring 13.4gm of iodine. Stir the mixture for 3hrs and then keep at room temperature for 2hrs. Remove the upper layer and extract the aqueous layer with ether. Combine the upper layer with ether layer. Distill and collect diethyl disulphide at b.p 151-152°C.
(5) 2, 4-dinitrophenyl thioether m.p 115°C
(6) 2, 4-dinitrophenyl sulphone m.p 160°C.
(7) 3, 5-dinitro thiobenzoate m.p 62°C.

b.p 67°C **Propanethiol** (n-Propyl Mercaptan) $CH_3CH_2CH_2SH$

(1) Liquid with highly repulsive odour, soluble in alkali.
(2) Gives purple colour with nitroprusside solution.
(3) Cool 9.9ml propanethiol, add with stirring 35ml of 15% NaOH solution. After all thiol has reacted add in portion with stirring 13.9gm of iodine. Stir the mixture for 3hrs and then keep at room temperature for 2hrs. Remove the upper layer and extract the aqueous layer with ether. Combine the upper layer with ether layer. Distill and collect dipropyl disulphide at b.p 190-192°C.
(4) 2, 4-dinitrophenyl thioether m.p 81°C
(5) 2, 4-dinitrophenyl sulphone m.p 128°C.
(6) 3, 5-dinitro thiobenzoate m.p 52°C.
(7) Mercury salt $Hg(CH_3\ CH_2CH_2S)_2$ m.p 70°C

b.p 97°C **Butanethiol** (n-Butyl Mercaptan) $CH_3\ CH_2CH_2CH_2SH$

(1) Colourless liquid with highly penetrating and disagreeable odour, insoluble in water, soluble in alcohol and ether.
(2) Gives purple colour with nitroprusside solution.
(3) Cool 10.5ml butanethiol, add with stirring 40ml of 15% NaOH solution. After all thiol has reacted add in portion with stirring

14.5gm of iodine. Stir the mixture for 3hrs and then keep at room temperature for 2hrs. Remove the upper layer and extract the aqueous layer with ether. Combine the upper layer with ether. Distill and collect dibutyl disulphide at b.p 223-226°C.

(4) 2, 4-dinitrophenyl thioether m.p 66°C
(5) 2, 4-dinitrophenyl sulphone m.p 92°C.
(6) 3, 5-dinitro thiobenzoate m.p 49°C.

b.p 118°C **3-Methyl Butane thiol** (Isoamyl Mercaptan)
$(CH_3)_2CHCH_2CH_2SH$

(1) Liquid with highly unpleasant odour.
(2) 2, 4-dinitrophenyl thioether m.p 59°C
(3) 2, 4-dinitrophenyl sulphone m.p 95°C.
(4) 3, 5-dinitro thiobenzoate m.p 43°C.

Aromatic Liquid

b.p 169°C **Thiophenol** (Phenyl Mercaptan)

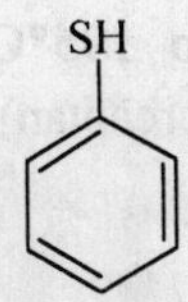

(1) Colourless liquid with unpleasant odour, insoluble in water, soluble in alcohol, ether and benzene.
(2) To 1ml of compound add conc H_2SO_4. A cherry red coloured solution which turns blue on warming is obtained.
(3) Yellow precipitate with lead acetate is obtained.
(4) Gives purple colour with nitroprusside solution.
(5) Cool a solution of 75ml of dichloromethane and 5.5ml of triethyamine. Add dropwise with stirring 4.1ml of thiophenol. Warm the mixture to 20°C and continue stirring for 3hrs. Filter off the triethylamine hydrochloride. Distill off the solvent and collect bis(phenylthio)methane, recrystallise from pet-ether, m.p 37°C.
(6) 2, 4-dinitrophenyl thioether m.p 121°C
(7) 2, 4-dinitrophenyl sulphone m.p 161°C. (8) 3, 5-dinitro thiobenzoate m.p 149°C.

b.p 194°C **Phenylmethane thiol** (Benzyl mercaptan)

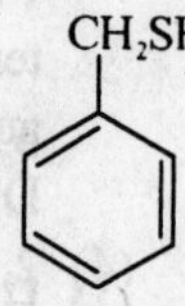

(1) Colourless liquid with highly unpleasant odour.
(2) Gives purple colour with sodium nitroprusside solution.
(3) Cool 11.5ml of compound, add with stirring 45ml of 15% NaOH solution. After all thiol has reacted add in portion with stirring 14.4gm of iodine. Stir the mixture for 3hrs and then keep at room temperature for 2hrs. Remove the upper layer and extract the aqueous layer with ether. Combine the upper layer with ether. Distill off the ether and collect dibenzyl disulphide, m.p 73°C.
(4) 2, 4-dinitrophenyl thioether m.p 130°C
(5) 2, 4-dinitrophenyl sulphone m.p 183°C.
(6) 3, 5-dinitro thiobenzoate m.p 120°C.

b.p 195°C **1-Methyl Thiophenol** (Thio-o cresol, o-Tolyl mercaptan)

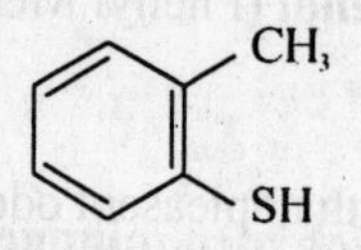

(1) Colourless liquid, soluble in alcohol and ether.
(2) 2, 4-dinitrophenyl thioether m.p 101°C
(3) 2, 4-dinitrophenyl sulphone m.p 155°C.

b.p 195°C **Thio-m-cresol** (m-Tolyl mercaptan)

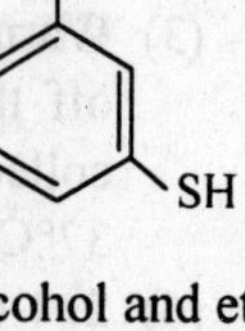

(1) Colourless liquid, insoluble in water, soluble in alcohol and ether.
(2) 2, 4-dinitrophenyl thioether m.p 91°C (3) 2, 4-dinitrophenyl sulphone m.p 145°C.

b.p 216°C **Phenol thiol** (Thiopyrocatechol)

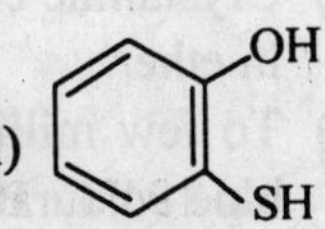

(1) Highly refractive liquid with penetrating odour, insoluble in water, soluble in ether.
(2) Gives yellow precipitate with lead acetate.
(3) To 1ml of compound add $FeCl_3$ solution. A violet colouration which changes to red on addition of NaOH.

b.p 285°C(d) **1-Thionaphthol**

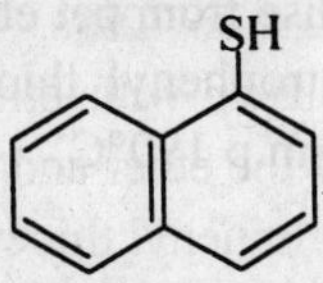

(1) Insoluble in water, soluble in alcohol and ether.
(2) Gives yellow precipitate with lead acetate solution.
(3) 2, 4-dinitrophenyl thioether m.p 176°C.

Aromatic Solid

m.p 30°C **Monothiohydroquinone**

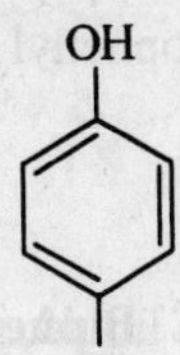

(1) Crystalline compound with highly penetrating odour, soluble in water.
(2) To few milligrams of compound add conc H_2SO_4. A blue green coloured solution is obtained.
(3) Gives yellow precipitate with lead acetate.

m.p 43°C **Thio-4-cresol** (p-Tolyl mercaptan)

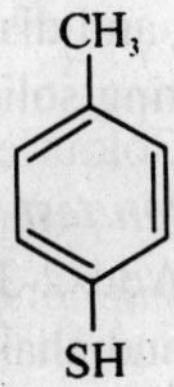

(1) Crystalline compound, sparingly soluble in alcohol, easily soluble in ether.
(2) To few milligrams of compound add conc H_2SO_4 and warm. A blue colouration is observed.
(3) Heat 19gm of compound with 75ml of acetic acid and excess of 40% ferric chloride solution (75gm) for 5-10min Cool and collect the precipitated di-p-tolyl disulphide, recrystallise from alcohol, m.p 43°C.
(4) To 6gm of compound in 10ml glacial acetic acid add dropwise with stirring 4ml of bromine. After addition is complete distill off the solvent and collect the precipitate p-toluenesulphonyl bromide, recrystallise from pet ether, m.p 93-94°C.
(5) 2, 4-dinitrophenyl thioether m.p 103°C (6) 2, 4-dinitrophenyl sulphone m.p 190°C.

m.p 81°C **2-Thionaphthol**

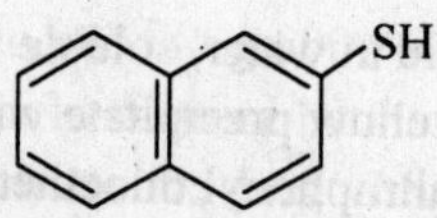

(1) Crystalline compound, insoluble in water.
(2) Gives yellow precipitate with lead acetate.
(3) 2, 4-dinitrophenyl thioether m.p 145°C.

m.p 111°C **Biphenylmercaptan**

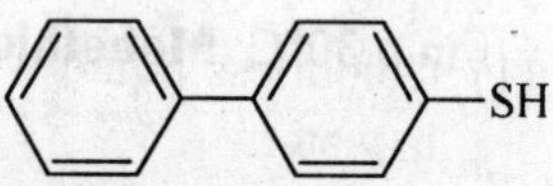

(1) Crystalline compound, soluble in alcohol, ether and benzene
(2) Gives dark red crystals with lead acetate.
(3) 2, 4-dinitrophenyl thioether m.p 146°C.

(iv) Sulphides and Disulphides

Sulphides are the sulphur analogues of ether in which the oxygen atom has been replaced by sulphur. (RSR). They are usually liquid with highly penetrating and disagreeable odour. Disulphides on the other hand are liquid or low melting solid with unpleasant odour (RSSR)

Identification test

(1) Add 2-3 drops of sulphide to 3ml of sodium plumbite solution and shake thoroughly. A light yellow precipitate is formed. Add

75mg of sulphur. The colour changes to orange

(2) Add 25mg of disulphide to 1ml of 1N hydroxylamine hydrochloride in methanol. Add few milligrams of zinc dust and shake the mixture vigorously for 1-2 minute. Decant the liquid and test for the presence of thiol derivative.

$$R\text{-}S\text{-}S\text{-}R + Zn\,[2H] \rightarrow 2RSH$$

Derivatives

(1) Oxidation of thiethers to sulphone:- Dissolve 5ml of compound in minimum amount of glacial acetic acid and then add in portion a slolution of 2gm of potassium permanganate dissolved in 30ml water. Shake after each addition. Remove excess of permanganate by adding sodium bisulphite. Filter, cool the filtrate and collect the sulphone derivative, recrystallise and determine its melting point.

(2) Dissolve 100mg of compound in 1ml ethanol and add this solution to a saturated solution of mercuric chloride in water/ethanol. Shake thoroughly and collect the crystalline adduct. Recrystallise and determine its melting point.

Aliphatic Liquid

b.p 37°C **Dimethyl sulphide** $(CH_3)_2S$

(1) Colourless liquid with highly alliaceous ethereal odour, insoluble in water.

(2) Stir 3.1gm of dimethyl sulphide with 5.8gm of 3-bromopropyne and 5ml of acetonitrile for 20hrs. Filter the white crytalline dimethyl-2-propynylsulfonium bromide, recrytallise from ethanol-ether, m.p 109-110°C.

(3) Sulfone m.p 109°C

(4) $HgCl_2$ adduct m.p 150°C.

b.p 46°C **Carbon disulphide** CS_2

(1) Colourless liquid with ethereal odour, miscible with alcohol, ether and benzene.

(2) Dissolve 2-3 drops of compound in 3ml NaOH solution. Acidify with acetic acid and then add 2-3 drops of 5% $CuSO_4$ solution.

Bright yellow coloured precipitate is obtained.

(3) Dissolve 2-3 drops of compound in 2ml alcohol and 1ml NH_4OH. Evaporate to dryness. Dissolve the residue in 1ml water, acidify with 1-2drop of dil HCl. Add 1-2 drop of $FeCl_3$ solution. Blood red colouration appears.

(4) To 2ml of compound add 2ml of phenylhydrazine solution (in ether). Shake thoroughly and cool. Collect the precipitated phenyhydrazine phenylthiocarbazinate, m.p 97°C.

b.p 66°C **Ethyl Methyl Sulphide** $CH_3CH_2SCH_3$

(1) Liquid with ethereal odour, insoluble in water, miscible with alcohol and ether.

(2) Sulphone m.p 36°C (3) $HgCl_2$ adduct m.p 128°C.

b.p 92°C **Diethyl sulphide** $(CH_3CH_2)_2S$

(1) Colourless liquid with highly alliaceous ethereal odour, insoluble in water.

(2) To 1ml of compound add solution of iodine in KI. A heavy brown oily layer due to the formation of iodine adduct is obtained.

(3) Sulphone m.p 73°C

(4) $HgCl_2$ adduct m.p 119°C.

b.p 110°C **Dimethyl disulphide** CH_3-S-S-CH_3

(1) Liquid with unpleasant alliaceous odour

(2) Cool a mixture of 3.7gm of 2-phenyl 1, 3-dithiane, 60ml THF and 25mg of n-butyl lithium in hexane to 20°C. Stir under these conditions for 2 hrs and then add a solution of 3ml dimethyl disulphide in 5ml THF. Remove the cooling source and stir at room temperature for 90min. Add 30ml saturated NaCl solution followed by 50ml $CHCl_3$. Remove the $CHCl_3$ layer, distill off $CHCl_3$ and cool, collect the precipitated 2-Methylthio-2-phenyl-1, 3-dithiane, m.p 78°C.

(3) Methyl mercaptan b.p 6°C.

(4) $HgCl_2$ adduct m.p 175°C.

b.p 138°C **Allyl sulphide** $(CH_2{=}CHCH_2)_2S$

(1) Liquid with strong garlic like odour, sparingly soluble in water, miscible with alcohol and ether.
(2) To 1ml of compound add alcoholic silver nitrate solution. Precipitation occurs due to the formation of adduct.

b.p 142°C **Di-n-propyl sulphide** $(CH_3CH_2CH_2)_2S$

(1) Liquid with unpleasant alliaceous odour, insoluble in water, soluble in alcohol and ether.
(2) Sulphone m.p 29°C (3) $HgCl_2$ adduct m.p 122°C.

b.p 153°C **Diethyl disulphide** CH_3CH_2-S-S- CH_2CH_3

(1) Liquid with garlic like odour, insoluble in water, miscible with alcohol and ether.
(2) Ethyl mercaptan b.p 36°C.

b.p 182°C (189°C) **Di-n-Butyl sulphide** $(CH_3CH_2CH_2CH_2)_2S$

(1) Insoluble in water, soluble in alcohol and ether.
(2) Sulphone m.p 44°C (3) $HgCl_2$ adduct m.p 111°C.

Aromatic Liquid

b.p 84°C **Thiophene** 

(1) Liquid with faint ethereal odour, insoluble in water, miscible with alcohol and ether.
(2) To few milligrams of isatin add 1ml conc H_2SO_4. Add a drop of $FeCl_3$ solution and then pour with stirring into dilute solution of compound in benzene (thiophen free). An intense blue colouration appears at once.
(3) Cool a mixture of 2ml thiophen, 2.4ml DMF and 8ml 1, 2 dichloroethane to 0°C. Add with stirring 3ml of phosphorous oxychloride and then reflux for 2hrs. Cool, pour in ice, neutralise

with sodium acetate. Extract with ether, distill off ether and collect 2-formylthiophen at 86°C/16mmHg. (b.p 198°C)

(4) Heat 8ml of compound with 5ml acetic anhydride at 70°C. Remove the heat source and add .5ml of orthophosphoric acid(exothermic reaction). After the reaction has subsided reflux for 2hrs. Add 20ml water. Distill and collect 2-acetyl thiophen at 90°C/10mmHg, on cooling it solidifies, m.p 10°C. (b.p 213°C)

(5) To 8ml of compound add 7.6gm of iodine and then add dropwise 11.2ml of HNO_3 solution (5.6ml HNO_3 in 5.6ml water) Heat the mixture under reflux for 30 min. Remove the organic layer and distill. Collect 2-iodothiophene at 90-93°C/36mmHg. (b.p 182°C)

(6) $HgCl_2$ adduct m.p 182°C.

b.p 188°C **Methyl Phenyl sulphide**

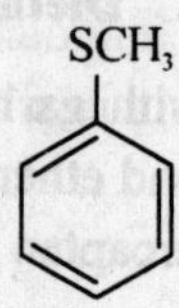

(1) Insoluble in water, soluble in alcohol.

(2) Sulphone m.p 88°C.

b.p 205°C **Ethyl Phenyl sulphide**

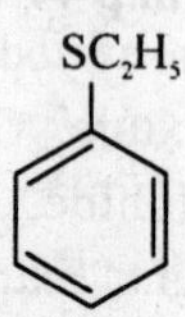

(1) Liquid with penetrating disagreeable odour.

(2) Sulphone m.p 42°C.

b.p 295°C **Diphenyl sulphide**

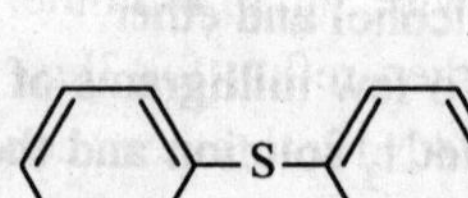

(1) Liquid with alliaceous odour, insoluble in water, soluble in ether and benzene.

(2) Stir 9.3gm of compound with 34.7gm of 1-chloro-3-iodopropane in 20ml nitromethane. Add in one portion 7.8 gm of silver

tetrafluoroborate and keep the mixture at room temperature for 16hrs. Add 20ml dichloromethane and filter. Remove the organic layer, distill off the solvent and collect 3-chloropropyldiphenylsulfonium tetrafluoroborate, m.p 103-105°C.

(3) Sulphone m.p 128°C.

Aromatic Solid

m.p 48°C **Di-p-tolyl disulphide**

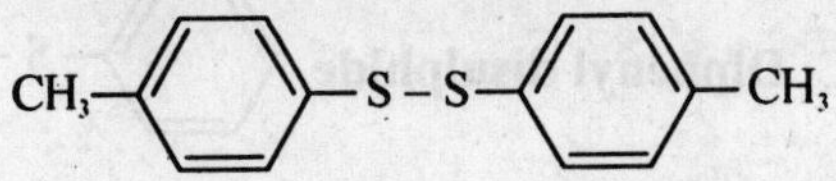

(1) Crystalline compound, soluble in alcohol and ether.

(2) Thio –p-cresol m.p 43°C (Zn/HCl)

m.p 49°C **Dibenzyl sulphide**

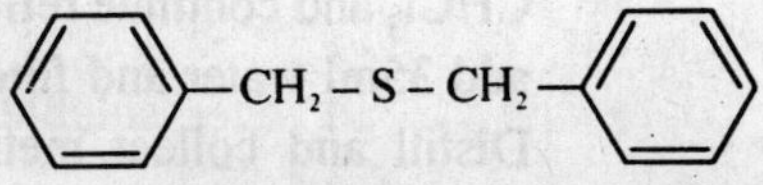

(1) Crystalline compound, insoluble in water, soluble in alcohol and ether.

(2) Add 1.175gm of sodium metaperiodate to 23ml of 1:1 water – methanol mixture. Cool to 0°C, add with stirring 1.07gm of dibenzyl sulphide. Continue stirring for 8-10hrs and then leave overnight. Extract with chloroform, distill off the solvent and collect dibenzyl sulphoxide, recrystallise from ethanol, m.p 135°C.

(3) Dissolve 5.65gm of compound in 75ml CCl_4 and then add dropwise 3ml of bromine. Focus a sunlamp on the reaction mixture and then reflux for 3hrs. Distill off the CCl_4 and to the residual oil add 12ml ether, cool and add a solution of 16gm of m-chloroperbenzoic acid in 75ml ether. Stir the mixture for 2 days. Distill off the solvent and add saturated sodium bicarbonate solition. Filter the precipitated α, α' dibromodibenzyl sulphone, mp 143-158°C.

(4) Sulphone m.p 151°C($KMnO_4$) (5) $HgCl_2$ adduct m.p 136°C.

m.p 57°C **Di-p-Tolyl Sulphide**

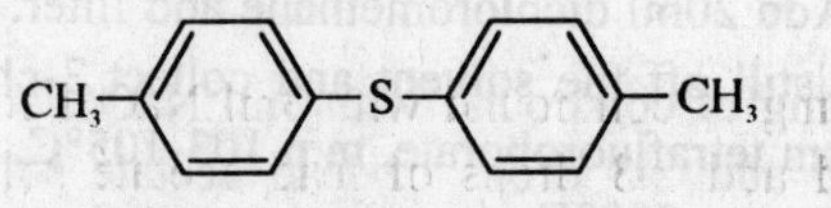

(1) **Crystalline compound, insoluble in water, soluble in hot alcohol, ether and benzene.**

(2) **Sulphone m.p 158°C($KMnO_4$).**

m.p 60°C **Diphenyl disulphide** 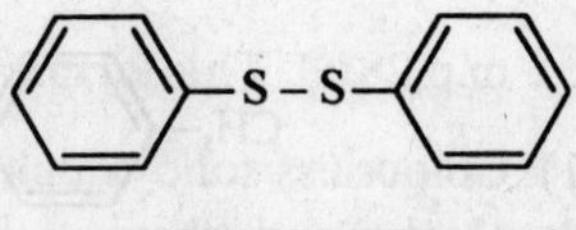

(1) **Colourless crystalline compound, insoluble in water, soluble in alcohol, ether and benzene.**

(2) **Reflux 5.5gm of compound with 45ml $CHCl_3$ and 45ml MeOH. Add in portion a solution of 45gm of lead tetracetate in 200ml $CHCl_3$ and continue refluxing for 12hrs. Distill off the chloroform, add 35ml water and filter. Concentrate, cool and leave overnight. Distill and collect methyl benzene sulphinate at b.p 59-60°C/.04mmHg.**

(3) **To a mixture of 4.5gm of silver difluoride and 5ml 1, 1, 2-trichloro-1, 2, 2-trifluoroethane add in portion 1gm of diphenyl disulphide (exothermic reaction, temp below 40°C). Heat the mixture under reflux for 5 min, filter hot. Distill the filtrate and collect the residue. Distill the residue under reduced pressure and collect phenylsulphur trifluoride at b.p 47-48°C/2.6mmHg as colourless liquid.**

(4) **Thiophenol b.p 169°C (Zn/HCl)**

m.p 72°C **Dibenzyl disulphide**

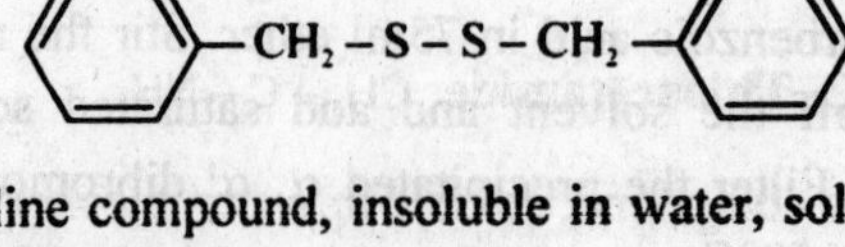

(1) **Crystalline compound, insoluble in water, soluble in hot alcohol, ether and benzene.**

(2) **Benzyl mercaptan b.p 194°C (Zn/HCl).**

(*v*) *Thioamides*

Identification test

(1) Boil 100mg of compound with 3ml NaOH solution for 1-2 min. Cool and add 2-3 drops of lead acetate solution. A brown or black precipitae is obtained.
(2) Dissolve 100mg of yellow mercuric oxide in alcohol, add 100mg of compound to it. Warm the mixture gently. A black precipitate is obtained.

m.p 28°C **Thioformamide** HCS. NH_2

(1) Colourless solid which changes to yellow syrupy mass, soluble in alcohol and ether.
(2) Reduces Fehling's solution.

m.p 76°C **Thioacetanilide** 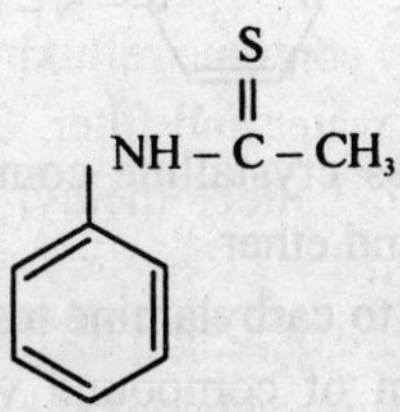

(1) Crystalline compound, soluble in alkali.

m.p 78°C **Allylthiourea** (thiosinamin) $CH_2=CHCH_2NHCS.NH_2$

(1) Crystalline compound, soluble in alcohol and ether.
(2) Reduces ammonical silver nitrate solution to give black precipitate.
(3) N-Acetate m.p 95°C

m.p 112°C **Thioacetamide** $CH_3-\overset{\overset{\displaystyle S}{\|}}{C}-NH_2$

(1) Crystalline compound, soluble in water, alcohol and ether.
(2) Reflux few milligrams of compound with dil H_2SO_4. Smell of vinegar is observed due to the formation of acetic acid.

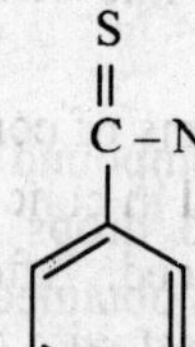

m.p 116°C **Thiobenzamide**

(1) Crystalline compound.
(2) Reflux with Zn and conc HCl in alcohol. Cool, extract with ether, distill off ether and collect benzyl amine, b.p 184°C.
(3) N Benzoate m.p 117°C.

m.p 154°C **Thiocarbanilide** (s-Diphenylurea)

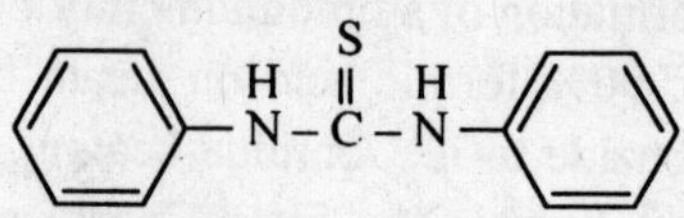

(1) Colourless crystalline compound, insoluble in water, soluble in alcohol and ether.
(2) Respond to carbylamine test.
(3) Heat 2gm of compound with 10ml conc HCl for 30 min. . A powerful irritating mustard like odour is observed due to the formation of phenyl mustard. Extract with ether, distill off ether and collect phenyl mustard at b.p 217-221°C.
(4) Boil 1gm of compound with 20ml alcohol and freshly prepared yellow mercuric oxide solution for 10-15 minutes. Filter hot, concentrate the filtrate, cool and collect carbanilide, m.p 236°C.

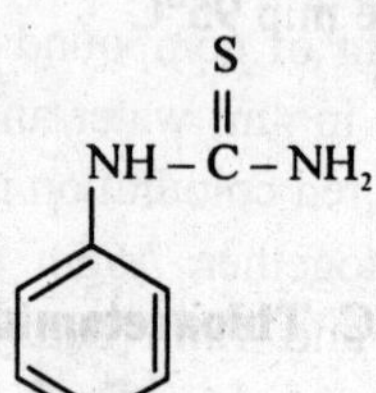

m.p 154°C **Phenylthiourea**

(1) White crystalline solid, soluble in alcohol.
(2) Respond to carbylamine test.

(3) Heat few milligrams of compound with conc HCl. A powerful irritating mustard like odour is observed due to the formation of phenyl mustard.

(4) Boil few milligrams of compound with dil NaOH solution. Bring a glass rod dipped in conc HCl near the mouth of test tube. White fumes are produced. After evolution of ammonia is complete perform the dye test with the resultant mixture.

(5) Reflux 200mg of compound with 2ml aniline. Cool and collect thiocarbanilide, m.p 153°C.

m.p 182°C **Thiosemicarbazide** NH_2-CS-$NHNH_2$

(1) Crystalline solid, soluble in water and alcohol.

(2) Heat few milligrams of compound with NaOH solution. Bring a glass rod dipped in HCl near the mouth of test tube. White fumes due to the formation of ammonium chloride are produced.

(3) Heat 8ml of 90% formic acid on steam bath and add 3.6gm of thiosemicarbazide to it. Continue heating for 30 min. Add about 10ml of boiling water and filter. Cool the filtrate in ice bath and collect 1-formyl 3-thiosemicarbazide, m.p 178°C.

(4) Hydrochloride m.p 188°C

(5) N-Acetate m.p 165°C.

m.p 182°C **Thiourea** NH_2-CS-NH_2

(1) White crystalline solid, soluble in water and alcohol, insoluble in ether.

(2) Dissolve 300mg of comound in 3ml acetic acid and heat. Add 3ml aqueous $K_4Fe(CN)_6$ solution. A green coloured solution changing to blue is obtained.

(3) Heat 1gm of compound in dry test tube until it melts. Dissolve the melt in 4ml water and then add few drops of $FeCl_3$ solution. A blood red colouration is produced.

(4) Reflux together 7.6gm of thiourea, 20ml warm water, 13.5ml ethanol and 12ml benzyl chloride on water bath until benzyl chloride dissolves. Cool, shake the mixture thoroughly and collect S-benzyl isothiouronium chloride, m.p 174°C.

(5) Heat together 3.8gm of thiourea with 6.5gm of ethyl aceto acetate,

6gm of sodium methoxide and 45ml methanol for 8hrs such that it is evaporated to dryness. Dissolve it in minimum amount of water and filter. Acidify the filtrate with glacial acetic acid and cool. Collect 2-thio 6-methyl uracil, m.p 280°C(d)

(6) Dissolve 3.8gm thiourea in 10ml water. Add with stirring 4ml of chloroacetone. Reflux for 2hrs. Cool and add 10gm of NaOH. Remove the oily layer and extract the aqueous layer with ether. Distill off ether and collect 2-amino 4-methyl thiazole, m.p 45°C. Repeat the same process with oily layer.

(7) Hydrochloride m.p 136°C (8) Xanthylthioamide m.p 136°C

(*vi*) *Alkyl Sulphates*

Alkyl sulphates are regarded as esters of sulphuric acid. But in these compounds the carbon is bonded to the oxygen and not to the sulphur. They are usually water insoluble liquids which are hydolysed to corresponding alcohols by water or dilute alkali.

$$R_2SO_4 + 2H_2O \rightarrow 2ROH + H_2SO_4$$

b.p 188°C (d) **Dimethyl sulphate** $(CH_3)_2SO_4$

(1) Colourless, odourless highly poisonous liquid, insoluble in water.

(2) To 15.2 gm of thiourea in 7ml water add 13.8gm of dimethyl sulphate (exothermix reaction). After the reaction has subsided reflux the mixture for one hour. Cool, add 20ml alcohol and filter. Collect the solid S-methyl isothiourea sulphate, m.p 235°C (d).

(3) Heat a solution of 5ml of N, N dimethylaniline in 10ml toluene to 40°C, remove the heat source and add 2ml of dimethyl sulphate (exothermic reaction). After the reaction has subsided heat on water bath for 60 min. Cool and collect phenyltrimethylammonium sulfomethylate. Dissolve it in 5ml 48% HBr acid and then dilute it with 5ml water. Add dropwise with stirring 1.3ml of bromine. Continue stirring for 5hrs. Cool and collect phenyltrimethylammonium tribromide, recrystallise from acetic acid, m.p 114-115°C.

(4) Stir a mixture of 12.4gm of potassium carbonate, 41ml acetone and 4.2gm of orcinol. Add dropwise 7.1ml of dimethyl sulphate. Mixture becomes warm and begins to reflux. After the reaction

has subsided reflux for 4hrs. Distill off acetone. Add 5ml aqueous ammonia, stir and heat for 10min. Remove the organic layer and extract the aqueous layer with ether. Combine the organic layers, distill off ether and collect orcinol dimethyl ether at b.p 134-135°C/40mmHg.

(5) To 5gm of m-hydroxy benzaldehyde in 16ml 2N NaOH solution add dropwise with stirring 3.3ml of dimethyl sulphate (temp 40-45°C). Add 9ml of NaOH solution and 1.6ml of dimethyl sulphate (temp 50°C). Continue stirring for 30min, cool, isolate with ether, distill off ether and collect m-methoxy benzaldehyde as pale yellow liquid at b.p 89-90°C(3mmHg).

(6) Cool a solution of 3.6gm of 2-naphthol in 15ml NaOH solution (1.1gm NaOH), add dropwise 2.4ml of dimethyl sulphate. Heat at 70-80°C for 60min. Cool and collect Methyl 2-naphthyl ether, m.p 72°C.

b.p 209°C (d) **Diethyl sulphate** $(CH_3CH_2)_2SO_4$

(1) Colourless liquid with peppermint like odour, toxic, insoluble in water.

(2) To 4.7gm of phenol add a solution of 2.1gm of NaOH in 20ml water. Cool the mixture and add dropwise 6.5ml of diethyl sulphate. After addition is complete reflux for 2hrs. Remove the organic layer and extract the aqueous layer with ether. Combine the two layer, wash with dil H_2SO_4 and water. Distill and collect ethyl phenyl ether at b.p 168-179°C.

(3) Cool a solution of 3.8gm of 2-naphthol in 18ml NaOH solution (1.3gm NaOH), add dropwise 3ml of diethyl sulphate. Heat at 70-80°C for 60min. Cool and collect Ethyl 2-naphthyl ether, m.p 37°C.

(vii) Sulphinic Acids

m.p 84°C **Benzenesulphinic acid** 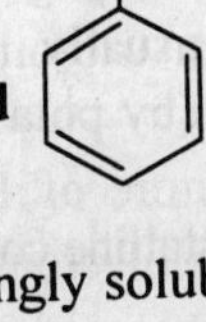

(1) Colourless crystalline solid, sparingly soluble in water, soluble in alcohol and ether.

(2) Heat together 2gm of compound with 2.25gm of o-nitrochlorobenzene and 1.25 gm of sodium acetate at 160°C for 3 hrs. Cool and collect 2-nitrodiphenyl sulphone, recrystallise from alcohol, m.p 147-148°C.

(3) Amide m.p 121°C (4) Anilide m.p 113°C

m.p 84°C **1-Naphthalene sulphinic acid**

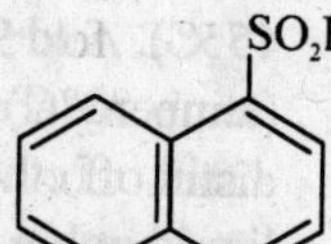

(1) Crystalline compound, sparingly soluble in water and alcohol.

(2) To few milligrams of compound add conc H_2SO_4. A blue coloured solution which slowly changes to olive green.

(3) Boil 500mg of compound with 5ml water. Cool and collect naphthalene 1-sulphonic acid, m.p 88°C.

(4) Boil 500mg of compound at 180°C with 5-7ml HCl. Sulphur dioxide gas is evolved, on cooling naphthalene is precipitated, m.p 80°C.

m.p 85°C **p-Toluene sulphinic acid**

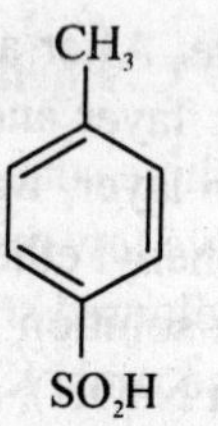

(1) Crystalline compound, sparingly soluble in water, soluble in alcohol and ether.

(2) Amide 120°C

(3) Anilide m.p 138°C.

(viii) Sulphoxides and Sulphones

Sulphoxides are usually low melting solids and can very easily be oxidised to sulphones (by potassium permanganate) or reduced to sulphides (Sn or Zn/HCl).

Sulphones are crystalline compounds which are resistant to most oxidising and reducing agents.

Liquid

b.p 189°C **Dimethysulphoxide** (DMSO) $(CH_3)_2SO$

(1) Colourless, odourless, hygroscopic liquid with bitter taste, miscible with water.
(2) To 15ml DMSO add 1.2gm of sodium hydride. Stir and heat to 75°C for 30min. Cool and add to it a solution of 2.2gm of diphenylacetylene in 10ml DMSO. Heat the mixture at 65°C for 2hrs. Cool and pour in ice. Extract with ether. Distill and collect 2, 3-diphenyl-1, 3-butadiene, m.p 47-48°C.
(2) Heat a mixture of 9ml of DMSO, 12ml t-butyl alcohol and 5.8gm of potassium t-butoxide to 80°C. When all the solid has dissolved remove the heat source and then add 7.2ml ethyl benzoate. Stir at room temperature for 4hrs and then remove the excess solvent. Pour in ice. Extract with ether. Discard the ether portion and acidify the aqueous layer with dil HCl. Keep the mixture at room temperature for 30hrs. Cool and collect phenylglyoxal hemimercaptal, m.p 103-105°C.

Solid

m.p 70°C **Phenyl Sulphoxide** (Thionyl benzene) $(C_6H_5)_2SO$

(1) Crystalline compound, easily soluble in alcohol, ether and benzene.
(2) Dissolve few milligrams of compound in conc H_2SO_4. A transient blue coloured solution is obtained.
(3) Oxidise with $KMnO_4$/AcOH and collect diphenylsulphone, m.p 128°C.

m.p 95°C **p-Tolyl Sulphoxide** $(CH_3C_6H_4)_2SO$

(1) Crystalline compound, soluble in alcohol, ether and chloroform.
(2) Sulphone m.p 158°C ($KMnO_4$/AcOH)
(3) Sulphide m.p 57°C (Zn/HCl)

m.p 102°C **Ethylbenzene Sulphone** $(C_2H_5C_6H_4)_2SO_2$

(1) Crystalline compound, soluble in ether, benzene and chloroform.

(2) Uneffected by oxidising and reducing agents.

m.p 127°C **Sulphonal** (Bisethylsulphone-dimethymethane) $(C_2H_5SO_2)_2C(CH_3)_2$

(1) Colourless, odourless solid, insoluble in water, moderately soluble in hot water, chloroform and benzene, soluble in conc H_2SO_4.
(2) Heat few milligrams of compound with sodium. Unpleasant odour of mercaptan is produced.

m.p 128°C **Phenyl Sulphone** (Sulphobenzid) $(C_6H_5)_2SO_2$

(1) Crystalline compound, insoluble in water, soluble in hot alcohol and benzene.
(2) Uneffected by oxidising and reducing agents.

m.p 135°C **Benzyl Sulphoxide** $(C_6H_5CH_2)_2SO$

(1) Crystalline compound, insoluble in water, soluble in hot water, alcohol and ether.
(2) Sulphone m.p 151°C ($KMnO_4$/AcOH)
(3) Sulphide m.p 49°C (Zn/HCl)

m.p 151°C **Benzyl Sulphone** $(C_6H_5CH_2)_2SO_2$

(1) Crystalline compound, insoluble in water, soluble in hot alcohol and benzene.
(2) Boil with conc HNO_3. Cool and collect benzoic acid m.p 121°C.

m.p 158°C **p-Tolyl Sulphone** (Sulphotoluid) $(CH_3C_6H_4)_2SO_2$

(1) Crystalline compound, insoluble in water, soluble in hot alcohol, chloroform and benzene.
(2) Phenyl sulphone dicarboxylic acid m.p >300°C ($KMnO_4$/AcOH).

m.p 239°C **Oxysulphobenzid** $(HOC_6H_4)_2SO_2$

(1) Crystalline compound, insoluble in water, soluble in hot water, alcohol and ether.

(*ix*) *Thiocarboxylic Acids*

Liquid

b.p 93°C **Thioacetic acid** CH_3COSH

(1) Yellow coloured liquid with unpleasant odour (similar to hydrogen sulphide), soluble in water and alcohol.
(2) Thioacetamide m.p 108°C
(3) Thioacetanilide m.p 76°C
(4) Thioacet-p-toluidide m.p 130°C

Solid

m.p 24°C **Thiobenzoic acid** 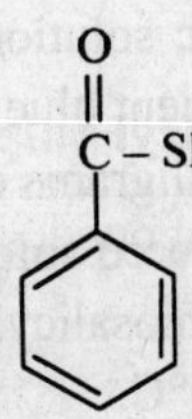

(1) Yellow coloured solid, insoluble in water, soluble in alcohol and ether.
(2) To few milligrams of compound add $CuSO_4$ solution. Greenish yellow coloured precipitate of copper salt is obtained.
(3) Thiobenzamide m.p 115°C (4) Thiobenz-p-toluidide m.p 129°C.

m.p 61°C **Phenylthioglycolic acid** 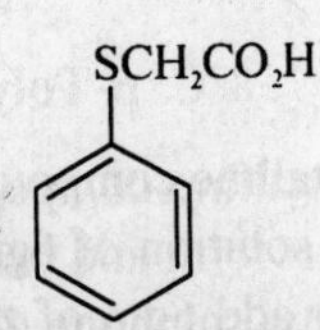

(1) Crystalline solid, insoluble in water, soluble in hot water, alcohol and ether.
(2) On oxidation with $KMnO_4$ gives phenylsulphonacetic acid, m.p 112°C.

m.p 129°C **1-Thiophenic acid** 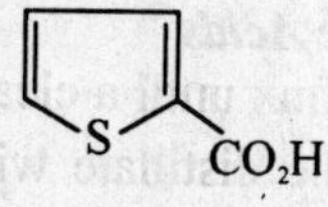

(1) Crystalline compound, with irritating odour, vapours cause coughing, soluble in hot water, alcohol and ether.
(2) To few milligrams of compound add isatin and H_2SO_4. A blue colouration is obtained.
(3) Amide, m.p 174°C (4) Anilide 140°C.

m.p 165°C **Thiosalicylic acid** 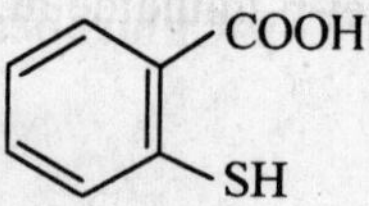

(1) Yellow crystalline solid insoluble in water, soluble in alcohol.
(2) To 1ml alcoholic solution of compound add few drops of $FeCl_3$ solution. A transient blue coloured solution is obtained.
(3) Dissolve few milligrams of compound in conc H_2SO_4 and heat. A carmine red coloured solution is obtained.
(4) On oxidation dithiosalicylic acid is obtained, m.p 289°C.
(5) Acetate, m.p 125°C.

(x) *Sulphonyl Chlorides*

Solid

m.p 14°C, b.p 246°C (d) **Benzenesulphonyl chloride** 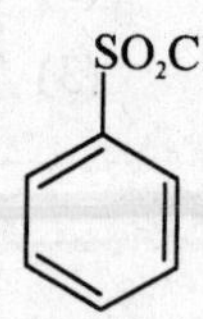

(1) Crystalline compound, insoluble in water, soluble in alcohol.
(2) To a solution of 6gm of compound in 30ml ether containing .5ml water add 6gm of zinc dust. Reflux for 1-2hrs. Cool, dissolve the precipitated solid in a solution of 6gm of sodium carbonate in 30ml water. Heat and then acidify with sulphuric acid. Cool, collect crystals of sulphinic acid, m.p 83-84°C.
(3) To 15ml conc H_2SO_4 containing 40gm ice add 7gm of compound. Add rapidly 15gm of zinc dust and stir for 1hr. Boil the mixture

under reflux until a clear solution is obtained (4-6hrs). Distill and extract the distillate with ether. Distill off the ether and collect thiophenol at b.p 167-169°C.

(4) Benzenesulphonamide 150°C.

m.p 55°C **p-Chlorobenzenesulphonyl chloride**

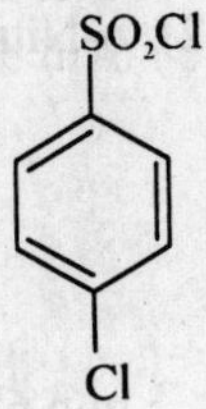

(1) Soluble in water, alcohol and ether.
(2) p-Chlorobenzenesulphonamide m.p 143°C.

m.p 63°C **Benzene 1, 3-disulphonyl chloride**

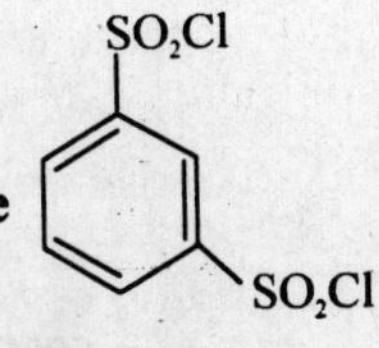

(1) Crystalline solid.
(2) m-Benzenedisulphonamide m.p 229°C.

m.p 63°C **m-Nitrobenzenesulphonyl chloride**

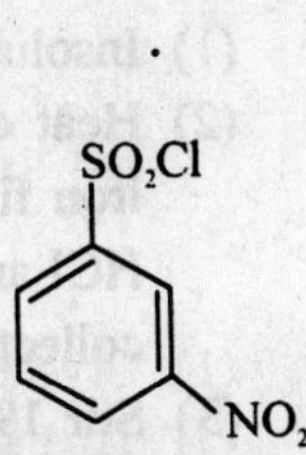

(1) Crystalline compound, insoluble in water, soluble in alcohol.
(2) To 6.66gm of compound add dropwise with stirring 21ml of 55% hydroiodic acid. Reflux on steam bath for 3hrs. Cool, add sodium bisulphite. Filter and collect m-nitrophenyl disulphide, recrystallise from acetone, m.p 82-83°C.
(3) m-Nitrobenzenesulphonamide, m.p 161°C.

m.p 67°C **o-Nitrobenzenesulphonyl chloride**

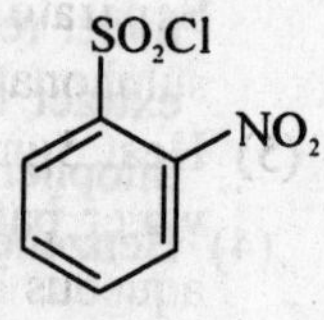

(1) Colourless crystalline solid, soluble in ether and hot alcohol.
(2) o-Nitrobenzenesulphonamide m.p 186°C

m.p 68°C **Naphthalene 1-sulphonyl chloride**

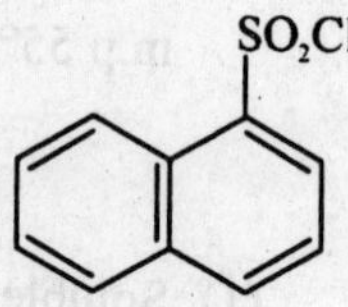

(1) Crystalline compound, soluble in alcohol, ether and benzene.
(2) Naphthalene 1-sulphonamide m.p 150°C

m.p 69°C **p-Toluenesulphonyl chloride**

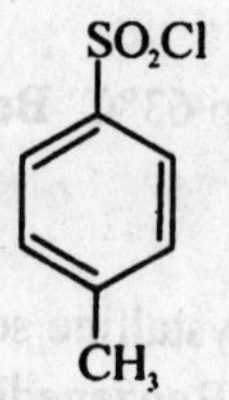

(1) Insoluble in water, soluble in alcohol and ether.
(2) Heat on water bath a mixture of 10gm of compound, 37gm of iron filings and 12ml acetic acid. Add gradually 100ml of 15% HCl and continue heating for 5hrs. Distill, cool the distillate and collect thiocresol, m.p 43°C by filtration.
(3) Stir 19gm of compound, 15ml alcohol and 17ml NaOH at 15°C for 6hrs. Cool and dilute the mixture with 15ml ice cold water. Filter the precipitated ethyl p-toluenesulphonate, m.p 33°C.
(4) To a cooled mixture of 11gm of compound and 9ml butanol add dropwise with stirring 17ml of 20% NaOH solution (temp below 15°C). Add another portion of 11gm of p-toluene sulphonyl chloride and 17ml NaOH solution. Continue stirring for 4hrs.

Separate the oily layer, distill and collect butyl toluene-p-sulphonate at b.p 132-133°C/3mmHg.

(5) Heat 5gm of compound with 10gm of ammonium carbonate on water bath for 1-2hrs. Cool, extract with water and discard the aqueous layer. Recrystallise the residue from hot water and collect crystals of toluene-p-sulphonate, m.p 138°C.

m.p 69°C **2-Chloronaphthalene –5-sulphonyl chloride**

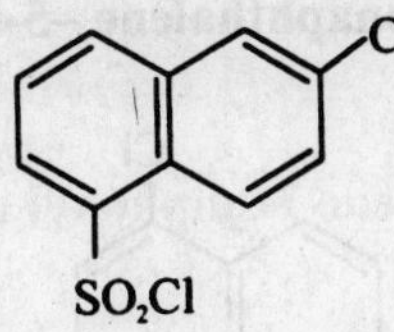

(1) Crystalline compound, soluble in chloroform and benzene.
(2) 2-Chloronaphthalene –5-sulphonamide m.p 214°C.

m.p 75°C **p-Bromobenzenesulphonyl chloride**

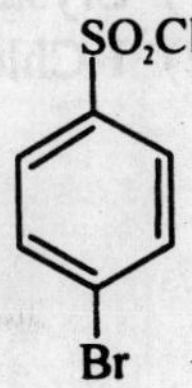

(1) Crystalline solid, insoluble in water, soluble in alcohol and ether.
(2) p-Bromobenzenesulphonamide m.p 160°C.

m.p 76°C **Naphthalene –2-sulphonyl chloride**

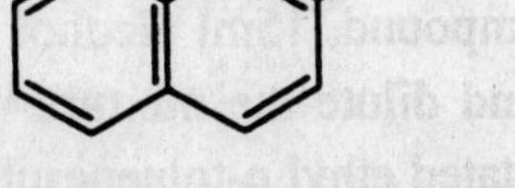

(1) Crystalline compound, soluble in alcohol, ether and benzene (comparitlvely less soluble than 1-isomer)
(2) Naphthalene –2-sulphonamide m.p 217°C.

m.p 80°C **1-Chloronaphthalene –2-sulphonyl chloride**

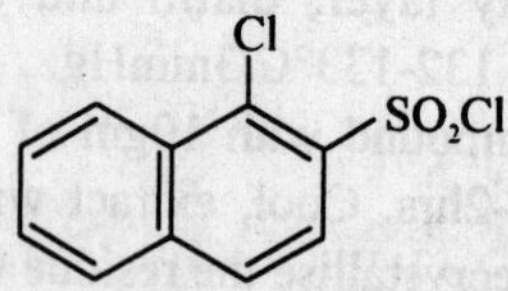

(1) Crystalline compound, soluble in ether and benzene.
(2) 1-Chloronaphthalene –2-sulphonamide m.p >250°C.

m.p 95°C **1-Chloronaphthalene –5-sulphonyl chloride**

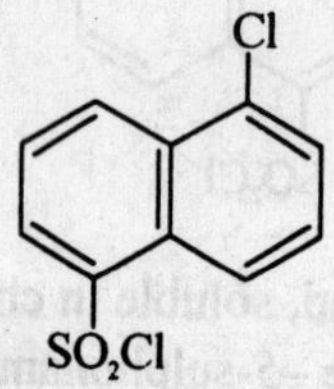

(1) Crystalline compound.
(2) 1-Chloronaphthalene –5-sulphonamide m.p 226°C.

SECTION II

SECTION II

1

Esterification

Esters are one of the important industrial and biological compounds that exist throughout nature. Waxes which are esters of higher fatty acids are used to coat the stems and leaves of the plants where they act as a protective layer thereby preventing dehydration. Waxes also find use in the production of cosmetics and medical supplies. A number of flowers, fruits and essential oils owe their fragrance to the presence of these esters : for eg isoamyl acetate has the odour of banana, amyl butyrate that of apricot and octyl acetate that of oranges. More recently a number of esters have been synthesized which find use as local anaesthetics eg ethyl p-amino benzoate and orthocaine.

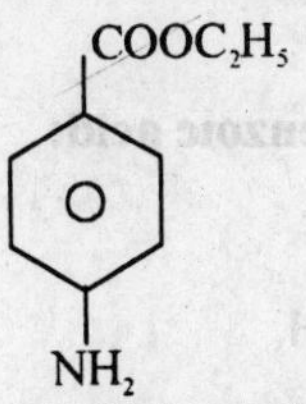

ethyl-p-aminobenzoate

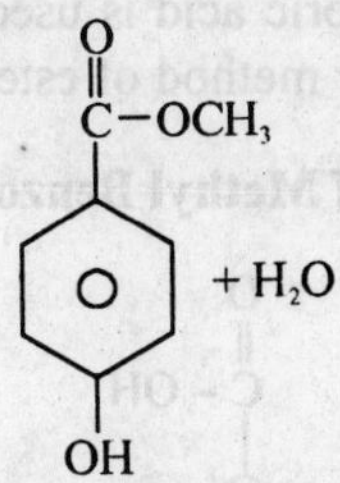

Orthocaine

Esters, which are derivatives of acids are usually prepared by the reaction of an alcohol and a carboxylic acid in the presence of an acid catalyst usually sulphuric acid. This reaction called esterification is reversible and involves the removal of a molecule of water. In these reactions the bonds are usually broken in two ways usually referred as acyl oxygen scission and alkyl oxygen scission.

$RCO - OH + H - OR' \rightleftharpoons RCOOR' + H_20$ acyl oxygen scission

$RCOO - H + HO - R' \rightleftharpoons RCOOR' + H_20$ alkyl oxygen scission

But it has been observed that the usual esterification of primary alcohols take place by acyl oxygen scission, secondary by either acyl or alkyl oxygen scission whereas tertiary by alkyl oxygen scission.

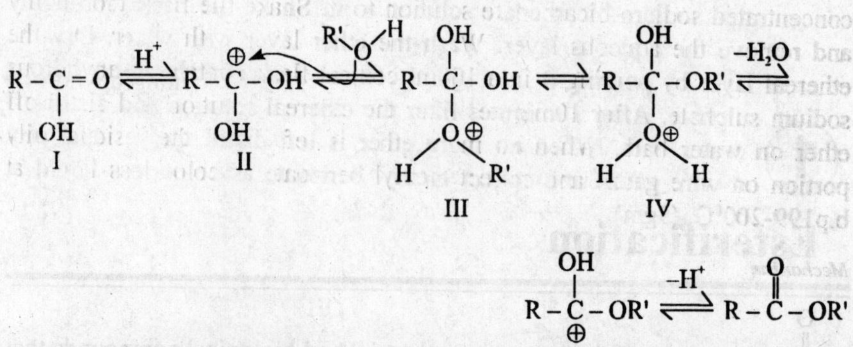

The carbonyl oxygen is first protonated which increases the electrophilicity of the carbonyl carbon because of the positive charge on oxygen. The weakly nucleophilic alcohol then attacks the electrophilic carbon leading to the generation of a tetrahedral intermediate *III* followed by proton transfer to give intermediate IV. Loss of a molecule of water and proton from the intermediate *IV* forms the ester.

If the carboxylic acid or alcohol contains an unsaturated group or either of them are aromatic then the sulphuric acid cannot be used as a catalyst because in the first case it will add on to the unsaturated group and in the second case sulphonation of the aromatic ring will result. Hence in such cases hydrochloric acid is used as an acid catalyst and the process is called Fischer –Speier method of esterification.

(i) Synthesis of Methyl Benzoate (Oil of Niobe) from benzoic acid:

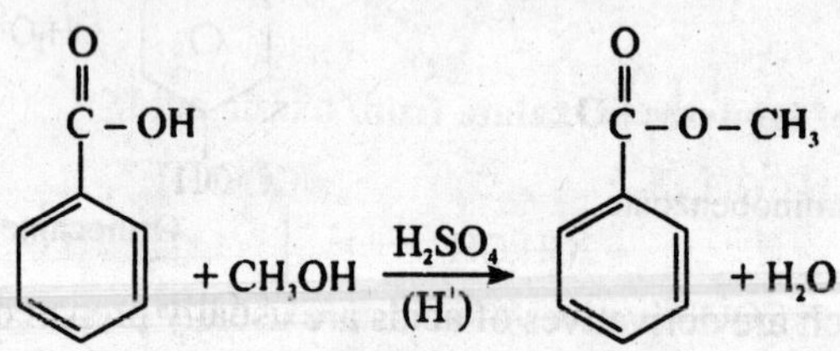

Place 5gm of benzoic acid and 15ml absolute methanol in a 100 ml round bottom flask. Cool the flask in ice and add slowly .5ml of concentrated sulphuric acid to it. Shake the flask so as to mix the components. Add a small piece of porcelain to the flask, attach a reflux condensor and reflux the mixture gently for 4-5hours. Set the apparatus for downward distillation, distill off the excess methanol on water bath and then cool the flask. Pour the concentrated oily layer in a 250 ml separating funnel and add 25 ml ether to it. Stopper the funnel and shake it thoroughly. Remove the upper ethereal layer. Repeat the process twice, each time using 25ml of ether. Combine the ethereal layers, transfer it to the separating funnel and add concentrated sodium bicarbonate solution to it. Shake the flask thoroughly and remove the aqueous layer. Wash the ether layer with water. Dry the ethereal layer by pouring it in a 100ml conical flask containing anyhdrous sodium sulphate. After 10minutes filter the ethereal solution and distill off ether on water bath. When no more ether is left distill the residual oily portion on wire gauze and collect methyl benzoate as colourless liquid at b.p199-200^{0}C. (5gm)

Mechanism

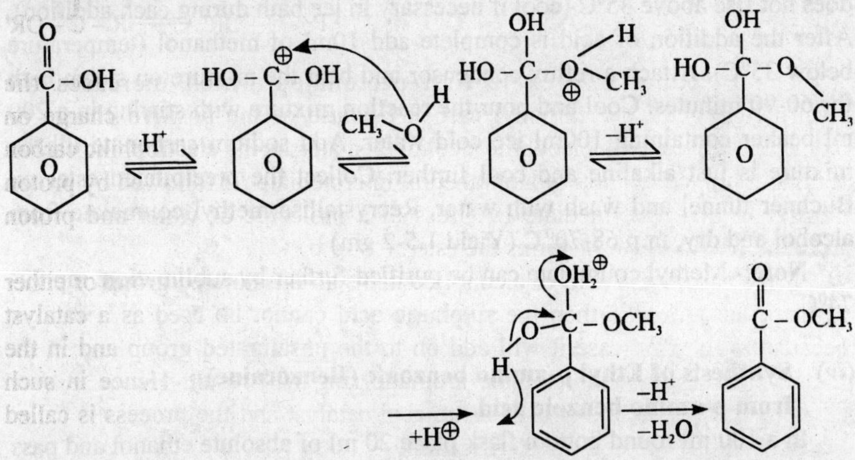

Note :-It is an important ester which is used in the formulation of many perfumes.

(*ii*) Synthesis of Dimethyl Oxalate from oxalic acid:-

$$\begin{array}{l} COOH \\ | \\ COOH \end{array} + 2CH_3OH \longrightarrow \begin{array}{l} COOCH_3 \\ | \\ COOCH_3 \end{array} + 2H_2O$$

Place 10 gm of powdered oxalic acid in a porcelain dish and heat on a wire gauze till no more water is given off. During the heating process the acid should be stirred and powdered. Transfer the anhydrous acid in a 100ml round bottom flask fitted with a reflux condensor. Add 25ml anhydrous methyl alcohol to the flask and reflux the resultant solution on water bath for 3hours. Set the apparatus for downward distillation and distill off the excess methanol on water bath. Distill the residual concentrate on wire gauze and collect the fraction boiling at 160-165^{0}C. Cool this distillate which soon solidifies and then filter it on a Buchner funnel. Recrystallise dimethyl oxalate from ethyl alcohol and dry, m.p 54^{0}C. (Yield 6-7 gm).

(*iii*) Synthesis of Methyl Coumalate from coumalic acid

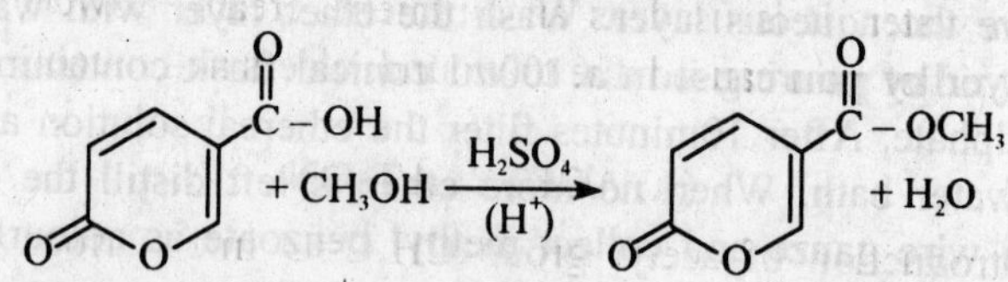

To 14ml of concentrated sulphuric acid placed in a 100 ml round bottom flask add in portions 5 gm of coumalic acid taking care that the temperature does not rise above 35^{0}C (cool if necessary in ice bath during each addition). After the addition of acid is complete add 10ml of methanol (temperature below 35^{0}C). Attach a reflux condensor and heat the mixture on steam bath for 60-90 minutes. Cool and pour the reaction mixture with stirring in a 200 ml beaker containing 100ml ice cold water. Add sodium carbonate till the mixture is just alkaline and cool further. Collect the precipitated ester on Buchner funnel and wash with water. Recrystallise methyl coumalate from alcohol and dry, m.p 68-70^0 C (Yield 1.5-2 gm)

Note:- Methyl coumalate can be purified further by sublimation. mp 72-74^{0}C

(*iv*) Synthesis of Ethyl p-amino benzoate (Benzocaine) from p-amino benzoic acid

In a 100 ml round bottom flask place 20 ml of absolute ethanol and pass dry HCl gas (prepared by the action of concentrated sulphuric acid on sodium chloride) until the alcohol is saturated and there is an increase in its weight by approximately 8-10gm. Now add 3 gm of p-amino benzoic acid to HCl saturated alcohol. Attach a reflux condensor and reflux the mixture for 90-120 minutes. Cool and pour the reaction mixture into a 200 ml beaker containing 50ml ice cold water. Add solid sodium carbonate with stirring until the solution is neutral and cool further. Collect the precipitated Ethyl p-amino benzoate on buchner funnel and recrystallise from alcohol, m.p 91°C (2gm).

$$NaCl + H_2SO_4 \rightarrow Na_2SO_4 + HCl$$

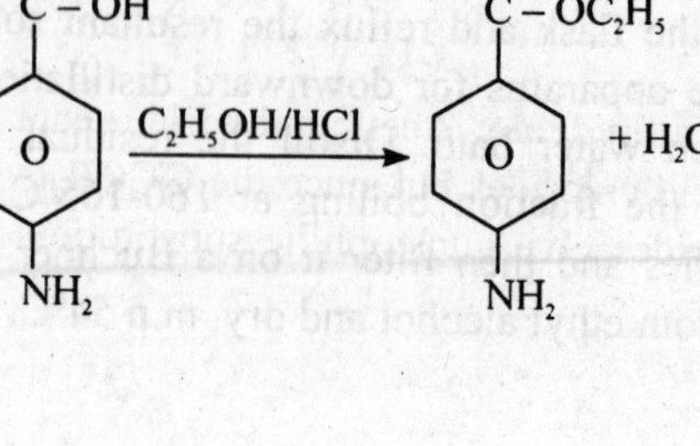

Note:- Ethyl p-amino benzoate is an important local anaesthetic. Being insoluble in water, it is used as an ointment, cream, powder and aerosol spray to get rid of pain caused due to wound, ulcers and mucous surfaces.

(A) ACETYLATION

The introduction of acetyl group $CH_3C{=}O$ into various nucleophilic agents such as alcohols, phenols and amines where the H atom of the functional group OH and NH_2 is replaced by $CH_3C{=}O$ group is known as acetylation. This reaction can very easily be carried out with acetyl chloride, acetic anhydride or glacial acetic acid with or without the aid of an inert solvent such as benzene or toluene.

$$CH_3COCl + HOR \rightarrow CH_3COOR + HCl$$
$$(CH_3CO)_2O + HOR \rightarrow CH_3COOR + CH_3COOH$$
$$(CH_3CO)_2O + H_2NR \rightarrow CH_3CONHR + CH_3COOH$$

Acetylation of alcohols, phenols or amines is in fact a better and faster method for the synthesis of esters than the direct esterification of the carboxylic acids with alcohol/ phenol. This method is often used for the quantitative determination of the hydroxyl or amino group in an organic compound and also for the qualitative analysis to determine the class of amine.

Mechanism

(i) By Acetyl Chloride: In acetyl chloride a highly electronegative chlorine atom is attached to the carbonyl group which decreases the electron density on the carbonyl carbon. As a result the carbonyl carbon is readily attacked by the nucleophile, leading to the generation of a tetrahedral intermediate (I). Cl⁻ being a weaker base is a good leaving group, departs readily leading to the generation of an ester.

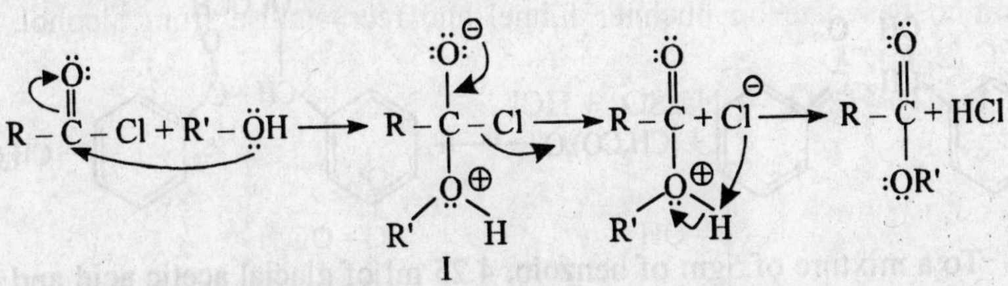

(*ii*) By Acetic anhydride: Acetic anhydride react with nucleophiles in the same way as acid chlorides but since the OCOR is not as elctronegative as Cl atom, anhydrides give nucleophilic substitution reaction product less vigorously.

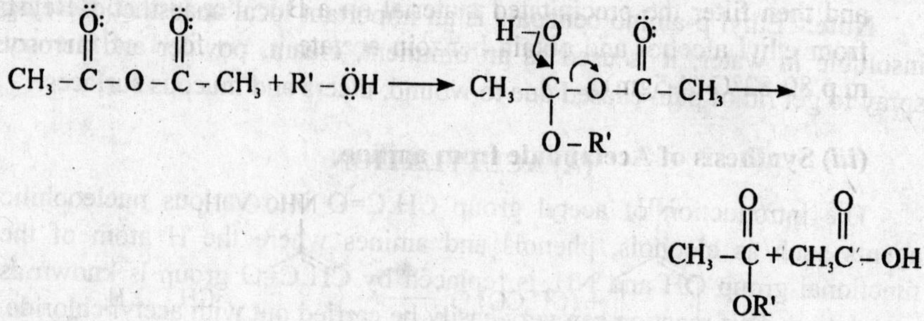

(i) Synthesis of p-acetanisidide from p-anisidine

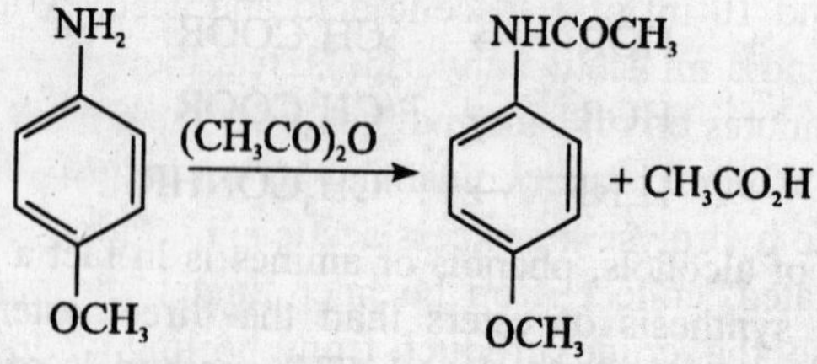

In a 100ml conical flask dissolve with stirring 6 gm of p-anisidine in 15ml of glacial acetic acid and 10 ml water. When p-anisidine has completely dissolved add 25gm of crushed ice to it. Add in one portion with stirring 6ml of acetic anhydride to the cold solution. Soon a crystalline mass will separate out. Heat the flask on water bath until the precipitated material completely dissolves and then cool the flask in ice bath. Collect the precipitated solid on Buchner funnel and wash with cold water. Recrystallise p-acetanisidide from hot water and dry, m.p 125-126°C. (Yield 6-7gm)

(*ii*) Synthesis of Benzoin Acetate from benzoin

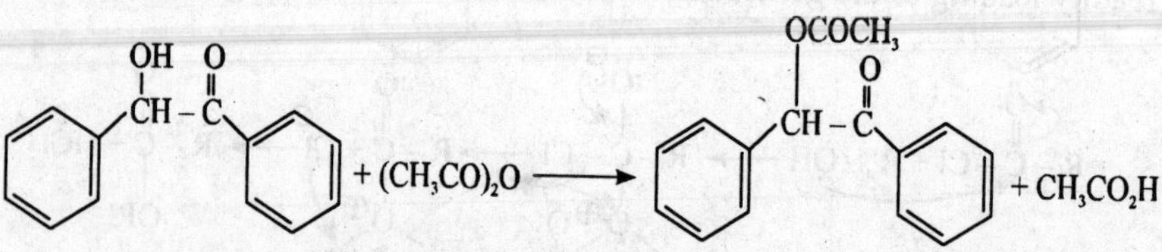

To a mixture of 5gm of benzoin, 4.25 ml of glacial acetic acid and 4.25 ml of acetic anhydride in a 100ml conical flask add with stirring .5ml of concentrated H_2SO_4. Heat the mixture on water bath for twenty minutes. Cool the hot mixture and then pour slowly with stirring in a 200ml beaker containing 100ml of ice water. Continue stirring for another 20-30minutes and then filter the precipitated material on a Buchner funnel. Recrystallise from ethyl alcohol and obtain benzoin acetate in the form of fine crystals, m.p 80-82°C (4-5gm)

(*iii*) Synthesis of Acetanilide from aniline

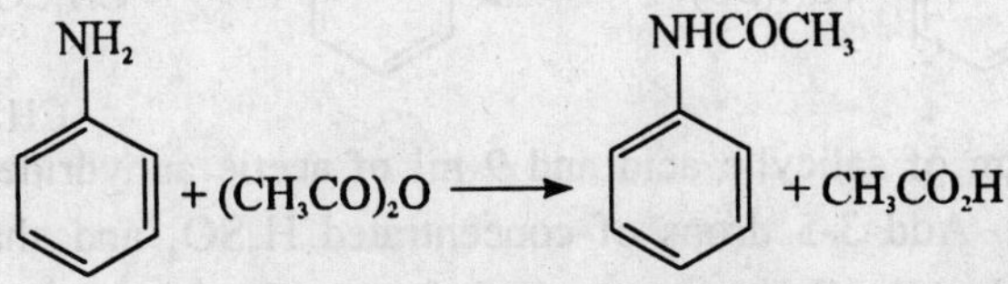

In a 100 ml round bottom flask fitted with a reflux condensor place 5ml of aniline and 10 ml of 1:1 acetic acid and acetic anhydride mixture (5ml acetic acid and 5 ml acetic anhydride). Heat the mixture gently under reflux for 15-20 minutes on oil bath and then pour the contents while still hot with stirring into a 200ml beaker containing 100ml ice cold water. Stir the mixture vigorously to hydrolyse the excess acetic anhydride. After all the acetanilide has precipitated, collect it on buchner funnel and wash with cold water. Recrystallise the crude product from boiling water. If the product is excessively coloured add a pinch of animal charcoal to hot water and filter hot through glass wool/ cotton plug. Pure colourless crystals of acetanilide melts at 114°C (5-5.5g)

Note : Acetanilide is medicinally important as it is used as febrifuge (antifibrin)

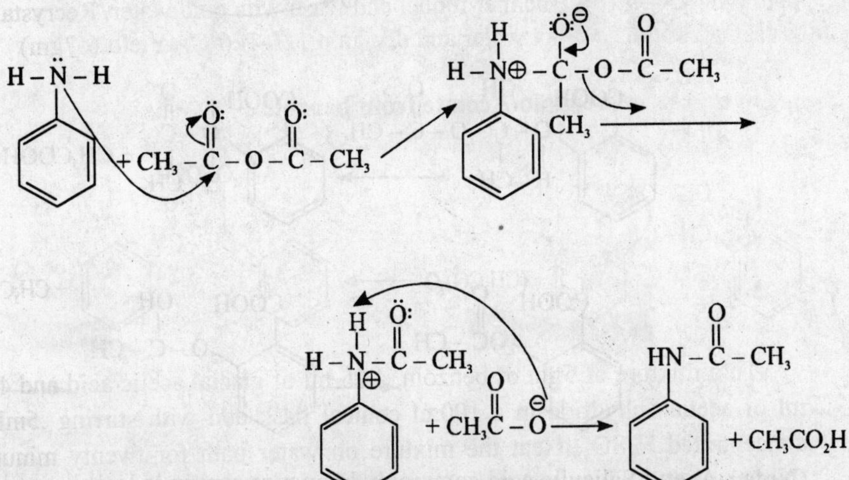

(*iv*) Synthesis of Acetyl Salicylic acid (Aspirin)from Salicylic acid

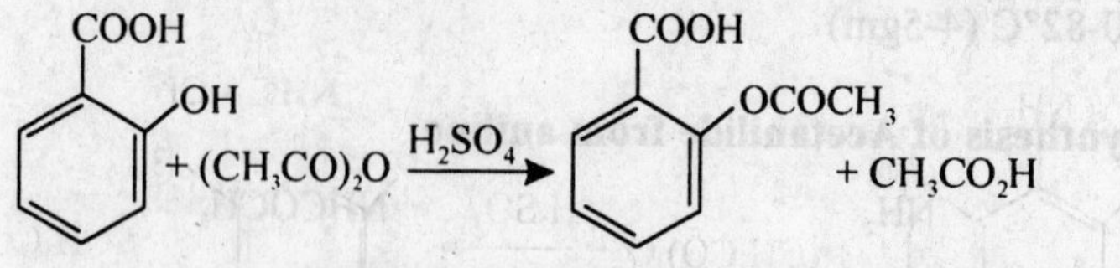

Place 5gm of salicylic acid and 9 ml of acetic anhydride in a 100 ml conical flask. Add 3-5 drops of concentrated H_2SO_4 and shake the flask thoroughly. Heat the flask on water bath for 10-15 minutes. Cool the flask at room temperature and then pour the reaction mixture with stirring into a 200ml beaker containing 50ml ice water. The crude acetyl salicylic acid either solidifies at once or separates out as oil and then rapidly crystallises as the stirring continues. Collect the crude product on buchner funnel by suction filtration and wash thoroughly with cold water. Recrystallise by dissolving the product in minimum amount of hot ethanol and then add equal amount of hot water and cool the clear solution. Acetyl salicylic acid is obtained as colourless needles, m.p 132-137°C d (5gm)

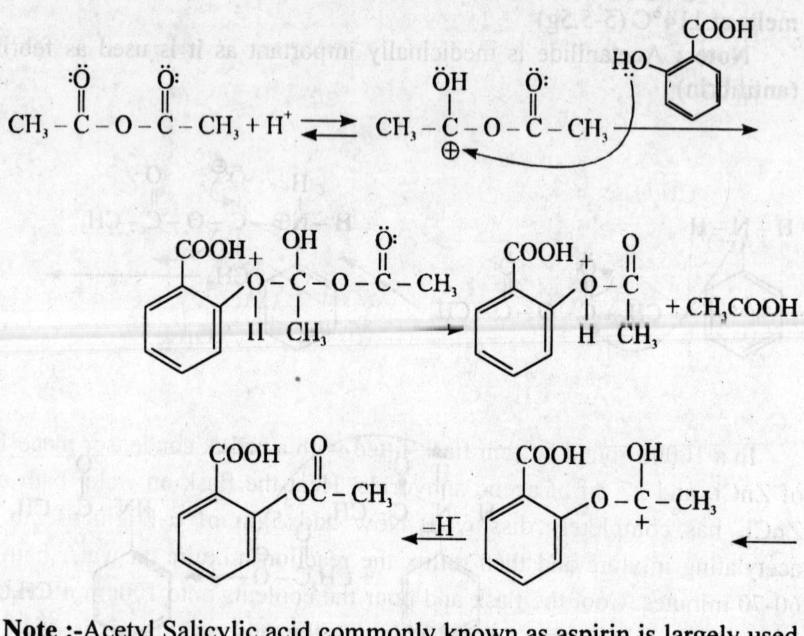

Note :-Acetyl Salicylic acid commonly known as aspirin is largely used in medicine as an analgesic and anti pyretic agent.

(*v*) Synthesis of 2, 4, 6 tribromoacetanilide from tribromo aniline

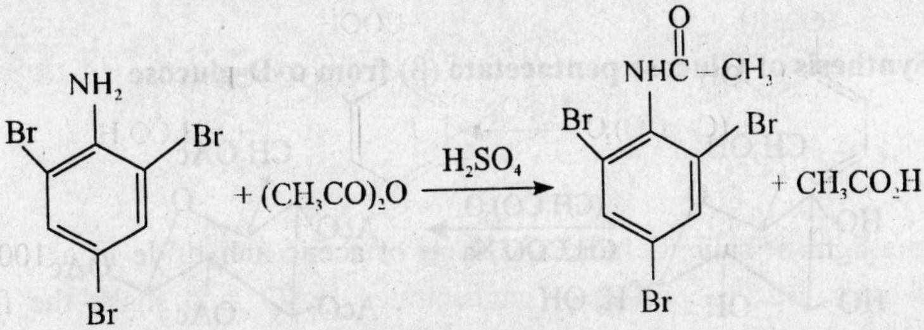

Heat a mixture of 2gm of tribromoaniline, 30ml of acetic anhydride and .5 ml of concentrated H_2SO_4 placed in a 100 ml conical flask for 10-15 minutes. Pour the reaction mixture into 100ml cold water. Filter the crude product on buchner funnel and wash with cold water. Recrystallise tribromoacetanilide from ethyl alcohol and dry, m.p 232°C (Yield 2.5gm)

(*vi*) Synthesis of Glucose pentacetate (α) from α-D-glucose

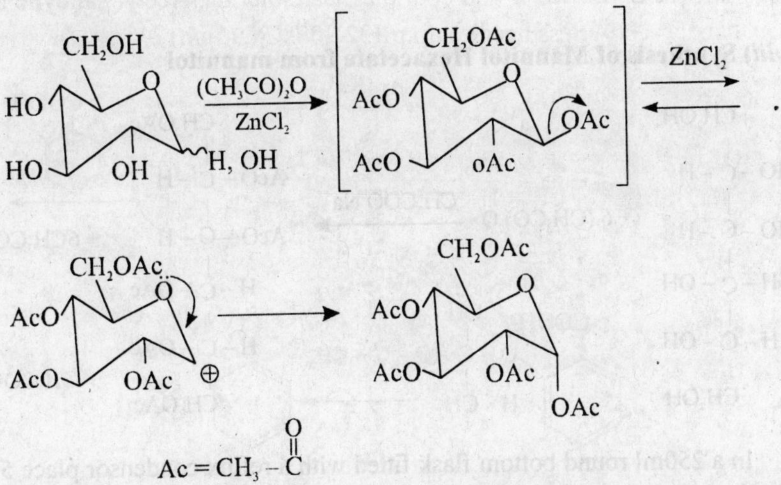

In a 100ml round bottom flask fitted with a reflux condensor place 1gm of $ZnCl_2$ and 27 ml of acetic anhydride. Heat the flask on water bath until $ZnCl_2$ has completely dissloved. Now add 5gm of α-D-glucose to the acetylating mixture and then reflux the reaction mixture on water bath for 60-70 minutes. Cool the flask and pour the contents onto 100gm of crushed ice and stir vigorously. After 10minutes the oily acetylated product will gradually solidify. Filter the pentacetate on buchner funnel and wash with cold water. Recrystallise the crude product from alcohol, m.p 109-111°C (5.5gm).

(*vii*) Synthesis of Glucose pentacetate (β) from α-D-glucose

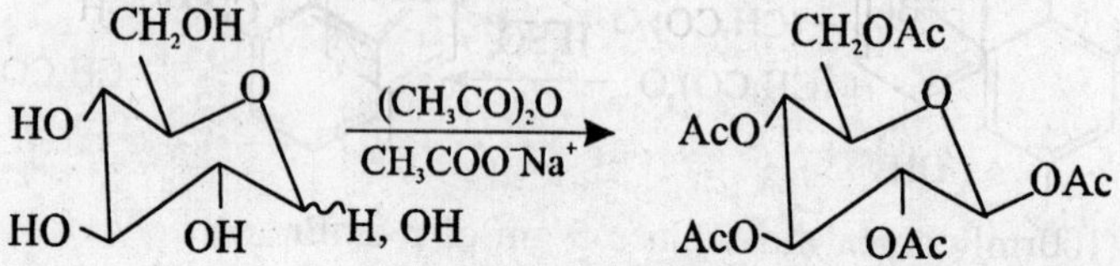

Add 6 gm of fused sodium acetate and 35ml of acetic anhydride to 7.5 gm of α-D-glucose placed in a 100 ml round bottom flask fitted with a reflux condensor. Heat the flask on water bath for 2- 2 ½ hour. Cool the flask at room temperature and then pour the contents onto 300 gm of crushed ice and stir vigorously. The oily layer will gradually solidfy. Filter the solid, wash with water and recrystallise from ethanol. Pure β pentacetate melts at 130-132°C (8gm).

(*viii*) Synthesis of Mannitol Hexacetate from mannitol

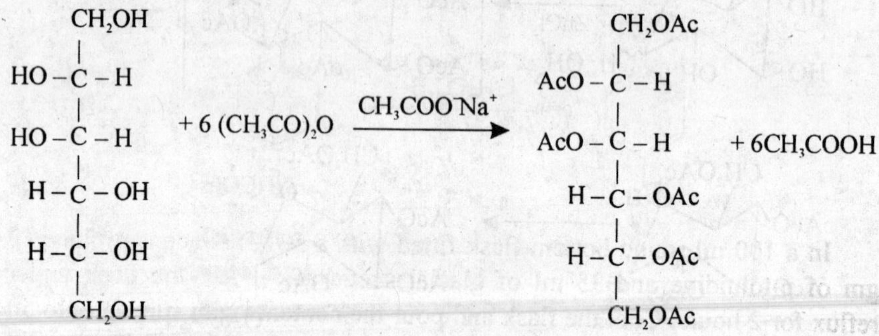

In a 250ml round bottom flask fitted with a reflux condensor place 5 gm of powdered mannitol, 30ml of acetic anhydride and 4gm of fused sodium acetate. Reflux the reaction mixture on wire gauze, first slowly over a low flame so that the sodium acetate and mannitol dissolves completely (If the reaction tends to become too vigorous, cool the flask in ice). Continue the reflux for further 20 minutes. Cool the flask and pour the reaction mixture with stirring onto 100 gm of crushed ice. Filter the precipitated product on buchner funnel and wash with water. Recrystallise mannitol hexacetate from ethanol and dry, m.p 120-122° C (7-8gm).

(*ix*) Synthesis of p-hydroxy acetanilide (Paracetamol) from p-amino phenol:

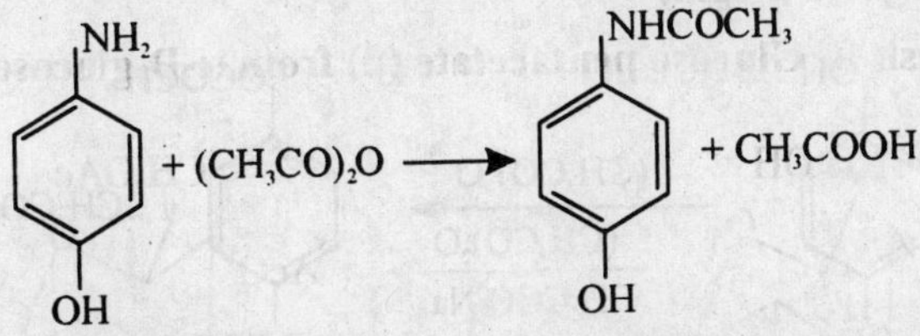

In a 100 ml conical flask place 5 gm of p-amino phenol, add 15 ml of water and 6 ml of acetic anhydride to it. Stir the mixture vigorously and heat the flask on water bath so that the solid dissolves completely. Cool the flask in ice bath and scratch the sides of flask to induce crystallization. Filter the solid acetyl derivative on buchner funnel and wash with cold water. Recrystallise the crude product from hot water and collect the crystalline p-hydroxy acetanilide, m.p 168-169°C (5.5gm)

Note : Paracetamol is an antipyretic and analgesic drug. It finds use mainly in arthritic and rheumatic conditions linked with muscular skeletal pain and is particularly useful for aspirin sensitive patients.

(*x*) Synthesis of p-acetotoluidide from p-toluidine

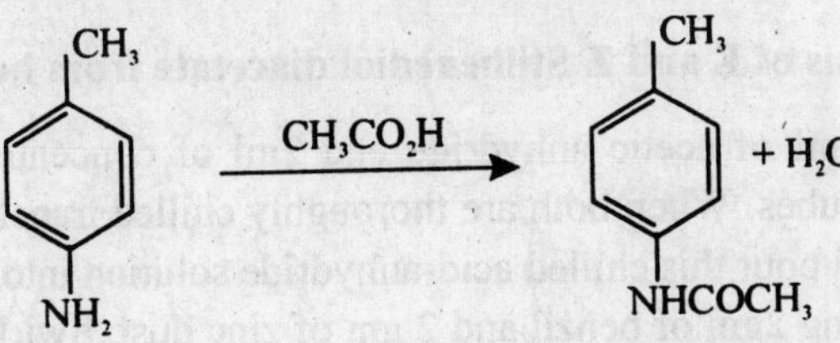

In a 100 ml round bottom flask fitted with a reflux condensor place 10 gm of p-toluidine and 35 ml of glacial acetic acid. Heat the flask under reflux for 2 hours. Cool the flask and pour the contents with stirring onto 50 gm of crushed ice contained in a 200 ml beaker. Filter the precipitated solid on buchner funnel, wash with cold water and recrystallise the crude acetylated product from ethanol. Pure crystallised p-acetotoluidide melts at 146°C (10-11 gm).

(*xi*) Synthesis of Hydroquinone diacetate from hydroquinone

In a 100ml conical flask dissolve with stirring 5gm of hydoquinone in 9ml of acetic anhydride containing a drop of concentrated sulphuric acid. After all the hydroquinone has dissolved and a clear solution is obtained, pour it in a 200ml beaker containing 50 gm of crushed ice. Filter the precipitated solid on buchner funnel and wash with cold water. Recrystallise from 50% aqueous alcohol and collect pure hydroquinone diacetate, m.p 121-122°C. (yield 5-6gm)

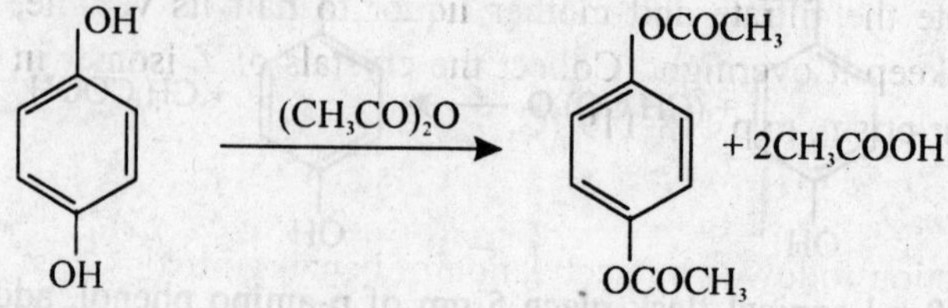

(*xii*) Synthesis of acetyl glycine from glycine

$$COOH - CH_2 - NH_2 + (CH_3CO_2)O \rightarrow$$
$$CO_2H - CH_2 - NHCOCH_3 + CH_3CO_2H$$

In a 100ml conical flask dissolve 10gm of glycine in 40ml water. Add 25ml of acetic anhydride in one portion to the glycine solution and stir vigorously for 20-30 minutes. Cool the conical flask in ice bath and leave overnight. Filter the precipitated solid on buchner funnel and wash with cold water. Recrystallise acetyl glycine from boiling water, m.p 207-208°C.

(*xiii*) Synthesis of E and Z Stilbenediol diacetate from benzil

Cool 14 ml of acetic anhydride and 2ml of concentrated HCl in two separate test tubes. When both are thoroughly chilled transfer the acid to the anhydride and pour this chilled acid-anhydride solution into a 100 m´ conical flask containing 2gm of benzil and 2 gm of zinc dust. Swirl the flask for 2-3 minutes in an ice bath. Remove the flask from ice bath and keep it at room temperature (if the reaction tends to be violent cool it again in ice bath). When the exothermic reaction has completely subsided keep it at room temperature for another 5 minutes and then add 50ml of ice water to it and stir vigorously. Collect the precipitated products on buchner funnel and wash with water. Dissolve the compounds in 50ml ether and dry over anhydrous sodium sulphate. Decant the ether and concentrate it to 10ml in 100ml conical flask. Cork the flask and leave it undisturbed for 5 minutes. The E isomer of stilbenediol diacetate seperates out as prismatic needles. Collect the crystals on buchner funnel, wash with ether and dry, m.p 153-156 °C.

Evaporate off the ether and dissolve the residue in 10ml methanol (heat if necessary). Keep the solution at room temperature for 10 minutes and then add one crystal of E isomer which helps in precipitation of second crop of E diacetate. Collect the crystals of E isomer and wash with methanol. Concentrate the filtrate and mother liquor to half its volume, stopper the flask and keep it overnight. Collect the crystals of Z isomer in the form of rectangular prism, m.p 118-119 °C.

Mechanism

Reduction followed by acetylation of benzil with zinc dust in a mixture of acetic anhydride –sulphuric acid involves1, 4 addition of hydrogen to the a diketone grouping and acetylation of the resulting enediol results in a mixture of E and Z isomers.

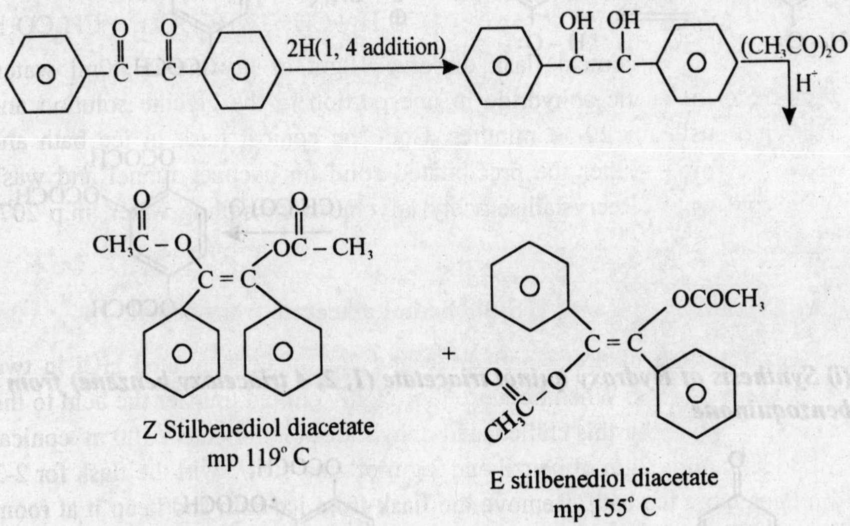

Z Stilbenediol diacetate
mp 119° C

E stilbenediol diacetate
mp 155° C

(b) Thiele Acetylation

The acid catalysed reaction of quinones with acetic anhydride gives triacetoxy derivatives, in which two acetoxy group are derived from two oxygen of the quinoid ring while the third is introduced during the reaction. This introduction of oxygen atom into aromatic nucleus is known as Thiele Winter acetoxylation of quinones or Thiele acetylation

The reaction is initiated by the 1, 4 addition of acetic anhydride across an αβ unsaturated carbonyl system.

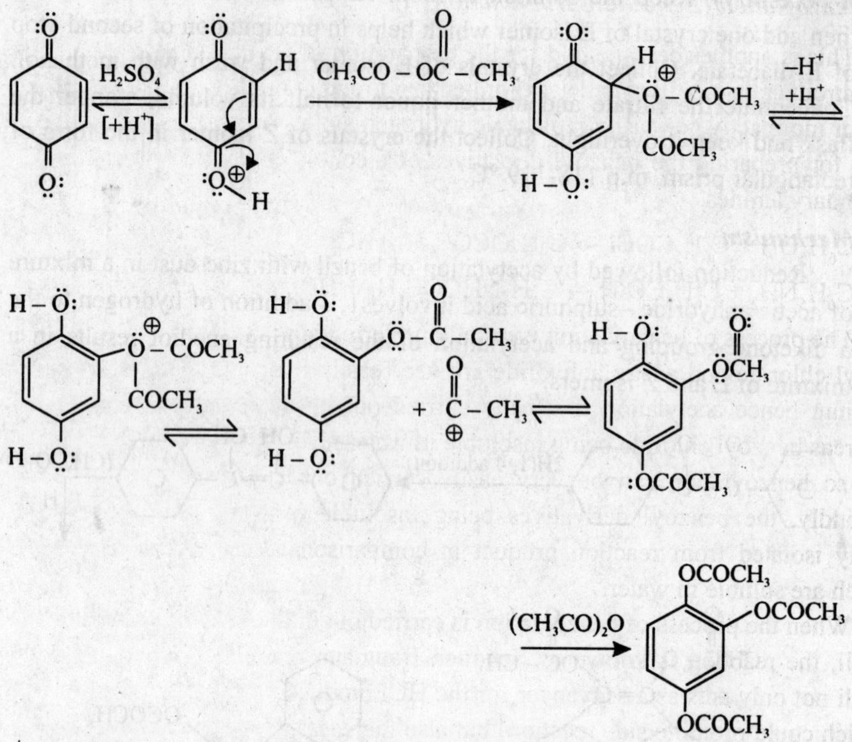

(i) Synthesis of Hydroxy quinol triacetate (1, 2, 4 triacetoxy benzene) from benzoquinone

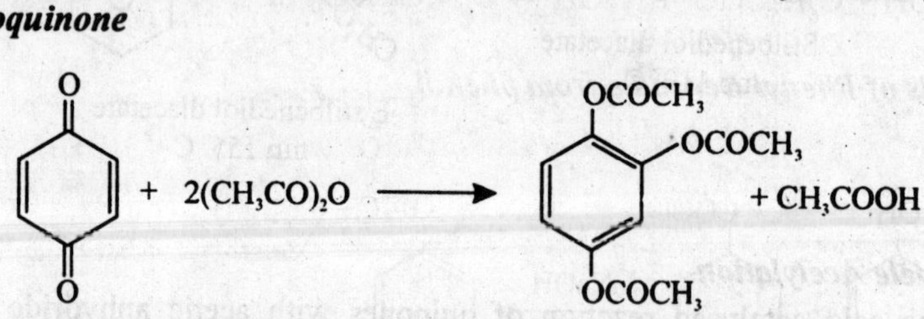

In a 250 ml beaker place 1.75 ml of concentrated sulphuric acid and 40 ml of acetic anhydride. Stir the mixture, add with stirring and in portion 15 gm of p-benzoquinone taking care that the temperature does not rise above 50°C. After the addition of quinone is complete continue stirring and maintain the temperature of reaction mixture between 40-50°C for 5-10 minutes. Allow the solution to cool at room temperature and then pour into 100ml ice cold water. Collect the precipitated triacetate on buchner funnel, wash with water and recrystallise from ethanol. The product practically white and rather granular melts at 97°C (25gm).

(c) Benzoylation (Benzene Carbonylation)

This is another best method for the preparation of an ester and involves the introduction of an aromatic acyl or benzoyl group in molecules containing one or more active or easily replacable hydrogen atom. This method is often used for preparing the benzoyl dervatives of alcohols, phenols, primary or secondary amines

$$C_6H_5OH + C_6H_5COCl \rightarrow C_6H_5OCOC_6H_5 + HCl$$

$$C_6H_5NH_2 + C_6H_5COCl \rightarrow C_6H_5NHCOC_6H_5 + HCl$$

The process of benzoylation is usually preferred over acetylation. Firstly, acetyl chloride and acetic anhydride are very easily hydrolysed in aqueous medium hence acetylation is usually carried out in non aqueous medium whereas benzoyl chloride being insoluble in water reacts with it very slowly, and so benzoylation can be very easily carried out in aqueous medium. Secondly, the benzoyl derivatives being insoluble in water can be very easily isolated from reaction product in comparison to acetyl derivatives which are soluble in water.

When the process of benzoylation is carried out in the presence of dilute alkali, the reaction is known as Schotten Baumann reaction. The aqueous alkali not only acts as a scavenger for the HCl produced during the reaction (which could promote side reaction) but also increases the nucleophilicity of the phenols and alcohols.

$$C_6H_5OH + C_6H_5COCl + NaOH \rightarrow C_6H_5OCOC_6H_5 + NaCl + H_2O$$

(i) Synthesis of Phenyl Benzoate from phenol

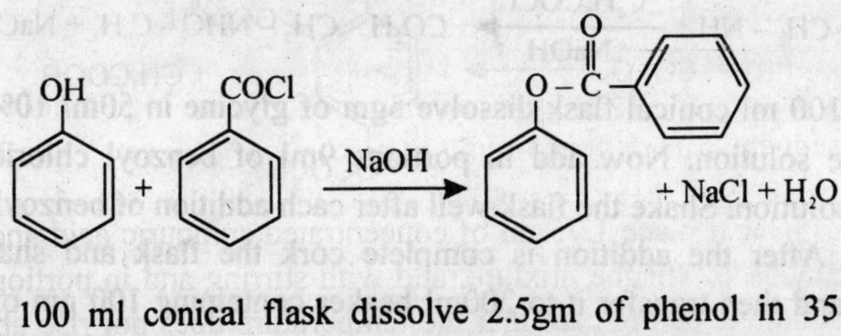

In a 100 ml conical flask dissolve 2.5gm of phenol in 35 ml 10% sodium hydroxide solution and then add 5ml benzoyl chloride to it. Cork the flask properly and shake the mixture vigorously for 20 minutes by which time the pungent smell of benzoyl chloride gets abated and a solid product is obtained. Filter off the benzoate at pump, wash with water and then recrystallise from rectified spirit. Phenyl benzoate is obtained as colourless needles, m.p 68-69°C (3-3.5 gm).

(ii) Synthesis of α-D-glucose pentabenzoate from α-D-glucose

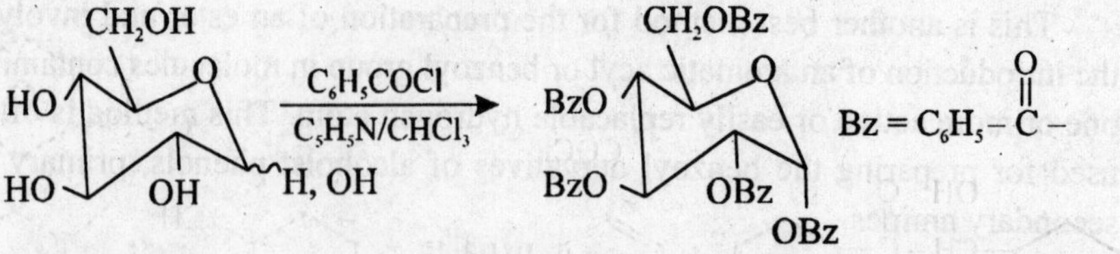

Cool a mixture of 13ml pyridine and 11 ml chloroform placed in a 100ml conical flask by keeping the flask in an ice bath. Add with stirring and in portions to the pyridine solution a well cooled solution of 11ml of benzoyl chloride in 11ml chloroform. Now add in portions with stirring 5 gm of α-D-glucose to the well cooled benzoylating reagent at such a rate that the temperature does not rise above 10°C during the addition process. After the addition is complete keep the resulting solution at 0°C for 24-30 hours. Transfer the solution to 250ml separating funnel, add 20ml chloroform to it and shake thoroughly. Remove the lower chloroform layer and repeat the process twice, each time using 20ml chloroform. Combine the chloroform layers and wash successively with dilute sulphuric acid, sodium hydrogen carbonate and water. Dry the organic layer over anhydrous sodium sulphate and decant. Distill off the chloroform and recover the crude product. Recrystallise from acetone-water and collect crystalline pentabenzoate, m.p 185-186°C (12-13gm).

(iii) Synthesis of Benzoyl glycine(Hippuric acid) from glycine

$$CO_2H - CH_2 - NH_2 \xrightarrow[NaOH]{C_6H_5COCl} CO_2H - CH_2 - NH\overset{O}{\overset{\|}{C}} - C_6H_5 + NaCl + H_2O$$

In a 100 ml conical flask dissolve 5gm of glycine in 50ml 10% sodium hydroxide solution. Now add in portions 9ml of benzoyl chloride to the alkaline solution. Shake the flask well after each addition of benzoyl chloride is made. After the addition is complete cork the flask and shake it for 1-2 min and then transfer it to 200ml beaker containing 100 gm of crushed ice. Add concentrated hydrochloric acid to the ice cooled solution until it is just acidic. Collect the precipitated benzoyl glycine on buchner funnel and wash well with cold water. Recrystallise from hot water, dry and collect crystals of Hippuric acid, m.p 185-187°C (8gm).

(iv) Synthesis of Benzoin benzoate from benzoin

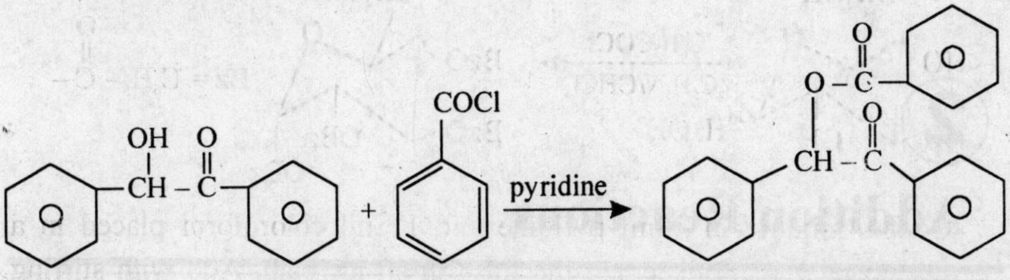

+ $C_5H_5N.HCl$

In a 100 ml conical flask place 2.6 gm of benzoin, 5ml of pyridine and 3ml of benzoyl chloride. Heat the mixture on water bath for 15-20 minutes and then finally on free flame for few minutes to complete the solution. Cool the flask and pour the contents onto 50 gm of crushed ice and stir vigorously until the oily layer solidifies. Filter the precipitated solid, wash with water and recrystallise from alcohol. Collect benzoin benzoate in the form of fine crystals and dry, m.p 127-128° C (2.7gm).

2

Addition Reactions

Addition reactions are those in which the attacking molecule adds to the substrate leading to the generation of a single species. The substrate on which addition takes place must possess a polar or non polar multiple bond. Alkenes and alkynes which fall in this category contain double bond and triple bonds and thus have high electron density in comparison to single bonded species. Excess electron density implies nucleophilic behaviour i.e. they react readily with electrophilies

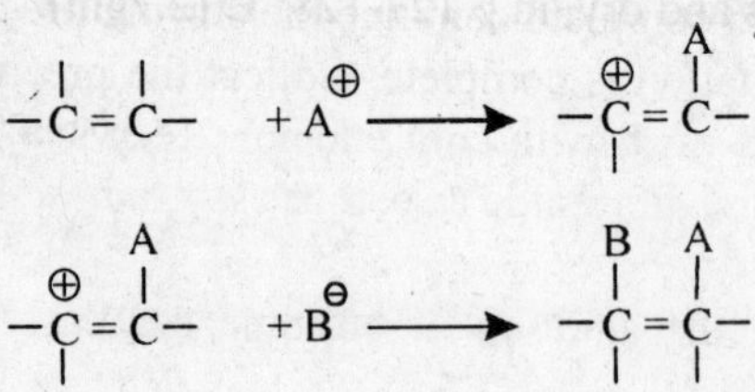

The addition of halogen across carbon-carbon double bond is a classical example of electrophilic addition reaction taking place in two steps with the formation of carbocation as intermediate. However in case of bromination, the carbocation though immediately cyclizes to a cyclic intermediate called bromonium ion. The bromide ion then attacks the intermediate from the opposite side resulting in the formation of the trans product. But when the alkene is conjugated with aryl group syn addition takes place.

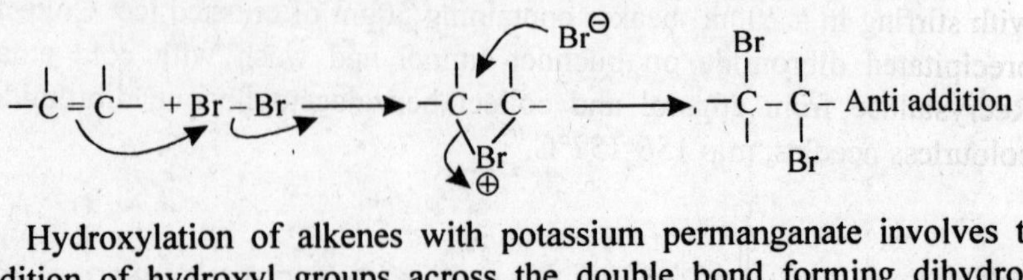

Hydroxylation of alkenes with potassium permanganate involves the addition of hydroxyl groups across the double bond forming dihydroxy compounds commonly known as glycols. The reaction involves the intermediate formation of cyclic manganese ester which is opened by the nucleophile (water) to give cis diols.

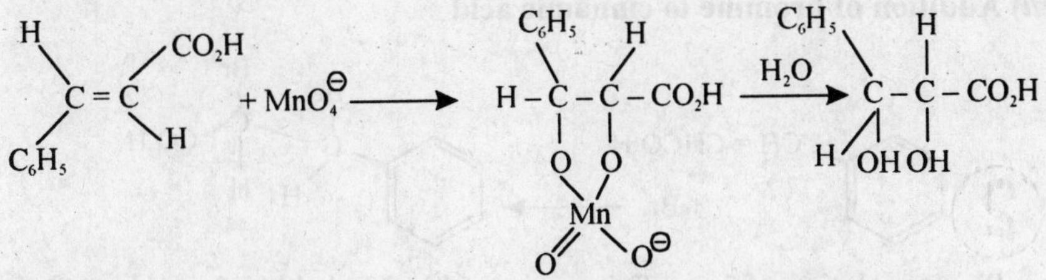

(i) Addition of bromine to benzalacetone

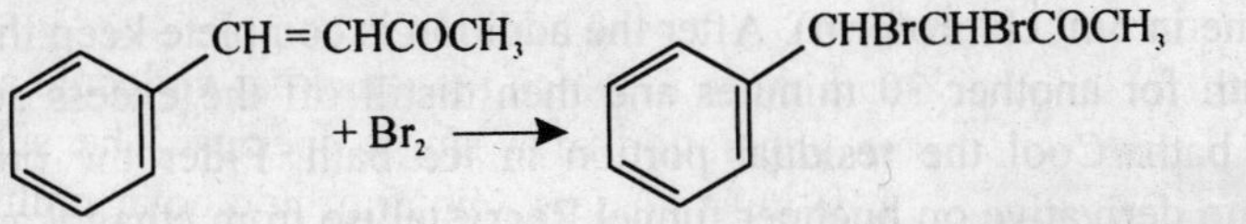

Place 5gm of benzalacetone and 15ml of carbon tetrachloride in a 100ml conical flask. Keep the flask in an ice-water bath so that the temperature inside the flask is in the range of 10-20°C. Add with stirring and in portions 4.7ml of bromine solution (1.7ml bromine in 3ml carbon tetrachloride). After the addition is complete continue the stirring process for 5-10 minutes or until the precipitation is complete. Collect the precipitated dibromide on buchner funnel and wash with cold ethanol. Recrystallise from ethanol and collect benzalacetone dibromide as colourless needles, m.p 124-125°C (yield 4.5-5.0gm).

Note:- During the addition of bromine the flask should be protected from sunlight

(*ii*) Addition of bromine to benzalacetophenone

In a 100ml conical flask dissolve 10gm of benzalacetophenone in 30ml carbon tetrachloride. Place the flask in ice bath, add dropwise with stirring 3ml of bromine. After the addition is complete remove the flask from ice bath, shake thoroughly and keep it at room temperature for 30 minutes. During this period shake the flask occasionally. Pour the reaction mixture with stirring in a 200ml beaker containing 50gm of crushed ice. Collect the precipitated dibromide on buchner funnel and wash with cold ethanol. Recrystallise from ethanol and collect benzalacetophenone dibromide as colourless needles, m.p 156-157°C.

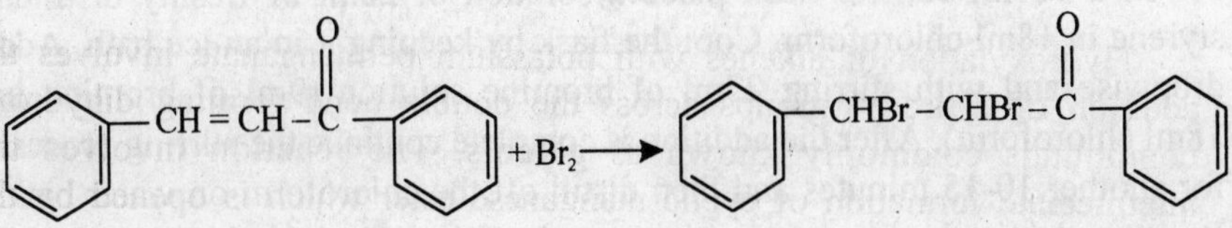

(*iii*) Addition of bromine to cinnamic acid

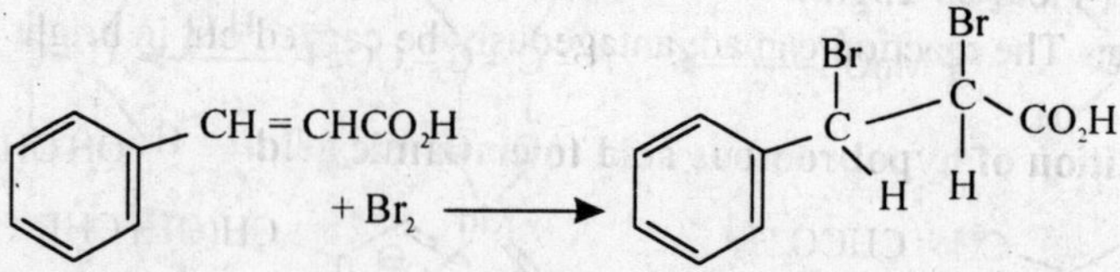

Place a solution of 5gm of cinnamic acid in 20ml chloroform in a 100ml conical flask. Cool the flask by keeping it in an ice bath (temperature below 5°C). Add with stirring and in portions 7.8ml of bromine solution (2.8ml of bromine in 5ml chloroform). After the addition is complete keep the flask in ice bath for another 30 minutes and then distill off the excess solvent on water bath. Cool the residual portion in ice bath. Filter the precipitated dibromo derivative on buchner funnel Recrystallise from ethanol and collect the pure specimen of erythro 2, 3-dibromo 3-phenyl propanoic acid, m.p 195°C (d).

Note:- During the addition of bromine the flask should be protected from sunlight

(*iv*) Addition of bromine to fumaric acid

$$HO_2C = CHCO_2H + Br_2 \rightarrow HO_2C\ CHBr\ CHBrCO_2H$$

In a 100ml conical flask dissolve by stirring 10gm of fumaric acid in 20ml water. Heat the flask on wire gauze and as soon as the solution begins to boil remove the flask from the flame. Now add dropwise with stirring 5.0 ml of bromine at such a rate that the mixture keeps on refluxing. After the addition is complete cool the flask in ice bath and stir the resulting solution for 30 minutes. Filter the precipitated solid on buchner funnel, wash with cold water. Dry the αβ dibromosuccinic acid at room temperature, m.p 167°C.

(*v*) Addition of bromine to styrene

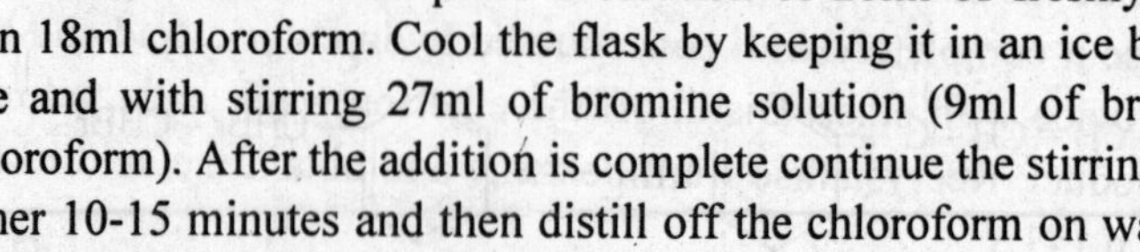

In a 100ml conical flask place a solution of 20ml of freshly distilled styrene in 18ml chloroform. Cool the flask by keeping it in an ice bath. Add dropwise and with stirring 27ml of bromine solution (9ml of bromine in 18ml chloroform). After the addition is complete continue the stirring process for another 10-15 minutes and then distill off the chloroform on water bath. Recover the crude dibromo product, wash with cold alcohol. Recrystallise from aqueous ethanol and collect crystals of 1, 2 dibromo phenylethane, m.p 73-74°C (Yield 20-25gm).

Note:- The reaction can advantageously be carried out in bright sunlight

(*vi*) Addition of hypobromous acid to cinnamic acid

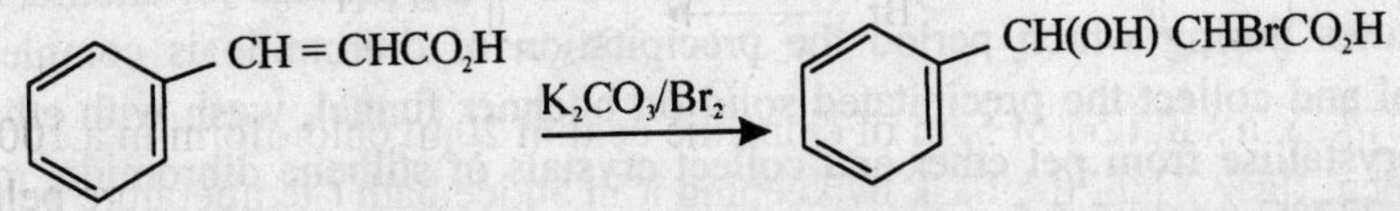

In a 100ml conical flask add 5gm of cinnamic acid to 50ml water and shake thoroughly. Neutralise the acidic solution by adding 4.5 gm of potassium carbonate and place it in an ice bath taking care that the temperature of the reaction mixture is well below 4°C. Now add dropwise with stirring 2ml of bromine. After the addition of bromine is complete add 20ml of sodium carbonate solution to the reaction mixture (9.5gm sodium carbonte in 20ml water). Remove the flask from ice bath, keep it at room temperature for 30-45 minutes with occasional shaking. Filter and acidify the filtrate with 8ml of concentrated hydrochloric acid. Unreacted cinnamic acid will be precipitated. Filter, transfer the filtrate to 250ml separating funnel, add 20ml ether to it. Shake the funnel thoroughly and remove the ethereal layer. Repeat the process twice, each time using 20ml ether. Combine the three organic layers and distill off the ether on water bath. Recover the crude acid, recrystallise from chloroform and collect crystals of α-bromo β-phenyl β-hydroxy propionic acid, m.p 124°C.

(*vii*) Hydroxylation of cinnamic acid

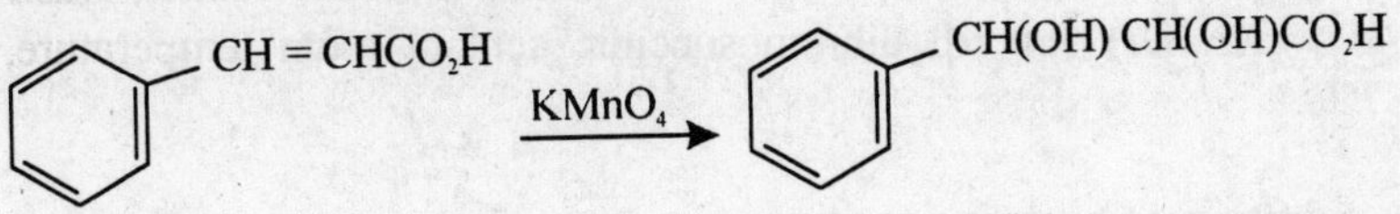

In a 100ml conical flask dissolve 5gm of cinnamic acid in 15ml of 5% aqueous caustic soda solution and add 12ml of 2% potassium permanganate solution to it. Keep the flask in ice bath, stir the mixture for 45 minutes. Filter the resulting mixture, neutralise with dilute hydrochloric acid. Concentrate the resulting mixture until the dissolved salt begins to precipitate. Neutralise it completely with concentrated hydrochloric acid and extract the product twice with ether. Distill off the ether on water bath and recover the crude product. Recrystallise from benzene and obtain the acid in crystalline form, m.p 141°C.

(*viii*) Addition of bromine to stilbene

$$C_6H_5CH = CHC_6H_5 + Br_2 \rightarrow C_6H_5CHBrCHBrC_6H_5$$

To a solution of 5gm of trans stilbene in 75ml ether placed in a 100ml conical flask, add dropwise and with stirring 1.5ml of bromine. After the addition of bromine is complete continue the stirring process for another 60 minutes during which period the precipitation of dibromide is complete. Cool and collect the precipitated solid on buchner funnel, wash with ether. Recrystallise from pet ether and collect crystals of stilbene dibromide, m.p 236-237°C (yield 5-5.5gm).

3

Elimination Reactions

Elimination reactions constitute one of the best general method for the preparation of multiple bonded compounds. It involves the removal of atoms or groups from two adjacent carbon atoms or from the same carbon atom. In these reactions two σ bonds are lost and a new π bond is generated. In fact by this method a saturated compound is converted to an unsaturated one.

These reactions are classified depending on the position from where the elimination has taken place. When both the atoms or groups are removed from the adjacent carbon atoms then the reaction is known as β or 1, 2 elimination and if the elimination takes place from the same carbon then it is known as α or 1, 1 elimination.

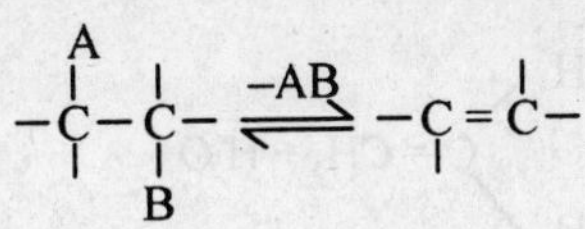

β elimination, or 1, 2 elimination

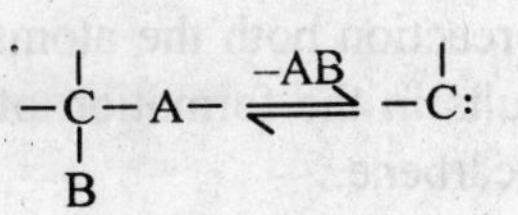

α elimination, or 1, 1 elimination

β or 1, 2 elimination reaction: These reactions are further classified into E1 and E2 reactions.

***(i)* Bimolecular Elimination or E2:-** This reaction follows second order kinetics and the rate of reaction is dependent on the concentration of alkyl halide as well as the base. The nucleophilic base abstracts the proton from the β carbon atom and the halide ion (leaving group) simultaneously departs from the α carbon atom along with its pair of electron. Thus the more concentrated the base is the more is a bimolecular reaction favoured. Some well known proton abstracting bases are R_3N, $C_6H_5O^-$, HO^-, RO^-, H_2N^-. This one step reaction which is also the rate determing step is highly stereospecific and takes place with ease when the two departing groups are trans.

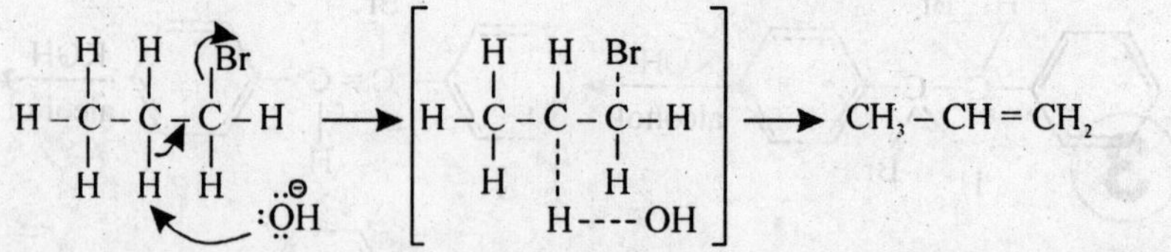

***(ii)* Monomolecular elimination or E1:-** This reaction follows first order kinetics and the rate is dependent only on the concentration of the substrate. In contrast to E2 process the bond breaking and bond formation process takes place in two steps. In the first step slow heterolysis of the substrate takes place leading to the formation of a carbocation. In the second step the carbocation rapidly loses a proton to the base forming the alkene.

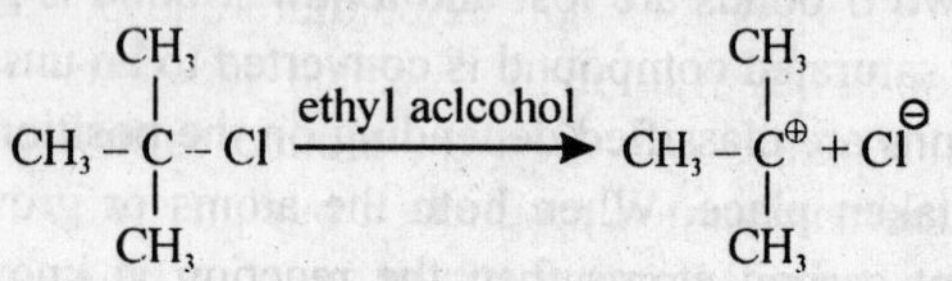

In the first step which is also the rate determining step only the substrate undergoes covalency change hence the rection is called monomolecular.

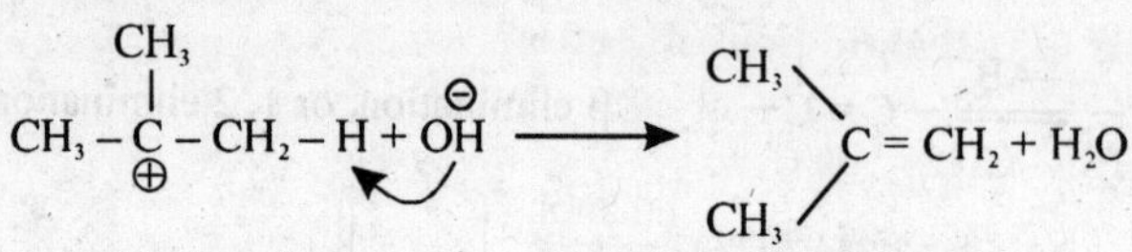

α or 1, 1 elimination reaction :-In this type of reaction both the atoms or groups are lost from the same carbon which results in the formation of an important divalent carbon intermediate known as carbene.

$$CHCl_3 + OH^{\ominus} \rightleftharpoons CCl_3^{\ominus} + H_2O$$

$$CCl_3^{\ominus} \rightleftharpoons Cl^{\ominus} + :CCl_2$$

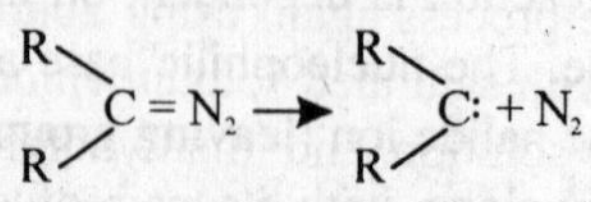

Carbenes usually exist in two forms, singlet and triplet carbenes. In the singlet carbene the carbon is sp^2 hybridized whereas in the triplet the carbon is sp hybridized.

(i) Synthesis of diphenyl acetylene (Tolan)

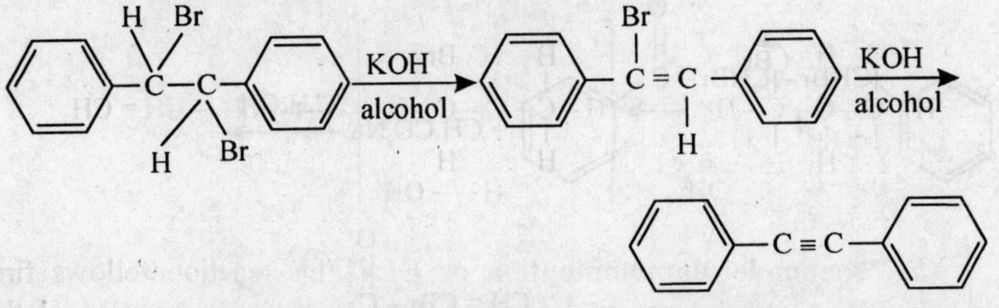

Dissolve 9gm of potassium hydroxide in 15ml absolute ethanol in a 500ml round bottom flask fitted with a reflux condensor. Cool the solution and add in portion 6.5 gm of stilbene dibromide. Reflux the reaction mixture in oil bath for 24 hours taking care that the solvent is not completely evaporated. Cool the flask to room temperature and pour the contents in a 200ml beaker containing 100gm of crushed ice whereby yellow-orange gummy mass will separate out. Allow the product to stand for 10-15 minutes and then collect the unsaturated product on buchner funnel. Wash with cold water, recrystallise the solid material from aqueous ethanol. Collect the pale yellow crystals of diphenyl acetylene (Tolan), m.p 60-61°C (Yield 3.5-4.0gm).

(ii) Synthesis of α bromo benzalacetone

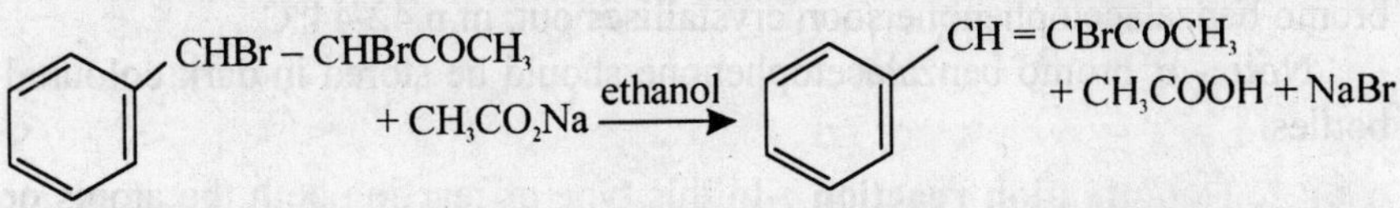

Reflux together 10gm of benzalacetone dibromide, 3gm of fused sodium acetate and 30ml of ethyl alcohol placed in a 100ml round bottom flask fitted with a reflux condensor for 4 hours. Cool the mixture at room temperature and filter. Distill the filtrate under reduced pressure to remove the excess alcohol. Transfer the residual oily portion to 100ml separating funnel and add 20ml ether to it. Shake the flask thoroughly and remove the organic layer. Repeat the process again using 20ml ether. Combine the two ether layers, wash first with saturated sodium chloride solution followed by 5% sodium bicarbonate solution and finally dry over anhydrous sodium sulphate. Distill off the ether on water bath, cool the residual oily portion in an ice bath, α bromo benzalacetone soon crystallises out, m.p 30-31°C.

Note:- α bromo benzalacetone should be stored in dark coloured bottles and kept in refrigerator.

(iii) Synthesis of α bromo benzalacetophenone

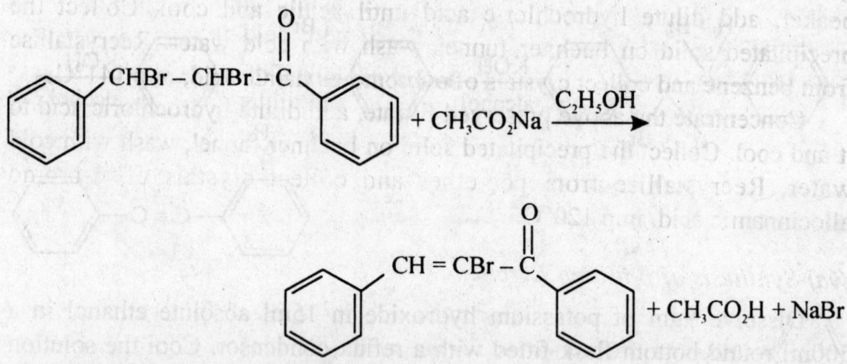

Reflux together 15gm of benzalacetophenone dibromide, 4.2gm of fused sodium acetate and 25ml of ethyl alcohol placed in a 100ml round bottom flask fitted with a reflux condensor for 5 hours. Shake the flask occassionally in between this period. Cool the mixture at room temperature and filter. Distill the filtrate under reduced pressure to remove the excess alcohol. Transfer the residual oily portion to 100ml separating funnel, add 20ml ether to it. Shake the flask thoroughly and remove the organic layer. Repeat the process again using 20ml ether. Combine the two ether layers, wash first with saturated sodium chloride solution followed by 5% sodium bicarbonate solution and finally dry over anhydrous sodium sulphate. Distill off the ether on steam bath. Cool the pale yellow oily layer in ice bath whereby α bromo benzalacetophenone soon crystallises out, m.p 42-44°C.

Note:- α bromo benzalacetophenone should be stored in dark coloured bottles.

(iv) Synthesis of α bromo cinnamic acid (cis) and α bromo allo cinnamic acid (trans)

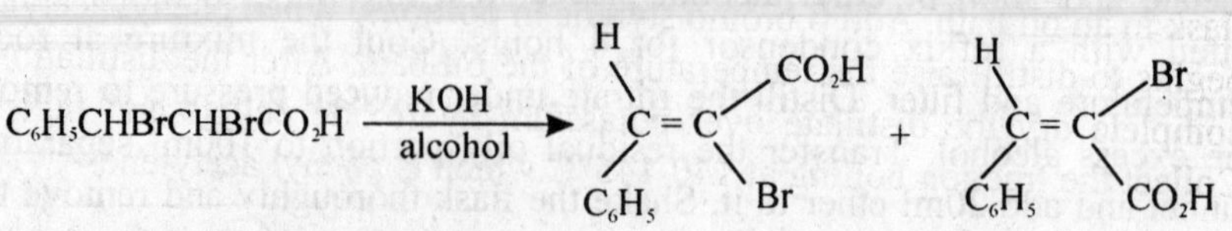

In a porcelain dish, to 5gm of cinnamic acid dibromide add 20ml of 10% alcoholic potassium hydroxide solution. Heat the porcelain dish on water bath and evaporate the mixture to dryness. Add about 5-10 ml of water, sufficient to dissolve the residue and then add excess of barium chloride solution. Barium salt of α bromo cinnamic acid is precipitated while α bromo allo cinnamic acid remains in solution. Filter the solid (preserve the filtrate), dissolve it in minimum amount of water in 100ml beaker, add dilute hydrochloric acid until acidic and cool. Collect the precipitated solid on buchner funnel, wash with cold water. Recrystallise from benzene and collect crystals of α-bromo cinnamic acid, m.p 141°C.

Concentrate the above preserved filtrate, add dilute hydrochloric acid to it and cool. Collect the precipitated solid on buchner funnel, wash with cold water. Recrystallise from pet ether and collect crystals of α-bromo allocinnamic acid, m.p 120°C.

(va) Synthesis of β-bromo styrene:

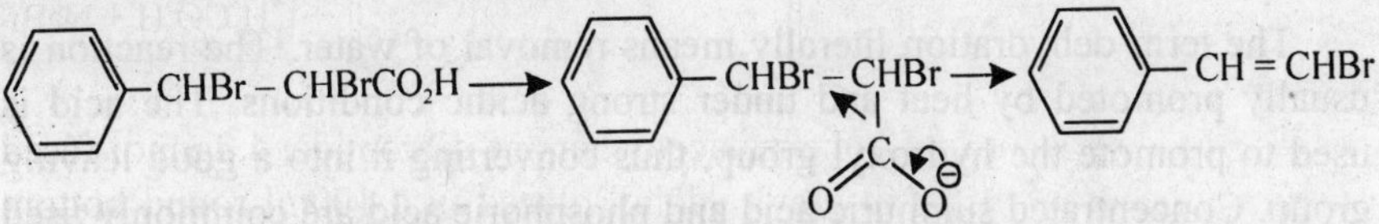

In a 100ml round bottom flask fitted with a reflux condensor place 5 gm of 2, 3-dibromo 3-phenyl propanoic acid and add 50ml of 10% aqueous sodium carbonate solution to it. Reflux the reaction mixture to boiling. Cool, transfer the reaction mixture to 200ml separating funnel, add 20ml ether to it. Shake the flask well and remove the organic layer. Repeat the process twice each time using 20ml of ether. Combine the three organic layers and then distill on water bath to remove the ether. Collect the residual portion which is β-bromo styrene, b.p 220°C.

(vb) Synthesis of Phenylacetylene :

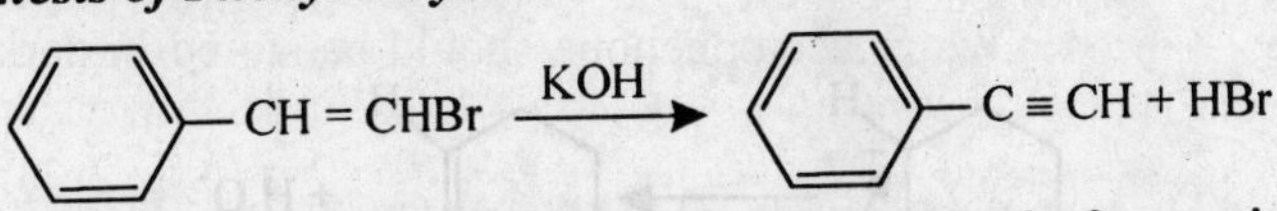

Place 10gm of potassium hydroxide pellets and 5ml of water in a 100ml round bottom flask. Set the assembly for downward distillation and heat the flask in an oil bath. Add β bromo styrene in portions. When phenyl acetylene begins to distill raise the temperature of the oil bath. After the distillation is complete dry the distillate over potassium hydroxide pellets and redistill. Collect the fraction boiling at 140-145°C which is phenyl acetylene.

4

Dehydration

The term dehydration literally means removal of water. The reaction is usually promoted by heat and under strong acidic conditions. The acid is used to promote the hydroxyl group, thus converting it into a good leaving group. Concentrated sulphuric acid and phosphoric acid are commonly used to catalyse these reactions.

Dehydration reactions generally proceed by E1 mechanism. Protonation of the hydroxyl group converts it into a good leaving group, water is lost forming a carbocation. Deprotonation of carbocation gives an alkene.

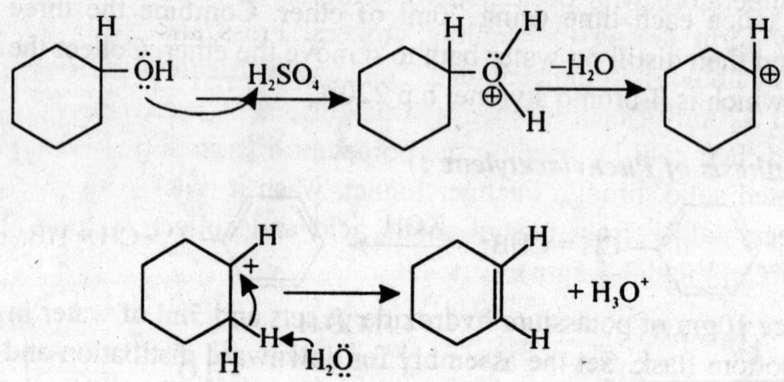

(i) Synthesis of Acetone dicarboxylic acid from citric acid

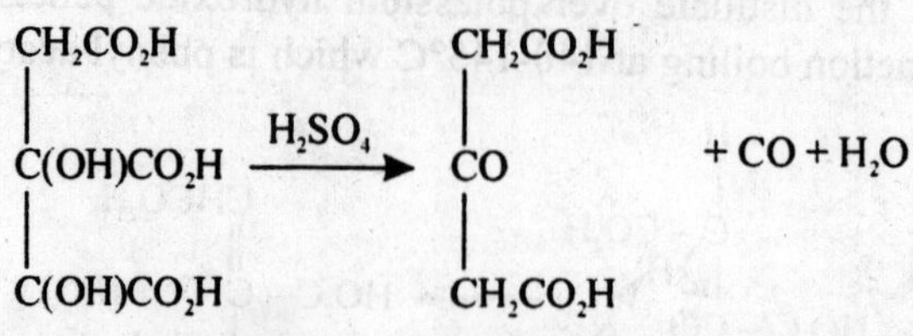

Cool 20ml of fuming sulphuric acid placed in a 100ml conical flask by placing the flask in an ice bath. Add with stirring 10gm of powdered citric acid. A vigorous reaction occurs, frothing takes place and carbon monoxide

is evolved. After the foaming has subsided remove the flask from the ice bath and place it at room temperature for five minutes. Cool it in a freezing mixture and add 25gm of crushed ice to it. A straw coloured solution soon crystallises out into a thick paste. Filter the crude product on bucher funnel and wash with cold ethyl acetate. Recrystallise from ether and collect acetone dicarboxylic acid as colourless crystals, m.p 135°C.

(*ii*) Synthesis of Aconitic acid from Citric acid

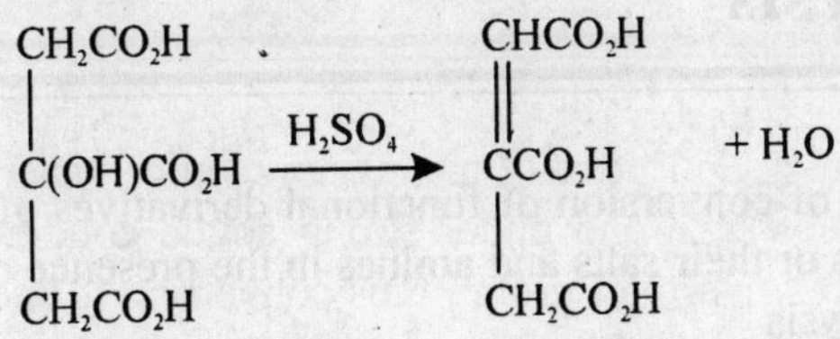

Place 10 gm of citric acid and a solution of 5.5ml of concentrated sulphuric acid in 5.5ml of water in a 100ml round bottom flask. Attach a reflux condensor to the flask and reflux the reaction mixture in an oil bath for 7-8hrs taking care that the temperature remains in the range of 140-145°C. Pour the light brown solution in a porcelain dish. Cool the liquid at room temperature with occassional stirring so as to break the solid lumps that may form during the cooling process. Press and drain the material thoroughly until it is completely dry. Transfer the dried material to a porcelain dish, add 3.5 ml of concentrated hydrochloric acid, stir it to a homogenous paste and then cool by keeping the porcelain dish in an ice bath. Filter the precipitated solid through buchner funnel, wash it twice with glacial acetic acid. Recrystallise from glacial acetic acid and collect aconitic acid, m.p 198-200°C. (Yield 3-3.5gm).

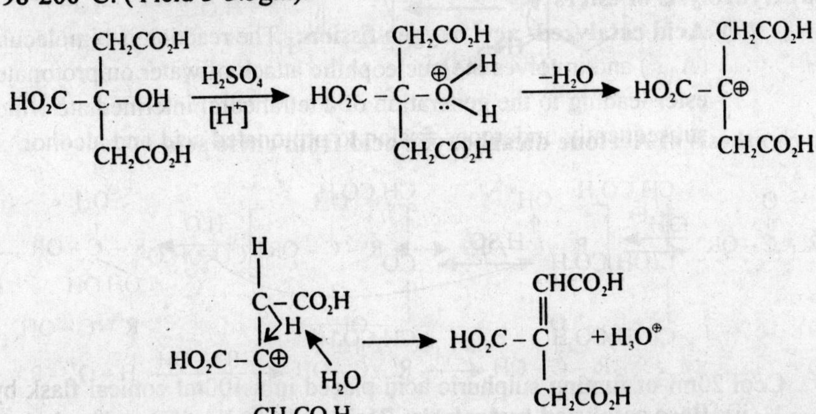

5

Hydrolysis

The process of conversion of functional derivatives of carboxylic acids to alcohols, acids or their salts and amines in the presence of acid or alkali is known as hydrolysis

The carboxylic acid derivatives are characterised by low carboxyl activity and hence hydrolysis takes place smoothly by heating them in the presence of acid or alkali.The hydrolysis can take place either by the cleavage of acyl oxygen bond or alkyl oxygen bond.

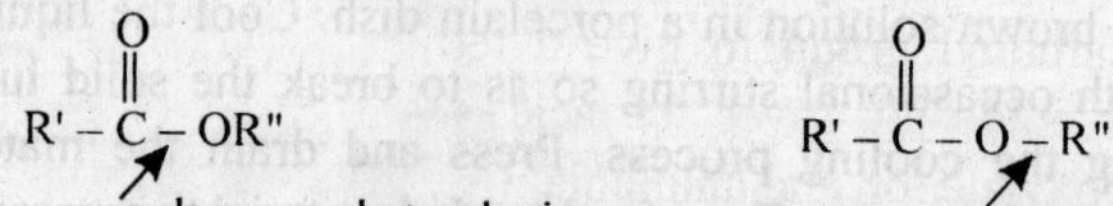

acyl oxygen heterolysis　　　　alkyl oxygen heterolysis

Acid hydrolysis is usually carried out only when the acid produced during the reaction is unstable to alkali.Hydrolysis reactions carried out by acid are reversible but can be driven to completion by using excess of water.

(a) Hydrolysis of esters

(i) **Acid catalyzed- acyl oxygen fission:-** The reaction is bimolecular ($A_{AC}2$) and involves the nucleophilic attack of water on protonated ester leading to the generation of a tetrahedral intermediate which subsequently undergoes fission to protonated acid and alcohol.

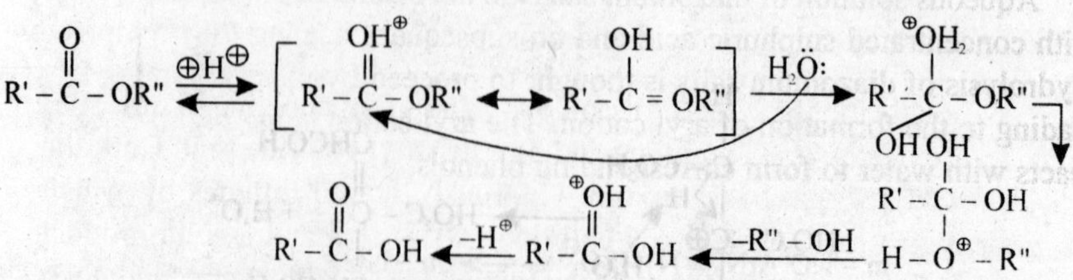

(ii) **Base catalyzed hydrolysis:** Basic hydrolysis of esters also known as saponification takes place more rapidly in comparison to acid hydrolysis. The hydroxyl ion which is a good nucleophile attacks the carboxyl group to give a tetrahedral intermediate. Elimination of alkoxide gives the acid,fast transfer of proton gives carboxylate ion and the alcohol.

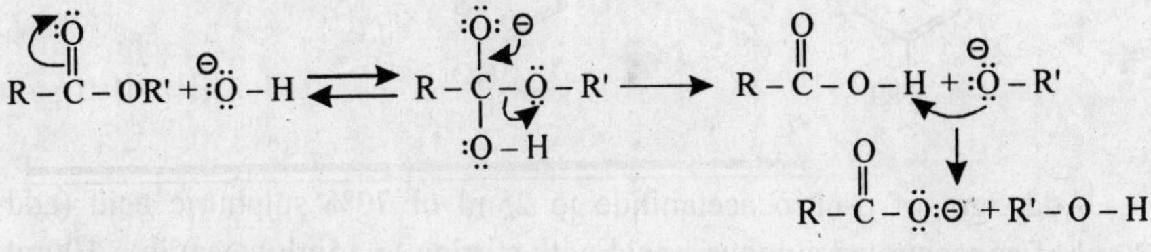

(b) Hydrolysis of Amides

Amides which are quite stable undergo hydrolysis much slower than esters and thus their hydrolysis to carboxylic acids and amines require more vigorous conditions.Amides are therefore hydrolysed in the presence of strong base and strong acid.

(i) **Basic hydrolysis :** The hydroxide which is a good nucleophile, attacks the carbonyl group to give a tetrahedral intermediate. Expulsion of amide ion gives the acid, fast transfer of proton gives the carboxylate ion and ammonia

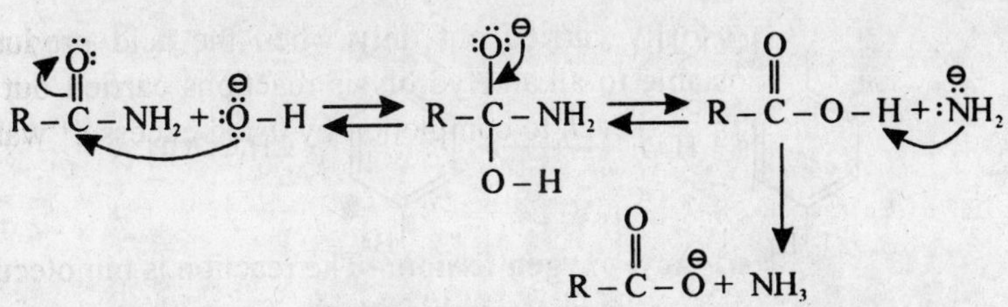

(c) Hydrolysis of diazonium salts:

Aqueous solution of diazonium salts are not stable, hence on acidification with concentrated sulphuric acid and on subsequent heating they hydrolyse. Hydrolysis of diazonium salts is thought to proceed by the SN^1 type process leading to the formation of aryl cation. The aryl cation being highly reactive reacts with water to form corresponding phenols.

$$m-NO_2ArN^{\oplus}\equiv N\}HSO_4^{\ominus} \longrightarrow m-NO_2Ar^{\oplus}+N_2$$

$$m-NO_2Ar^{+}+H_2O \longrightarrow m-NO_2ArOH$$

(i) Hydrolysis of p-nitro acetanilide to p-nitro aniline

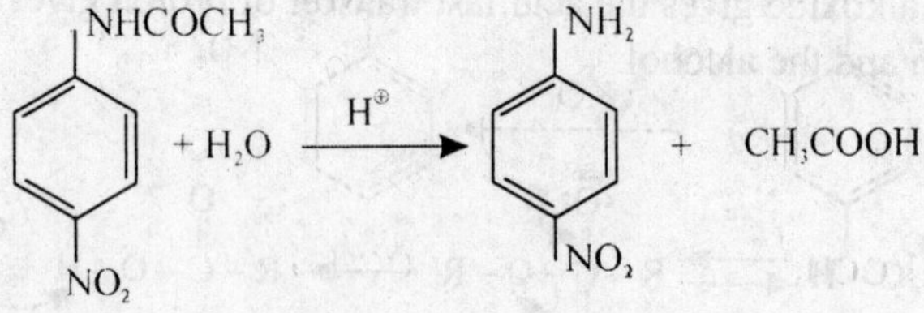

Add 5gm of p-nitro acetanilide to 25ml of 70% sulphuric acid (add 20ml of concentrated sulphuric acid with stirring to 15ml water) in a 100ml round bottom flask fitted with a reflux condensor. Boil the mixture gently for 25-35 minutes. Pour the hot solution in a 200ml beaker containing 100ml of chilled ice water and then add excess of dilute sodium hydroxide solution or until the precipitation of yellow p-nitro aniline is complete.Filter the precipitated solid on buchner funnel, wash with cold water. Recrystallise from hot water and collect crystals of p-nitro aniline, mp 147-148°C. (Yield 3.6gm).

(ii) Hydrolysis of p-bromo acetanilide to p-bromo aniline

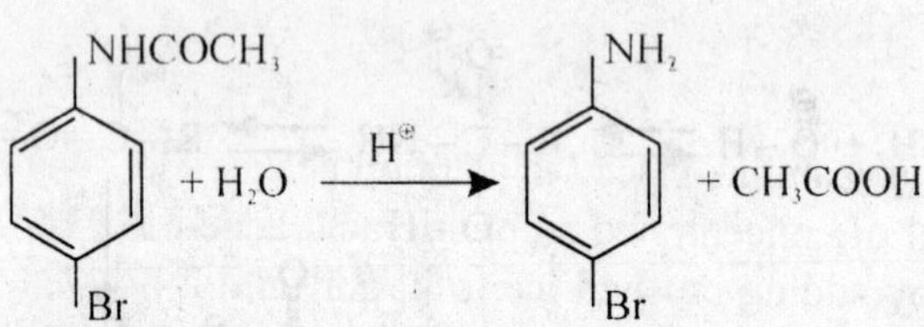

In a 100ml round bottom flask dissolve by heating on water bath 5gm of p-bromo acetanilide in 10ml of ethanol. After a clear solution is obtained add in portions 6ml of concentrated hydrochloric acid to the hot solution and then reflux for 40minutes. Cool the flask, add 20ml of water to it.Set the assembly for downward distillation and distill the mixture on wire gauze until 10ml of distillate is collected. Pour the residual concentrate with stirring in a 200ml beaker containing 50gm of crushed ice.Add with stirring dilute sodium hydroxide solution until the solution is alkaline and cool further.Filter the precipitated solid on buchner funnel, wash with cold water.Recrystallise from dilute alcohol and collect crystals of p-bromo aniline, m.p 66°C. (Yield 3-3.5gm).

(iii) Hydrolysis of 2-nitro 4-methoxy acetanilide to 2-nitro 4-methoxy aniline

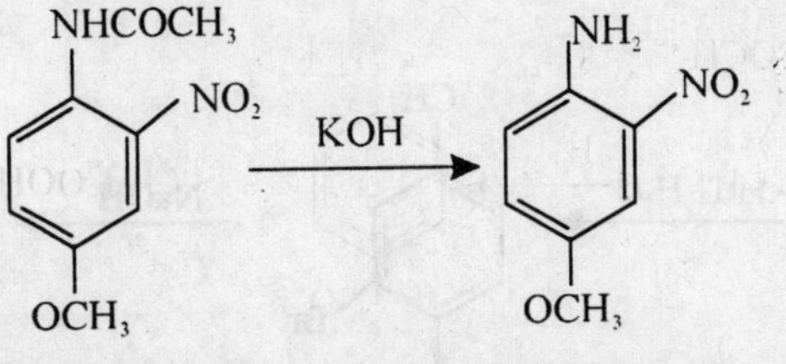

In a 100ml conical flask place 5gm of 2-nitro 4-methoxy acetanilide and add 15 ml of potassium hydroxide solution (3gm in 15ml water) to it.Heat the flask on steam bath for 20minutes. First the solid will completely dissolve and then a thick paste will be formed in the flask.Add 30 ml of hot water and again heat the flask on steam bath for 20 minutes with stirring.Cool the flask in ice bath and allow precipitation to take place.Filter the precipitated solid on buchner funnel, wash with water.Recrystallise from ethanol and collect crystals of 2-nitro 4-methoxy aniline, m.p 122-124°C (Yield 4-4.5gm).

(iv) Diazotization of m-nitro aniline and its hydrolysis to m-nitro phenol

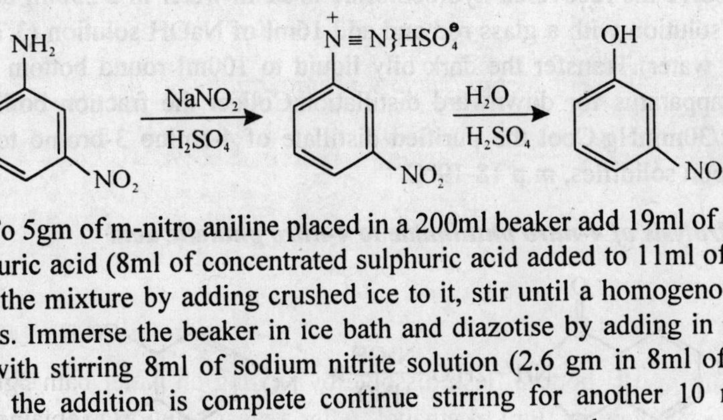

To 5gm of m-nitro aniline placed in a 200ml beaker add 19ml of aqueous sulphuric acid (8ml of concentrated sulphuric acid added to 11ml of water). Cool the mixture by adding crushed ice to it, stir until a homogenous paste results. Immerse the beaker in ice bath and diazotise by adding in portions and with stirring 8ml of sodium nitrite solution (2.6 gm in 8ml of water). After the addition is complete continue stirring for another 10 minutes. Decant and preserve the solid and liquid portion separately.

In a 250ml round bottom flask add 25ml of concentrated sulphuric acid to 21ml of water and then heat it to boiling.Add first the supernatant liquid (diazonium solution) in portion at such a rate that the mixture boils vigorously. After the addition of the liquid portion is complete add the residual solid, taking care that excessive frothing does not take place.Boil the mixture for another five minutes after the addition is complete and then pour with stirring in a 500ml beaker containing 200gm of crushed ice.Filter the crude product through buchner funnel, wash with cold water. Recrystallise from dilute hydrochloric acid and collect light yellow crystals of m-nitro phenol, m.p 96°C. (yield 2-2.5 gm)

(v) Hydrolysis of 3-bromo 4-acetamido toluene to 4-amino 3-bromo toluene

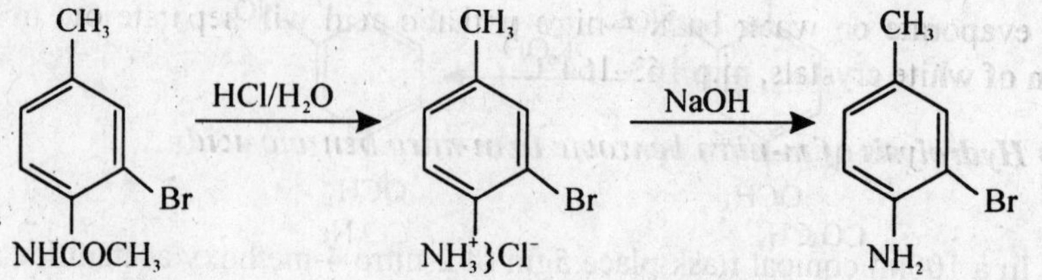

In a 100ml round bottom flask fitted with a reflux condensor dissolve by heating 10gm of 3-bromo 4-acetamido toluene in 15 ml of ethanol. After the solid portion has completely dissolved introduce 15 ml of concentrated hydrochloric acid through the condensor and reflux the reaction mixture for 3 hours.Pour the hot reaction mixture in a 200ml beaker and cool thoroughly.Filter the crystals of hydrochloride through buchner funnel and wash with cold alcohol.

Dissolve the recovered hydrochloride in 22ml water in a 250ml beaker. Stir the solution with a glass rod and add 16ml of NaOH solution (3.75 gm in 16ml water).Transfer the dark oily liquid to 100ml round bottom flask, set the apparatus for downward distillation.Collect the fraction boiling at 120-122/30mmHg.Cool the purified distillate of 4-amino 3-bromo toluene which soon solidifies, m.p 18-19°C.

(vi) Hydrolysis of 4-nitro phthlimide to 4-nitro phthalic acid

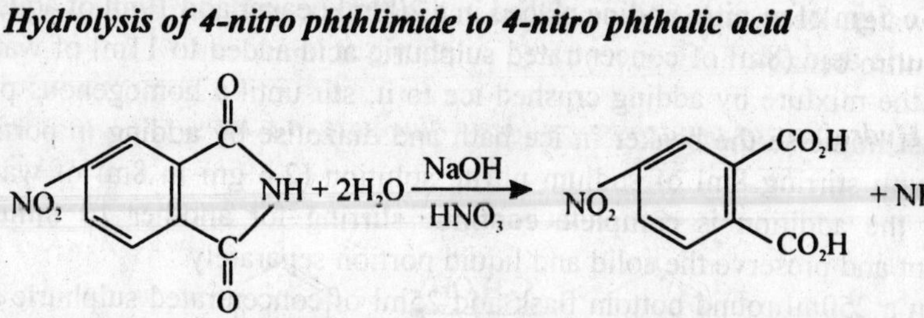

In a 100ml round bottom flask fitted with a reflux condensor place 4gm of 4-nitro phthalimide and 12ml of sodium hydroxide solution (1.4gm in 12ml of water). Heat the mixture to boiling for 10 minutes. Cool the flask and add concentrated nitric acid until the solution is just acidic. Swirl the flask gently and then reflux the resultant mixture for 5 minutes.Cool the flask and transfer the contents to a 500ml separating funnel, add 20ml of ether to it.Shake the flask thoroughly and remove the organic layer.Repeat the process twice each time using 20ml ether.Combine the three organic layers, dry over anhydrous sodium sulphate and then filter. Distill off the ether on water bath, transfer the concentrated solution in a porcelain dish and evaporate on water bath. 4-nitro phthalic acid will separate out in the form of white crystals, m.p 163-164°C.

(vii) Hydrolysis of m-nitro benzoate to m-nitro benzoic acid

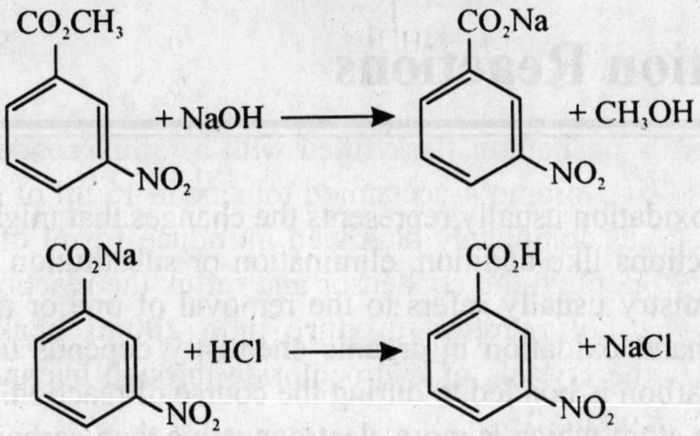

Heat to boiling 9gm of m nitro benzoate with 16ml of sodium hydroxide solution (4gm in 16ml water) placed in a 100ml round bottom flask fitted with a reflux condensor for 10-15 minutes.Cool the flask, add 10ml of water to it.Pour the resultant dilute solution in a 200ml beaker containing 13ml of concentrated hydrochloric acid.Cool the resultant mixture. Filter the light brownish coloured precipitated solid through buchner funnel and wash with cold water. Recrystallise from dilute hydrochloric acid and collect crystals of m-nitro benzoic acid, m.p 141°C. (yield 3.5-4.0gm)

(viii) Hydrolysis of p-nitro benzylidene diacetate to p nitro benzaldehyde

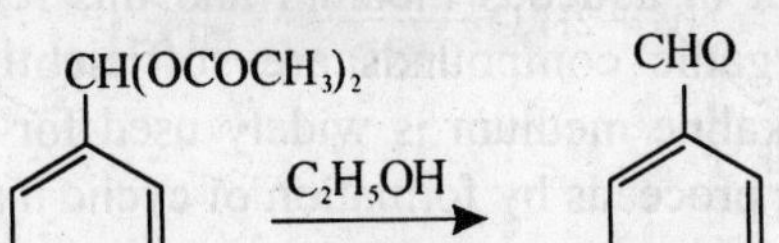

Reflux together 5gm of benzylidene diacetate with 14ml of ethanol, 14ml of water and 1.2ml of concentrated sulphuric acid placed in a 100ml round bottom flask for 30-40 minutes.Filter the solution hot, cool the filtrate in an ice bath. Filter the precipitated solid through buchner funnel and wash with cold water. Recrystallise from ethanol and collect crystals of p-nitro benzaldehyde, m.p 106°C.

6 Oxidation Reactions

The term oxidation usually represents the changes that might accompany during the reactions like addition, elimination or substitution. Oxidation in inorganic chemistry usually refers to the removal of one or more electron. On the other hand oxidation in organic chemistry depends upon to which type of atom carbon is bonded to during the course of reaction. Formation of a bond with an atom which is more electronegative than carbon is said to be oxidation. But in simpler terms addition of oxygen or removal of hydrogen from an organic molecule is almost always regarded as an oxidation process.

Oxidation of organic compounds can be brought about by a variety of oxidising agents like hexavalent chromium, potassium permanganate, lead tetraacetate, hydrogen peroxide, nitric acid and manganese dioxide.

Oxidation by permanganate depends upon whether it is used in acidic, basic or neutral medium. In acidic medium heptavalent manganese is reduced to divalent manganese due to transfer of five electrons, whereas in basic or neutral medium there is net transfer of three electrons and there is also the formation of manganese dioxide. Permanganate oxidation reactions are generally carried out in aqueous medium and this restricts its usefulness, since not many organic compounds are sufficiently soluble in water. Permanganate in alkaline medium is widely used for cis hydroxylation of alkene. The reaction proceeds by formation of cyclic manganese ester.

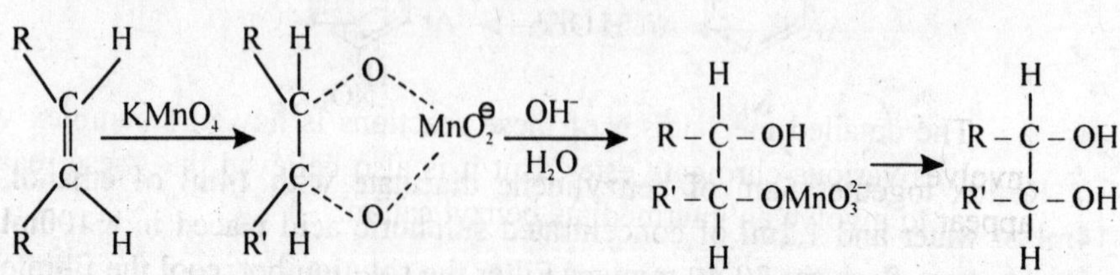

Oxidation by hexavalent chromium is usually brought about by one of its versatile derivative, chromic acid. During the oxidation process the chromium is reduced from hexavalent to trivalent state.

$$K_2Cr_2O_7 + H_2O + H_2SO_4 \longrightarrow 2HO-\overset{\overset{\displaystyle O}{\|}}{Cr}-OH + 2HSO_4^-$$

$$CrO_3 + H_2O \longrightarrow HO-\underset{\underset{\displaystyle O}{\|}}{\overset{\overset{\displaystyle O}{\|}}{Cr}}-OH \rightleftharpoons H^+ + O^- -\underset{\underset{\displaystyle O}{\|}}{\overset{\overset{\displaystyle O}{\|}}{Cr}}-OH$$

Chromium VI oxide can also be used by dissolving in reagents such as acetic anhydride, t-butanol, pyridine or hydrochloric acid. In these solutions the reactive species are chromyl acetate, t-butyl chromiteVI, pyridine-chromium VI oxide complex and chromyl chloride respectively.

$$(CrO_3)_n + CH_3-\overset{\overset{\displaystyle O}{\|}}{C}-O-\overset{\overset{\displaystyle O}{\|}}{C}-CH_3 \longrightarrow CH_3-\overset{\overset{\displaystyle O}{\|}}{C}-O-\underset{\underset{\displaystyle O}{\|}}{\overset{\overset{\displaystyle O}{\|}}{Cr}}-O-\overset{\overset{\displaystyle O}{\|}}{C}-CH_3$$

$$(CrO_3)_n + (CH_3)_3COH \longrightarrow (CH_3)_3CO-\underset{\underset{\displaystyle O}{\|}}{\overset{\overset{\displaystyle O}{\|}}{Cr}}-O-C(CH_3)_3$$

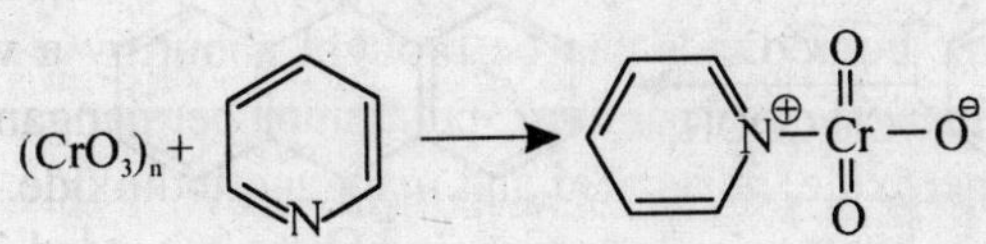

$$(CrO_3)_n + HCl \longrightarrow Cl-\underset{\underset{\displaystyle O}{\|}}{\overset{\overset{\displaystyle O}{\|}}{Cr}}-Cl$$

The acidic chromic acid solution readily attacks the alkyl side chain of aromatic ring forming aromatic carboxylic acid. The reaction is believed to proceed by the following pathway.

$$ArCH_3 \longrightarrow ArCH_2OH \longrightarrow Ar-\underset{\underset{\displaystyle H}{|}}{\overset{\overset{\displaystyle O}{\|}}{C}} \longrightarrow Ar-\overset{\overset{\displaystyle O}{\|}}{C}-OH$$

The detailed mechanism of these reactions is however complex which involves various chromate ester. But it is also believed that these reactions appear to involve an intermediate benzyl cation.

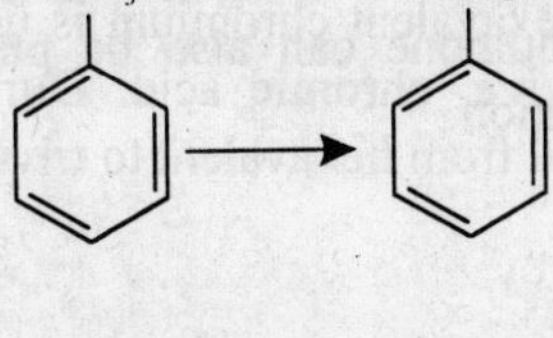

Chromic acid oxidation is however useful in conversion of secondary alcohols to ketones. The reaction proceeds by the initial formation of chromate ester. Elimination of the chromate ester gives the ketone. The carbinol carbon retains the oxygen but loses the hydrogen during the elimination process.

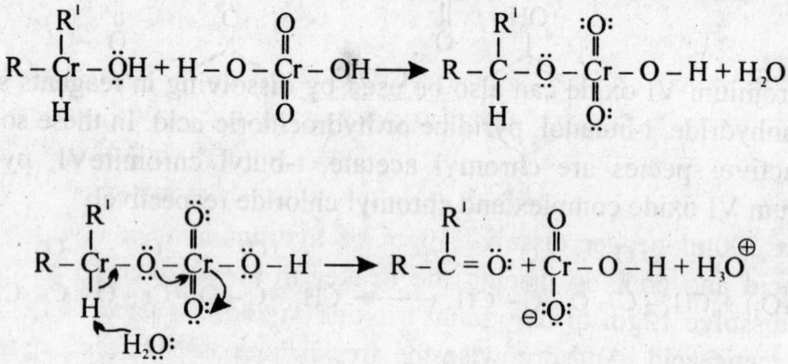

(i) Oxidation of anthracene to anthraquinone (9, 10 anthraquinone)

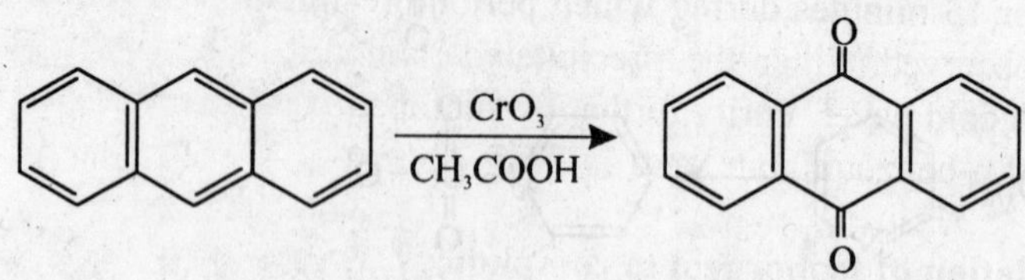

In a 100ml round bottom flask fitted with reflux condensor place 2.5gm of anthracene and 13ml of glacial acetic acid. Heat the flask on wire gauze so that most of the anthracene dissolves. In the meantime prepare chromium trioxide solution by dissolving 5gm of chromium trioxide in 5ml water and 13ml of glacial acetic acid. Remove the flask from the heat source and then add this freshly prepared chromium trioxide solution in portions such that the mixture continues to reflux. As the oxidation proceeds the anthracene will completely dissolve in the boiling acetic acid during the addition of chromium trioxide solution. After the addition is complete reflux the reaction mixture for another 15 minutes. Cool and pour the contents with stirring in a 200ml beaker containing 100ml of ice cold water. Collect the precipitated anthraquinone on buchner funnel, wash it first with hot water then with dilute sodium hydroxide and finally with cold water. Recrystallise the crude product with hot glacial acetic acid and obtain yellow crystals of anthraquinone, m.p 273°C (Yield 2-2.5gm)

Note: (i) Anthraquinone can also be purified by the process of sublimation.

(*ii*) Similarly 1, 4 naphthaquinone can be prepared by using 6.4gm of naphthalene, 12gm chromium trioxide and 15 ml glacial acetic acid.

(*ii*) Oxidation of hydroquinone to p-benzoquinone

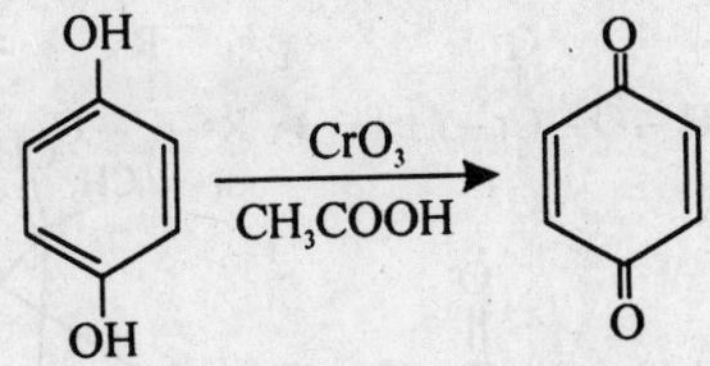

In a 500ml beaker dissolve 10gm of hydroquinone in 45ml of 60% acetic acid and cool by placing the beaker in an ice bath. In a separate beaker dissolve 13gm of chromium trioxide in 24ml of water and 11ml of glacial acetic acid. Add dropwise the freshly prepared chromium trioxide solution to well stirred hydroquinone solution at such a rate that the temperature does not rise above 10°C. After the addition is complete continue stirring for 15 minutes during which period precipitation of yellow quinone will be observed. Filter the precipitated quinone on buchner funnel and wash with cold water. Purify further by sublimation and collect bright yellow crystals of p-benzoquinone, m.p 116°C.

(*iii*) Oxidation of Isoborneol to camphor

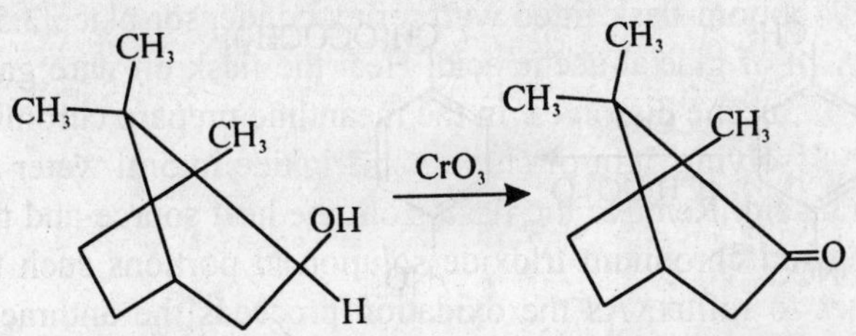

In a 250ml conical flask dissolve 2.5gm of isoborneol in 25ml of acetone. Cool by keeping the flask in ice bath. Pre cool 10ml of chromic acid solution (dissolve 27gm of chromium trioxide in 50ml water and then add 23ml concentrated sulphuric acid to it) and add in portion with stirring to the isoborneol solution. After all chromic acid has been added swirl the flask for 10 minutes and filter the reaction mixture through a thin layer of celite filter aid. Distill off the acetone on water bath. After the acetone has been distilled off add 25ml of distilled water to the residue and transfer the resultant solution to a 100ml separating funnel. Add 20ml ether to the funnel and shake thoroughly. Remove the organic layer and repeat the process again using 20ml ether. Combine the two organic layers and wash it first with dilute sodium carbonate solution followed by water. Dry the ether solution over anhydrous sodium sulphate and then decant. Distill off the ether on water and recover the residue which is camphor. Camphor can be further purified by sublimation, m.p 173-175°C (Yield 1.5-1.7gm).

Mechanism

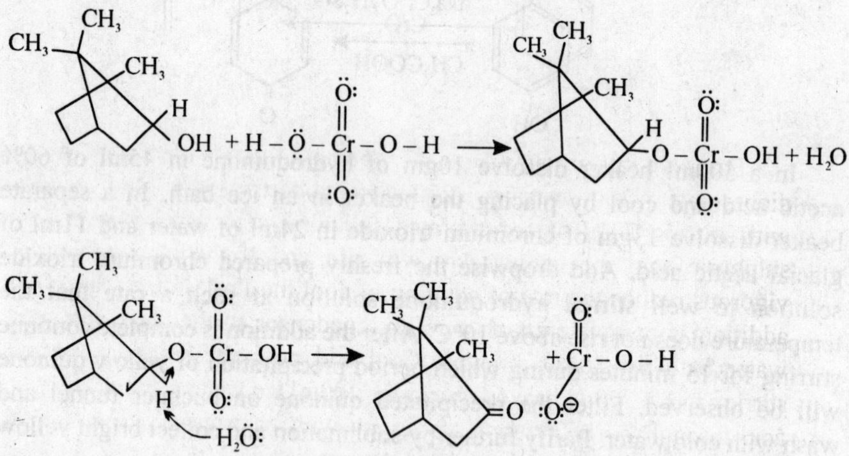

(*iv*) Oxidation of p-nitro toluene to p-nitro benzaldiacetate.

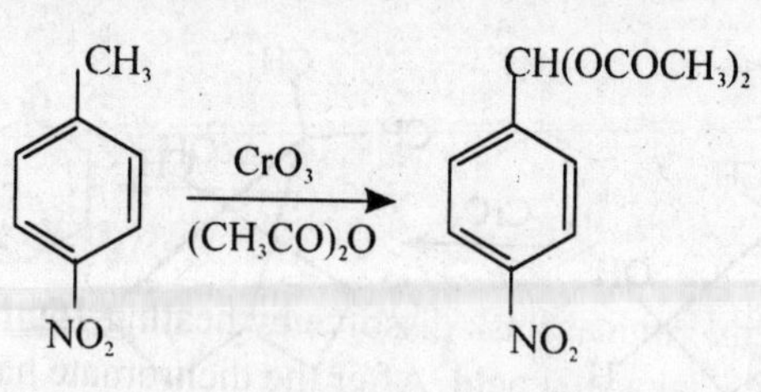

In a 200ml conical flask place 40ml of acetic anhydride and 5gm of p-nitro toluene. Cool by keeping the flask in an ice bath. Now add slowly and with stirring 8ml of concentrated sulphuric acid to the cooled mixture. After the addition of sulphuric acid is complete and the temperature of the mixture is 0°C, add slowly and with stirring a solution of 10gm of chromium trioxide in 45ml acetic anhydride such that the temperature of the reaction mixture does not rise above 8°C. After the addition is complete continue stirring the resultant mixture for another 2 hours. Pour the mixture in a 250ml beaker containing 50gm of crushed ice. Filter the precipitated p-nitro benzaldiacetate on buchner funnel and wash with cold water. Recrystallise from hot alcohol and collect crystals of p-nitro benzaldiacetate, m.p 125-126°C. (yield 4-4.5gm).

(*v*) Oxidation of p-nitro toluene to p-nitro benzoic acid

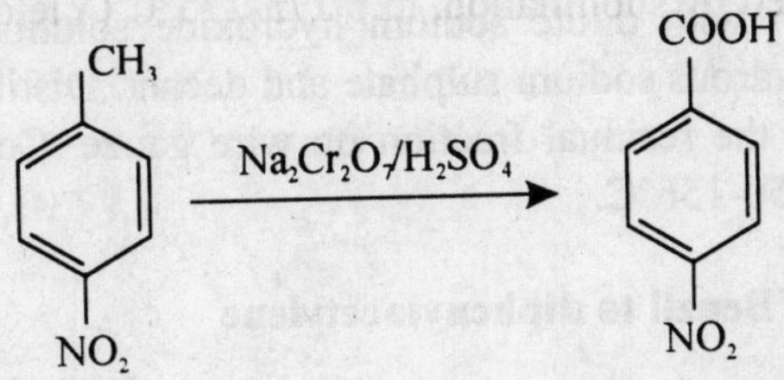

Dissolve by stirring 5gm of p-nitro toluene and 15gm of sodium dichromate in 30ml water in a 250ml round bottom flask. Now add dropwise with stirring 20ml of concentrated sulphuric acid. During the addition of sulphuric acid the temperature will rise and if the reaction tends to be vigorous reduce the rate of addition or cool the flask in tap water. After the addition is complete attach a reflux condensor to the flask and heat it on water bath for 30-40 minutes. Cool and pour the reaction mixture with stirring into a 200ml beaker containing 100ml ice water. Filter the precipitated acid through buchner funnel and wash with cold water. Recrystallise from benzene and collect crystals of p-nitro benzoic acid, m.p 239°C. (yield 4gm)

(*vi*) Oxidation of cyclohexanol to cyclohexanone

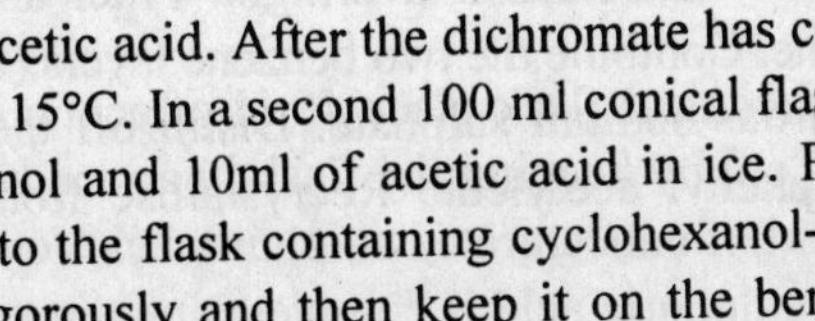

In a 100ml conical flask dissolve by heating 15gm of sodium dichromate in 25ml of glacial acetic acid. After the dichromate has completely dissolved cool the solution to 15°C. In a second 100 ml conical flask chill a mixture of 16ml of cyclohexanol and 10ml of acetic acid in ice. Pour the dichromate solution in one lot to the flask containing cyclohexanol-acetic acid mixture. Shake the flask vigorously and then keep it on the bench. The reaction is exothermic and if the temperature rises above 60°C cool it in ice bath just enough to prevent a further rise in temperature and then by intermittent brief cooling keep the temperature close to 60°C for 15 minutes. When the temperature stops rising allow the flask to stand for 15 minutes at room temperature and then pour the solution in a 250ml round bottom flask containing 100ml water. Set the apparatus for downward distillation and collect about 80ml of distillate. Transfer the distillate to a 250ml separating funnel and add a saturated solution of sodium chloride to it. Add 25ml of ether to the separating funnel and shake thoroughly. Remove the ethereal layer and repeat the process again using 25ml ether. Combine the two ether extracts and wash with dilute sodium hydroxide solution. Dry the ether solution over anhydrous sodium sulphate and decant. Distill off the ether on water bath, distill the residual fraction on wire gauze. Collect the fraction boiling between 150-156°C.

(*vii*) Oxidation of Benzil to diphenylacetylene

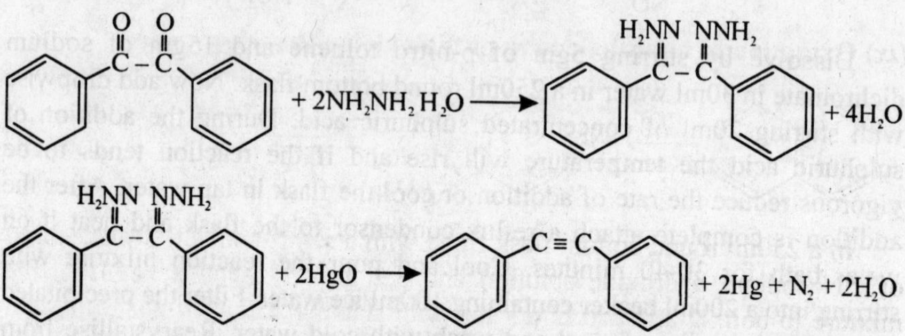

In a 100ml round bottom flask fitted with a reflux condensor place 5gm of benzil, 17ml of n-propyl alcohol and 3.8 gm of hydrazine hydrate. Reflux the mixture for 60 hours. Cool the flask in ice bath and collect the precipitated benzil dihydrazone on buchner funnel, wash with cold alcohol and dry, m.p 150-151°C.

Dissolve benzil hydrazone (4.5gm) in 25ml benzene in 100ml round bottom flask and add 200mg of yellow mercuric oxide to it. Attach a reflux condensor to the flask and warm the mixture on steam bath. Add 12gm of mercuric oxide in small instalments. After the addition is complete stir the mixture for 1hour and then keep it overnight. Filter and wash the residue with 10-15ml benzene. Combine the two benzene layers (filtrate and washing) and dry over anhydrous sodium sulphate. Distill off the benzene on water bath and collect diphenyl acetylene. Recrystallise from ethanol, m.p 60-61°C.

(*viii*) Oxidation of phthalic anhydride to monoper phthalic acid

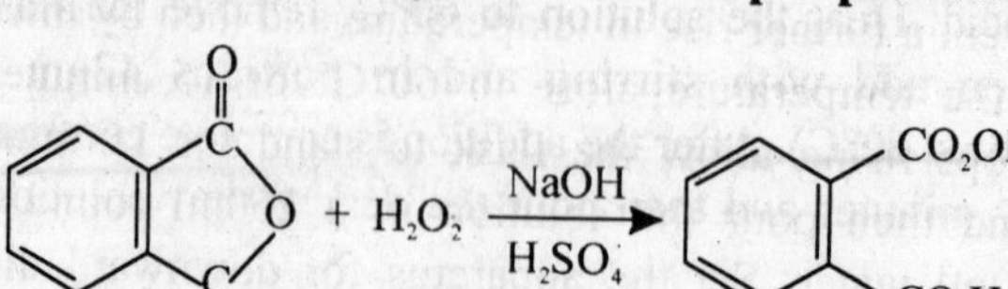

Cool 25ml of 15% sodium hydroxide solution in 250ml conical flask by placing it in an ice bath. Add in portions 11ml of 30% hydrogen peroxide to this cold alkali solution. When the temperature inside the flask reaches 10°C add 7.5 gm of powdered phthalic anhydride with stirring. Continue the stirring process until all the anhydride dissolves. Add 25ml of 20% cold sulphuric acid to the anhydride solution. Filter the acidic solution through glass wool. Transfer the filtrate to 250ml separating funnel and add 20ml ether to it. Shake the funnel thoroughly and remove the ethereal layer. Repeat the process again using 20ml ether. Combine the two ether layers and shake it thoroughly with ammonium sulphate and leave overnight. Distill off the ether on water bath and collect crystalline monoperphthallic acid.

(*ix*) Oxidation of naphthalene to phthalaldehydic acid

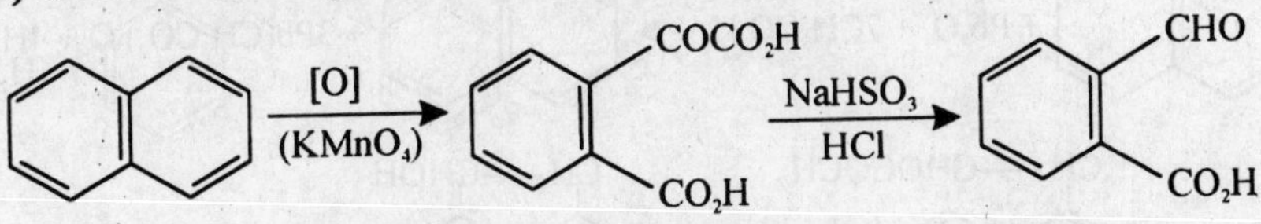

In a 250ml round bottom flask fitted with a reflux condensor place 50ml of .5N sodium hydroxide solution and 3.2 gm of naphthalene. Heat the mixture to boiling and add in portion with shaking 150ml of hot potassium permanganate solution (21gm dissolved in 150ml water). After the addition is complete continue heating the reaction mixture for 45-60 minutes. Add 5ml of alcohol, cool the resultant mixture and filter. Acidify the filtrate with 15ml concentrated hydrochloric acid and concentrate the filtrate to approximately 50ml. Cool and filter. Neutralise the filtrate by adding 15ml of 30% sodium hydroxide solution. Transfer the alkaline solution to porcelain dish, add 5gm of sodium bisulphite to it and evaporate it to dryness. Add 5-10ml of concentrated hydrochloric acid to the residue and again evaporate it to dryness. Add 10ml of benzene to the residue and heat it on water bath. Remove the benzene layer and repeat the process again using 10ml benzene. Combine the two benzene layers and distill off benzene on water bath. Collect the crude phthalaldehydic acid and recrystallise it from hot water, m.p 96°C.

(*x*) Oxidation of acenapthene to acenaphthenol

In a 250ml conical flask place 7.7gm of acenaphthene and 55ml of glacial acetic acid. Heat the solution to 60°C, remove the flask from heat source and then add with stirring and in portions 42gm of red lead (temperature below 60°C). After the addition is complete continue the stirring process for few minutes and then pour the dark syrupy solution into 500ml separating funnel containing 200ml water. Add 40ml ether to the seperating funnel and shake it thoroughly. Remove the organic layer and repeat the extraction twice each time using 40ml ether. Wash the combined ethereal layer with water and dry over anhydrous sodium sulphate. Distill off the ether on water bath and recover acenaphthenol acetate as yellow oil.

Dissolve the acetate in 15ml methanol contained in 100ml round bottom flask. Add 20ml of sodium hydroxide solution (2gm in 20ml water). Attach a reflux condensor and reflux the resultant mixture for 2hours. Cool the flask in ice bath and collect yellow acenaphthenol on buchner funnel, wash with cold water. Recrystallise from benzene, m.p 144-145°C.

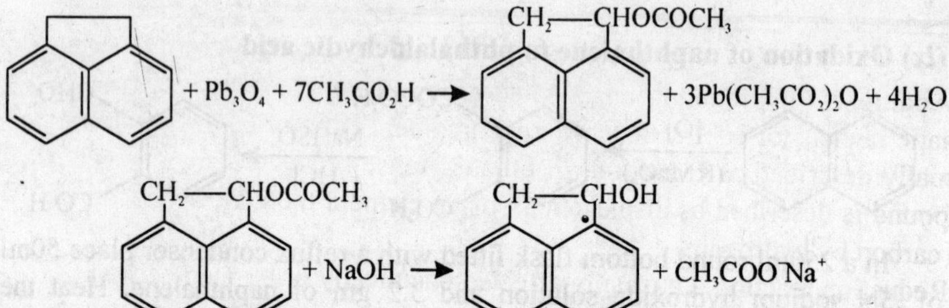

(xi) Oxidation of Benzoin to benzil

In a 100 ml round bottom flask fitted with a condensor place 4 gm of benzoin and 20 ml of concentrated nitric acid. Heat the flask in boiling water bath for 80-90 minutes or until the fumes of the oxides of nitrogen have subsided. Pour the reaction mixture in 200 ml beaker containing 40gm crushed ice. Stir the mixture well so that the oil crystallises in the form of solid. Filter the crude benzil through brehner funnel and recrystallise from alcohol, m.p 95-96°C (2-3gm)

$$C_6H_5\underset{|}{\overset{OH}{C}}H-\overset{O}{\overset{\|}{C}}-C_6H_5 \xrightarrow{[O]} C_6H_5-\overset{O}{\overset{\|}{C}}-\overset{O}{\overset{\|}{C}}-C_6H_5$$

7

Reduction Reaction

Addition of hydrogen to unsaturated groups such as olefins, carbonyls, aromatic nucleus or subtraction of oxygen content from organic compound is usually described as reduction. In broader sense reduction of an organic compound is described as displacement of an element more electronegative than carbon by hydrogen.

Reduction reaction are usually effected chemically or by the addition of molecular hydrogen to one or more unsaturated groups under the influence of a catalyst. The catalysts more commonly used for hydrogenation process are platinum, palladium, nickel and copper chromite.

$$R-\overset{\overset{O}{\|}}{C}-R' \xrightarrow{[2H]} R-\underset{\underset{OH}{|}}{\overset{\overset{H}{|}}{C}}-R' \quad \text{Increase in number of hydrogen}$$

$$R-\overset{\overset{O}{\|}}{C}-OH \xrightarrow{[H]} R-\overset{\overset{O}{\|}}{C}-H \quad \text{Decrease in number of oxygen}$$

$$R-X \xrightarrow{[H]} R-H \quad \text{Removal of more electronegative element halogen by hydrogen}$$

Chemical methods of reduction are usually classified into two groups :

(i) Some reactions take place first by addition of electron to the unsaturated compound followed or accompanied by transfer of proton. In these reactions alkali metals like calcium, magnesium, zinc, tin and iron are used as a source of electron whereas the proton donor may be either an alcohol, acid or water.

$$(C_6H_5)_2C=O \xrightarrow[C_2H_5OH]{Zn,\ NaOH} C_6H_5-\underset{\underset{OH}{|}}{CH}-C_6H_5$$

$$C_6H_{13}CHO \xrightarrow{Fe,\ CH_3COOH} C_6H_{13}CH_2OH$$

$$C_6H_5NO_2 \xrightarrow{Sn,\ HCl} C_6H_5NH_2$$

$$C_6H_{11}COCH_3 \longrightarrow C_6H_{11}\underset{\displaystyle OH}{\underset{|}{C}H} - CH_3$$

Mechanism :

Reduction with metal and acid of nitro compounds :- In acidic medium the final product is amine. The metal furnishes electron whereas acid provides proton.

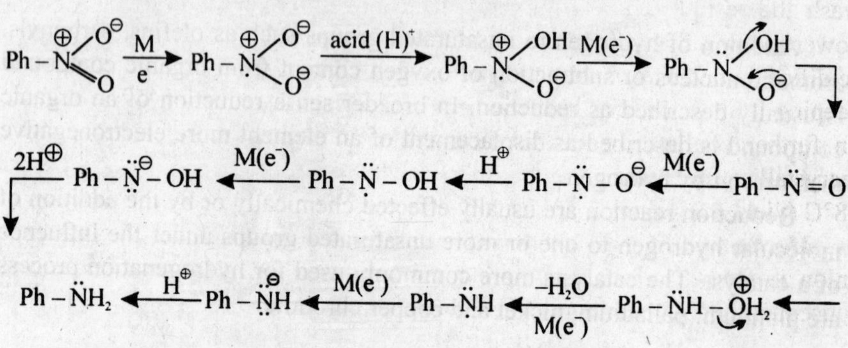

(*ii*) Included in the second category are reactions which proceed by transfer of hydride. Lithium aluminium hydride, sodium boro hydride and aluminium isopropoxide are some of the metal hydrides which are commonly employed as reducing agents or hydride transfer agents. In these reactions a hydride ion (a proton with a pair of electron) H:⁻ usually add across the carbonyl group generating an enolate ion. Protonation of the enolate ion gives the product.

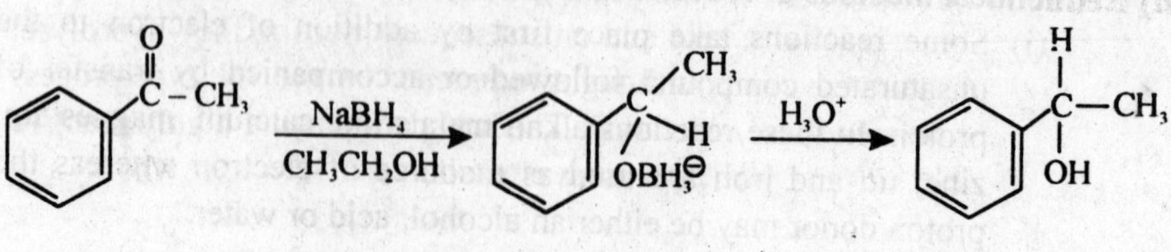

$$CH_2 = CH - CH = CHO \xrightarrow{LiAlH_4} CH_2 = CH - CH = CH - CH_2OH$$

(i) Reduction of benzophenone to benzhydrol (diphenyl methanol)

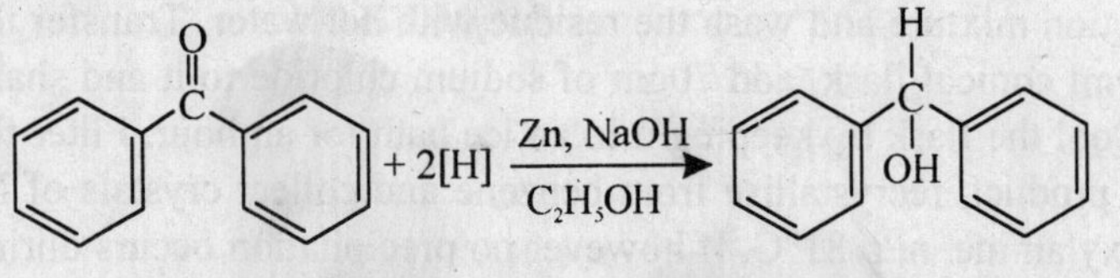

Place 5gm of benzophenone, 55ml of 95% ethanol and 5gm of sodium hydroxide in a 100ml round bottom flask fitted with a reflux condensor. Stir and heat the mixture to boiling for 20-25 minutes or until it turns dark brown. To the dark brown solution add in portions 5gm of zinc dust. After the addition is complete heat the solution until the brown colour disappears (usually 2-3hrs). Cool the flask, filter the reaction mixture by suction and wash the residue twice with 5ml of ethanol (do not allow residual zinc power to dry). Pour the filtrate in a 200ml beaker containing ice cold water acidified with 25ml of concentrated hydrochloric acid, cool the beaker by keeping it in an ice bath. Benzhydrol first separates out as a viscous oil but on further cooling it soon solidifies. Filter the solid on buchner funnel, recrystallise from ethanol and collect benzhydrol as colourless crystals, m.p 68°C (yield 4.0-4.5gm)

Mechanism: In the reduction of benzophenone first a resonance stabilised anion is formed which on protonation forms a radical A. The radical is subsequently converted to anion followed by the formation of benzhydrol.

$$C_6H_5-\overset{\overset{O}{\|}}{C}-C_6H_5 \xrightarrow{Zn[2e^-]} C_6H_5-\overset{\overset{\ddot{O}:}{|}}{\underset{\ominus}{\ddot{C}}}-C_6H_5 \xrightarrow[{[H^+]}]{C_2H_5OH} C_6H_5-\overset{\overset{:\dot{O}:}{|}}{\underset{A}{CH}}-C_6H_5 \xrightarrow{Zn[2e^-]}$$

$$C_6H_5-\overset{\overset{:\ddot{O}:^{\ominus}}{|}}{CH}-C_6H_5 \xrightarrow{H_3O^{\oplus}} C_6H_5-\overset{\overset{OH}{|}}{CH}-C_6H_5$$

(*ii*) Reduction of nitrobenzene to N-phenyl hydroxylamine

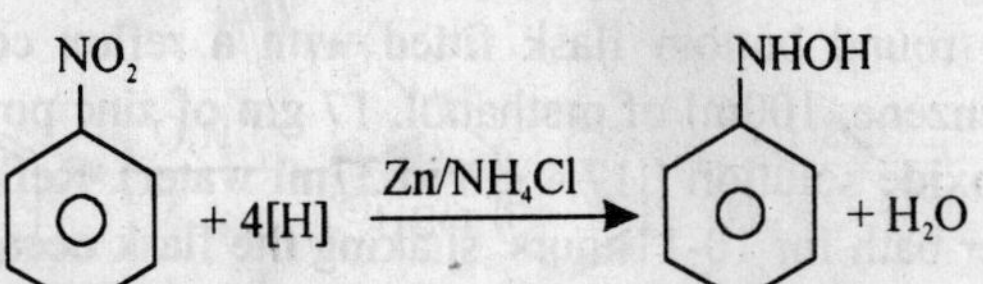

In a 500ml beaker place 10ml of nitrobenzene, 6gm of ammonium chloride and 200ml of water. Stir the mixture and add in portions 15gm of zinc powder taking care that the temperature does not rise above 60°C.

After the addition is complete continue stirring for another 20 minutes. Filter the reaction mixture and wash the residue with hot water. Transfer the filtrate to 500ml conical flask, add 70gm of sodium chloride to it and shake thoroughly. Cool the flask by keeping it in an ice bath for an hour. Filter the yellow crude product, recrystallise from benzene and collect crystals of N-phenyl hydroxylamine, m.p 81°C. If however no precipitation occurs during the cooling process transfer the filtrate to 500ml separating funnel and add 20ml ether to it. Shake the funnel and remove the ethereal layer. Repeat the process again using 20ml ether. Combine the two organic layers and evaporate off the ether on water bath by simple distillation and collect N-phenyl hydroxylamine.

(*iii*) Reduction of 2-nitro fluorene to 2-amino fluorene

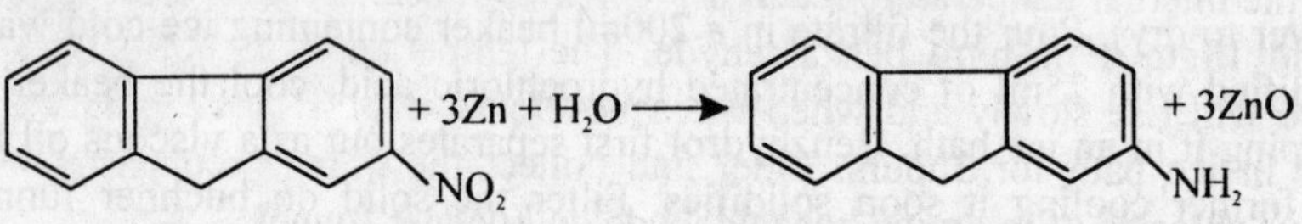

Reflux together a mixture of 3gm of 2-nitro fluorene, 100ml of 78% alcohol (82ml alcohol and 18ml water), 2ml of calcium chloride solution (1gm in 2ml water) and 30 gm of zinc dust placed in a 250ml round bottom flask fitted with a reflux condensor for 2 hours. Filter the solution while still hot and wash the residue with 10ml boiling 78% alcohol. Combine the two filtrates and pour it into a 500ml beaker containing 100gm of crushed ice and then leave for 20-30minutes. Collect the precipitated 2-amino fluorene, recrystallise from ethanol, m.p 127°C (yield 2-2.2gm)

(*iv*) Reduction of nitrobenzene to hydrazobenzene (sym diphenyl hydrazine)

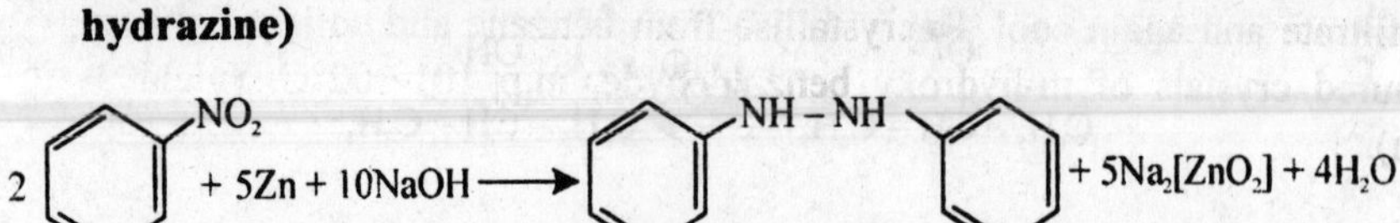

In a 250ml round bottom flask fitted with a reflux condensor place 8.5ml of nitrobenzene, 100ml of methanol, 17 gm of zinc powder and 37ml of sodium hydroxide solution (17.4 gm in 37ml water). Reflux the reaction mixture on water bath for 10-11hours, shaking the flask occasionally during this period. The colour of solution gradually changes from reddish to pale yellow. Filter the solution while still hot through preheated metallic funnel and wash the precipitate with 10-15ml of hot methanol. Transfer the filtrate to 250ml conical flask, cork it and place it in an ice bath for 60-90 minutes.

After the crystallisation is complete, collect the colourless crystals of hydrazobenzene on buchner funnel, wash with 50% methanol and dry, m.p 125-126°C.

Note : Crystals of hydrazobenzene should be placed in a vaccuum desicator

(*v*) Reduction of m-nitro benzaldehyde to m-hydroxy benzaldehyde

$$m-NO_2C_6H_4CHO \xrightarrow{SnCl_2,\ HCl} m-NH_2C_6H_4CHO\ .\ (Sn\ Cl_4)_n \xrightarrow{NaNO_2,\ HCl}$$

$$(m-ClC_6H_4CHO)_2\ .\ SnCl_4 \xrightarrow[heat]{H_2O} m-HOC_6H_4CHO$$

Cool 60ml of stannous chloride solution (45gm in 60ml concentrated hydrochloric acid) placed in 250ml beaker by keeping it in an ice bath. When the internal temperature reaches 2°C remove the beaker from ice bath and add 10gm of m-nitro benzaldehyde. The temperature of the reaction mixture will rise slowly and when it reaches 95°C, cool it by keeping the beaker in ice bath for 3hours. Filter and collect the orange red paste of stannous chloride of m-amino benzaldehyde on buchner funnel.

In a 250ml beaker dissolve the salt of m-amino benzaldehyde in 60ml concentrated hydrochloric acid and place the beaker in ice bath. Now add dropwise with stirring a 15ml solution of sodium nitrite (4.6gm in 15ml water) taking care that the temperature does not rise above 5°C. After the addition is complete continue stirring for 60 minutes and then collect the diazonium salt on buchner funnel.

Add immediately and in portions the diazonium salt to 160ml of boiling water placed in 500ml beaker. After the addition is complete filter and keep the filtrate for 14-15hours in ice cold water. If the crystals of m-hydroxy benzaldehyde appear then collect it on buchner funnel otherwise concentrate the filtrate and again cool. Recrystallise from benzene and collect light tan coloured crystals of m-hydroxy benzaldehyde, m.p 101-102°C. (yield 3-4gm).

(*vi*) Reduction of anthraquinone to anthrone

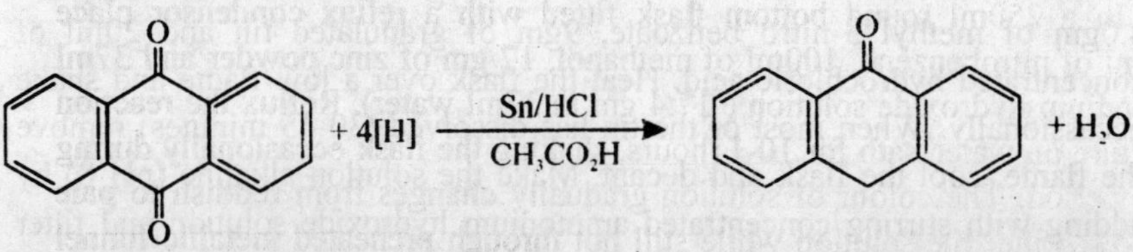

Reflux together a mixture of 5gm of anthraquinone, 5gm of granulated tin and 38ml of glacial acetic acid placed in a 200ml round bottom flask fitted with a reflux condensor. Add slowly and in portion 13ml of concentrated hydrochloric acid. After the addition is complete continue refluxing for another 45 minutes. Filter the reaction mixture and add 10ml of ice cold water to the filtrate. Cool the filtrate in an ice bath whereby anthrone will be precipitated. Filter the crude product on buchner funnel and wash with cold water. Recrystallise from a mixture of benzene: hexane (3:1) and obtain pure anthrone, m.p 155°C. (Yield 3-3.5gm).

Mechanism : The reaction proceeds leading to the generation of 10 hydroxy anthracene also known as anthranol, but it is unstable and soon isomerises to more stable anthrone (keto-enol tautomerism)

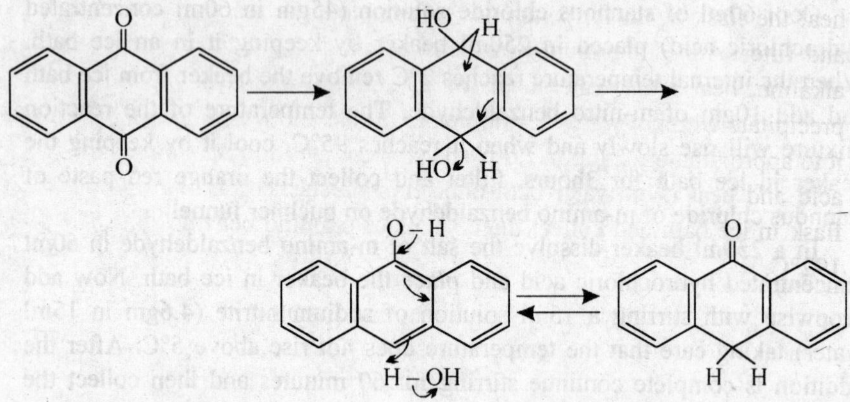

(*vii*) Reduction and hydrolysis of 3-nitro methyl benzoate to 3-amino benzoic acid

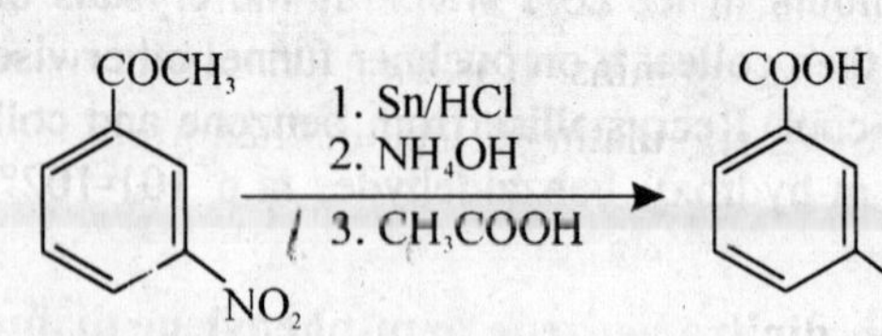

In a 100ml round bottom flask fitted with a reflux condensor place 4.0gm of methyl 3-nitro benzoate, 9gm of granulated tin and 20ml of concentrated hydrochloric acid. Heat the flask over a low flame and shake occassionally. When most of the tin has dissolved (30-45 minutes) remove the flame, cool the flask and decant. Make the solution alkaline (pH 8) by adding with stirring concentrated ammonium hydroxide solution and filter again. Acidify the clear filtrate with glacial acetic acid to pH 5, cool in ice and allow the acid to precipitate. If no product separates concentrate the solution and then cool. Collect the precipitated solid by suction filteration and wash with cold water. Recrystallise from alcohol and obtain 3-amino benzoic acid as white granular crystals, m.p 174°C. (yield 2-2.5gm)

(*viii*) Reduction of p-nitro benzoic acid to p-amino benzoic acid

In a 500ml round bottom flask place 5gm of p-nitro benzoic acid, 12 gm of powdered tin and 25ml of concentrated hydrochloric acid. Attach the condensor to the flask and heat it over a low flame and when the reaction commences remove the heat source. Shake the flask gently for 20 minutes so that all the tin dissolves and a clear solution is obtained. But if not then heat the flask again for 5 minutes to complete the reaction. Cool the flask and filter. Add ammonium hydroxide solution to the filtrate until it is just alkaline, heat it on water bath for 25 minutes and then filter. Wash the precipitate with hot water. Combine the filtrate and washing and concentrate it to approximately 25ml. Acidify the concentrated filtrate with glacial acetic acid and heat it on water bath until crystals start separating out. Cool the flask in ice bath and collect the crystals of p-amino benzoic acid, m.p 190-192°C.

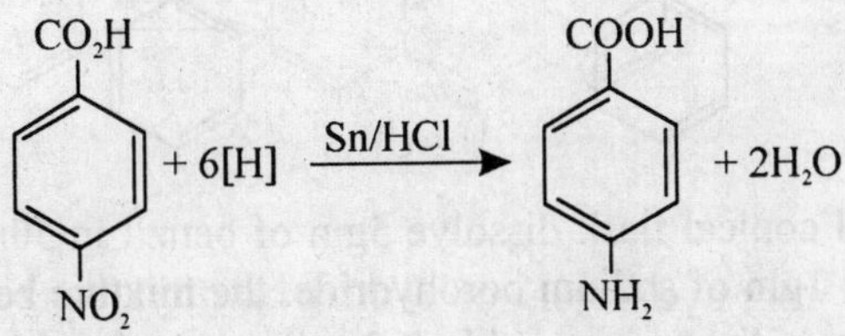

Note :- p-Amino benzoic acid is nutritionally very important and is often classed among the vitamins. It is thought to play an essential role in microbial nutrition by its stimulating effect on vitamin synthesis by intestinal microflora.

(*ix*) Reduction of m-dinitro benzene to m-phenylene diamine

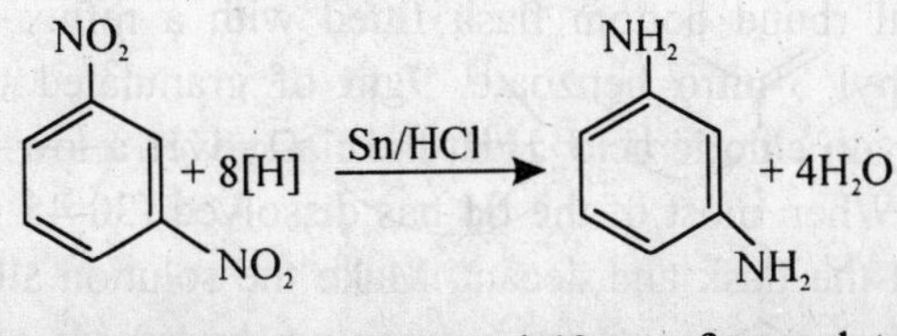

Place 10gm of m-dinitro benzene and 40gm of granulated tin in a 250ml round bottom flask. Add in portions 80ml of concentrated hydrochloric acid to the flask. When all the acid has been added attach a condensor to the flask and heat it for 60-70 minutes. Dilute the reaction mixture with 100ml of water and pass H_2S gas in order to precipitate tin as tin sulphide. Keep the reaction mixture overnight and filter the precipitated residue. Add sodium hydroxide solution to the filtrate until alkaline and transfer the alkaline solution to 250ml separating funnel. Add 20ml ether to the funnel and shake vigorously. Remove the ether extract and repeat the process twice each time using 20ml ether. Combine the three ether extracts and dry over anhydrous sodium sulphate. Distill off the ether on water bath and cool the residual solution in porcelain dish. Crystals of m-phenylene diamine soon precipitate out, m.p 63°C. (Yield 5-5.5gm)

(*x*) Reduction of benzil to mesohydrobenzoin

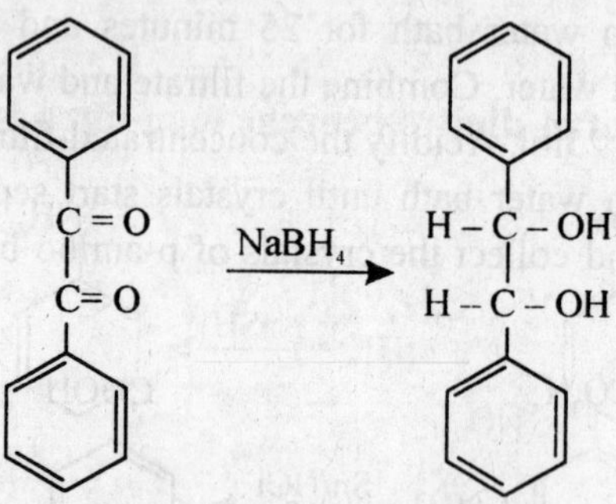

In a 100ml conical flask dissolve 5gm of benzil in 50ml of ethanol and cool. Then add 1gm of sodium borohydride, the mixture becomes warm and the yellow colour disappears within 2-3 minutes. After 10 minutes add 5ml of water to the reaction mixture and warm the conical flask on water bath so that the mixture begins to boil. Cool the flask in ice bath and collect mesohydrobenzoin on buchner funnel. Recrystallise from alcohol, m.p 136-137°C (Yield 3.0-3.5gm)

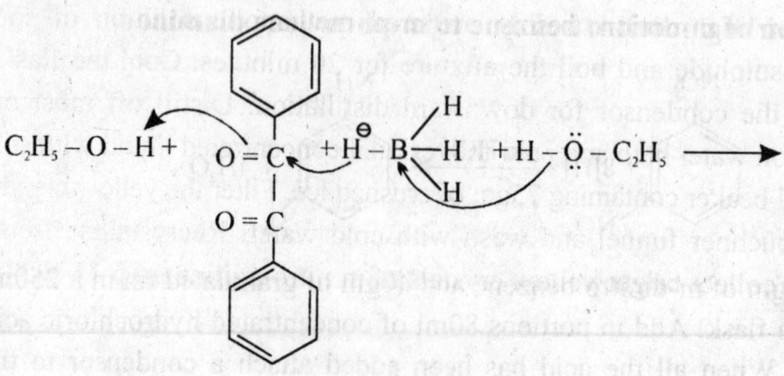

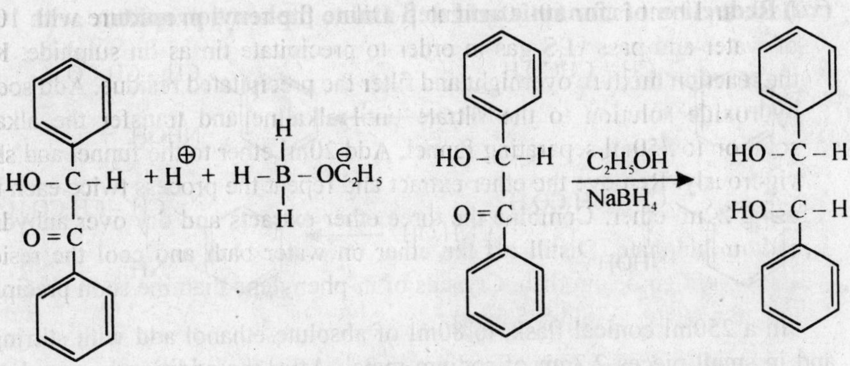

(*xi*) Reduction of m-dinitro benzene to m-nitro aniline

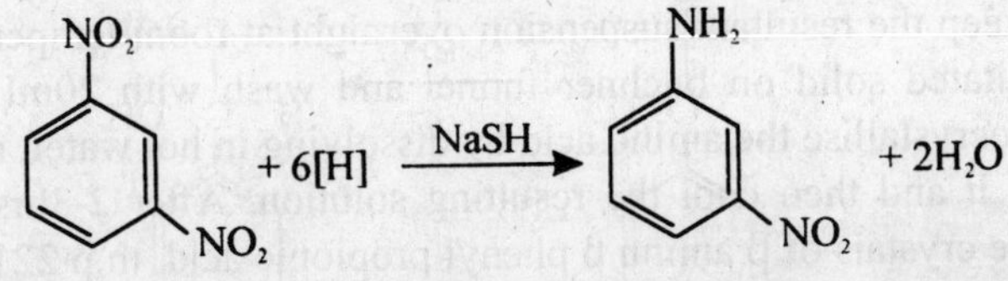

In a 200ml beaker dissolve 10gm of sodium sulphide in 25ml of water and add in portions 3gm of sodium bicarbonate to it. After the addition of bicarbonate is complete add 30ml of methanol and cool the resulting solution to 15°C. Filter off the precipated sodium carbonate and wash with 5ml methanol. Retain the filtrate and washings as it contains the reducing agent sodium hydrogen sulphide.

Dissolve 5gm of m-dinitro benzene in 30ml of methanol in a 250ml round bottom flask fitted with a reflux condensor, heat if necessary. Add with a swirling motion freshly prepared methanolic solution of sodium hydrogen sulphide and boil the mixture for 20 minutes. Cool the flask and rearrange the condensor for downward distillation. Distill off most of the methanol on water bath and pour the residual concentrated liquid with stirring into 200ml beaker containing 75gm of crushed ice. Filter the yellow crystalline solid on buchner funnel and wash with cold water. Recrystallise from hot water and collect bright yellow crystals of m nitro-aniline, m.p 114°C (yield 2gm).

(*xii*) Reduction of cinnamic acid to β amino β phenyl propionic acid

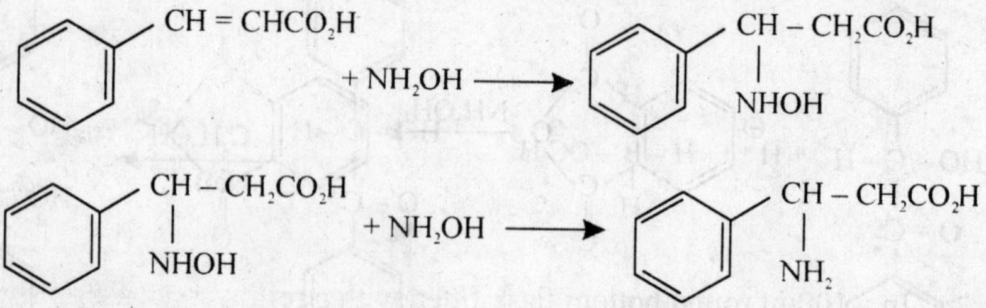

In a 250ml conical flask, to 80ml of absolute ethanol add with stirring and in small pieces 2.3gm of sodium metal. After the addition is complete continue stirring for 10 minutes and then add with stirring a solution of 7gm of hydroxylamine hydrochloride in 5ml water. Cool the resulting suspension by keeping the flask in ice bath. Filter and transfer the filtrate to a 250ml round bottom flask fitted with a reflux condensor and add 7.5 gm of cinnamic acid to it. Reflux the reaction mixture on steam bath/water bath for 9 hours and then keep the resultant suspension overnight at room temperature. Filter the precipitated solid on buchner funnel and wash with 20ml of absolute ethanol. Recrystallise the amino acid by dissolving in hot water, add absolute ethanol to it and then cool the resulting solution. After 2-3hrs collect the snow white crystals of β amino β phenyl propionic acid, m.p 221°C(d).

(*xiii*) Reduction of Benzilic acid to diphenyl acetic acid

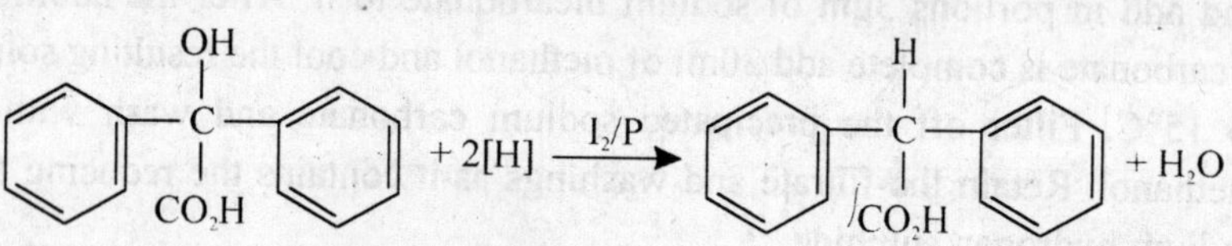

In a 250 round bottom flask fitted with a reflux condensor place 25ml of glacial acetic acid, 2 gm of phosphorous and 1.0 gm of iodine. Keep the mixture at room temperature for 20-25 minutes so that the iodine completely dissolves and then add 1ml of water and 10gm of benzilic acid to it. Reflux the mixture for 2 ½ hour and then filter while still hot. Pour the hot filtrate with stirring into 200ml beaker containing 100ml of cold sodium bisulphite solution. Diphenyl acetic acid gets precipitated as white or yellow coloured powder. Filter the product on [illegible]ner [illegible]unnel and wash with cold water. Recrystallise from 50% ethanol [illegible]nd coll[illegible]t crystals of diphenyl acetic acid, m.p 144-145°C. (yield 7-7.5gm)

Note :- Diphenyl acetic acid is used as an intermediate in the synthesis of powerful analgesic methadone

(*xiv*) Reduction of phthalic anhydride to phthalimide

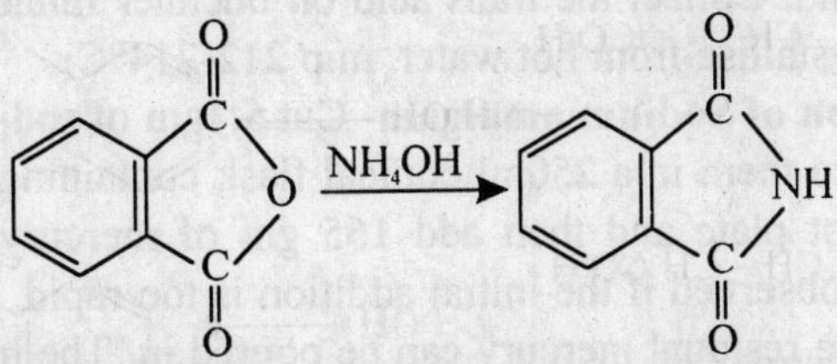

In a 100ml round bottom flask fitted with a reflux condensor place 10gm of phthalic anhydride and add to it 12ml of concentrated ammonia solution. Heat the flask on wire gauze for 2 hours with occasional shaking. During this period the mixture forms a homogenous paste as most of the water is evaporated. Pour the contents of the flask while still hot in a porcelain dish. Cool and grind the solidified mass in form of fine powder. Recrystallise from ethanol and collect pure crystals of phthalimide, m.p 233-234°C.

Mechanism

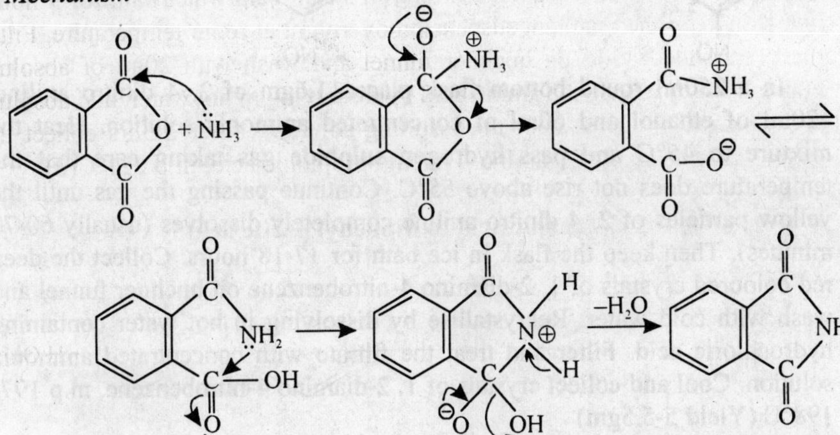

(*xv*) Reduction of phthalic acid to trans 1, 2 dihydro phthalic acid

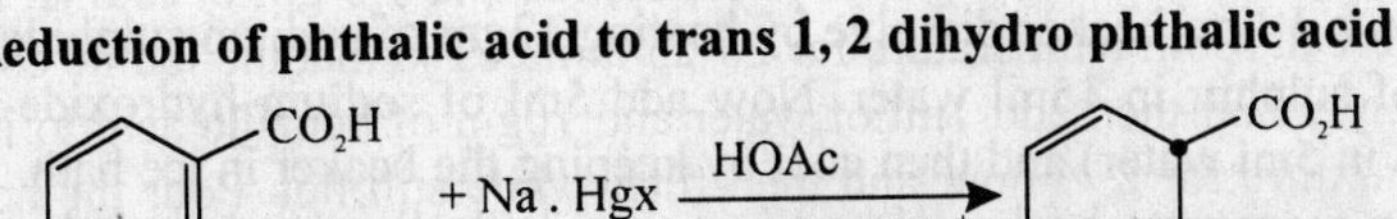

In a 500ml beaker dissolve by stirring 17gm of phthalic acid and 28gm of sodium acetate in 170ml of water. After a clear solution is obtained place the beaker in an ice bath. Add in portion 340 gm of sodium amalgam. After each addition of sodium amalgam add 2-3ml of 50% acetic acid. After the addition is complete decant the solution from mercury through celite (silica gel can also be used). Cool the filtrate and add 170ml of cold 20% sulphuric acid. Allow the crystallisation to take place by keeping the solution at 10-15°C for 5 hours. Collect the trans acid on buchner funnel, wash with cold water and recrystallise from hot water, m.p 212-214°C.

Preparation of Sodium amalgam- Cut 5.1gm of sodium metal in small pieces and place them in a 250ml conical flask containing mineral oil. Heat the flask on hot plate and then add 165 gm of mercury in portions. Fire flashes can be observed if the initial addition is too rapid. Once the reaction has subsided the residual mercury can be poured in. The molten amalgam is then poured in a shallow metallic pan. Once cool the amalgam is washed with pet ether to remove the mineral oil.

(*xvi*) Reduction of 2, 4-dinitro aniline to 1, 2-diamino 4-nitrobenzene

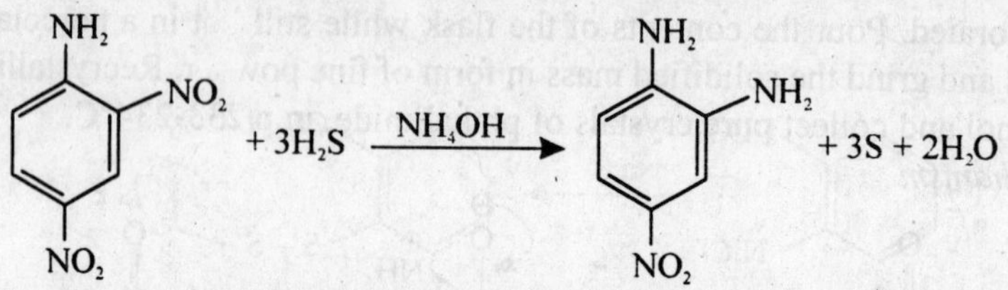

In a 250ml round bottom flask place 11.6gm of 2, 4 dinitro aniline, 120ml of ethanol and 60ml of concentrated ammonia solution. Heat the mixture to 45°C and pass hydrogen sulphide gas taking care that the temperature does not rise above 55°C. Continue passing the gas until the yellow particles of 2, 4 dinitro aniline completely dissolves (usually 60-70 minutes). Then keep the flask in ice bath for 17-18 hours. Collect the deep red coloured crystals of 1, 2-diamino 4-nitrobenzene on buchner funnel and wash with cold water. Recrystallise by dissolving in hot water containing hydrochloric acid. Filter and treat the filtrate with concentrated ammonia solution. Cool and collect crystals of 1, 2-diamino 4-nitrobenzene, m.p 197-198°C (Yield 5-5.5gm)

(*xvii*) Reduction of Anthranilic acid to Thio Salicylic acid

In a 1 litre beaker dissolve by heating 13gm of sodium sulphide and 1.7 gm of sulphur in 15ml water. Now add 5ml of sodium hydroxide solution (2gm in 5ml water) and then cool by keeping the beaker in ice bath.

In a separate beaker placed in an ice bath dissolve by stirring 7gm of anthranilic acid in 25ml water and 10ml concentrated hydrochloric acid. Now add with stirring 14ml of ice cooled sodium nitrite solution (3.5gm of sodium nitrite in 14ml water) taking care that the temperature does not rise above 5°C. Now add this cold diazonium salt solution to alkaline cold sulphide solution in portions taking care that the temperature does not rise above 5°C. After the addition is complete remove the beaker from the ice bath and keep it at room temperature for 2 hours. Add 9ml of concentrated hydrochloric acid or until the solution is just acidic. Cool the solution and collect precipated dithiosalicylic acid on buchner funnel and wash with water.

In a 100ml round bottom flask fitted with a reflux condensor place the dithiosalicylic acid, 1.35 gm of zinc dust and 15ml of glacial acetic acid. Reflux the mixture for 4 hours. Cool and filter the precipitated solid. Dissolve the solid in 10ml water contained in 100ml beaker and heat it to boiling. Add sodium hydroxide solution to the hot solution till alkaline and continue heating for 20 minutes. Filter the solution hot. Cool and add concentrated hydrochloric acid to the filtrate until it is acidic. Collect thiosalicylic acid on buchner funnel and recrystallise from alcohol, m.p 163-164°C (Yield 4-5gm)

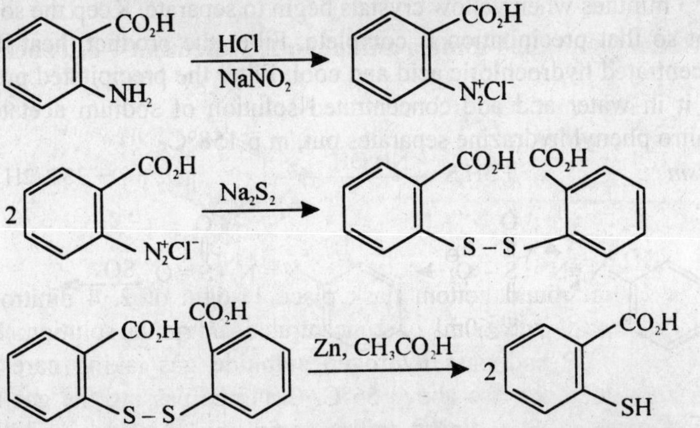

(*xviii*) Reduction of p-nitro aniline to p-nitro phenyl hydrazine

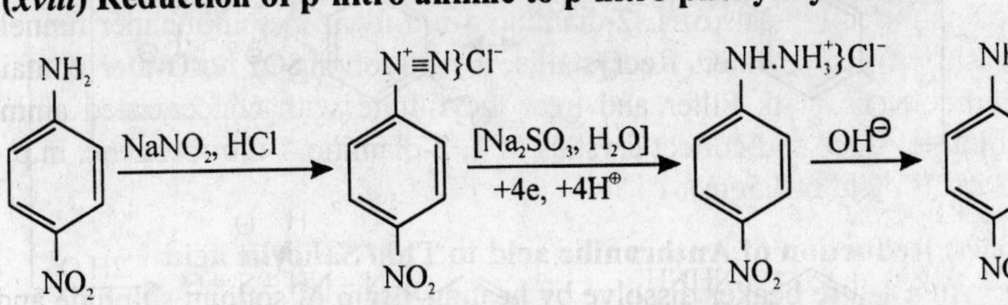

In a 250ml beaker dissolve 5gm of p-nitro aniline in a mixture of 10ml concentrated hydrochloric acid and 10ml water. Cool the beaker by keeping it in an ice bath (the hydrochloride of the base may be precipitated). When the temperature has fallen to 3°C slowly add with stirring and in portion 6ml of cold sodium nitrite solution (2.6gm of sodium nitrite in 6ml water) taking care that the temperature does not rise above 7°C during the diazotization process. After the addition is complete continue stirring for another 10 minutes and filter.

In the meantime prepare sodium sulphite solution by first dissolving 15gm of sodium hydroxide in 150ml water in a 500ml beaker. To the sodium hydroxide solution add 35 gm of sodium metabisulphite solution and stir so that the solid completely dissolves. Cool the solution to 25°C and add few drops of phenolphthalein indicator. Add a small portion of sodium meta bisulphite so that the pink colour just disappears and then add 3.5 gm of sodium meta bisulphite. Continue stirring and cool the solution to 5°C by first keeping the beaker in an ice bath and then add 25gm of crushed ice to it.

After the sodium sulphite has been synthesized add to it the cold filtered diazonium salt solution slowly and with stirring. After the addition is complete keep the solution at room temperature for 10 minutes and then acidify with concentrated hydrochloric acid (7ml). Heat the beaker on water bath at 25-30°C for 3 minutes when yellow crystals begin to separate. Keep the solution overnight so that precipitation is complete. Filter the product, heat it with 5ml concentrated hydrochloric acid and cool. Filter the precipitated product, dissolve it in water and add concentrated solution of sodium acetate and cool, p-nitro phenyl hydrazine separates out, m.p 158°C.

Mechanism

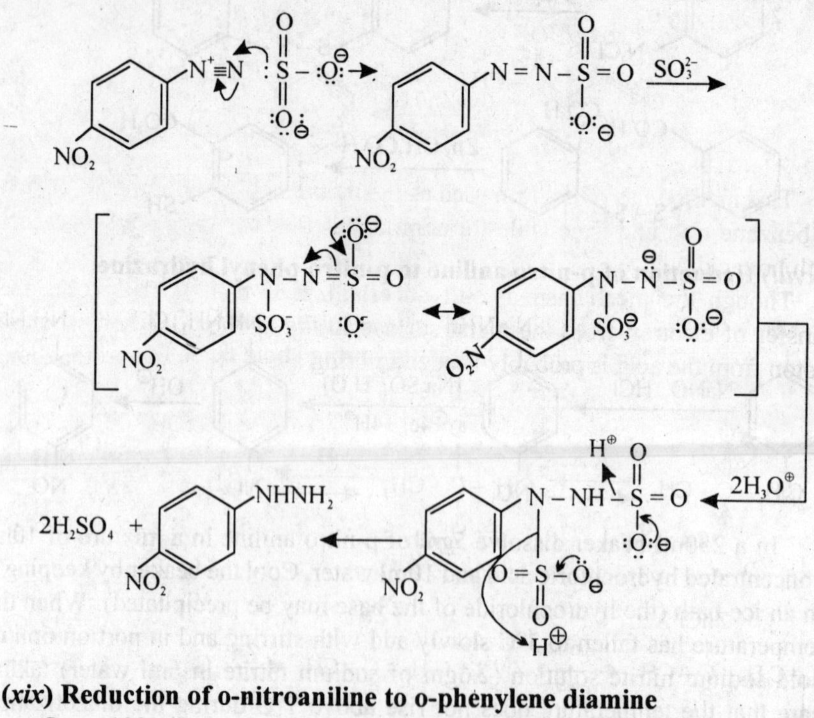

(*xix*) Reduction of o-nitroaniline to o-phenylene diamine

In a 100ml round bottom flask fitted with a reflux condensor place 9.2gm of o-nitroaniline, 5.5 ml of 20% sodium hydroxide solution and 35ml of ethanol. Heat the reaction mixture on water bath and as soon as it begins to boil remove the heat source and add in portion 19gm of zinc powder at such a rate that the mixture continues to reflux. After the addition of zinc dust is complete reflux the mixture for another one hour. Filter the mixture hot, wash the zinc residue with hot ethanol. Combine the filtrate and washing and concentrate under reduced pressure. Cool the concentrate in an ice bath. Filter the precipitated o-phenylene diamine, wash with cold water and dry, m.p 97-100°C. Recrystallise from hot water and obtain fine yellow crystals of o-phenytene diamine, m.p 99-101°C.

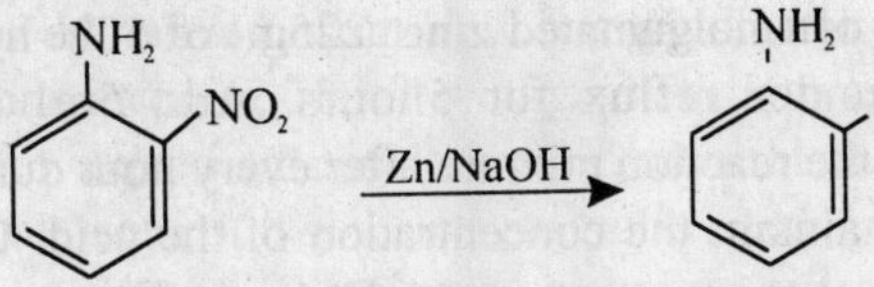

(A) CLEMMENSEN REDUCTION

Reduction of the carbonyl group of aldehydes and ketones to methyl and methylene in the presence of amalgamated zinc and hydrochloric acid is known as clemmensen reduction.

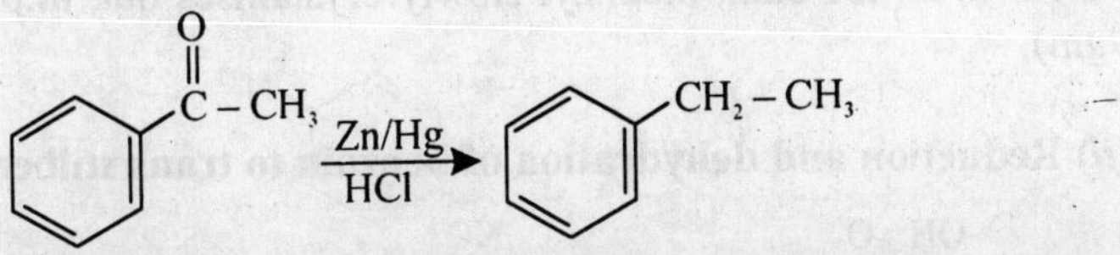

This method has often been used as the indirect method for the alkylation of benzene ring and is suitable for compounds that can withstand treatment with hot acid.

Though the mechanism is still uncertain, it is thought to involve the transfer of electron from the metal surface to the carbonyl carbon whereas proton from the acid is probably needed to bring about the initial protonation.

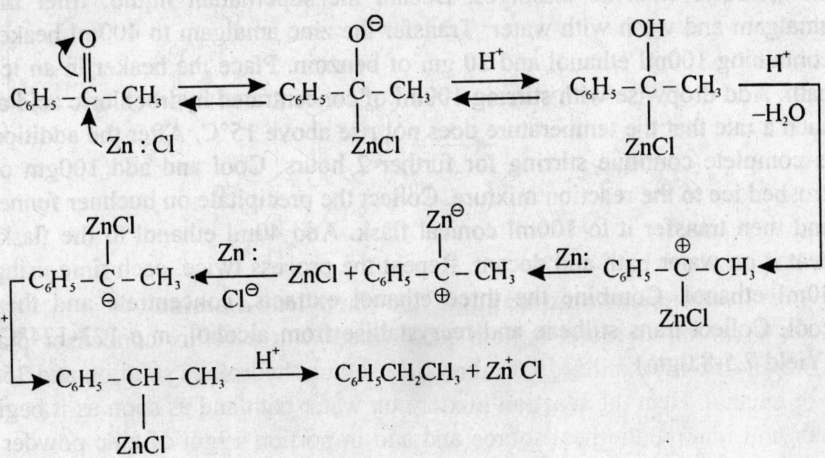

(i) Reduction of benzil to dibenzyl

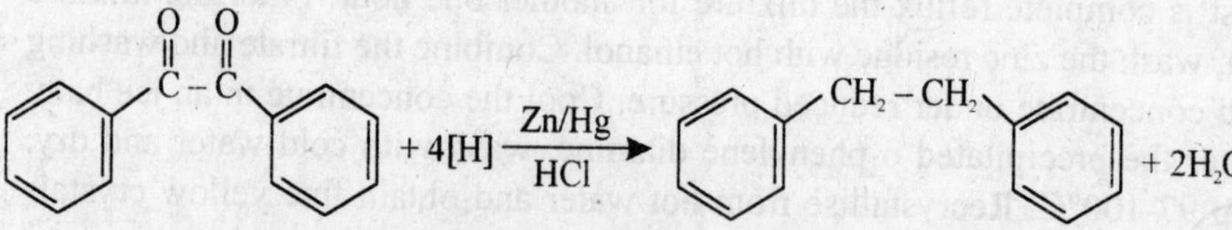

In a 250ml round bottom flask fitted with a reflux condensor place 15 gm of benzil, 60gm of amalgamated zinc, 225ml of 50% hydrochloric acid. Boil the mixture under reflux for 5hours. Add 5ml of concentrated hydrochloric acid to the reaction mixture after every hour during the refluxing period in order to maintain the concentration of the acid. Cool the mixture and allow it to settle at room temperature. Decant off the unreacted zinc and transfer the filtrate to 250ml separating funnel. Add 25ml ether to the funnel and shake vigorously. Remove the ethereal layer and repeat the process twice, each time using 25ml of ether. Combine the three ethereal layers and dry over anhydrous sodium sulphate. Concentrate the dried ether extract and cool it in an ice bath, bibenzyl slowly crystallises out, m.p 52°C (Yield, 8-9gm)

(*ii*) Reduction and dehydration of benzoin to trans stilbene

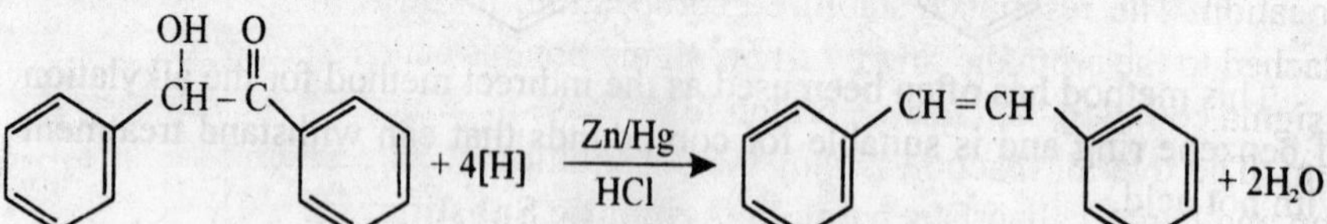

In a 200ml beaker place 10gm of mercuric chloride and 100ml of water. Now add with stirring 40gm of zinc dust. Continue the stirring process until all mercuric chloride dissolves. Decant the supernatant liquid, filter the amalgam and wash with water. Transfer the zinc amalgam to 400ml beaker containing 100ml ethanol and 20 gm of benzoin. Place the beaker in an ice bath. Add dropwise with stirring 100ml of concentrated hydrochloric acid at such a rate that the temperature does not rise above 15°C. After the addition is complete continue stirring for further 2 hours. Cool and add 100gm of crushed ice to the reaction mixture. Collect the precipitate on buchner funnel and then transfer it to 100ml conical flask. Add 40ml ethanol to the flask, heat it on water bath and decant. Repeat the process twice, each time using 40ml ethanol. Combine the three ethanol extracts, concentrate and then cool. Collect trans stilbene and recrystallise from alcohol, m.p 123-124°C. (Yield 7.5-8.0gm)

8

Electrophilic Aromatic Substitution

Aromatic system such as benzene, naphthalene, anthracene, thiophene and others that contain double bonds and in analogous with the alkenes are basic and hence react with electrophiles.

The π electrons of the alkenes are less tightly held and they attack an electrophile to give a carbocation. The electrophilic carbocation is then attacked by nucleophile to give addition product. In aromatic system though the π electron cloud are involved in the stable aromatic system, they are still available to attack electrophilic species to give a resonance stabilized carbocation. The resonance stabilized carbocation in which the electrophile is attached to the aromatic ring by a new sigma bond is called sigma complex. The sigma complex then loses a proton to the base to regenerate the aromatic system. The overall reaction is thus the substitution of an electrophile for a proton and hence called **Electrophilic Aromatic Substitution.**

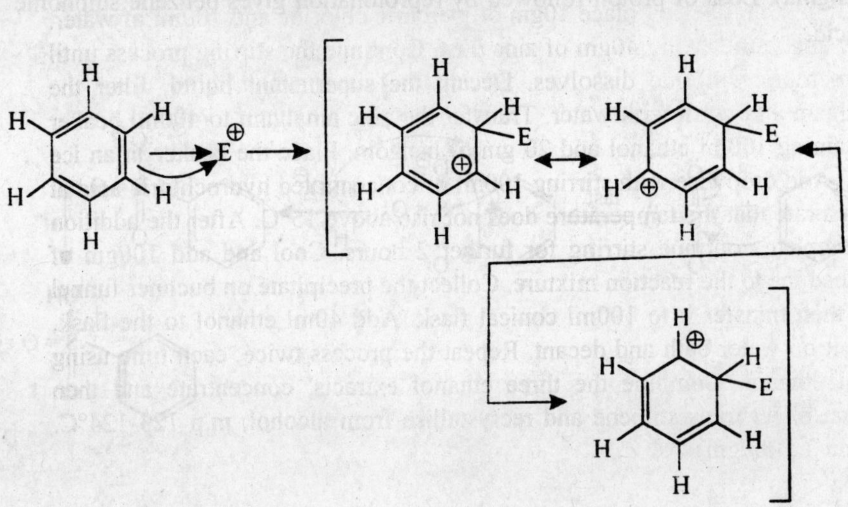

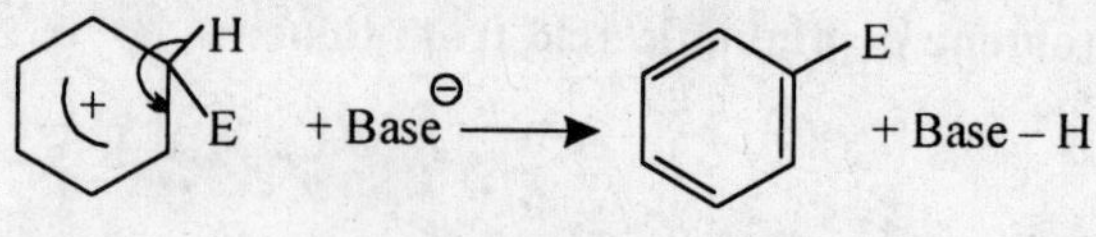

(A) SULPHONATION

Sulphonic acids which are strong acid catalyst are derived by the replacement of hydrogen from a hydrocarbon by univalent sulphonic group (-SO_3H). Alkane sulphonic acids which are usually colourless liquids are less common in comparison to aromatic sulphonic acids. The sodium salts of alkyl aryl sulphonic acids are more commonly being used as detergents. Arylsulphonated compounds often find use in the preparation of dyes and drugs.

Sulphonation is usually accomplished by using fuming sulphuric acid, concentrated sulphuric acid or chloro sulphonic acid. The electrophile in sulphonation is believed to be sulphur trioxide which has three sulphonyl bonds withdrawing electron density away from sulphur atom.

$$2H_2SO_4 \rightleftharpoons SO_3 + H_3O^+ + HSO_4^{\ominus}$$

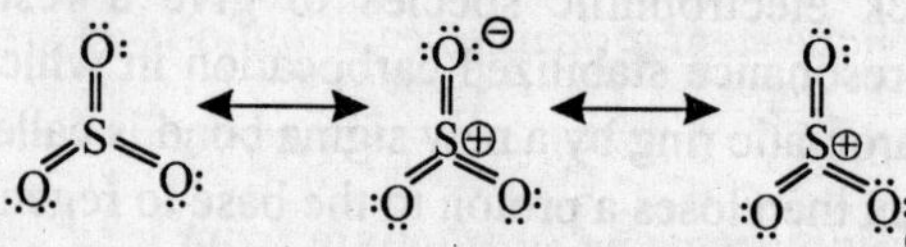

Aromatic ring attacks the sulphur trioxide forming intermediate complex (sigma). Loss of proton followed by reprotonation gives benzene sulphonic acid.

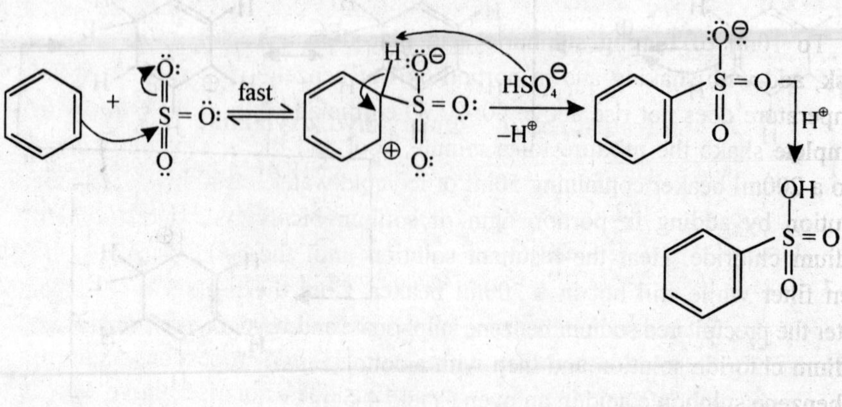

(*i*) Synthesis of toluene p-sulphonic acid from toluene

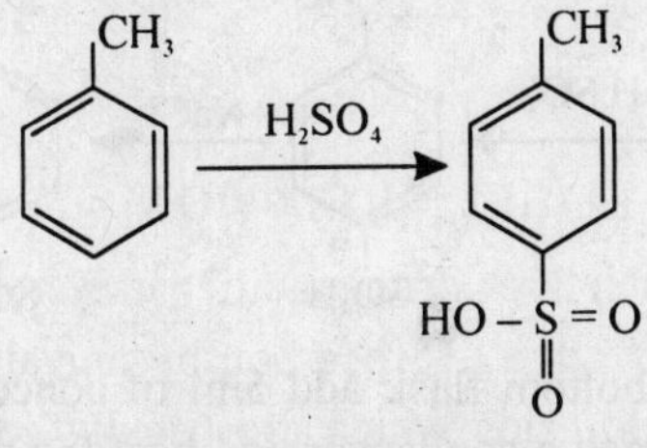

Reflux 20ml of toluene with 5ml of concentrated sulphuric acid placed in a 50ml round bottom flask with frequent shaking (to mix the two layers) for 2-3 minutes. Cool the flask in ice whereby the lower layer of the crude sulphonic acid will solidify. Decant off the excess toluene. Now add 20ml of concentrated hydrochloric acid and heat the flask gently until a clear solution is just obtained. Cool the flask in ice whereby toluene p-sulphonic acid along with deliquescent o-sulphonic acid gets precipitated. Filter off the crystals quickly on suction pump and repeat the crystallisation process again with 10ml of concentrated hydrochloric acid (keep overnight). Filter rapidly through filter pump (suction pump) which now gives sulphonic acid almost devoid of deliquescent properties (without o-isomer), m.p 106-107°C (Yield 7- 7.5gms)

(ii) Synthesis of sodium benzene sulphonate from benzene

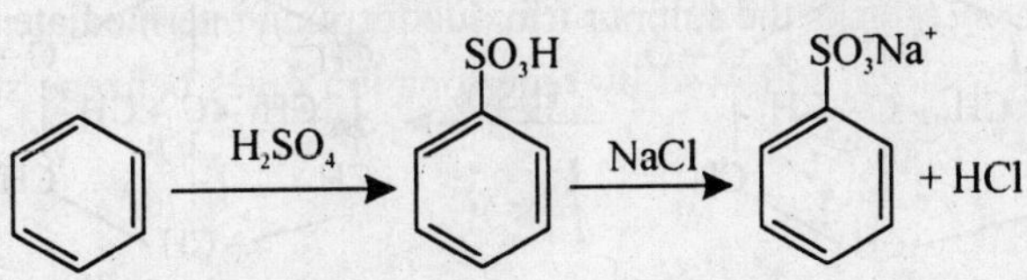

To 10ml of fuming sulphuric acid placed in a 100ml round bottom flask, add with shaking and in portion 6ml of benzene taking care that the temperature does not rise above 50°C. When the addition of the benzene is complete shake the mixture for a minute, cool and then pour with stirring into a 200ml beaker containing 50ml of ice cold water. Neutralise the acidic solution by adding in portion 6gm of sodium bicarbonate and 10gm of sodium chloride. Heat the resultant solution until the solid dissolves and then filter while still hot in a 200ml beaker. Cool the beaker in ice bath. Filter the precipitated sodium benzene sulphonate and wash first with saturated sodium chloride solution and then with alcohol. Finally dry the sodium salt of benzene sulphonic acid in an oven (Yield 4-5gm).

(iii) Synthesis of sodium toluene p-sulphonate from toluene

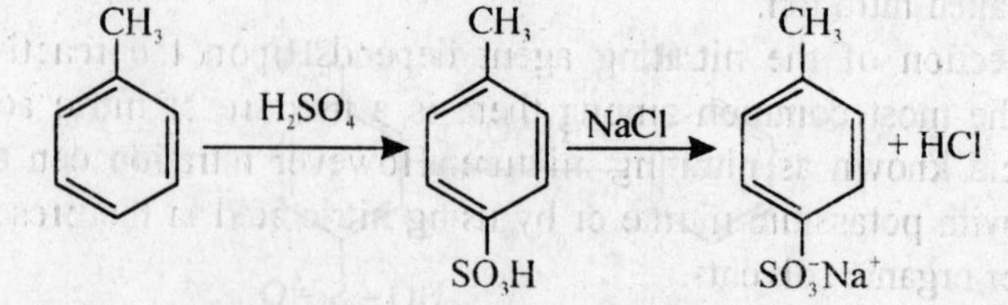

In a 100ml round bottom flask add 5ml of concentrated sulphuric acid to 10ml of toluene. Attach a condensor and heat the mixture with stirring in an oil bath for 60minutes (temperature between 110- 120°C). When the toluene layer has completely disappeared cool it to laboratory temperature and then pour with stirring into a 200ml beaker containing 30ml of ice cold water. Neutralise the acidic solution by adding in portion 4.5gm of sodium bicarbonate and 14gm of sodium chloride. Heat the solution to boiling and filter while still hot into a 200ml beaker. Cool the beaker by keeping it in an ice bath. Filter the precipitated sodium toluene p-sulphonate on buchner funnel, first wash with saturated sodium chloride solution and then with water. Dry the hydrated crystals in an oven (Yield 6 – 7gm)

(*iv*) Synthesis of D,L-10 Camphor Sulphonic acid (Reychler's Acid) from camphor

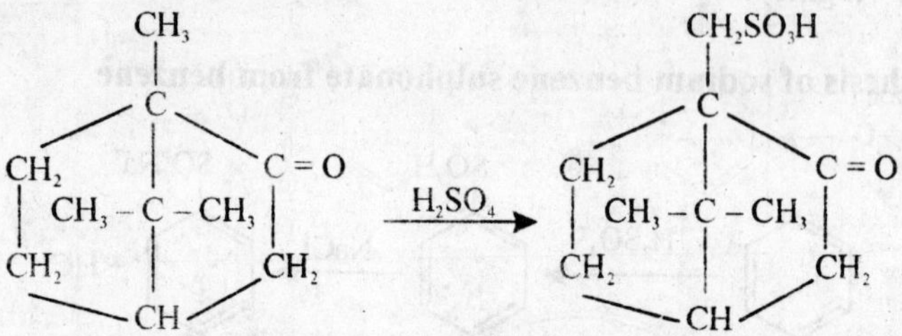

In a 100ml conical flask containing 7ml concentrated sulphuric acid (cool by placing the flask in an ice bath) add with stirring 22ml of acetic anhydride taking care that the temperature does not rise above 18°C. After the addition of acetic anhydride is complete, add 18gm of powdered camphor. Stir the resulting mixture until camphor is completely dissolved. Remove the flask from ice bath and keep it at room temperature for 36 hours. Collect the precipitated camphor sulphonic acid on buchner funnel, wash with ether and dry, m.p 202-203°C.

(B) NITRATION

Aromatic hydrocarbons react with concentrated nitric acid directly or in the presence of strong acid catalyst namely sulphuric acid to form nitro derivatives. Since a hydrogen is replaced by monovalent nitro group, this reaction is called nitration.

The selection of the nitrating agent depends upon the reactivity of substrate. The most common among them is a mixture of nitric acid and sulphuric acid known as nitrating mixture. However nitration can also be carried out with potassium nitrate or by using nitric acid in the presence of acetic acid or organic solvents.

The nitration reaction is initiated by acid catalyzed dehydration of nitric acid which acts as a base to form the electrophilic nitronium ion (NO_2^+) which is the active species in the reaction and the one responsible for the incorporation of nitro group onto the aromatic ring.

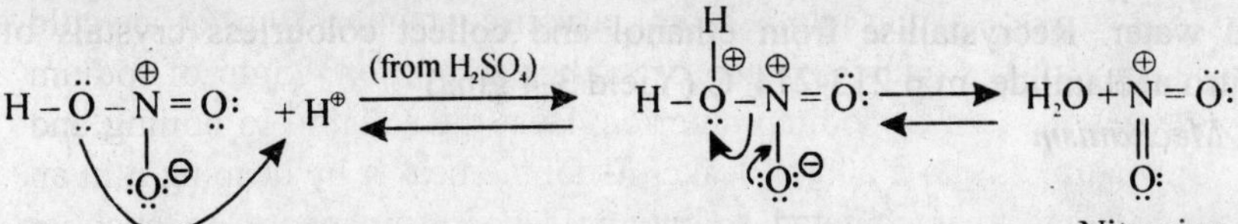

Nitronium ion

The mechanism involves the reaction of the electrophile nitronium ion with the aromatic ring to form the intermediate sigma complex or arenium ion (I), followed by removal of a proton by the basic hydrogen sulphate ion to form the nitro derivative.

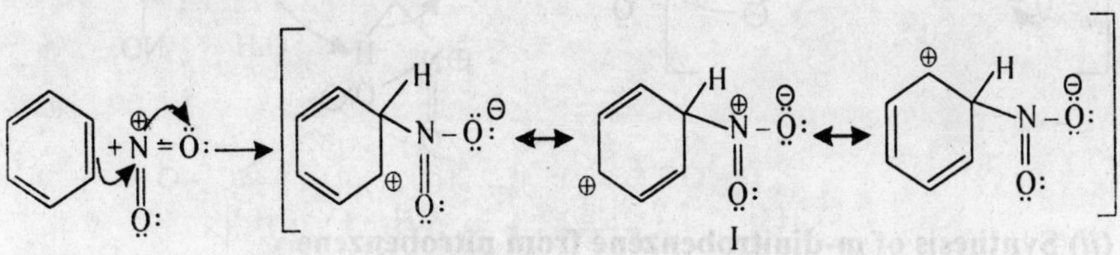

I

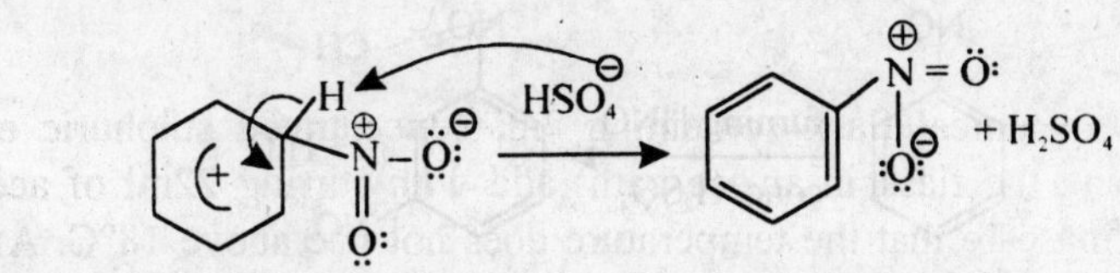

(*i*) Synthesis of p-nitro acetanilide from acetanilide

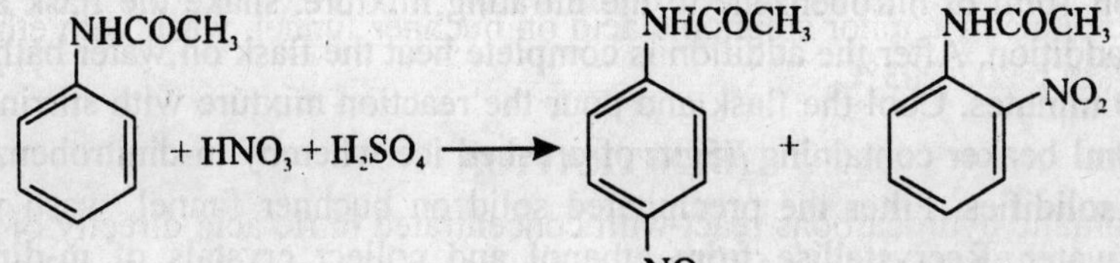

Soluble in water

In a 100ml conical flask dissolve by stirring with a glass rod 5gm of acetanilide in 5ml of glacial acetic acid and then add 10ml of concentrated sulphuric acid so that a clear solution is obtained. Place the flask in an ice bath. When the temperature of the reaction mixture falls to 0°C add dropwise with stirring 3.9ml of cold nitrating mixture (2.5ml of concentrated nitric acid and 1.4ml of concentrated sulphuric acid) taking care that the temperature of the mixture does not rise above 10°C during each addition. After all the nitrating mixture has been added remove the flask from the ice bath and keep it at room temperature for an hour. Pour the contents of the flask with stirring in a 200ml beaker containing 50gm of crushed ice. Filter the crude pale yellow product with suction on buchner funnel and wash with cold water. Recrystallise from ethanol and collect colourless crystals of p-nitro acetanilide, m.p 213-214°C (Yield 3-4 gms)

Mechanism

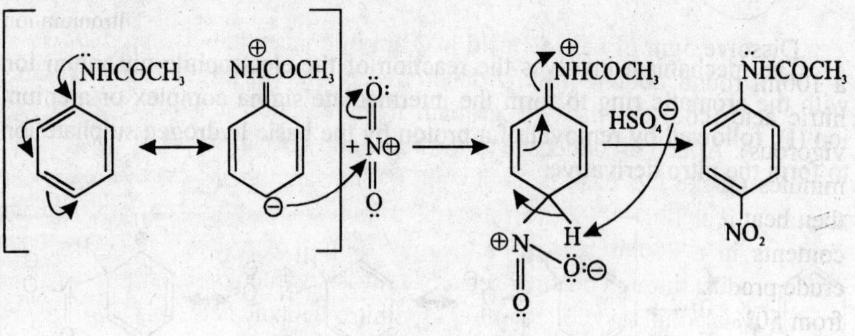

(ii) Synthesis of m-dinitrobenzene from nitrobenzene

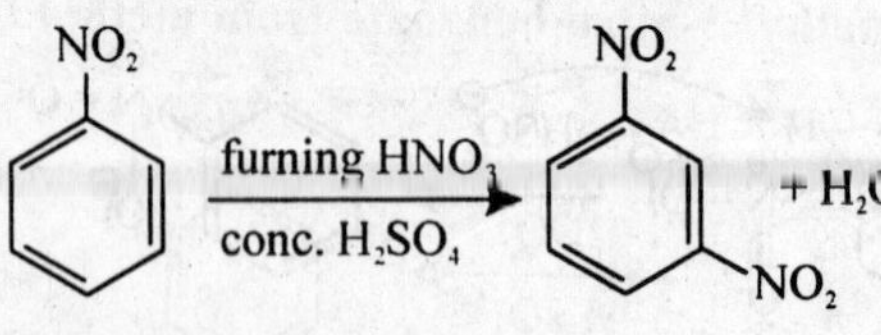

Place 17ml of concentrated sulphuric acid and 12ml of fuming nitric acid in a 100ml round bottom flask fitted with a reflux condensor. Add in portion 10ml of nitrobenzene to the nitrating mixture, shake the flask after each addition. After the addition is complete heat the flask on water bath for 30-40 minutes. Cool the flask and pour the reaction mixture with stirring in a 200ml beaker containing 75gm of crushed ice whereby m-dinitrobenzene soon solidifies. Filter the precipitated solid on buchner funnel, wash with cold water. Recrystallise from ethanol and collect crystals of m-dinitro benzene, m.p 89 – 90°C (Yield 10 – 11gm)

Mechanism

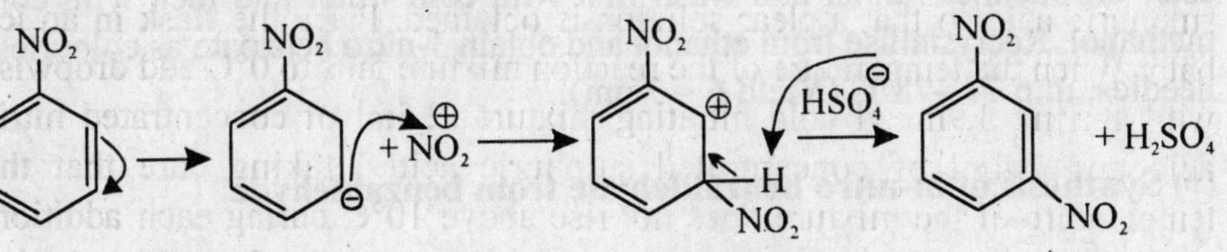

(*iii*) Synthesis of 3, 5-dinitro benzoic acid from benzoic acid

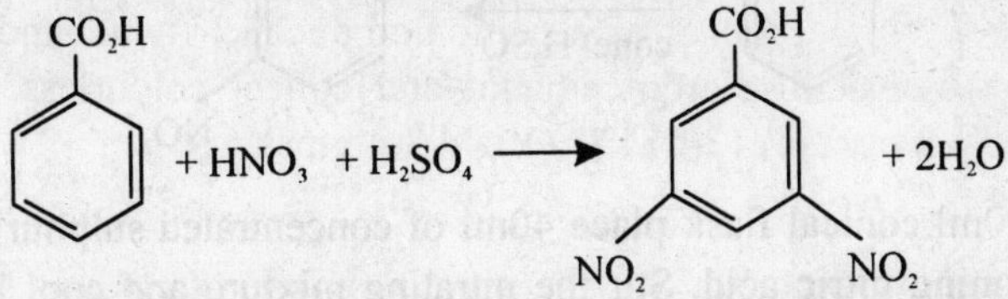

Dissolve 5gm of benzoic acid in 23ml of concentrated sulphuric acid in a 100ml round bottom flask. Add dropwise with stirring 7.3ml of fuming nitric acid (cool the flask in ice bath if the reaction tends to become too vigorous). Attach a condensor to the flask and heat it on water bath for 45 minutes (shake the flask occassionally). Transfer the flask to oil bath and then heat it at 130 – 140°C for 70 – 80 minutes. Cool the flask and pour the contents in a 200ml beaker containing 100gm of crushed ice. Filter the crude product through buchner funnel and wash with cold water. Recrystallise from 50% alcohol and collect pure 3,5 dinitro benzoic acid, m.p 206 -207°C (Yield 4- 4.5 gm)

(*iv*) Synthesis of methyl 3-nitro benzoate from methyl benzoate

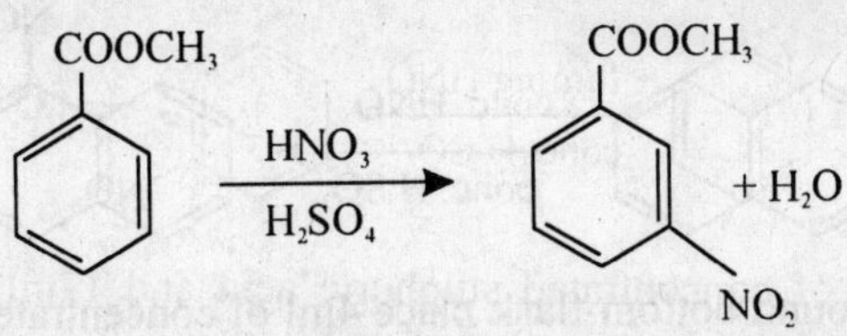

Cool the conical flask containing 10ml of pure methyl benzoate in ice bath so that the temperature is in the range of 0–10°C. Now add dropwise a cooled mixture of 6.5ml of concentrated sulphuric acid and 6.5ml of concentrated nitric acid. During the addition process swirl the flask occassionally and also take care that the temperature does not rise above 15°C. When all the nitrating mixture has been added continue the stirring process for another fifteen minutes and then pour the mixture with stirring in a 200ml beaker containing 75gm of crushed ice. Filter off the crude nitro ester on buchner funnel and wash first with cold water and then with cold methanol. Recrystallise from ethanol and obtain 3-nitro benzoate as colourless needles, m.p 77 –78°C (Yield 6 – 7 gm)

(*v*) Synthesis of m-nitro benzaldehyde from benzaldehyde

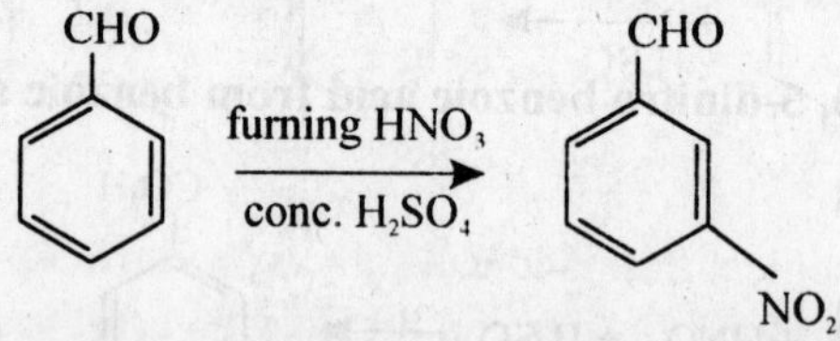

In a 100ml conical flask place 40ml of concentrated sulphuric acid and 3.5ml of fuming nitric acid. Stir the nitrating mixture and cool by keeping the flask in an ice bath. Add dropwise 10ml of freshly distilled benzaldehyde taking care that the temperature does not rise above 7°C during each addition. When the addition is complete remove the flask from ice bath and then warm it to 40°C for 1-2 minutes. Cool and pour the reaction mixture with stirring in a 200ml beaker containing 50gm of crushed ice. Filter the precipitated solid on buchner funnel and wash with cold water. Recrystallise the crude nitro derivative by first dissolving in hot benzene and then add pet ether to it until slight turbidity results. Cool and collect crystals of m-nitro benzaldehyde, m.p 56-58°C (Yield 6-7gm).

(*vi*) Synthesis of 1-nitro naphthalene from naphthalene

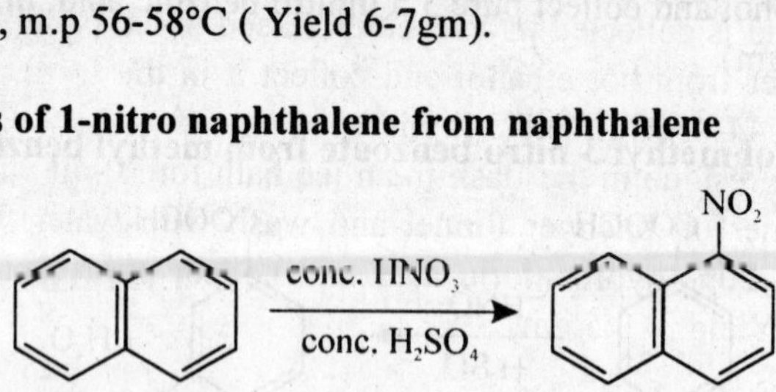

In a 100ml round bottom flask place 4ml of concentrated nitric acid and add in portion with shaking 4ml of concentrated sulphuric acid to it. Cool the nitrating mixture by keeping the flask in ice-water bath. Add in portion with stirring 5gm of finely powdered naphthalene while maintaining the temperature between 45-50°C. When the addition is complete attach a condensor to the flask and heat it on water bath at 50-60°C for 45 minutes. Pour the reaction mixture with stirring onto 100ml of ice cold water. Filter off the nitro derivative at pump and wash with water. Recrystallise from dilute alcohol and collect crystals of 1-nitro naphthalene, m.p 60- 61°C (Yield 5.5-6.0 gm)

(*vii*) Synthesis and separation of o and p-nitro phenol from phenol

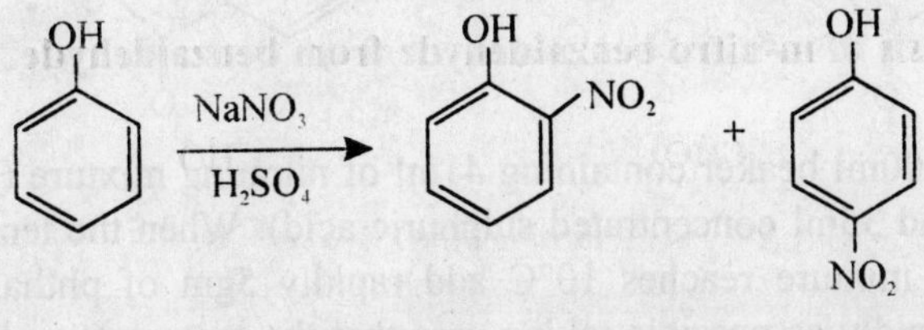

In a 250ml conical flask add carefully with stirring 14ml of concentrated sulphuric acid to 42ml of water and then dissolve 16gm of sodium nitrate in acid solution. Cool the nitrating mixture by keeping the flask in ice bath. Add dropwise with stirring phenolic solution to the cold nitrating mixture (10gm of phenol in 3ml of water). Care should be taken that the temperature does not rise above 15°C during each addition. After all the phenolic solution has been added continue the stirring process for another 2-2 ¼ hours. Decant off the mother liquor from the mixture of nitro compounds. Add 50ml of water to the conical flask and shake thoroughly. Allow the contents to settle and then decant off the wash liquid. Repeat the washing process again in order to remove the acid from the mixture. Transfer the solid portion in a 100ml round bottom flask and set the apparatus for downward distillation. Pass steam into the mixture of nitro phenols. o-nitro phenol being volatile distills off and is collected in a receiver placed in ice bath. Recrystallise the solid o-isomer from hot ethanol and collect it in the form of bright yellow crystals, m.p 44-45°C (Yield 3-4 gm)

Cool the residue in the flask in an ice bath for 30-40 minutes. Filter the crude p-isomer on buchner funnel and wash with water. Recrystallise by boiling with 20% hydrochloric acid and collect pure p-nitro phenol, m.p 110-112°C (Yield 3- 3.5 gm)

Mechanism

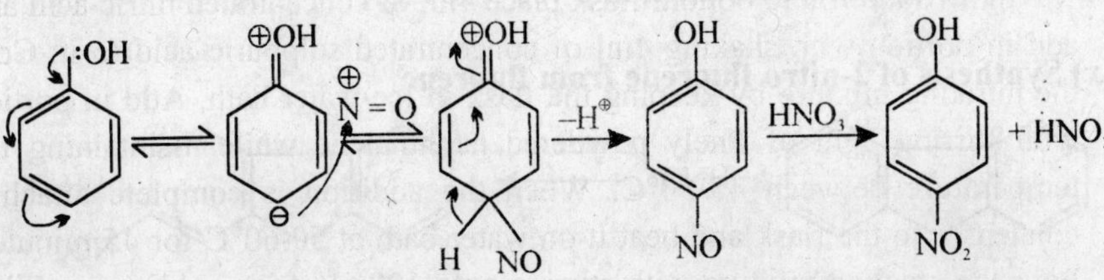

(*viii*) Synthesis of 4-nitro phthalimide from phthalimide

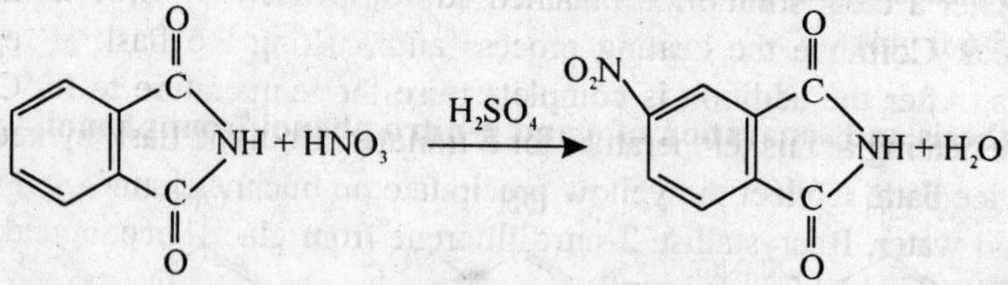

Cool a 200ml beaker containing 41ml of nitrating mixture (6ml fuming nitric acid and 35ml concentrated sulphuric acid). When the temperature of the nitrating mixture reaches 10°C add rapidly 5gm of phthalimide with stirring as rapidly as possible taking care that the temperature does not rise above 15°C. After the addition is complete keep the reaction mixture at room temperature for 6 hours (preferably overnight). Pour the clear pale yellow solution slowly and with vigorous stirring into a 200ml beaker containing 100gm of crushed ice. Filter the crude product by suction on buchner funnel and wash twice with cold water. Recrystallise 4-nitro phthalimide from alcohol, m.p 196 – 198°C (Yield 2 -3 gm)

(*ix*) Synthesis of 1,5 dichloro 2, 4-dinitro benzene from m-dichloro benzene

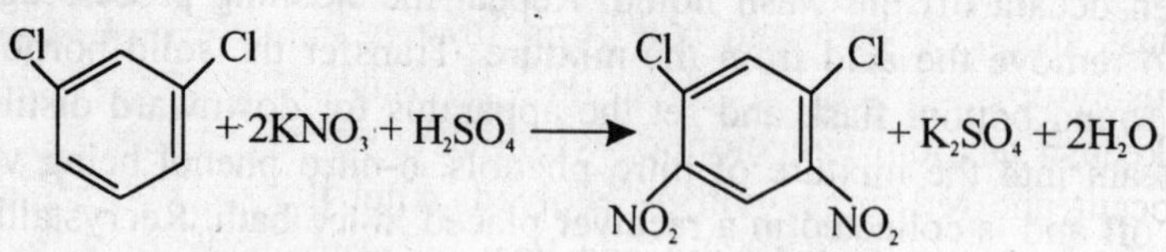

In a 250ml conical flask add 5gm of m-dichloro benzene to a well stirred solution of 7gm of potassium nitrate in 25ml of concentrated sulphuric acid. The temperature of the reaction mixture will rise to 135–145°C. Continue the stirring process and maintain the temperature between 125–135°C for 1 hour. Cool the mixture and pour with stirring into a 200ml beaker containing 50gm of crushed ice. Collect the precipitated solid on buchner funnel and wash with cold water. Recrystallise from hot water and collect crystals of 1,5 dichloro 2,4 dinitro benzene, m.p 103 – 104°C.

(*x*) Synthesis of 2-nitro fluorene from fluorene

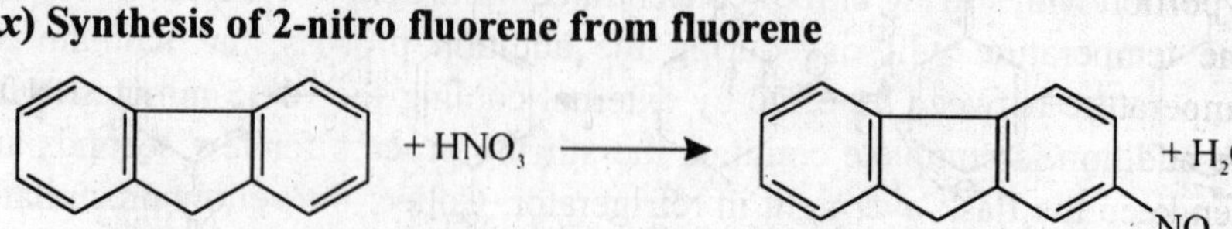

In a 100ml round bottom flask fitted with a reflux condensor dissolve

6gm of fluorene in 50ml of glacial acetic acid by heating on water bath to 50°C. After a clear solution is obtained add dropwise 8ml of concentrated nitric acid. Continue the heating process and shaking the flask at regular intervals. After the addition is complete raise the temperature to 80°C and continue heating at this temperature for 5 minutes. Cool the flask by keeping it in an ice bath. Collect the yellow precipitate on buchner funnel and wash with cold water. Recrystallise 2-nitro fluorene from glacial acetic acid, m.p 156-157°C (Yield 4.75 - 5.5 gm)

(*xi*) Synthesis of 2,4,6-trinitro phenol from phenol

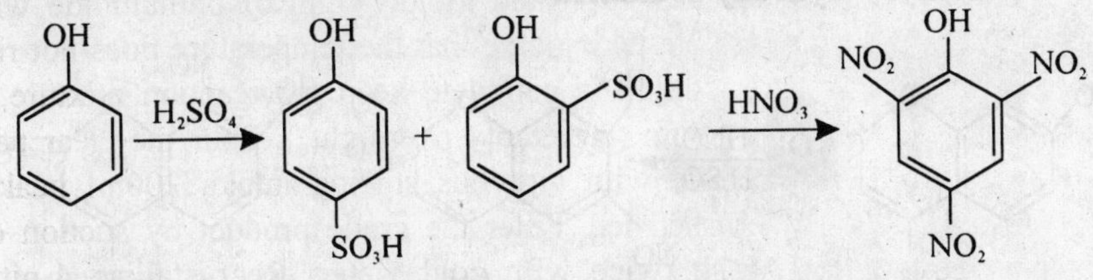

Weigh about 5gm of phenol in 250ml round bottom flask and add to it 6.5ml of concentrated sulphuric acid. First shake the flask thoroughly and then heat it on water bath for 35 minutes. Cool the flask by keeping it in an ice bath and allow precipitation of sulphonic acids to take place. Remove the flask from ice bath and add 19ml of concentrated nitric acid to the viscous syrupy solution. Shake the flask thoroughly for 20-30 seconds and then keep it at room temperature. Within few seconds a vigorous reaction will occur and red fumes will be evolved. After the reaction has subsided attach a condensor to the flask and heat on water bath for 2 hours with occassional shaking. Cool the flask in ice bath, collect the yellow precipitate on buchner funnel and wash with water. Recrystallise from dilute alcohol and collect pale yellow crystals of picric acid (2,4,6 trinitro phenol), m.p 120-122°C (Yield 6-7 gm).

(*xii*) Synthesis of 2-nitro 4-methoxy acetanilide from p-aceto anisidide

In a 100ml conical flask dissolve 6gm of p-aceto anisidide in 15ml glacial acetic acid and then cool the flask by keeping it in an ice bath. Add in portion with stirring 6ml of concentrated nitric acid to the cooled solution. The temperature will rise during the addition process but maintain the temperature between 55-65°C by external cooling for 10-15 minutes. After the addition is complete continue the stirring process for few seconds and then keep the flask overnight in refrigerator. Collect the yellow precipitated solid on buchner funnel and wash with water. Recrystallise 2-nitro 4-methoxy acetanilide from ethanol, m.p 116-117°C (Yield 4.0-4.5 gms)

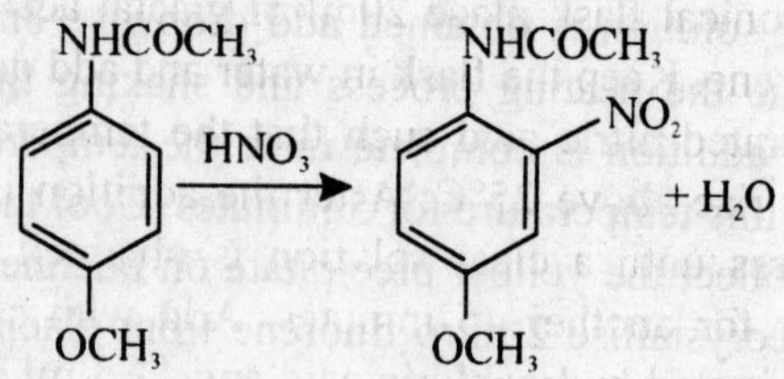

(*xiii*) Synthesis of 6 and 8 nitro-4-methyl-7-hydroxy coumarin from 4-methyl 7-hydroxy coumarin

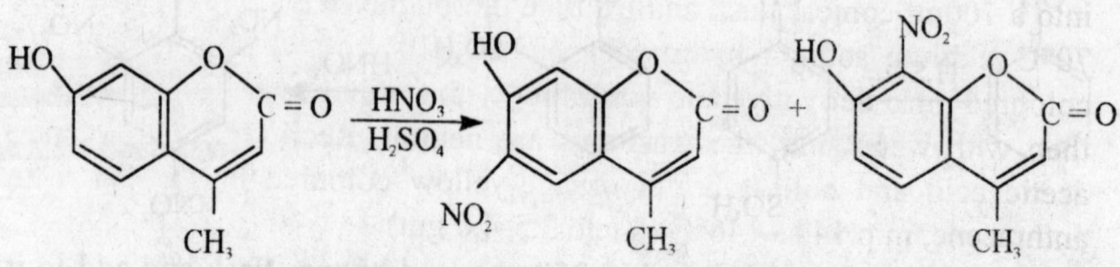

In a 100ml conical flask dissolve 5 gm of 4-methyl 7-hydroxy coumarin in 40ml of concentrated sulphuric acid and then keep the flask in an ice bath. When the temperature inside the flask is below 1°C, add dropwise with stirring 7ml of nitrating mixture (2ml of concentrated nitric acid and 5ml of concentrated sulphuric acid) taking care that the temperature does not rise above 10°C. After the addition is complete remove the flask from the ice bath and keep it at room temperature for an hour. Shake the flask occassionally during this period and then pour with stirring in a 200ml beaker containing 100gm of crushed ice. Filter the crude product which is a mixture of 6 and 8 nitro derivatives on buchner funnel and wash with cold water. Transfer the crude mixture in a 250ml conical flask containing 100ml of ethanol and boil. Filter the solution while still hot. The residue is 6-nitro derivative. Crystallise from benzene and obtain pure crystalline 6-nitro 4-methyl 7-hydroxy coumarin, m.p 260- 262°C (Yield 1.5-2.00 gm)

Concentrate the filtrate, cool in an ice bath, 8-nitro derivative soon crystallises out. Recrystallise from ethanol and collect 8-nitro 4-methyl 7-hydroxy coumarin, m.p 255 - 256°C (Yield 1.0 - 1.5 gm)

(*xiv*) Synthesis of 9-nitro anthracene from anthracene

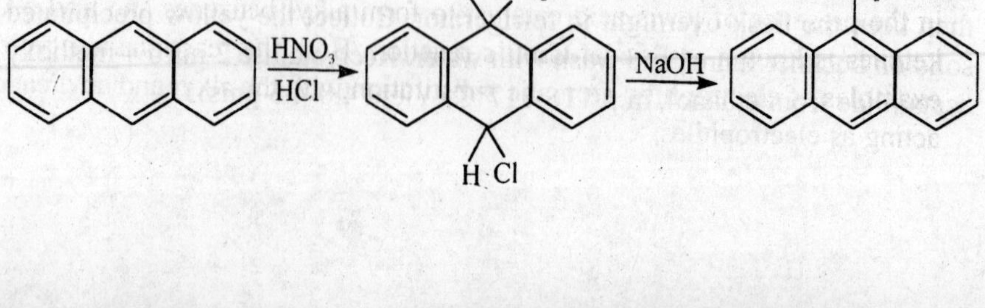

In a 100ml conical flask place 20ml of glacial acetic acid and 5 gm of powdered anthracene. Keep the flask in water and add dropwise with stirring 2.5ml of concentrated nitric acid such that the temperature of the reaction mixture does not rise above 25°C. After the addition is complete continue the stirring process until a clear solution is obtained and then stirring is continued further for another 45 minutes. Add with stirring, a mixture of 12.5 ml of concentrated hydrochloric acid and 12.5 ml of glacial acetic acid to the clear solution. Cool and collect 9-nitro 10-chloro 9,10 dihydroanthracene on buchner funnel, wash first with glacial acetic acid and then with water until the washings are neutral. Transfer the crude product into a 100ml conical flask and triturate thoroughly with 15ml of warm (65-70°C) dilute sodium hydroxide solution (10%). Filter the crude orange coloured nitro derivative and wash first with sodium hydroxide solution and then with water until the washings are neutral. Recrystallise from glacial acetic acid and collect bright orange-yellow coloured needles of 9-nitro anthracene, m.p 144 - 146°C (Yield 3.5-4.0 gm)

(*xv*) Synthesis of 2,4 dinitrochlorobenzene from chlorobenzene

In a 100ml round bottom flask fitted with a reflux condensor place 10ml of chlorobenzene and 16ml of concentrated sulphuric acid. Heat the mixture on water bath, add in portion 34 ml of nitrating mixture (20ml concentrated nitric acid and 14ml concentrated sulphuric acid). Continue heating on water bath for another 2 hours. Transfer the assembly to sand bath and heat for 20-30 minutes in order to complete the reaction. Cool, pour with stirring in 400ml beaker containing 100gm of crushed ice. Filter the precipitated solid, wash with cold water, recrystallise from ethanol and collect 2, 4-dinitrochlorobenzene, m.p 52- 53°C.

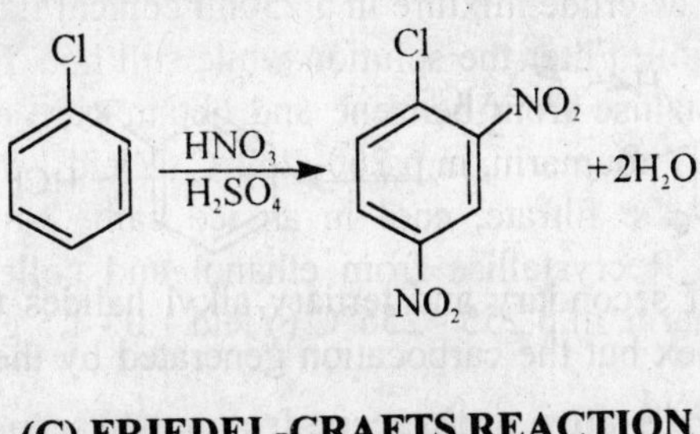

(C) FRIEDEL-CRAFTS REACTION

The introduction of an alkyl or acyl group into the aromatic compounds in the presence of lewis acid catalyst to form alkyl benzenes and aryl alkyl ketones is known as Friedel-Craft's reaction. Both the reactions are typical examples of electrophilic aromatic substitution with the alkyl and acyl cation acting as electrophiles.

(a) Friedel-Crafts alkylation :- Alkylation is usually carried out by the reaction of alkyl halide, alkenes or alcohols with aromatic compounds in the presence of lewis acid catalyst. The most common lewis acid catalyst are Aluminium Chloride, Boron Triflouride, Zinc Chloride and Aluminium Bromide etc. The function of the lewis acid catalyst is to provide an electrophilic agent.

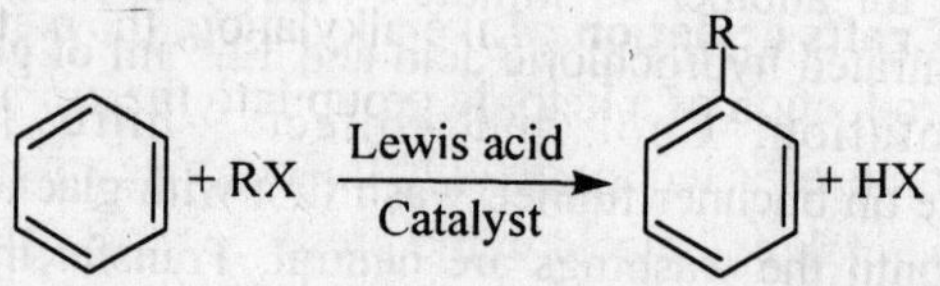

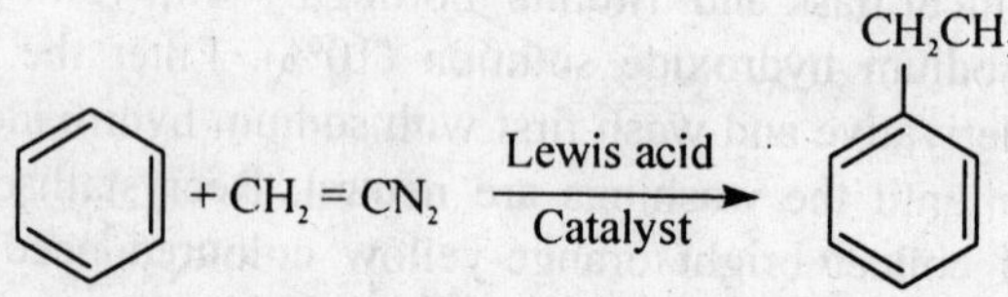

In case of primary alkyl halides the electrophile is infact a complex of lewis acid with the alkyl halide. The carbon-hydrogen bond in this complex is highly weak and there is considerable positive charge on the carbon atom. The benzene ring then reacts with the electrophile to form sigma complex. Loss of proton gives the product.

$$R - Cl + AlCl_3 \rightleftharpoons R^{\delta-} \text{----} Cl \text{----} AlCl_3^{\delta-}$$

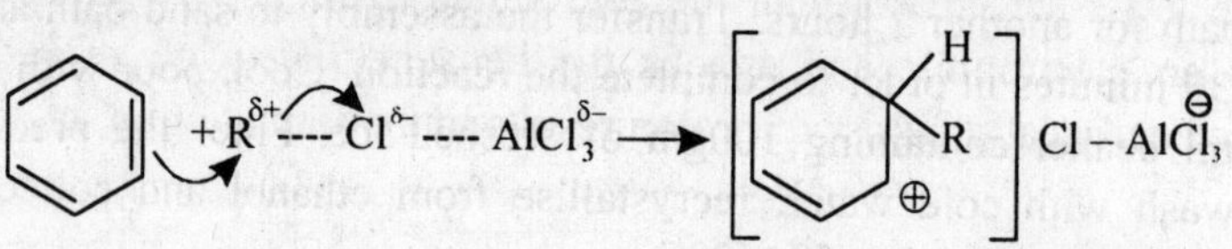

Sigma Complex

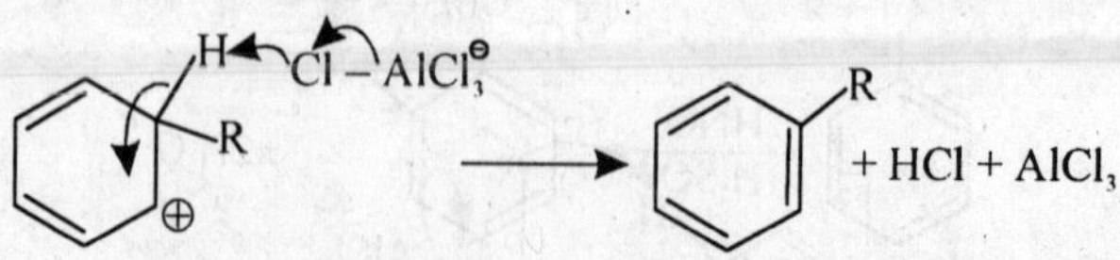

But in case of secondary and tertiary alkyl halides the electrophile is infact not a complex but the carbocation generated by the reaction of alkyl halide with lewis acid.

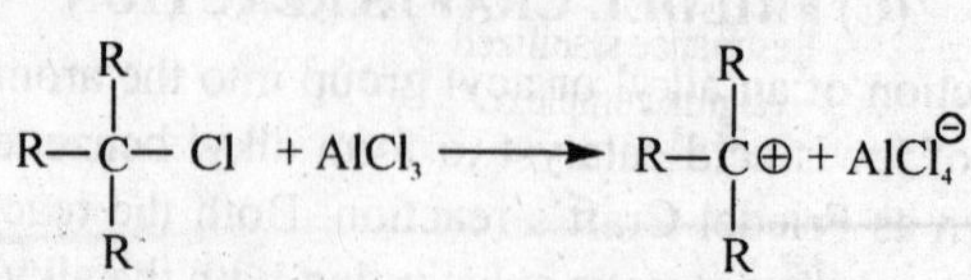

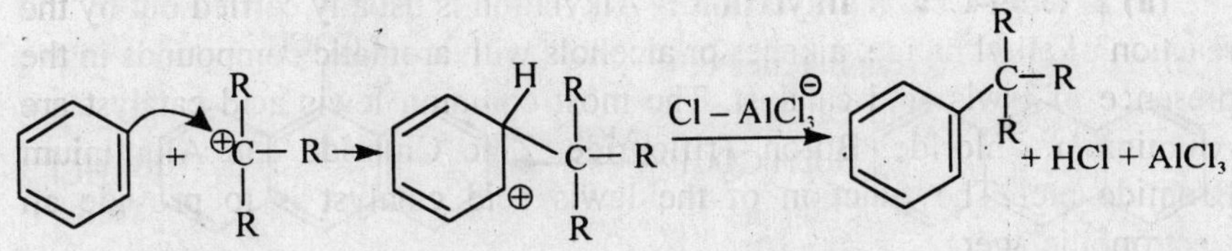

(*b*) Friedel-Crafts acylation : Like alkylation, the method of acylation is used for the introduction of a ketonic group into the aromatic ring system. Generally acylation is effected by treating the benzene ring with acyl halide, acid anhydride and carboxylic acid in the presence of lewis acid, catalyst.

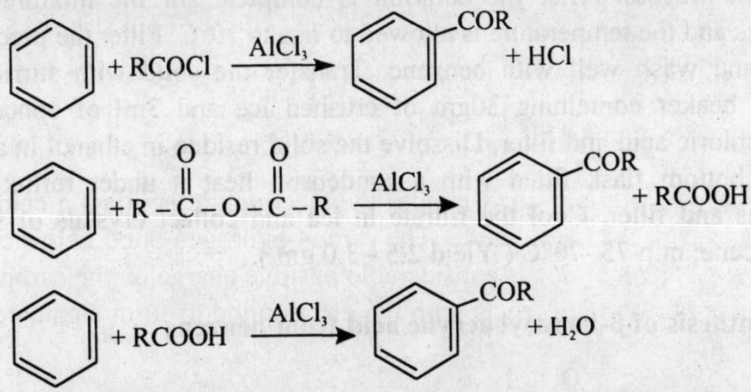

The lewis acid reacts with the acid halide to form acylium ion which acts as an electrophile and attacks the benzene ring to form resonance stabilised carbocation. Loss of proton results in the formation of an aromatic ketone.

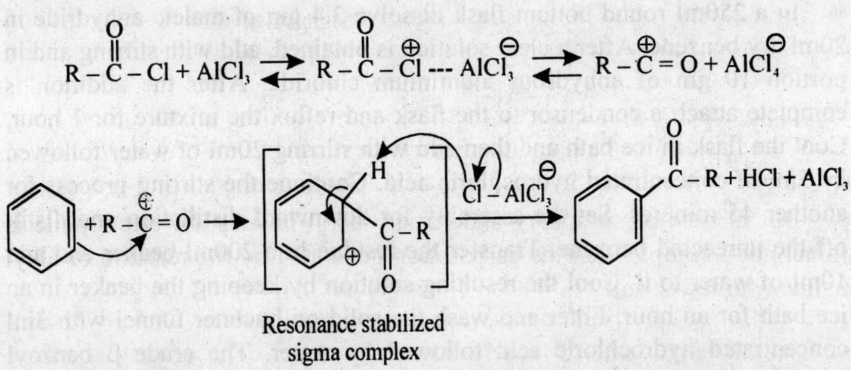

Resonance stabilized sigma complex

(i) Synthesis of 9-acetyl anthracene from anthracene

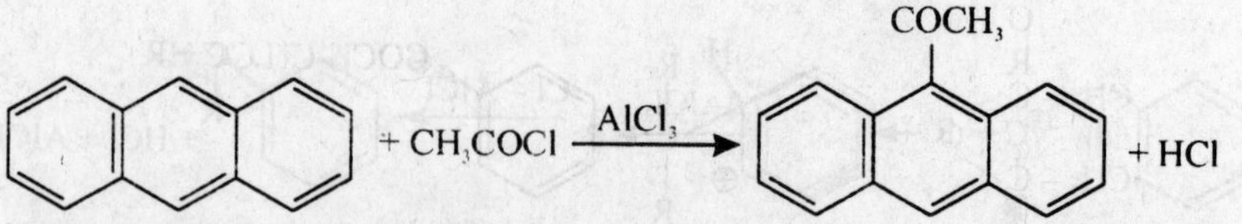

In a 100ml conical flask place 5gm of finely powdered anthracene, 35ml of sodium dried thiophene free benzene and 12ml of acetyl chloride. Shake the flask thoroughly to ensure thorough mixing and then keep it in an ice bath. Add with stirring and in portions 7.5 gms of anhydrous aluminium chloride taking care that the temperature does not rise above 0°C during the addition process. After the addition is complete stir the mixture for 30 minutes and the temperature is allowed to rise to 10°C. Filter the precipitated solid and wash well with benzene. Transfer the solid with stirring in a 100ml beaker containing 30gm of crushed ice and 3ml of concentrated hydrochloric acid and filter. Dissolve the solid residue in ethanol in a 100ml round bottom flask fitted with a condensor, heat it under reflux for 10 minutes and filter. Cool the filtrate in ice and collect crystals of 9-acetyl anthracene, m.p 75 -76°C (Yield 2.5 - 3.0 gm).

(ii) Synthesis of β-benzoyl acrylic acid from benzene

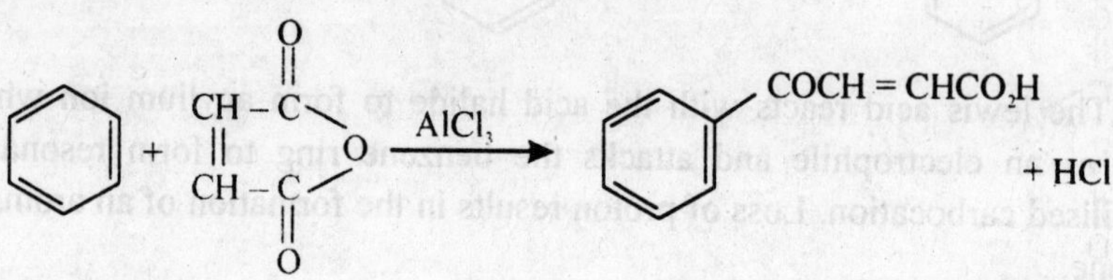

In a 250ml round bottom flask dissolve 3.4 gm of maleic anhydride in 20ml dry benzene. After a clear solution is obtained, add with stirring and in portion 10 gm of anhydrous aluminium chloride. After the addition is complete attach a condensor to the flask and reflux the mixture for 1 hour. Cool the flask in ice bath and then add with stirring 20ml of water followed by 5ml of concentrated hydrochloric acid. Continue the stirring process for another 45 minutes. Set the assembly for downward distillation and distill off the unreacted benzene. Transfer the residue in a 200ml beaker and add 10ml of water to it. Cool the resulting solution by keeping the beaker in an ice bath for an hour. Filter and wash the solid on buchner funnel with 3ml concentrated hydrochloric acid followed by water. The crude β benzoyl acrylic acid may be recrystallised from benzene, m.p 94 - 96°C (Yield 4.5 - 5.0 gm)

(*iii*) Synthesis of β-benzoyl propionic acid from benzene

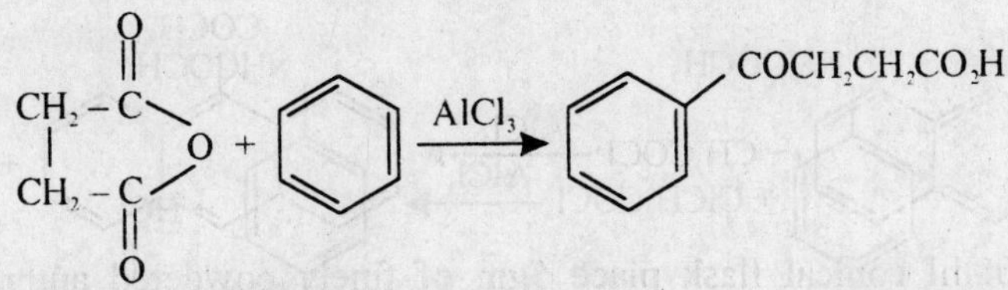

In a 200ml round bottom flask place 20ml of sodium dried thiophene free benzene and 4 gm of succinic anhydride. Swirl the flask and add in portion 10gm of anhydrous aluminium chloride. The mixture gets hot and hydrochloric acid is evolved. Attach a condensor to the flask and heat in on oil bath for 30–35 minutes. Swirl the flask occassionally during this period. Cool the flask by keeping it in an ice bath and add 20ml of water followed by 5ml of concentrated hydrochloric acid to it. Rearrange the condensor, set the assembly for downward distillation and distill off the unreacted benzene. Pour the residue in a 200ml beaker containing 50gm of crushed ice. Collect the precipitated acid on buchner funnel and wash with cold water. Recrystallise from hot water and collect pure β benzoyl propionic acid, m.p 115 - 116°C (Yield 5 - 6 gm)

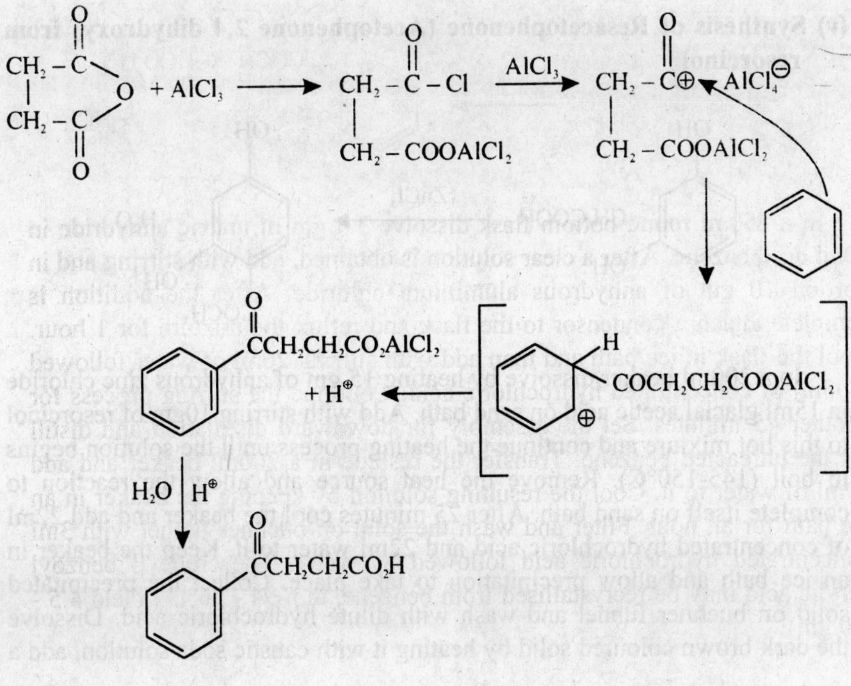

(*iv*) Synthesis of p-chloro acetyl acetanilide from acetanilide

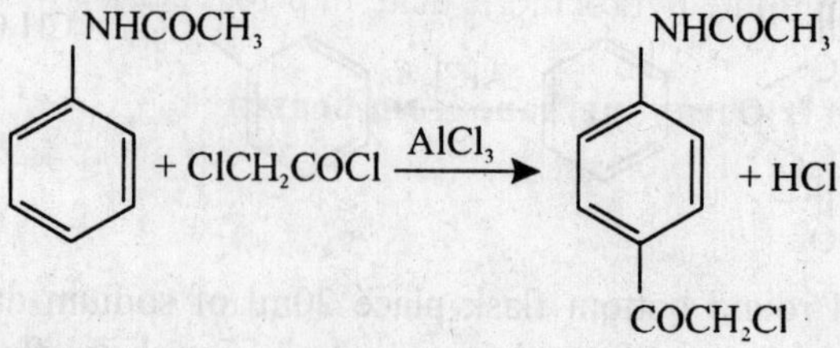

In a 250ml round bottom flask place 55ml of carbon disulphide, 10gm of acetanilide and 15 gm of chloro acetyl chloride. Stir the mixture and add in portion 30 gm of anhydrous aluminium chloride. After the addition is complete attach a condensor to the flask and reflux the resulting mixture for 30 minutes. Keep the flask at room temperature for 2 hours during which period the mixture separates out into two layers. The upper carbon disulphide layer is decanted and the red brown lower layer is poured with stirring into a 200ml beaker containing 50 gm of crushed ice and 5ml of concentrated hydrochloric acid. Collect the precipitated solid on buchner funnel and wash with cold water. Recrystallise p-chloro acetyl acetanilide from ethanol, m.p 214 - 216°C (Yield 6.5 - 7.0 gm)

(*v*) Synthesis of Resacetophenone (Acetophenone 2,4 dihydroxy) from resorcinol

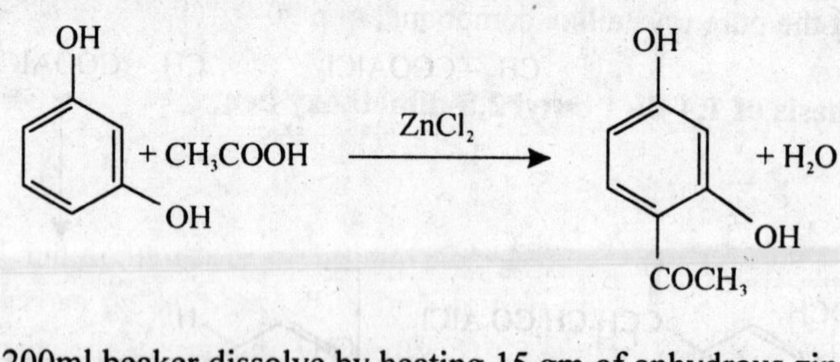

In a 200ml beaker dissolve by heating 15 gm of anhydrous zinc chloride in 15ml glacial acetic acid on sand bath. Add with stirring 10gm of resorcinol to this hot mixture and continue the heating process until the solution begins to boil (145-150°C). Remove the heat source and allow the reaction to complete itself on sand bath. After 25 minutes cool the beaker and add 22ml of concentrated hydrochloric acid and 22ml water to it. Keep the beaker in an ice bath and allow precipitation to take place. Collect the precipitated solid on buchner funnel and wash with dilute hydrochloric acid. Dissolve the dark brown coloured solid by heating it with caustic soda solution, add a pinch of animal charcoal to it and filter hot. Cool the filtrate and acidify it with dilute hydrochloric acid. Collect the precipitated resacetophenone, recrystallise from dilute hydrochloric acid, m.p 142°C (d) (Yield 4 - 5 gms)

(*vi*) Synthesis of triphenylmethane from benzene

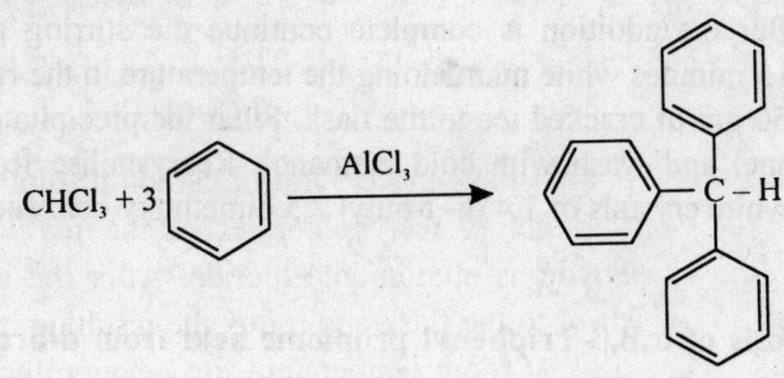

In a 100ml round bottom flask place 25ml of sodium dried thiophene free benzene and 3.5 ml of chloroform. Add with stirring and in portions 3.2 gm of anhydrous aluminium chloride. After the addition is complete immediately attach a condensor to the flask and heat it on water bath for 30-35 minutes. Cool the flask, pour the contents in a beaker containing 50gm of crushed ice and 1ml of concentrated hydrochloric acid. Remove the benzene layer and distill off the benzene. Continue the distillation process of the residual portion and collect fraction boiling at 190-215°C.This is triphenylmethane which soon solidifies on cooling. Recrystallise from ethanol and collect the pure crystalline compound, m.p 90-92°C (Yield 2-2.5 gm).

(*vii*) Synthesis of 1,4-di- t butyl 2,5-dimethoxy benzene

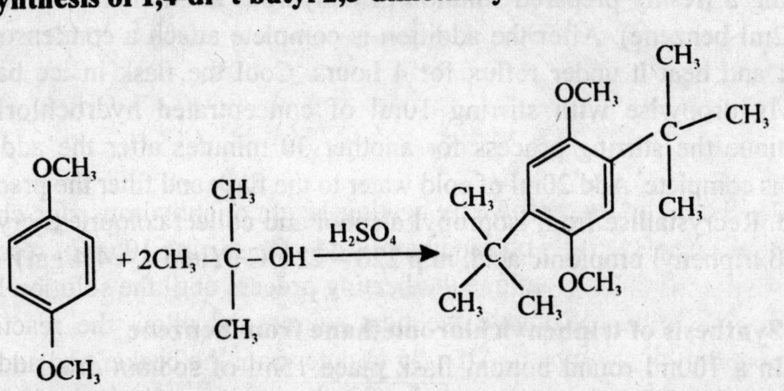

Place 6gm of 1, 4 dimethoxy benzene, 10ml of dry t-butyl alcohol and 20ml of glacial acetic acid in a 100ml conical flask. Swirl the flask thoroughly and then cool by keeping it in an ice bath. In a separate 50ml conical flask place 10ml of concentrated sulphuric acid and add 10ml of fuming sulphuric acid to it. Cool this acid containing flask by keeping it in an ice bath. When the temperature of the first flask (containing the 1,4 dimethoxy benzene solution) is in the range of 0-3°C add the chilled sulphuric acid solution dropwise. Agitate the flask gently during the addition of sulphuric acid solution. After the addition is complete continue the stirring process for another 10-15 minutes while maintaining the temperature in the range of 20-25°C. Add 50 gm of cracked ice to the flask. Filter the precipitated solid on buchner funnel and wash with cold methanol. Recrystallise from ethanol and collect white crystals of 1,4 di- t-butyl 2,5 dimethoxy benzene, m.p 104-105°C.

(*viii*) Synthesis of α,β,β-Triphenyl propionic acid from dibromohydrocinnamic acid

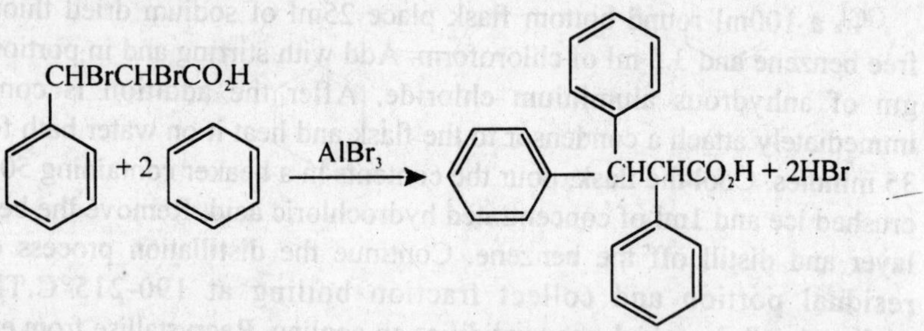

In a 100ml round bottom flask place 9.18 gm of dibromohydrocinnamic acid and 18ml of sodium dried thiophene free benzene. Add dropwise with stirring a freshly prepared solution of anhydrous aluminium bromide (9gm in 12ml benzene). After the addition is complete attach a condensor to the flask and heat it under reflux for 4 hours. Cool the flask in ice bath, add slowly dropwise with stirring 10ml of concentrated hydrochloric acid. Continue the stirring process for another 30 minutes after the addition of acid is complete. Add 20ml of cold water to the flask and filter the precipitated solid. Recrystallise from isopropyl alcohol and collect colourless crystals of α,β,β triphenyl propionic acid, m.p 220 – 221°C (Yield 4 – 4.4 gm)

(*ix*) Synthesis of triphenylchloromethane from benzene

In a 100ml round bottom flask place 15ml of sodium dried thiophene free benzene and 3.3 ml of carbon tetrachloride. Keep the flask in ice bath and add in portion 3.5gm of anhydrous aluminium chloride. The addition should be regulated so that the mixture does not reflux. After the addition is complete remove the flask from the ice bath, attach a condensor to it and then heat on steam bath until the evolution of hydrogen chloride ceases (30 - 40 minutes). Cool and pour the contents into a beaker, with stirring containing a mixture of 20 gm of crushed ice and 20ml of concentrated hydrochloric acid. Separate the benzene layer and extract the aqueous layer with 10ml benzene. Combine the two organic layers and wash with concentrated hydrochloric acid. Dry the organic extract over anhydrous calcium chloride. Distill to remove the benzene. Add 1ml benzene and .5ml of acetyl chloride to the residue, heat the mixture to boiling. Cool by keeping the flask in an ice bath for 1-2 hours. Filter the precipitated triphenylchloromethane, wash with pet-ether and dry, m.p 111-112°C.

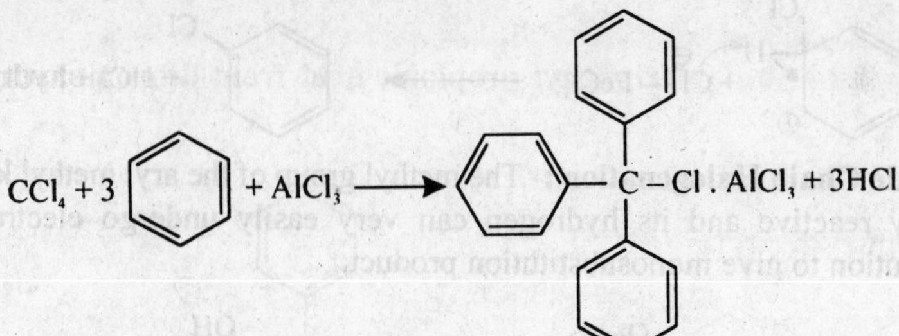

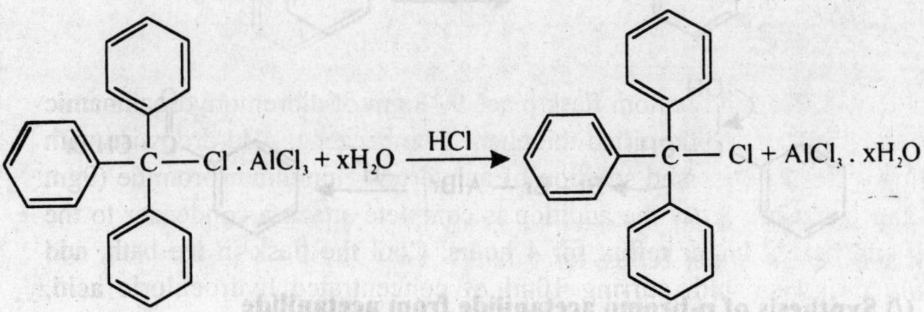

(D) HALOGENATION

Halogenation is usually carried out with molecular halogens (except fluorine). Halogens themselves are not strong electrophiles to be attacked by the aromatic system. However in the presence of polar solvents or so called halogen carriers like Aluminium Chloride, Boron Triflouride, Zinc Chloride and Aluminium Bromide polarises the halogens and impart some properties of lewis acid to it. The aromatic system now attack the newly created strong electrophilic species to form the sigma complex.

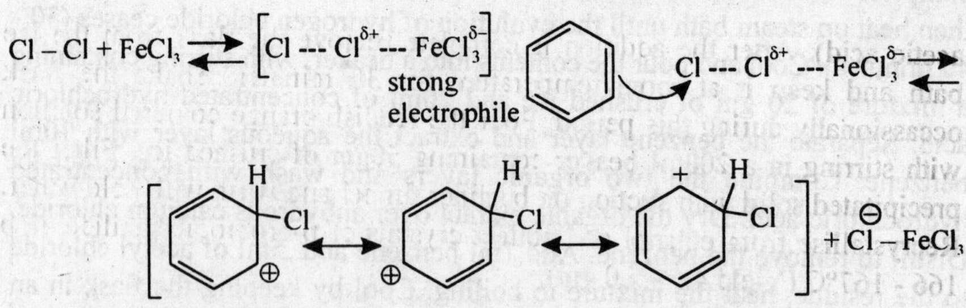

The sigma complex is not aromatic but it regains aromaticity by the loss of proton on the tetrahedral carbon leading to the substitution product.

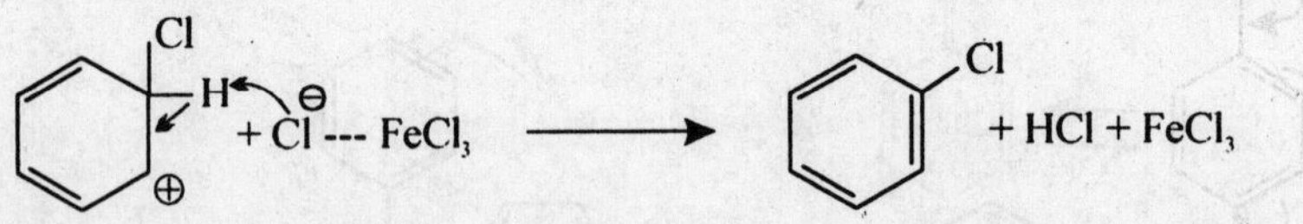

Side Chain Halogenation :- The methyl group of the aryl methyl ketones is very reactive and its hydrogen can very easily undergo electrophilic substitution to give monosubstitution product.

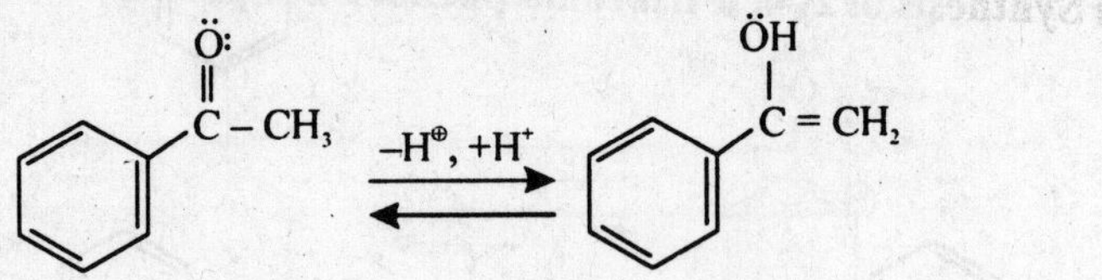

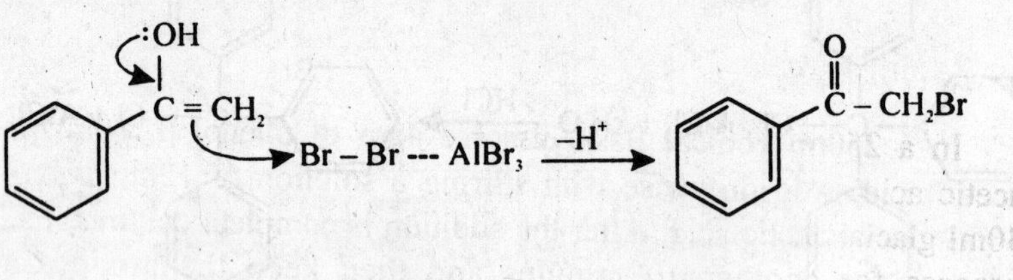

(*i*) Synthesis of p-bromo acetanilide from acetanilide

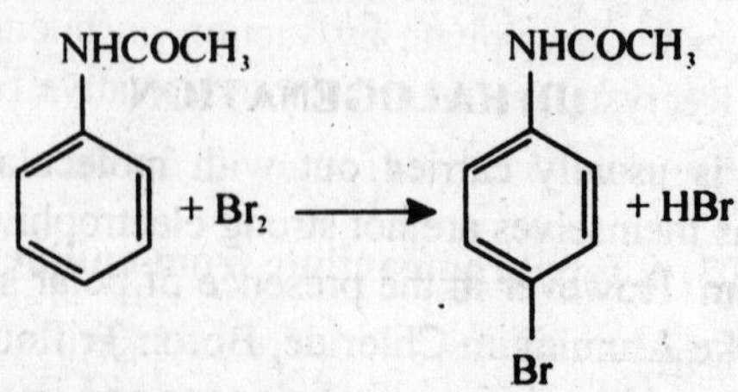

In a 100ml conical flask dissolve 5gm of acetanilide in 17ml of glacial acetic acid and then cool the flask by keeping it in an ice bath. Add slowly and with stirring 12.1ml of bromine solution (2.1ml bromine in 10ml glacial

acetic acid). After the addition is complete remove the flask from the ice bath and keep it at room temperature for 30 minutes. Shake the flask occassionally during this period. Pour the reddish orange coloured solution with stirring in a 200ml beaker containing 50gm of crushed ice. Filter the precipitated solid with suction on buchner funnel and wash with cold water. Recrystallise from ethanol and collect crystals of p-bromo acetanilide, m.p 166 - 167°C (Yield 4.5 - 5.0 gm).

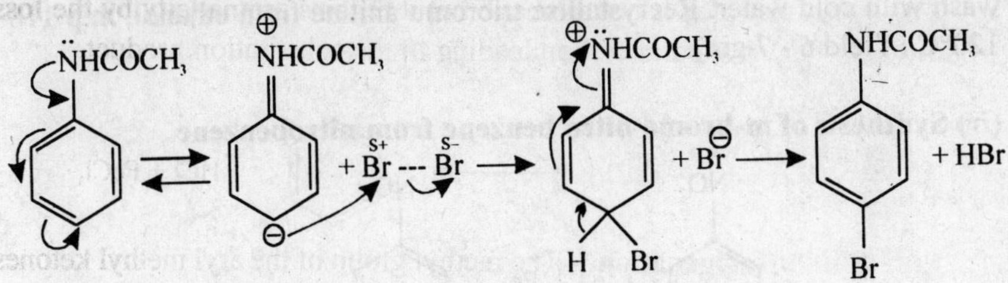

(*ii*) Synthesis of 2, 4, 6-tribromo phenol from phenol

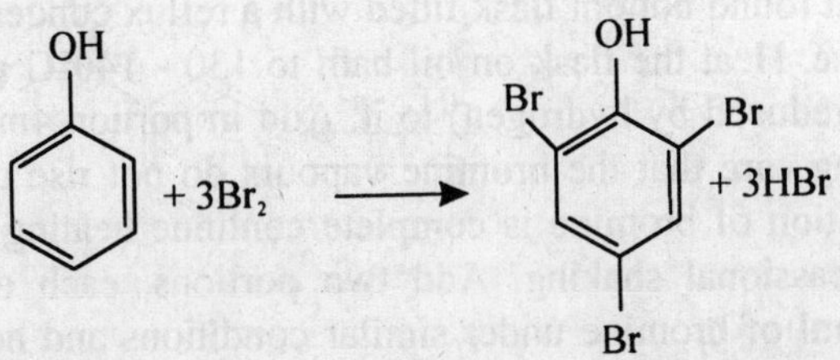

In a 250ml conical flask dissolve 4gm of phenol in 30ml of glacial acetic acid. Add dropwise with stirring a solution of 12ml of bromine in 30ml glacial acetic acid. After the addition is complete continue the stirring process for another 10 minutes and then keep the mixture at room temperataure for 20 minutes. Cool the flask in ice bath and add 100ml of ice water to it. Collect the precipitate by suction on buchner funnel and wash with cold water. Recrystallise the tribromo derivative from ethanol, m.p 93-95°C.

(*iii*) Synthesis of 2, 4, 6-tribromoaniline from aniline

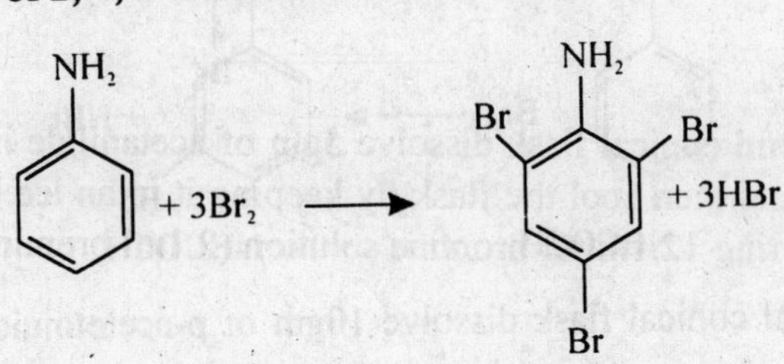

Place 5ml of aniline and 20ml of glacial acetic acid in a 200ml beaker, cool by keeping the beaker in an ice bath. Add dropwise with stirring 27ml of bromine solution (9ml of bromine in 18ml of glacial acetic acid). After the addition is complete remove the beaker from ice bath and keep it at room temperature for 30 minutes. Shake the beaker occassionally during this period and then pour the contents in another 200ml beaker containing 100ml of ice cold water. Filter the crude product on buchner funnel and wash with cold water. Recrystallise tribromo aniline from ethanol, m.p 118-120°C (Yield 6 - 7 gm).

(*iv*) Synthesis of m-bromo nitro benzene from nitrobenzene

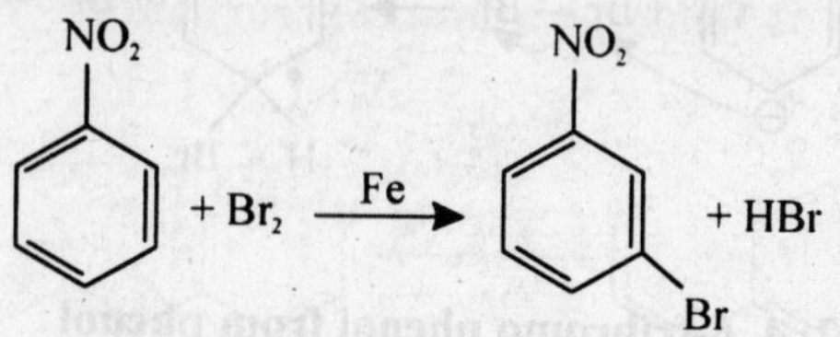

In a 100ml round bottom flask fitted with a reflux condensor place 15ml of nitrobenzene. Heat the flask on oil bath to 130 - 140°C and add 1gm of iron powder (reduced by hydrogen) to it. Add in portion 4ml of bromine to the flask taking care that the bromine vapours do not rise upto condensor. After the addition of bromine is complete continue heating the flask for 1 hour with occassional shaking. Add two portions, each of 1gm of iron powder and 4ml of bromine under similar conditions and heating the flask for 1 hour after each addition. Pour the dark coloured liquid into a 250ml round bottom flask containing 100ml of water and 10ml of saturated sodium meta bisulphite solution. Attach a condensor to the flask and set the assembly for downward distillation. Steam distill the resulting mixture and collect about 100ml of distillate. Cool the residual portion and collect yellow coloured m-bromo nitro benzene on buchner funnel and dry, m.p 51-52°C (yield 12-13 gm)

(*v*) Synthesis of 4-acetamido 3-bromo toluene from p-acetotoluidide

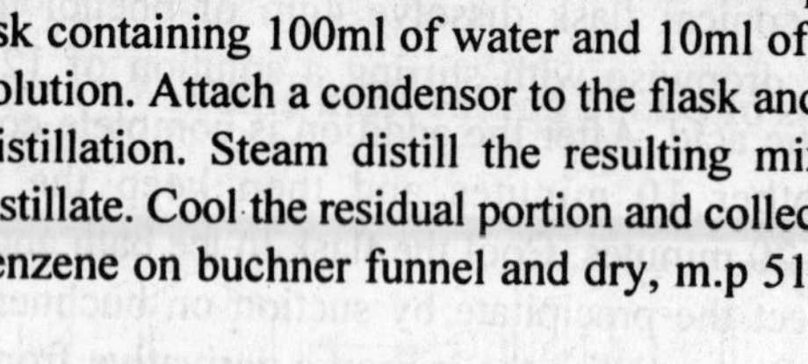

In a 100ml conical flask dissolve 10gm of p-acetotoluidide in 35ml of glacial acetic acid. Heat the flask on water bath taking care that the temperature of the solution is in the range of 45-50°C. Remove the flask from the water bath and add dropwise with stirring 5.5ml of bromine taking care that the temperature does not rise above 55°C during the addition process. After the addition is complete continue stirring for another 30 minutes and then pour the reaction mixture in a 200ml beaker containing 50gm of crushed ice. Filter the bromo derivative on buchner funnel and wash well with cold water. Recrystallise the crude product from 50% aqueous ethanol and collect 4-acetamido 3-bromo toluene as colourless needles, m.p 115 - 117°C (Yield 10 -12 gm).

(*vi*) Synthesis of 9-chloro anthracene from anthracene

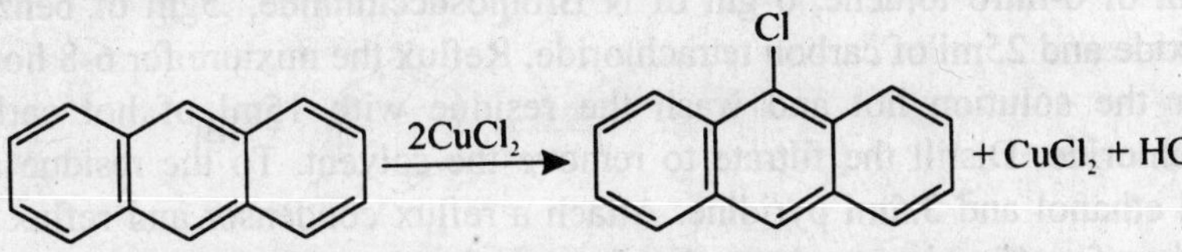

In a 500ml round bottom flask place 9gm of anthracene, 14gm of anhydrous cupric chloride and 250ml of carbon tetrachloride. Attach a condensor to the flask and reflux the mixture for 20- 24 hours. Shake the flask occassionally during this period. Cool the flask at room temperature and filter. Set the assembly for downward distillation and distill the filtrate to remove carbon tetrachloride. Cool the residue, recrystallise from pet-ether and collect 9-chloro anthracene as yellow needles, m.p 104 - 106°C (Yield 6 -7gm)

(*vii*) Synthesis of p-iodo aniline from aniline

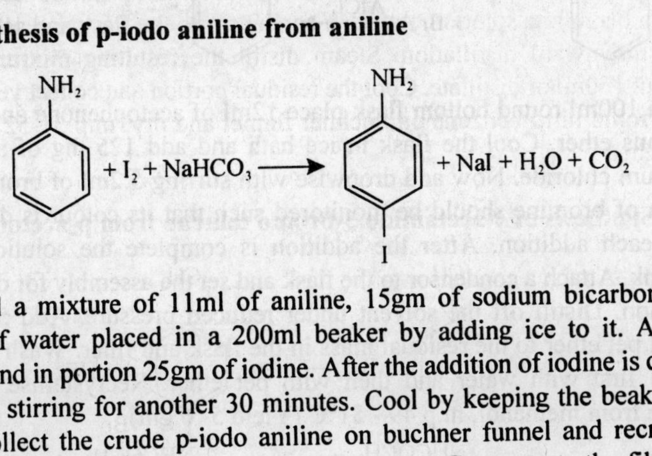

Cool a mixture of 11ml of aniline, 15gm of sodium bicarbonate and 100ml of water placed in a 200ml beaker by adding ice to it. Add with stirring and in portion 25gm of iodine. After the addition of iodine is complete continue stirring for another 30 minutes. Cool by keeping the beaker in ice bath. Collect the crude p-iodo aniline on buchner funnel and recrystallise from pet-ether, m.p 62-63°C (Yield 7-7.5 gm). Concentrate the filtrate and collect the iodine.

SIDE CHAIN HALOGENATAION

(i) Synthesis of o-nitrobenzyl pyridinium bromide from o-nitro toluene

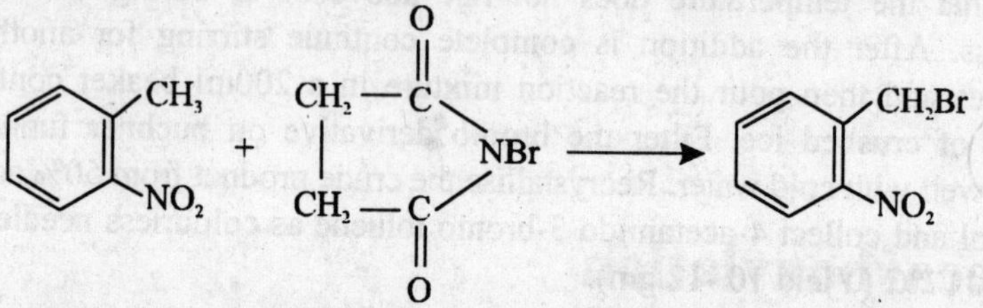

In a 100ml round bottom flask fitted with a reflux condensor place 5.1gm of o-nitro toluene, 6 gm of N Bromosuccinimide, .5gm of benzoyl peroxide and 25ml of carbon tetrachloride. Reflux the mixture for 6-8 hours. Filter the solution hot and wash the residue with 15ml of hot carbon tetrachloride. Distill the filtrate to remove the solvent. To the residue add 20ml ethanol and 3.5ml pyridine. Attach a reflux condensor and reflux the solution for 45 minutes. Immediately transfer the resultant solution to a 200ml beaker and cool by keeping it in an ice bath. Collect the crystals of o-nitrobenzyl pyridinium bromide, wash with cold ethanol and dry, m.p 128-135°C (Yield 3 - 3.5gm)

(ii) Synthesis of phenacyl bromide from acetophenone

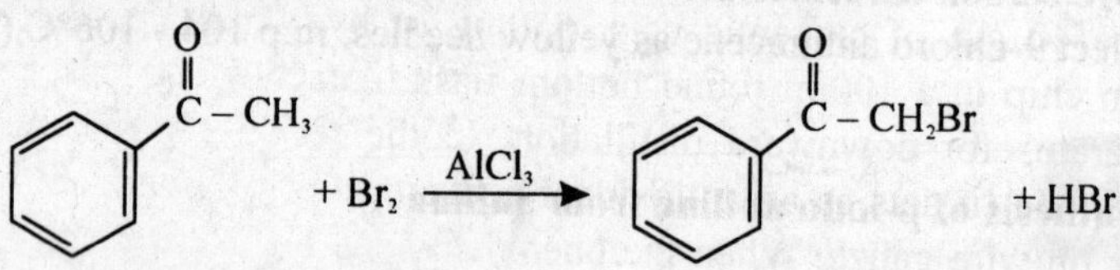

In a 100ml round bottom flask place 12ml of acetophenone and 15ml of anhydrous ether. Cool the flask in ice bath and add 125 mg of anhydrous aluminium chloride. Now add dropwise with stirring 5.2ml of bromine. The addition of bromine should be monitored such that its colour is discharged during each addition. After the addition is complete the solution finally turns pink. Attach a condensor to the flask and set the assembly for downward distillation. Distill off the solvent under reduced pressure. Add 5ml water and 5ml pet ether to the residual mass in the flask and filter. Wash the crude material first with water and then with pet ether. Recrystallise phenacyl bromide from methanol, m.p 49 - 51°C (Yield 5- 6 gm)

9

Decarboxylation

Decarboxylation is the process in which a molecule of carbon dioxide is eliminated from carboxylic acid. Decarboxylation can very easily be carried out by using quicklime or by heating the carboxylic acid at higher temperatures.

(*i*) Synthesis of phenylethylene (Styrene) from cinnamic acid

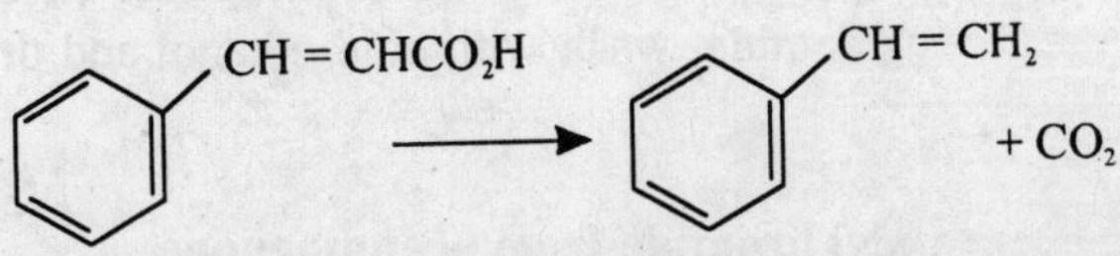

Place 40 gm of cinnamic acid, 550mg of hydroquinone and a small porcelain chip in a 100ml round bottom flask. Attach a condensor and set the apparatus for downward distillation. In the receiver place 125mg of hydroquinone (it acts as an antioxidant and prevents polymerization). Heat the flask on wire gauze. When the phenylethylene begins to distill regulate the heating in such a manner that the temperature of the vapours at the top of the distillation head does not exceed 130°C. After the distillation is complete add 30ml water to the oily distillate. Distill off the aqueous solution preferably with the help of steam. Phenylethylene being volatile is collected as a colourless oil, b.p 146°C.

(ii) Synthesis of Imidazole from Imidazole –4,5 dicarboxylic acid

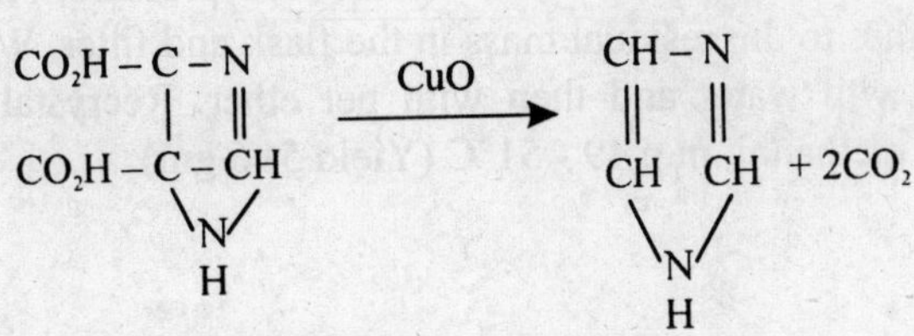

In a 100ml round bottom flask place 10 gm of Imidazole –4,5 dicarboxylic acid and .2gm of copper-chromium oxide catalyst or powdered copper oxide.Attach a condensor and set the assembly for downward distillation. Heat the flask on free flame and collect imidazole at 262-264°C. Cool and dissolve the product in 10ml benzene, add a pinch of animal charcoal.Heat the solution to boiling and filter hot.Cool the filtrate in ice bath and collect pure white crystals of imidazole, m.p 88-90°C.

10

Deamination

Replacement of an amino group by hydrogen is known as deamination. This method offers a useful method for the synthesis of trihalogen arenes which otherwise cannot be synthesized by the direct halogenation of arenes.Direct halogenation of benzene would give the monohalogenated product, however halogenation of aniline which has an activating amino group directs halogenation to its ortho and para position. Removal of the amino group (deamination) gives the desired product.

The mechanism involves the intermediate formation of aryl cation which reacts with ethanol to give the desired product.

$$2NaNO_2 + H_2SO_4 \longrightarrow Na_2SO_4 + 2HNO_2$$

$$2HNO_2 \rightleftharpoons O=N-O-N=O + H_2O \rightleftharpoons N=O + NO_2^{\ominus}$$

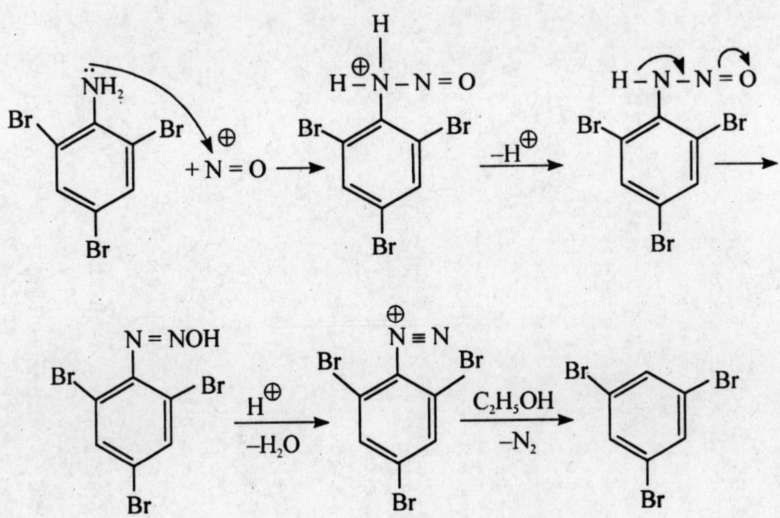

(i) Synthesis of sym tribromo benzene from sym tribromo aniline

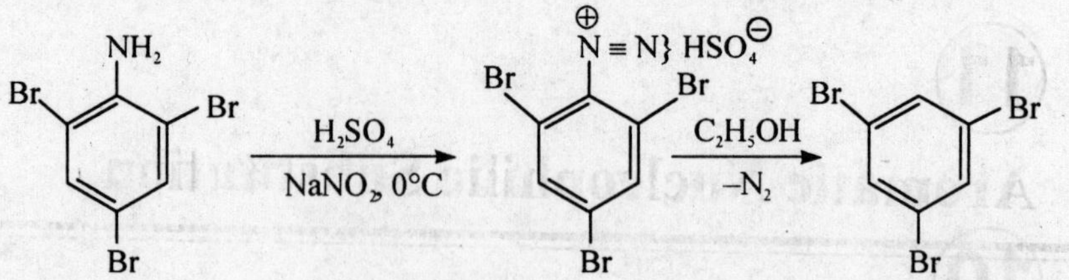

In a 100ml round bottom flask dissolve 5gm of sym tribromo aniline in 15ml of ethanol, add 4ml of benzene to it. Heat the flask on water bath so that the halogenated amine completely dissolves. Now add 1ml of concentrated sulphuric acid to the hot solution and swirl the resulting solution. Attach a condensor to the flask and heat it on water bath. When the solution begins to boil remove the flask from the water bath, add 1gm of sodiun nitrite in two instalments.After each addition fit the condensor to the flask and shake vigorously.The heat of the reaction will make the reactants to boil.Add the second instalment of sodium nitrite only after the first reaction has subsided. After the addition is complete heat the flask on water bath for 75-90 minutes.Remove the flask from water bath, keep it at room temperature for 10 minutes and then immerse it in an ice bath.Filter the precipitated solid on buchner funnel, wash with water so as to remove sodium sulphate which is also precipitated along with tribromobenzene. Recrystallise from methanol and collect crystals of sym tribromo benzene, m.p 120-122°C (Yield 1.5-2.00gm).

11

Aromatic Nucleophilic Substitution

In the hydrolysis of alkyl halide by aqueous sodium hydroxide, alcohol and sodium halide are obtained. This is a substitution reaction in which the halide group directly attached to the carbon in the substrate has been replaced by a hydroxyl group. Since this substitution has been brought about by a basic electron rich species called nucleophile, the reaction is referred to as Nucleophilic Substitution (SN).

A nucleophilic substitution reaction can take place either by a single step mechanism in which the bond making and bond breaking occur in a single step (SN^2) or by a two step mechanism when in the first step one group from the substrate (leaving group) departs along with its bonding electron and the intermediate carbocation so formed reacts with the nucleophile in the second step (SN^1)

Aromatic compounds being rich source of π electrons rarely undergo nucleophilic substitution reaction. However when electron withdrawing substituent like nitro, cyano and carboxyl group are present they markedly activate the aryl halide towards nucleophilic substitution reaction.Nucleophilic aromatic substitution as in the case of SN^2 follow second order kinetics i.e bimolecular but the reaction normally takes place in two steps. In the first step the nucleophilic reagent attacks the aromatic ring leading to the generation of a carbanion or σ complex which is usually represented by a number of resonating structures.

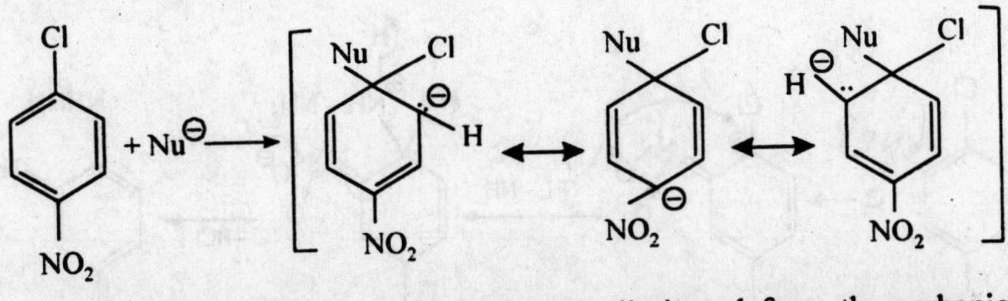

In the second step the halide ion gets eliminated from the carbanion thus yielding the product.

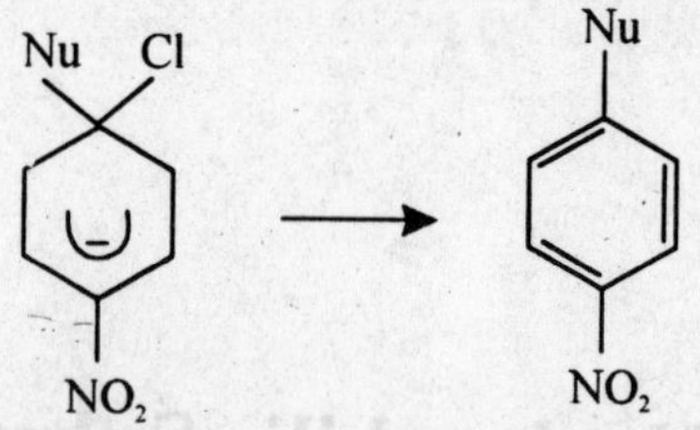

(I) Synthesis of 2,4 dinitrophenyl hydrazine from 1-chloro 2,4 dinitro benzene

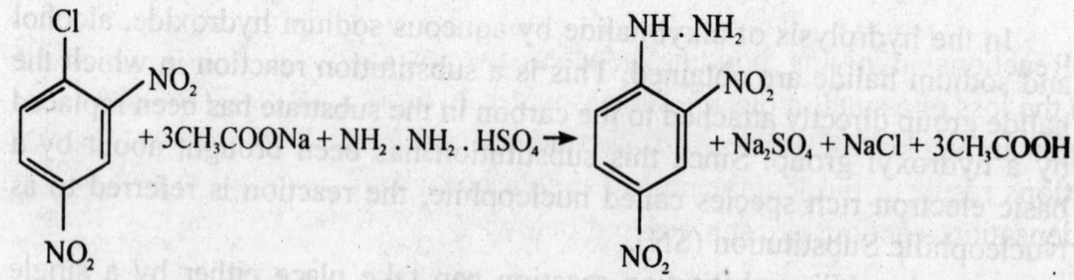

In a 250ml beaker dissolve 7gm of hydrazine sulphate in 25ml of hot water and then add with stirring 24 gm of sodium acetate.Heat the mixture to boiling for 5 minutes, cool to 70°C and then add 15ml of ethanol to it.Filter the solution at pump, wash with 20ml of ethanol and keep the filtered hydrazine solution for the preparation.

Dissolve 10gm of 1 chloro 2,4 dinitro benzene in 50ml ethanol contained in a 250ml round bottom flask; add the freshly prepared hydrazine solution to it.Attach a condensor to the flask and reflux the mixture with occasional shaking for an hour.Cool the resulting reaction mixture, allow precipitation to take place and then filter on a buchner funnel. Wash the precipitated solid with ethanol. Recrystallise 2,4 dinitrophenyl hydrazine from butanol, m.p 200°C (d) (Yield 5-6 gms)

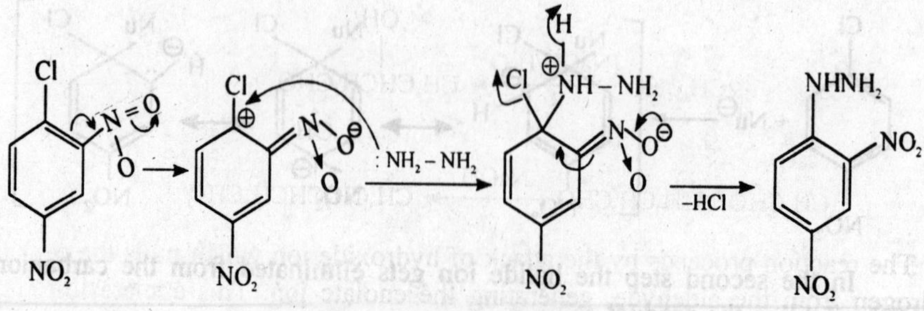

12

Condensation Reactions

Reactions involving bonding between two or more molecules, often with the loss of a small molecule such as water, hydrochloric or hydrobromic acid or an alcohol are referred to as condensation reactions. Condensation reactions result in the lengthening of the carbon chain. Some examples of condensation reactions are summarised below.

$$2CH_3COCH_3 \xrightarrow{Ba(OH)_2} (CH_3)_2C(OH)CH_2COCH_3$$

$$CH_3CHO + CH_3COCH_3 \xrightarrow{NaOH} CH_3CH(OH)CH_2COCH_3$$

(A) ALDOL CONDENSATION

The electrophilicity of the carbonyl carbon atom and nucleophilicity of the enolate carbanion leads to interesting carbonyl condensation reactions. When an enolate ion generated by the deprotonation of an aldehyde containing a hydrogen, attacks the carbonyl group of another aldehyde, addition takes place across carbonyl double bond. The product, a β-hydroxy aldehyde, is also known as aldol, as it contains both an aldehyde and an alcoholic group.

$$2CH_3CHO \xrightarrow{NaOH/H_2O} CH_3CH(OH)CH_2CHO$$

$$CH_3CHO + CH_3CH_2CHO \xrightarrow{NaOH} CH_3CH_2CH(OH)CH_2CHO$$

The reaction proceeds by the attack of hydroxide ion which pulls the α hydrogen from the aldehyde, generating the enolate ion. This enolate ion which now acts as a nucleophile attacks the carbonyl carbon of the other aldehyde to form an anion. The anion then exchanges proton with water to form aldol.

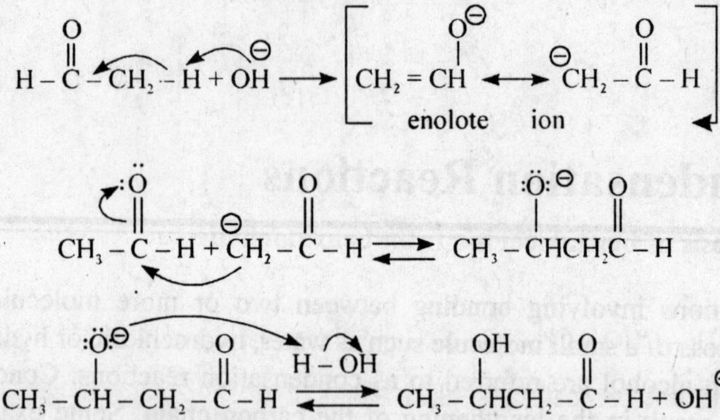

Aldehydes undergo self-condensation because of the electron withdrawing nature of the carbonyl group renders the α hydrogen susceptible to base attack forming a carbanion. Aldehydes are good acceptors because of low degree of steric hinderance whereas ketones are poor acceptors and their ability to undergo self condensation is not quite favourable.

(B) CLAISEN SCHMIDT REACTION

Condensation of aromatic aldehydes lacking α hydrogen with aliphatic aldehydes or ketones in the presence of dilute mineral alkali leads to the formation of aldol, elimination of molecule of water results in the formation of αβ unsaturated aldehyde or ketone. In this condensation reaction known as Claisen Schmidt reaction, the aromatic aldehyde acts as an electrophile whereas the aliphatic aldehyde or ketone acts as a neuclophile.

$$C_6H_5CHO + CH_3CHO \xrightarrow{NaOH} C_6H_5CH = CHCHO$$

$$C_6H_5CHO + CH_3COOC_2H_5 \xrightarrow{NaOH} C_6H_5CH = CHCOOC_2H_5$$

The reaction proceeds by the formation of enolate ion which attacks the carbonyl carbon of the aldehyde to form an intermediate anion. Addition of a proton to the anionic oxygen gives the hydroxy compound. Elimination of a molecule of water results in the formation of αβ unsaturated ketonic compound.

$$CH_3-\overset{O}{\overset{\|}{C}}-CH_2H + \overset{\ominus}{O}H \longrightarrow \overset{\ominus}{C}H_2-\overset{O}{\overset{\|}{C}}-CH_3 + H_2O$$

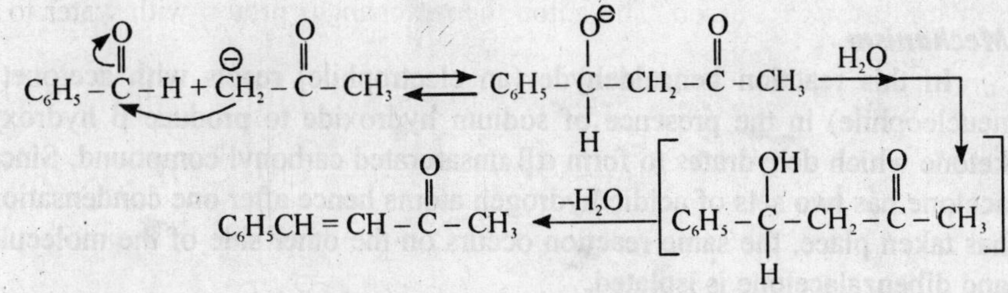

(j) Synthesis of benzalacetone from benzaldehyde

$$C_6H_5-CHO+CH_3-\overset{\displaystyle O}{\overset{\|}{C}}-CH_3 \xrightarrow{NaOH} C_6H_5CH=CH-\overset{\displaystyle O}{\overset{\|}{C}}-CH_3$$

In a 100ml conical flask place 10ml of freshly distilled benzaldehyde and 20ml of acetone. Keep the flask in cold water bath and then add dropwise with stirring 2.5ml of 10% sodium hydroxide solution such that the temperature does not rise above 30°C. After the addition is complete stir the mixture for 2 hours. Add dilute hydrochloric acid to the reaction mixture until it is just acidic and then transfer it to a 250ml separating funnel. Add 20ml chloroform/ether to the separating funnel and shake thoroughly. Remove the organic layer and repeat the process twice, each time using 20ml of chloroform/ether. Combine the three organic layers, wash with water and then distill on water bath to remove the solvent. Distill the residual portion under reduced pressure and collect the fraction boiling at 150-155°C/ 25mmHg. Cool the distillate which soon solidifies, collect benzalacetone, m.p 42°C.

(*ii*) Synthesis of dibenzalacetone from benzaldehyde

$$2C_6H_5CHO+CH_3-\overset{\displaystyle O}{\overset{\|}{C}}-CH_3 \xrightarrow{NaOH} C_6H_5CH=CH-\overset{\displaystyle O}{\overset{\|}{C}}-CH=CH-C_6H_5$$

In a 100ml conical flask dissolve 5gm of sodium hydroxide in 30ml of water and add 25 ml of 95% ethyl alcohol to it. Cork the flask and then shake it thoroughly taking care that the temperature does not rise above 25°C. Remove the cork add 4ml of acetone and 11ml of benzaldehyde. The colour of the solution immediately changes from yellow to orange. Cork the flask and shake it thoroughly for 30-45minutes. Cool the flask and collect the precipitated product by suction filtration on buchner funnel. Wash the product with water so as to remove all the traces of sodium hydroxide. Recrystallise from ethanol and collect yellow crystals of dibenzalacetone, m.p 110-112°C.

Mechanism

In this reaction benzaldehyde (an electrophile) reacts with acetone(a neucleophile) in the presence of sodium hydroxide to produce β hydroxy ketone which dehydrates to form αβ unsaturated carbonyl compound. Since acetone has two sets of acidic hydrogen atoms hence after one condensation has taken place, the same reaction occurs on the other side of the molecule and dibenzalacetone is isolated.

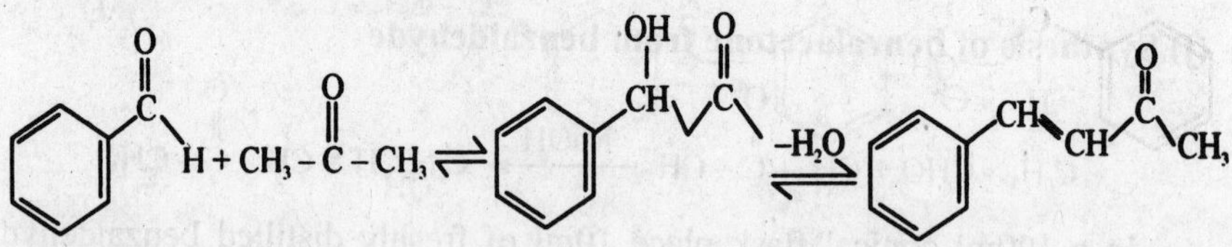

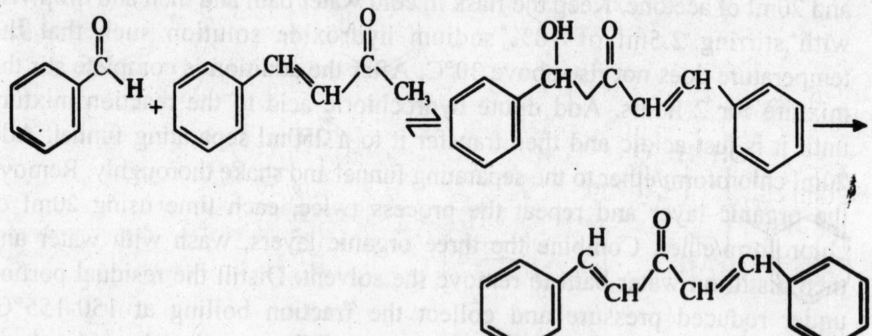

(*iii*) Synthesis of Benzalacetophenone (Chalcone) from acetophenone

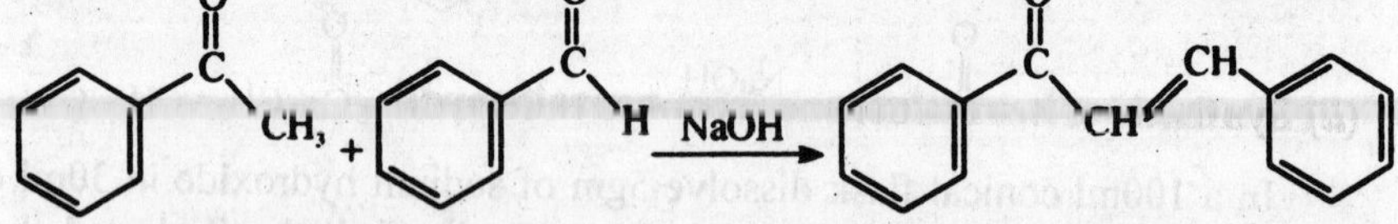

In a 100ml conical flask dissolve 5gm of sodium hydroxide in 40ml of water. Add 30ml of 95% ethanol and swirl the flask so as to effect mixing of the two liquids thoroughly. Cool the flask in ice water so that the temperature falls to about 20°C. Add 12ml of acetophenone and after briefly swirling the flask add 11ml of benzaldehyde. Stir the reaction mixture at room temperature for 2 hrs and then keep it overnight in a refrigerator. Filter the yellow precipitated benzalacetophenone through buchner funnel and wash it first with water and then with cold ethanol. Recrystallise from 95% ethanol and collect pure yellow crystals of benzalacetophenone, m.p 56-58°C (yield 5-6gm).

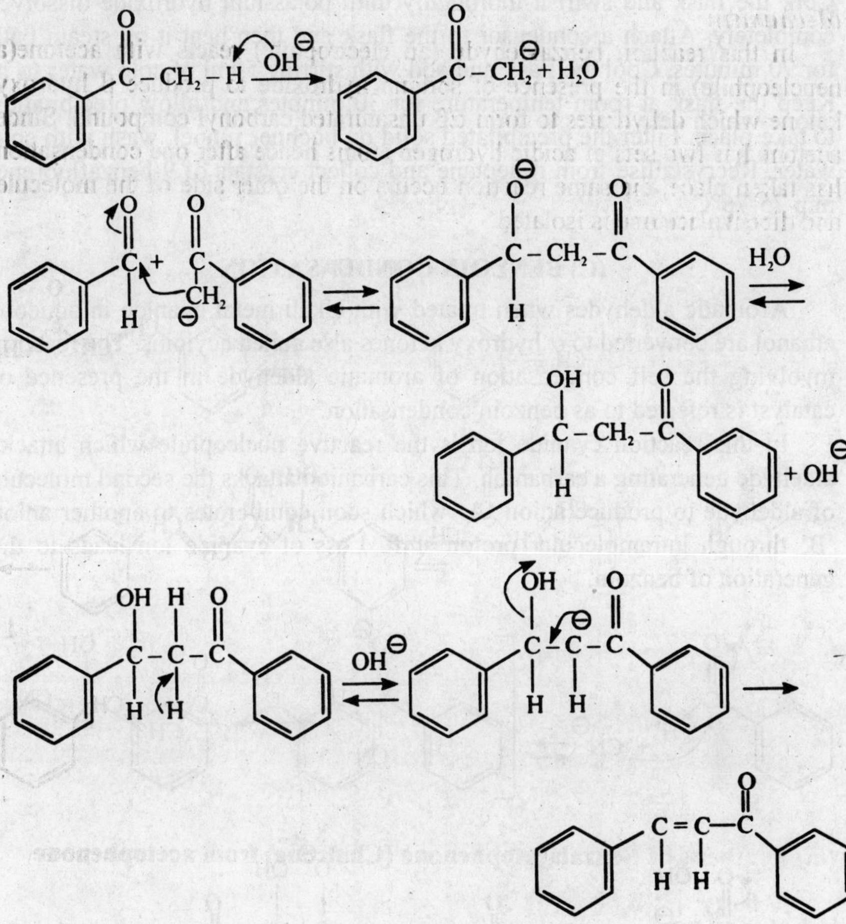

(*ii*) Synthesis of benzafluorene from benzaldehyde

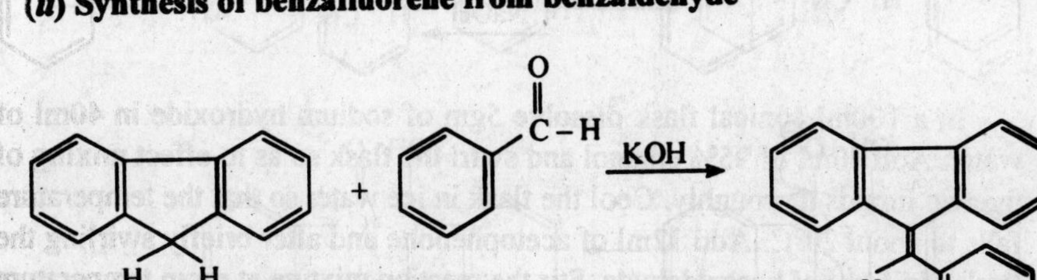

In a 100ml round bottom flask place 3gm of fluorene, 3.5 ml of benzaldehyde, 16ml of benzyl alcohol and 1.5gm of potassium hydroxide.

Cork the flask and swirl it thoroughly until potassium hydroxide dissolves completely. Attach a condensor to the flask and then heat it on steam bath for 90 minutes. Cool the flask and add with stirring 50ml of cold water to it. Keep the flask at room temperature for 30 minutes and allow precipitation to take place. Filter the precipitated solid on buchner funnel, wash with cold water. Recrystallise from n-heptane and collect crystals of 9-benzafluorene, m.p 75-76°C.

(C) BENZOIN CONDENSATION

Aromatic aldehydes when treated with alkali metal cyanide in aqueous ethanol are converted to α hydroxy ketones also called acyloins. The reaction involving the self condensation of aromatic aldehyde in the presence of catalyst is referred to as benzoin condensation.

In this reaction cyanide ion is the reactive nucleophile which attacks aldehyde generating a carbanion. This carbanion attacks the second molecule of aldehyde to produce anion 'A' which soon equilibrates to another anion 'B' through intramolecular proton shift. Loss of cyanide ion leads to the generation of benzoin.

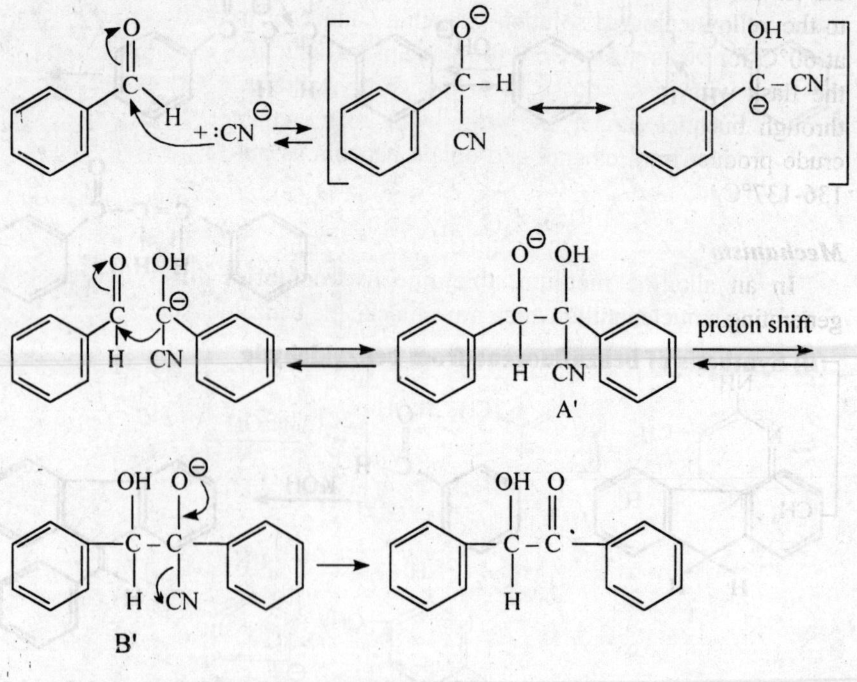

(i) Synthesis of benzoin from benzaldehyde

$$2C_6H_5CHO \xrightarrow[\text{aqueous ethanol}]{\text{NaCN}} C_6H_5-\overset{\overset{\displaystyle OH}{|}}{C}H-\overset{\overset{\displaystyle O}{\|}}{C}-C_6H_5$$

In a 100ml round bottom flask dissolve 1gm of sodium cyanide in 10ml of water. Add to the cyanide solution 15ml of ethanol and 10ml of freshly distilled benzaldehyde. Attach a reflux condensor to the flask, reflux the mixture for 30-35 minutes. Allow the contents of the flask to cool, first at room temperature and then cool in an ice bath. Filter the precipitated benzoin through buchner funnel and wash with cold water. Recrystallise the crude product from ethanol and obtain benzoin as pure white crystals, m.p 136-137°C. (Yield (7-7.5 gm).

(*ii*) Synthesis of benzoin from Vitamin B_1 (thiamine hydrochloride)

To a solution of 2.5gm of thiamine hydrochloride in 7ml of water in a 100ml round bottom flask add 25ml of 95% ethanol. Keep the flask in ice bath, add dropwise 4ml of 3N sodium hydroxide solution taking care that the temperature does not rise above 20°C. Now add 15ml of benzaldehyde to the yellow coloured solution. Attach a condensor to the flask and heat it at 60°C for 90 minutes. Cool the flask in ice bath and scratch the sides of the flask with glass rod so as to induce crystallization. Filter the product through buchner funnel and wash with cold methanol. Recrystallise the crude product from ethanol and obtain benzoin as pure white crystals, m.p 136-137°C.

Mechanism

In an alkaline medium, thiamine hydrochloride loses two proton generating a nucleophilie which now attacks the benzaldehyde

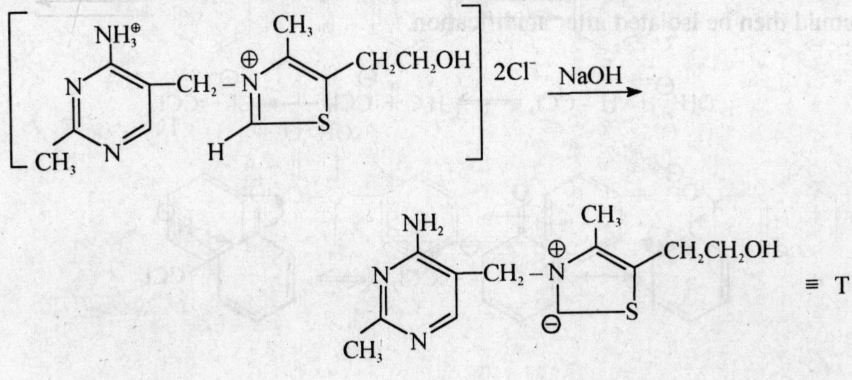

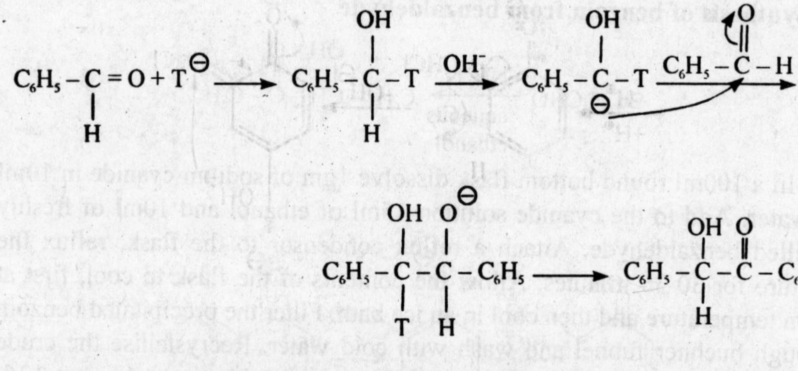

(D) REIMER –TIEMANN REACTION

Phenols on treatment with chloroform in presence of a base introduce an aldehydic group on to the aromatic ring generally ortho to the phenolic group. This method of generation of phenolic aldehydes is known as Reiman –Tiemann reaction

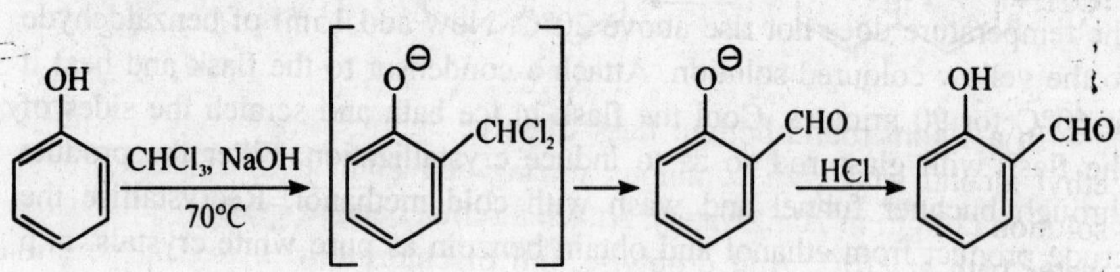

The reaction involves electrophilic substitution on the highly reactive phenoxide ring by dichlorocarbene [:CCl_2 (I)] generated from chloroform by the action of base. This results in the generation of benzylidene dichloride (*II*) which is then hydrolysed to hydroxyaldehyde. The phenolic aldehyde could then be isolated after acidification.

$$\overset{\ominus}{OH}{:} + H-CCl_3 \rightleftharpoons H_2O + :\overset{\ominus}{C}Cl_3 \longrightarrow \overset{\ominus}{Cl} + \underset{I}{:CCl_2}$$

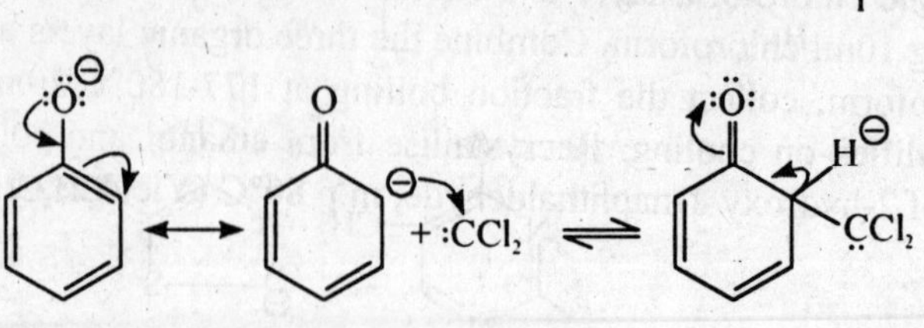

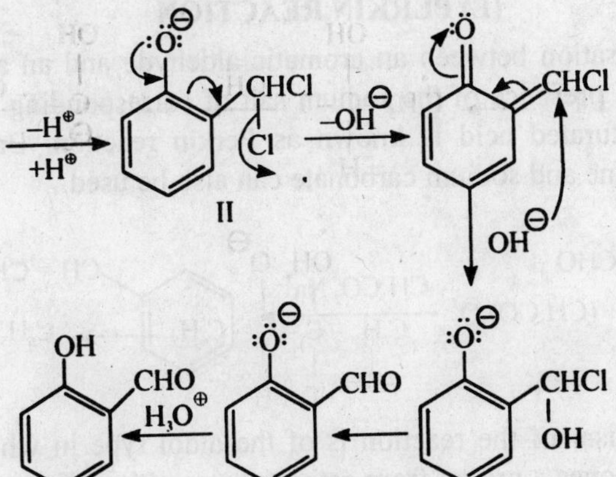

(i) Synthesis of 2-hydroxy-1-naphthaldehyde :-

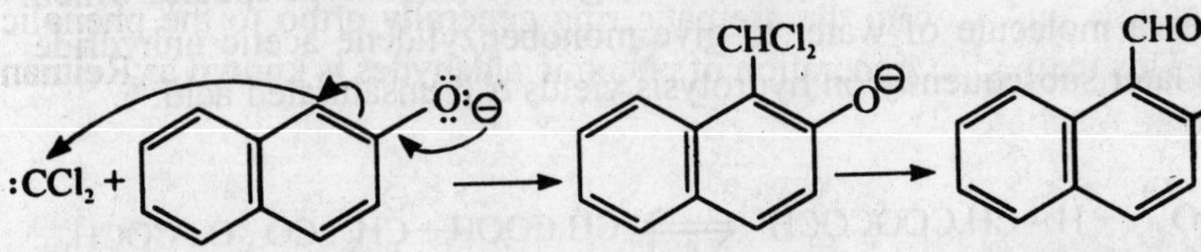

In a 100ml round bottom flask dissolve 5gm of 2-naphthol in 15ml of ethyl alcohol and with continous stirring add 20ml of sodium hydroxide solution (10 gm in 20ml water). Attach a reflux condensor, heat the flask on water bath at 80°C. Add dropwise 5ml of chloroform. After the reaction commenses (indicated by the formation of blue colour) remove the flask from the water bath and continue the addition of chloroform with stirring at such a rate that the mixture refluxes gently. Continue stirring till there is complete precipitation of the sodium salt of phenolic aldehyde. Distill off the excess chloroform and alcohol. Add dropwise with stirring concentrated hydrochloric acid to the residue until it is acidic to litmus. A dark oily liquid accompanied by considerable amount of sodium chloride separates out. Add water to dissolve the salt and then add 10ml chloroform and shake thoroughly. Remove the chloroform layer and repeat the extraction process twice each time using 10ml chloroform. Combine the three organic layers and distill off the chloroform, collect the fraction boiling at 177-180°C/20mmHg which soon solidifies on cooling. Recrystallise from ethanol and collect the fine crystals of 2-hydroxy-1-naphthaldehyde, m.p 80°C (Yield 2.5-2.8gm)

(E) PERKIN REACTION

The condensation between an aromatic aldehyde and an aliphatic acid anhydride in the presence of the sodium salt of corresponding anhydride to yield α β unsaturated acid is known as Perkin reaction. Bases such as pyridine, quinoline and sodium carbonate can also be used.

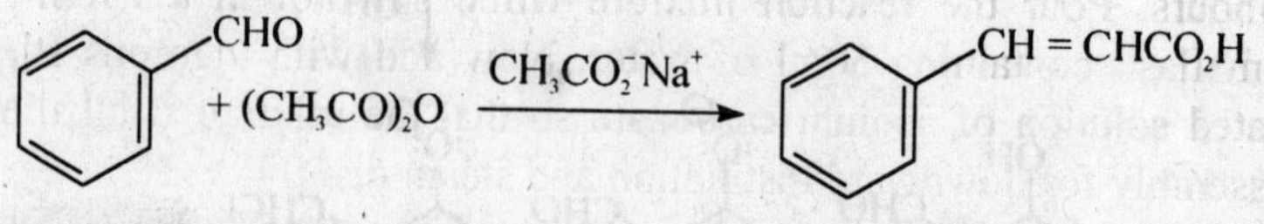

The mechanism of the reaction is of the aldol type in which the base (acetate ion) removes a proton from activated α position of the anhydride to yield the carbanion (A). The carbanion then adds to the carbonyl group of the aldehyde, the product after protonation gives an aldol type species which then loses a molecule of water to give monobenzylidene acetic anhydride (B). The later subsequently on hydrolysis yields α β unsaturated acid.

$$CH_3COO^{\ominus} + H-CH_2COOCOCH_3 \rightleftharpoons CH_3COOH + \underset{A}{\overset{\ominus}{C}H_2} . CO . O . COCH_3$$

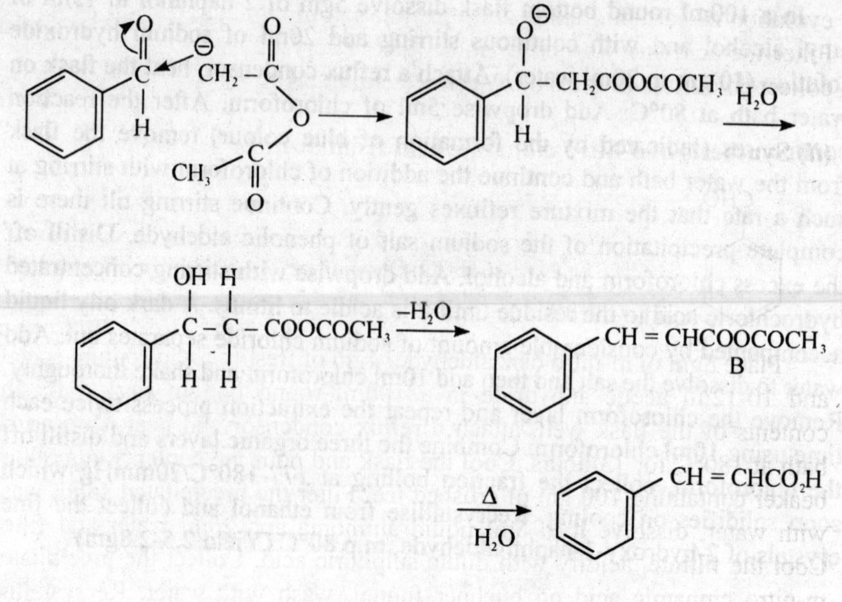

(i) Synthesis of cinnamic acid from benzaldehyde

Method I :-

In a 100ml round bottom flask fitted with a reflux condensor place 10ml of benzaldehyde, 15ml of acetic anhydride and 5gm of finely powdered anhydrous sodium acetate. Reflux the mixture in an oil bath at 175-180°C for 4hours. Pour the reaction mixture while still hot in a 250ml round bottom flask containing 50ml of water. Now add with vigorous stirring a saturated solution of sodium carbonate so that the mixture is alkaline. Set the assembly for downward distillation and steam distill the solution until all the unreacted benzaldehyde has been removed and the solution is no longer turbid. Cool the residual solution and filter. Cool the filtrate, acidify with concentrated hydrochloric acid and stir vigorously until there is no more evolution of carbon dioxide and precipitation of cinnamic acid is complete. Filter the acid, wash with cold water. Recrystallise from hot water and collect colourless crystals of cinnamic acid, m.p 132°C. (Yield 8-9gm).

Method *II* :- After refluxing the contents for 4hrs, cool the flask and add 50ml of water to it followed by addition of a saturated solution of sodium carbonate till alkaline. Extract the unreacted benzaldehyde with two 25ml portions of chloroform. Cool the aqueous layer, acidify with concentrated hydrochloric acid and stir vigorously until there is no more evolution of carbon dioxide and precipitation of cinnamic acid is complete. Filter the acid and wash with cold water. Recrystallise from hot water and collect colourless crystals of cinnamic acid, m.p 132°C. (Yield 8-9gm).

(*ii*) Synthesis of 3-nitro cinnamic acid from m-nitro benzaldehyde:-

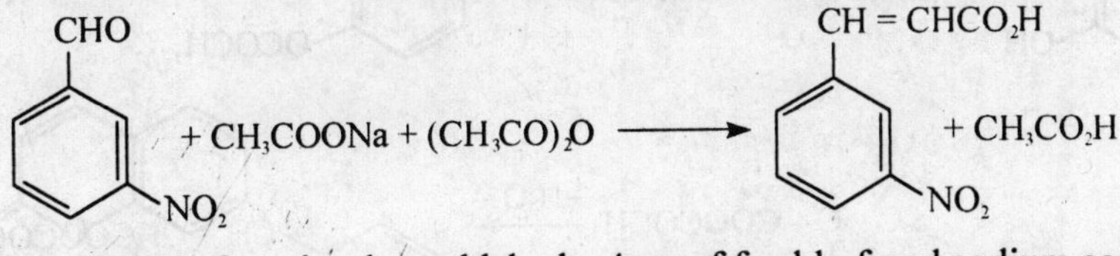

Place 5gm of m-nitro benzaldehyde, 4gm of freshly fused sodium acetate and 10-12ml acetic anhydride in a 100ml round bottom flask. Mix the contents of the flask well, attach a reflux condensor and heat it on an oil bath at 180°C for 13 hours. Cool the flask and pour the contents in a 500ml beaker containing 100 gm of crushed ice. Filter the precipitated solid, wash with water, dissolve it in 5ml liquid ammonia and 20ml water and filter. Cool the filtrate, acidify with dilute sulphuric acid. Collect the precipitated m-nitro cinnamic acid on buchner funnel, wash with water. Recrystallise from benzene and collect m-nitro cinnamic acid in the form of yellow crystals, m.p 192-194°C.

$$CH_3COO^{\ominus} + H-CH_2COOCOCH_3 \rightleftharpoons CH_3COOH + \overset{\ominus}{C}H_2CO.O.COCH_3$$

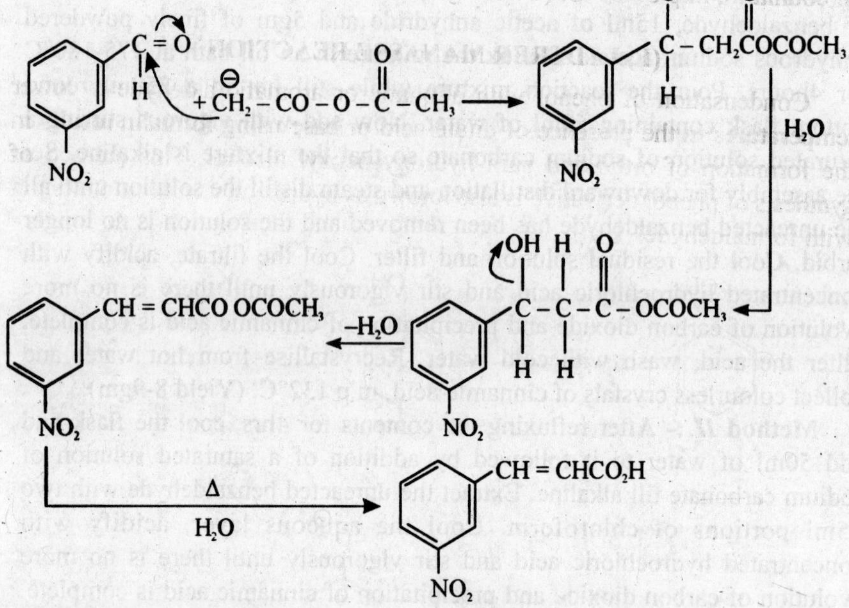

(*iii*) Synthesis of Coumarin from Salicylaldehyde

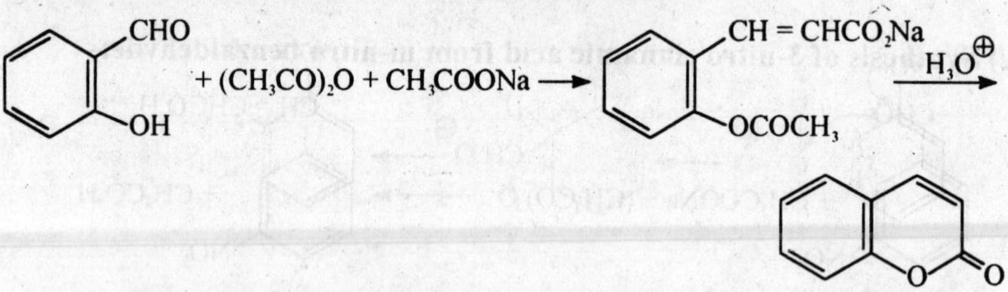

In a 100ml round bottom flask attached with a reflux condensor place 9ml of salicylaldehyde, 26 ml of acetic anhydride, 16 gm of fused sodium acetate and 250mg of iodine. Reflux the reaction mixture, first gently at 120°C for 2 hours and then further for 5 hrs at 180-190°C. Pour the reaction mixture in a 500ml beaker containing 100ml of ice cold water. Transfer the aqueous portion to a separating funnel and add 25ml ether to it. Shake the funnel thoroughly and remove the organic layer. Repeat the process twice each time using 25ml of ether. Combine the ethereal layers and add to it concentrated sodium bisulphite solution and filter. This procedure removes the unreacted salicylaldehyde. Wash the ether solution with water and dry it over calcium chloride. Distill off the solvent and recover the crude coumarin. Recrystallise the product from pet-ether or 40% methanol and collect crystals of coumarin, m.p 67-69°C. (Yield 4gm).

(F) LEDERER MANASSE REACTION

Condensation of phenol with aliphatic or aromatic aldehyde at lower temperatures in the presence of dilute acid or base using formalin results in the formation of ortho and para hydroxy benzyl alcohol. This method of synthesis of phenol derivative is also known as hydroxymethylation of phenol with formaldehyde.

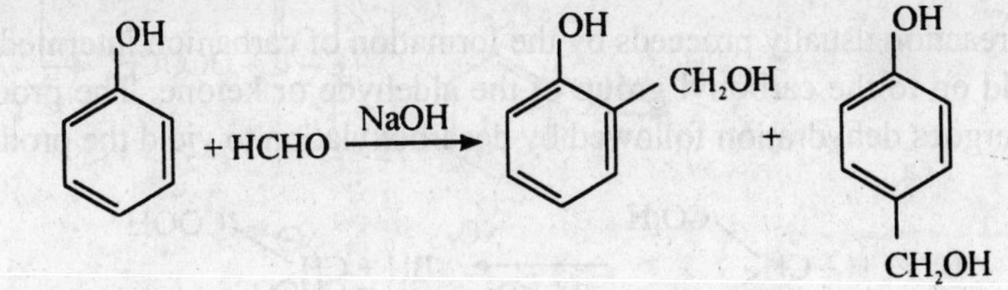

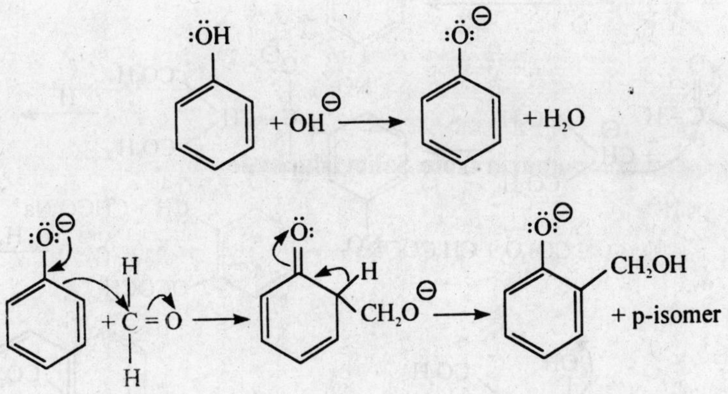

(i) Synthesis of ortho and para hydroxy benzyl alcohol from phenol

In a 250ml conical flask dissolve 10gm of phenol in 55ml of 10% sodium hydroxide solution and then add 30ml of 40% formaldehyde solution to it. Cork the flask and keep the reaction mixture at room temperature for six days. Neutralise the reaction mixture by adding hydrochloric acid to it. Transfer the resultant solution to a separating funnel and add 20ml ether to it. Shake the funnel thoroughly, remove the ethereal layer. Repeat the process twice each time using 20ml of ether. Combine the three organic layers and distill off the ether on steam bath. Dissolve the residue in benzene and filter.

Cool the filtrate, para isomer is precipitated first and is collected, m.p 112°C. Concentrate the mother liquor and cool, collect the ortho isomer, m.p 82°C.

(G) KNOEVENAGEL REACTION

The reaction of carbonyl compounds with compounds containing active methylene group like malonic ester, malonic acid, ethyl acetoacetate etc in the presence of base resulting in the formation of unsaturated compounds is known as Knoevenagel reaction. The base commonly used are piperidine, pyridine, ammonia, primary or secondary amines.

$$CH_3CHO + CH_2(CO_2H)_2 \longrightarrow CH_3CH = C(CO_2H)_2 \xrightarrow{-CO_2} CH_3CH = CHCO_2H$$

The reaction usually proceeds by the formation of carbanion intermediate which add on to the carbonyl group of the aldehyde or ketone. The product then undergoes dehydration followed by decarboxylation to yield the product.

$$B: + H-CH(CO_2H)_2 \longrightarrow BH + \overset{\ominus}{C}H(COOH)_2$$

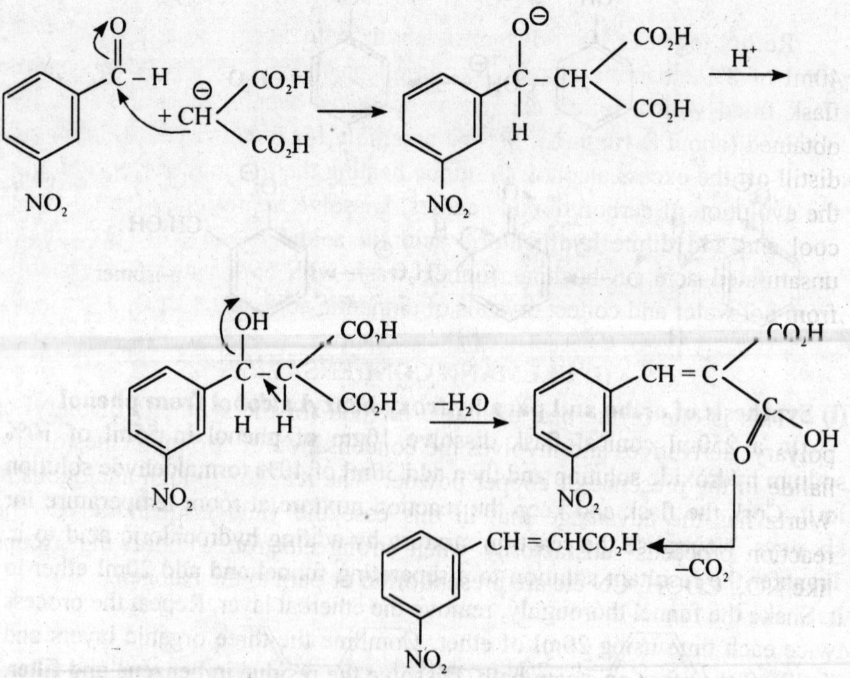

(*i*) Synthesis of m-nitro cinnamic acid from m-nitro benzaldehyde

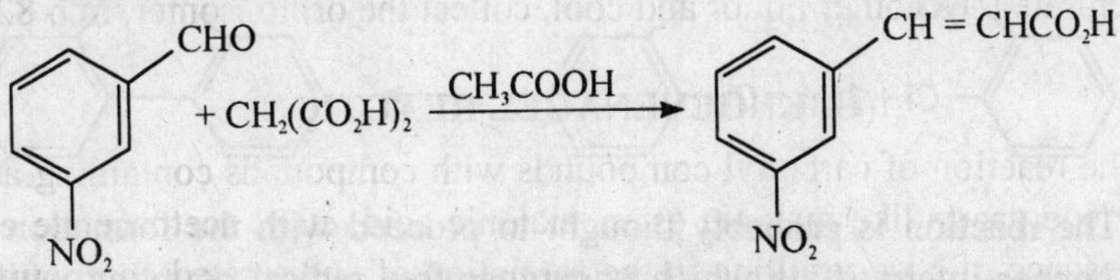

Reflux together 3gm of m-nitro benzaldehyde with 2.3gm of malonic acid, 7ml of ethanol and .5ml of pyridine placed in a 100ml round bottom flask fitted with a reflux condensor on steam bath for 7-8 hours. Cool and freeze the mixture in ice bath. Filter and collect the precipitated solid on buchner funnel, wash with cold water. Recrystallise the crude product from ethanol and collect yellow crystals of m-nitro cinnamic acid, m.p 200-201°C.

(*ii*) Synthesis of cinnamic acid from benzaldehyde.

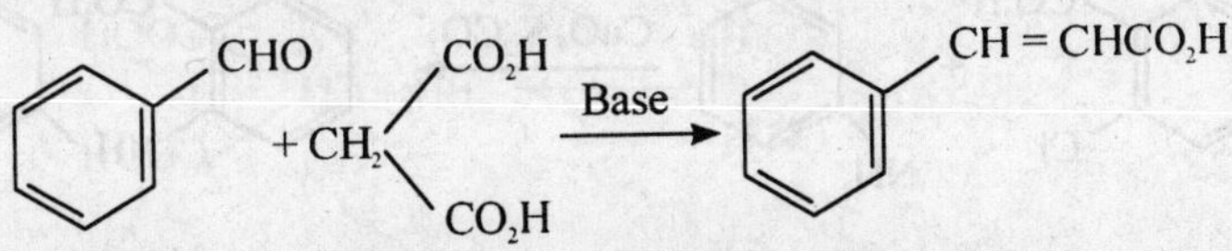

Reflux together 10ml of benzaldehyde with 10gm of malonic acid and 40ml of 8% ethanolic ammonia solution placed in a 100ml round bottom flask fitted with a reflux condensor on water bath till a clear solution is obtained (about 8-10hours). Set the assembly for downward distillation and distill off the excess alcohol. Continue heating the residual oily portion until the evolution of carbon dioxide ceases. Dissolve the residue in 20ml water, cool and add dilute hydrochloric acid till acidic. Collect the precipitated unsaturated acid on buchner funnel, wash with cold water. Recrystallise from hot water and collect crystals of cinnamic acid, m.p 132°C.

(I) ULLMANN CONDENSATION

This is one of the best available method for the synthesis of biaryl or polyaryl derivatives and involves the condensation of two molecules of aryl halide in the presence of copper powder. The reaction though analogous to Wurtz has the advantage that in this case the yield is quantitative. The reaction proceeds satisfactorily when strong electron withdrawing groups like NO_2, COOH, CN etc are present ortho or para to the halogen.

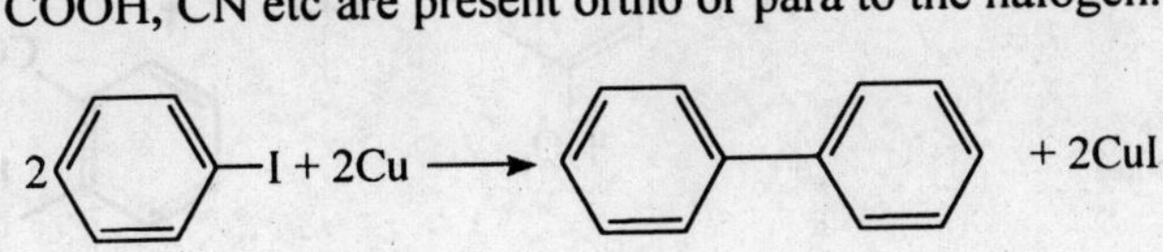

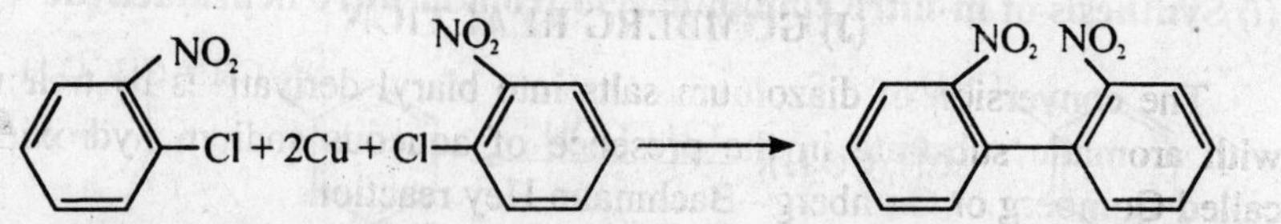

The reaction is probably thought to proceed with the formation of an aryl copper intermediate which undergoes free radical coupling while still coordinated to copper.

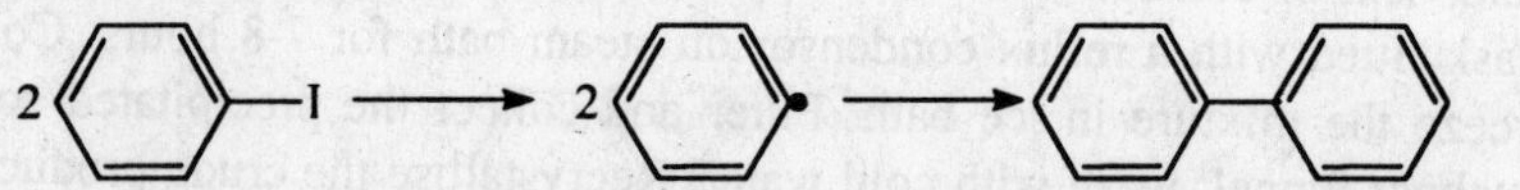

(i) Synthesis of N Phenyl Anthranilic acid from o-chloro benzoic acid

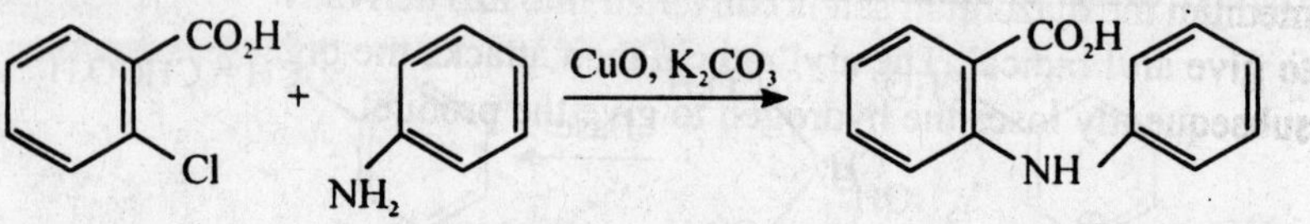

In a 100ml round bottom flask fitted with an air condensor place 5gm of o-chloro benzoic acid, 20ml of aniline, 5gm of potassium carbonate and 3gm of copper oxide. Reflux the mixture on an oil bath for 2 hours. Cool the flask at room temperature, set the assembly for downward distillation and distill off the excess aniline. To the residual brown solution containing potassium N phenyl anthranilic acid add a pinch of animal charcoal, boil for 10 minutes and filter hot through glass wool or through cotton plug. Cool the filtrate, add dilute hydrochloric acid until no more precipitation occurs and then cool again. Filter the precipitated acid by suction filteration and wash with water. Recrystallise from alcohol and collect colourless crystals of N Phenyl Anthranilic acid, m.p 185-186°C.

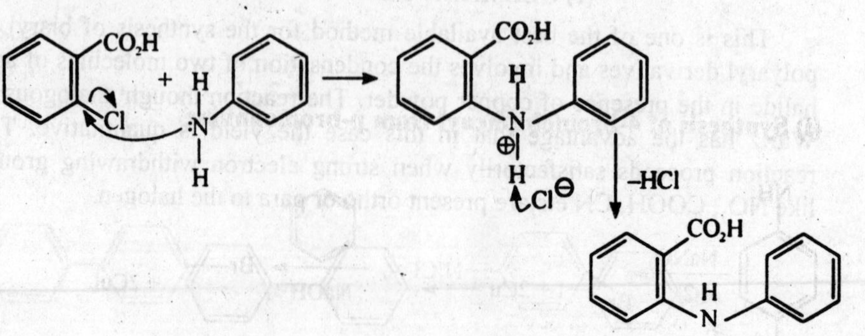

(J) GOMBERG REACTION

The conversion of diazonium salts into biaryl derivatives by treatment with aromatic substrate in the presence of aqueous sodium hydroxide is called Gomberg or Gomberg –Bachmann Hey reaction.

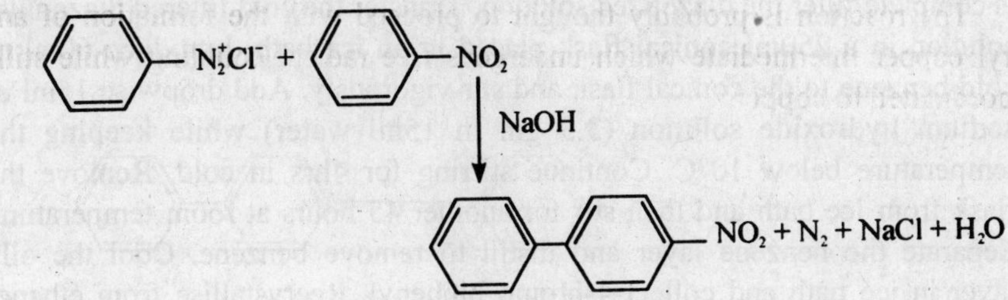

The reaction is believed to proceed by free radical mechanism. In alkaline medium the diazonium salt is converted into azo derivative which decomposes to give aryl radical. The aryl radical then attacks the organic substrate which subsequently loses the hydrogen to give the product.

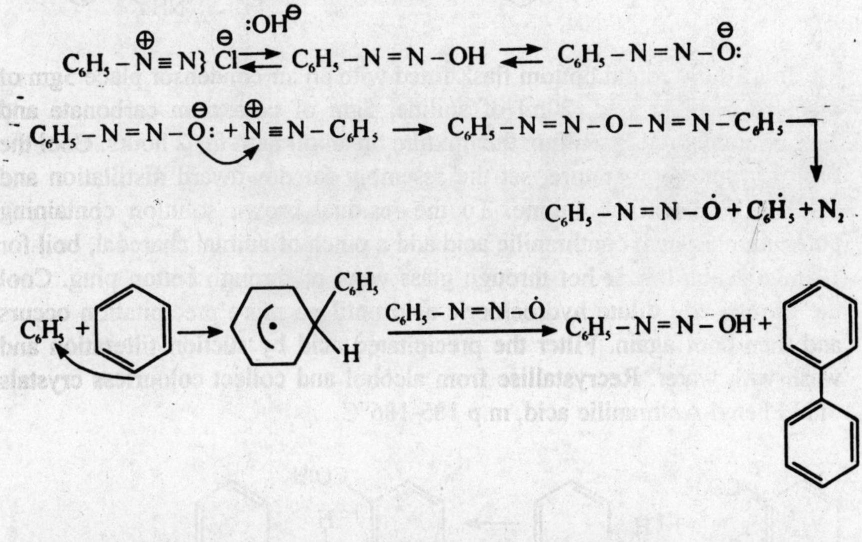

(i) Synthesis of 4-bromobiphenyl from p-bromoaniline

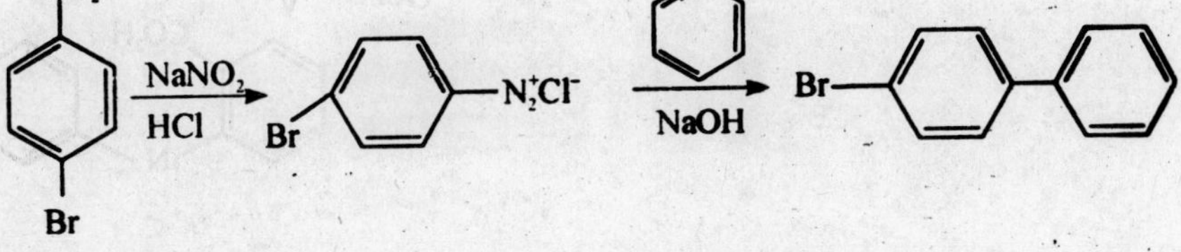

In a 100ml conical flask dissolve 5gm of p-bromoaniline in a mixture of 5ml of concentrated hydrochloric acid and 5ml of water. Place the flask in an ice bath. When the temperature of the reaction mixture falls below 5°C, then add cold solution of sodium nitrite (2.5 gm in 10 ml of water) dropwise taking care that the temperature does not rise above 10°C. After the addition is complete filter the diazotized solution. Transfer the cold filtered diazonium solution in a 250ml conical flask placed in an ice bath. Introduce 60ml of cold benzene to the conical flask and stir vigorously. Add dropwise 15ml of sodium hydroxide solution (3.3 gm in 15ml water) while keeping the temperature below 10°C. Continue stirring for 4hrs in cold. Remove the flask from ice bath and then stir for another 45 hours at room temperature. Separate the benzene layer and distill to remove benzene. Cool the oily layer in ice bath and collect 4-bromo biphenyl. Recrystallise from ethanol and collect crystals of 4-bromobiphenyl, m.p 90°C.

13
Rearrangement Reactions

During most organic reactions the carbon skeleton of the reactant normally remains unchanged and also there is no change in the position of the functional group. But there are reactions during which the functional group migrate from one atom to another within the same molecule or there is change in the carbon skeleton. Such reactions are referred to as rearrangement.

Rearrangement reactions usually take place if during the course of reaction carbon or any hetro atom develops a sextet of electron.This intermediate being very rich in energy can be stabilised by the migration of substituent present at α position along with its bonding electron to the electron deficient atom.The new centre generated by the migration also has a sextet of electron which can be stabilised in subsequent reaction either by nucleophilic substitution or elimination.

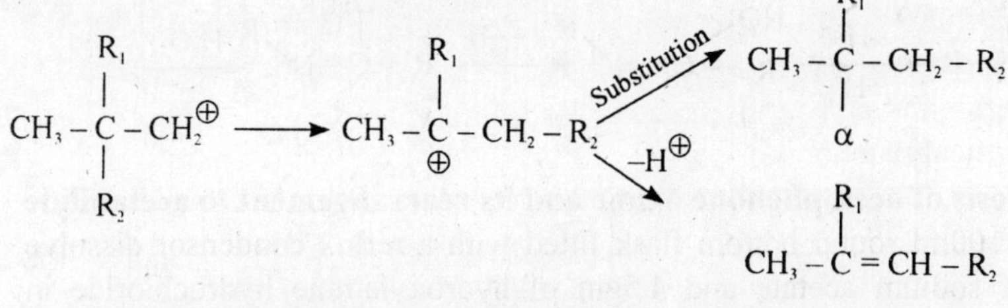

(a) BECKMANN REARRANGEMENT

Beckmann rearrangement consists of the conversion of ketoxime, the oxime of ketone to N substituted amide.The reaction is catalysed by a number of acidic reagents such as concentrated sulphuric acid,phosphorous pentachloride, p-toluene sulphonyl chloride and trifluoroacetic acid. These acids only assist in the departure of the hydroxyl group.

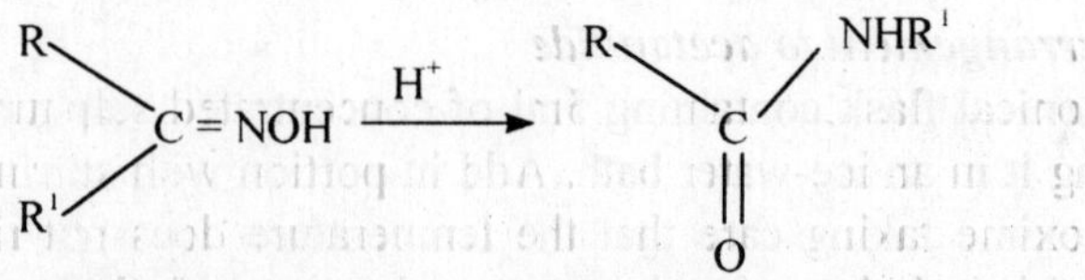

The rearrangement involves the migration of the alkyl or aryl group from carbon to electron deficient nitrogen. But this reaction is highly stereospecific and it has been observed that only the group which is trans (anti) to the leaving group migrates.

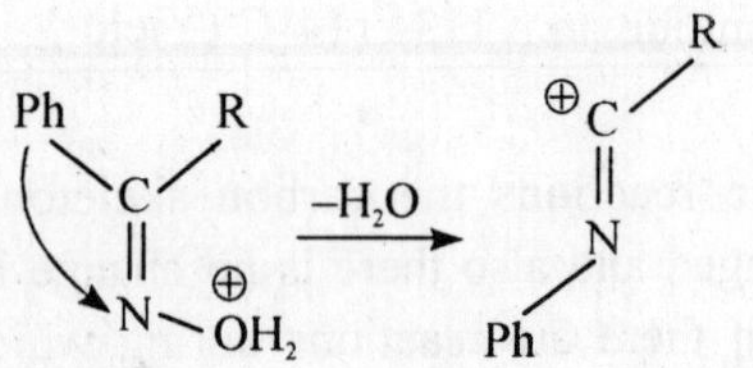

The reaction usually involves the protonation of the oxime followed by heterolysis of the leaving group (H_2O) yielding electron deficient nitrogen. Immediate migration of the group R takes place followed by attack of water on the carbocation which loses a proton yielding enolic form of amide which finally reverts to amide.

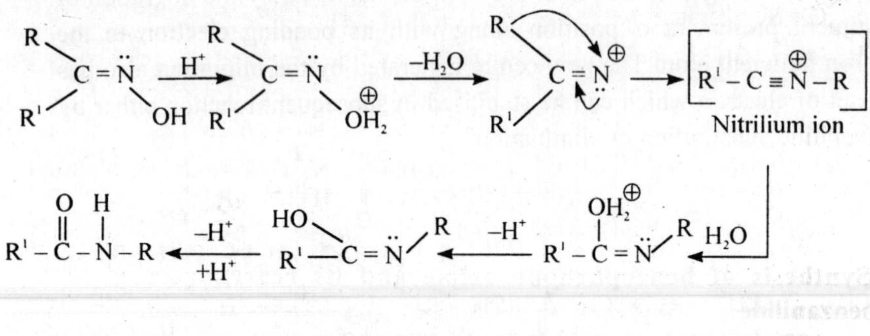

(i) Synthesis of acetophenone oxime and its rearrangement to acetanilide

In a 100ml round bottom flask fitted with a reflux condensor dissolve 5.0gm of sodium acetate and 4.5gm of hydroxylamine hydrochloride in 20ml water (heat on water bath if necessary). Now with stirring add slowly a solution of 7ml of acetophenone in 6ml of 95% ethanol. Heat the flask on oil bath (100-110°C) or on steam bath for 20 minutes. Cool the flask in ice bath and if necessary scratch the walls of the flask with a glass rod to induce crystallization. Filter the product through buchner funnel and wash with water. Recrystallise acetophenone oxime from water or petroleum ether, m.p 59°C (Yield 5-7gms)

Beckmann rearrangement to acetanilide

Cool the conical flask containing 5ml of concentrated sulphuric acid to 15°C by placing it in an ice-water bath. Add in portion with stirring 5gm of acetophenone oxime taking care that the temperature does not rise above 25°C.Transfer this solution of oxime to another conical flask containing 2.5ml of concentrated sulphuric acid. Heat the flask on steam bath for 30 minutes. Cool and pour the contents of the flask in a 400ml beaker containing 100ml of ice cold water. Filter the precipitated product on buchner funnel and wash with cold water. Recrystallise from hot water and collect almost colourless crystals of acetanilide, m.p 113- 114°C (Yield 3.5-4 gm)

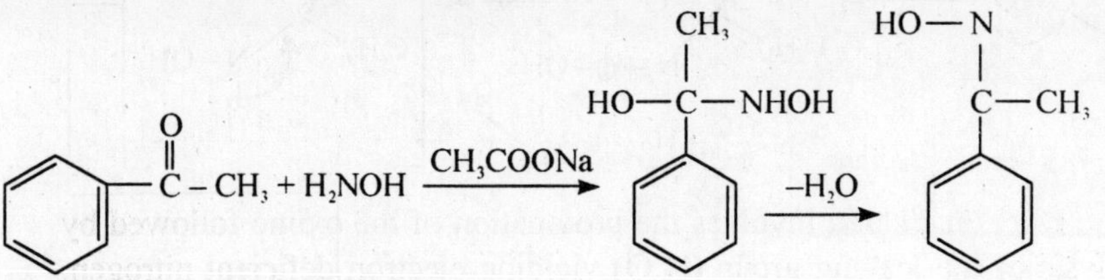

Mechanism

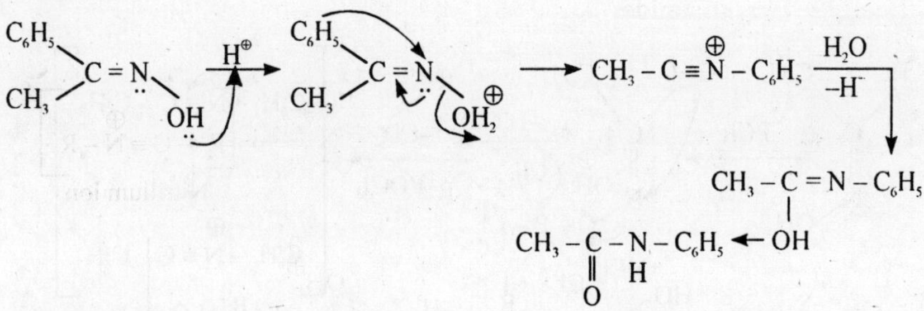

(ii) Synthesis of benzophenone oxime and its rearrangement to benzanilide

In a 100ml round bottom flask fitted with a reflux condensor place 5gm of benzophenone,3gm of hydroxylamine hydrochloride, 10ml of ethyl alcohol and 2ml water. To the resulting mixture now add with shaking and in portions 5.6gm of solid sodium hydroxide taking care that the reaction does not become too vigorous (if so cool the flask under tap water). After the addition of sodium hydroxide is complete heat the flask under reflux for 5-10 minutes. Cool and pour the contents of the flask in a 400ml beaker containing a solution of 15ml concentrated hydrochloric acid in 100ml water. Filter the precipitated oxime through buchner funnel and wash with cold water.Recrystallise the product from methanol, m.p 140- 142°C (Yield 4-5gms).

Beckmann rearrangement to benzanilide

In a 50ml round bottom flask dissolve 3.0gm of benzophenone oxime in 30ml ether and add 4.5gm of phosphorous pentachloride to it. Set the assembly for downward distillation, heat the flask on water bath and distill off most of the solvent. Add 40ml water to the residue, boil the solution for 10-15 minutes and break the lumps. Filter the product through buchner funnel, recrystallise from ethanol and collect pure crystalline benzanilide, m.p 161- 163°C (Yield 2 – 2.5gms).

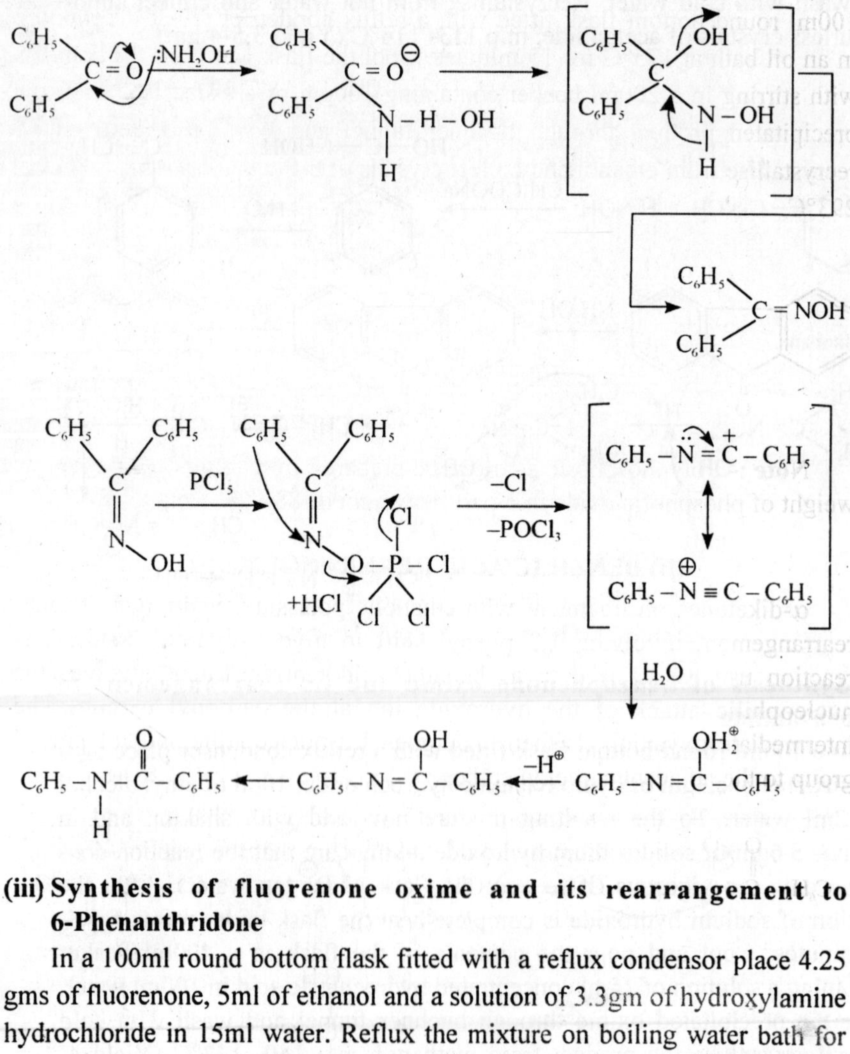

(iii) Synthesis of fluorenone oxime and its rearrangement to 6-Phenanthridone

In a 100ml round bottom flask fitted with a reflux condensor place 4.25 gms of fluorenone, 5ml of ethanol and a solution of 3.3gm of hydroxylamine hydrochloride in 15ml water. Reflux the mixture on boiling water bath for

30-40 minutes. Remove the flask from hot water, cool at room temperature and then add 6gms of sodium acetate to it. Heat the resultant mixture at 100°C for 60 minutes. Cool and collect the precipitated pale yellow coloured derivative on buchner funnel, recrystallise from ethanol and collect fluorenone oxime, m.p 195-197°C.

Beckmann rearrangement to 6-Phenanthridone

Place 1 gm of fluorenone oxime and 60gm of polyphosphoric acid in a 100ml round bottom flask fitted with a reflux condensor. Heat the mixture in an oil bath at 175°C for 15 minutes. Cool the flask and pour the contents with stirring in a 200ml beaker containing 100gm of crushed ice. Filter the precipitated product through buchner funnel and wash with cold water, recrystallise from ethanol and collect crystals of 6-Phenanthridone, m.p 292-293°C.

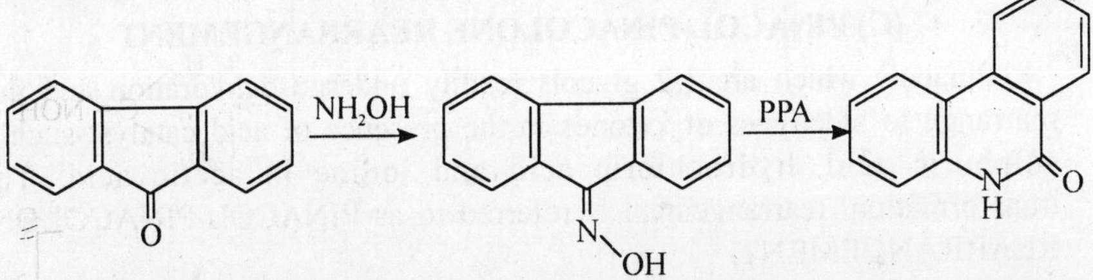

Note :- Polyphosphoric acid can be prepared by dissolving 1.8 parts by weight of phosphoric oxide in 1 part by weight of 88-90% H_3PO_4

(B) BENZILIC ACID REARRANGEMENT

α-diketones on treatment with ethanolic potassium hydroxide undergo rearrangement involving 1,2 phenyl shift to give α-hydroxy acids. The reaction usually referred to as benzilic acid rearrangement involves the nucleophilic attack of the hydroxide ion at the carbonyl carbon. The intermediate oxyanion (I) then undergoes 1,2 nucleophilic shift of phenyl group to the adjacent carbonyl carbon.

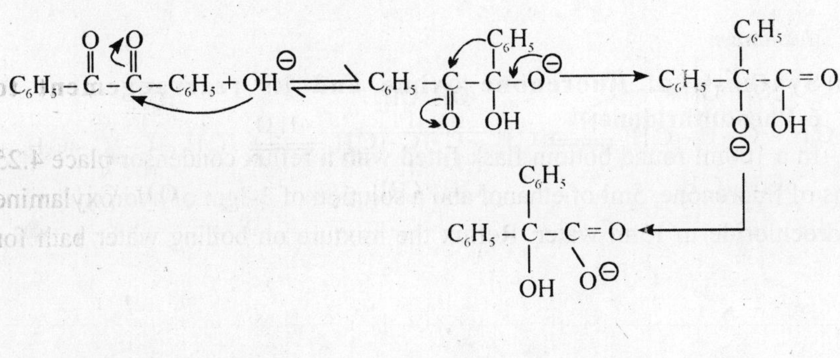

(i) Synthesis of benzilic acid from benzil

In a 100ml round bottom flask dissolve 5gm of potassium hydroxide in 10ml water and then add 15ml of ethanol and 5gm of benzil to it. Reflux the mixture on water bath for 20-30 minutes. Transfer the contents of the flask into a porcelain dish and keep overnight at room temperature. Collect crystals of potassium salt of benzilic acid by suction filtration and wash with ice cold water.

Dissolve potassium benzilate in 20ml water and then add dropwise with stirring concentrated hydrochloric acid until the solution is acidic and then cool. Benzilic acid will precipitate as a red brown sticky mass. Filter the precipitated product through buchner funnel and wash with cold water. Dissolve the coloured sticky mass in minimum amount of hot water, add a pinch of decolourising carbon to it, heat on water bath for few seconds and filter hot. Cool the filtrate and collect crystals of benzilic acid, m.p 148-150°C.

(C) PINACOL-PINACOLONE REARRANGEMENT

Pinacols which are 1,2 glycols readily undergo dehydration and then rearrange to aldehydes or ketones in the presence of acid catalyst such as sulphuric acid, hydrochloric acid and iodine in acetic acid. This transformation/ rearrangement is referred to as PINACOL-PINACOLONE REARRANGEMENT.

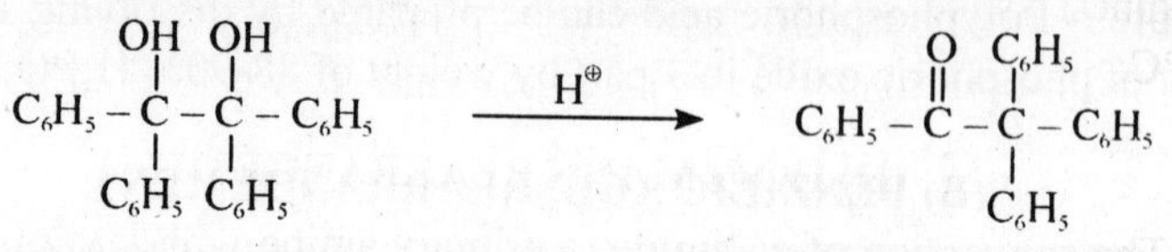

In the first step protonation of one of the hydroxyl groups takes place. This loses a molecule of water to give a tertiary carbocation. The migration of alkyl/aryl group with its bonding electron pair from one carbinyl position to adjacent electron deficient centre results in the formation of carbocation which is resonance stabilised. Deprotonation gives the desired product, pinacolone.

Mechanism

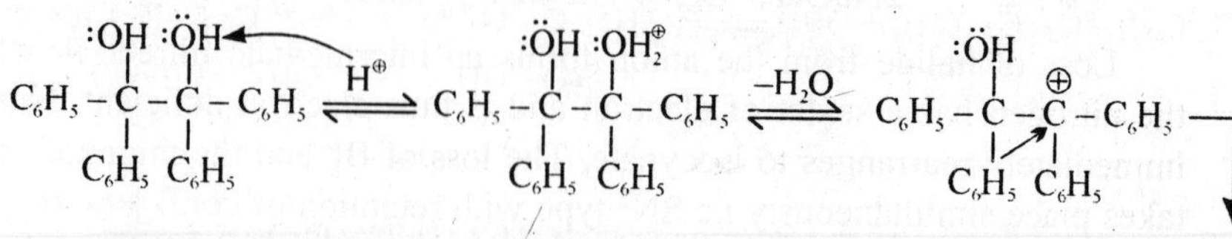

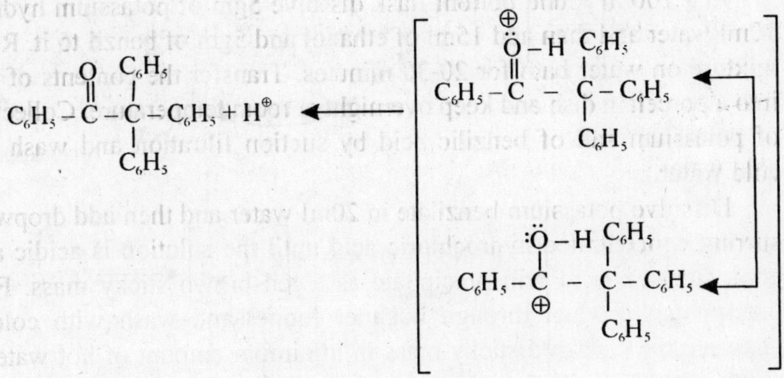

(i) Synthesis of Benzopinacolone from Benzopinacol

In a 100ml round bottom flask fitted with a reflux condensor place 5gm of benzopinacol,20ml of glacial acetic acid and two crystals of iodine. Heat the flask in an oil bath until all the crystals dissolve. Continue heating for another 5 minutes and then cool it to room temperature whereby pinacolone separates out as a stiff paste. Filter the precipitated benzopinacolone on buchner funnel, wash with 10-15ml of ethanol to make it free from iodine. Recrystallise by first dissolving in benzene and then add pet-ether until cloudiness appears. Collect colourless crystals of benzopinacolone, m.p 179-180°C.

(D) HOFFMANN REARRANGEMENT

The conversion of an amide to primary amine with loss of one carbonyl carbon atom by the action of alkaline hypobromite is known as HOFFMANN REARRANGEMENT.

$$RCONH_2 + Br_2 + 4NaOH \rightarrow RNH_2 + 2NaBr + Na_2CO_3 + H_2O$$

The reaction involves the treatment of amide with hypobromite (formed by treatment of bromine with alkali) to form N bromamide (I) which is subsequently converted to its anion under alkaline conditions.

$$2NaOH + Br_2 \rightarrow Na^+OBr^- + NaBr + H_2O$$

Loss of halide from the anion forms an intermediate nitrene in which the nitrogen has a sextet of electron and is thus electron deficient. Nitrene immediately rearranges to isocynate. The loss of Br^- and the migration of R takes place simultaneously i.e SN^2 type with retention of configuration. The reaction is completed by hydration of the isocynate to yield the carbamic acid which undergoes spontaneous decarboxylation to the amine.

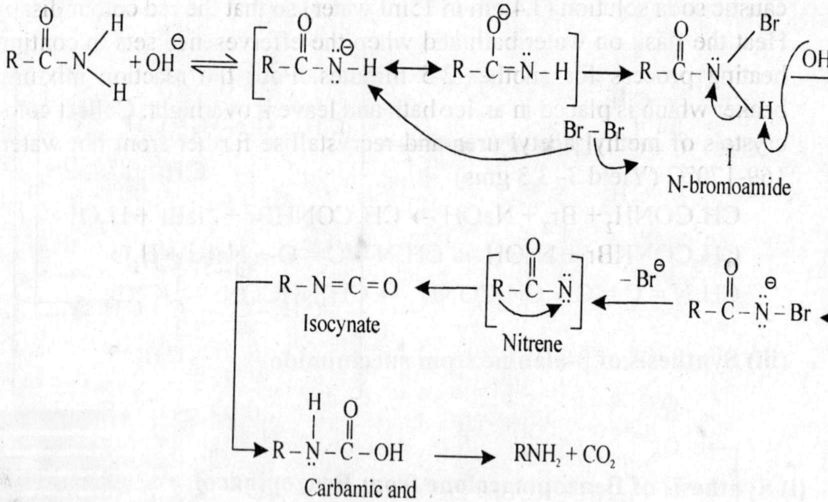

(i) Synthesis of anthranilic acid from phthalimide

To 24ml of sodium hydroxide solution (6gm in 24ml water) placed in a 100ml conical flask (cooled by keeping the flask in an ice bath), add 1.7ml of bromine with stirring, taking care that the temperature is maintained at 0°C. Now add with stirring 5gm of finely powdered phthalimide (preferably convert it into a thick paste by treating with water) to the cold hypobromite solution. Remove the flask from ice bath and shake vigorously to get a clear solution. Add 16ml of sodium hydroxide solution (4.5gm in 16ml water) and heat the flask to 80°C for 2 – 5 minutes. Filter, cool the filtrate in ice bath and add dilute hydrochloric acid with stirring until the solution is just acidic. Filter the precipitated anthranilic acid on buchner funnel and wash with cold water. Recrystallise from hot water and collect crystals of anthranilic acid, m.p 144- 145°C (Yield 2.5 – 3gms)

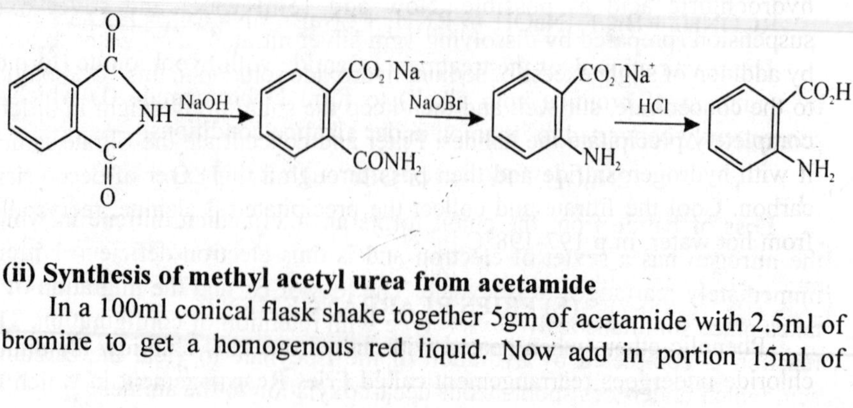

(ii) Synthesis of methyl acetyl urea from acetamide

In a 100ml conical flask shake together 5gm of acetamide with 2.5ml of bromine to get a homogenous red liquid. Now add in portion 15ml of caustic soda solution (3.4 gm in 15ml water) so that the red colour disappears. Heat the flask on water bath and when the effervesence sets in continue the heating process for another 2-5 minutes. Pour the reaction mixture in a beaker which is placed in an ice bath and leave it overnight. Collect colourless crystals of methyl acetyl urea and recrystallise further from hot water, m.p 169-170°C (Yield 3- 3.3 gms)

$$CH_3CONH_2 + Br_2 + NaOH \rightarrow CH_3CONHBr + NaBr + H_2O$$

$$CH_3CONHBr + NaOH \rightarrow CH_3N = C = O + NaBr + H_2O$$

$$CH_3N = C = O + CH_3CONH_2 \rightarrow CH_3NHCONHCOCH_3$$

(iii) Synthesis of β-alanine from succinimide

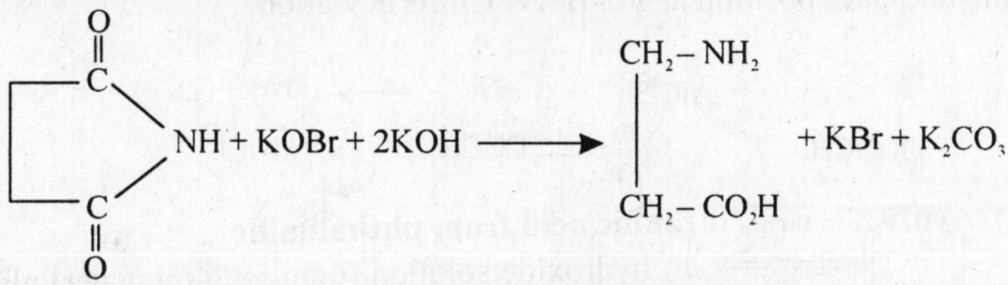

Cool 130 ml of potassium hydroxide solution (15gm in 130ml water) contained in a 250ml conical flask by placing it in an ice bath. Now add with stirring 1.6ml of bromine to it. After the addition of bromine is complete add with stirring 3gm of succinimide. Heat the resultant mixture on water bath at 55-60°C for 2 hours. Leave the mixture overnight and then acidify with concentrated hydrochloric acid. Evaporate the mixture on water bath. Treat the residue with 50ml hot alcohol and filter. Evaporate the filtrate to dryness. Extract the residue with 50ml of 95% alcohol.Concentrate the alcoholic extract,dilute it with 30ml water and add 10ml of ether to it. Remove the ethereal layer and reflux the aqueous layer for 90minutes. Set the apparatus for downward distillation and distill so as to remove as much hydrochloric acid as possible. Now add 15ml water and silver oxide suspension (prepared by dissolving 1gm silver nitrate in 5ml water followed by addition of slight excess of sodium hydroxide solution in this 10% solution) to the concentrate, stir well and then keep the solution overnight in order to completely precipitate the halides. Filter and concentrate the filtrate,saturate it with hydrogen sulfide and then pass through a thin layer of decolorising carbon. Cool the filtrate and collect the precipitated β alanine, recrystallise from hot water, m.p 197-198°C.

(E) FRIES REARRANGEMENT

Phenolic ethers when heated with anhydrous zinc chloride/ aluminium chloride undergoes rearrangement called Fries Rearrangement in which the acyl group migrates to ortho and para position to form phenolic ketones. However if para position is occupied then only ortho isomer is formed.

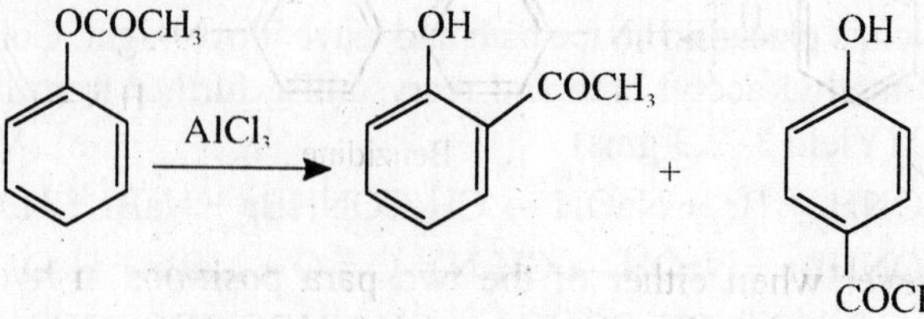

The reaction proceeds by the formation of complex between the ester and the lewis acid. An acylium ion is eliminated which gets substituted at ortho and para position as in Friedel Crafts acylation.

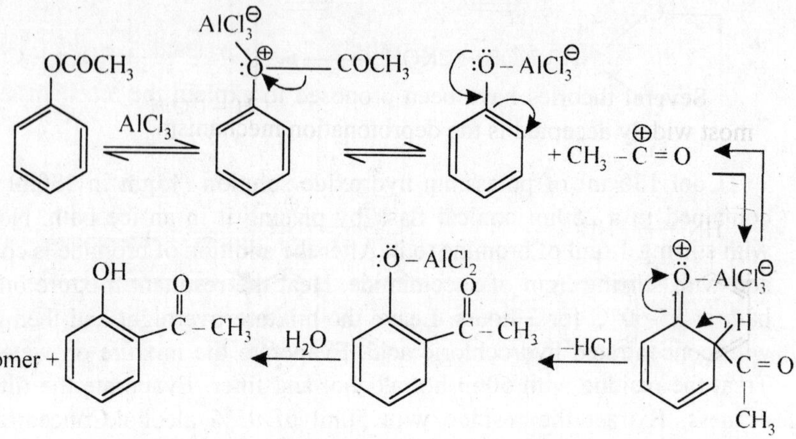

(i) Synthesis of 2,5 dihydroxy acetophenone from hydroquinone diacetate

In a 100ml round bottom flask fitted with a reflux condensor place 5gm of hydroquinone diacetate and 12gm of aluminium chloride. Heat the flask in an oil bath at 155-165°C for 3 ½ hours. Remove the heat source, cool the flask at room temperature and add 30gm of crushed ice followed by 3ml of concentrated hydrochloric acid to it.Collect the precipitated solid on buchner funnel and wash with cold water. Recrystallise 2, 5 dihydroxy acetophenone from hot water and collect it in the form of silky green crystals, m.p 201-203°C (Yield 2.5 – 3.5gms)

(F) BENZIDINE REARRANGEMENT

The transformation of hydrazoarenes into diaminobiaryls under strong acidic conditions is termed as Benzidine Rearrangement.

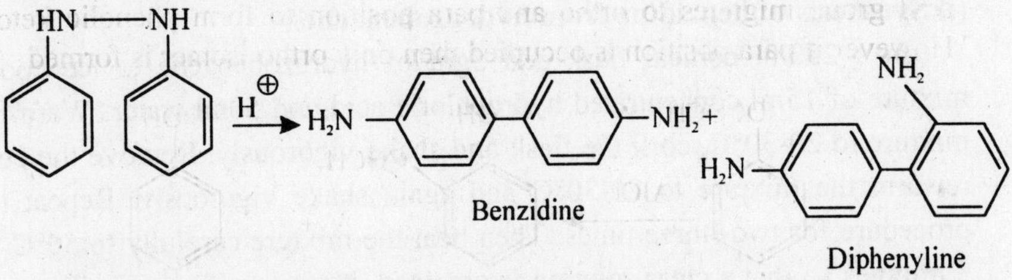

However when either of the two para positions in hydrazoarenes are occupied then the product formed may be either ortho semidine or diphenyline depending upon the nature of the substituent.

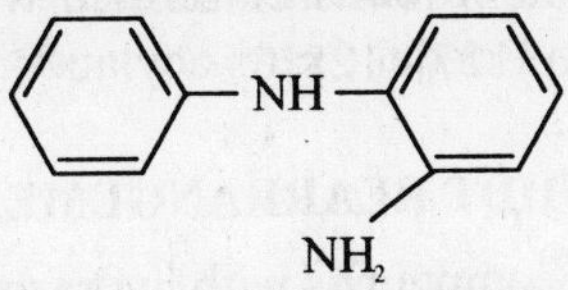

Several theories have been proposed to explain the mechanism but the most widely accepted is the deprotonation mechanism.

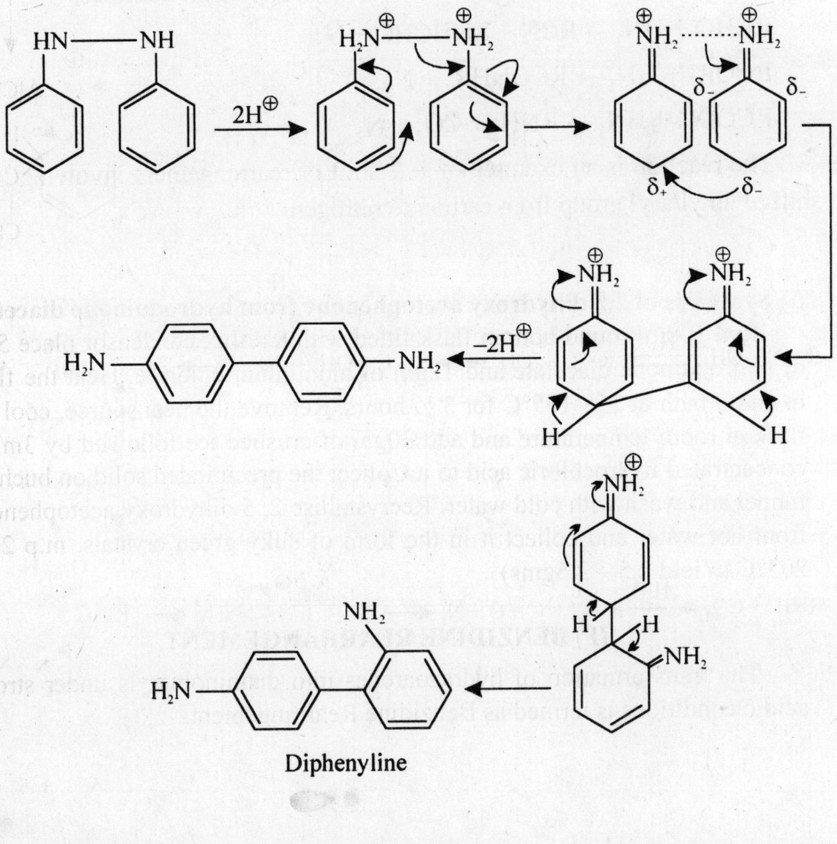

(i) Synthesis of benzidine from hydrazobenzene

In a 250ml conical flask add 5gm of hydrazobenzene to ice cooled mixture of 15ml concentrated hydrochloric acid and 50ml water. Warm the mixture to 20-30°C, cork the flask and shake vigorously. Remove the cork, rewarm the mixture to 20-30°C and again shake vigorously. Repeat this procedure for two- three times. Then heat the mixture carefully to 50°C for 5 minutes so that a clear solution is obtained. However if some undissolved material remains then filter hot. Cool the filtrate and add with stirring aqueous sodium hydroxide solution until the solution is alkaline. Filter the greyish white precipitate of benzidine on buchner funnel, wash it with water and recrystallise from ethanol, m.p 127 - 128°C.

(G) SCHMIDT REARRANGEMENT

The reaction of carbonyl compounds with hydrazoic acid in the presence of concentrated sulphuric acid is known as Schmidt reaction or rearrangement. Aldehydes give a mixture of cyanide and formyl derivatives of primary amines, ketones give amides whereas carboxylic acids give amines.

$$RCHO + HN_3 \rightarrow RCN + RNHCHO + N_2$$

$$RCOR' + HN_3 \rightarrow RCONHR' + N_2$$

$$RCOOH + HN_3 \rightarrow RNH_2 + CO_2 + N_2$$

The reaction is an example of 1, 2 shift or rearrangement involving the shift of alkyl/aryl group from carbon to nitrogen.

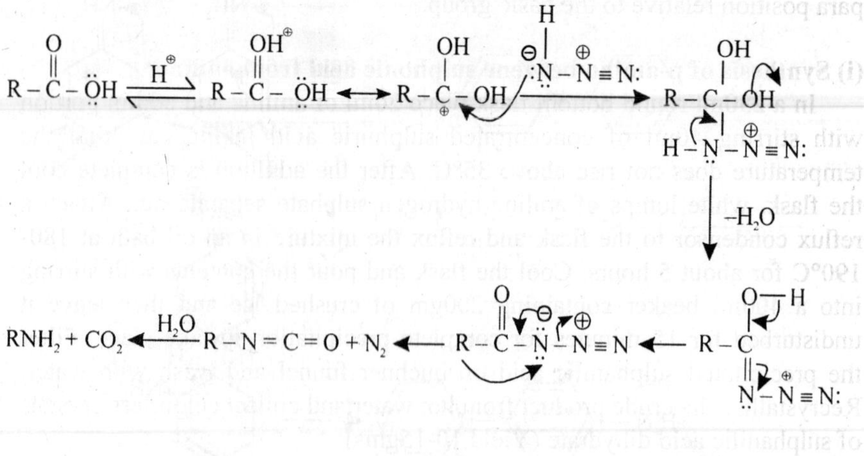

(i) Synthesis of 3, 5 dinitroaniline from 3, 5 dinitrobenzoic acid

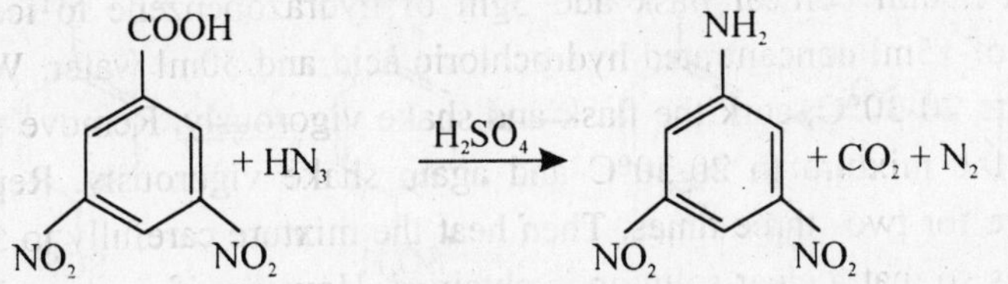

In a 250ml round bottom flask fitted with a reflux condensor place 10gm of 3, 5 dinitrobenzoic acid, 18ml of 10% oleum and 4ml of concentrated sulphuric acid. Add 20ml of chloroform and heat the flask on water bath at 45°C. Add in portion 3.5gm of sodium azide to the reaction mixture while maintaining the temperature in the range of 35 to 45°C. Addition of sodium azide should be regulated as foaming takes place during its addition. After all the sodium azide has been added raise the temperature so that the chloroform refluxes vigorously and maintain it at this temperature for 3 hours. Cool, pour the reaction mixture in a 500ml beaker containing 100gm of crushed ice and leave it for 1hour. Filter the yellow precipitated 3,5 dinitroaniline through buchner funnel, wash with cold water and dry, m.p 162-163°C.

(H) INTRAMOLECULAR REARRANGEMENT OF AROMATIC AMINE SULPHATES

When an amine is treated with sulphuric acid, a sulphate is usually formed. If this sulphate is heated either alone or with excess of concentrated sulphuric acid, rearrangement takes place, the sulphonic group entering the para position relative to the basic group.

(i) Synthesis of p-amino benzene sulphonic acid from aniline

In a 250ml round bottom flask place 20ml of aniline and add in portion with stirring 40ml of concentrated sulphuric acid taking care that the temperature does not rise above 35°C. After the addition is complete cool the flask, white lumps of aniline hydrogen sulphate separate out. Attach a reflux condensor to the flask and reflux the mixture in an oil bath at 180-190°C for about 5 hours. Cool the flask and pour the contents with stirring into a 400ml beaker containing 200gm of crushed ice and then leave it undisturbed for 15 minutes for complete precipitation to take place. Filter the precipitated sulphanilic acid on buchner funnel and wash with water. Recrystallise the crude product from hot water and collect colourless crystals of sulphanilic acid dihydrate (Yield 10-15gms)

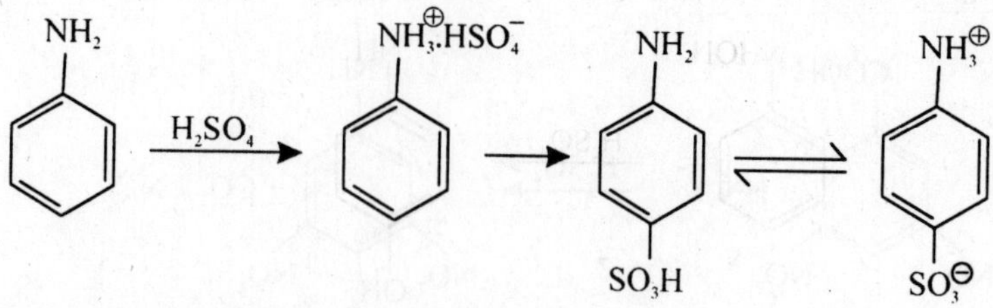

(i) Rearrangement of Salicylic acid to p-hydroxy benzoic acid

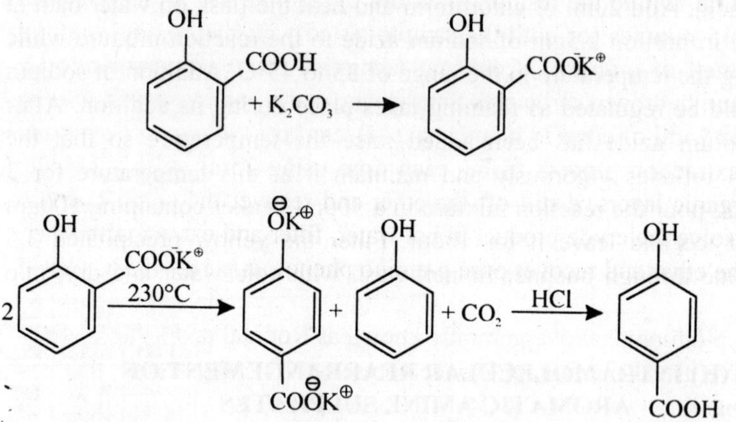

In a porcelain dish place 20gm of salicylic acid and 30ml of water. Add in portion and with stirring 12gm of potassium carbonate. Heat the porcelain dish on steam bath and stir the solution with glass rod. After sometime the solution slowly gets converted into a thick pasty solid. Break the solid into small pieces, dry and then finally grind it into a fine powder. Transfer the powder to 100ml round bottom flask and set the assembly for downward distillation. Heat the flask in an oil bath at 240°C for 60-90 minutes. Phenol formed during the reaction is distilled off. When no more phenol is distilled transfer the contents of the flask to a 500ml beaker containing 100ml hot water. Acidify the salt solution with concentrated hydrochloric acid, heat nearly to boiling and filter. Concentrate the filtrate, cool and collect brown coloured solid on buchner funnel. Recrystallise the crude product from hot water and collect crystals of p-hydroxy benzoic acid, m.p 210-212°C (Yield 7-8gm).

(j) Rearrangement of N-Phenyl hydroxylamine to p-aminophenol

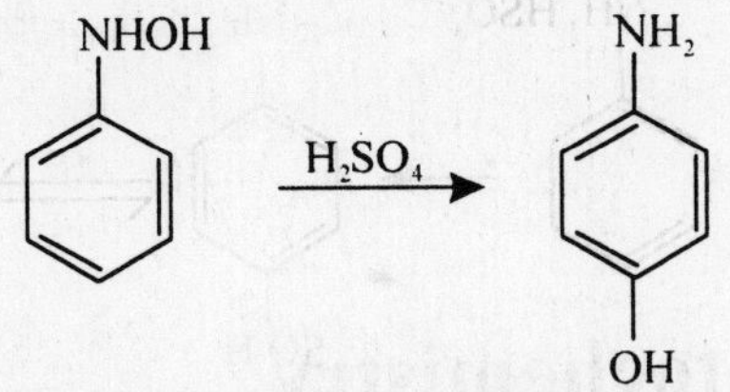

In a 1 litre beaker placed in an ice bath containing a mixture of 23ml of concentrated sulphuric acid and 70gm of crushed ice add 5gm of N phenyl-hydroxylamine. Dilute the resulting mixture by adding 425ml of water and boil for 10-15 minutes (or until the sample tested with dichromate solution gives the smell of quinone). Cool and neutralise the reaction mixture by adding sodium hydrogen carbonate. Saturate the neutral solution with sodium chloride, add 25ml of ether to it and shake. Decant off the organic layer and repeat the extraction process twice, each time using 25ml ether. Combine the three organic layers, distill off the ether and recover the crude p-amino phenol. Dissolve the crude product in hot water, filter and extract with ether. Distill off the ether and recover pure p-amino phenol, m.p 183-185°C (Yield 3.5 –4 gm)

Note :- p-amino phenol commonly known as Rodinal and its derivatives find extensive use as photographic developers.

Mechanism

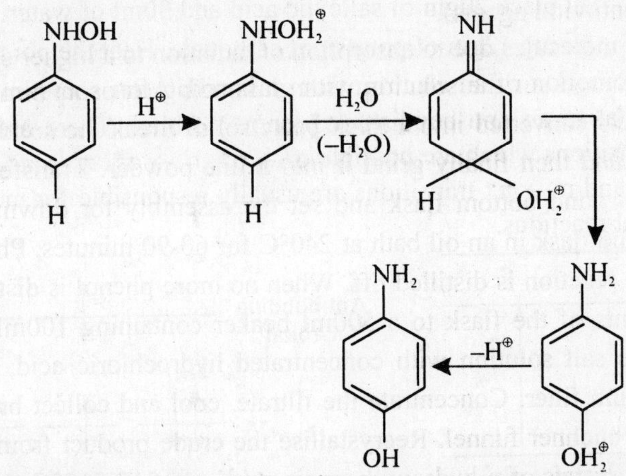

14

Organic Photochemistry

Photochemistry is the chemistry of reactions initiated by electronically excited molecules produced by the absorption of suitable radiation. Many important synthetically useful reactions like cycloaddition, oxidation, reduction, substitution, dimerization and rearrangement take place due to absorption of light by organic molecules. Synthesis of desired product which may require several steps by general chemical procedures besides giving a number of side products can however be carried out in a single step and in good yield by photochemical method.

Photochemical transformations find relevance not only in synthetic chemistry but also in the field of biology and physics. Sun which acts as a central figure is responsible for supplying solar energy for all photobiological processes, like the transformation of carbon dioxide to sugar under mediation of chlorophyll in green plants. Some well known photochemical synthesis are—synthesis of vitamin D_2 from ergosterol and synthesis of cubanes (which are well known anti viral agents).

Excitation of molecules due to absorption of radiation to a higher energy level involves promotion of an electron from a bonding (σ or π) or a non-bonding (n) orbital to an anti-bonding (σ^* or π^*) orbital. There are four such types of transitions which are possible $\sigma \rightarrow \sigma^*$, $n \rightarrow \sigma^*$, $\pi \rightarrow \pi^*$, $n \rightarrow \pi^*$. The $\pi \rightarrow \pi^*$ and $n \rightarrow \pi^*$ transitions are usually responsible for most of the photochemical reactions.

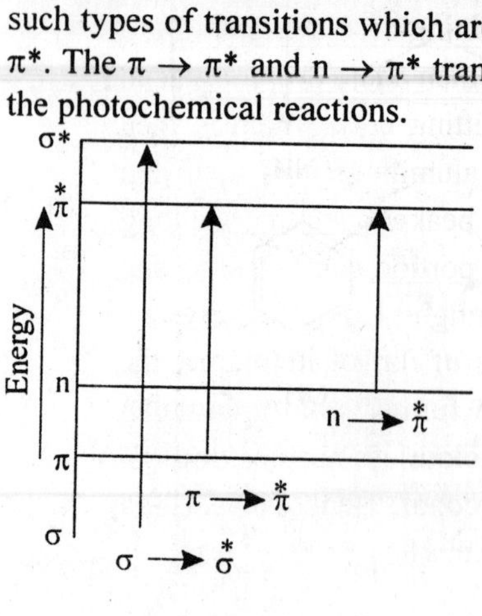

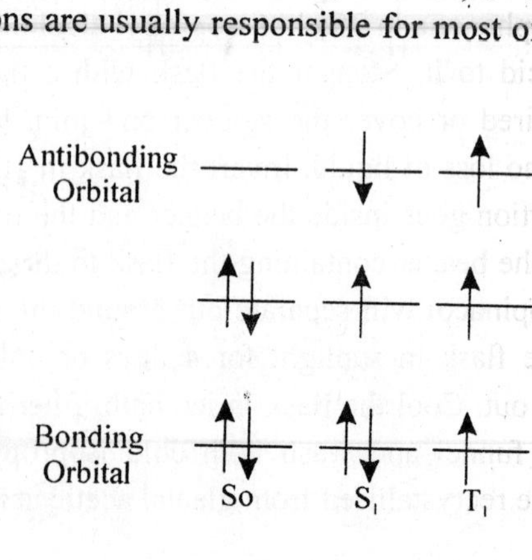

Molecules having even number of electrons are paired (spin in opposite direction) and the energy states associated with them is called singlet (S) states. Inversion of the spin of one of the electron leads to the formation of different electronic state in which the two unpaired electrons have the same spin and this state is referred to as triplet (T).

The absorption of radiation by the molecule results in the promotion of an electron to the vibrationally excited singlet state S_1 which lose excess vibrational energy by collision with solvent molecules to give thermally equilibrated excited singlet state S_1. The excited molecule may return to the ground state with emission of light i.e flourescence is observed, or it may get converted to triplet (T_1) state by spin inversion (intersystem crossing). The half life of triplet state is longer than that of singlet excited state and hence enough time for the chemical reaction to take place.

(i) Synthesis of benzopinacol from benzophenone

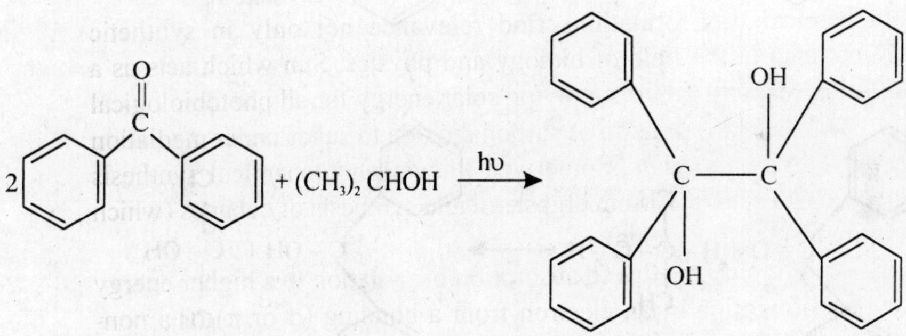

In a 50ml round bottom flask dissolve by warming on water bath 5 gm of benzophenone in 35ml of isopropyl alcohol and then add a drop of glacial acetic acid to it. Stopper the flask with a tight fitting cork which is then either wired or cover the stopper and joint with aluminium foil such that there is no loss of liquid. Invert the flask in 100ml beaker such that the long neck portion goes inside the beaker and the round portion comes to the top. Expose the beaker containing the flask to direct sunlight. Colourless crystals of benzopinacol will separate out around the walls of flask within 24hours. Keep the flask in sunlight for 4 days or until no further solid appears to separate out. Cool the flask in ice bath, filter the colourless needles through buchner funnel and wash with cold isopropyl alcohol. Benzopinacol can further be recrystallised from glacial acetic acid, m.p. 184-186°C.

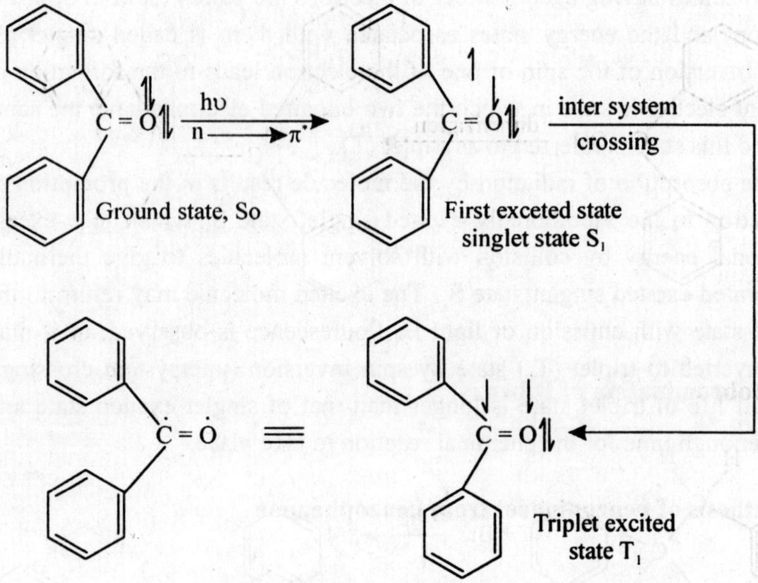
hυ
n → π*
inter system
crossing
Ground state, So
First exceted state
singlet state S_1
Triplet excited
state T_1

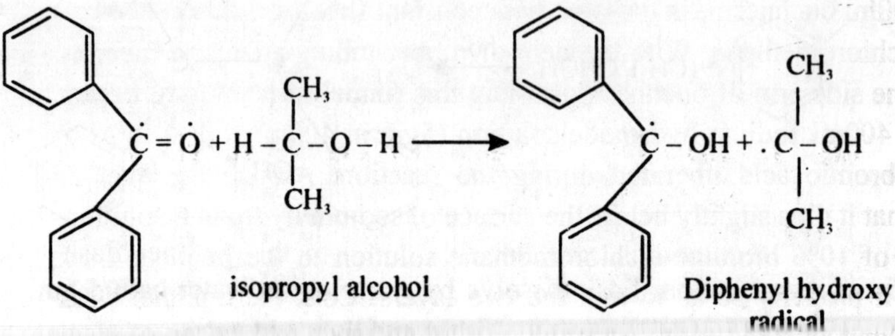
isopropyl alcohol
Diphenyl hydroxy
radical

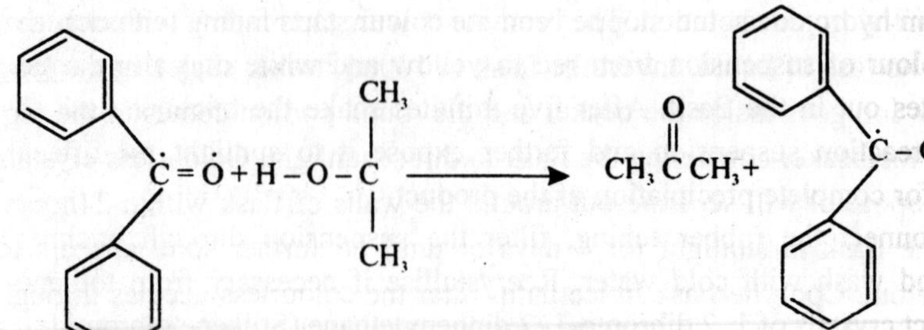

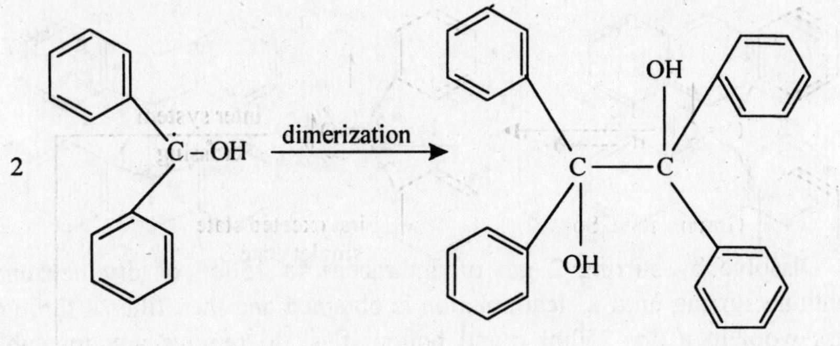

(ii) Photobromination of Dibenzyl

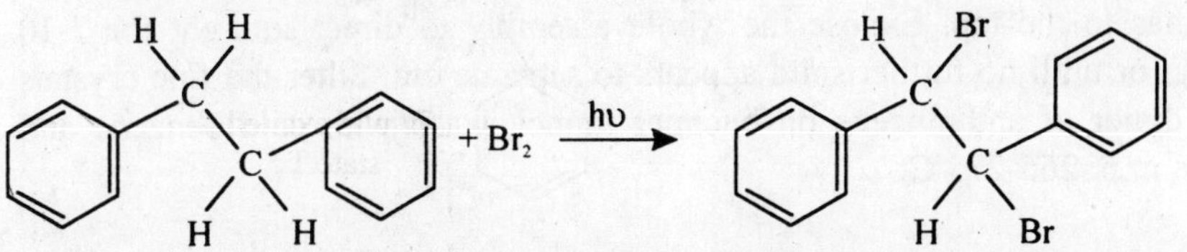

In a 500ml buchner flask dissolve with constant shaking 5gm of dibenzyl in 40ml dichloromethane. With the help of rubber tubing attach an inverted funnel to the side arm of buchner flask. Dip this funnel into a 500ml beaker containing 400ml sodium hydroxide solution (5gm in 400ml water) so as to trap hydrobromic acid liberated during the reaction. Adjust the level of funnel so that it dips slightly below the surface of sodium hydroxide solution. Add 30ml of 10% bromine-dichloromethane solution to the buchner flask and shake it properly so as to mix the two layers. Cork the buchner flask thoroughly and shift the entire assembly out into direct sunlight. A vigorous reaction occurs with evolution of HBr gas which is immediately trapped by the sodium hydroxide solution. The bromine colour starts fading with change of the colour of suspension from red to yellow and white crystalline solid precipitates out in the flask. After five minutes shake the buchner flask to mix the reaction suspension and further expose it to sunlight for fifteen minutes for complete precipiation of the product.

Disconnect the rubber tubing, filter the suspension through buchner funnel and wash with cold water. Recrystallise if necessary from toluene and collect crystals of 1, 2 dibromo-1, 2 diphenylethane (Stilbene dibromide), m.p 243-245°C. (Yield 2-3 gm)

(iii) Dimerization of anthracene

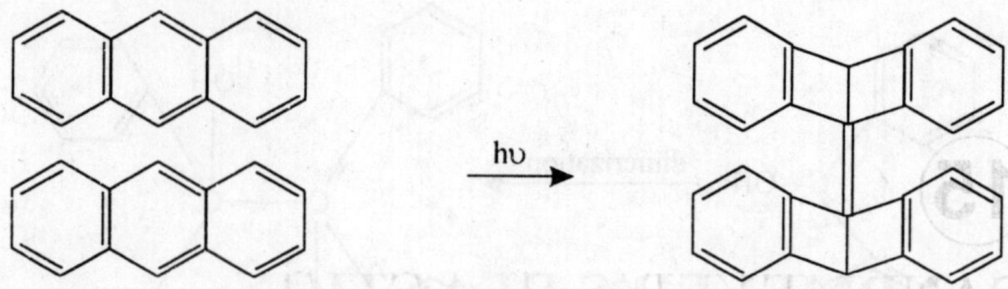

Dissolve by stirring 2 gm of anthracene in 150ml of dry benzene. Continue stirring until a clear solution is obtained and then filter it through glass wool in a dry 250ml round bottom flask to remove any insoluble material that may be present. Cork the flask and seal it properly or cover the stopper and joint with aluminium foil. Invert the flask in a 25ml breaker such that the long neck portion goes inside the breaker and round portion comes to the top. Expose the whole assembly to direct sunlight for 7-10 days or until no further solid appears to separate out. Filter the fine crystals of dimer of anthrancene on buchner funnel, wash with cold benzene and dry, m.p. 265-275°C.

15

SANDMEYER'S REACTION

Sandmeyer's reaction is a versatile method by which cuprous chloride, cuprous bromide and cuprous cyanide react with arenediazonium salts to give aryl chloride, aryl bromide and aryl cyanide respectively.

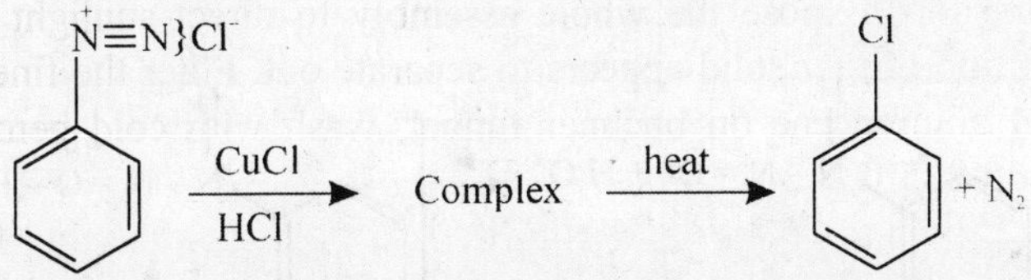

The reaction mechanism is not certain but it is believed to proceed by free radical pathway.Cuprous halide acts as a catalyst in this free radical reaction.Cuprous ion reduces the diazonium ion resulting in the formation of aryl free radical which then abstracts halogen from cupric halide reducing it to cuprous halide with the formation of aryl halide.

$$CuCl + C_6H_5N_2^+Cl^- \longrightarrow CuCl_2^{\ominus} + C_6H_5N_2^{\oplus}$$

$$C_6H_5N_2^+ + CuCl_2^{\ominus} \longrightarrow C_6H_5^{\cdot} + N_2 + CuCl_2^{\cdot}$$

$$C_6H_5^{\cdot} + CuCl_2^{\cdot} \longrightarrow C_6H_5Cl + CuCl$$

Diazotisation: When aromatic primary amine is treated with nitrous acid in the presence of hydrochloric acid under cooled conditions,aromatic diazonium salt results.This process is known as diazotization.The diazonium salts are important as they are used as starting material for the synthesis of a number of dyes and medicinally important compounds

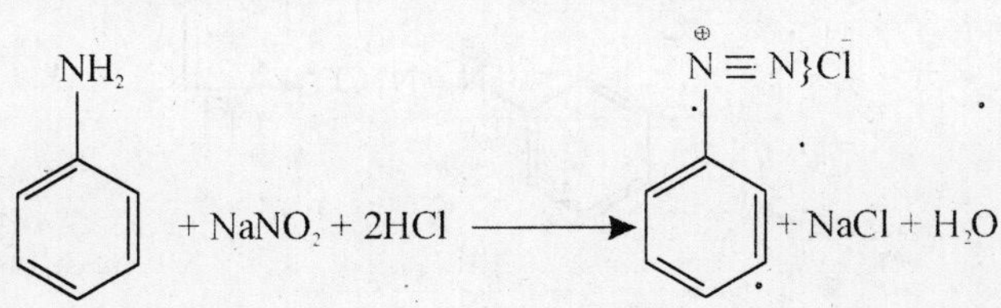

Nitrous acid is generated in the reaction mixture by mixing sodium

nitrite with cold dilute hydrochloric acid. Under acidic conditions nitrous acid protonate and lose water molecule to give nitrosonium ion. The amine then attacks (nucleophile) nitrosonium ion to form N-nitroso derivative which tautomerises by transfer of a proton from nitrogen to oxygen resulting in the formation of hydroxyl group.

$$Na^{+}O-N=O+HCl \longrightarrow H-O-N=O+NaCl$$

$$H-O-N=O+H^{\oplus} \rightleftharpoons H-\overset{H}{\underset{\oplus}{O}}-N=O \rightleftharpoons H_2O+\left[\overset{\oplus}{N}=\ddot{O}: \longleftrightarrow N\equiv\overset{\oplus}{O}:\right]$$

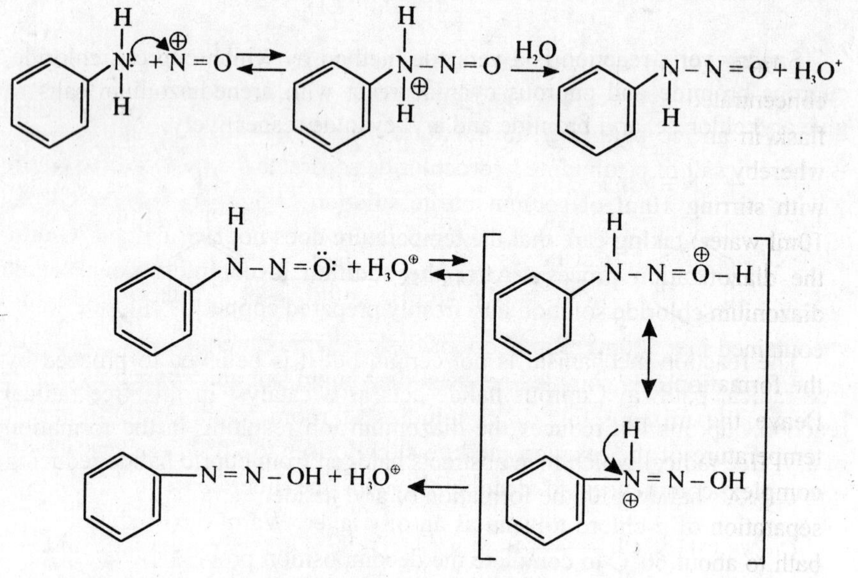

Protonation of hydroxyl group followed by loss of water molecule gives diazonium cation.The cation then takes up Cl⁻ to yield diazonium salt.

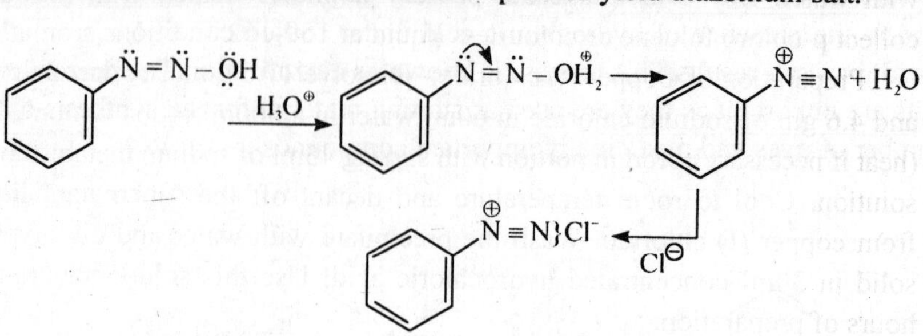

(i) Synthesis of p-chloro toluene from p-toluidine

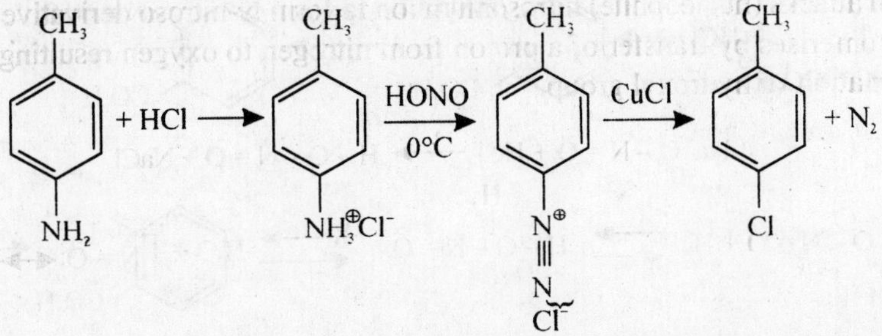

In a 100ml conical flask dissolve 5gm of p-toluidine in 12ml of concentrated hydrochloric acid and 12ml of water. Cool by keeping the flask in an ice bath. Stir the mixture vigorously and add a little ice to it whereby salt of p-toluidine hydrochloride separates out. Add in portions and with stirring 10ml of sodium nitrite solution (3.5gm of sodium nitrite in 10ml water) taking care that the temperature does not rise above 5°C during the diazotization process. After the addition is complete pour the cold diazonium chloride solution into freshly prepared copper (I) chloride solution contained in a 500ml round bottom flask. The mixture becomes thick due to the formation of a complex between diazonium salt and copper (I) chloride. Leave the mixture for 10-15 minutes at room temperature so that the temperature of the reaction mixture slowly rises to 20°C whereby the solid complex commences to decompose with the liberation of nitrogen and separation of p-chloro toluene as an oily layer. Warm the mixture on water bath to about 60°C to complete the decomposition process. Steam distill the mixture and transfer the distillate to separating funnel. Remove the p-chloro toluene layer, wash it first with dilute sodium hydroxide solution and then with water. Dry over anhydrous sodium sulphate and decant. Distill and collect p-chloro toluene as colourless liquid at 159-162°C.

Preparation of Copper (I) chloride:- Dissolve 17.5 gm of copper sulphate and 4.6 gm of sodium chloride in 60ml water in a 500ml round bottom flask (heat if necessary). Add in portion with stirring 45ml of sodium metabisulphite solution. Cool to room temperature and decant off the supernatant liquor from copper (I) chloride. Wash the precipitate with water and dissolve the solid in 30ml concentrated hydrochloric acid. Use this solution within 18 hours of preparation.

$$4CuSO_4 + 4NaCl + Na_2S_2O_5 + 3H_2O \rightarrow 4CuCl + 6NaHSO_4$$

(ii) Preparation of o-chloro benzoic acid from anthranilic acid

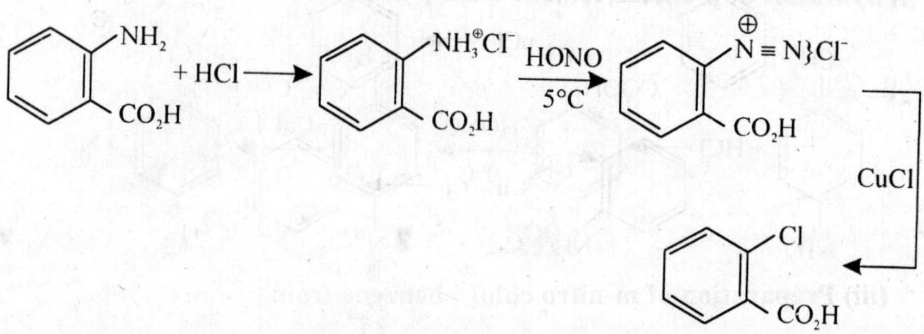

In a 100ml round bottom flask dissolve 5gm of anthranilic acid in a solution of 7ml of concentrated hydrochloric acid and 35 ml of water. Cool by keeping the flask in an ice bath and when the temperature inside the flask reaches below 5°C then add in portions with stirring 10ml of cold sodium nitrite solution (2.3 gm in 10ml water). After the addition is complete continue keeping the flask in ice until the preparation of diazonium salt is complete.

In a separate 250ml round bottom flask prepare copper (I) chloride by dissolving 8.5gm of crystalline copper (II) sulphate and 4gm of sodium chloride in 15ml water. Boil the resultant solution, add 27ml of concentrated hydrochloric acid and 5gm of copper turning. Continue refluxing until the solution is practically colourless. Cool the flask in ice and then add with stirring and in portions cold diazonium chloride solution whereby frothing occurs. After the addition is complete keep the mixture at room temperature for 2-3hrs and shake frequently. Filter the precipitated solid on buchner funnel, wash with cold water. Recrystallise from hot water and collect colourless crystals of o-chloro benzoic acid, m.p 138-139°C. (Yield 4.5-5.0gms).

Mechanism

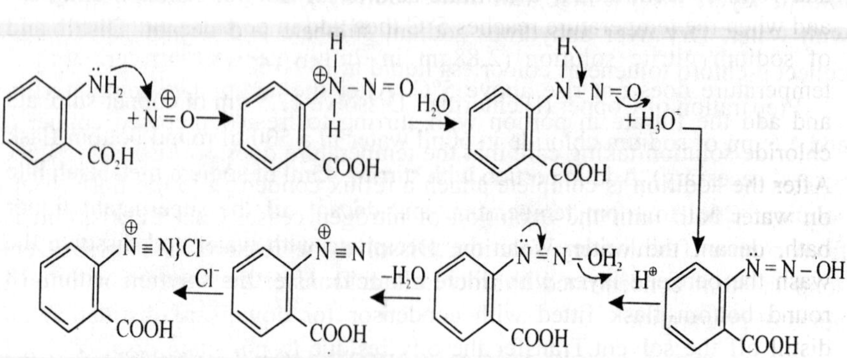

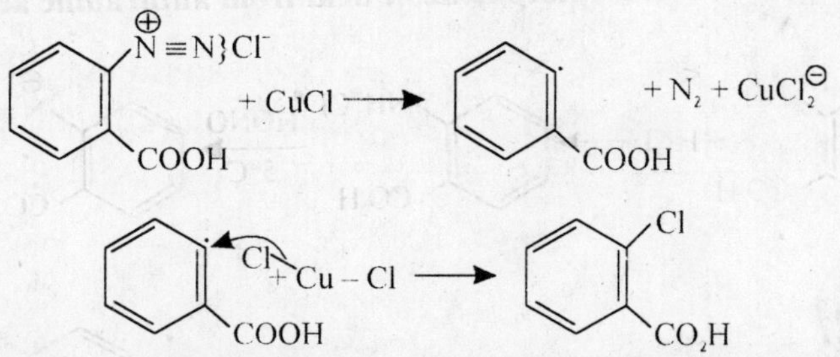

(iii) Preparation of m-nitro chloro benzene from m-nitro aniline

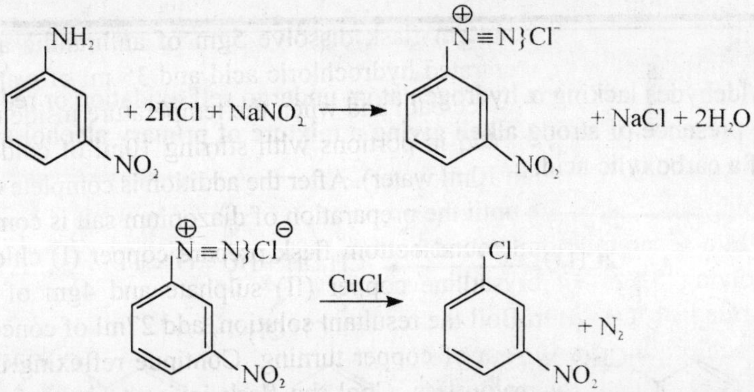

In a 1 litre beaker dissolve 12gm of copper (II) sulphate and 4gm of sodium chloride in 40ml water.Add a saturated solution of 2gm of sodium bisulphite, heat the resultant mixture to 60-70°C.Cool, filter the precipitated copper (I) chloride, dry it and immediately add it to a solution of 20ml water and 15ml concentrated hydrochloric acid placed in a 250ml conical flask.

In a 250ml round bottom flask dissolve 5.52gm of m-nitro aniline in a mixture of 5ml concentrated hydrochloric acid and 10ml hot water and then add 10ml concentrated hydrochloric acid to it.Place the flask in an ice bath and when the temperature reaches 5°C then add in portions with stirring 7ml of sodium nitrite solution (2.88gm in 7ml water) taking care that the temperature does not rise above 5°C. After the addition is complete filter and add the filtrate in portion with stirring to freshly prepared copper (I) chloride solution taking care that the temperature does not rise above 25°C. After the addition is complete attach a reflux condensor to the flask, heat it on water bath until the evolution of nitrogen ceases.Cool the flask in ice bath, decant the watery solution. Dissolve the residue in 20ml benzene, wash the benzene layer with dilute alkali.Transfer the benzene solution in round bottom flask fitted with condensor for downward distillation and distill off the solvent.Transfer the oily residue to porcelain dish, cool it in ice bath and collect crystals of m-nitro chloro benzene, m.p 44-45°C. (Yield 2.5-3.0gm).

16

Cannizaro Reaction

Aldehydes lacking α hydrogen atom undergo self oxidation or reduction in the presence of strong alkali giving a mixture of primary alcohol and the salt of a carboxylic acid.

$$2CH_2O \xrightarrow[RT]{50\%\ NaOH} CH_3OH + HC(=O)-O^-Na^+$$

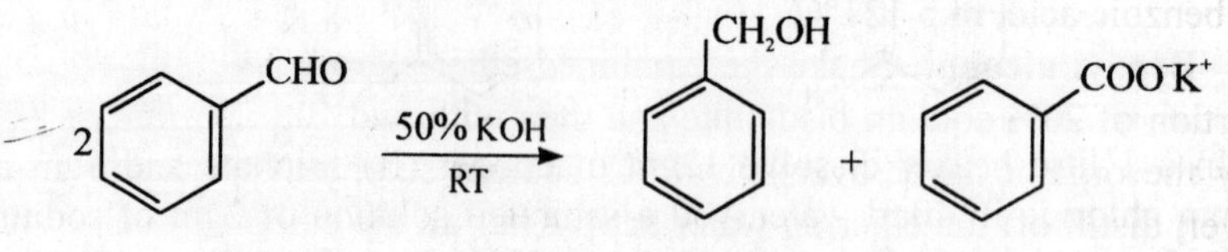

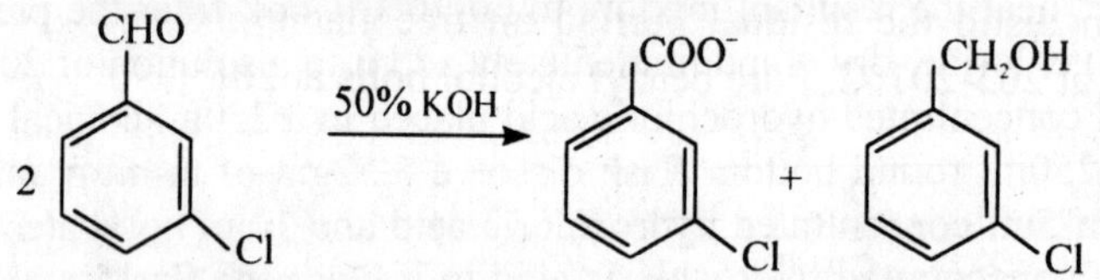

Evidence chiefly from kinetics and experiments with isotopically labelled compounds indicate that the carbonyl compounds follow the nucleophilic addition pathway. In this reaction the nucleophile is hydroxide ion which attacks the carbonyl group of the aldehyde. The resulting anion which is a good hydride donor transfer its hydride to the second aldehyde. This is infact an oxidation-reduction step. The hydride transfer results in the formation of an acid and alkoxide ion which then undergo a proton exchange among themselves to yield the more stable pair, alcohol and the acid anion.

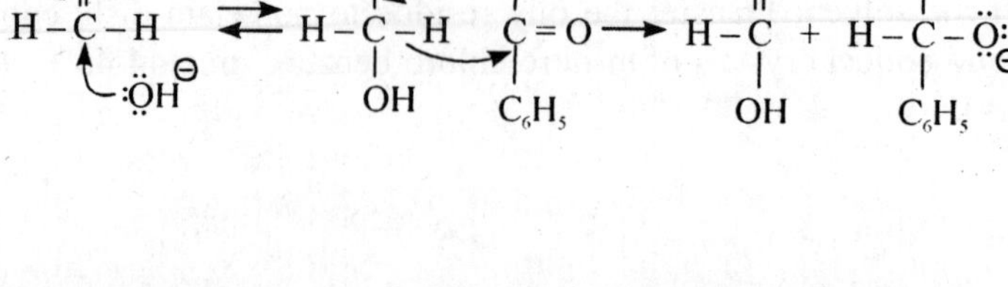

(i) Synthesis of Benzoic acid and Benzyl alcohol

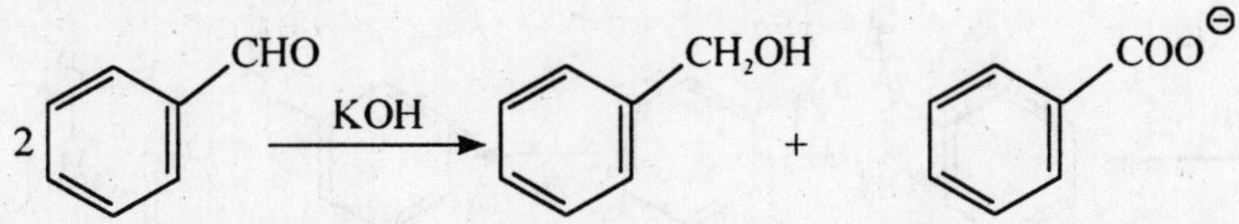

In a 250ml round bottom flask dissolve 25gm of potassium hydroxide in 25ml water. Add 24ml of benzaldehyde to the alkali solution, shake the mixture for 45 minutes and then keep it overnight. Add water in order to dissolve the precipitated potassium benzoate.Transfer the resultant mixture to a 250ml separating funnel and add 25ml ether to it. Shake the funnel thoroughly and remove the organic layer. Repeat the process twice each time using 25ml of ether.Combine the three organic fractions and use it for the isolation of benzyl alcohol.

Benzoic acid: Transfer the aqueous layer from the separating funnel into a 400 ml beaker containing 50ml concentrated hydrochloric acid and 100ml of ice water. Stir the mixture thoroughly whereby benzoic acid gets precipitated.Collect the precipitated acid by suction filteration and wash with cold water.Recrystallise from hot water and collect pure white crystals of benzoic acid, m.p 121°C.

Benzyl alcohol: Shake the combined ether extract first with two 10ml portion of 20% sodium bisulphite and then with two 10ml portion of water, dry the organic layer over anhydrous sodium sulphate.Decant the organic layer, distill off the ether on water bath. After all the ether has been distilled off then distill the residual portion on free flame and collect the fraction boiling at 203-207°C. Pure benzyl alcohol boils at 205°C.

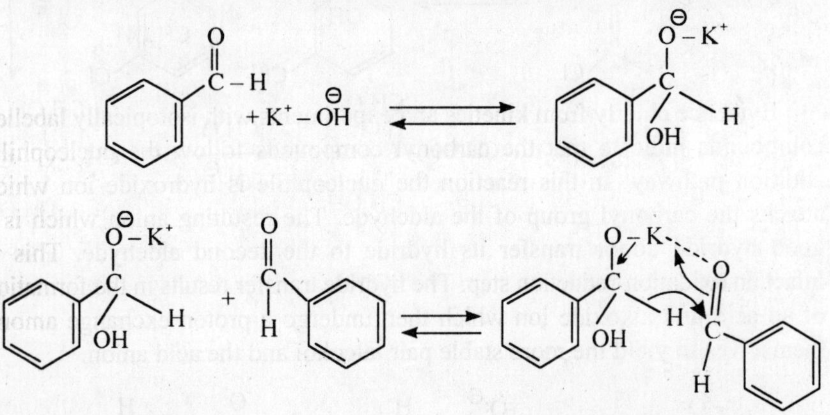

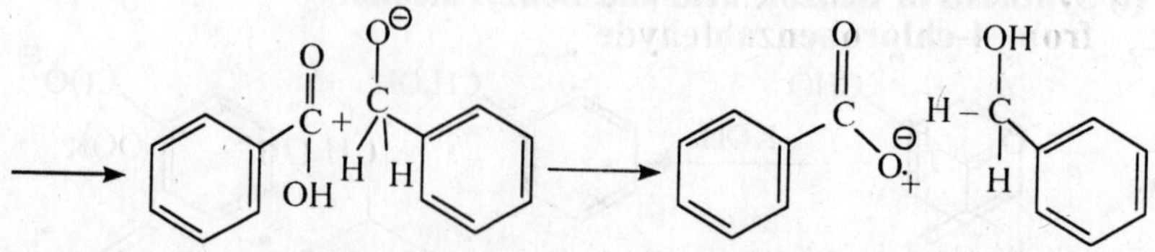

(ii) Synthesis of Pentaerythritol (Tetrahydroxy methyl methane) from acetaldehyde

$$8CH_2O + 2CH_3CHO + Ca(OH)_2 \rightarrow 2(CH_2OH)_4 + (HCO_2)_2Ca$$

Mix together 10gm of paraformadehyde, 50ml of water and 3gm of acetaldehyde in a 250ml conical flask.To this suspension add in portion with stirring 2.2 gm of finely powdered quicklime taking care that the temperature does not rise above 55°C. After the addition is complete continue the stirring process for another 3 hours and then filter. Acidify the yellow coloured filtrate with concentrated hydrochloric acid,add a pinch of animal charcoal, heat the resultant mixture on water bath and filter. Concentrate the colourless filtrate to one fourth its volume either by heating on water bath or by simple distillation. Cool the concentrated liquid in ice bath and leave overnight. Collect the precipitated product on buchner funnel. Recrystallise from a mixture of hot water and concentrated hydrochloric acid and collect pure pentaerythritol, m.p 258-260°C (Yield 3-4gm)

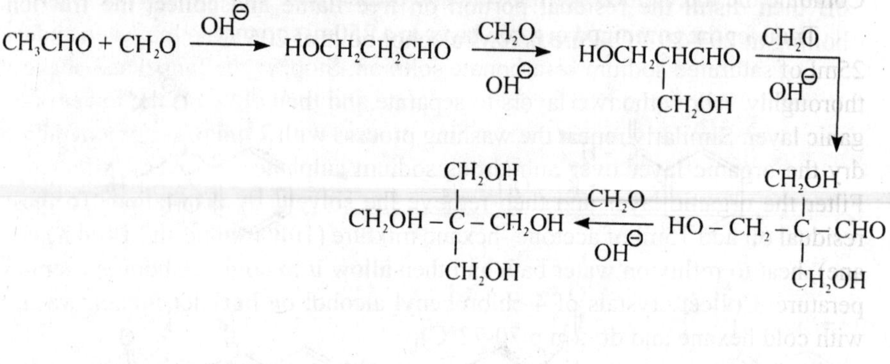

(iii) Synthesis of 4-chlorobenzyl alcohol and 4-chlorobenzoic acid from 4-chlorobenzaldehyde

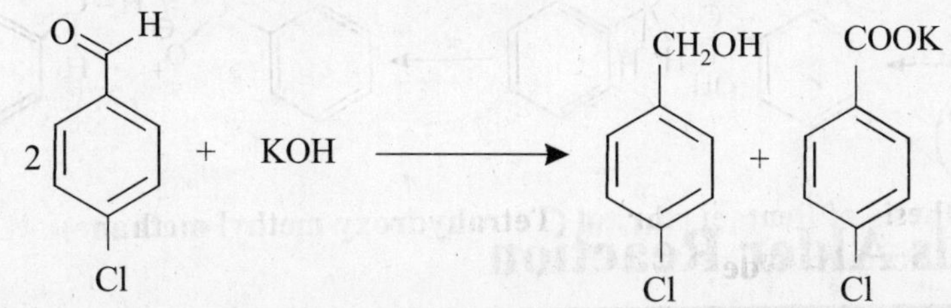

In a 100ml conical flask dissolve 10gm of 4-chlorobenzaldehyde in 30ml of methanol. In a separate 250ml conical flask dissolve 21 gms of KOH in 30ml of distilled water. After KOH has completely dissolved, allow the aqueous KOH solution to cool at laboratory temperature. Pour the organic solution into aqueous KOH solution. Swirl the flask vigorously so as to mix the two solutions. The solution becomes warm, as the reaction is exothermic. After the reaction has subsided, heat the flask on steam bath at 50-60°C for one hour (swirl the flask regularly during this period). After the heating process allow the reaction to cool at laboratory temperature and then transfer it into a 500ml separating funnel. Wash the conical flask with 25ml of chloroform and transfer this washing into the separating funnel. Stopper the separating funnel and shake thoroughly. Allow the two layers to separate and then draw off the lower organic layer. Repeat the process twice each time using 25ml of chloroform. Combine the three chloroform layers and preserve the aqueous layer.

Transfer the combined organic layer in a 250ml separating funnel and add 25ml of saturated sodium bicarbonate solution. Stopper the funnel and shake thoroughly. Allow the two layers to separate and then draw off the lower organic layer. Similarly repeat the washing process with 25ml of water and then dry the organic layer over anhydrous sodium sulphate for several minutes. Filter the organic layer and then remove the solvent by distillation. To the residual oil add 15ml of acetone -hexane mixture (1ml acetone and 14ml hexane), heat to reflux on water bath and then allow it to cool at laboratory temperature. Collect crystals of 4-chlorobenyl alcohol on buchner funnel, wash with cold hexane and dry (m.p 70-72°C).

Transfer the preserved aqueous layer to 100ml conical flask; add 1 gm of charcoal to it. Heat the mixture on water bath for several minutes, cool and then filter through elite/ silica/cotton wool. To this clear aqueous solution add dropwise with stirring dilute HCl until the solution is just acidic and then cool in an ice bath. Filter the precipitated white solid through buchner funnel. Dry and preserve the 4-chlorobenzoic acid (m.p 237-239°C).

17

Diels Alder Reaction

Diels Alder reaction involves the formation of a six membered ring by the reaction of an alkene or alkyne containing electron withdrawing group with a conjugated diene. The alkene also known as the ene component is ordinarily electron deficient hence is known as dienophile (lover of diene) whereas diene is electron rich. In general the reactivity of the dienophile is increased if electron withdrawing groups like CO, CO_2R, CN and NO_2 are present in conjugation with the double bond or the triple bond. It is for this reason αβ unsaturated carbonyl compounds and quinones are most commonly used as dienophiles in diels alder synthesis.

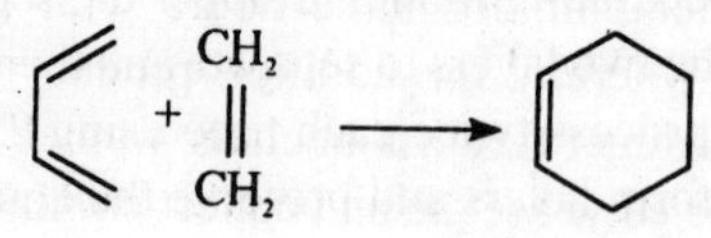

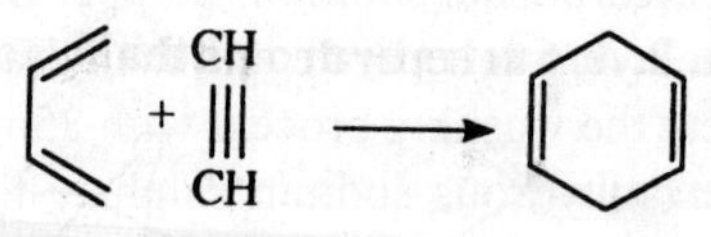

The reaction is also often called (4 + 2) cycloaddition because a ring is formed by the interaction of 4π electron of the diene with 2π electron of alkene or alkyne.In this reaction two new σ bonds are formed at the expense of two π bonds.

The reaction mechanism involves the simultaneous cyclic movement of 4π electron of the diene and 2π electron of the dienophile and is also called concerted reaction because the bond breaking and bond formation occur simultaneously.For a more effective reaction to occur the transition state must have a geometry that favours the overlap of the two end π orbitals of the diene with those of the dienophile. This geometry is only possible when the two conjugated double bonds of the diene must be in the S configuration.

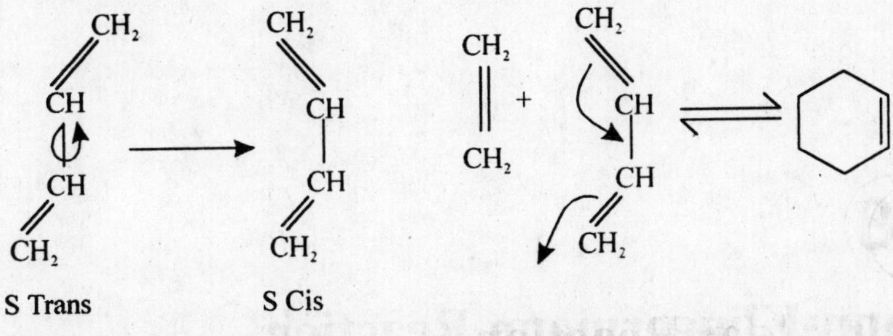

(i) Synthesis of 9, 10-dihydro anthracene endo α β succinic anhydride from anthracene

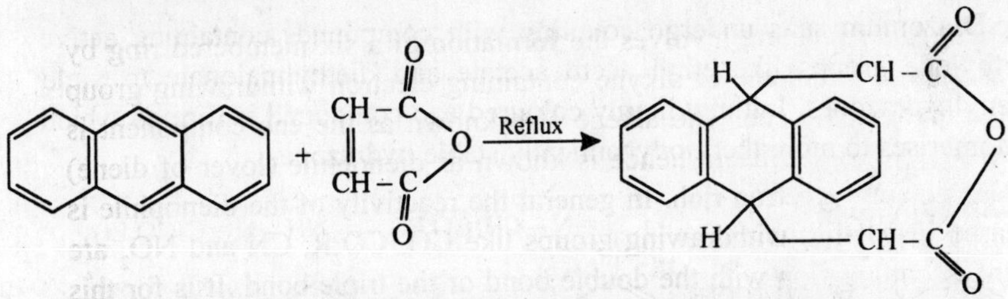

In a 100ml round bottom flask fitted with a reflux condensor place 4gm of anthracene, 50ml of xylene or benzene and 11.6 gm of maleic anhydride. Reflux the mixture on water bath for three hours with frequent shaking. Cool the flask in ice bath whereby the addition product will immediately crystallise. Filter the solid adduct through buchner funnel. Recrystallise the crude product from xylene or ethyl acetate and collect crystals of 9, 10 dihydro anthracene endo α β succinic anhydride, m.p 262-263°C. (Yield 5-5.5gm).

(ii) Synthesis of 3, 6 oxo 1, 2, 3, 6 tetrahydro phthalic anhydride

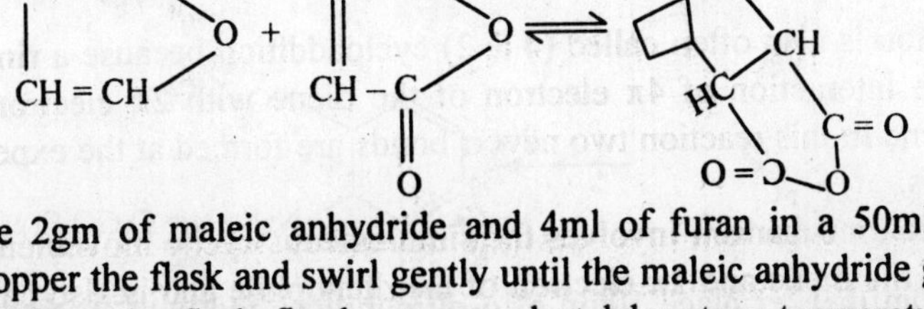

Place 2gm of maleic anhydride and 4ml of furan in a 50ml conical flask. Stopper the flask and swirl gently until the maleic anhydride goes into solution. Keep the flask firmly stoppered at laboratory temperature for 2 hours with occassional shaking. Add 2ml of benzene and cool the mixture in ice. Collect the precipitate by suction filtration, wash with cold benzene. Recrystallise the solid product from chloroform-pet ether solution and obtain crystals of 3, 6 oxo 1, 2, 3, 6 tetrahydro phthalic anhydride, m.p 125°C.

18

Japp-Klingermann Reaction

Diazonium salts undergo coupling with compounds containing active methylene groups like ethyl aceto acetate and diethylmalonate to yield phenylhydrazones. Initially highly coloured azo compound is formed which tautomerises to more thermodynamically stable hydrazone.

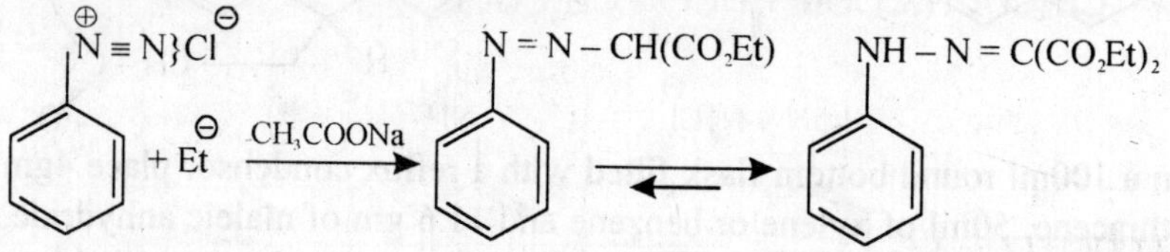

(i) Synthesis of Diethyl mesoxalate o-nitrophenylhydrazone from o-nitroaniline

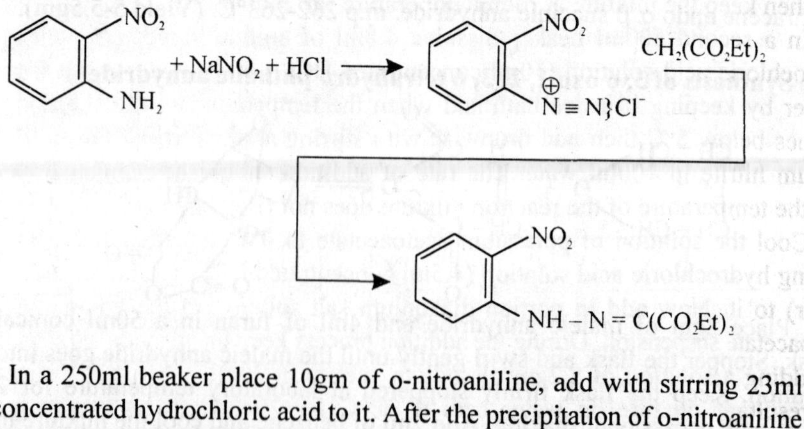

In a 250ml beaker place 10gm of o-nitroaniline, add with stirring 23ml of concentrated hydrochloric acid to it. After the precipitation of o-nitroaniline hydrochloride is complete cool the beaker by keeping it in an ice bath. Now add dropwise with stirring a solution of 6gm of sodium nitrite in 50ml water. The addition of sodium nitrite should be regulated such that the temperature does not rise above 5°C during the addition process.

In a separate 250ml beaker containing 50ml water add with stirring 12ml of diethylmalonate. Continue the stirring process for 10-15 minutes so that an emulsion results and then cool by keeping the beaker in ice bath. Now add dropwise the o-nitrobenzenediazonium chloride solution to the diethylmalonate suspension taking care that the temperature remains below 5°C. During the addition process keep the reaction alkaline by portionwise addition of sodium carbonate. After the addition is complete keep the mixture for 5 minutes and then filter the precipitated yellow diethyl mesoxalate o-nitrophenylhydrazone. Wash with water and recrystallise from ethanol, m.p 72-73°C

(ii) Synthesis of methylglyoxal-ω-phenylhydrazone from ethyl acetoacetate

$$CH_3COCH_2CO_2C_2H_5 + KOH \rightarrow CH_3COCH_2CO_2K + C_2H_5OH$$

$$CH_3COCH_2COOK + HCl \rightarrow CH_3COCH_2COOH + KCl$$

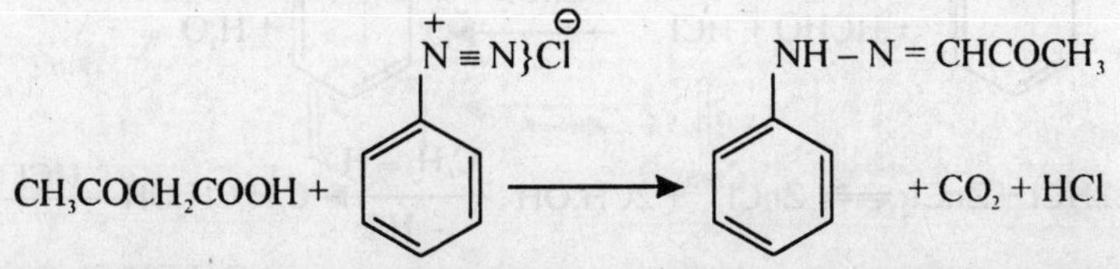

In a 500ml beaker place a solution of 3.5gm of potassium hydroxide in 110ml water.To this solution add with stirring 6.5ml of ethyl acetoacetate and then keep the mixture at room temperature for 24hours.

In a second 500ml beaker dissolve 4.8ml of aniline in 20ml aqueous hydrochloric acid solution (10ml concentrated HCl + 10ml water).Cool the beaker by keeping it in ice bath and when the temperature of the solution reaches below 5°C then add dropwise with stirring a solution of 3.6 gm of sodium nitrite in 100ml water.The rate of addition should be regulated so that the temperature of the reaction mixture does not rise above 10°C.

Cool the solution of potassium acetoacetate to 0°C and then add with stirring hydrochloric acid solution (4.5ml concentrated HCl in 15ml ice cold water) to it. Now add in portion diazonium salt solution to the potassium acetoacetate suspension. During the addition process keep the reaction alkaline by adding a solution of 8.2gm of sodium acetate in 30ml water to it. Heat the resultant mixture on water bath at 50°C for 2hours. Cool, filter the precipitated solid. Recrystallise from toluene and collect methylglyoxal-ω-phenylhydrazone, m.p 148-150°C.

19

Chloromethylation Reaction

The replacement of hydrogen atom of an aromatic compound by a chloromethyl group ($-CH_2Cl$) is known as chloromethylation reaction or Blanc chloromethylation reaction.

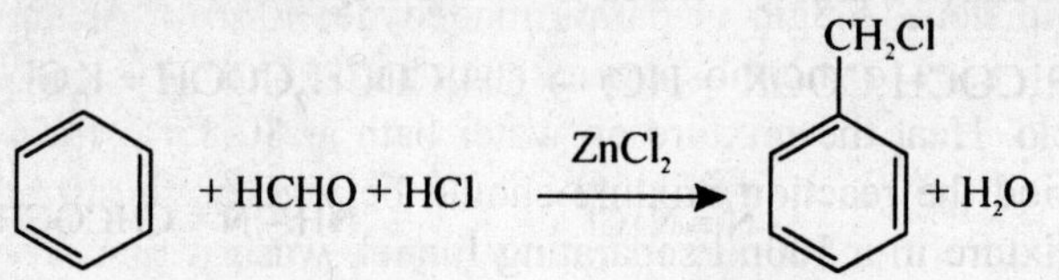

$$2CH_2O + 2HCl + ZnCl_2 \rightleftharpoons ZnCl_4^{2\ominus} + 2CH_2OH \xrightarrow[-H\oplus]{C_6H_5-H} C_6H_5CH_2OH \xrightarrow{HCl} C_6H_5CH_2Cl$$

The reaction consists essentially of the interaction of the aromatic compound with formalin or formaledhyde and hydrochloric acid in the presence of zinc chloride as catalyst. The formaldehyde may be replaced by methylal or by chloromethyl ether.

$$(CH_2O)_3 + 3CH_3OH + HCl \xrightarrow{ZnCl_2} 3CH_3OCH_2Cl + 3H_2O$$

(i) Synthesis of 2-hydroxy-5-nitrobenzyl chloride from p-nitro phenol

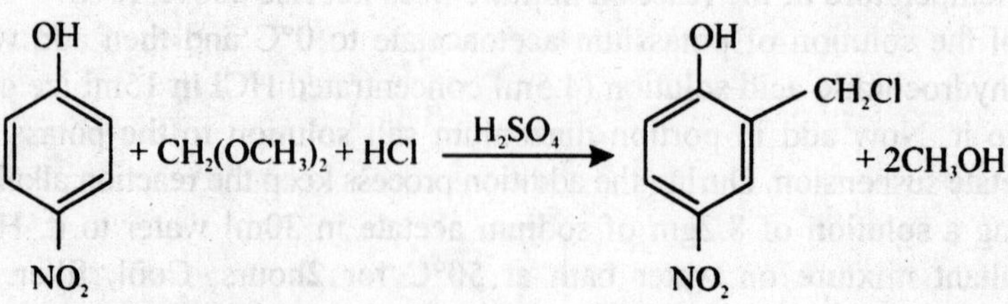

In a 100ml round bottom flask place 5gm of p-nitrophenol, 65ml of concentrated sulphuric acid and 7.6gm of methylal. Stir and heat the mixture on water bath at 70°C for 4-5 hours. During this period pass hydrogen chloride gas into the reaction mixture. The temperature of the reaction mixture should not be allowed to rise above 70°C as it would result in the formation of oily resinous material. Cool by keeping the flask in ice bath for 1hour. Filter the precipitated solid, recryatallise from hot benzene and collect 2 hydroxy-5-nitrobenzyl chloride as white crystals, m.p 129-130°C.

(ii) Synthesis of 1-chloromethylnaphthalene from naphthalene

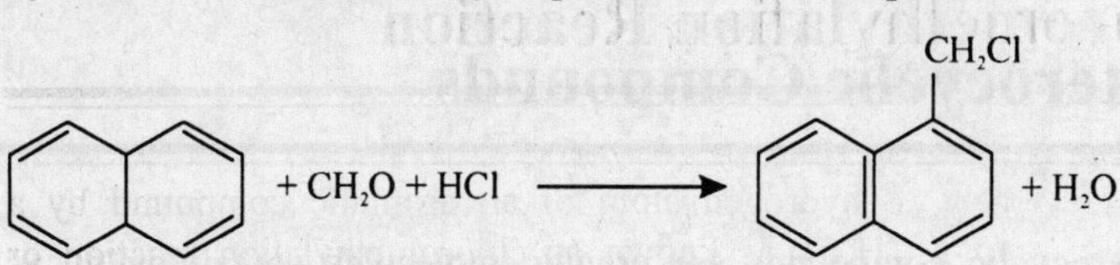

In a 250ml round bottom flask fitted with a reflux condensor place 12.6gm of naphthalene, 5.5gm of paraformadehyde, 13ml of glacial acetic acid, 8.2 ml of 85% orthophosphoric acid and 18ml of concentrated hydrochloric acid. Heat the mixture on water bath at 80-85°C for 6 hours. During this period the reaction mixture should be vigorously stirred. Cool and pour the mixture in a 500ml separating funnel, wash it first with water then with dilute sodium carbonate solution and finally with water. The product is the lower layer in all the three washings Add 20ml ether to the separating funnel and shake thoroughly. Add anhydrous potassium carbonate to the separating funnel, shake thoroughly and leave for 1 hour. Remove the ethereal layer, dry again over anhydrous potassium carbonate for 6-7 hours. Repeat the extraction process again using 20ml ether. Combine the two ethereal extracts. Distill under reduced pressure and collect 1-chloromethylnaphthalene at b.p 148-152°C/13mmHg.

20

Heterocyclic Compounds

Heterocyclic compounds are organic molecules containing a cyclic structure in which one or more atoms is an element other than carbon.However the most common other atoms (hetero atoms) are nitrogen, oxygen and sulphur.

The field of heterocyclic chemistry includes any ring system starting from a three membered ring system.The heterocyclic ring may either be aromatic or non aromatic. Some common heterocyclic structures are:

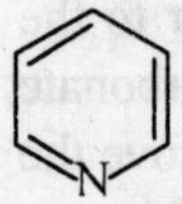
pyridine

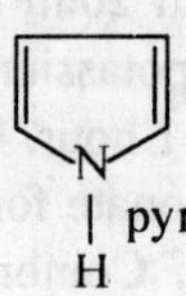
pyrole

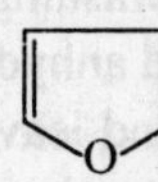
Furan

Thiophene

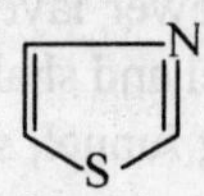
Thiazole

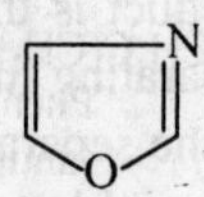
Oxazole

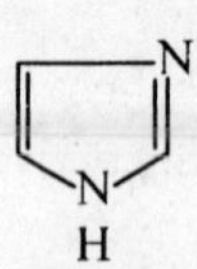
Imidazole

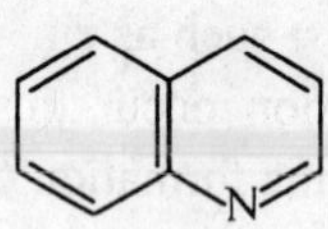
Quinoline

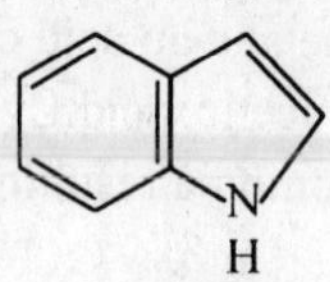
Indole

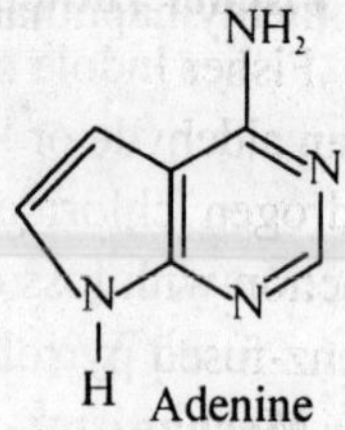
Adenine

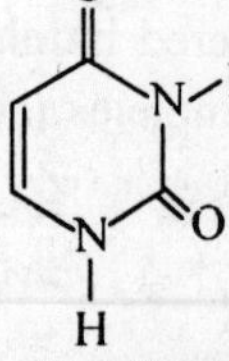
Uracil (RNA)

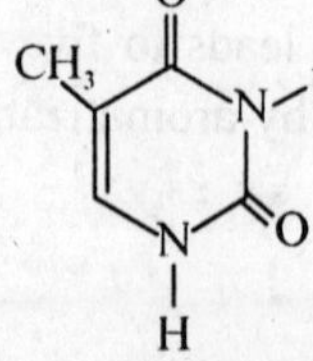
Thymine (DNA)

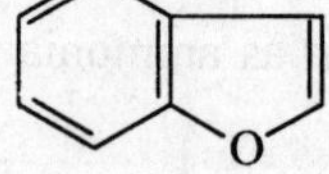
Benzofuran

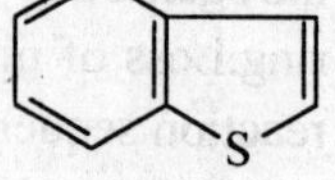
Benzothiophene

A survey of naturally occuring biologically active materials reveal the biological importance of heterocyclic ring system.Many of the most effective drugs used to treat diseases contain one or more hetrocyclic ring. Some of the biologically active heterocyclic ring system are

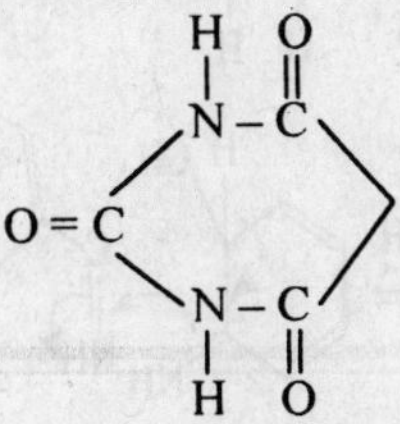

Bartituric acid
(sedative)

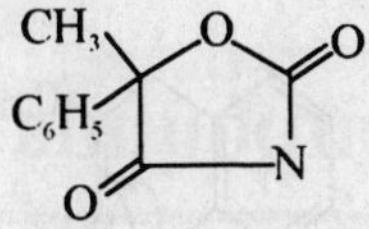

Paramethadione
(anticonvulant)

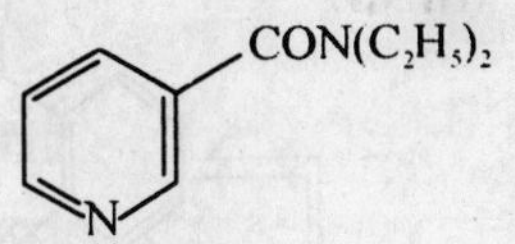

Nikethamide
(analeptic)

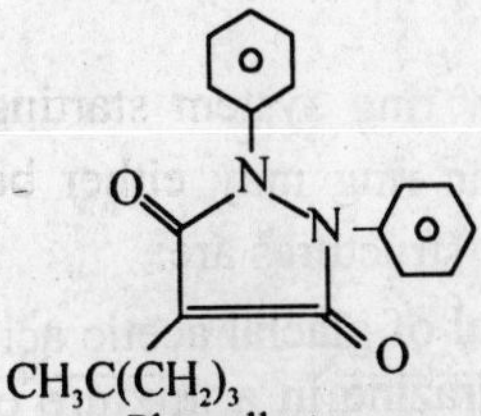

Phenylbutazone
(antipyretic, analgesic)

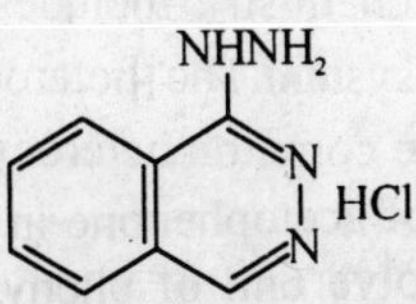

Hydralazine hydrochloride
(antihypertensive)

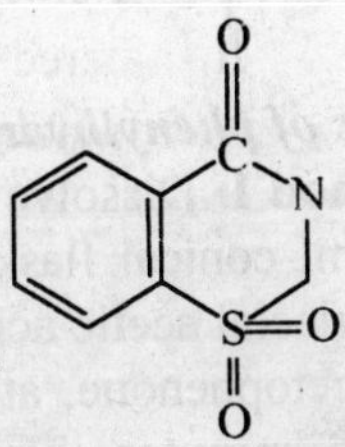

Saccharin
(artificial sweetner)

(i) Fischer Indole Synthesis

Fisher Indole synthesis involve the reaction between the phenylhydrazone of an aldehyde or ketone and an acid catalyst such as zinc chloride, ethanolic hydrogen chloride or acetic acid.Cyclization occur during the course of reaction with loss of ammonia resulting in the formation of indole derivative (benz-fused pyrrole)

Mechanism:- Conversion of the carbonyl group into a phenylhydrazone is the first step. When protonated under the reaction conditions the phenylhydrazone equilibrates to its enamine tautomer.Cleavage of nitrogen-nitrogen bond and simultaneous cyclization leads to five membered indole ring.Loss of nitrogen as ammonia followed by aromatization completes the reaction sequence.

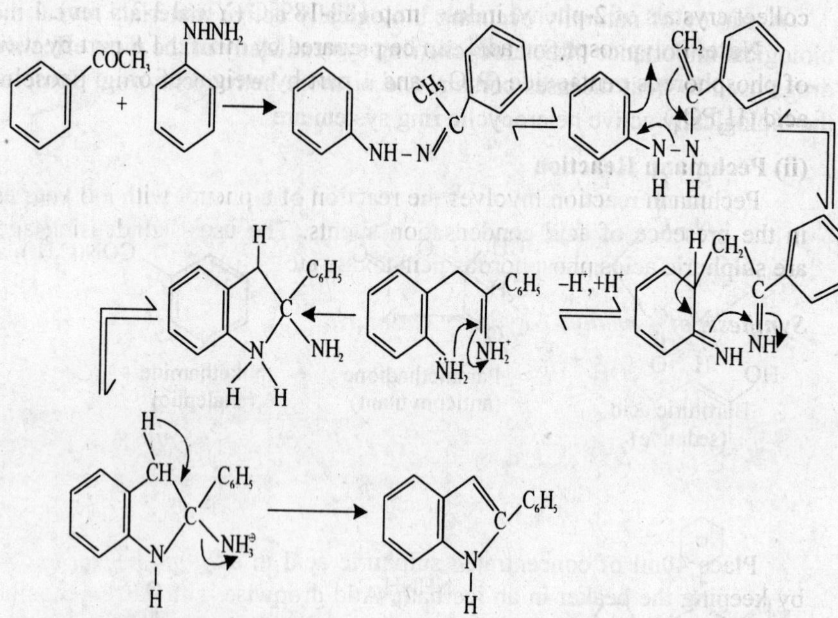

Synthesis of phenylhydrazone

Method I: Dissolve 5ml of acetophenone in 15ml of glacial acetic acid in a 100ml conical flask.Dissolve 6ml of phenylhydrazine in a mixture of 10ml of glacial acetic acid and 10ml of water. Add the hydrazine solution to that of acetophenone, at once cork and shake the conical flask vigorously for 10-15 minutes. The mixture becomes warm and phenylhydrazone seperates out (heating may sometimes be required for 10-15minutes).Cool the flask and filter the contents at pump, wash first with dilute acetic acid and then with water. Recrystallise the phenylhydrazone from ethanol, m.p 105°C (Yield 5-6gm)

Method II: Heat a mixture of 6.6ml of acetophenone with 6ml of phenylhydrazine contained in 50ml round bottom flask fitted with a reflux condensor on water bath for 60 minutes. Cool the flask and stir with glass rod until the mixture solidifies. Filter the crystalline acetophenone phenylhydrazone and recrystallise from ethanol, m.p 105°C (Yield 6-7gm)

Synthesis of 2-Phenyl Indole

Place 5gm of phenylhydrazone and 30gm of polyphosphoric acid in a 100ml round bottom flask. Heat the flask on water bath with stirring for 10-15 minutes. Cool, add 50ml of cold water to the flask and stir. Filter the precipitated solid at pump and wash with water. Reflux the crude product with 50ml of ethanol, add a little animal charcoal and filter the solution while still hot through a pre-heated metallic funnel. Cool the filtrate and collect crystals of 2-phenyl indole, m.p 188-189°C (Yield 3-3.5 gm).

Note: Polyphosphoric acid can be prepared by mixing 1.8 part by weight of phosphorous pentaoxide (P_2O_5) and 1 part by weight of ortho phosphoric acid (H_3PO_4)

(ii) Pechmann Reaction

Pechmann reaction involves the reaction of a phenol with a β keto ester in the presence of acid condensation agents. The usual condensing agents are sulphuric acids,phosphorous pentaoxide etc.

Synthesis of 4-methyl-7-hydroxy coumarin

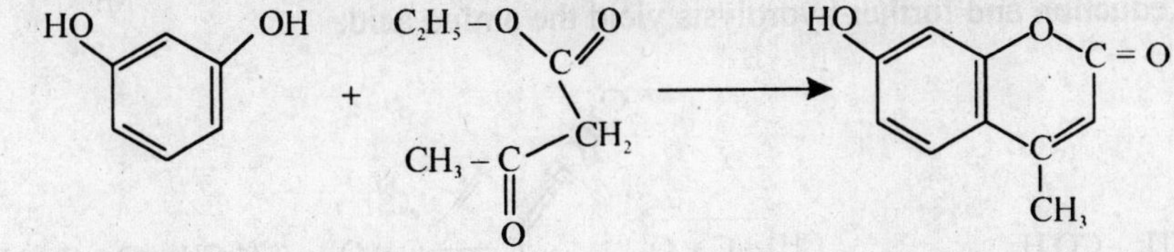

Place 40ml of concentrated sulphuric acid in a 200ml beaker and cool by keeping the beaker in an ice bath. Add dropwise with stirring a solution of 4gm of resorcinol in 5.5 ml of ethyl acetoacetate. Maintain the temperature of the beaker below 10°C during the addition process. After the addition is complete continue the stirring process for 90 minutes. Keep the reaction mixture overnight at room temperature and then pour with stirring in a 200ml beaker containing 50gm of crushed ice. Filter the pecipitated solid at pump and purify by first dissolving it in dilute sodium hydroxide solution and reprecipitating it with dilute hydrochloric acid. Recrystallise from ethanol and collect 4-methyl-7-hydroxy coumarin as colourless crystals, m.p 184-185°C (Yield 5-6gm).

$$C_2H_5O-\overset{O}{\overset{\|}{C}}-CH_2-\overset{O}{\overset{\|}{C}}-CH_3 + H^{\oplus} \longrightarrow C_2H_5-O-\overset{O}{\overset{\|}{C}}-CH_2-\overset{\overset{\oplus}{OH}}{\overset{\|}{C}}-CH_3$$

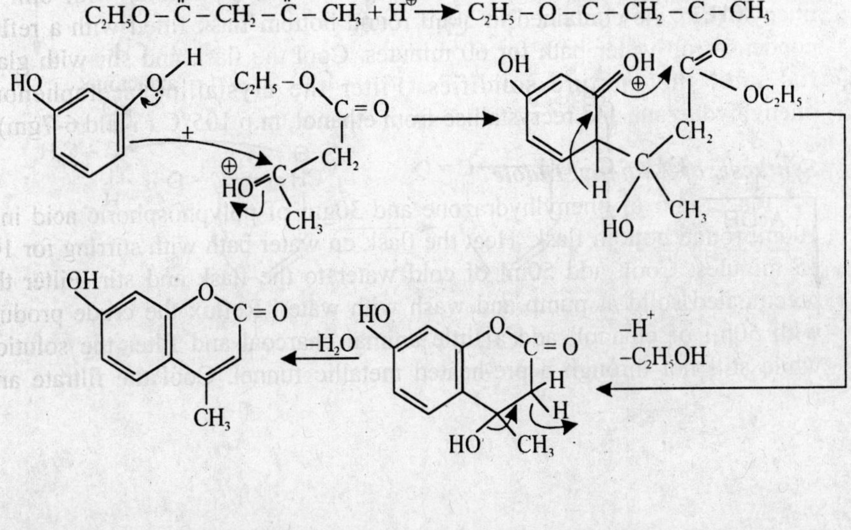

COUMARIN is a natural perfume and is also used as an artificial flavouring agent.

(iii) Erlenmeyer azlactone synthesis

Erlenmeyer azlactone synthesis is one of the most convenient method for the synthesis of α amino acids with aromatic side chain. The procedure involves the initial preparation of azlactone formed by the dehydration of benzoyl glycine when the latter is heated with acetic anhydride in the presence of sodium acetate. The methylene group in azlactone is reactive which condenses with benzaldehyde to form benzylidene derivative.The azlactone ring is readily cleaved by boiling water to yield acylamino acrylic acid. Reduction and further hydrolysis yield the amino acid.

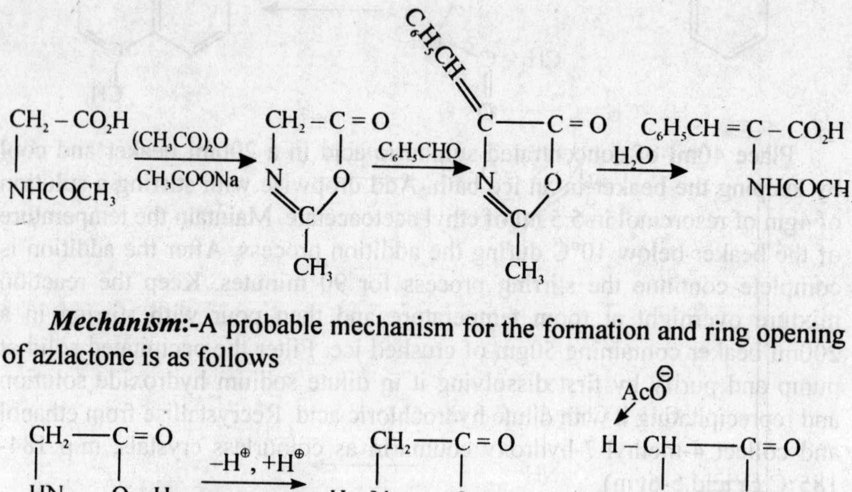

***Mechanism*:-**A probable mechanism for the formation and ring opening of azlactone is as follows

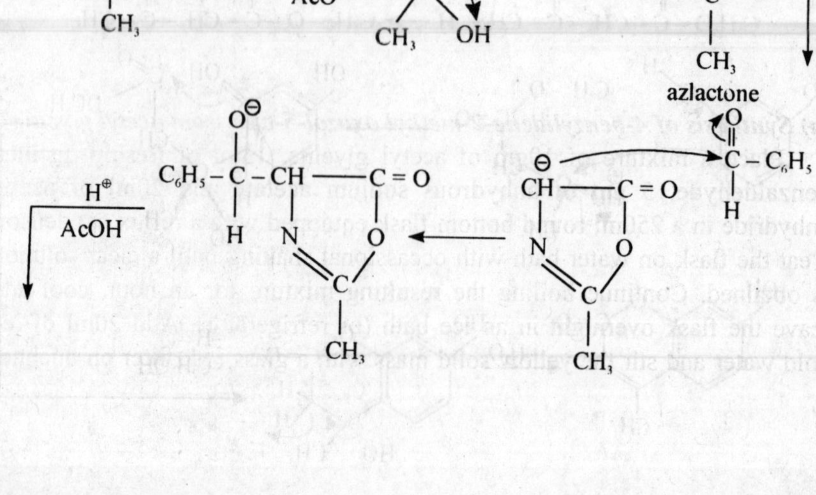

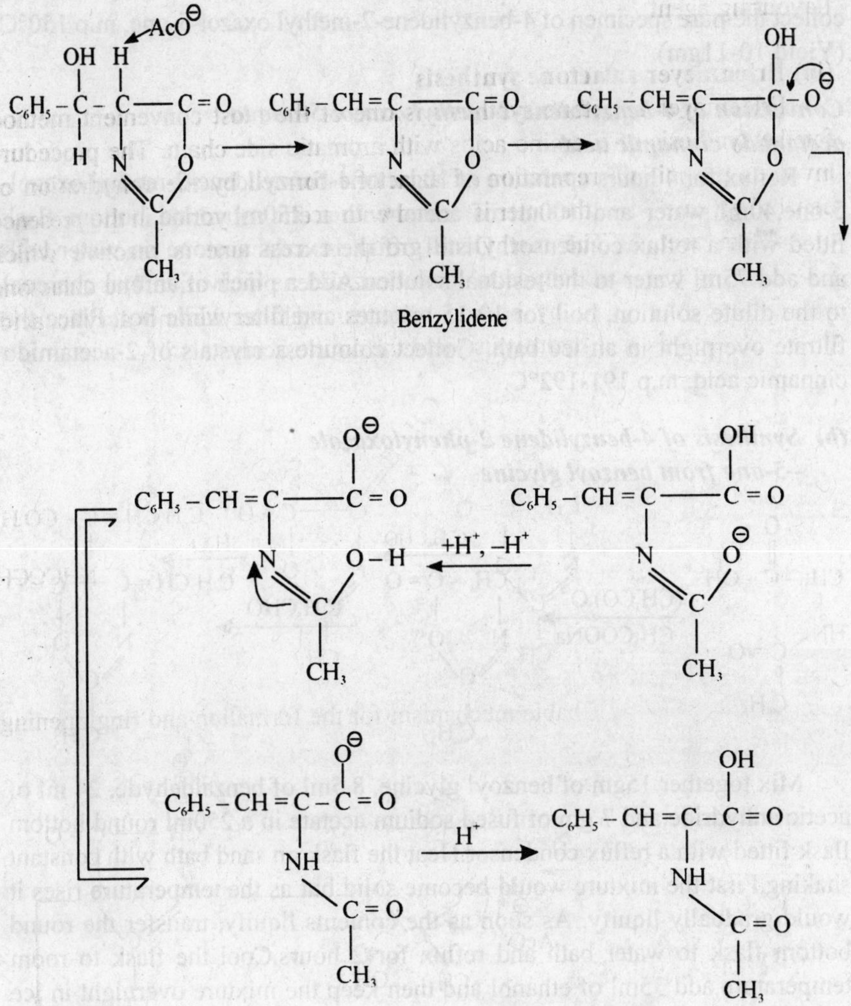

(a) Synthesis of 4-benzylidene-2-methyl oxazol-5-one from acetyl glycine

Place a mixture of 10gm of acetyl glycine, 13ml of freshly distilled benzaldehyde, 5 gm of anhydrous sodium acetate and 20ml of acetic anhydride in a 250ml round bottom flask equipped with a reflux condensor. Heat the flask on water bath with occassional shaking until a clear solution is obtained. Continue boiling the resulting mixture for an hour, cool and leave the flask overnight in an ice bath (or refrigerator). Add 20ml of ice cold water and stir the yellow solid mass with a glass rod, filter on buchner funnel and wash with cold water.Recrystallise from carbon tetrachloride and collect the pure specimen of 4-benzylidene-2-methyl oxazol-5-one, m.p 150°C (Yield 10-11gm)

Conversion of 4-benzylidene-2-methyl oxazol-5-one to acetamido cinnamic acid

Reflux for 4 hours a mixture of 10gm of 4-benzylidene-2-methyl oxazol-5-one,40ml water and 100ml of acetone in a 250ml round bottom flask fitted with a reflux condensor. Distill off the excess acetone on water bath and add 75ml water to the residual solution.Add a pinch of animal charcoal to the dilute solution, boil for 10-15 minutes and filter while hot. Place the filtrate overnight in an ice bath. Collect colourless crystals of 2-acetamido cinnamic acid, m.p 191-192°C.

(b) Synthesis of 4-benzylidene 2-phenyloxazole –5-one from benzoyl glycine

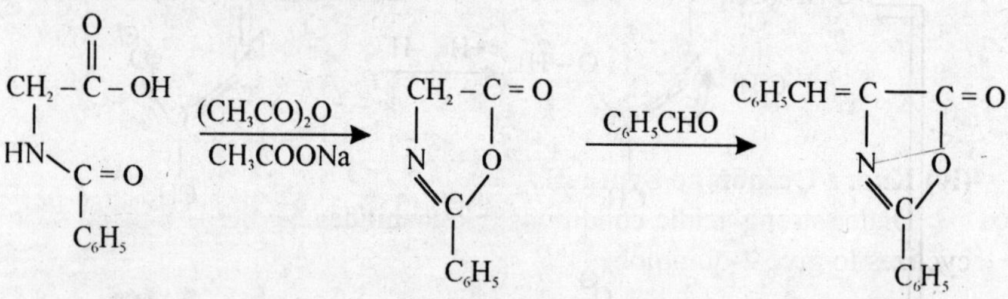

Mix together 15gm of benzoyl glycine, 8.5ml of benzaldehyde, 24 ml of acetic anhydride and 7 gm of fused sodium acetate in a 250ml round bottom flask fitted with a reflux condensor.Heat the flask on sand bath with constant shaking.First the mixture would become solid but as the temperature rises it would gradually liquify. As soon as the contents liquify, transfer the round bottom flask to water bath and reflux for 2 hours.Cool the flask to room temperature,add 35ml of ethanol and then keep the mixture overnight in ice bath. Collect the yellow crystalline product by suction filtration on buchner funnel and wash first with cold ethanol and then with boiling water.Recrystallise from benzene and collect pure 4-benzylidene 2-phenyloxazole –5-one, m.p 167-168°C. (Yield 10-11gm)

Conversion of 4-benzylidene 2-phenyloxazole –5-one to DL Phenylalanine

Place 10gm of 4-benzylidene-2-phenyloxazole –5-one, 8gm of red phosphorous and 50ml of acetic anhydride in a 250ml round bottom flask. Add with stirring and in portion 50ml of hydroiodic acid.After the addition is complete attach a condensor to the flask and reflux the mixture for 3-4 hours. Cool the flask and filter the contents with suction, wash the unreacted phosphorous with two 5ml portion of glacial acetic acid. Transfer the residue in a 250ml conical flask containing 50ml water and evaporate it to dryness on water bath (rotatory evaporator). Shake the reisdue in the flask with 60ml water and add 50ml ether to it.Shake the flask thoroughly and discard the ether layer. Repeat the process twice, each time using 20ml ether. Add a pinch of animal charcoal and few milligrams of sodium sulphite to the aqueous phase, heat it to boiling and filter hot. Neuralise the hot filtrate with ammonia solution and cool. Collect the colourless crystals of DL Phenylalanine, wash with ethanol and dry, m.p 285-288°C (d) (Yield 4-4.5gm).

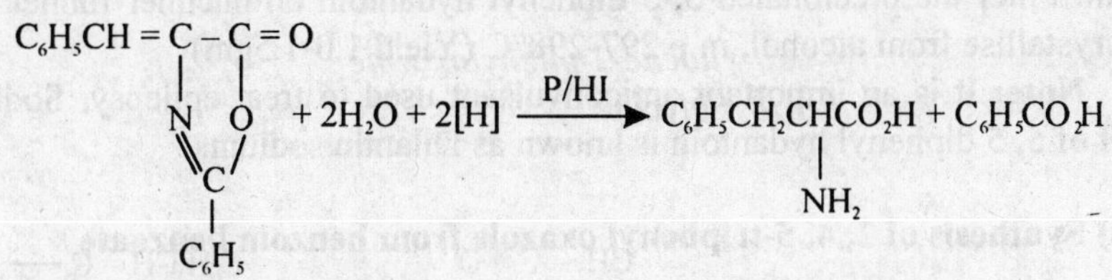

(iv) Knorr Quinolone Synthesis

Under strong acidic conditions β ketoanilides having free ortho position cyclizes to give 2-quinolones

Synthesis of 4-methyl –2-quinolone from acetoacetanilide

In a 100ml conical flask place 30ml of concentrated sulphuric acid and heat it on water bath to 75°C. Now add in portion 13.2 gm of acetoacetanilide to the hot sulphuric acid such that the temperature remains between 70-75°C. (if the reaction tends to be vigorous, cool under tap water). After the addition is complete heat the mixture at 100°C for 1hour. Cool and pour with stirring into a 500ml beaker containing 200ml ice cold water. Filter the precipitated solid, recrystallise from ethanol and collect 4-methy –2-quinolone as colourless crystals, m.p 220-222°C.

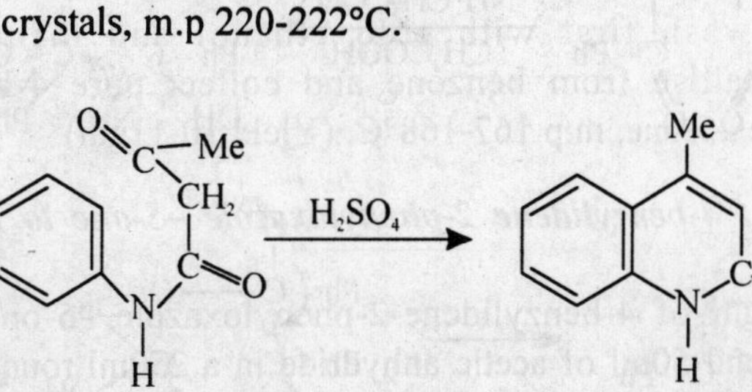

(v) Synthesis of 5, 5-diphenyl hydantoin from benzil

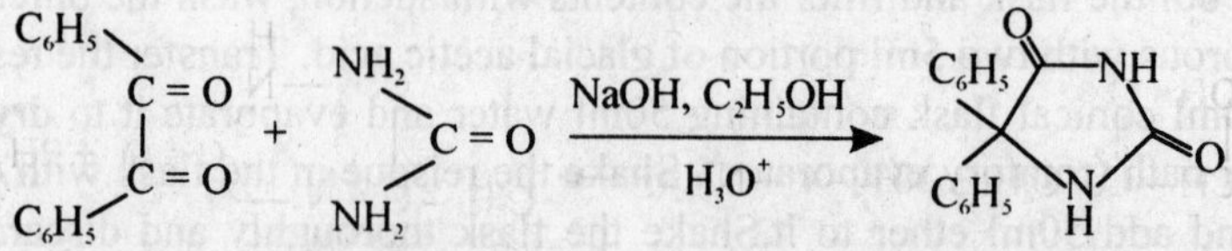

In a 100ml round bottom flask fitted with a reflux condensor place 2.65 gm of benzil, 1.5 gm of urea, 8ml of 30% aqueous sodium hydroxide solution and 40ml of ethanol. Reflux the mixture on sand bath for 2 hours. Cool the mixture at laboratory temperature and then pour with stirring in a 200ml beaker containing 50ml of water. Filter and acidify the filtrate with concentrated hydrochloric acid and then cool the acidic solution in an ice bath. Filter the precipitated 5, 5 diphenyl hydantoin on buchner funnel and recrystallise from alcohol, m.p 297-298°C (Yield 1.0-1.5gm)

Note: It is an important anticonvulsant used to treat epilepsy. Sodium salt of 5, 5 diphenyl hydantoin is known as Dilantin sodium.

(vi) Synthesis of 2, 4, 5-triphenyl oxazole from benzoin benzoate

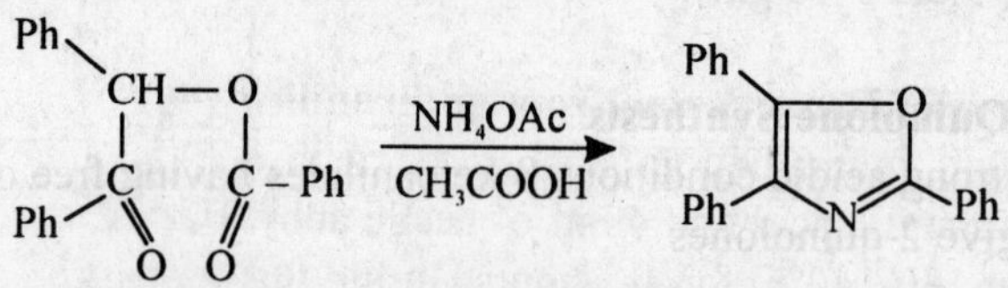

Reflux together 1.8gm of benzoin benzoate with 3gm of ammonium acetate and 25 ml of glacial acetic acid in a 100ml round bottom flask fitted with a reflux condensor for 60 minutes. Còol the flask, add 20ml of ice cold water to it and stir the mixture well. Filter the precipitated solid on buchner funnel and wash with 60% aqueous acetic acid. Recrystallise from ethanol and obtain pure specimen of 2, 4, 5-triphenyl oxazole, m.p 115-116°C.

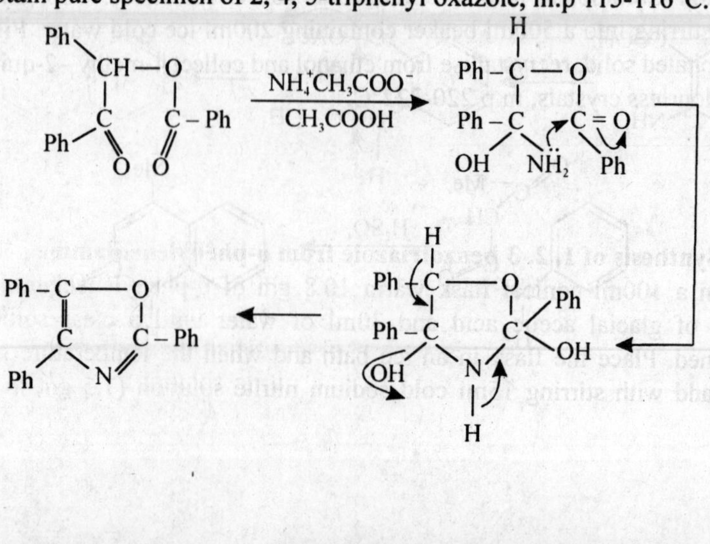

(vii) Synthesis of 4, 5 diphenylglyoxalone from benzoin

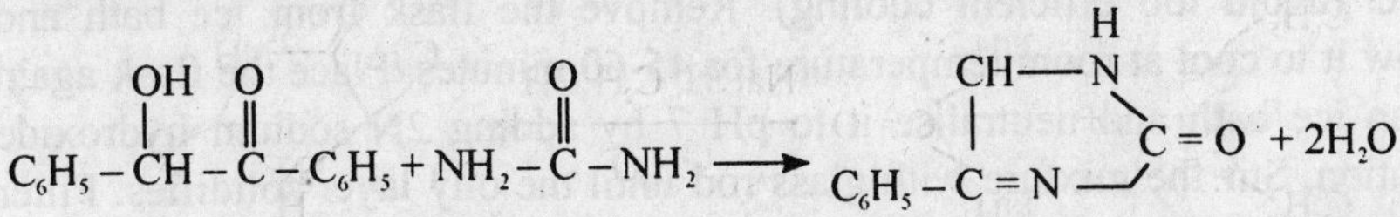

Reflux together 10.5gm of benzoin, 5.5 gm of urea and 40ml of glacial acetic acid placed in a 100ml round bottom flask fitted with a reflux condensor for 7 hours. Pour the hot mixture in a 200ml beaker, allow it to stand at room temperature for 3 hours and then finally cool in an ice bath. Filter the precipitated solid on buchner funnel Transfer the dried solid in a 200 ml beaker, add 25ml of ether to it and stir well for 20 minutes and again filter. Dry the solid and then dissolve by heating with 20ml of glacial acetic acid in a 100ml round bottom flask fitted with a reflux condensor until a clear solution is obtained. Pour the hot solution with stirring in a 200ml beaker containing 50 gm of crushed ice/100ml of ice water.Continue the stirring for 30 minutes and then filter the solid on buchner funnel.Wash with water and then dry, 4, 5 diphenylglyoxalone is obtained as white fluffy crystals, m.p 331-335°C. (Yield 9-10 gm)

(viii) Synthesis of Benzoxazine from anthranilic acid

In a 100ml round bottom flask fitted with a reflux condensor place 4.6gm of anthranilic acid and 40ml of acetic anhydride. Reflux the mixture on oil bath at 130-140°C for 4-5 hours.During this period the colour of the solution gradually changes from yellow to red. Evaporate off the excess acetic anhydride under reduced pressure. Recrystallise the resulting gummy mass from anhydrous ether and collect colourless crystals of 2 methyl –4H-3, 1-benzoxazine –4-one, m.p 80-81°C.

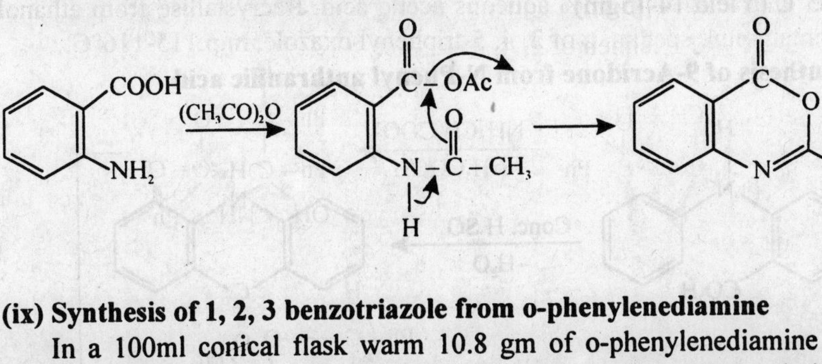

(ix) Synthesis of 1, 2, 3 benzotriazole from o-phenylenediamine

In a 100ml conical flask warm 10.8 gm of o-phenylenediamine with 15ml of glacial acetic acid and 30ml of water until a clear solution is obtained. Place the flask in an ice bath and when the temperature reaches 5°C add with stirring 15ml cold sodium nitrite solution (7.5 gm in 15 ml water).The mixture will turn dark green,allow the temperature to rise to 70°C (avoid too efficient cooling). Remove the flask from ice bath and allow it to cool at room temperature for 45-60 minutes. Place the flask again in an ice bath and neutralise it to pH 7 by adding 2N sodium hydroxide solution. Stir the mixture with glass rod until the oily layer solidifies. Filter the precipitated solid on buchner funnel and wash with cold water.Recrystallise benzotriazole from benzene, m.p 100-101°C. (Yield 7-8gm). A white crystalline product can also be obtained by sublimation.

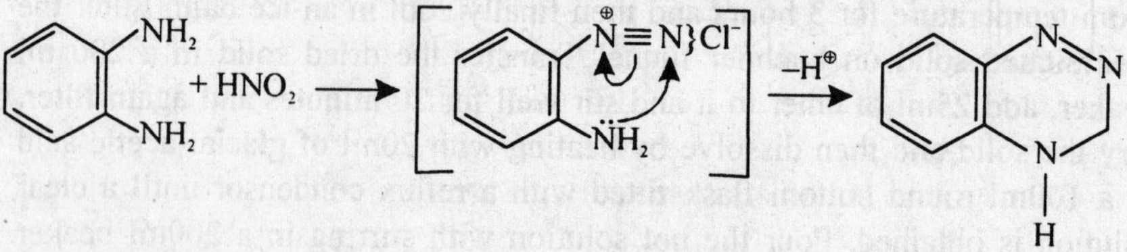

(x) Synthesis of Phenothiazine from diphenylamine

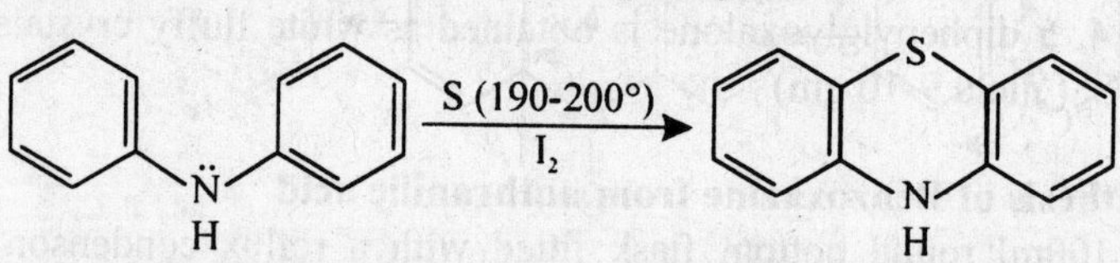

In a 100ml round bottom flask fitted with a reflux condensor place 6.4 gm of sulphur, 16.8 gm of diphenylamine and 300mg of iodine. Reflux the mixture on oil bath at 190-200°C for 30-40 minutes. Cool and grind the residual mass. Add 5ml of ethanol and a pinch of animal charcoal to it, heat it on water bath until the solid mass dissolves.Filter the solution hot and cool the filtrate in an ice bath. Collect yellow crystals of Phenothiazine, m.p 184-185°C (Yield 14-15gm)

(xi) Synthesis of 9-Acridone from N-Phenyl anthranilic acid

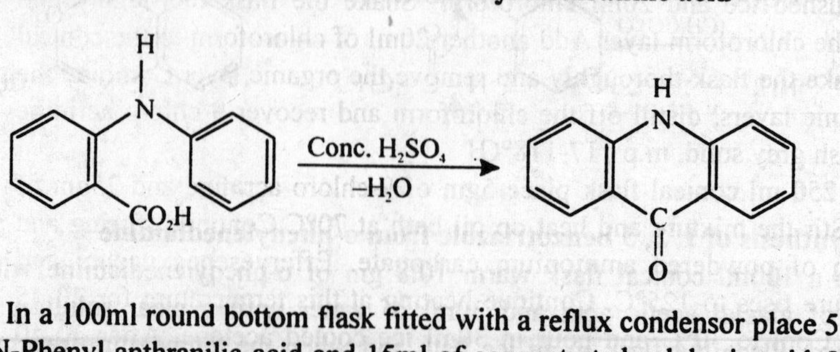

In a 100ml round bottom flask fitted with a reflux condensor place 5 gm of N-Phenyl anthranilic acid and 15ml of concentrated sulphuric acid. Heat the flask on steam bath for 3-4 hours. Shake the flask occassionally during the refluxing period. After the reaction is complete pour the coloured solution with stirring in a 200ml beaker containing 100ml boiling water. Heat the mixture to boiling for 5 minutes and filter hot. Cool the filtrate in an ice bath and collect the yellow precipitated solid on buchner funnel. Wash with cold water, recrystallise from glacial acetic acid or from isoamyl alcohol, 9-acridone melts at 348-350°C.

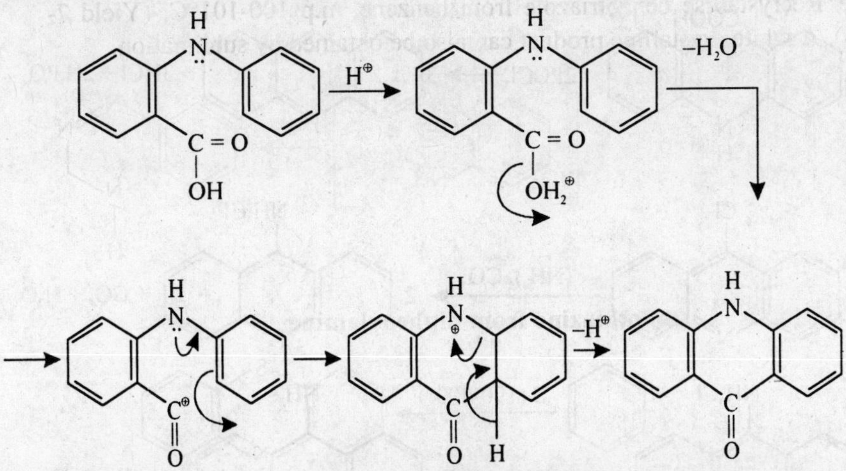

(xii) Synthesis of 9-amino acridine from N phenyl anthranilic acid

Heat a mixture of 5gm of N phenyl anthranilic acid with 16ml of phosphorous oxychloride in a 100ml round bottom flask fitted with a reflux condensor for 15 minutes on water bath. As soon as the vigorous reaction sets in remove the flask from water bath and keep it at room temperature until the reaction subsides. Heat the mixture in an oil bath at 130-140°C for 2 hours. Distill off the excess phosphorous oxychloride. Cool the residue and pour it in a 250ml conical flask containing 20ml ammonium hydroxide, 50gm crushed ice and 20ml chloroform. Shake the flask thoroughly and remove the chloroform layer. Add another 20ml of chloroform to the conical flask, shake the flask thoroughly and remove the organic layer. Combine the two organic layers, distill off the chloroform and recover 9-chloro acridine as greenish grey solid, m.p 117-118°C.

In a 250 ml conical flask place 5gm of 9-chloro acridine and 25gm of phenol. Stir the mixture and heat on oil bath at 70°C. Continue stirring and add 3gm of powdered ammonium carbonate. Effervesence occurs and temperature rises to 125°C. Continue heating at this temperature for 40-45 minutes. Cool to 30°C and pour in 30ml ice cooled acetone. After 30-40 minutes, filter the precipitated 9-amino hydrochloride. Dissolve it in 100ml boiling water containing 1-2 drops of concentrated hydrochloric acid and filter hot. To the filtrate add 30ml of sodium hydroxide solution and cool (6gm sodium hydroxide in 30ml water).Collect the bright yellow coloured solid on buchner funnel and dry. 9-amino acridine can be further purified from acetone, m.p 232-233°C.

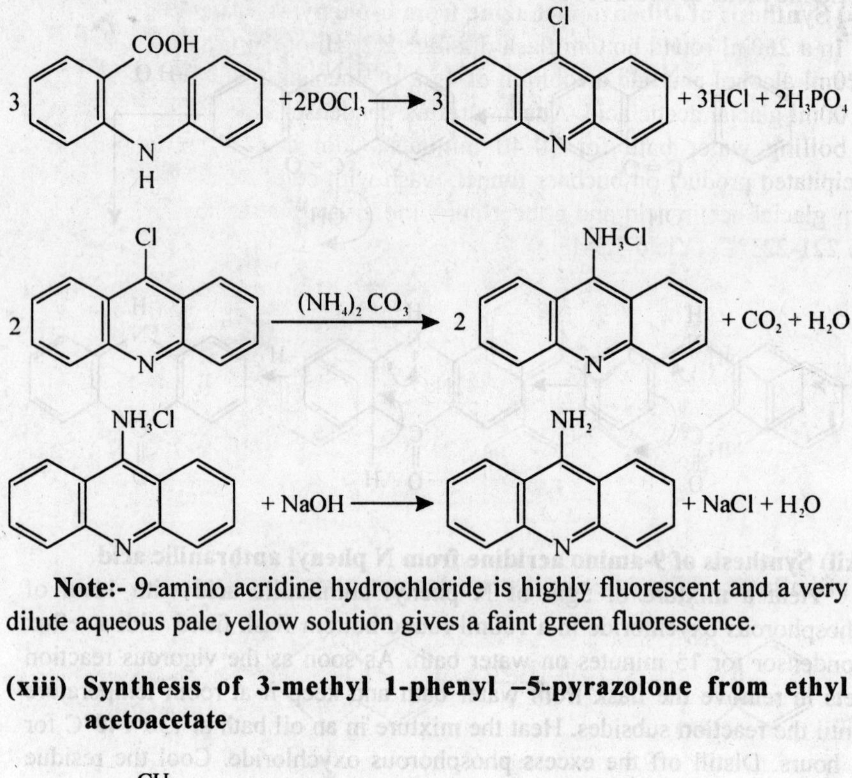

Note:- 9-amino acridine hydrochloride is highly fluorescent and a very dilute aqueous pale yellow solution gives a faint green fluorescence.

(xiii) Synthesis of 3-methyl 1-phenyl –5-pyrazolone from ethyl acetoacetate

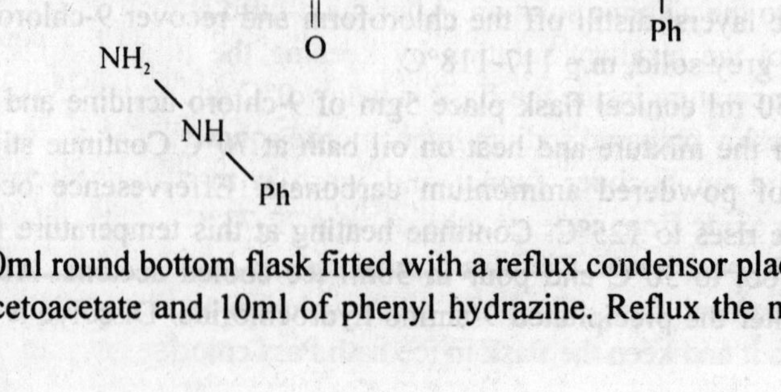

In a 50ml round bottom flask fitted with a reflux condensor place 13.5ml of ethyl acetoacetate and 10ml of phenyl hydrazine. Reflux the mixture in oil bath at 120-130°C for 60 minutes. Cool the oily liquid and stir with a glass rod until it solidifies.Filter the precipitated 3-methyl 1-phenyl –5-pyrazolone and recrystallise from 50% ethanol, m.p 125-127°C.

(xiv) Synthesis of Dibenzophenazine from o-phenylene diamine

In a 250ml round bottom flask dissolve 2.2 gm of o-phenylene diamine in 20ml alcohol and add a solution of 4gm of phenanthraquinone dissolved in 100ml glacial acetic acid. Attach a reflux condensor and heat the mixture on boiling water bath for 30-40 minutes. Cool and filter the yellow precipitated product on buchner funnel, wash with cold water. Recrystallise from glacial acetic acid and collect pure and crystalline dibenzophenazine, m.p 221-223°C. (Yield 3.5-4gm)

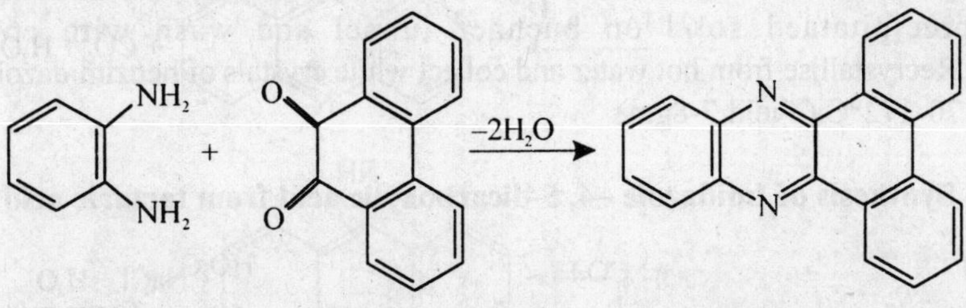

(xv) Synthesis of Benzofurazan oxide from o-nitro aniline

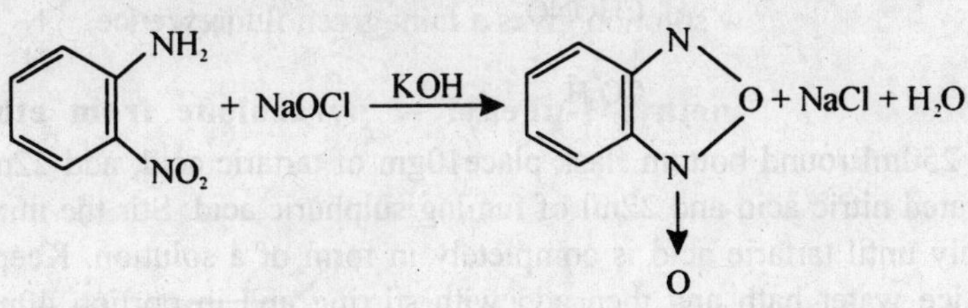

Dissolve 2.1gm of potassium hydroxide in 25ml ethanol placed in a 100ml conical flask by warming on water bath. Add with stirring 4 gm of o-nitro aniline to the alkaline solution. After the o-nitro aniline has completely dissolved cool the resultant solution by keeping the flask in an ice bath. When the temperature inside the flask is below 0°C add in portion and with stirring the freshly prepared sodium hypochlorite solution. Collect the yellow coloured solid on buchner funnel and wash with water. Recrystallise benzofurazan oxide from aqueous ethanol, m.p 72-73°C.

Sodium hypochlorite: Dissolve 5gm of sodium hydroxide in 20ml water in a 100ml conical flask. Cool the alkali solution to 0°C, add 10gm of crushed ice to it and keep the flask in ice bath.Pass chlorine gas into the cold alkali solution until there is an increase in weight by 4-6gm. Preserve this sodium hypochlorite solution in a dark coloured bottle at 0°C.

(xvi) Synthesis of benzimidazole from o-phenylene diamine

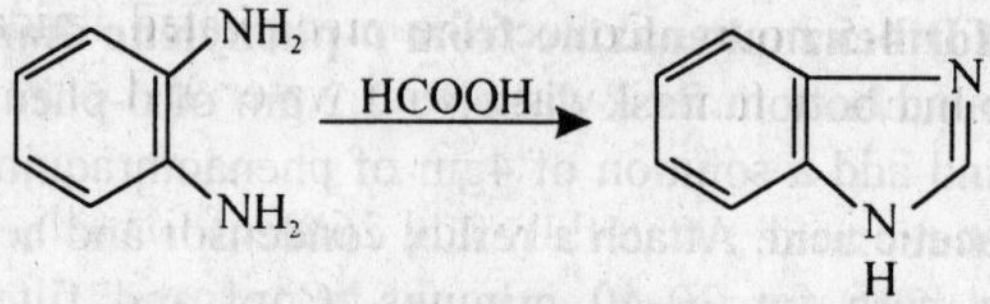

Heat together 10gm of o-phenylene diamine with 40ml of 90% formic acid placed in a 100ml round bottom flask at 100°C on water bath for 1-2hours. Cool, dilute the reaction mixture with 20ml cold water and add with stirring, dilute sodium hydroxide solution until the mixture is alkaline. Collect the precipitataed solid on buchner funnel and wash with cold water.Recrystallise from hot water and collect white crystals of benzimidazole, m.p 170-172°C.(Yield 7-8gm)

(xvii) Synthesis of Imidazole –4, 5-dicarboxylic acid from tartaric acid

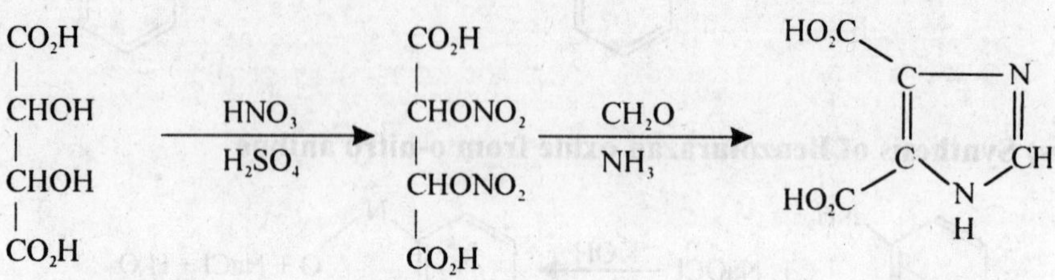

In a 250ml round bottom flask place10gm of tartaric acid, add 22ml of concentrated nitric acid and 22ml of fuming sulphuric acid. Stir the mixture thoroughly until tartaric acid is completely in form of a solution. Keep the flask in ice water bath and then add with stirring and in portion 40ml of concentrated sulphuric acid taking care that the temperature does not rise above 45°C. After the addition is complete keep the flask at room temperature for 3hours. Filter the precipitated solid on buchner funnel and press to remove as much liquid as possible. The solid nitro derivative is then transferred in a 200ml beaker containing 50gm of crushed ice and stirred thoroughly. Pour this cold solution in a 250ml conical flask placed in an ice bath. Neutralise the cold solution by adding 40ml concentrated ammonium hydroxide solution. The addition should be regulated such that the temperature does not rise above –5°C. Now add dropwise freshly prepared chilled hexamethylene- tetramine solution to the ammonical solution of tartaric acid dinitrate (temp below 2°C). Continue the stirring process for another 10 minutes after the addition is complete and then keep the mixture overnight at room temperature.

Filter and add 5ml of ethanol to the filtrate,which is then acidified by adding 20ml of concentrated hydrochloric acid to it. Cool the acidified mixture in ice bath for 4-5 hours. Collect the precipitated Imidazole–4, 5-dicarboxylic acid on buchner funnel, wash with water and dry, m.p 280°C (d) (Yield 4-5gm).

Hexamethylenetetramine : Carefully add 26ml of formalin to 25ml of ammonium hydroxide.The temperature should be below 20°C during the addition process.Finally chill the solution to 0°C.

21

DYES

Dyes are coloured organic compounds (natural or synthetic) which impart colour to substances like fibres, papers, food stuffs, leathers, plastics, medicines etc. A good dye not only has the ability to bind itself to the fabric and retain colour even on prolonged exposure to sunlight (ultraviolet light) but it also resists the action of dilute acids and dilute alkalies particularly the latter due to alkaline nature of soaps and washing powders.

The synthetic dyes are mainly aromatic hydrocarbon derivatives. These aromatic compounds are converted into dyes by introducing polar substituents like nitro, sulphonic and alkoxy group in parent hydrocarbon. The colour of these dyes is due to the presence of unsaturated group known as chromophores in the moiety; and greater the number of chromophores greater is the intensity of colour. The compounds containing these chromophores are known as chromogen. Some important chromophores are

$-N=N-$ azo

$-N=O$ nitroso

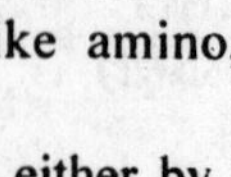

nitro

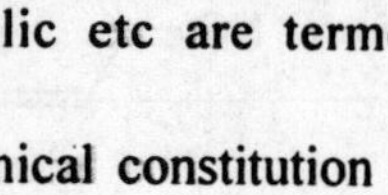

quinonoid

In addition these organic compounds also contain groups which are themselves unable to produce colour but can deepen it in the presence of chromophore. Such groups like amino, carboxylic etc are termed as auxochromes.

Dyes are usually classified either by their chemical constitution or by their application to textile fibre. The chemical constitution method is being used here to describe different dyes in this book.

(a) AZO DYES

These compounds are characterised by the presence of $-N=N-$ linkage and are formed by the direct coupling between diazonium compound and a phenol or aromatic base The coupling reaction is infact an electrophilic aromatic substitution reaction in which diazonium ion being a weak electrophile reacts only with strongly activated ring.

$$Ar-\overset{\oplus}{N}\equiv N+H-Ar^{1}\longrightarrow Ar-N=N-Ar^{-1}+H^{+}$$

These coupling reactions often take place in weakly acidic medium when amines are used as coupling compounds and in weakly alkaline medium when phenols are used.

Most of these azo dyes possess at least one sulfonate ($-SO_3^-$) or carboxylate (COO^-) group in the molecule which not only promote solubility in water but also helps to bind the dye to the polar surface of common fibre such as cotton and wool.

(i) Synthesis of Para Red (p-nitrobenzene azo 2-naphthol) from p-nitro aniline

In a 200ml beaker dissolve 5gm of p-nitro aniline in 12ml of concentrated hydrochloric acid and 15ml of water.Cool the beaker by placing it in an ice bath and stir vigorously whereby the hydrochloride of the amine crystallises out. Add dropwise with stirring the nitrite solution (prepared by dissolving 4gm of sodium nitrite in 9ml water) but taking care that the temperature does not rise above 10°C.The solid hydrochloride will dissolve to a clear solution of diazonium salt.

In a second 200ml beaker dissolve 5.2gm of 2-naphthol in 25ml sodium hydroxide solution (7gm sodium hydroxide in 25ml water), cool by keeping the beaker in ice bath and add the cold diazonium salt solution slowly and with stirring to it. The solution slowly turns red and red crystal of p-nitrobenzene azo 2-naphthol seperates out. After the addition is complete keep the beaker in ice bath for another 10-15 minutes with occasional stirring. Filter the solid through buchner funnel and wash with cold water. Recrystallise from toluene and collect red crystals of p-nitrobenzene azo 2-naphthol, m.p 244-246°C (Yield 6-7gm).

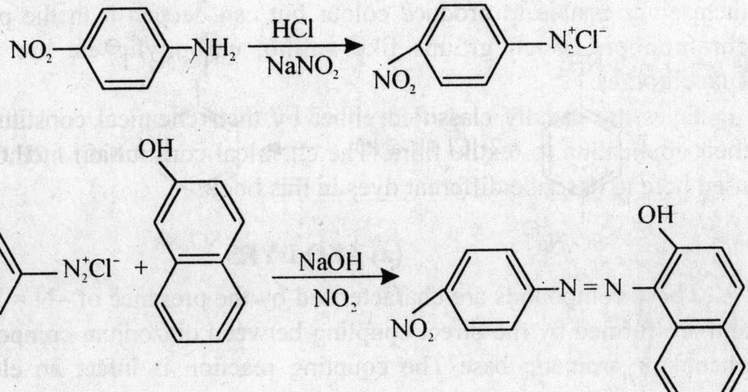

(ii) Synthesis of Methyl Orange (Helianthin, Orange III, Sodium p-dimethyl amino phenyl azobenzene-4-sulphonate) from Sulfanilic acid

In a 100ml conical flask dissolve by warming 5gm of sulfanilic acid dihydrate in 50ml of sodium carbonate solution (1.33gm of sodium carbonate dissolved in 50ml water). Cool by keeping the flask in ice water and when the temperature inside the flask reaches 10°C add in portion 5ml of cold sodium nitrite solution (1.8gm of sodium nitrite in 5ml water). Pour this solution with stirring in a 200ml beaker containing 5ml of concentrated hydrochloric acid and 25gm of crushed ice. In a minute or two crystals of diazobenzene sulphonate will separate out.

In a second 100ml conical flask thoroughly mix 3.2ml of dimethyl aniline with 2ml of glacial acetic acid and add this solution of dimethyl aniline acetate with vigorous stirring to the above synthesized suspension of diazotized sulfanilic acid. Keep the reaction mixture at room temperature for 10-15 minutes and then add slowly with stirring 20ml of dilute sodium hydroxide solution (20%). The solution turns red due to the formation of sodium salt of methyl orange. Add 5gm of sodium chloride to the red solution and heat it to 90°C so that all the salt dissolves. Cool the solution first at room temperature and then in an ice bath, the dye separates out in the form of orange crystals. Filter the crystalline solid at pump and wash with cold ethanol. Recrystallise from hot water and collect reddish orange crystals of methyl orange (Yield 5-6gm)

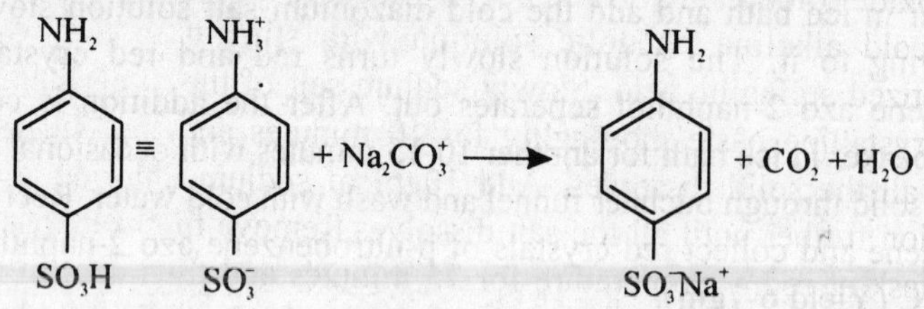

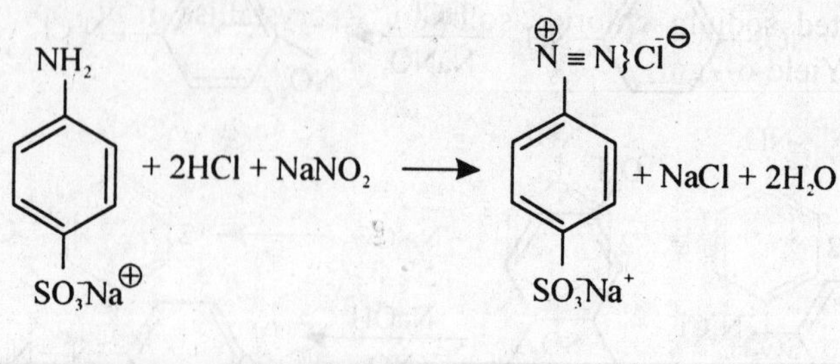

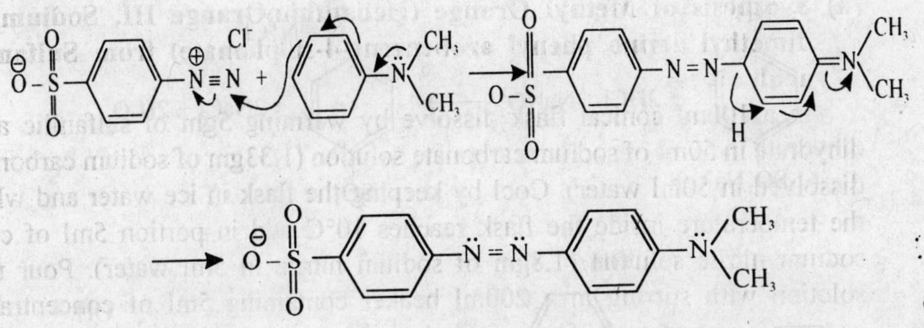

methyl orange

(iii) Synthesis of Orange II (2-naphthol orange, Sodium p-[2-hydroxy-1-naphthylazo] benzene sulphonate) from Sulfanilic acid

In a 100ml conical flask dissolve by warming 5gm of sulfanilic acid dihydrate in 50ml of sodium carbonate solution (1.33gm of sodium carbonate dissolved in 50ml water). Cool by keeping the flask in ice water and when the temperature inside the flask reaches 10°C add in portion 5ml of cold sodium nitrite solution (1.8gm of sodium nitrite in 5ml water).Pour this solution with stirring in a 200ml beaker containing 5ml of concentrated hydrochloric acid and 25gm of crushed ice. In a minute or two crystals of diazobenzene sulphonate will separate out.

In a 200ml beaker dissolve 3.6gm of 2-naphthol in 20ml of dilute sodium hydroxide solution (10%) and cool by keeping the beaker in an ice bath.Pour this cold alkaline phenolic solution with stirring to the suspension of diazotized sulfanilic acid. Slowly sodium salt of the dye separates out. Stir the crystalline paste thoroughly for 10 minutes and then heat the mixture until all the solid dissolves. Add 10gm of sodium chloride and warm the solution further until all the salt dissolves.Remove the heat source and keep the flask at room temperature for 75 minutes and then cool in ice until the crystallization is complete. Filter the dye on buchner funnel and wash with saturated sodium chloride solution. Recrystallise from boiling water and dry. (Yield 6-7gm)

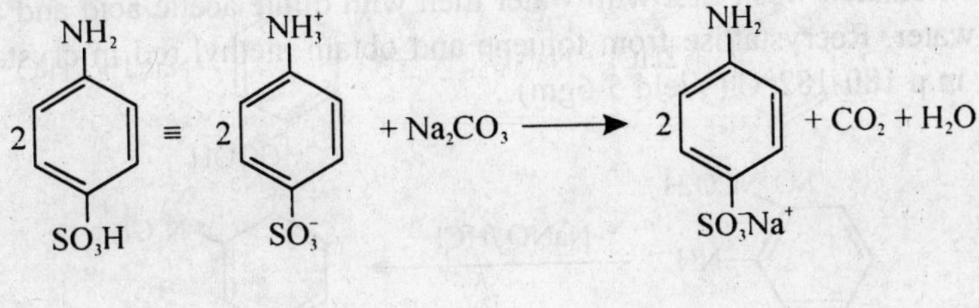

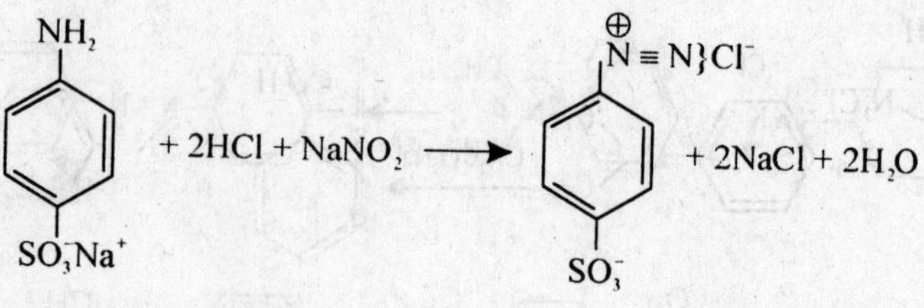

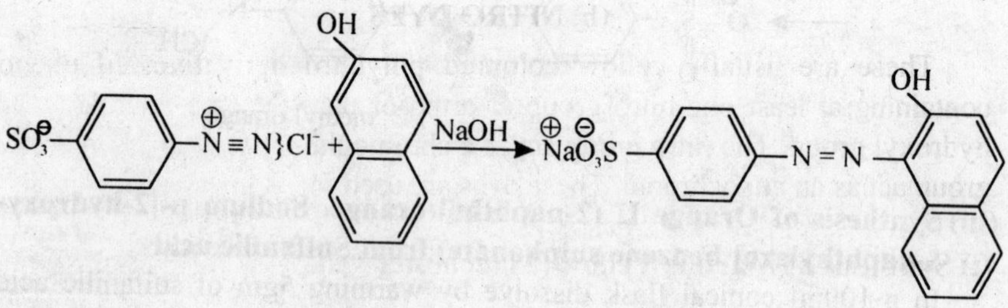

(iv) Synthesis of Methyl Red from anthranilic acid

Dissolve by heating 5gm of anthranilic acid in a mixture of 4ml concentrated hydrochloric acid and 12ml water in a 200ml beaker. Place the beaker in an ice bath and add with stirring 20gm of crushed ice and 6ml of concentrated hydrochloric acid. When the temperature inside the beaker is in the range of 0-5°C add slowly with stirring 6ml of cold sodium nitrite solution (3gm of sodium nitrite in 6ml water) taking care that the temperature does not rise above 6°C. To this cold diazonium salt solution add with stirring 7ml of dimethyl aniline and continue the stirring process for another 15-20 minutes (temperature below 6°C)

In a test tube dissolve 5gm of sodium acetate in 8ml water and add 4ml of this solution to the cold diazonium salt-amine solution. Continue keeping the beaker in ice bath for another 75 minutes and then add the remainder of the sodium acetate solution. Remove the beaker from the ice bath, keep it at room temperature for 20-30minutes. To the resulting reaction mixture add with stirring 4ml of dilute sodium hydroxide solution (20%) and leave the mixture undisturbed for 60-70 minutes. Cool, filter the precipitated solid on buchner funnel, wash first with water then with dilute acetic acid and again with water. Recrystallise from toluene and obtain methyl red in crystalline form, m.p 180-182°C (Yield 5-6gm)

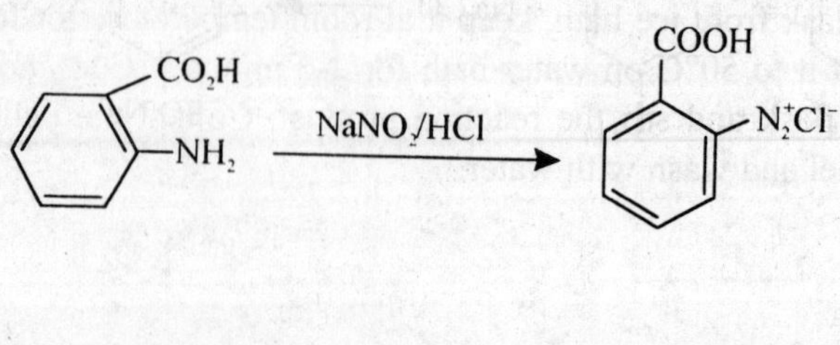

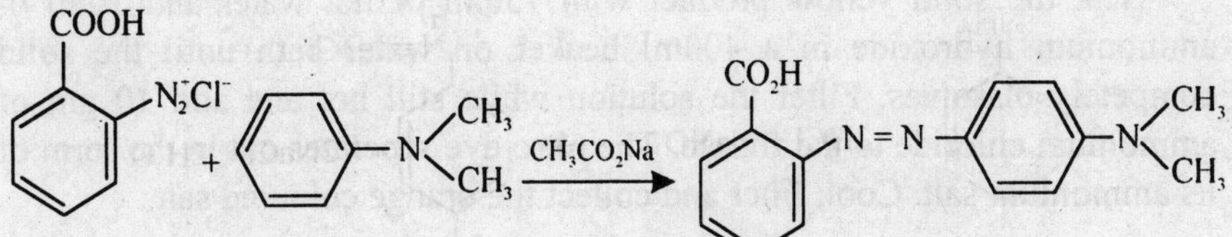

(b) NITRO DYES

These are usually yellow coloured polynitro derivatives of phenol containing at least one nitro group at ortho or para position relative to the hydroxyl group. The nitro group act as a chromophore whereas the hydroxyl group act as an auxochrome. These dyes are used for dying silk and wool.

(i) Synthesis of Martius Yellow (Manchester yellow) from 1-naphthol

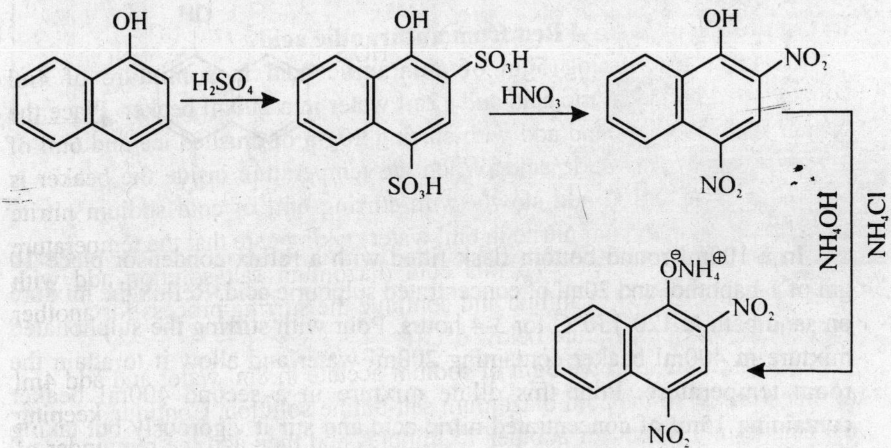

In a 100ml conical flask dissolve 5gm of 1-naphthol in 10ml of concentrated sulphuric acid by heating on steam bath for 5-10 minutes. Cool the flask by keeping it in an ice bath and then add 20ml water to it. When the temperature inside the flask is below 10°C add in portion with stirring 6ml of concentrated nitric acid taking care that the temperature does not rise above 10°C during the addition process. After the addition is complete remove the flask from ice bath, keep it at room temperature for few minutes and then heat it to 50°C on water bath for 2-5 minutes. Cool, add 25ml ice water to the flask and stir the reaction product. Collect the yellow dye on buchner funnel and wash with water.

Heat the solid yellow product with 150ml of hot water and 10ml of ammonium hydroxide in a 400ml beaker on water bath until the solid competely dissolves. Filter the solution while still hot and add 10 gm of ammonium chloride to the filtrate. The nitro dye seperates out in the form of its ammonium salt. Cool, filter and collect the orange coloured salt.

(ii) Synthesis of Naphthol Yellow from 1-naphthol

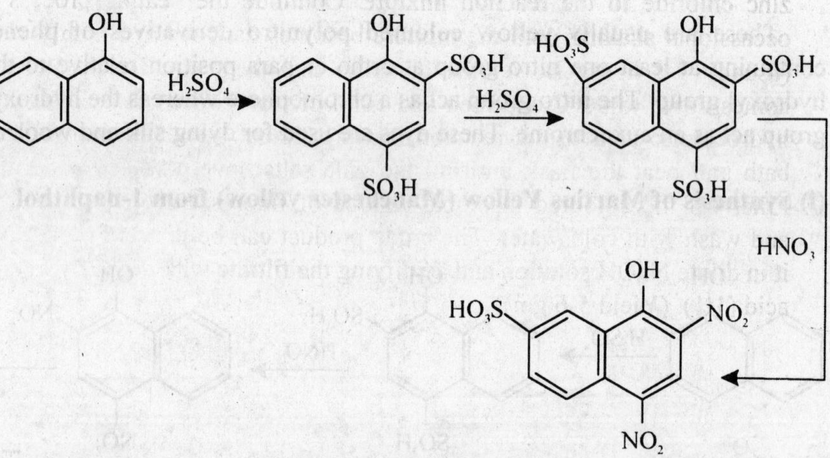

In a 100ml round bottom flask fitted with a reflux condensor place 10 gm of 1-naphthol and 20ml of concentrated sulphuric acid. Reflux the mixture on sand bath at 120-130°C for 3-4 hours. Pour with stirring the sulphonated mixture in 400ml beaker containing 200ml water and allow it to attain the room temperature. Pour this dilute mixture in a second 400ml beaker containing 15ml of concentrated nitric acid and stir it vigorously but taking care that the temperature does not rise above 35°C. Add with stirring and in portion 10ml of concentrated nitric acid (keep temperature below 35°C). Keep the beaker at room temperature for 10-15 minutes and then cool. Filter the nitrated product through buchner funnel and wash with 10% sodium chloride. Dissolve the crude precipitate in hot water and add sodium carbonate until the solution is neutral. Precipitate the dye by adding 4gm of potassium chloride. Collect the orange yellow coloured dye and dry.

(c) XANTHENE DYES

These compounds are so classified because of the presence of xanthene ring (dibenzo-1, 4 pyran) as the basic skeleton.The dyes are obtained by the introduction of auxochromes mainly amino or hydroxyl group at para position with respect to the carbon atom linking the two benzene rings, i.e., at 3 and 6 position.

(i) Synthesis of Fluorescein from resorcinol

Grind together 3.75gm of phthalic anhydride with 5.5gm of resorcinol and heat in a 100ml round bottom flask fitted with a reflux condensor to 180°C in an oil bath. Add in portions 1.75 gm of finely powdered anhydrous zinc chloride to the reaction mixture. Continue the heating process with occassional shaking until the mixture becomes dark red and highly viscous (usually 1-2 hours).Lower the flame so that the temperature of the flask comes down to 90°C and then add 50ml of water and 3ml of concentrated hydrochloric acid to the reaction mixture. Raise the temperatue of the oil bath and heat the flask until all the zinc salts have completely dissolved. Filter off the coloured insoluble residue of fluorescein on buchner funnel and wash with cold water. The crude product can be purified by dissolving it in dilute NaOH solution and acidifying the filtrate with dilute hydrochloric acid (1:1). (Yield 5-6 gm)

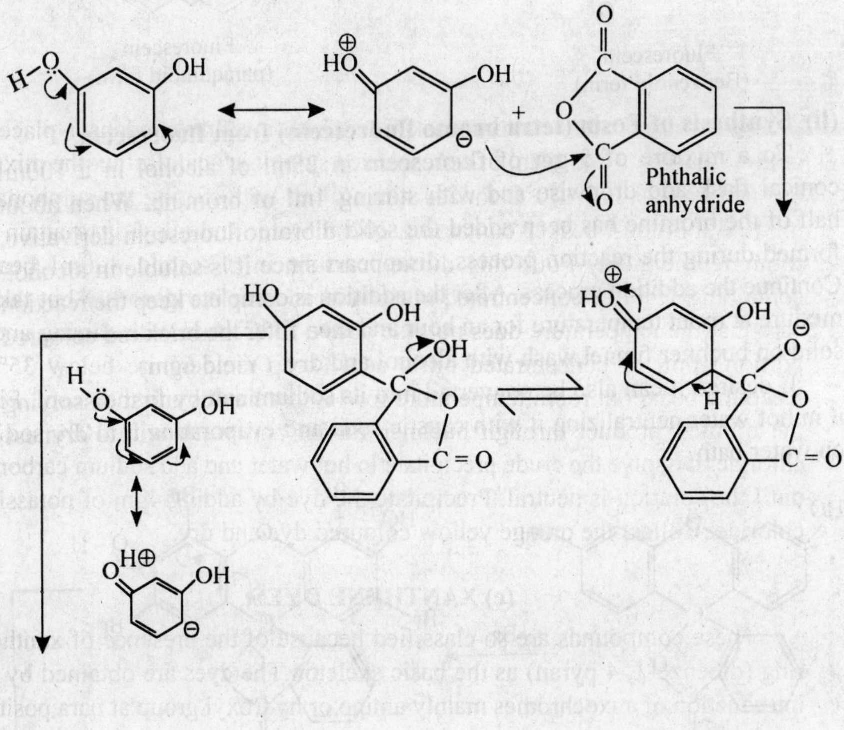

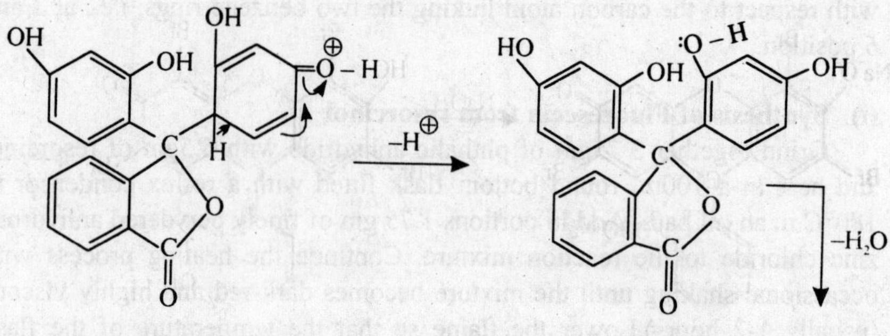

Phthalein

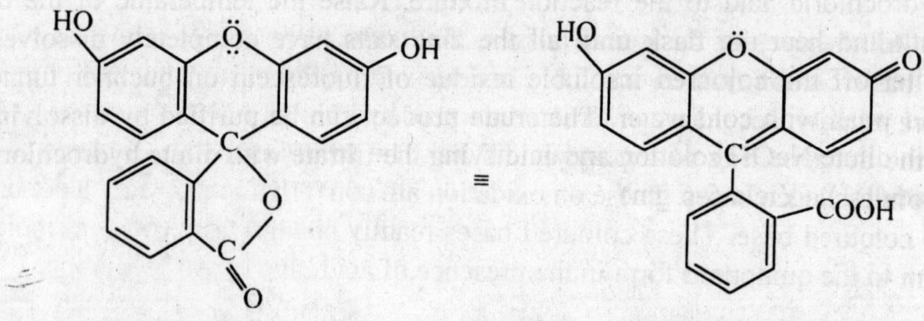

Fluorescein
(Benzenoid form)

Fluorescein
(paraquinoid form)

(ii) Synthesis of Eosin (tetra bromo fluorescein) from fluorescein

To a mixture of 5 gm of fluorescein in 25ml of alcohol in a 100ml conical flask add dropwise and with stirring 4ml of bromine. When about half of the bromine has been added the solid dibromofluorescein derivative, formed during the reaction process, disappears since it is soluble in alcohol. Continue the addition process. After the addition is complete keep the reaction mixture at room temperature for an hour and then filter the brick red coloured solid on buchner funnel,wash with alcohol and dry. (Yield 6gm)

If desired it can also be converted into its sodium salt by first dissolving it in hot water,neutralizing it with caustic soda and evaporating it to dryness on water bath.

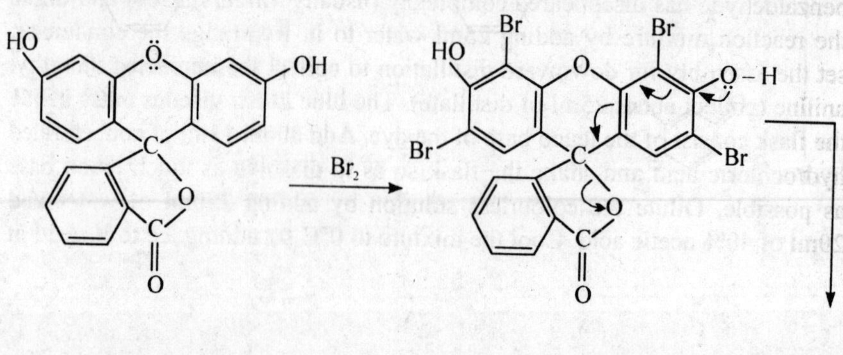

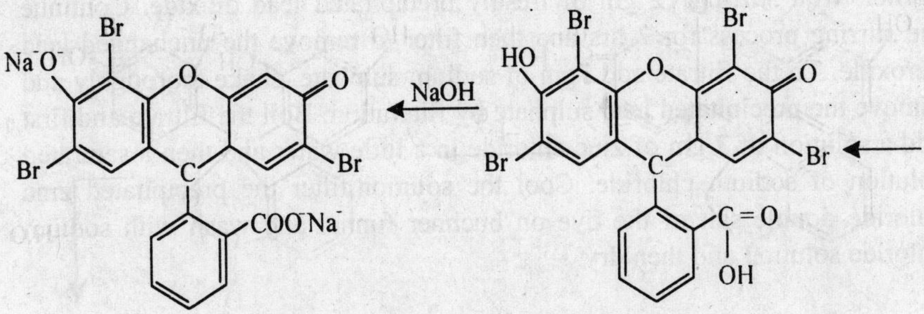

(d) TRIPHENYLMETHANE DYES

All the dyes of this series possess the hydrocarbon triphenylmethane as the basic carbon framework. They are derived by the introduction of one or more primary, secondary or tertiary amino or hydroxy group into para position of the benzene ring of triphenylmethane. The compound thus obtained are colourless leuco bases. These on oxidation are converted into tertiary alcohol, the coloured base. These coloured bases readily change from the benzenoid form to the quinonoid form in the presence of acid, due to salt formation.

$$\underset{\text{(colourless)}}{\text{Leucobase}} \xrightarrow{\text{oxidation}} \text{Coloured base} \xrightarrow{\text{acid}} \text{dye (coloured)}$$

Triphenylmethane dyes are notable for their brilliant intense colours and the typical shades are red, violet,blue and green.But these dyes suffer from one disadvantage that their colour fades gradually and thus finds less use in textile industry. They are usually used for colouring papers, typewriter ribbons etc.

(i) Synthesis of Malachite green (Benzaldehyde green) from benzaldehyde

Place 9.5 ml of benzaldehyde and 25ml of dimethyl aniline along with 10gm of fused powdered zinc chloride in a 500ml round bottom flask fitted with a reflux condensor. Heat the flask on water bath until the smell of benzaldehyde has disappeared completely (usually 4 hours). Cool and dilute the reaction mixture by adding 25ml water to it. Rearrange the condensor, set the assembly for downward distillation to collect the unreacted dimethyl aniline (collect about 25ml of distillate). The blue green viscous mass inside the flask consist of the leuco base of the dye. Add about 13ml of concentrated hydrochloric acid and shake the flask so as to dissolve as much leuco base as possible. Dilute the colourless solution by adding 250ml of water and 20ml of 40% acetic acid. Cool the mixture to 0°C by adding ice to it. Add in portion with stirring 12 gm of freshly precipitated lead dioxide. Continue the stirring process for 2 hrs and then filter to remove the uncharged lead peroxide. To the filtrate add 7gm of sodium sulphate, shake thoroughly and remove the precipitated lead sulphate by filteration. Boil the filtrate and first add a solution of 3 gm of zinc chloride in a little water and then a saturated solution of sodium chloride. Cool the solution,filter the precipitated zinc chloride double salt of the dye on buchner funnel and wash with sodium chloride solution and then dry.

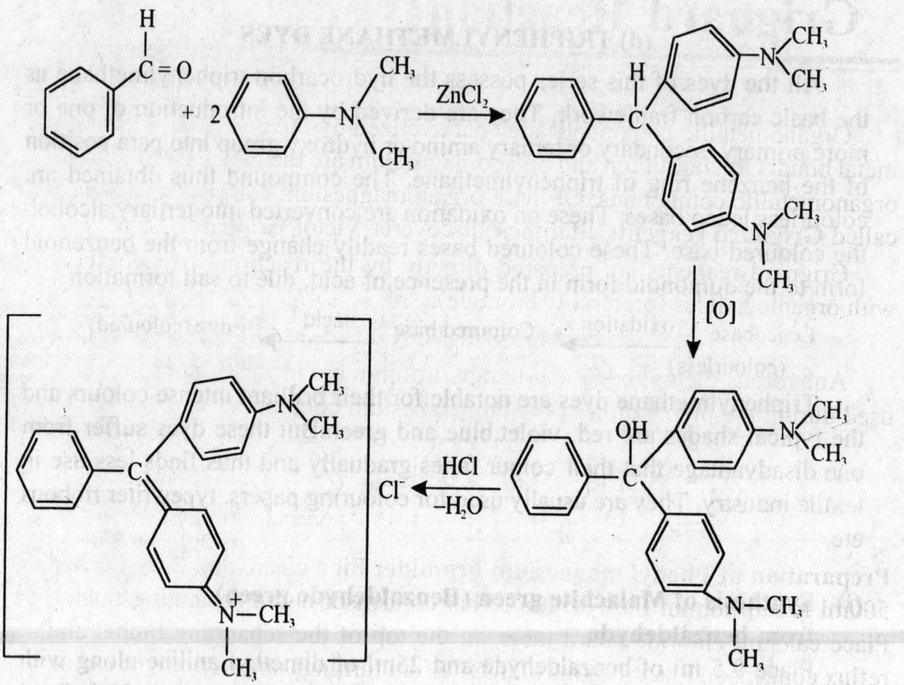

malachite green hydrochloride

22

Grignard Reaction

Organometallic compounds are characterized by the presence of carbon-metal bond. Organolithium and organomagnesium are the two most important organometallic compounds. Of these organomagnesium compounds are also called **Grignard reagents** after their discoverer Victor Grignard.

Grignard reagents are prepared directly by the reaction of magnesium with organic halides (preferably bromide).

$$RX + Mg \rightarrow RMgX$$

Anhydrous diethyl ether and tetrahydrofuran are most common solvents used during the preparation of organomagnesium compounds.

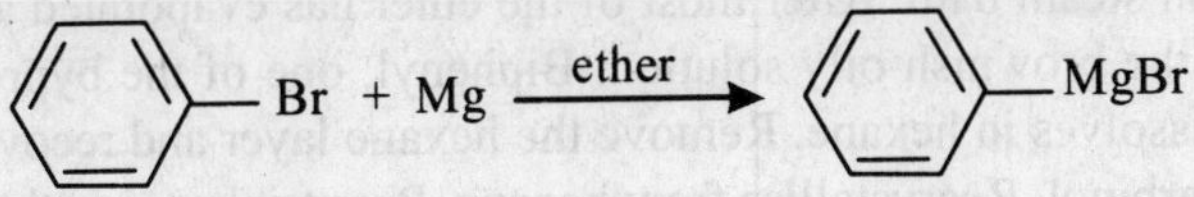

Preparation of Phenyl magnesium bromide: Fit a clean, dry, three necked 500ml round bottom flask with a reflux condensor and a separating funnel. Place calcium chloride guard tubes on the top of the separating funnel and reflux condensor. Place 1.8gms of magnesium turnings and a crystal of iodine in the flask. Measure out in a separate vessel 7ml of bromobenzene and 10-15ml of anhydrous ether. Transfer this bromobenzene solution to the separating funnel attached to the flask. Introduce about 15-20ml of anhydrous ether and 1.5ml of bromobenzene into the flask. Heat the flask on water bath so that ether just begins to boil. As soon as the reaction commences, remove the water bath and set the stirrer in motion. Evidence of the reaction taking place will be the formation of brownish grey cloudy material in the solution, ether will continue to reflux even though no heat is being supplied and formation of small bubbles at the surface of magnesium will also be noticeable. Add the remainder of the bromobenzene solution to the reaction mixture at such a rate that a gentle reflux is maintained throughout the addition. As the reaction is exothermic, moderate it with an ice bath if it tends to reflux too vigorously. After all the bromobenzene solution has been added, swirl the flask and then allow it to stand until the reaction mixture stops refluxing. Grignard reagent is now ready for further reaction.

(i) Synthesis of triphenylmethanol from benzophenone

Weigh about 8gms of benzophenone and dissolve it in 20-25ml of anhydrous ether. Transfer this benzophenone solution to the separating funnel attached to a 500ml three necked round bottom, containing freshly formed Grignard reagent. Add the benzophenone solution dropwise to the Grignard reagent. During the addition vigorous reaction will take place and the solution will boil. After all the benzophenone solution has been added, swirl the flask and then heat it on steam bath for few minutes. Allow the flask to cool at laboratory temperature. Meanwhile prepare 50ml of 10% sulphuric acid solution. Cool this acidic solution by adding crushed ice to it. Increase the volume of the reaction mixture by adding 5-10ml ether to it. Add with stirring the cold aqueous acidic solution to the reaction mixture until the aqueous layer is acidic to litmus. Swirl the flask so that the organic and aqueous layer mix intimately and all the salts hydrolyze. Add more acidic solution so as to complete the hydrolysis. Transfer this combined solution to the separating funnel. Stopper the flask and shake thoroughly. Withdraw the lower aqueous layer. Wash the ether layer in the separating funnel, first with water and then with sodium carbonate solution and again with water. After all these washings, evaporate the ether on steam bath. After most of the ether has evaporated add 10-15ml hexane to the brownish oily solution. Biphenyl, one of the byproduct of the reaction dissolves in hexane. Remove the hexane layer and recover the crude triphenylcarbinol. Recrystallise from hexane. Pure triphenylmethanol melts at 162°C.

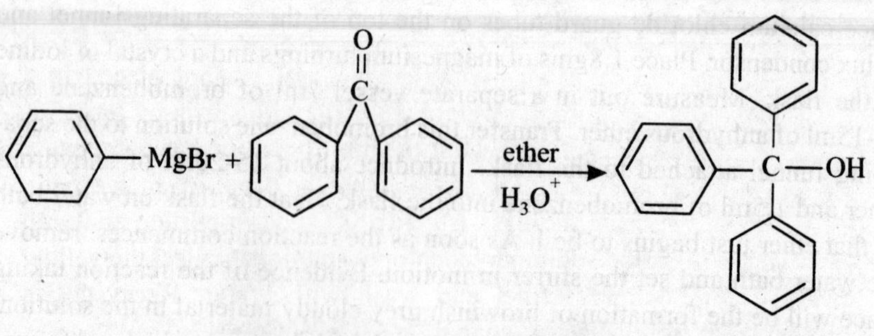

(ii) Synthesis of triphenylcarbinol from methyl benzoate

Measure about 8 ml of methyl benzoate and dissolve it in 20-25ml of anhydrous ether. Transfer this ester solution to the separating funnel attached to 500ml three necked round bottom flask, containing freshly prepared Grignard reagent (prepared from about 15ml of bromobènzene). Add the methyl benzoate solution dropwise to the Grignard reagent. During the addition, vigorous reaction will take place and the solution will boil. After all the ester solution has been added, swirl the flask and then heat it on steam bath for 30-45 minutes. Allow the flask to cool at laboratory temperature. Meanwhile prepare 50ml of 10% sulphuric acid solution. Cool this acidic solution by adding crushed ice to it. Increase the volume of the reaction mixture by adding 5-10ml ether to it. Add with stirring the cold aqueous acidic solution to the reaction mixture until the aqueous layer is acidic to litmus. Swirl the flask so that the organic and aqueous layer mix intimately and all the salts hydrolyze. Add more acidic solution so as to complete the hydrolysis. Transfer this combined solution to the separating funnel. Stopper the flask and shake thoroughly. Withdraw the lower aqueous layer. Wash the ether layer in the separating funnel first with water and then with sodium carbonate solution and again with water. After all these washings, evaporate the ether on steam bath. After most of the ether has evaporated add 10-15ml hexane to the brownish oily solution. Biphenyl, one of the byproduct of the reaction dissolves in hexane. Remove the hexane layer and recover the crude triphenylcarbinol. Recrystallise from hexane. Pure triphenylmethanol melts at 162°C.

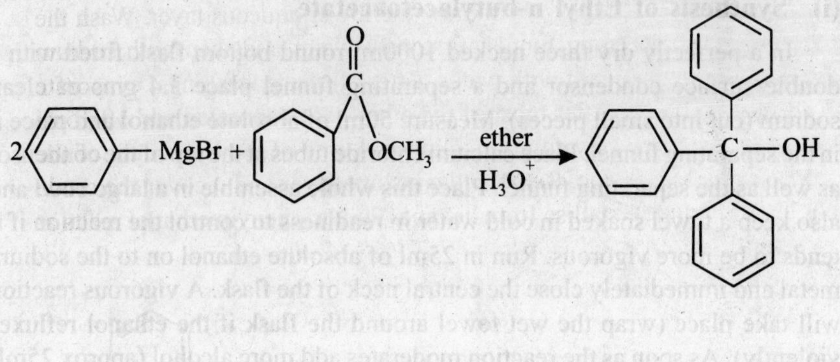

23 Alkylation Reaction

A general method for the synthesis of alkylated ketones involves alkylation of the sodio-derivative of ethyl acetoacetate with alkyl halide under anhydrous conditions.

$$CH_3COCH_2COOC_2H_5 + Na^{+-}OC_2H_5 \xrightarrow{-C_2H_5OH} [CH_3COCH^-COOC_2H_5]Na^+$$

$$[CH_3COCH^-COOC_2H_5]Na^+ \xrightarrow{RX} CH_3COCH(R)COOC_2H_5$$

(i) Synthesis of Ethyl n-butylacetoacetate

In a perfectly dry three necked 1000ml round bottom flask fitted with a double surface condensor and a separating funnel place 3.4 gms of clean sodium (cut into small pieces). Measure 50ml of absolute ethanol and place it in the separating funnel. Place calcium chloride tubes at the top of the condensor as well as the separating funnel. Place this whole assemble in a large bowl and also keep a towel soaked in cold water in readiness to control the reaction if it tends to be more vigorous. Run in 25ml of absolute ethanol on to the sodium metal and immediately close the central neck of the flask. A vigorous reaction will take place (wrap the wet towel around the flask if the ethanol refluxes violently). As soon as the reaction moderates add more alcohol (approx 25ml) in order to maintain the refluxing. In this manner most of the sodium will react to produce sodium ethoxide. Now add the remainder of the ethanol from the separating funnel at such a rate that a gentle reflux is maintained throughout the addition. After all the alcohol has been added, swirl the flask and then heat it on water bath until all the sodium has reacted. Remove the stopper from the central neck of the flask and add 20gms (19.5 ml) of redistilled ethyl acetoacetate. Stir the solution and then heat it to gentle boiling. Add in instalments 18ml of dry butyl bromide. Continue the refluxing until the solution is neutral to moist litmus paper (usually 4-7 hrs). Cool the solution at laboratory temperature. Decant to remove the precipitated sodium bromide. Wash the salt with absolute ethanol and mix this washing with the main solution. Distill off ethanol under reduced pressure and collect the fraction distilling at 112-116°C/16mmHg, which is pure ethyl n-butylacetoacetate.

$$CH_3COCH_2COOC_2H_5 + Na^{+-}OC_2H_5 \xrightarrow{-C_2H_2OH} [CH_3COCH^-COOC_2H_5]Na^+$$

$$[CH_3COCH^-COOC_2H_5]Na^+ \xrightarrow{CH_3(CH_2)_2CH_2Br} \begin{array}{l} CH_3CO\underset{|}{C}HCOOC_2H_5 + NaBr \\ \qquad CH_2CH_2CH_2CH_3 \end{array}$$

24

Drug and Drug Intermediates

(a) Barbituric Acid (4-hydroxyuracil)

Barbituric acid is the parent compound of family called barbiturates. These compounds posses both hypnotic as well as sedative properties and as a result they are chiefly used for the suppression of anxiety, induction of sleep and control of seizures. 5,5-diethylbarbituric acid commonly called Barbitone or barbital or veronol belongs to the category of long acting barbiturates, is a powerful hypnotic drug whose duration of action is usually 7-10 hours.

Barbituric acid itself is synthesized by the condensation of diethyl malonate with urea in the presence of sodium ethoxide

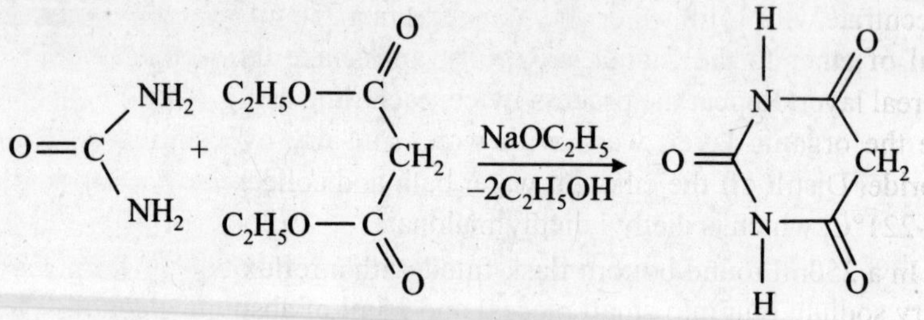

Urea   Diethyl malonate   Barbituric acid

(i) Synthesis of Barbituric acid from diethyl malonate

In a perfectly dry 250ml round bottom flask fitted with a reflux condensor place 2 gms of thoroughly clean and dry sodium metal. Add in one portion 50 ml of absolute ethanol to it (If the reaction between sodium and alcohol tend to be more vigorous keep the flask in ice bath). After all the sodium metal has reacted (if not heat the flask gently on water in order to complete the reaction) add 16 ml of diethyl malonate to this sodium ethoxide solution followed by 50ml of hot urea solution (6gms urea in 50ml of hot absolute ethanol). Attach a calcium chloride tube at the top of the condensor and reflux the resulting reaction mixture in an oil bath at 105-115°C for seven hours. After the refluxing period white solid slowly separates out in the flask. Treat the resulting mixture first with 80 ml of hot water (50°C) and then add with stirring 9ml of concentrated hydrochloric acid or until the solution is just acidic. Filter the resulting acidic solution and allow it to cool in ice bath (preferably keep it in a refrigerator overnight). Filter the solid on buchner funnel and then dry at 100°C, m.p 245°C (d).

(ii) Synthesis of 5,5-Diethyl Barbituric acid (Barbitone barbital or veronol) from diethylmalonate

In 250ml round bottom flask fitted with a reflux condensor place 2 gms of thoroughly clean and dry sodium metal. Add in one portion 60 ml of absolute ethanol to it (If the reaction between sodium and alcohol tend to be more vigorous keep the flask in ice bath). After all the sodium metal has reacted (if not heat the flask gently on water in order to complete the reaction) allow the sodium ethoxide solution to cool and when sodium ethoxide starts separating out add in portions 15 ml of dry diethyl malonate to it (during a period of 30 minutes). After the addition is complete, heat the flask on water bath for 15 minutes and then allow it to cool. When diethyl sodiomalonate starts precipitating out, add in portions about 16ml of ethyl iodide to it (15-20 minutes). After the addition is complete heat the flask on water bath for 3 hours. Rearrange the assembly and set it for downward distillation and distill off as much alcohol on water bath (cover the receiver with cloth or towel). Dilute the concentrate with 15ml water and transfer it in a 250 ml separating funnel. Add 20ml of ether to the funnel, stopper it and shake thoroughly. Remove the ethereal layer. Repeat the process twice, each time using 20 ml of ether. Combine the organic layer, wash with water and dry over anhydrous calcium chloride. Distill off the ether on water bath and collect the fraction boiling at 217-221°C, which is diethyl diethylmalonate.

In a 250ml round bottom flask fitted with a reflux condensor place 1 gm of dry sodium (cut into small pieces) and 25ml of absolute alcohol. When all the sodium has reacted heat the flask on water bath to complete the reaction. Cool the flask and add 5 gm of diethyl diethylmalonate and 1.2 gms of dry urea to it. Heat the flask on oil bath at 110-115°C for 4 hours. Cool the flask, add 20ml of water to it and heat on water bath so as to dissolve the precipitated solid. Pour the solution in a beaker and add dropwise concentrated sulphuric acid. As soon as the mixture becomes acidic veronal crystallizes out. Heat the beaker on water bath until all the veronal dissolves. Cool the hot solution and collect crystals of veronal on buchner funnel, m.p 188-190°C.

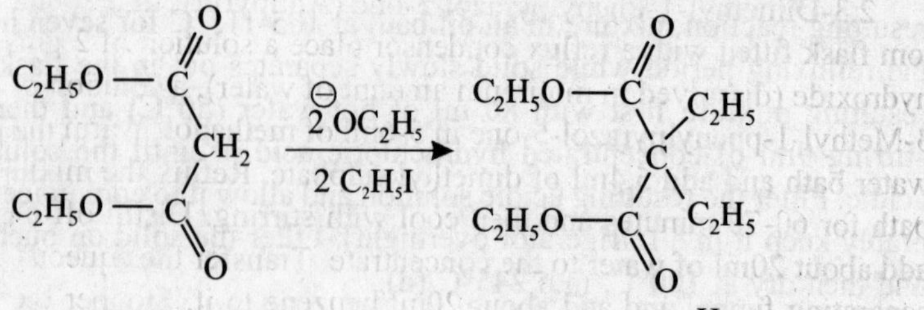

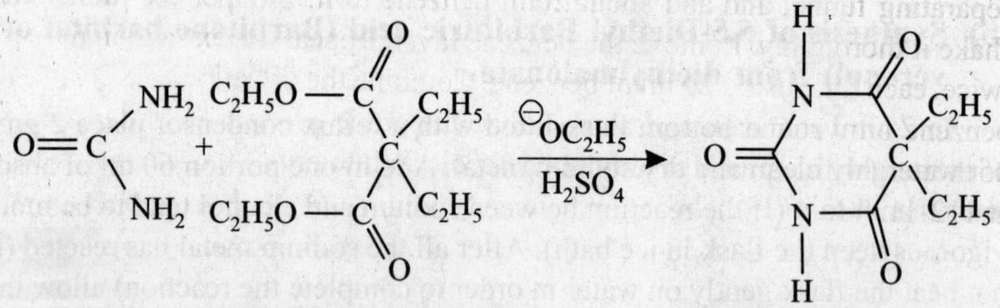

Urea Diethyl diethylmalonate 5,5 diethylbarbituric acid

(b) Antipyrine

One of the earliest antipyritic drugs, which also possess local anesthetic and styptic properties. Nowadays, used in combination with benzocaine for treating ottis media, a middle ear infection.

(i) Synthesis of 2,3-Dimethyl-1-Phenylpyrazol-5-one (Antipyrin) from ethylacetoacetae

3-Methyl-l-phenylpyrazol-5-one: In a large evaporating dish place 10ml of freshly distilled ethyl acetoacetate and 7.5ml of phenylhydrazine. Heat the dish on boiling water bath for two hours in a fume cupboard (or continue heating until the solid just starts precipitating out). During this period stir the reaction mixture occasionally with glass rod. Cool the heavy reddish syrup and add about 20ml of ether to it. Place the evaporating dish on ice bath and allow crystals to separate out. Filter the precipitated solid on buchner funnel and recrystallise from hot water or from 50% alcohol, m.p 126-127°C (Yield 10gms)

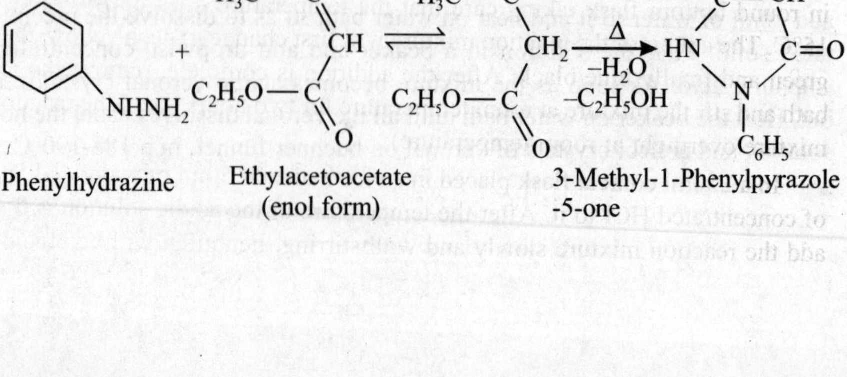

Phenylhydrazine Ethylacetoacetate (enol form) 3-Methyl-1-Phenylpyrazole -5-one

2,3-Dimethyl-1-Phenylpyrazol-5-one (Antipyrin): In a 100ml round bottom flask fitted with a reflux condensor place a solution of 2 gms of sodium hydroxide (dissolved in minimum amount of water), a solution of 8.7 gms of 3-Methyl-1-phenylpyrazol-5-one in 5-6ml of methanol. Warm the mixture on water bath and add 5.4ml of dimethyl sulphate. Reflux the mixture on water bath for 60-70 minutes and then cool with stirring. Distill off methanol and add about 20ml of water to the concentrate. Transfer the aqueous extract into separating funnel and add about 20ml benzene to it. Stopper the funnel and shake it thoroughly. Remove the benzene layer. Repeat the extraction process twice, each time using 20 ml of benzene. Combine the organic layer. Distill off benzene on water bath and recrystallize the crude product from benzene or hot water (with a pinch of animal charcoal). White crystals of antipyrin melt at 112-113°C.

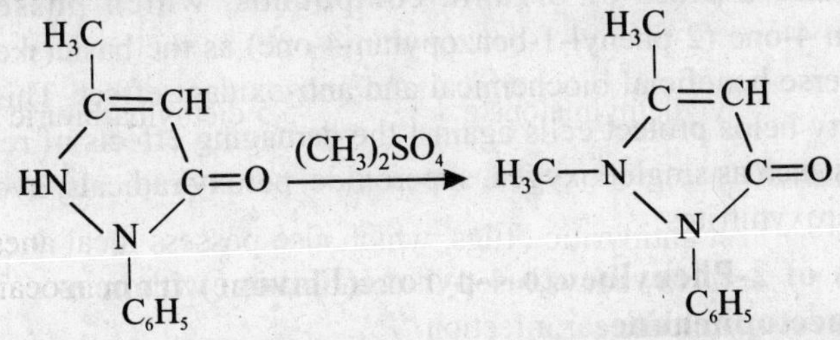

3-Methyl-1-Phenylpyrazole-5-one → 2,3-Dimethyl-1-Phenylpyrazol-5-one (Antipyrin)

(c) Fenbufen

One of the non-steroidal anti-inflammatory drug used chiefly to treat inflammation in spondylitis, osteorthritis and tendonitis.

(i) Synthesis of 4-Biphenyl-4-yl-4-oxo-butyric acid (Fenbufen)

In a 100 ml round bottom flask containing a stirrer bar, flush it with nitrogen and then add 2 gms of succinic anhydride and 35 ml of dichloromethane. Cool the resulting solution to 0°C by keeping the flask in an ice bath, and then add 3.0 gms of aluminium chloride.

In a 100 ml conical flask, dissolve 3.0 gms of biphenyl in 10-15ml of dichloromethane. Add this biphenyl solution dropwise to the reaction mixture in round bottom flask taking care that the temperature does not rise above 15°C. The colour of the reaction mixture will first change to deep yellow, then green and finally blue/black. After the addition is complete, remove the ice bath and stir the mixture at room temperature for two hours (keep the reaction mixture overnight at room temperature).

In a 250ml conical flask placed in an ice bath add 50ml of water and 5ml of concentrated HCl to it. After the temperature of the acidic solution is 0°C add the reaction mixture slowly and with stirring. Fenbufen will be precipitated as white solid. Continue the stirring for few more minutes. Filter the solid on buchner funnel, dry and recrystallize from ethanol, m.p 186-187°C.

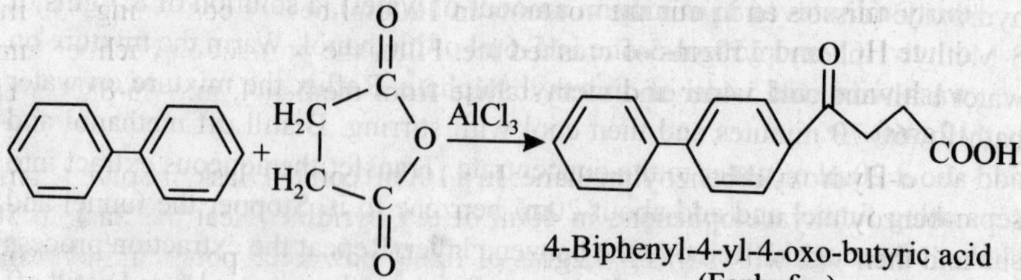

biphenyl   Succinic anhydride   4-Biphenyl-4-yl-4-oxo-butyric acid (Fenbufen)

(d) Flavones

Flavones are a class of organic compounds, which possess 2-phenylchromen-4-one (2-phenyl-1-benzopyran-4-one) as the basic skeleton. They have diverse beneficial biochemical and anti-oxidant effect. This anti-oxidant property helps protect cells against the damaging effects of reactive oxygen species such as singlet oxygen, superoxide, peroxy radicals, hydroxyl radicals and peroxynitrites.

(i) Synthesis of 2-Phenylbenzo-4-pyrone(Flavone) from o-hydroxyacetophenone

o and p-Hydroxyacetophenone: In a 250ml round bottom flask fitted with a reflux condensor and a calcium chloride tube at the top, place 11.2 gms of anhydrous aluminium chloride and 12ml of carbon disulphide. Stir the suspension and add slowly in portions 9.5ml of phenyl acetate. The rate of addition should be such that the reaction mixture begins to boil. After the addition is complete, reflux the reaction mixture on water bath for 90-120 minutes (or when the evolution of hydrogen chloride gas ceases). Turn the condensor downwards and distill off as much solvent as possible on water bath. Cool the flask and decompose aluminium chloride by adding 15ml of dilute hydrochloric acid and 25 ml of water. Much heat will be evolved and oily liquid collects on the surface. Keep this acidified oily solution overnight. Transfer it into a separating funnel and extract with three-20ml portion of toluene. Combine the organic layers, dry over magnesium sulphate. Distill off toluene under reduced pressure. Steam distill the residue and collect o-hydroxy acetophenone (b.p 218°C, m.p 28°C) while the p-isomer, which is non-volatile in steam, remains in the residue (Alternatively distill the residue under reduced pressure to collect the o-isomer). Recrystallize the residue from dil alcohol and collect crystals of p-hydroxy acetophenone, m.p 108-109°C.

o-Benzoyloxyacetophenone: In a 100ml conical flask place 10ml of o-hydroxy acetophenone, add 12ml of benzoyl chloride and 15ml of redistilled pyridine. Stopper the flask and shake it thoroughly so as to mix the contents, remove the stopper and release the pressure. Continue this process for at least thirty minutes and pour the contents in 1000ml beaker containing 350 ml of dilute HCI and 150gms of crushed ice. Filter the product on buchner funnel, wash with cold water and recrystallise from methanol, m.p 86-88°C (Yield 12gms)

o-Hydroxy dibenzoylmethane: In a 100ml conical flask dissolve 12gms of o-benzoyloxyacetophenone in 40ml of dry pyridine. Heat the flask to 50°C and then add with stirring 4.2gms of finely powdered potassium hydroxide. Continue the stirring process for another 15-20 minutes, cool the flask to laboratory temperature and then acidify by adding 60ml of 10% aqueous acetic acid. Filter the precipitated solid on buchner funnel and recrystallise from methanol, m.p 120-121°C (Yield 9.2 gms).

2-Phenylbenzo-4-pyrone (Flavone): Dissolve 9gms of o-hydroxy dibenzoylmethane in 50ml of glacial acetic acid in a 100ml round bottom flask fitted with a reflux condensor. Add with shaking 2ml of concentrated sulphuric acid and then heat the resultant mixture on boiling water bath for 60-70 minutes (shake the flask occasionally during this period). Pour the reaction mixture with stirring onto 250gms of crushed ice placed in a 500ml beaker. Filter the precipitated flavone on buchner funnel and wash with cold water. Recrystallise from pet-ether and collect pure white crystals of flavone, m.p 97-98°C (Yield 7.5gms).

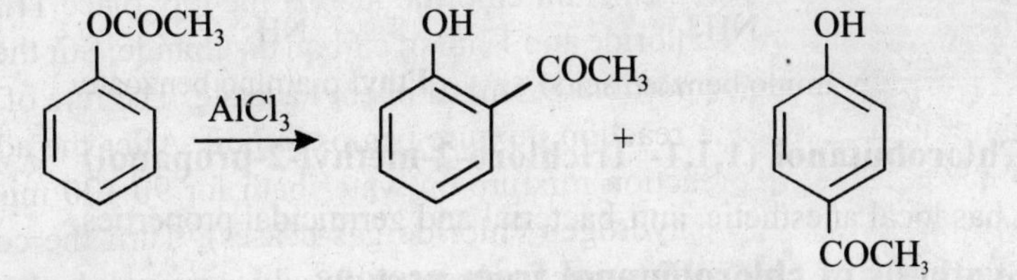

Phenyl acetate o-hydroxyacetophenone p-hydroxyacetophenone

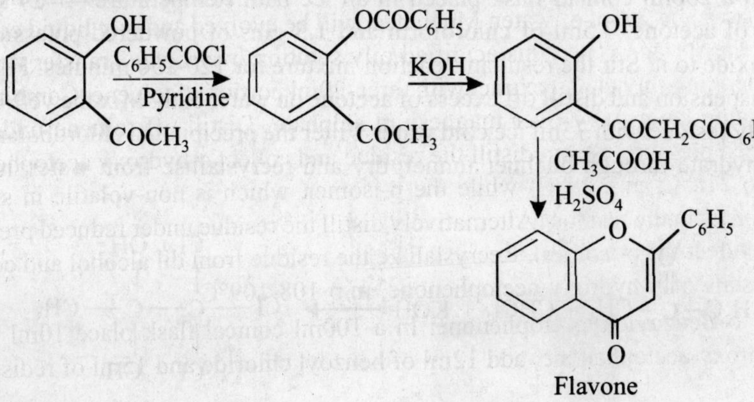

Flavone

(e) Benzocaine (Ethyl p-amino benzoate)

It is an important local anesthetic used chiefly as a topical pain reliever. Being insoluble in water, it is used as an ointment, cream, powder and aerosol spray to get rid of pain caused due to wound, ulcers and mucous surgaces. In combination with antipyrine it is used to form eardrops to relieve earpain and to remove cerumen.

(i) Synthesis of benzocaine from p-aminobenzoic acid

In a 100 ml round bottom flask place 20 ml of 95% ethanol and add 3 gm of p-aminobenzoic acid to it. Shake the flask thoroughly such that the benzoic acid derivative completely dissolves. Cool the flask in ice bath and then add 3ml of conc H_2SO_4 to it. Attach a reflux condensor and reflux the mixture for 60-90 minutes. Cool and pour the reaction mixture into a 200 ml beaker containing 50ml ice-cold water. Add solid sodium carbonate with stirring until the solution is neutral and cool further. Collect the precipitated Ethyl p-aminobenzoate on buchner funnel and recrystallise from alcohol, m.p 91°C (Yield 2gms)

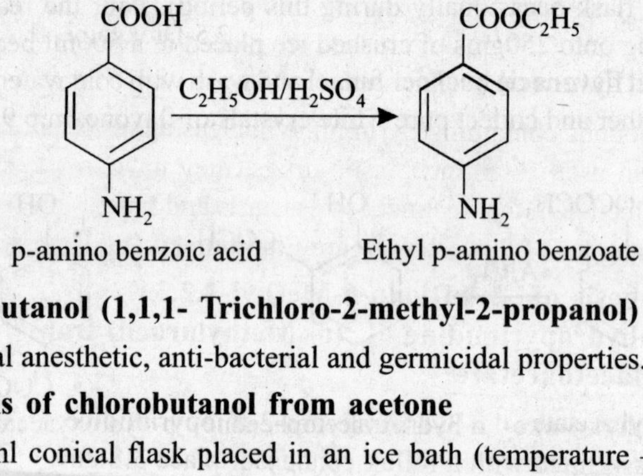

p-amino benzoic acid   Ethyl p-amino benzoate

(f) Chlorobutanol (1,1,1- Trichloro-2-methyl-2-propanol)

It has local anesthetic, anti-bacterial and germicidal properties.

(i) Synthesis of chlorobutanol from acetone

In a 250ml conical flask placed in an ice bath (temperature –5°C) add 70ml of acetone, 7.5ml of chloroform and 1.5 gms of powdered potassium hydroxide to it. Stir the resultant reaction mixture for 120-150 minutes. Filter the suspension and distill off excess of acetone on water bath. Mix the yellowish oily residue with 75ml ice cold water. Filter the precipitated chlorobutanol hemihydrate through buchner funnel, dry and recrystallise from water, m.p 79°C.

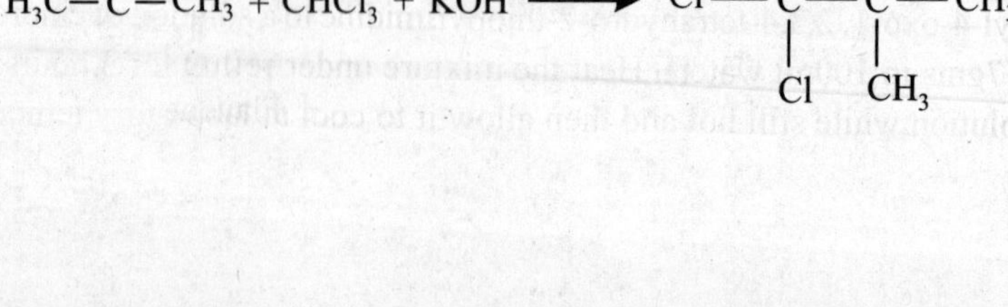

(g) Piperazines

A number of piperazines and its salts have commonly been used for the treatment of roundworm and pinworm infections. Besides a number of piperazine derivatives are moderately potent antihistaminics.

(i) Synthesis of 2,5-dioxopiperazine from glycine

In a 100ml round bottom flask fitted with a reflux condensor place 7.5 gms of glycine and 45ml of ethylene glycol. Reflux the resultant mixture for 30-40 minutes. Cool the dark brown solution in an ice bath and filter the precipitated solid through buchner funnel. Crystallize 2,5-Dioxopiperazine from water, m.p > 300°C (Yield 2 gms)

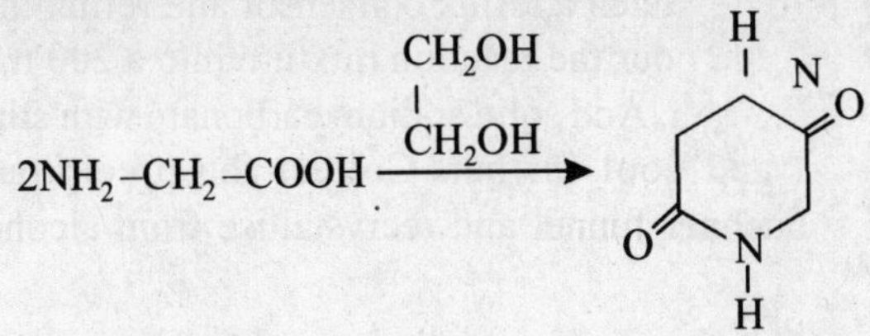

2,5-Dioxypiperazine

(ii) 6-Methyluracil

Compounds containing pyrimidine nucleus apart from being more commonly been used as antineoplastic drugs, they have also been found to be equally effective in the treatment of psoriasis and fungal infections. The well-known examples of this category are Fluorouracil and Cytarabine.

(h) Synthesis of 2,4-Dioxo-6-Methyl-1,2,3,4-tetrahydropyrimidine (6- Methyluracil) from ethylacetoacetate

6-Methyl-4-oxo-l,2,3,4-tetrahydro-2-thiopyrimidine: In a 100ml round bottom flask fitted with a reflux condensor place 3.8 gms of thiourea, 7.4ml of ethylacetoacetate and 6ml of ethanol. To this reaction mixture add with stirring 3ml of potassium hydroxide solution (3.25 gms in 3ml water). Reflux the resultant mixture for 90-120 minutes. Acidify carefully with hot hydrochloric acid (10ml conc HCl in 5ml water). Cool and collect 6-Methyl-4-oxo-1, 2,3,4-tetrahydro-2-thiopyrimidine as white crystalline product on buchner funnel, m.p > 290°C (Yield 5gms).

2,4-Dioxo-6-Methyl-1, 2,3,4-tetrahydropyrimidine (6-Methyluracil): In a 250ml round bottom flask fitted with a reflux condensor add 5gms of 6-Methyl-4-oxo-1, 2,3,4-tetrahydro-2-thiopyrimidine to a solution of chloroacetic acid (7gms in 100ml water). Heat the mixture under reflux for 3 hours. Filter the solution while still hot and then allow it to cool at laboratory temperature and collect 2,4-Dioxo-6-Methyl-1, 2,3,4-tetrahydropyrimidine as colourless crystals on buchner funnel, m.p > 300°C (yield 2.5gms)

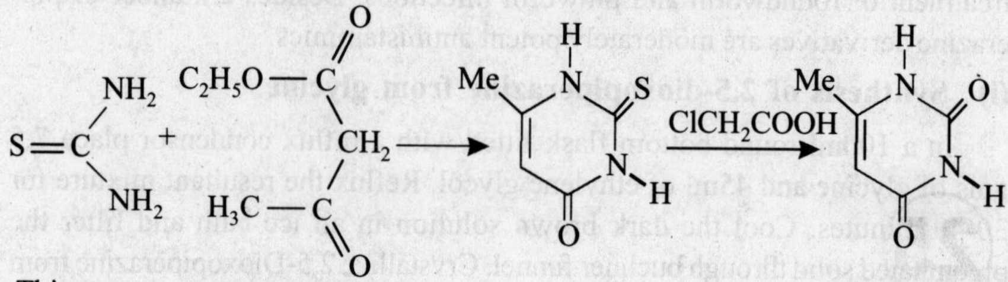

Thiourea Ethylacetoacetate 6-Methyluracil

25 Industrial Preparations

(a) Polymers

Polystyrene belongs to the category of addition polymers, which are formed by the reaction in which monomer units are simply added to one another to form a long chain. The monomer units usually contain carbon-carbon double bonds.

In this experiment polystyrene is prepared by free radical mechanism in which benzoyl chloride acts as an initiator

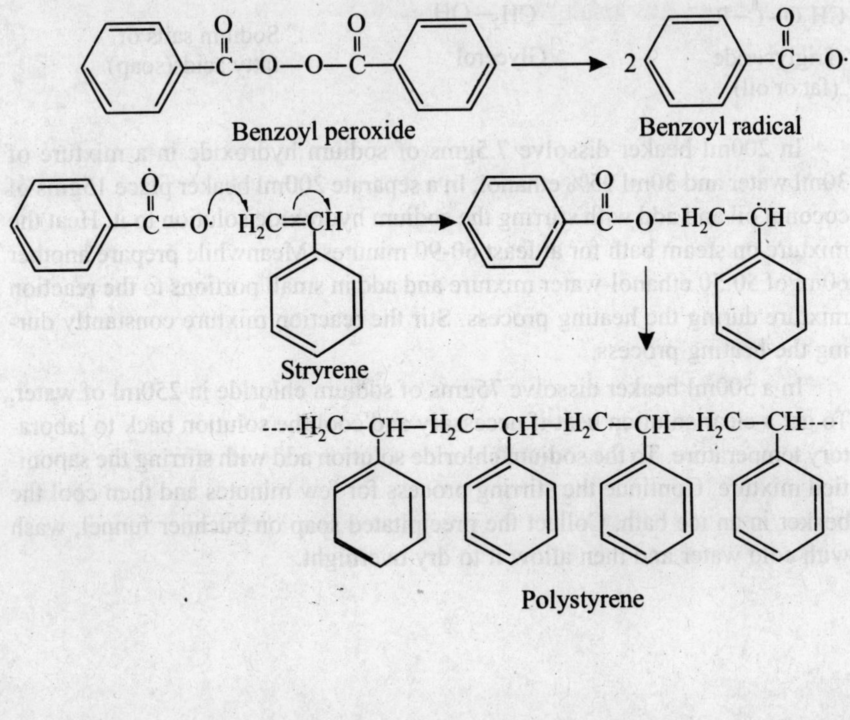

In a 200ml beaker place 30ml of styrene and add 1gm of benzoyl peroxide to it. Place the beaker on hot plate and heat until the mixture turns yellow. When the colour of the reaction mixture disappears and bubbles begin to appear, immediately remove the beaker from hot plate (this is an exothermic reaction, if it tends to become too violent then cool in ice bath). After the reaction has subsided put the beaker again on the hot plate and continue the heating process until the reaction mixture turns syrupy. With the help of a glass rod draw out the reaction mixture from the beaker, if the filament can be broken after few seconds of cooling, polyester is ready to be poured in a watch glass. If however the filament does not breaks then continue the heating process until the filament breaks. Pour the syrupy liquid on watch glass and after cooling at laboratory temperature for few minutes, lift the polystyrene from watch glass by gently prying with a metallic spatula.

(b) Soaps

Soaps well known for their cleansing action, are sodium or potassium salt of fatty acids, obtained by hydrolysis of fats or oils with aqueous alkali. Soaps obtained from coconut oil contain salts of lauric and myristic acid.

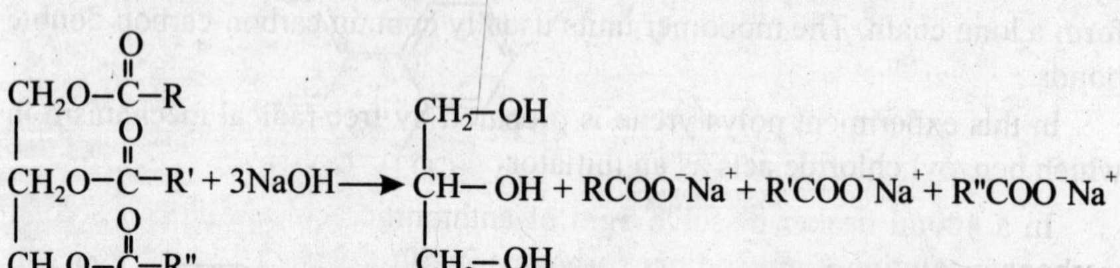

Triglyceride (fat or oil) Glycerol Sodium salts of fatty acids(soap)

In 200ml beaker dissolve 7.5gms of sodium hydroxide in a mixture of 30ml water and 30ml 95% ethanol. In a separate 200ml beaker place 15gms of coconut oil and add with stirring the sodium hydroxide solution to it. Heat the mixture on steam bath for at least 60-90 minutes. Meanwhile prepare another 60ml of 50:50 ethanol-water mixture and add in small portions to the reaction mixture during the heating process. Stir the reaction mixture constantly during the heating process.

In a 500ml beaker dissolve 75gms of sodium chloride in 250ml of water. To get a clear solution heat if necessary and cool the solution back to laboratory temperature. To the sodium chloride solution add with stirring the saponified mixture. Continue the stirring process for few minutes and then cool the beaker in an ice bath. Collect the precipitated soap on buchner funnel, wash with cold water and then allow it to dry overnight.

26

Miscellaneous Preparations

(i) Synthesis of biphenic acid from anthranilic acid

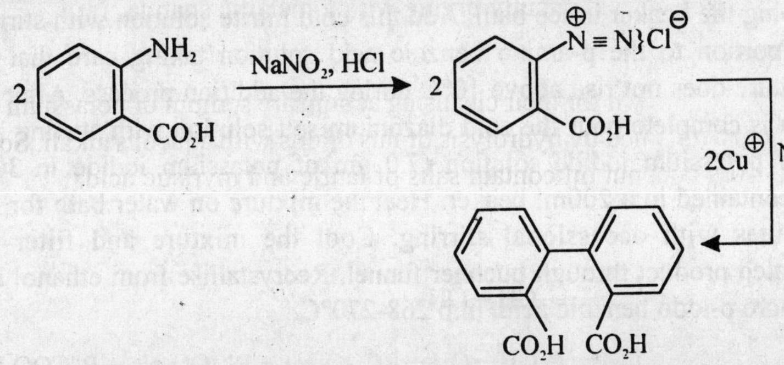

In a 400ml beaker dissolve 5gm of anthranilic acid in 50ml of sodium carbonate solution (2gm sodium carbonate in 50ml water). Slowly add with stirring 3 gm of sodium nitrite, cool the beaker by keeping it in an ice bath. Pour this well cooled solution to a well stirred mixture of 10ml of concentrated hydrochloric acid and 30gm of crushed ice in a 400ml beaker.

Prepare separately ammonical cuprous solution in a 250ml conical flask by adding 20ml ammonia solution to 50ml of copper sulphate solution (12gm in 50ml water) and cool by keeping the conical flask in ice bath. Cool a solution of 3.8 gm of hydroxylamine hydrochloride in 13ml water placed in 100ml beaker and add 7ml of sodium hydroxide solution to it (2.2 gm in 7ml water). Immediately add this hydroxylamine solution to the copper solution.

Transfer the cuprous solution in a 250ml round bottom flask, cool it by keeping it in an ice bath. Now add in portion with stirring the diazotized solution taking care that the temperature does not rise above 5°C. After the addition is complete attach a condensor to the flask and heat the mixture to boiling for 2-5 minutes. Cool, slowly add 35ml of concentrated hydrochloric acid and keep the mixture overnight. Filter the precipitated solid on buchner funnel, wash with cold water. Recrystallise from hot water containing acetic acid and collect crystals of biphenic acid, m.p 227-229°C (Yield 2-2.5 gm)

(ii) Synthesis of p-iodo benzoic acid from p-amino benzoic acid

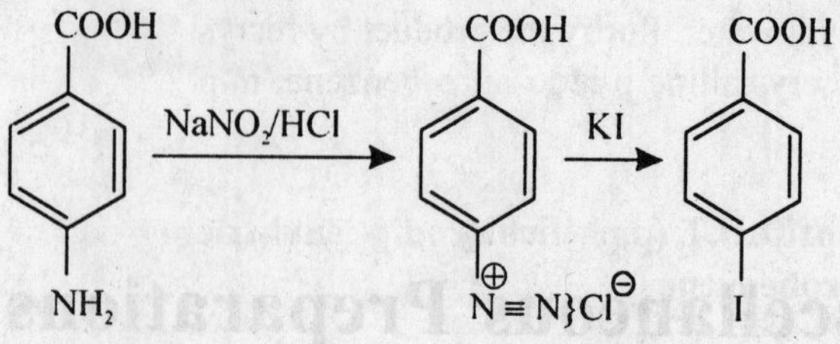

Dissolve 5 gm of p-amino benzoic acid in 30ml of 3M hydrochloric acid in a 100ml conical flask. Cool the flask by keeping it in an ice bath. In a separate 100ml beaker dissolve 3gm of sodium nitrite in 10ml water, cool by keeping the beaker in ice bath. Add this cold nitrite solution with stirring and in portion to the p-amino benzoic acid solution taking care that the temperature does not rise above 10°C during the addition process. After the addition is complete pour the cold diazonium salt solution with stirring into 30ml of potassium iodide solution (7.0 gm of potassium iodide in 30ml water) contained in a 200ml beaker. Heat the mixture on water bath for 20-25 minutes with occassional stirring. Cool the mixture and filter the precipitated product through buchner funnel. Recrystallise from ethanol and obtain pure p-iodo benzoic acid, m.p 268-270°C.

(iii) Synthesis of p-iodo nitro benzene from p-nitro aniline

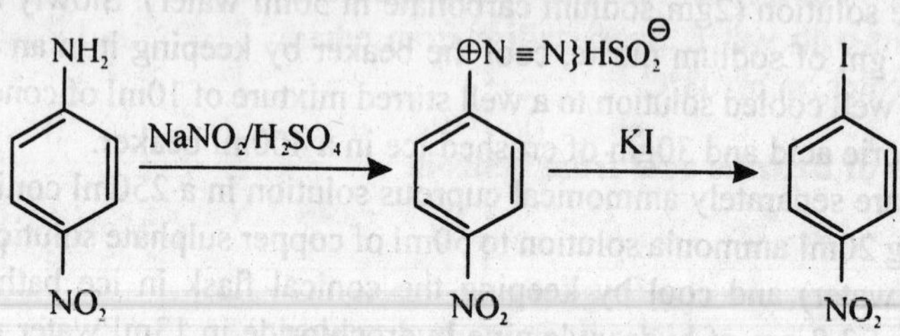

In a 100ml conical flask mix together 5 gm of p-nitro aniline, 4.5 ml of concentrated sulphuric acid and 30ml of water, stir the resulting mixture for an hour. Cool the flask by keeping it in an ice bath. In a separate 100ml beaker dissolve 2.5gm of sodium nitrite in 7.5 ml water, cool by keeping the beaker in an ice bath. Add this cold nitrite solution with stirring and in portion to the cold p-nitro aniline solution taking care that the temperature does not rise above 8°C. After the addition is complete filter the cold diazonium salt solution and add the filtrate with stirring to 30ml of potassium iodide solution (10 gm of potassium iodide dissolved in 30ml water) contained in 200ml beaker. Keep the beaker at room temperature for 40-45 minutes and then heat it on water bath for 10 minutes with occassional stirring. Cool by keeping the beaker in ice bath and filter the precipitated solid on buchner funnel, wash with water. Purify the product by recrystallisation from ethanol and collect the crystalline p-iodo nitro benzene, m.p 170-171°C (Yield 5-6 gm)

(iv) Synthesis of D.D.T (p,p'-dichlorodiphenyl-trichloroethane) from chlorobenzene

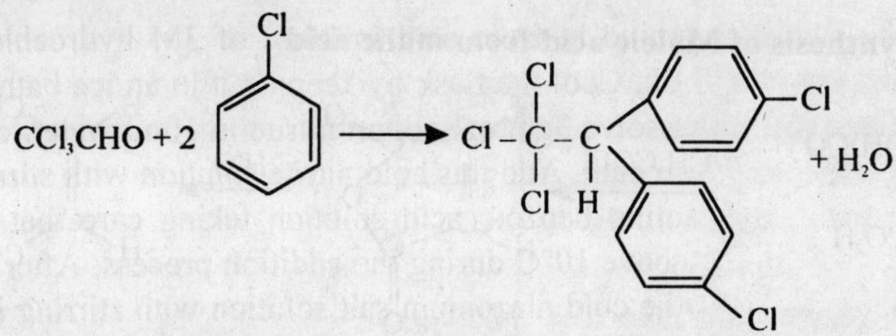

In a 250ml conical flask provided with with a stopper dissolve 13gm of chloralhydrate crystals in 18ml chlorobenzene by heating on water bath. Cool the flask and add dropwise with stirring 135ml of concentrated sulphuric acid (temperature between 40-50°C). After the addition is complete stopper the flask and shake thoroughly for 60-90 min. Pour the contents of flask with stirring into a 500ml beaker containing 100gm crushed ice. Filter the precipitated solid on bucher funnel, wash with cold water, sodium carbonate and then again with water. Recrystallise from ethanol and collect DDT as colourless needles, m.p 108°C.

(v) Synthesis of Benzilic acid from benzoin

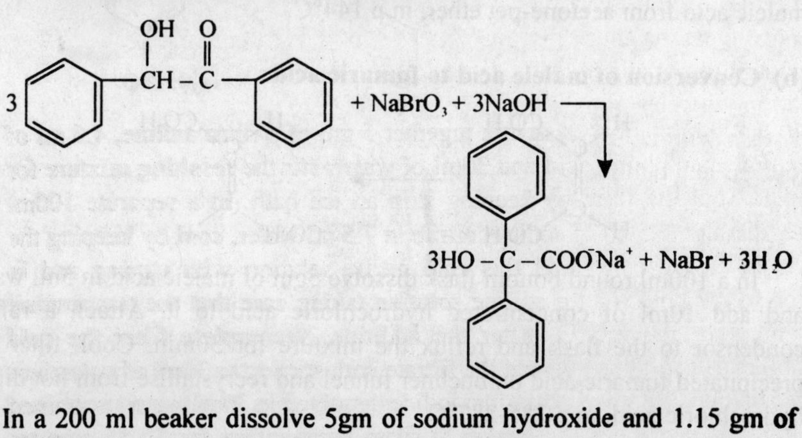

In a 200 ml beaker dissolve 5gm of sodium hydroxide and 1.15 gm of sodium bromate (or 1.25 gm of potassium bromate) in 10ml water. Heat the solution on steam bath /water bath and then add with stirring 5gm of benzoin to it.Continue the heating process on steam bath/water bath for 5-6 hrs taking care that the solution does not get completely dry (add 5-8ml water during this period). Dilute the resultant mixture by adding 10ml water to it and then leave overnight. Filter and to the cooled filtrate add dropwise with stirring dilute sulphuric acid until the solution is acidic. Filter the precipitated benzilic acid through buchner funnel, wash with water and recrystallise from hot water, m.p 148-150°C.

(vi).(a) Synthesis of Maleic acid from malic acid

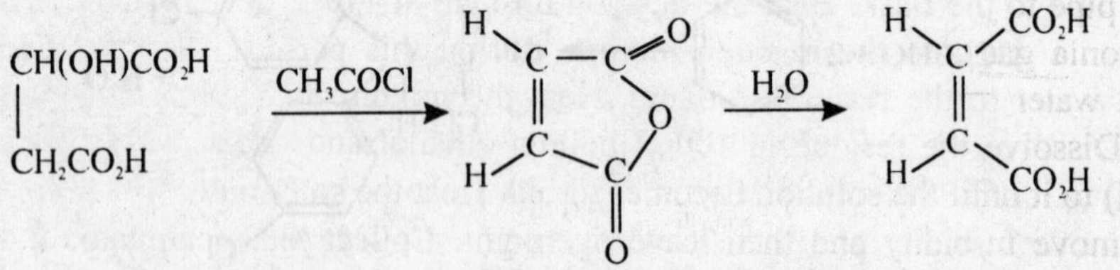

In a 100ml round bottom flask place 9gm of malic acid and add slowly 11.4ml of acetyl chloride to it.Attach a reflux condensor to the flask and heat the reaction mixture on water bath. As soon as the reaction starts remove the heat source and allow the reaction to proceed (exothermic reaction). After the reaction has subsided heat the mixture on water bath for 1 ½ - 2hours. Set the apparatus for downward distillation and distill. When the temperature of the mixture rises above 185°C remove the water source from the condensor and collect maleic anhydride at 190-200°C.On cooling the anhydride soon solidifies,recrystallise from chloroform, m.p 54°C.

In a porcelain dish heat maleic anhydride with twice its weight of water on water bath. Evaporate off the water completely. Recrystallise the crude maleic acid from acetone-pet ether, m.p 144°C.

(b) Conversion of maleic acid to fumaric acid

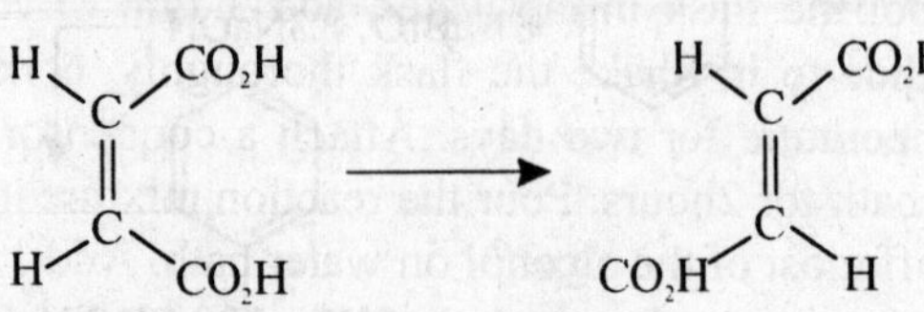

In a 100ml round bottom flask dissolve 5gm of maleic acid in 5ml water and add 10ml of concentrated hydrochloric acid to it. Attach a reflux condensor to the flask and reflux the mixture for 30min. Cool, filter the precipitated fumaric acid on buchner funnel and recrystallise from hot dilute hydrochloric acid, m.p 285-287°C.

(vii) Synthesis of 2, 4-dinitroaniline from 1-chloro 2, 4-dinitrobenzene

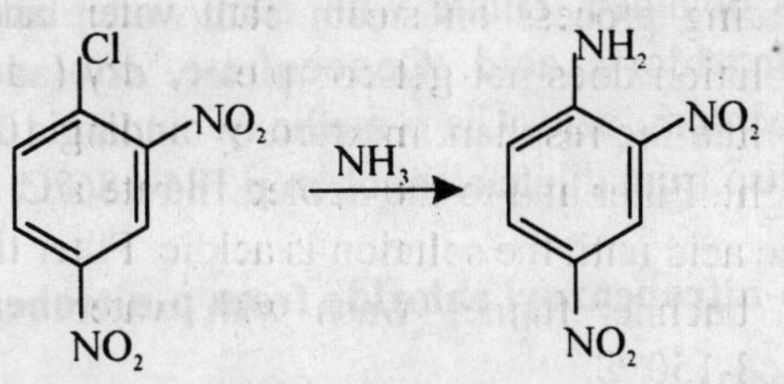

In a two necked 100ml round bottom flask place 3.6 gm of ammonium acetate and 10gm of 1 chloro 2,4 dinitrobenzene. Fit a condensor and gas inlet pipe to the flask. Heat the flask on oil bath at 170°C for 6hrs and pass ammonia gas into the reaction mixture during this period. Cool and add 20ml water to the reaction mixture. Heat the mixture to boiling and filter hot. Dissolve the residue in 100ml boiling ethanol and add water (approx 30ml) to it until the solution becomes turbid. Heat the solution on water bath to remove turbidity and then leave overnight. Collect the precipitated 2,4 dinitroaniline on buchner funnel, recrystallise from dilute ethanol, m.p 180°C.

(viii) Synthesis of cis-methoxy cinnamic acid from coumarin

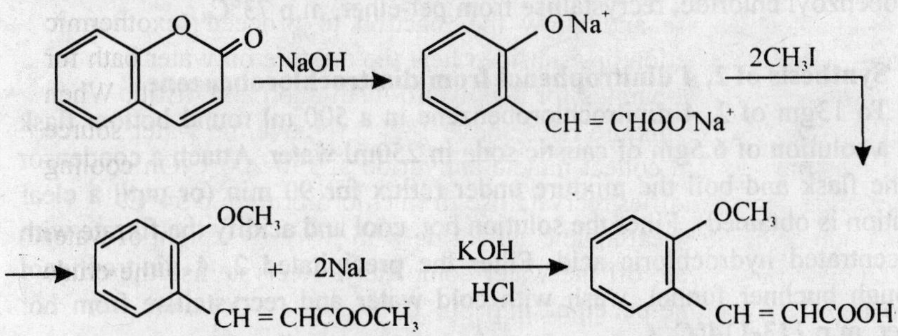

In a 250ml round bottom flask dissolve 1.25gm of sodium in 17.5ml of absolute ethanol. Cool the flask in tap water, add 3.7gm of coumarin and 7.5gm of methyl iodide to it. Shake the flask thoroughly, cork it and then leave it at room temperature for two days. Attach a condensor to the flask and heat it on water bath for 2hours. Pour the reaction mixture in a porcelain dish and evaporate off most of the alcohol on water bath. Add 15ml water to the residue and transfer the resultant solution to a 100ml separating funnel. Add 15ml ether and shake the funnel thoroughly. Remove the ether layer. Repeat the extraction process again using 15ml ether. Combine the two organic layers, distill off the solvent and collect cis o-methoxy cinnamic methyl ester as yellow oil.

To 2gm of ester add a solution of 1gm of caustic potash dissolved in

3ml water. Add 15ml absolute alcohol and reflux the reaction mixture until a clear solution is obtained. Dilute with 10ml water and neutralise with concentrated hydrochloric acid. Concentrate, cool and acidify with concentrated hydrochloric acid. Cis o-methoxy cinnamic acid separate out, filter and recrystallise from dilute alcohol, m.p 184-186°C.

(ix) Synthesis of p-nitrobenzoyl chloride from p-nitrobenzoic acid

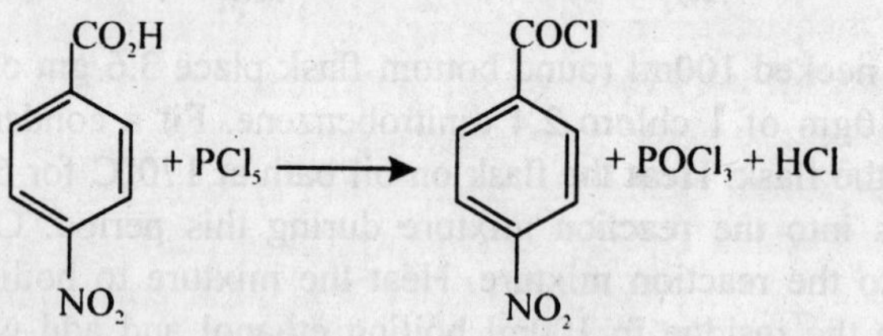

To a 100ml round bottom flask fitted with a reflux condensor place 5gm of p-nitro benzoic acid and 6.5gm of phosphorous pentachloride. Heat the mixture on water bath with occasional shaking until a clear solution is obtained. Set the apparatus for downward distillation and distill off the phosphorous oxychloride on oil bath. Cool the residue and collect p-nitrobenzoyl chloride, recrystallise from pet-ether, m.p 73°C.

(x) Synthesis of 2, 4 dinitrophenol from dinitrochlorobenzene

To 13gm of 2, 4 dinitrochlorobenzene in a 500 ml round bottom flask add a solution of 6.5gm of caustic soda in 250ml water. Attach a condensor to the flask and boil the mixture under reflux for 90 min (or until a clear solution is obtained). Filter the solution hot, cool and acidify the filtrate with concentrated hydrochloric acid. Filter the precipitated 2, 4 dinitrophenol through buchner funnel, wash with cold water and recrystallise from hot water, m.p 113-114°C.

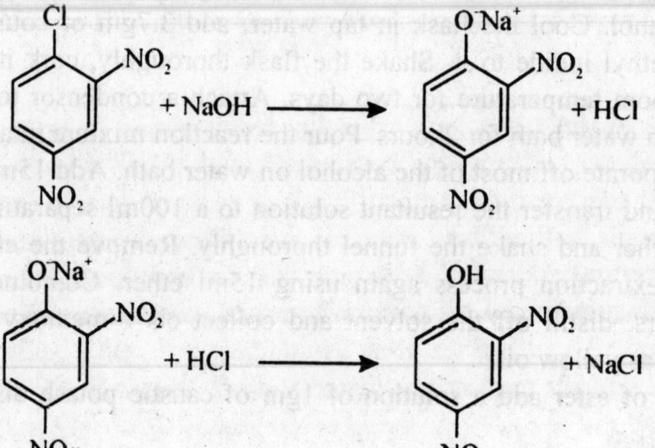

SECTION III

1

Flavonoids

Flavonoids also known as anthoxanthins, represent a group of naturally occuring plant pigments. They usually occur in all the parts of the plant and are isolated in the free state or in the form of their glycosides. All flavonoids are believed to be derived by substituting benzo γ pyrone (chromone) with a phenyl ring. These plant pigments are thought to be related biosynthetically to essential amino acid phenylalanine and the C_{15} skeleton is probably generated by the coupling of C_6 –C_3 unit produced by shikimic acid pathway with a six carbon unit produced by the combination of three acetate units

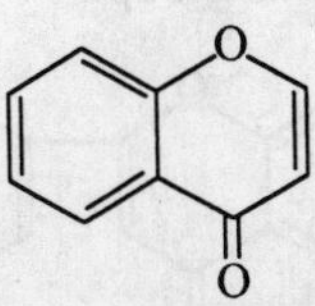

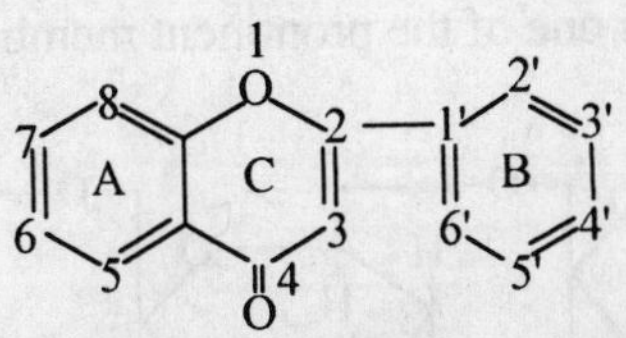

Flavonoids are usually classified into large number of categories chiefly on the basis of oxidation level of oxygen ring or carbon chain bridging the two benzene rings.

1. Flavones: are the parent flavonoid with the phenyl ring substituted at position 2 of the chromone unit. They contain hydroxyl groups normally at position 5 and 7 in ring A and one or more hydroxy groups at position 3', 4' and 5' in ring B. The ring A and B exhibit all the reactions typical of a benzenoid ring whereas it is the ring C which is responsible for most of the reactions typical of a flavone. Flavones are biologically very important and some of them have been found to exhibit anti-inflammatory and anti-tumor activity. Some prominent members of this group are chrysin and apigenin.

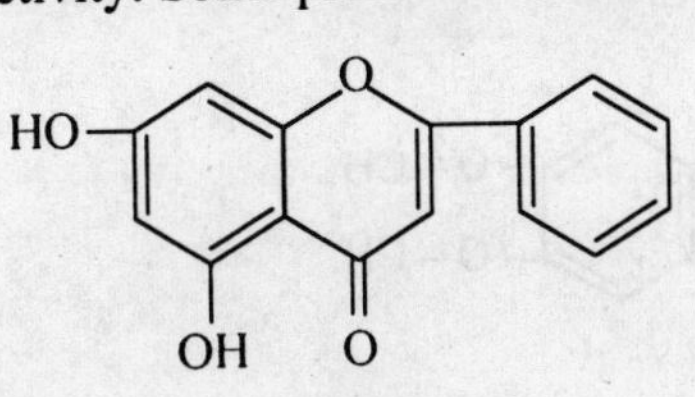

Chrysin

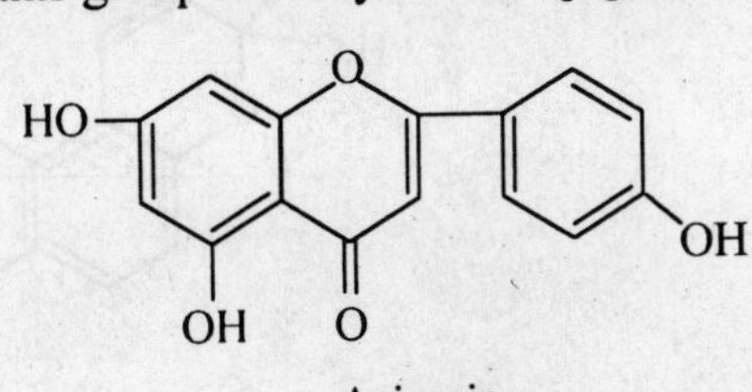

Apigenin

2. Flavonols: These are derivatives of flavones in which the position 3 is substituted by a hydroxyl group. The most potent member of this group is quercetin which is physiologically very active and plays an indirect role in stimulating germination and plant growth. Recent studies have shown that quercetin has a number of other physiological actions associated with it like anti-inflammatory, anti-fungal and anti-bacterial activity.

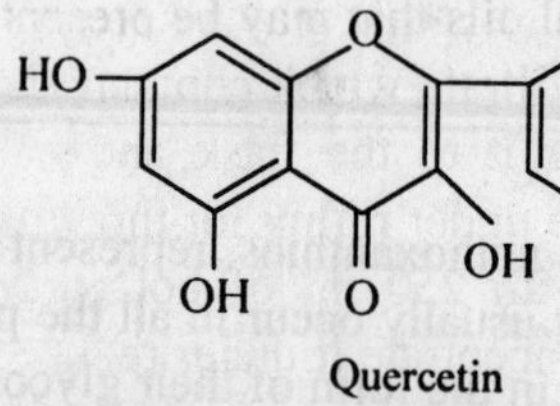

Quercetin

3. Flavonones (dihydroflavone): are simple reduction product of flavones and are normally isolated in its hydroxylated form either in the free state or in the form of its glycoside. The carbonyl group at position 4 of the flavones does not normally react with some carbonyl group reagent such as hydorxylamine but the flavonones show all the reactions of carbonyl function. Hesperdin is one of the prominent member of this group.

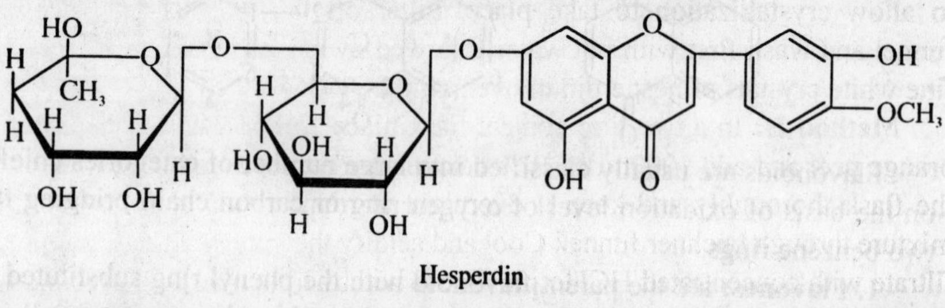

Hesperdin

4. Isoflavones : are isomers of flavones derived biosynthetically by the migration of aryl group from 2 to 3 position. They are colourless compounds and are capable of forming colours with metals and are thus used as mordant dyes. Recently Toxicarol isoflavone isolated from Legumenoceae family plant showed anti-cholesterolemic effect.

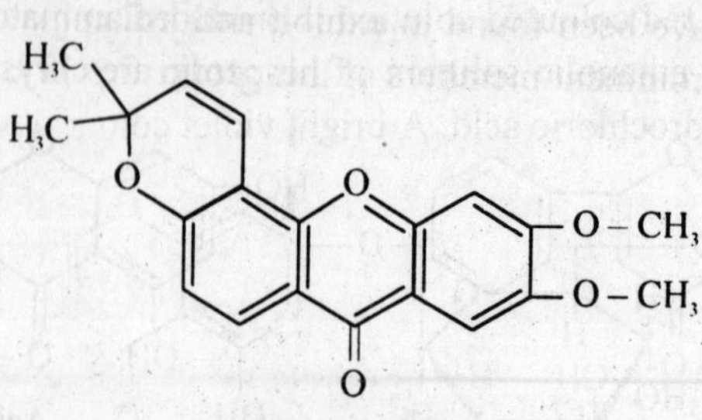

Toxicarol Isoflavone

Isolation of Hesperdin from Orange Peel

Hesperdin is a glycoside of flavonone hespertin and it has been found to strengthen the weak capillary blood vessels.

Method I : Powder 250 gm of dried orange peel in a pencil mortar and then place it in a two litre round bottom flask fitted with a reflux condensor. Add 1250ml pet ether to the flask and reflux on water bath for 60 minutes so as to remove all the essential oils that may be present. Filter the contents while still hot and discard the filtrate which contains the essential oils. Dry the powder and again transfer it to the flask and now add 1250 ml of methanol to it. Boil the mixture under reflux for three hours and filter while still hot. Wash the powder with 250 ml of hot methanol. Combine the washing and the filtrate and concentrate it under reduced pressure. Cool the syrupy residue, crystallise from dilute acetic acid and obtain Hesperdin in form of fine white needles, m.p 250-254° C.

Purification of Hesperdin: Dissolve hesperdin in minimum amount of 10% formamide solution and add a pinch of animal charcoal to it. Heat the contents for 30 minutes and add a little glacial acetic acid so that the aqueous formamide solution is slightly acidic. Filter the solution through celite and dilute the filtrate with a little water, allow it to stand for few hours in order to allow crystallization to take place. Filter off the crystals on Buchner funnel and wash first with hot water followed by isopropyl alcohol. Dry the fine white crystals of hesperdin in oven, m.p 260 -263°C.

Method II: In a two litre conical flask place 250gm of dried powdered orange peel and add 1000ml of 10% calcium hydroxide solution to it. Shake the flask thoroughly and keep it overnight at room temperature. Filter the mixture through buchner funnel. Cool and acidify the yellow orange coloured filtrate with concentrated HCl to pH 4 - 5 whereby hesperdin will precipitate out. Collect the precipitated hesperdin on buchner funnel, wash it thoroughly with water and recrystallise from formamide, m.p 260-263°C.

Colour Reactions of Hesperdin

(1) To few milligrams of hesperdin in a test tube add $FeCl_3$ solution. A wine red coloured solution is produced.

(2) To 2ml ethanolic solution of hesperdin first add magnesium and then hydrochloric acid. A bright violet colour develops.

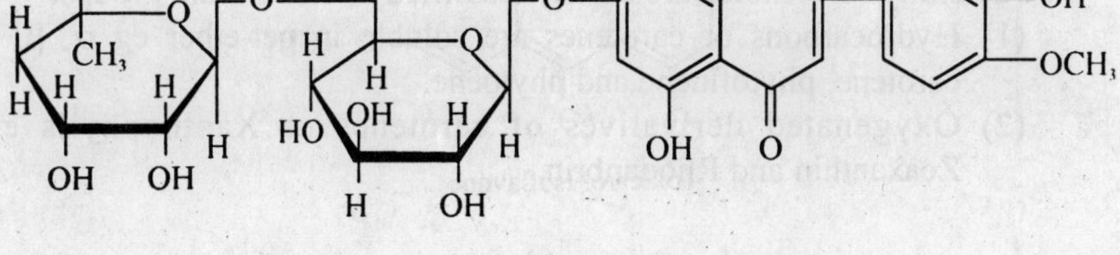

Hesperdin

2

Carotenoids

Carotenoids are yellow, orange or red nitrogen free polyene pigments, widely distributed in nature from bacteria and fungi to vegetable and animal life. Most of the carotenoids found in animals are similar to those isolated from plants and enter the animal with its food. In plants they normally occur as colloidal suspension in the form of their esters or as protein complex whereas in animals they are either dissolved in fat or combined with protein in aqueous phase. Due to their behaviour being fat soluble they are also known as lipochrome pigments.

Carotenoids are highly unsaturated acyclic, monocyclic,bicyclic hydrocarbons or oxygenated compounds consisting chiefly of a long acyclic carbon chain united by uninterrupted function of conjugated double bonds. This system of conjugation functions as a chromophore and is responsible for the colour of these hydrocarbons due to $\pi \rightarrow \pi^*$ transition. The colour of these pigments vary from bright yellow to deep red, violet or even dark blue, the intensity of the shade increases with increase in the number of conjugation in consecutive union and decrease with the saturation of the double bonds.

Almost all naturally occuring carotenoids are composed of eight isoprene units because of which they are known as tetraterpenes and the configuration of the double bond in these pigments has been found to be trans.

In plants, carotenoids normally act as light filters for chlorophyll and also as protector of enzymes in cell. But the great importance which has been attached to the carotenoid group from the viewpoint of medicine and nutrition is due to the fact that about 10 of the naturally occuring carotenoids (derived from dietary sources) are provitamin 'A' active and are capable of conversion to Vitamin 'A' inside the human body.

Structurally carotenoids are chiefly classified into two main groups:-

(1) Hydrocarbons or carotenes are soluble in pet-ether eg α, β, γ carotene, phytofluene and phytoene.

(2) Oxygenated derivatives of carotenes or Xanthophylls eg Zeaxanthin and Rhodanbrin.

Carotenes: are polyene hydrocarbon of general formula $C_{40}H_{50}$. In almost all carotenes the central molecule is composed of a long conjugated chain joined at the two ends by two open chain structures, one open chain structure and one ring or by two rings. There are several types of carotenes which have been designated as α, β, γ, δ, φ, κ, and ϕ carotenes.

Xanthophylls: These are the oxygenated derivatives of carotenoids. and are normally isolated along with carotene and chlorophyll in all the green parts of the plant. The oxygen atom may either be present in the form of ester, methoxy, oxo or carboxy group eg Rhodovibrin and crocetin.

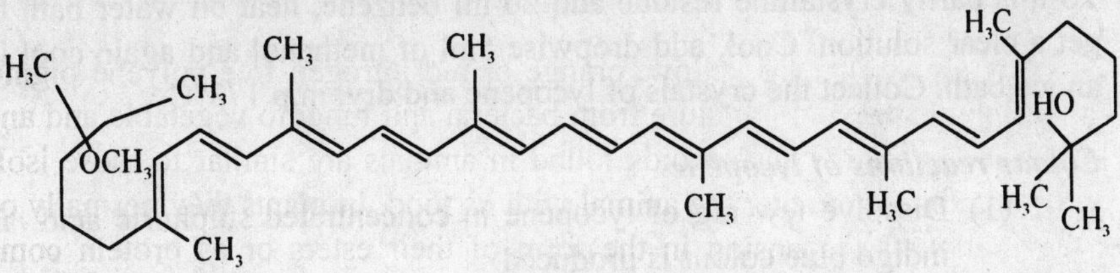

Rhodovibrin

Isolation of Lycopene from tomatoes

Lycopene a C_{40} carotenoid is the red pigment of tomatoes and is made up of eight isoprene units whereas β carotene the yellow pigment of carrot is an isomer of lycopene in which the double bonds at C_1-C_2 and $C_{1'}$ –$C_{2'}$ are replaced by bonds extending from C_1-C_6 and from $C_{1'}$- $C_{6'}$ to form rings

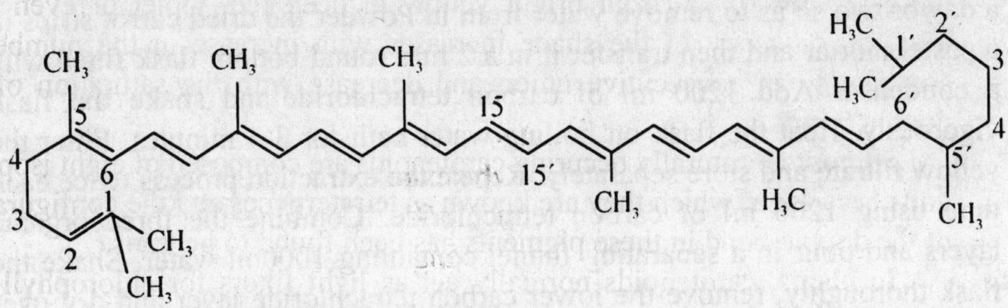

Lycopene

Place 1000 gm of tomato paste in a 2 litre conical flask provided with a stopper and add 1200 ml of methanol in order to dehydrate it. Shake the mixture vigorously (taking care that the formation of lumps does not take place) and then allow the mixture to stand for 1-2 hours. Filter the suspension on a Buchner funnel. Transfer the dark red cake to the 2 litre round bottom flask fitted with a condensor while discarding the yellow filtrate. Add 600 ml carbon tetrachloride and 600ml methanol and shake the flask vigorously.

Heat the flask on boiling water bath for 3-4 minutes. Filter the yellow filtrate and store separately. Repeat the extraction process twice each time using 600 ml carbon tetrachloride and 600ml methanol. Combine the three organic layers and pour in a separating funnel containing 1000ml water. Shake the flask thoroughly, remove the lower carbon tetrachloride layer and dry over anhydrous sodium sulphate. Filter the solution, transfer it in a 5 litre round bottom flask and distill off the solvent on steam bath. Transfer the dark oily residue to porcelain dish, add 10 ml of benzene to it and evaporate it off on water bath to remove the carbon tetrachloride completely. To this partly crystalline residue add 20 ml benzene, heat on water bath to get a clear solution. Cool, add dropwise 5ml of methanol and again cool in an ice bath. Collect the crystals of lycopene and dry, m.p 173°C.

Colour reactions of lycopene

(1) Dissolve few mg of lycopene in concentrated sulphuric acid, an indigo blue colour is produced.

(2) Add a solution of antimony trichloride in chloroform to a solution of lycopene in chloroform, an intense blue colour is produced.

(3) Dissolve few mg of compound in chloroform, add a drop of fuming nitric acid, a purple colouration develops which soon disappears

Isolation of β-carotene from carrot

Cut about 1kg of carrot into thin strips and then leave it overnight or for a day or two so as to remove water from it. Powder the dried carrot strips in a pastel mortar and then transfer it in a 2 litre round bottom flask fitted with a condensor Add 1200 ml of carbon tetrachloride and shake the flask vigorously. Heat the flask on boiling water bath for 3-4 minutes. Filter the yellow filtrate and store separately. Repeat the extraction process twice each time using 1200 ml of carbon tetrachloride. Combine the three organic layers and pour in a separating funnel containing 1000ml water. Shake the flask thoroughly, remove the lower carbon tetrachloride layer and dry over anhydrous sodium sulphate. Filter the solution, transfer it in a 5 litre round bottom flask and distill off the solvent on steam bath. Transfer the dark oily residue to porcelain dish, add 10 ml of benzene to it and evaporate it off on water bath to remove the carbon tetrachloride completely. To this partly crystalline residue add 25 ml benzene, heat on water bath to get a clear solution. Cool, add dropwise 5ml methanol and again cool in an ice bath. Collect the crystals of β-carotene and dry, m.p 183°C.

Note:- If the carrots are not dried then carry out the dehydration process as in case of lycopene.

Colour Reactions of β-carotene:-

(1) Dissolve 1-2mg of β-carotene in 2 ml chloroform and add concentrated H_2SO_4 to it. The acid layer turns blue.

(2) Dissolve 1-2 mg of β-carotene in chloroform, add a drop of fuming HNO_3. An immediate blue colouration is produced which then turns green and finally yellow.

(3) Dissolve 1-2 mg of β-carotene in chloroform, add a solution of antimony trichloride in chloroform. A dark blue colouration is produced.

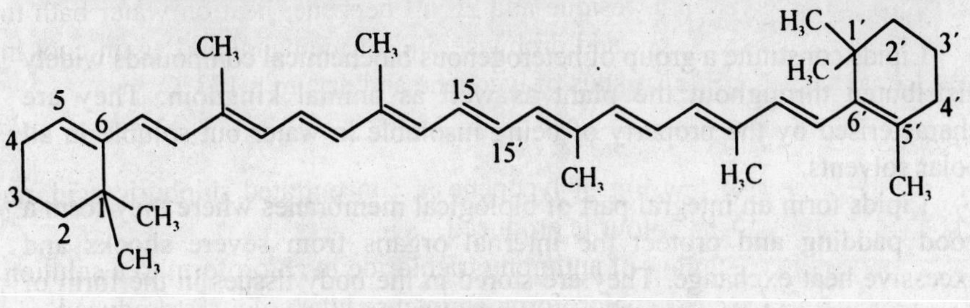

β-Carotene

Note: Since these carotenoids are very sensitive to photochemical air oxidation hence the solution as well the solid should be protected from undue exposure of light

3
Lipids

Lipids constitute a group of heterogenous biochemical compounds widely distributed throughout the plant as well as animal kingdom. They are characterised by the property of being insoluble in water but soluble in all polar solvents.

Lipids form an integral part of biological membranes where they form a good padding and protect the internal organs from severe shocks and excessive heat exchange. They are stored in the body tissues in the form of fat which on oxidation provide energy for life processes.

Lipids are chiefly classified into two main groups

(1) Simple lipids

(2) Compound or conjugate lipids

Simple Lipids: are those which are easily hydrolysed to simpler constituents. Most simple lipids are esters of long chain carboxylic acids which on hydrolysis yield fatty acids and alcohols. Included in this category are waxes and triglycerides.

(a) **Waxes:** are esters of higher fatty acid with alcohols usually containing even number of carbon atoms. They occur widely in plants and animals where they serve as protecting agents. Waxes are low melting solids which in contrast to fats and oils are resistant to saponification and atmospheric oxidation. They have a satiny waxy touch because of which they find wide application in manufacture of furniture polishing, boot polishes, cosmetics, electrical insulators, candles etc. Some of the well known waxes are beewax, carnuba wax and spermaceti.

(b) **Triglycerides:** also known as triacyl glycerols, are triesters of glycerols with a long chain of fatty acid. When all the three hydroxyl groups of glycerol are esterified with the same acid then the triester is known as simple glyceride, eg. tristearin (all the three OH groups are esterified with stearic acid) and if the two esters are different then it is known as mixed glyceride eg αα' distearo β palmitin

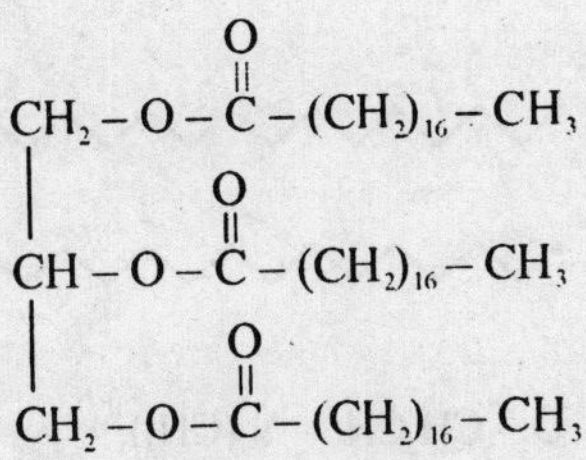

Tristearin

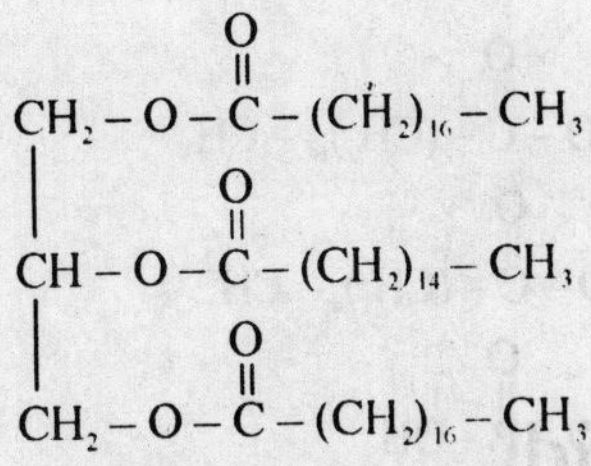

αα′ distearo β palmitin

These triesters of glycerol are commonly called fats if they are solid and oils if they are liquid at room temperature. These triglycerides isolated from plants and cold blooded animals are normally oils as corn oil, peanut oil or fish oil whereas those isolated from warm blooded animals are fats such as beef, lard etc.

Compounds or Conjugate lipids: These lipids on hydrolysis yield other substances in addition to fatty acids and alcohols like phosphoric acid, sugars etc, Some important members of this category are glycolipids and phospholipids.

(a) **Glycolipids or Cerebrosides:** are lipids containing a sugar, fatty acid and sphingosine. They differ from other complex lipids in sense that they do not contain phosphoric acid or glycerol. The sugars normally present are D-glucose and D-galactose. They are found in brain and nerve cells especially in the myelin sheath. Many hereditary fat metabolised diseases have been found to be due to the errors in the metabolism of glycolipids.

(b) **Phospholipids:** occur in the membrane of the cell and are the basic structural unit of living organisms. They are regarded as derivatives of triglycerides in which one of the acyl group has been replaced by a phosphate group. Hence the name diacyl glycerol phosphate or phospho glycerides. or phosphatidic. A number of phospholipids contain an additional alcohol esterified to the phosphoric acid group. If the base is choline then the phosphatide is known as lecithin and if the base is ethanolamine then the phosphatide is called cephalin. Lecithin is an important constituent of bile and hence plays an important role in fat metabolism. It also serves as an excellent emulsifing agent. Cephalin is found in blood platelets and is responsible for the clotting of blood.

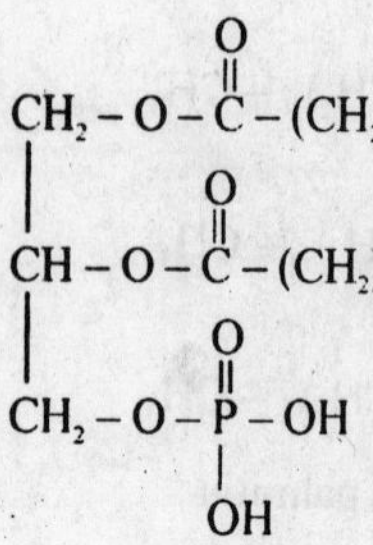

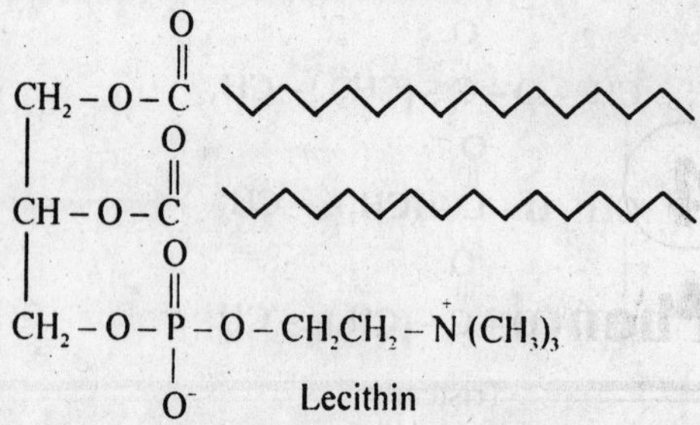

Lecithin

Isolation of Lecithin from Egg Yolk

In a porcelain dish place yolk of two hard boiled eggs, add 75 ml of ether and grind properly. Decant the organic layer and repeat the process again using 75 ml of ether. Discard the remaining solid material. Combine the two ethereal extracts, transfer it to a 250 ml round bottom flask and distill on steam bath. Add 20ml of hot ethyl alcohol to the heavy oil, shake properly and remove the alcoholic layer by filtration. Repeat the process again using 20ml of hot ethyl alcohol. Combine the two alcoholic layers and evaporate off the alcohol. Dissolve the alcoholic residue in a porcelain dish in 15 ml cold ether and add 20 ml acetone to it. Stir the resulting organic solution until particles of lecithin starts precipitating and finally adhere together in the form of a ball.

Isolation of Trimyristin from Nutmeg

Crush about 50gm of nutmeg in the form of powder in a pencil mortar and then transfer it into a 500ml round bottom flask fitted with a reflux condensor. Add 350 ml of chloroform to the flask and reflux it on water bath for 90 minutes.and filter. Set the apparatus for downward distillation and distill off the chloroform on water bath. Transfer the resultant semi solid residue in a porcelain dish and add 20 ml ethanol to it. Cool the ethanolic solution whereby trimyristrin starts precipitating. Filter the crystalline lipid and wash with cold ethanol on buchner funnel. Crystals of trimyristrin are colourless and odourless and melts at 54-55°C.

$$
\begin{array}{l}
CH_2OCOC_{13}H_{27} \\
| \\
CHOCOC_{13}H_{27} \\
| \\
CH_2OCOC_{13}H_{27}
\end{array}
$$

4

Phenols

Phenols are derivatives of aromatic hydrocarbon containing hydroxyl group directly attached to the nucleus. They are mostly liquid or low melting solids, because of hydrogen bonding they have quite high boiling points. Phenols are fairly acidic compounds and differ from alcohols which are weakly acidic. Since phenols are acidic in nature hence they can be separated from non acidic compounds by means of its solubility in a base whereas it can be separated from carboxylic acids by means of its insolubility in bicarbonate.

A number of phenols and its esters which occur in nature have been found to exert physiological effects. Salicylic acid which occurs in the willow tree has been used as febrifuge. Salicylic acid derivative like methyl salicylate or oil of winter green is a common ingredient of liniment. Thymol is a flavouring ingredient of thymene and is mainly used as a mouthwash because of its antiseptic property. Clove oil used by dentist as an antiseptic also contains a phenol eugenol. Acetyl salicylic acid commonly known as asprin has analgesic as well as antipyretic properties.

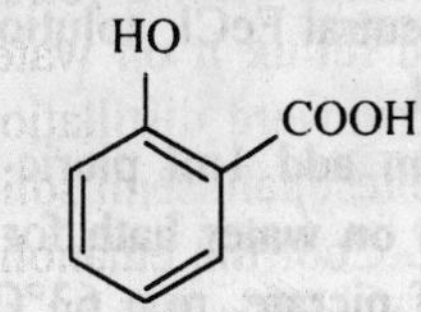

Salicylic Acid

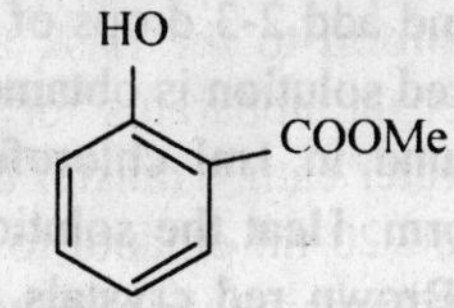

Methyl Salicylate

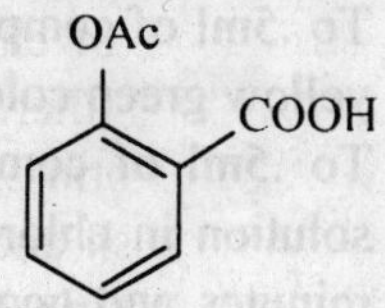

Aspirin

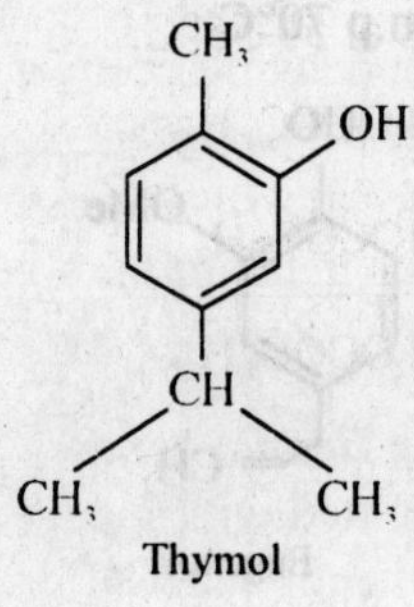

Thymol

Isolation of Eugenol from Cloves

Place 50gm of cloves in a 500ml round bottom flask and add 250 ml water to it. Set the apparatus for downward distillation. Heat the flask first strongly and when the boiling starts reduce the flame in order to prevent the foam from being carried over to the receiver. Collect about 150ml of the distillate and further add 250ml of water to the flask. Repeat the distillation process again and collect additional 150ml of distillate. Transfer the combined distillate to 500ml separating funnel and add 75ml dichloromethane to it. Shake the flask first gently and then vigorously and remove the lower organic layer. Repeat the process twice each time using 75ml of dichloromethane. Combine the organic layer and tranfer it to the 500ml separating funnel and add 50ml of 5% potassium hydroxide solution. Shake the flask thoroughly and remove the aqueous basic layer. Repeat the process twice each time using 50ml of 5% potassium hydroxide solution. Combine the three basic aqueous layers. Dry the dichloromethane layer over anhydrous sodium sulphate, decant the solution in a 250ml conical flask and evaporate off the solvent. The residue consists chiefly of acetyl eugenol and other steam volatile compounds from cloves.

Pour the basic aqueous layer in a 500ml beaker and add dilute HCl until the solution is acidic (check with litmus paper). Transfer the aqueous solution back to the separating funnel and extract with two 50ml portion of dichloromethane. Wash the combined organic layer with water and dry over anhydrous sodium sulphate. Decant off the dichloromethane and evaporate it off on a steam bath. The practically pure eugenol is obtained as a pale yellow oil, b.p 250°C, with an overpowering odour of cloves.

Colour reactions and derivatives:

(1) To .5ml of compound add 2-3 drops of neutral $FeCl_3$ solution. A yellow green coloured solution is obtained.

(2) To .5ml of compound in 1ml chloroform add 1ml picric acid solution in chloroform. Heat the solution on water bath for few minutes and cool. Brown red crystals of picrate, m.p 63°C are obtained.

(3) Eugenol benzoate, m.p 70°C.

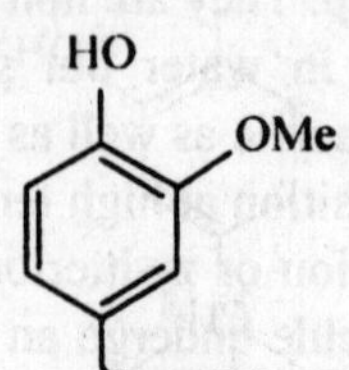

Eugenol

5

Amino Acids

Carbohydrates, lipids and proteins are a major source of fuel for the human beings. Protein in diet not only functions as an energy supplier but it also furnishes certain important components which are essential for the tissues. Proteins are polymers and the building block units (monomers) from which they are derived are called amino acids. Thus the physical and chemical properties of a protein are determined by its constituent amino acids which are joined to one another by amide linkages called peptide bonds. Amino acids are one of the essential constituent needed for the growth of the body. There are some amino acids which are not synthesized by the body but are responsible for the maintenance of proper nitrogen balance are known as essential amino acids, eg. lysine, trptophan, arginine, serine etc.

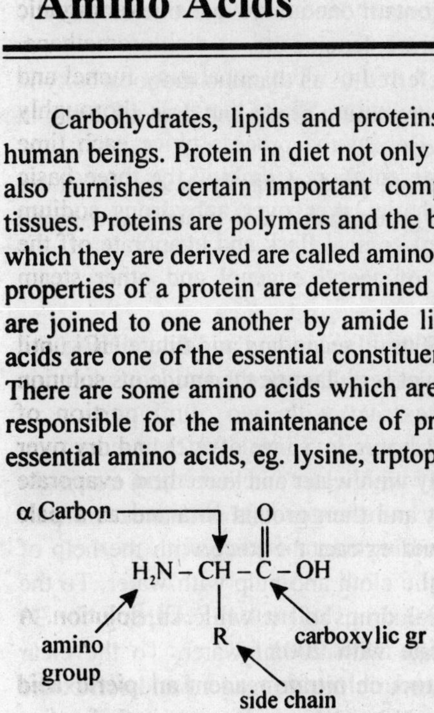

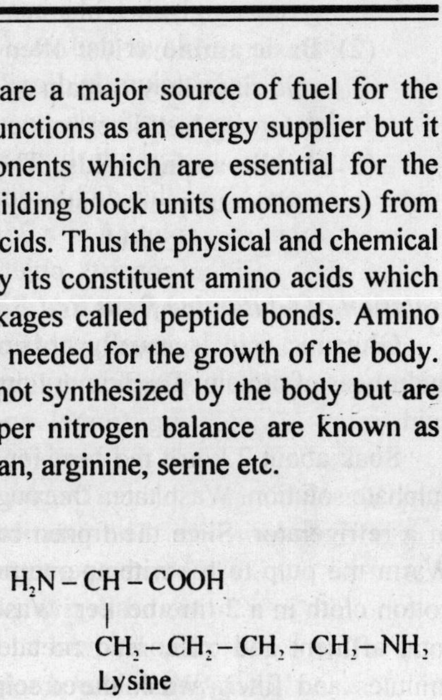

Lysine

Amino acids are organic compounds that contain both an amino group and a carboxylic group. In fact they are regarded as derivatives of carboxylic acids in which one or more hydrogen atom of the alkyl group are replaced by equal number of amino group. They are non volatile colourless crystalline compounds generally soluble in water but sparingly soluble in organic solvents. Since they contain an acidic as well as basic group they are regarded as salts and melt with decomposition at high temperatures. In fact in solution amino acids exist as a dipolar ion or zwitterion, in which the carboxyl and amino group of the same molecule undergo an internal acid base reaction in which the carboxyl group loses a proton to become carboxylate ion whereas the amino group gets protonated to form ammonium ion.

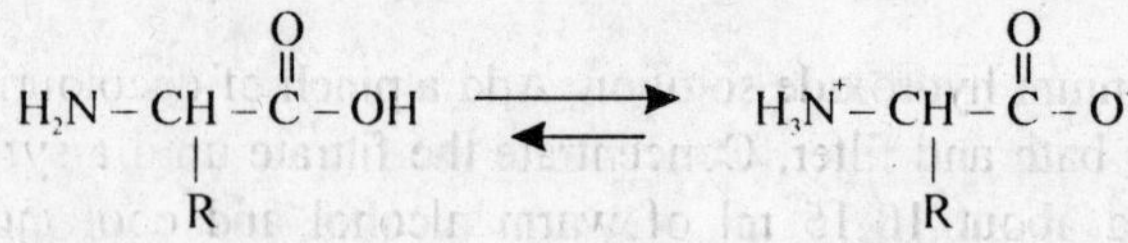

Zwitterion

The amino acids are conveniently classified into three broad categories depending upon the number of their amino and carboxyl groups.

(1) **Neutral amino acids:** often referred to as mono amino and mono carboxylic acid since they contain one amino and one carboxylic group eg glycine, alanine, serine, leucine etc.

(2) **Basic amino acids:** often referred to as diamino monocarboxylic acid since they contain an extra amino group and thus are basic in nature eg histidine, lysine etc.

(3) **Acidic amino acids:** These are predominantly acidic in nature and contain more carboxylic group than amino group, eg. aspartic and glutamic acid.

Isolation of Glutamine from Red Beet

Glutamic acid is usually obtained as a secondary product during the hydrolysis of protein. The primary product is glutamine the amide of glutamic acid.

Soak about 2 kg of red beet for 24 hours in 5 litre of 0.2M ammonium sulphate solution. Wash them thoroughly with water and keep them overnight in a refrigerator. Slice the frozen beet and then ground in a meat mincer. Warm the pulp to room temperature and extract the juice with the help of cotton cloth in a 2 litre beaker. Wash the cloth and pulp with water. To the total affluent add basic lead acetate solution, boil it while stirring for 30 minutes and filter. Wash the precipitate with 200ml water. To the clear yellow filtrate add freshly prepared mercuric nitrate reagent and reflux the mixture for 3 to 4 hours. Cool and add 1N NaOH solution until a permanent turbidity appears. Dilute it with 200 ml water and filter. To the suspension again add 10% NaOH solution so that the pH of the suspension is between 5 and 6 and then leave the suspension overnight.

Collect the precipitate on buchner funnel and wash it with water. Transfer the precipitate into a 500ml conical flask, add minimum amount of water to it so that the precipitate is now in the form of suspension. Add .1 ml of 4N H_2SO_4 to the suspension and pass H_2S gas for 2 hours with continuous agitation. Filter the precipitate on a buchner funnel and wash with water. Heat the clear pale yellow filtrate for 20 minutes on water bath in order to remove last traces of H_2S. Neutralise the solution to approximately 6 pH by adding ammonium hydroxide solution. Add a pinch of decolourising carbon heat on water bath and filter. Concentrate the filtrate until a syrupy mass is obtained. Add about 10-15 ml of warm alcohol and cool the flask in a refrigerator and keep it overnight. Filter the crystals of glutamine, wash with cold alcohol and dry, m.p 254-256°C.

Reagents

(1) **Lead Acetate Solution:** Dissolve 36gm of lead acetate in 140ml of hot water and then add 22gm of lead oxide (litharge) to it.

(2) **Mercuric nitrate reagent:** Add with stirring 44gm of mercuric oxide to 32ml of concentrated nitric acid.

Colour reaction :

(1) Heat few milligrams of compound, it turns yellow at 240°C.

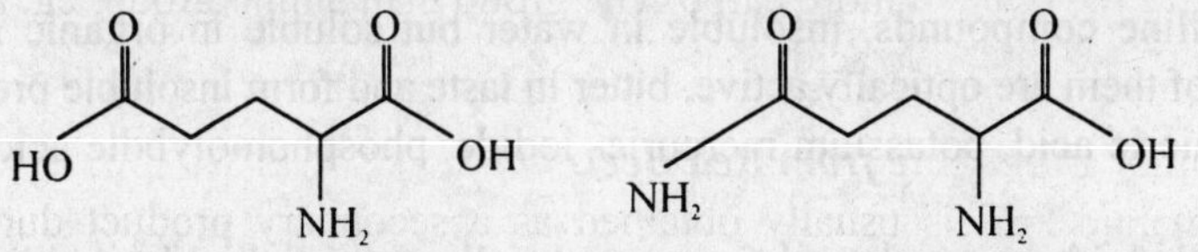

L(+) Glutamic acid L(+) Glutamine

6

Alkaloids

Alkaloids are a class of organic compounds of plant origin which are basic in nature. They posseses one or more nitrogen heterocyclic rings as their structural unit. Except for few exception, alkaloids are usually colourless, crystalline compounds, insoluble in water but soluble in organic solvents. Most of them are optically active, bitter in taste and form insoluble precipitates with picric acid, potassium mercuric- iodide, phosphomolybdic acid reagent etc.

Alkaloids are isolataed from nearly all parts of the plant chiefly from Apocynaceae, Papaveraceae, Papilionaceae and Rubiaceae family plants. Since they are basic in nature they occur largely as salts of malic, tartaric, citric and aromatic acid etc. They are also known to occur in the form of their glycosides, esters or as acid amides.

Though alkaloids have complex structures but they exhibit pronounced pharmacological activity which makes them medicinally very important. For eg quinine is an important anti-malarial agent, morphine is a powerful analgesic, reserpine is a tranquilizer, atropine finds use in the dilation of the pupil of the eye, strychnine serves as a nerve stimulant whereas emetine is used to cure amoebic dysentry.

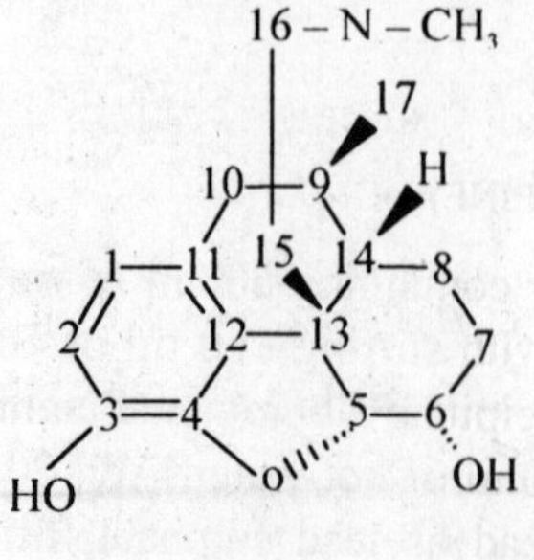

Morphine

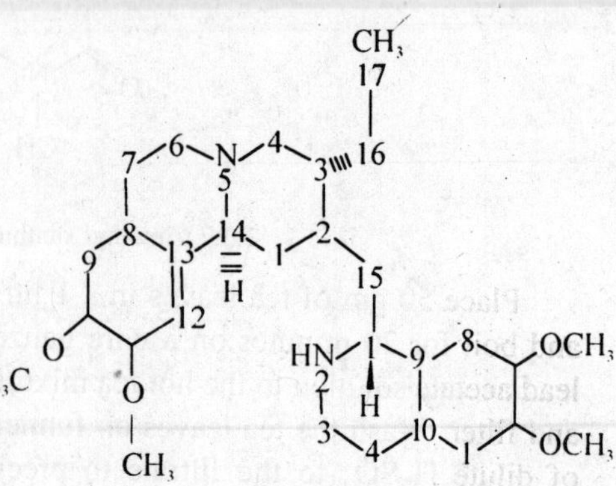

Emetine

Though a great development has been made in the field of alkaloids but it is still a subject of speculation regarding its function in plants. They are thought to act as plant stimulants, regulators that controls reproduction, growth and other metabolic activities.

Alkaloids are normally classified according to the type of nucleus (chiefly heterocyclic) that is present in the molecule.

(1) Phenylethylamine group
(2) Pyrrolidine group
(3) Pyridine and piperidine group
(4) Pyridine and pyrolidine group
(5) Quinoline group
(6) Isoquinoline group
(7) Phenanthrene group
(8) Indole group
(9) Tropane group
(10) Tropolone group
(11) Steroid group

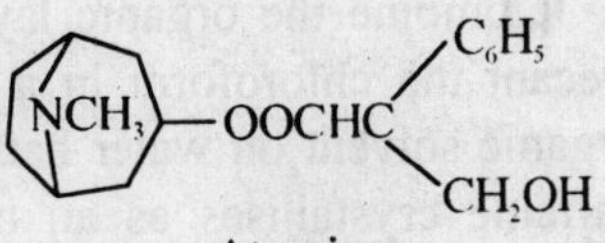

Atropine

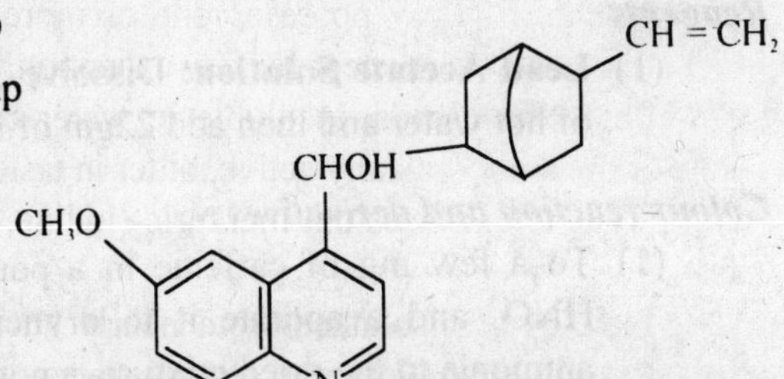

Quinine

Isolation of Caffeine from Tea

Caffeine is a methyl derivative of xanthine and acts as a stimulant of the central nervous system. It has the ability to counteract the hynotic effect of other drugs.

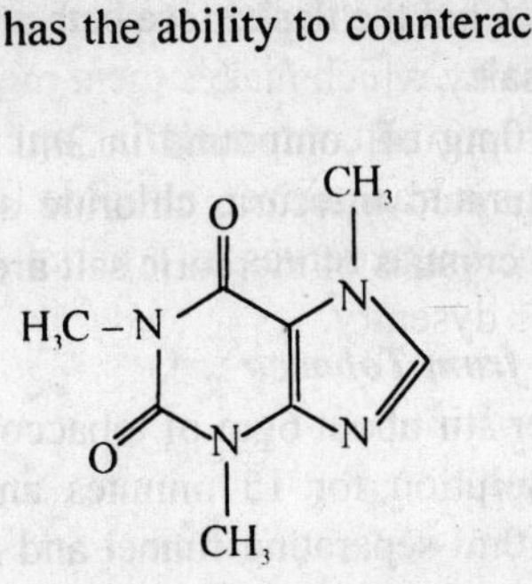

1:3:7 trimethyl xanthine (CAFFEINE)

Place 50 gm of tea leaves in a 1 litre beaker containing 600ml of water and boil for 20 minutes on a wire gauze. Add with stirring 160 ml of 10% lead acetate solution to the hot tea mixture to precipitate albumin and tannins and filter. Wash the tea leaves on funnel with 50ml of hot water. Add 20ml of dilute H_2SO_4 to the filtrate to precipitate lead as lead sulphate, filter.

Concentrate the filtrate to 100 ml and then cool it to room temperature. Transfer the concentrated liquid to 500ml separating funnel and add 50ml of chloroform to it. Shake the funnel slowly and drain off the lower organic layer (avoid shaking too vigorously as the formation of emulsion takes place). Repeat the extraction process twice each time using 50ml of chloroform

Combine the organic layers and dry over anhydrous sodium sulphate. Decant the chloroform in a 250ml round bottom flask and distill off the organic solvent on water bath. As the last traces of solvent are distilled off caffeine crystallises as an off white or cream coloured solid. It can be further purified by sublimation or by recrystallisation from benzene, m.p 233-235°C.

Reagents

(1) **Lead Acetate Solution:** Dissolve 36gm of lead acetate in 140ml of hot water and then add 22gm of lead oxide (litharge) to it.

Colour reaction and derivatives :-

(1) To a few mg of caffeine in a porcelain dish add 3-5 drops of HNO_3 and evaporate it to dryness. Add 2-3 drops of liquid ammonia to the dried mixture, a purple colour is developed.
(2) In a 100ml conical flask dissolve 300mg of caffeine and 225mg of salicylic acid in 15ml benzene. Heat the flask on water bath until benzene begins to boil. Add pet-ether just enough to create cloudiness. Cool the flask in ice bath and collect crystals of caffeine salicylate salt.
(3) Dissolve 10mg of compound in 2ml warm water. Cool and add 1ml of saturated mercuric chloride solution. After few minutes long white crystals of mercuric salt are obtained, m.p 213°C

Isolation of Nicotine from Tobacco

In a 200ml beaker stir about 6gm of tobacco (from 6 cigarettes) with 75 ml of 25% NaOH solution for 15 minutes and filter. Transfer the dark brown filtrate to a 250ml separating funnel and add 40ml ether to it. Shake the flask thoroughly and leave it for 5 minutes and allow the layers to separate. Remove the ethereal layer. Repeat the extraction process again using 40ml of ether. Combine the two organic layers and distill off the ether on steam bath. Transfer the residual oily portion in a porcelain dish and evaporate off the residual ether. The thick oily residue in the porcelain dish is nicotine, b.p 247°C.

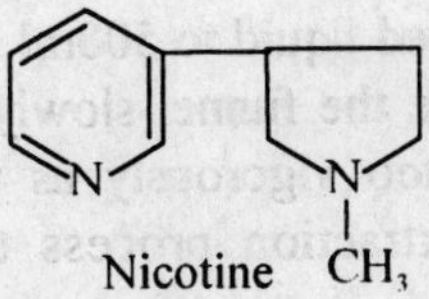

Nicotine

Colour reaction and derivative

(1) To 1ml of compound add concentrated HCl. A light violet or brown colouration is obtained on warming.

(2) To 1ml of nicotine in 5ml methanol add 5ml of saturated methanolic picric acid solution. Heat it on water bath for few seconds. Cool and collect crystals of picrate, m.p 219°C.

Isolation of Piperine from Black Pepper

In a 1litre round bottom flask fitted with reflux condensor place 50gm of powdered black pepper and 250ml of 95% ethanol. Reflux the mixture for 3 hours. Filter the resulting solution and transfer the filtrate to 500ml round bottom flask and set the condensor for downward distillation. Distill off most of the alcohol and in the end only 25-30ml should remain in the flask. Transfer the concentrated extract to a porcelain dish, add 20ml of 10% alcoholic potassium hydroxide solution to it and then keep it at room temperature for 5-10 minutes. Decant off the alcohol from the insoluble residue and leave it overnight in a refrigerator. Piperine crystallises out in form of yellow crystals, m.p 124-126°C (Yield 1.5gm)

Colour reaction:-

(1) To a few mg of compound add concentrated H_2SO_4 yellow colouration which soon changes to greenish brown is observed.

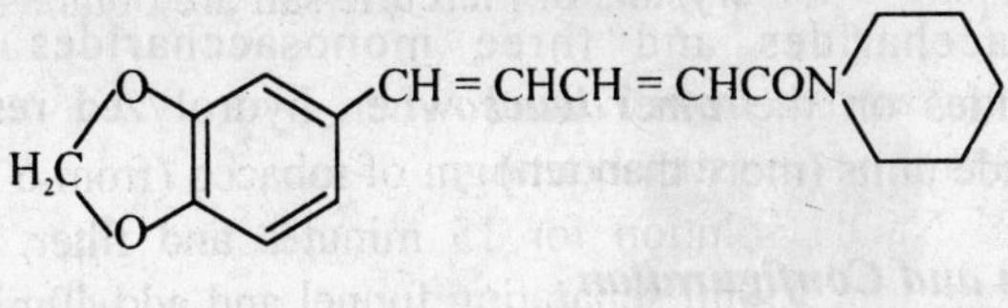

7

Carbohydrates

Carbohydrates are a class of organic compounds which are distributed universally in plants and animals. They are the chief structural unit of plants comprising more than 50% of its dry weight principally in the form of cellulose. In higher animals glucose is the vital constituent of the blood. Carbohydrates are also biologically very important since they are an important source of energy for animals. They also occur in bound form in ATP which plays an important role in metabolic process.

Chemically carbohydrates are compounds with molecular formula $C_n(H_2O)_m$ thereby indicating the relationship between carbon and the water molecule,hence called hydrates of carbon or carbohydrates. But according to the modern definition all polyhydroxy compounds containing aldehydic and ketonic carbonyl group (usually in the acetal or hemiacetal form) or compounds that are easily hydrolysed to them are included into the category of carbohydrates.

Sugars also called saccharides are the simplest type of carbohydraes. Monosaccharides are the simplest sugar that cannot be hydrolysed into simpler compounds whereas the hydrolysis of a disaccharide and trisaccharide gives two monosaccharides and three monosaccharides respectively. Oligosaccharides on the other hand when hydrolyzed result in several monosaccharide units (more than ten)

Classification and Configuration

Although most sugars have their own specific common names but they are not systematic and it is difficult to remember their structures. Thus they are classified according to the number of carbon and the carbonyl group they contain.

Sugars with aldehyde group are called aldoses and those with ketonic group are called ketoses. The number of carbon atom present in a sugar are designated by the term triose (for three carbon), tetrose (four carbon), Pentose (five carbon), hexose (six carbon) and heptose (seven carbon). The two group of terms are then combined to represent a particular monosaccharide for, eg. erythrose, a four carbon sugar with aldehydic group is called aldotetrose, glucose contains an aldehyde group and six carbon hence called aldohexose, similarly fructose is called keto hexose

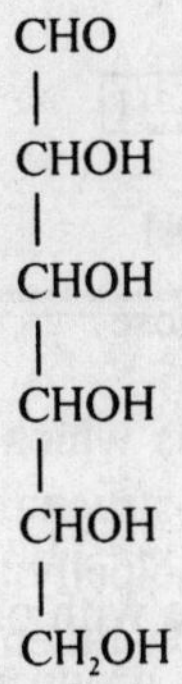

an aldohexose

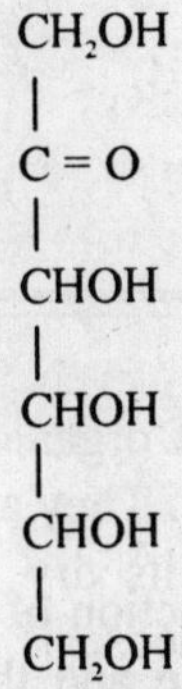

ketohexose

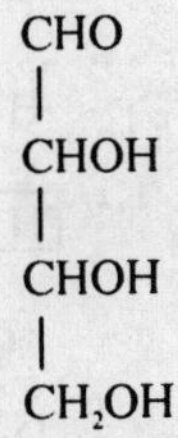

aldotetrose

Most sugars are closely related to one another but differ only in the stereochemical arrangement at a single carbon atom. Such sugars are called epimers for ex glucose and mannose differ only at C-2, glucose and allose differ at C-3 and in ketohexoses fructose and psicose differ at C-3.

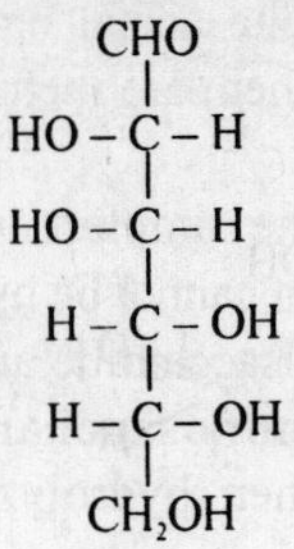

D-mannose

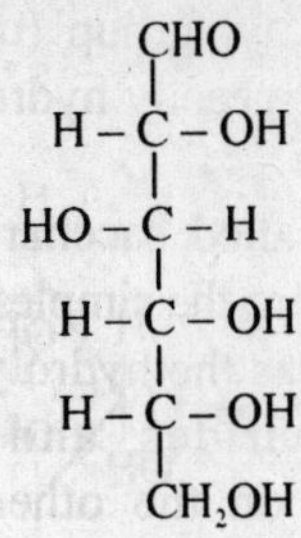

D-Glucose

According to convention the configuration of all the sugars are decided by taking into account the stereochemistry of the chiral centre most distant from the carbonyl group. They are designated D when the hydroxyl group of this distant chiral carbon lies on the right side of the Fischer projection formulae,while those with the opposite stereochemistry at this centre are members of the L family. Thus all D series of monosaccharides are related to D(+) glyceraldehyde and mirror images of the D series ie the L series are equated with L(-) glyceraldehyde.

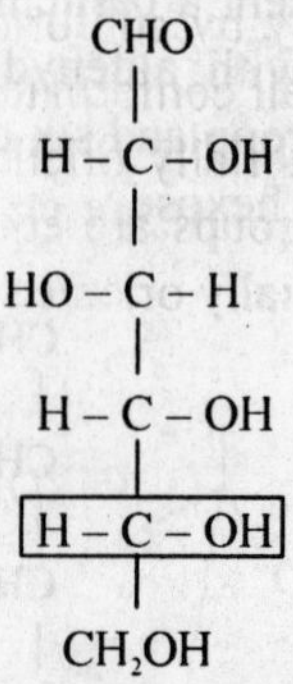

D-Glucose

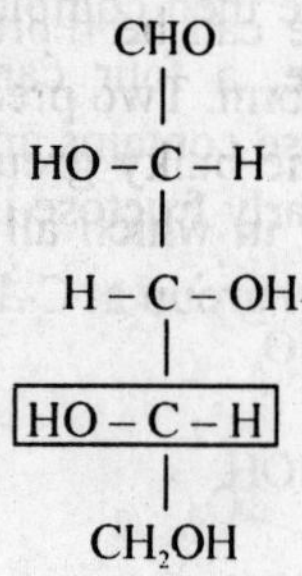

L-Xylose

Cyclic structures

Hemiacetal results by the reaction of an aldehyde with one molecule of alcohol. However if the aldehyde and the alcoholic group are part of the same molecule than a cyclic hemiacetal results. These cyclic hemiacetals are more stable than their open chain structures. Glucose which is an aldohexose contains both an aldehydic and several alcoholic groups. Thus the solid crystalline form of the glucose is normally a cyclic hemiacetal which arises due to the reaction of carbonyl function at C-1 and hydroxyl group at C-5 of the open chain structure to give pyranose structure.

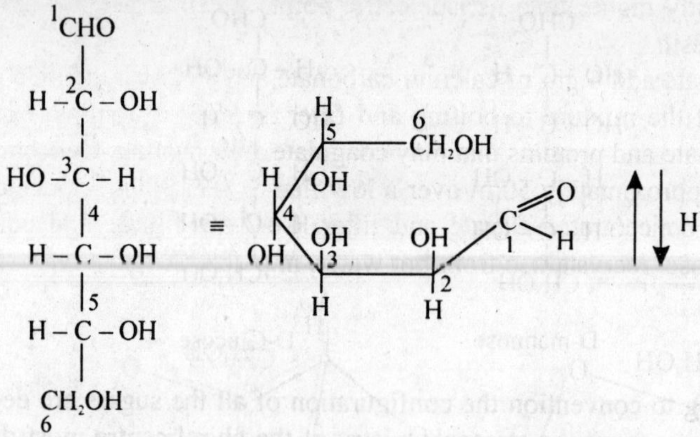

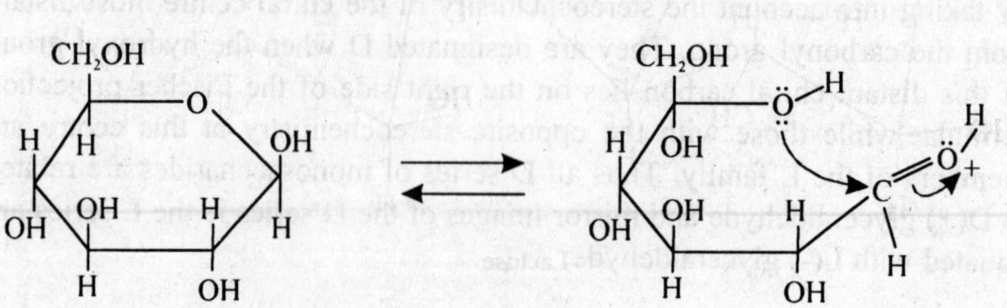

Glucose being six membered ring bear quite resemblance with cyclohexane, hence can be represented easily by conformational structures in the stable chair form. Two predominant chair conformations are possible :- one in which all the bulky groups are equatorially oriented is known as β form and secondly in which all the bulky groups are equatorially oriented except the hydroxyl group at C-1 which is axially oriented is known as α.

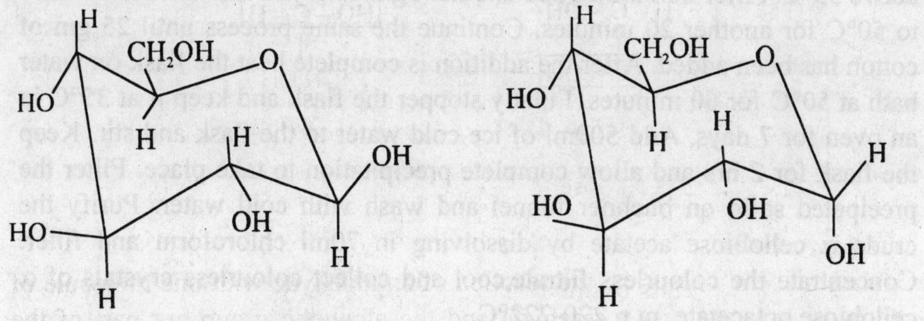

β-form α-form

Isolation of lactose and Caesin from milk

In a 500ml beaker add 300ml of skimmed milk and heat it on water bath to 60°C. Add dropwise and with stirring 10% acetic acid solution until a flocculent precipitate forms (precipitation of caesin takes place more readily at a point slightly higher than its isoelectric point of pH 4.5). Filter off the precipitated caesin.

To the filtrate add 6 gm of calcium carbonate and stir with the help of a glass rod. Heat the mixture to boiling and filter in order to remove excess calcium carbonate and proteins that may coagulate after heating. Concentrate the filtrate to approximately 50ml over a low flame. Add 300ml of acetone/ ethanol to the concentrated filtrate and filter. Cool the filtrate and collect crystals of lactose, recrystallise from hot water, m.p 203°C.

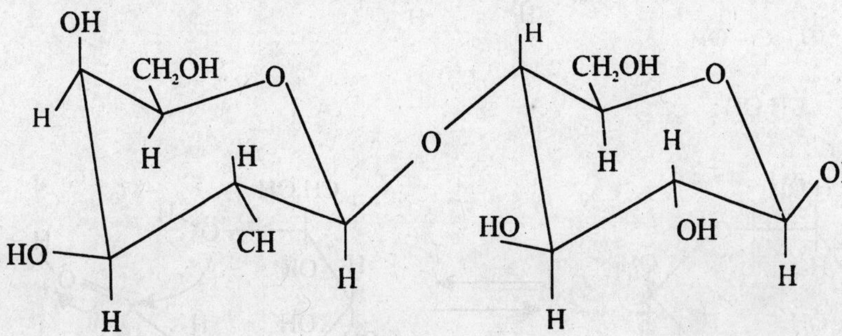

Lactose

Isolation of α-cellobiose acetate

In a 2 litre conical flask provided with a stopper add 100ml of acetic anhydride and keep the flask in an ice bath. Now add in one portion with stirring 9 ml of concentrated H_2SO_4. Remove the flask from ice bath and then add 5 gm of absorbent cotton to it. Heat the flask on water bath to 50°C and stir the mixture. Care should be taken that the temperature does not rise above 55°C. After 20minutes add another 5gm of cotton and stir the mixture to 50°C for another 20 minutes. Continue the same process until 25 gm of cotton has been added. After the addition is complete heat the flask on water bath at 50°C for 60 minutes. Finally stopper the flask and keep it at 35°C in an oven for 7 days. Add 500ml of ice cold water to the flask and stir. Keep the flask for 2 hrs and allow complete precipitation to take place. Filter the precipated solid on buchner funnel and wash with cold water. Purify the crude α cellobiose acetate by dissolving in 70ml chloroform and filter. Concentrate the colourless filtrate,cool and collect colourless crystals of α cellobiose octacetate, m.p 220-222°C.

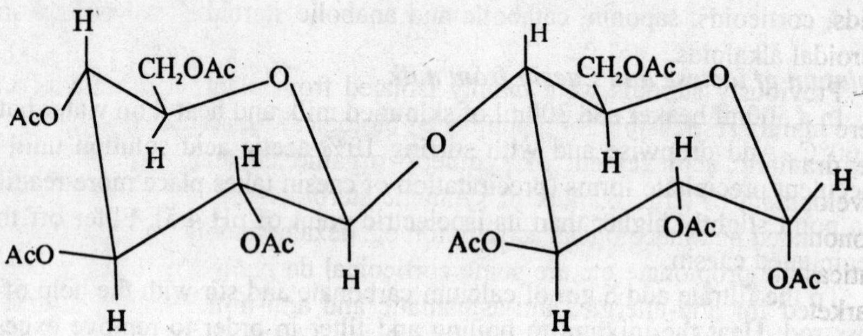

8

Steroids

Steroids encompasses a wide variety and an important class of compound which possess tetracyclic perhydro cyclopenteno phenanthrene ring system as the basic skeleton. Biogenetically they are thought to belong to triterpenoid class of compound embodying six isopentyl units.

Included into this class are compounds which exhibit remarkable pharmacodynamic properties like cardiotonic glycosides, sex hormones, bile acids, corticoids, saponin, catabolic and anabolic steroids, toad poison and steroidal alkaloids.

Previously steroids were mainly isolated from plants and animals and were mainly responsible for a number of biological function in them. One of the dramatic achievement of synthetic organic chemistry has been the development of large number of synthetic steroidal derivatives which have pronounded pharmacological activity for, eg. dexamethasone, betamethasone, fluticasone propionate etc are some corticoidal derivatives which are being marketed for anti-allergic, anti-asthamatic and anti-inflammatory diseases. Formestane and androgenic hormone are being used for curing breast cancer. The development of birth control oral contraceptive pills have further revolutionised the steroidal field.

All steroids have been found to posseses the following basic skeleton and depending upon the nature of substituent R at C-17 they are classified into following categories.

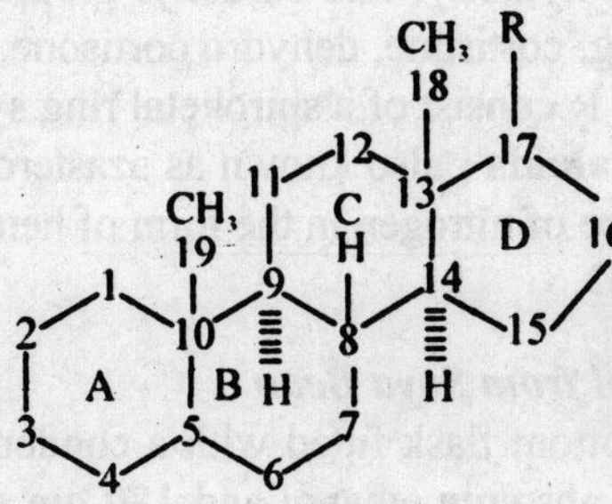

(1) **Sterols:** If R consists of an eight or nine carbon, aliphatic saturated or unsaturated side chain and with one or more hydroxyl group attached to the alicyclic rings, eg. cholesterol and stigmasterol.

(2) **Bile acids:** the group R consists of a five membered carbon chain terminating with a carboxylic group. The hydroxyl groups attached to the alicyclic rings are generally α oriented, eg. lithocholic acid and deoxy cholic acid.

(3) **Sex hormones :**

(a) If the group R is either a ketonic or hydroxy group with hydroxy or keto group at C-3, then this is categorised as a male hormone or androgens, eg. testosterone.

(b) If the group R is either a ketonic or hydroxyl group and ring A is aromatic then it is a female hormone or estrogens, eg. estrone and estradiol.

(c) If R is a two carbon side chain with ketonic group at C-3 then the steroidal hormone is a gestogen, eg. progesterone.

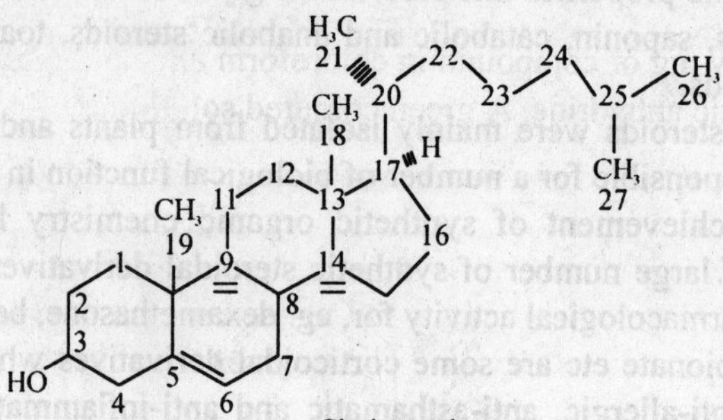

Cholesterol

(4) **Cardiac aglycones :** are characterised by the presence of αβ unsaturated lactone in the form of R at C-17. These compounds commonly called cardenolides occur in the form of their glycosides or esters, eg. digitoxigenin.

(5) **Corticoids :** if R consist of α ketol side chain at C-17 (CO-CH_2OH) with hydroxyl and carbonyl groups at 3, 9, 11, 12, 14, 17 position, eg. cortisone, dehydro cortisone.

(6) **Sapogenins :** R consist of a spiroketal ring system, eg. diosgenin

(7) **Steroidal alkaloids :** also known as azasteroids are characterised by the presence of nitrogen in the form of heterocyclic ring usually fused to ring D.

Isolation of Stigmasterol from Soya Bean

In a 3 litre round bottom flask fitted with a condensor place 200 gm of soya bean oil, 1 litre of absolute ethanol and 180 gm of KOH dissolved in minimum amount of water. Reflux the flask on water bath for 2½ hours. Cool the flask and dilute the solution by adding 1 litre of water to it. Transfer the resultant solution to a separating funnel and add 500ml of ether to it. Shake the flask well, remove the organic layer which contains the saponified material. Repeat the same process four times, each time using 500ml of ether. Combine the organic layer and concentrate it to one fourth its volume on water bath. Transfer the concentrated liquid to 1 litre separating funnel and add 200ml of water to it. Shake the flask and remove the lower aqueous layer from it. Repeat the process thrice or until the organic layer is free from alkali. Transfer the alkali free concentrate to 1 litre round bottom flask and concentrate further on steam bath. Dissolve the concentrated residue containing the unsaponifiable material in 250 ml pet-ether, pass steam into the solution until it is nearly saturated and then leave it overnight. Collect colourless crystals of stigmasterol on buchner funnel, wash with cold water and dry, m.p 140-144°C.

Colour reaction

(1) To a few mg of compound in chloroform add concentrated H_2SO_4 and acetic anhydride. A green coloured solution is obtained.

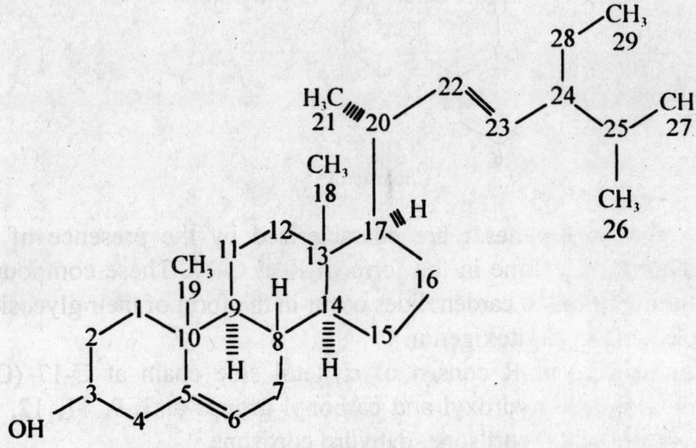

Stigmasterol

SECTION IV

1

Estimation of Sulphur (Messenger's Method)

Principle: Sulphur of the organic compound is oxidised by alkaline $KMnO_4$ to acid sulphate. The latter is then precipitated as barium sulphate by treating with barium chloride. From the weight of barium sulphate, the percentage of sulphur can thus be computed.

$$H_2NCSNH_2 + 8KMnO_4 + 8KOH \rightarrow H_2SO_4 + H_2NCONH_2 + 8K_2MnO_4 + 3H_2O$$

$$H_2SO_4 + BaCl_2 \rightarrow BaSO_4 + 2HCl$$

Procedure: Weigh accurately about .2 – .3 gm of sulphur compound (thiourea) and place it in a 250ml round bottom flask Now add 50ml of distilled water, 2 gm potassium permangnate and 1gm of KOH to it. Place the flask on sand bath, attach the condensor to it and arrange the assembly such that it is inclined at an angle of 60°. Add few chips of porcelain to the flask and reflux the contents for 5-6 hrs. Cool the flask, add about 10-20ml concentrated HCl to it. Heat the flask again until a colourless solution results and then filter the contents into a beaker. Wash the flask and filter paper thoroughly with distilled water and transfer all the washing to the beaker. Concentrate the solution to about 40-45ml. Add 2% barium chloride solution until the precipitation is complete. Digest the precipitate on water bath for 30min and then allow it to stand, preferably overnight. Filter the precipitate on a gravimetric filter paper (Whatman filter paper) and wash with distilled water to remove chloride ions (check by treating the filtrate with silver nitrate solution). Fold the filter paper properly, place it in a weight constant silica crucible. Heat the crucible first gently and then strongly until the black matter has completely disappeared. Cool the crucible to room temperature and then weigh it to a constant weight.

Calculations

Weight of sulphur containing compound = M

Weight of $BaSO_4$ precipitate	=	m
Molecular weight of sulphur compound (thiourea)	=	233
233 gm of $BaSO_4$	=	32 gm of Sulphur
$\therefore$ m gm of $BaSO_4$	=	$\frac{m \times 32 \text{ gm of Sulphur}}{233}$
% of Sulphur	=	$\frac{m \times 32 \times 100}{233 \times M}$

2

Estimation of Nitrogen (Kjeldahl's Method)

Principle : A known weight of nitrogen containing compound is digested in kjeldahl flask with concentrated sulphuric acid in the presence of a catalyst until the organic matter has been completely oxidised and the nitrogen converted quantitatively to ammonium sulphate. The acidic reaction mixture is then treated with excess of sodium hydroxide solution. The liberated ammonia is steam distilled and absorbed into saturated boric acid solution. Ammonium borate so formed is titrated with standard hydrochloric acid solution using methyl red or methylene blue as indicator. From the amount of ammonia evolved, the percentage of nitrogen can be calculated.

$$\text{Nitrogenous compound} + H_2SO_4 \rightarrow (NH_4)_2SO_4$$

$$(NH_4)_2SO_4 + 2NaOH \rightarrow Na_2SO_4 + 2H_2O + 2NH_3$$

$$NH_3 + H_3BO_3 \rightarrow NH_4^+ + H_2BO_3^-$$

$$H_2BO_3^- + HCl \rightarrow H_3BO_3$$

Reagents

(i) Sodium hydroxide solution (.1N)	Dissolve 4gm of pellets in 1000ml distilled water, allow the solution to settle and then standardise with .1N oxalic acid solution.
(ii) Catalyst	a mixture of 1gm selenium powder, 20gm potassium sulphate and 1gm copper sulphate.
(iii) Hydrochloric acid	.02 N

Procedure : Weigh accurately 30-50mg of nitrogenous compound and then transfer into a clean and dried kjeldahl flask. Add 1gm of catalyst and 5-10ml of concentrated sulphuric acid to it. Keep the flask loosely stoppered and then heat it in an inclined position first slowly (until the frothing subsides) over the micro burner and then vigorously for 1-2 hrs. (or until the reaction mixture is colourless). Transfer the reaction mixture to distillation flask, wash the kjeldahi flask 4-5 times with distilled water and transfer each washing to the distillation flask. Fit the distillation flask with dropping funnel and kjeldahl trap. Connect the kjeldahl trap with condensor with the lower end dipping in 25ml saturated boric acid solution contained in 100ml conical flask. Introduce 30-40ml of standard sodium hydroxide solution from the dropping funnel and heat the dilute reaction mixture inside the distillation flask by passing steam. Ammonia evolved is absorbed into boric acid solution. The amount of ammonia evolved is obtained by titrating with standard hydrochloric acid using methyl red or methylene blue as indicator.

Calculations

(i) Weight of nitrogen sample = w_1 gms

(ii) Normality of HCl (N_1) = .02N

(iii) Volume of .02N HCl used to neutralise ammonia = v_1 ml

$$HCl \equiv NH_3 \equiv N$$

$$\frac{w \times 1000}{14} = (N_1 \times v_1)\ HCl$$

$$\text{Weight of nitrogen (w)} = \frac{14 \times N_1 \times v_1}{1000}$$

$$\%\text{ of Nitrogen} = \frac{w \times 100}{w_1}$$

3

Estimation of Molecular Weight of Acid

The molecular weight of carboxylic acids can be estimated by :

(a) Silver salt method

(b) Titration with standard alkali solution.

Silver Salt Method

Principle : Silver salts of most carboxylic acids are sparingly soluble in water and they also do not contain any water of crystallisation. Thus when ignited in a weighed crucible the organic matter is driven off, leaving only the residue of pure metallic silver.

Procedure: Dissolve 1-1.5gm of accurately weighed organic acid in 20-25ml distilled water and then add dilute ammonia solution until the well stirred solution gives smell of ammonia. Make the solution neutral by boiling the alkaline solution so as to drive off excess ammonia (test with the red litmus paper). Cool and add dilute aqueous silver nitrate solution until no further precipitation of silver salt as silver acetate takes place. Filter the precipitated silver salt through buchner funnel, wash with cold water to remove the excess silver nitrate and then drain thoroughly. Transfer the precipitate in a small porcelain dish cover it with a watch glass and then heat it on steam oven or water bath (protect from light). Take about .5gm of this well dried silver salt in a weighed silica crucible (weight constant) and then heat the crucible first gently over a low flame and then after 5-10 minutes over a free flame in order to completely decompose the silver salt. Continue the heating process until the crucible becomes red hot. Remove the heating source, allow the crucible to cool and then keep it in a dessicator. Weigh it, repeat the heating process and again weigh until a constant weight is obatined.

Calculations;

Weight of crucible	=	W^1
Weight of crucible + silver salt	=	W^2
Weight of crucible + silver (after heating)	=	W^3
Weight of silver salt	=	$W^2 - W^1$

Weight of silver = $W^3 - W^1$

Atomic weight of silver = 108

Molecular weight of silver salt = $\dfrac{(W^2 - W^1) \times 108}{(W^3 - W^1)}$

One hydrogen atom (atomic weight 1) is replaced by one silver atom (atomic weight 108)

Equivalent weight of acid = $\dfrac{(W^2 - W^1) - (W^3 - W^1) \times 108}{W^3 - W^1} + 1$

Molecular weight of acid = Equivalent weight × basicity

Titration with Alkali Solution Method

Procedure : Weight accurately about 2 - 2.5gm of organic acid, dissolve it in distilled water in a 100ml graduated flask and then make the solution upto the graduated mark. Pipette out 25ml of the acid solution in a 100ml conical flask, add 2-3 drops of phenolphthalein indicator and then titrate with N/2 NaOH solution. Perform the titration again till at least two consistent readings are obtained.

Calculations

Weight of acid = W^1

25ml of this solution requires = V^1 ml of N/2 NaOH solution

100ml of this solution requires = $(V^1 \times 4)$ ml of N/2 NaOH solution

= $\dfrac{(V^1 \times 4)}{2}$ ml of N NaOH solution

W^1 gm of organic acid (monobasic) are equivalent to $\dfrac{(V^1 \times 4)}{2}$ ml of NNaOH solution

$\dfrac{W^1 \times 1000}{\dfrac{V^1 \times 4}{2}}$ gm of organic acid (monobasic) are equivalent to 1000ml of NNaOH

Molecular weight of acid (monobasic) = $\dfrac{W^1 \times 1000}{\dfrac{V^1 \times 4}{2}}$

Molecular weight of acid (dibasic) = $\dfrac{W^1 \times 2000}{\dfrac{V^1 \times 4}{2}}$

4

To determine the Saponification Value of Oils and Fats

Principle: Fats and oils, one of the major constituents of our food are triesters of glycerol with higher aliphatic monocarboxylic acids. Those, which are solid at ordinary temperature, are called fat while those, which are liquid, are called oils.

Theory: The term saponification originates from the alkaline hydrolysis of oils that lead to the formation of soaps. Saponification value or number of an oil or fat is the amount of KOH required to neutralize the free fatty acid present in 1 gm of oil or fat. The process involves hydrolyzing the weighed amount of an oil or fat by boiling with a known volume (excess) of standard caustic potash solution. The residual unreacted alkali is then treated with standard acid solution using phenolphthalein as an indicator.

The measurement of saponification value helps in the identification of given fatty oil, extent of adulteration and also to distinguish between fatty and mineral oils.

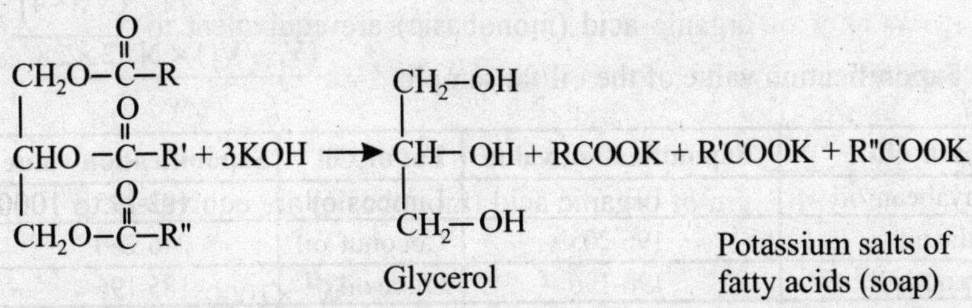

Preparation of Reagent:

N/2 Potassium hydroxide solution (Eq wt 56): Weigh (56/2=28/4=6.5) 6.5 gms of potassium hydroxide and transfer it in a 250 ml measuring flask with the help of 95% ethanol. Add more alcohol, stopper the flask and shake thoroughly so that the alkali completely dissolves. Make up the volume by adding more alcohol and then leave overnight. Decant or filter to remove any insoluble potassium carbonate. Standardize by titrating with N/2 HCl using phenolphthalein as indicator.

Phenolphthalein indicator: See Experiment 12.

Procedure: Transfer 1 gm of accurately weighed fat or oil in 100ml round bottom flask. Add about 20-30ml of N/2 standard alcoholic KOH solution and a small piece of porcelain chip to the flask and then reflux on sand bath/oil bath till a clear solution is obtained (usually 90-120 minutes). Add the same amount of solvent to another flask for blank determination. Cool and disconnect the condensor, wash it thoroughly with distilled water and transfer the washing in the flask. Transfer the saponified solution in a 250ml conical flask. Wash the round bottom flask with minimum distilled water and transfer it into the conical flask. Now add 10-12 drops of phenolphthalein as indicator to each flask and titrate against standard N/2 HCl solution, until the pink colour just disappears.

Perform the blank titration (using all the above procedure and reagent, except for the use of oil or fat sample.

Observation and Calculation:

Weight of fat/oil sample taken = w gms

Volume of N/2 HCl solution used for blank = V_1 ml

Volume of N/2 HCl solution used for sample = V_2 ml

Volume of N/2 HCl equivalent to KOH used in saponification of w gms of sample = $(V_1 - V_2)$ ml

1000ml of 1N HCl = 1000ml of 1N KOH = 56 gms of KOH

$$(V_1 - V_2)\text{ml of N/2 HCl} = \frac{(V_1 - V_2) \times N/2 \times 56}{1000} \text{ gms KOH}$$

$$\text{Saponification value of the oil/fat sample} = \frac{(V_1 - V_2) \times N/2 \times 28}{w}$$

Fat or Oil	Saponification value	Fat or Oil	Saponification value
Soyabean oil	193	Linseed oil	192-195
Palm oil	196-205	Coconut oil	246-260
Peanut oil	190-196	Olive oil	185-196
Butter	220-233	Rapeseed oil	170-179
Cottonseed oil	193-195	Corn Oil	188-195

5

Estimation of Hydroxyl Groups (Alcohols)

Principle: The procedure generally adopted for the estimation of hydroxyl groups include esterification and oxidation. Esterification process is applicable to nearly all compounds whereas oxidation method is generally applicable to compounds containing vicinal hydroxyl groups.

The most common esterification method is acetylation and it involves heating the known volume of alcohol with known volume of acetic anhydride (used in excess) and pyridine until acetylation is complete. During this process the hydroxyl group is replaced by acetyl group. Excess amount of unreacted acetic anhydride present in the reaction mixture is hydrolysed to acetic acid by adding water to it. The total free acetic acid is then estimated by titrating with standard sodium hydroxide solution using phenolphthalein as an indicator.

$$R(OH)_m + m\ CH_3COOCOCH_3 \rightarrow R(OCOCH_3)_m + mCH_3COOH$$

$$(CH_3CO)_2O + H_2O \rightarrow 2CH_3COOH$$

A blank titration (without addition of hydroxy compound) is also performed under similar conditions. The difference in the volume of sodium hydroxide used in the two titrations is equivalent to the amount of acetic acid formed. If the molecular weight of hydroxy compound is known then the number of hydroxyl groups present in the the compound can be calculated.

Reagents

1. **Acetylating reagent:** Mix together one volume of acetic anhydride with three volumes of dry pyridine and shake thoroughly.
2. **Indicator:** Phenolphthalein

Procedure: Place accurately weighed 1gm of the alcoholic substance (mannitol, benzyl alcohol, cyclohexanol) in a 100ml round bottom flask fitted with a reflux condensor. Add 20ml of acetylating mixture to the flask and heat it on water bath for 30-40 minutes. Remove the heat source and pour about 20ml distilled water down the condensor and shake the flask thoroughly to ensure complete hydrolysis of the residual acetic anhydride. Remove the condensor, cool the flask and keep it at room temperature for 5-10 minutes. Titrate the contents of the flask with standard (1N) sodium hydroxide solution using phenolphthalein as an indicator. Perform the blank titration (using all the above procedure and reagent except for the use of the hydroxyl sample).

Calculations :-

Weight of sample taken	=	a gms
Volume of NaOH used in blank titration	=	b ml
Volume of NaOH used with sample	=	c ml
Molecular weight of the hydroxy compound	=	M
Volume of NaOH used which is equivalent to the acetic acid used	=	(b – c) ml

1000ml N NaOH ≡ 1 gm mol NaOH ≡ 1 gm mol CH_3COOH ≡ 1 hydroxyl group

$$\therefore (b - c) \text{ ml NaOH} \equiv \frac{b-c}{1000} \text{ hydroxyl group}$$

As 'a' gm of the sample contain $\frac{b-c}{1000}$ hydroxyl group

$$\therefore 1 \text{ gm mol (M) of the sample will have } \frac{b-c}{1000} \times \frac{M}{a} \text{ hydroxyl group}$$

$$\therefore \text{Number of hydroxyl group} = \frac{b-c}{1000} \times \frac{M}{a}$$

If the normality of NaOH is other than 1 N say N/n where n is any integer, then Number of hydroxyl groups = $\frac{b-c}{1000} \times \frac{M}{a} \times N/n$

6

Estimation of Phenols

The number of phenolic groups in any compound can very easily be determined by the acetylation or bromination methods.

(1) Acetylation method:-

Acetylation of phenolic compounds by acetic anhydride is as follows :-

$$Ar(OH)_m + m\,(CH_3CO)_2O \rightarrow Ar(OCOCH_3)_m + mCH_3COOH$$

The principle and procedure is similar to that described for the acetylation of alcoholic compounds.

(2) Bromination Method :-

Principle: Weighed quantity of phenol (having free ortho and para position) is brominated with excess of standard bromate –bromide solution in presence of hydrochloric acid forming 2,4,6 tribromophenol in quantitative yield. The excess of bromine (unreacted) is then treated with potassium iodide solution when quantitative amount of iodine is liberated. The liberated iodine is then estimated by titrating with sodium thiosulphate solution using starch as an indicator.

$$KBrO_3 + 5KBr + 6HCl \rightarrow 3Br_2 + 6KCl + 3H_2O$$

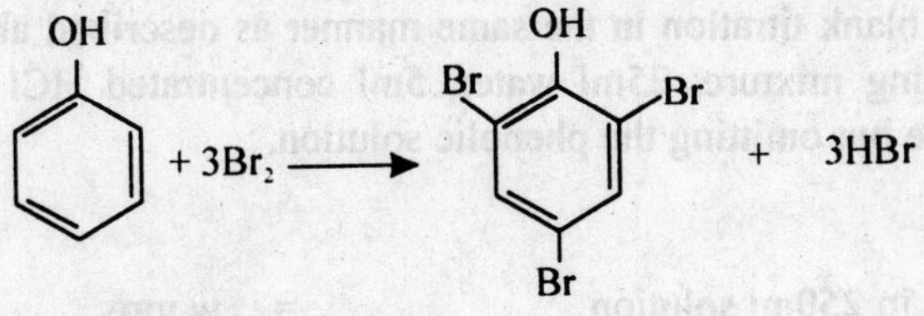

$$3Br_2 + 2KI \rightarrow I_2 + 2KBr$$

$$I_2 + 2Na_2S_2O_3 \rightarrow 2NaI + Na_2S_4O_6$$

Reagents :

(1) Brominating mixture (0.2N) : Dissolve 2.7835 gm of AR potassium

bromate and 37.5 gm of A.R potassium iodide in distilled water and then dilute upto 500ml

(2) Sodium thiosulphate (0.1N): Dissolve 12.5gm of A.R sodium thiosulphate in 500ml of distilled water.

(3) Potassium iodide : Dissolve 15gm of potassium iodide in 100ml distilled water.

(4) Starch indicator solution : Make a paste of 1gm of starch with few ml of water and then pour this suspension into 100ml water, boil for 2-3 minutes,cool and add 3gm of potassium iodide.

(5) Standard phenol solution : Dissolve accurately weighed .25gm of phenol in 5ml of 10% NaOH in a 250ml measuring flask and then make it upto 250ml with distilled water.

Procedure: Pipette out 25ml of phenolic solution into 250ml iodine flask and then add to it 25ml of brominating mixture followed by 25ml of water and 5ml concentrated HCl. Stopper the flask immediately and shake it for a minute. Keep the flask at room temperature for 25-30 minutes, shaking it occassionally during this period. Slightly dislodge the stopper and immediately add 10ml of potassium iodide solution. Stopper the flask immediately, shake it and then leave it for another 5 minutes. Remove the stopper and titrate the liberated free iodine (equivalent to excess of bromine taken) with standard sodium thiosulphate solution using starch as an indicator. The end point is marked by the disappearence of blue colour.

Perform the titration with 25ml of unknown phenolic solution in a similar manner as described for the standard solution of phenol.

Perform also the blank titration in the same manner as described above using 25ml brominating mixture, 25ml water, 5ml concentrated HCl and 10ml potassium iodide but omitting the phenolic solution.

Calculation :-

Weight of phenol in 250ml solution = w gms

Volume of thiosulphate used for sample = v ml

Volume of thiosulphate used for unknown sample = v^1 ml

Volume of thiosulphate used for blank = v^2 ml

Normality of sodium thiosulphate solution = N^1

Number of bromine atoms substituted in phenol = S

Molecular weight of phenol = M

25ml of brominating mixture = v^2 ml of thiosulphate

Brominating mixture consumed by phenol = $v^2 - v$

Brominating mixture consumed by unknown phenol = $v^2 - v^1$

% Purity of phenolic sample = $\dfrac{\left(v^2 - v\right) \times N^1 \times M \times 100}{w \times S \times 2000}$

v^2 ml of sodium thiosulphate solution = 25ml bromination mixture

1ml of sodium thiosulphate solution = $\dfrac{25}{v^2}$ ml bromination mixture

v ml of sodium thiosulphte solution = $\dfrac{25}{v^2} \times$ Vml brominating mixture

Volume of bromination mixture used for 25 ml standard phenolic solution = $v^2 - \dfrac{25}{v^2} \times v$ ml

Volume of bromination mixture used for 25ml unknown phenolic solution = $v^2 - \dfrac{25}{v^2} \times v^1$ ml

Weight of phenol in unknown solution = $\dfrac{4 \times w \times v^2 - \dfrac{25}{v^2} \times v^1}{v^2 - \dfrac{25}{v^2} \times v}$ gms/litre

7

Estimation of Number of Acetyl Groups (Mannitol Hexa-Acetate)

Principle: A known weight of acetyl compound is hydrolysed by boiling with excess known volume of 1N H_2SO_4. The excess of H_2SO_4 is then estimated by titrating with 1N NaOH solution and also the volume of H_2SO_4 actually used to hydrolyse the acetyl compound thus determined. As the molecular weight of the compound is known,the number of acetyl groups can thus be estimated.

Procedure: Weigh two conical flasks A and B, add about 750mg of acetyl compound in flask A and weigh again. Add about 20ml of 1N H_2SO_4 to each flask. Attach a reflux condensor to each flask and boil the solution under reflux for 30min. Transfer the contents in two separate beakers A and B. Wash the condensor and the flask with 10ml water and then pour in their respective beakers. Titrate the contents of each beaker with 1N NaOH solution using phenolphthalein as an indicator.

Calculations

Weight of acetyl compound = .750gm

Volume of NaOH solution used by beaker A = D ml

Volume of NaOH solution used by beaker B (blank) = E ml

Difference in the volume of NaOH solution = D – E

1000ml of NaOH solution = 1 acetyl group

D – E ml of NaOH solution $\doteq \frac{D-E}{1000}$ acetyl group

.750gm of compound contain $\frac{D-E}{1000}$ acetyl group

1 gm mol wt (434) of compound contain $\frac{D-E}{1000} \times \frac{1\text{ gm mol wt }(434)}{.750}$

= acetyl group

8

Estimation of Amino Acids

Principle: Amino acids cannot be estimated by the direct titration procedure due to the presence of two opposing groups viz :acidic as well as basic group in the molecule. The amino acid is therefore first treated with aqueous formaldehyde solution followed by titration with sodium hydroxide solution. But amino acids like glycine are known to exist as 'zwitterion' and are therefore incapable to react with formaldehyde. However glycine on treatment with sodium hydroxide is converted into its sodium salt which then reacts with formaldehyde to give stable cation.

$$NH_3^+CH_2COO^- + OH^- \rightarrow NH_2CH_2COO^- + H_2O$$

$$CH_2O + NH_2CH_2COO^- \rightarrow CH_2 = NCH_2COO^- + H_2O$$

Procedure: Weigh about 750mg – 1gm of glycine, transfer it to a 250ml graduated flask. Dissolve glycine in distilled water and make the solution upto 250ml mark.

In a 100ml conical flask place 50ml of aqueous formaldehyde solution (40% formalin) and add 4-5 drops of phenolphthalein indicator. Now add dropwise N/10 sodium hydroxide solution until the solution is faintly pink.

Pipette out 25ml of glycine solution and transfer it to a 100ml conical flask. Add 2-3 drops of phenolphthalein indicator followed by dropwise addition of N/10 NaOH solution until the solution is faintly pink. To this pink coloured solution add 10ml of neutral formaldehyde solution. The pink colour disappears and the solution becomes acidic. Titrate with N/10 NaOH solution until the pink colour of the solution is restored. Note down the amount of alkali consumed during this process. Repeat the process again and obtain at least two consistent readings.

Calculations

Molecular weight of amino acid (glycine)	=	75
Weight of glycine	=	m gms

25ml glycine requires $= v^1$ ml of .1N NaOH solution

250ml glycine requires $= \dfrac{v \times 250}{25}$ ml of .1N NaOH solution

$= v^1$ ml of N NaOH

1000ml N NaOH $=$ 75 gm glycine

v^1 ml of N NaOH $= \dfrac{75 \times v^1}{1000}$ gm glycine

Sample used contain $= \dfrac{75 \times v^1 \times 100}{1000 \times m}$ % glycine

9

Estimation of Sugars (Glucose)

Carbohydrates (sugars) in solution are normally estimated by chemical, polarimetric and fermentation method. Although polarimetric method is more accurate but chemical method is of great value for the estimation of sugars in biological fluids. Thus in this section estimation by chemical method is discussed.

Principle: Fehling's solution which is alkaline solution of copper salts is reduced by aldose sugars (glucose) of known and unknown strength to cuprous oxide. The strength of the given aldose sugar is then determined from the volume of sugar solution consumed.

Reagents :-

(1) **Fehling's Solution A:** Dissolve 17.32gm of A.R copper sulphate in distilled water and then make up the solution to 250ml

(2) **Fehling's Solution B:** Dissolve 86.5 gm of A.R Sodium Potassium tartarate (Rochelle salt) and 25gm of sodium hydroxide in distilled water and make the solution upto 250ml.

Procedure: Weigh accurately about 1.25 gm of A.R glucose, dissolve it in distilled water and make the solution upto 250ml.

Pippette out 12.5ml each of freshly prepared fehling's solution A and B in a 250ml conical flask and add to it 20ml of distilled water. Boil the fehling's solution gently over a gauze. Now titrate the warm solution with the glucose solution from the burette until the blue colour of the solution has disappeared. Continue the heating during the titration process. The disappearence of the blue colour can be noticed more accurately by allowing the solution to cool at intervals so that the precipitated cuprous oxide can settle. Now tilt the flask on one side and view the supernatant liquid against the white background to observe any blue colour still present in the solution. Repeat the titration until at least two consistent readings are obtained.

The Fehling's solution can be used for titration with unknown solution of glucose in the same manner as described above.

Calculation :

Weight of glucose in
250ml standard solution = W gms

Volume of standard glucose solution
used for 25ml fehling's solution = v^1ml

Volume of unknown glucose solution
used for 25ml fehling's solution = v ml

Strength of unknown glucose solution = $\frac{4 \times W \times v^1}{v}$ gms/litre

1ml of Fehling's solution = .0050 gm of glucose

v^1 ml of glucose solution contain 25 × .005 gm of pure glucose

250ml of glucose solution contain $\frac{25 \times .005 \times 250}{v^1}$ gm pure glucose

$\therefore$ sample contains $\frac{25 \times .005 \times 250}{v^1} \times 100\%$ of pure glucose

10

Estimation of Amino Groups

The number of amino groups can be estimated by the acetylation as well as the bromination method. The acetylation method involves heating the known volume of amine with known volume of acetic anhydride (used in excess) and pyridine until acetylation is complete. Excess amount of unreacted acetic anhydride present in the reaction mixture is hydrolysed to acetic acid by adding water to it. The total free acetic acid is then estimated by titrating with standard sodium hydroxide solution using phenolphthalein as an indicator.

$$RNH_2 + CH_3COOCOCH_3 \rightarrow RNHCOCH_3 + CH_3COOH$$

$$(CH_3CO)_2O + H_2O \rightarrow 2CH_3COOH$$

A blank titration (without addition of amino compound) is also performed under similar conditions. The difference in the volume of sodium hydroxide used in the two titrations is equivalent to the amount of acetic acid formed. If the molecular weight of amino compound is known then the number of amino groups present in the the compound can be calculated.

Reagents

1. **Acetylating reagent:** Mix together one volume of acetic anhydride with three volumes of dry pyridine and shake thoroughly.
2. **Indicator:** Phenolphthalein

Procedure: Place accurately weighed 1gm of the amino compound in a 100ml round bottom flask fitted with a reflux condensor. Add 20ml of acetylating mixture to the flask and heat it on water bath for 30-40 minutes. Remove the heat source and pour about 20ml distilled water down the condensor and shake the flask thoroughly to ensure complete hydrolysis of the residual acetic anhydride. Remove the condensor, cool the flask and keep it at room temperature for 5-10 minutes. Titrate the contents of the flask with standard (1N) sodium hydroxide solution using phenolphthalein as an indicator. Perform the blank titration (using all the above procedure and reagent except for the use of the amino sample).

Calculations :-

Weight of amino compound taken	=	a gms
Volume of NaOH used in blank titration	=	b ml
Volume of NaOH used with sample	=	c ml
Volume of NaOH used which is equivalent to the acetic acid used	=	(b – c) ml
Molecular weight of the amino compound	=	M

1000ml N NaOH ≡ 1 gm mol NaOH ≡

1 gm mol CH_3COOH ≡ 1 amino group

$\therefore$ (b – c) ml N. NaOH ≡ $\frac{b-c}{1000}$ amino group

As 'a' gm of the sample contain $\frac{b-c}{1000}$ hydroxyl group

$\therefore$ 1 gm mol (M) of the sample will have $\frac{b-c}{1000} \times \frac{M}{a}$

$$\therefore \text{Number of amino groups} = \frac{b-c}{1000} \times \frac{M}{a}$$

If the normality of NaOH is other than 1 N say $\frac{N}{n}$ where n is any integer, then

$$\text{Number of amino groups} = \frac{b-c}{1000} \times \frac{M}{a} \times \frac{N}{n}$$

(b – c) ml N. NaOH is equivalent to 'a' gm of amino group

$$1000\text{ml of N. NaOH is equivalent to } \frac{a \times 1000}{(b-c)}$$

$$\text{Molecular weight of amino compound} = \frac{a \times 1000}{(b-c)}$$

Bromination method

Principle: Weighed quantity of amino compound (say aniline) is brominated with excess of standard bromate –bromide solution in presence of hydrochloric acid forming 2, 4, 6 tribromoaniline in quantitative yield. The excess of bromine (unreacted) is then treated with potassium iodide solution when quantitative amount of iodine is liberated. The liberated iodine is then estimated by titrating with sodium thiosulphate solution using starch as an indicator.

$$KBrO_3 + 5KBr + 6HCl \rightarrow 3Br_2 + 6KCl + 3H_2O$$

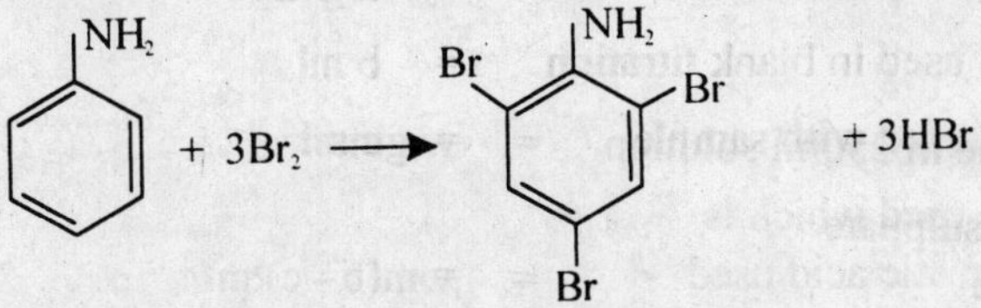

$$3Br_2 + 2KI \rightarrow I_2 + 2KBr$$
$$I_2 + 2Na_2S_2O_3 \rightarrow 2NaI + Na_2S_4O_6$$

Reagents:

(1) Brominating mixture (0.2N):	Dissolve 2.7835 gms of AR potassium bromate and 37.5 gms of potassium bromide in distilled water and then dilute up to 500ml.
(2) Sodium thiosulphate (0.1N):	Dissolve 12.5gm of A.R sodium thiosulphate in 500ml of distilled water.
(3) Potassium iodide:	Dissolve 15gm of potassium iodide in 100ml distilled water.
(4) Indicator:	1% starch solution.
(5) Standard amino solution:	Dissolve accurately weighed .25gm of amino compound in 5ml of dilute HCl in 250ml measuring flask and then make it upto 250ml with distilled water.

Procedure: Pipette out 25ml of amino solution (aniline) into 250ml iodine flask and then add to it 25ml of brominating mixture followed by 25ml of water and 7.5ml concentrated HCl. Stopper the flask immediately and shake it for a minute. Keep the flask at room temperature for 25-30 minutes, shaking it occassionally during this period. Slightly dislodge the stopper and immediately add 10ml of potassium iodide solution. Stopper the flask immediately, shake it and then leave it for another 5 minutes. Remove the stopper and titrate the liberated free iodine (equivalent to excess of bromine taken) with standard sodium thiosulphate solution using starch as an indicator. The end point is marked by the disappearnce of blue colour.

Perform the titration with 25ml of unknown amino solution in a similar manner as described for the standard solution of amine.

Perform also the blank titration in the same manner as describe above using 25ml brominating mixture, 25ml water, 5ml concentrated HCl and 10ml potassium iodide but omiting the amino solution.

Calculation :-

Weight of amine in 250ml solution	=	**w gms**
Volume of thiosulphate used for sample	=	**v ml**
Volume of thiosulphate used for unknown sample	=	v^1 **ml**
Volume of thiosulphate used for blank	=	v^2 **ml**
Normality of sodium thiosulphate solution	=	N^1
Number of bromine atoms substituted in amino compound	=	**S**
Molecular weight of amino compound	=	**M**
25ml of brominating mixture	≡	v^2 **ml of thiosulphate**
Brominating mixture consumed by amine	=	$v^2 - v$
Brominating mixture consumed by unknown amine	=	$v^2 - v^1$
% Purity of amino sample	=	$\dfrac{\left(v^2 - v\right) \times N^1 \times M \times 100}{w \times S \times 2000}$
Strength of given amino solution	=	$\dfrac{\left(v^2 - v^1\right)}{\left(v^2 - v\right)} \times w \times 4$ **gm/litre.**

11

Estimation of Carbonyl Group

There are a number of methods which are employed for the determination of percentage purity of carbonyl compounds. But the method applicable to both the aldehydes and ketones is hydroxylamine hydrochloride–pyridine procedure as well as hydrazone formation method.

Principle (Hydroxylamine hydrochloride method): A known quantity of carbonyl compound is treated with excess of hydroxylamine hydrochloride in the presence of pyridine to yield oxime and pyridine hydrochloride.

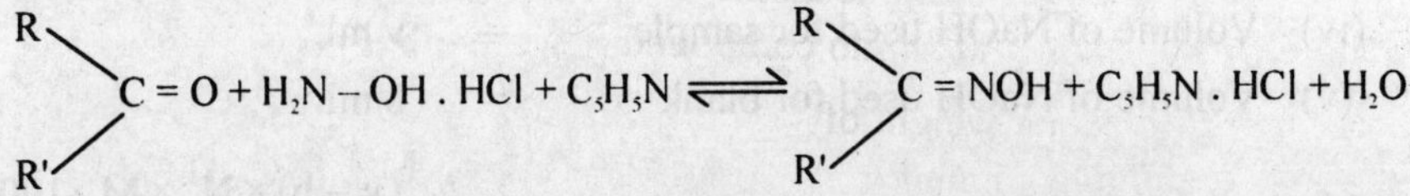

The equilibrium of the reversible reaction is displaced to the right due to the presence of pyridine and hydroxylamine hydrochloride. Pyridine hydrochloride formed during the reaction is acidic enough to be titrated with sodium hydroxide solution using bromophenol blue as an indicator.

$$C_5H_5N.\ HCl + NaOH \rightarrow C_5H_5N + NaCl + H_2O$$

Reagents

(i) Hydroxylamine hydrochloride (0.5N)	Dissolve 17.5gm of hydroxylamine hydrochloride in 80ml distilled water and then make the solution upto 500ml with ethanol.
(ii) Sodium hydroxide solution (.5N)	Dissolve 10gm of sodium hydroxide in 50ml distilled water and then make the solution upto 500ml with methanol.
(iii) Bromophenol indicator	Mix .2ml,4% alcoholic solution of bromophenol blue with 16ml pyridine and then make the solution upto 800ml with ethanol.

Procedure: To 30ml of hydroxylamine hydrochloride in a 250ml conical flask add 1gm of accurately weighed carbonyl compound (benzaldehyde or cyclohexanone) followed by 100ml of bromophenol blue indicator. Stopper the flask and keep it at laboratory temperature for 30 minutes (in case of benzil,acetophenone reflux on water bath).

Meanwhile carry out the blank titration using 30ml of of hydroxylamine hydrochloride and 100ml bromophenol blue indicator with methanolic sodium hydroxide solution to blue green colour. Keep the coloured solution aside for colour comparison. Now titrate the reaction mixture for HCl liberated during the reaction with methanolic sodium hydroxide solution until the colour matches with that of the blank.

Calculations :

(i) Weight of sample = w^1 gms

(ii) Molecular weight of sample = M

(iii) Normality of NaOH = N^1

(iv) Volume of NaOH used for sample = v ml

(v) Volume of NaOH used for blank = b ml

$$\% \text{ of Carbonyl compound} = \frac{(v-b) \times N^1 \times M \times 100}{w^1 \times 1000}$$

12 To Determine the Iodine Value of Vegetable Oil

Principle: Iodine value indicates the amount of unsaturation present in oils and fats. It is defined as the number of grams of iodine consumed by 100gms of oil or fat. It is determined by treating excess of iodine monochloride solution (Wig's solution) with fat or oil and then titrating the residual unreacted iodine with standard sodium thiosulphate solution.

$$CH_2 = CH_2 + ICl \rightarrow \underset{I}{\underset{|}{CH}} — \underset{Cl}{\underset{|}{CH}}$$

$$ICl + KI \rightarrow KCl + I_2$$

$$I_2 + 2Na_2S_2O_3 \rightarrow 2NaI + Na_2S_4O_6$$

Preparation of reagents: Iodine Monochloride (Wig's solution): Dissolve 6.50gms of resublimed iodine in 500ml of glacial acetic acid and then pass chlorine gas to the iodine solution until a permanent orange yellow coloured solution is obtained.

N/10 Sodium thiosulphate (Eq wt 248.20) solution: A standard solution of sodium thiosulphate cannot be prepared by direct weighing because it decomposes in solution due to bacterial action, sulphur gets deposited as a result the strength of solution gets changed. However a solution of approximate N/10 normality is prepared by dissolving (248.2/10 × 4 = 6.204) 6.2 gms of sodium thiosulphate in 250ml of distilled water. This solution is then kept overnight to make it homogenous. It is then standardized by titrating against N/10 copper sulphate solution.

Starch indicator (1%): Make slurry of 1 gm of starch with cold water. Pour this suspension in 100ml of boiling water with constant stirring. Boil the solution for few minutes and then cool it at laboratory temperature. If the solution is not clear then decant to get a clear solution. To the clear solution add 2-3 gms of potassium iodide.

Potassium iodide solution (10%): Transfer 10gms of potassium iodide in a 100ml measuring flask. Add 50ml distilled water, stopper the flask and shake thoroughly so that potassium iodide dissolves completely. Make up the volume to 100ml mark by adding distilled water.

Procedure: Dissolve 1 gm of accurately weighed fat or oil in 20ml of carbon tetrachloride in a 250ml conical flask. Now add 25ml of iodine monochloride solution to it and shake thoroughly. Stopper the flask and keep it in a dark place for an hour with occasional shaking during this period. Add 20ml of potassium iodide solution followed by 100ml of distilled water. Titrate the liberated iodine with standard sodium thiosulphate solution until a straw coloured solution is obtained. Add few drops of starch indicator. The solution turns blue. Titrate again until the blue colour becomes colourless. Similarly perform the blank titration (without sample).

Observation:

Weight of fat or oil sample = M gms

Normality of standardized hypo solution = N/Y

Volume of sodium thiosulphate solution used with sample = V_1 ml

Volume of sodium thiosulphate solution used for blank = V_2 ml

$$\text{Iodine Value} = \frac{(V_2 - V_1) \times N/Y \times 127 \times 100}{M \times 1000}$$

Fat or Oil	Iodine number	Fat or Oil	Iodine number
Soyabean oil	137-143	Sunflower oil	119-135
Palm oil	51.5-57	Coconut oil	8-10
Peanut oil	83-100	Olive oil	79-88
Sesame oil	103-108	Rapeseed oil	94-102
Milk fat	26-50	Corn oil	111-130

13

Chemical Oxygen Demand (COD)

Carbonaceous organic matter present in waste-water has adverse effect on aquatic plants and animals, as they tend to deplete the concentration of oxygen in waste water. In order to overcome this problem all organic matter present in water can be oxidized to carbon dioxide and water by treatment with a strong chemical oxidizing agent. The amount of oxygen required to completely oxidize the organic matter in the water sample is called Chemical Oxygen Demand (COD).

Preparation of Reagents

Standard N/4 Potassium dichromate solution (Eq wt 49.04): Weigh accurately (49.04/4 × 4 = 0.065) 3.1 gms of potassium dichromate and transfer it in a 250ml measuring flask containing about 50-100ml of distilled water. Stopper and shake the flask thoroughly so that potassium dichromate completely dissolves. Now make the solution up to 250 ml mark by adding more distilled water.

N/10 Ferrous Ammonium Sulphate solution (Eq wt 392.10): Weigh (392.10/l0 × 4 = 9.8) 9.8 gms of Ferrous ammonium sulphate and transfer it in a 250ml measuring flask containing 50-100ml of distilled water and 15-20ml of dil H_2SO_4. Stopper and shake the flask thoroughly so that ferrous ammonium sulphate completely dissolves. Now make the solution up to 250ml mark by adding more distilled water.

Ferroin indicator: Dissolve 1.485 gms of 1,10 Phenonthraoline and 0.695 gms of ferrous sulphate in 100 ml distilled water.

Procedure: In a 500ml round bottom flask fitted with a reflux condensor place 50ml of waste water sample and then add 1 gm of mercuric sulphate to it. Shake the mixture thoroughly and then add 25ml N/4 standard potassium dichromate solution followed by 75ml of sulphuric acid. Add a small piece of porcelain chip and reflux the resultant mixture for 2 hours. Cool at laboratory temperature and wash the inside portion of the condensor with 10-15ml distilled water. Dilute the reaction mixture by adding 150ml of distilled water to it. Pipette out 20ml of this solution in a 100ml conical flask and titrate against ferrous ammonium sulphate using 2-3 drops of ferroin indicator. At the end point the colour of solution changes from blue green to reddish brown. Repeat the titration in order to get at least two concordant readings.Perform the blank titration using 20ml of distilled water.

Observation and Calculation

Volume of water sample taken (in ml)	Volume of Ferrous ammonium sulphate used (in ml)
20	Say V_1 ml
20	Say V_1 ml
20	Say V_1 ml

Blank titration

Volume of water sample taken (in ml)	Volume of Ferrous ammonium sulphate used (in ml)
20	Say V_2 ml
20	Say V_2ml
20	Say V_2ml

Volume of N/10 Ferrous ammonium sulphate equivalent to Potassium dichromate = $(V_2 - V_1)$ml used in satisfying chemical oxygen demand.

Volume × Normality(of water sample) = Volume × Normality (of FAS)

20 × Normality (of water sample) = $(V_2 - V_1) \times 1/10$

$$\text{Normality} = \frac{(V_2 - V_1)}{20 \times 10}$$

$$\text{C.O.D} = \frac{(V_2 - V_1)}{20 \times 10} \times 8 \text{ gm/litre}$$

$$= \frac{(V_2 - V_1) \times 8 \times 1000}{20 \times 10} \text{ mg / litre.}$$

14

Dissolved Oxygen

Dissolved oxygen is one of the essential requirement for the survival of aquatic life. The amount of dissolved oxygen in water depends upon (a) respiration and photosynthesis. (b) oxidation of wastes

During pollution of water the Dissolved Oxygen concentration drops and which is quite harmful for the aquatic life. The estimation of Dissolved Oxygen can be easily be carried out either by Winkler's method or its Rideal Stewart modification.

Winkler method is one of the best method for measuring the Dissolved Oxygen. It is based on the reaction of manganous sulphate with potassium hydroxide in alkaline potassium iodide mixture to produce a white precipitate of manganous hydroxide.

$$MnSO_4 + KOH \rightarrow Mn(OH)_2 + K_2SO_4$$

If white precipitate is obtained then it signifies that there is no Dissolved Oxygen in the water sample. However if brown precipitate is obtained then it indicates that Oxygen was present which reacted with manganous hydroxide to give brown precipitate of manganic basic oxide.

$$2Mn(OH)_2 + O_2 \rightarrow 2MnO(OH)_2$$

Addition of sulphuric acid dissolves the precipitate forming manganic sulphate

$$2MnO(OH)_2 + 2H_2SO_4 \rightarrow Mn(SO_4)_2 + 3H_2O$$

Manganic sulphate immediately reacts with potassium iodide, liberating iodine. The amount of iodine liberated is equivalent to the amount of oxygen present in the sample. The amount of iodine is estimated by titrating a portion of the solution with standard sodium thiosulphate solution.

$$Mn(SO_4)_2 + 2KI \rightarrow MnSO_4 + K_2SO_4 + I_2$$

$$2Na_2S_2O_3 + I_2 \rightarrow Na_2S_4O_6 + 2NaI$$

Preparation of Reagents

Manganous sulphate solution: Dissolve 24gms of manganous sulphate crystals ($MnSO_4.4H_2O$) in 500ml of distilled water.

Alkaline potassium iodide solution: Dissolve 250gms of KOH and 75gms of potassium iodide in 500ml of distilled water (l0gms of sodium azide can be added to this solution if nitrite is present in the water sample).

N/40 Sodium thiosulphate (Eq wt 248.20) solution: A standard solution of sodium thiosulphate cannot be prepared by direct weighing because it decomposes in solution due to bacterial action, sulphur gets deposited as a result the strength of solution gets changed. However, a solution of approximate N/40 normality is prepared by dissolving (248.2/40 = 6.2) 6.2 gms of sodium thiosulphate in 1 litre of distilled water. This solution is then kept overnight to make it homogenous. Standardize this solution by titrating against potassium dichromate solution.

Procedure: Transfer the water sample from 300ml capacity glass stoppered bottle to 500ml conical flask. With the aid of pipette add 2ml of manganous sulphate solution by dipping the pipette just below the surface of water. Add 2ml of alkaline potassium iodide in a similar manner. Stopper the flask and shake thoroughly. Keep the flask aside for few minutes and allow the precipitate to settle. Add 1-2ml of conc. H_2SO_4. Stopper the flask and shake thoroughly. The precipitate will dissolve. Keep the flask at laboratory temperature for five minutes. Transfer 100ml of this solution in a 250ml conical flask and titrate against standard sodium thiosulphate solution. When the colour of the solution is slightly yellow add 2ml of freshly prepared starch solution. The solution turns blue. Continue the titration process until the blue coloured solution turns colourless. Repeat the titration process in order to get at least two concordant readings.

Observation and Calculations:

Amount of water sample = 300ml = Vml

Volume of sample titrated = V_1ml

Volume of $MnSO_4$ + KI added (2 + 2 = 4) = V_2ml

Normality of hypo solution = N/40

If whole sample is used

Volume of water sample taken (in ml)	Volume of hypo solution used (in ml)
300	Say V_3 ml
300	Say V_3ml
300	Say V_3ml

$$\text{Dissolved oxygen} = \frac{V_3 \times 1/40}{V \times V_2} \times 8 \text{ gm/litre}$$

$$\text{Putting values} = \frac{V_3 \times 8}{300 \times 4 \times 40}$$

$$= \frac{V_3 \times 8 \times 1000}{300 \times 4 \times 40} \text{ mg/litre}$$

When part of water sample is used

Volume of water sample taken (in ml)	Volume of hypo solution used (in ml)
100	Say V_4 ml
100	Say V_4ml
100	Say V_4ml

$$\text{Dissolved oxygen} = \frac{V_4 \times 1/40}{V_1 \times \left(\frac{V - V_2}{V}\right)} \times 8 \text{gms / litre}$$

$$= \frac{V_4 \times 8}{100 \times \left(\frac{300-4}{300}\right) \times 40} \text{ Putting values}$$

$$\text{Dissolved oxygen} = \frac{V_4 \times 8 \times 1000}{100 \times 0.98 \times 40} \text{ mg/litre}$$

$$= \frac{V_4 \times 8}{0.98 \times 4} \text{ ppm.}$$

15

Biochemical Oxygen Demand(BOD)

All naturally occurring compounds can be degraded by microbial activity in the presence of oxygen. Biochemical oxygen demand (BOD) is a measure of the oxygen utilized/required by microorganism to oxidize the organic material. As the amount of oxygen required to convert the given quantity of biologically oxidizable organic matter to carbon dioxide and water is fixed, the BOD data can be interpreted in terms of concentration of organic matter, as well as oxygen consuming tendency.

Procedure: Take 300ml of polluted water sample and determine its Dissolved Oxygen (DO) as given in previous experiment. Take another 300ml sample of polluted water; incubate it at 20°C for 5 days. Now make Dissolved Oxygen determination on this incubated sample.

$$BOD = \frac{DO_1 - DO_2}{E} \times F$$

Where DO_1 = Initial DO of the sample

DO_2 = DO of the sample after 5 days of incubation.

E = Volume of sample before dilution

F = Volume of sample after dilution.

Abbreviations and Symbols Used

Ac_2O	Acetic anhydride
AgI	Silver Iodide
$AgNO_3$	Silver nitrate
$AlCl_3$	Aluminium chloride
BF_3	Boron trifluoride
b.p	Boiling point
Br_2	Bromine
Cl_2	Chlorine
$CHCl_3$	Chloroform
conc	Concentrated
CCl_4	Carbon Tetrachloride
CrO_3	Chromium trioxide
$CuSO_4$	Copper Sulphate
CS_2	Carbon disulphide
2,4-DNP	2,4 -dinitro phenylhydrazine
DMF	Dimethyl formamide
dil	Dilute
$FeCl_3$	Ferric chloride
H_2SO_4	Sulphuric acid
HNO_3	Nitric acid
HCl	Hydrochloric acid
$HgCl_2$	Mercuric chloride
HI	Hydroiodic acid
hr/hrs	hour/hours
$KMnO_4$	Potassium permanganate
KOH	Potassium hydroxide
$K_2Cr_2O_7$	Potassium dichromate
$KClO_3$	Potassium chlorate
KI	Potassium Iodide
KNO_3	Potassium Nitrate

I_2	Iodine
mgs	Milligrams
MeOH	Methanol
m.p	melting point
NaOH	Sodium hydroxide
$NaNO_2$	Sodium nitrite
NH_4OH	Ammonium hydroxide
NaCl	Sodium chloride
NH_4Cl	Ammonium chloride
$NaBH_4$	Sodium borohydride
PCl_5	Phosphorous pentachloride
temp	Temperature
Pet-ether	Petroleum Ether
THF	Tetrahydrofuran

Bibliography

1. Systematic identification of organic compounds By R.L. Shriner; R.C. Fuson; D.Y. Curtin
2. Identification of Organic compounds By N.D. Cheronis and J.B. Entrikin
3. Organicum - Practical handbook of Organic Chemistry By B.J. Hassard
4. Organic Experiments By Fieser Williamson
5. A hand book of organic analysis By H.T. Clarke
6. Introduction to Organic Chemistry By S. Heathcock
7. Experimental Organic Chemistry By H.Dupont Durst and George W. Gokal
8. Operational Organic Chemistry By John. W. Lehman
9. Natural Product Chemistry By Raphael Ikan
10. Natural Product chemistry. Edited by Atta-ur-Rahman
11. Phytochemistry Vol. I-The process and product of photosynthesis Edited by Lawrence P. Miller
12. Organic synthesis Collective Volume I,II,III,IV,V and VI
13. Organic Chemistry By L.G. Wade
14. Practical Heterocyclic Chemistry By A.O. Fitton and R.K. Smalley
15. Reactions of Organic Chemistry By Hickinbotton
16. Practical Organic Chemistry By F.G. Mann and B.C. Saunders
17. Textbook of Practical Organic chemistry By A.I.Vogel
18. Unitized Experiment in Organic Chemistry By R.Q. Brewster, C.A. Vanderwerf and W.E. Mcewen
19. Systematic Organic Chemistry By W.M. Cumming, I.U. Hopper and T.S. Wheeler
20. Practical chemistry-an integrated course By J.W. Buttle and D.J. Daniels

Some two, three and multistep reactions included in this book

(1) Acenaphthene → acenaphthene diacetate $\xrightarrow{\text{Pg 647}}$ acenaphthenenol

(2) Acetophenone $\xrightarrow{\text{Pg 716}}$ acetophenone oxime $\xrightarrow{\text{Pg 716}}$ acetanilide

(3) Acetophenone $\xrightarrow{\text{Pg 752}}$ Acetophenone phenylhydrazone $\xrightarrow{\text{Pg 752}}$ 2- phenyl indole

(4) Aniline $\xrightarrow{\text{Pg 727}}$ Aniline sulphate $\xrightarrow{\text{Pg 727}}$ p amino benzene suphonic acid $\xrightarrow{\text{Pg 767}}$ methyl orange

(5) Aniline $\xrightarrow{\text{Pg 611}}$ acetanilide $\xrightarrow{\text{Pg 682}}$ p- chloroacetyl acetanilide

(6) Aniline $\xrightarrow{\text{Pg 611}}$ acetanilide $\xrightarrow{\text{Pg 669}}$ p- nitro acetanilide $\xrightarrow{\text{Pg 636}}$ p- nitro aniline $\xrightarrow{\text{Pg 661}}$ p- nitro phenylhydrazine

(7) Aniline $\xrightarrow{\text{Pg 687}}$ 2,4,6- tribromoaniline → 2,4,6- tribromoacetanilide

(8) Aniline $\xrightarrow{\text{Pg 687}}$ 2,4,6- tribromoaniline $\xrightarrow{\text{Pg 693}}$ 2,4,6- tribromobenzene

(9) Aniline $\xrightarrow{\text{Pg 727}}$ Aniline hydrogen sulphate $\xrightarrow{\text{Pg 727}}$ sulphanilic acid $\xrightarrow{\text{Pg 768}}$ Orange II

(10) Acetanilide $\xrightarrow{\text{Pg 686}}$ p- bromoacetanilide $\xrightarrow{\text{Pg 636}}$ p- bromo aniline $\xrightarrow{\text{Pg 713}}$ 4 -bromo biphenyl

(11) Aniline $\xrightarrow{\text{Pg 611}}$ acetanilide $\xrightarrow{\text{Pg 669}}$ p-nitro acetanilide $\xrightarrow{\text{Pg 636}}$ p- nitro aniline $\xrightarrow{\text{Pg 766}}$ Para red

(12) Aniline $\xrightarrow{\text{Pg 611}}$ acetanilide $\xrightarrow{\text{Pg 669}}$ p- nitro acetanilide $\xrightarrow{\text{Pg 636}}$ p- nitro aniline $\xrightarrow{\text{Pg 777}}$ p- iodonitrobenzene

(13) p- Anisidine $\xrightarrow{\text{Pg 610}}$ p- acetoanisidide $\xrightarrow{\text{Pg 675}}$ 4 -methoxy 2 - nitro acetanilide $\xrightarrow{\text{Pg 637}}$ 2- nitro 4 -methoxy aniline

(14) Anthracene $\xrightarrow{\text{Pg 642}}$ Anthraquinone $\xrightarrow{\text{Pg 653}}$ anthrone

(15) Benzaldehyde $\xrightarrow{\text{Pg 699}}$ benzalacetone $\xrightarrow{\text{Pg 623}}$ benzalacetone dibromide $\xrightarrow{\text{Pg 629}}$ α- bromo benzalacetone

(16) Benzaldehyde $\xrightarrow{\text{Pg 707}}$ Cinnamic acid $\xrightarrow{\text{Pg 624}}$ 2,3 -dibromo 3 phenyl propionic acid $\xrightarrow{\text{Pg 630}}$ α-bromo cinnamic acid and α- bromo allo cinnami acid

↓

(17) Cinnamic acid $\xrightarrow{\text{Pg 625}}$ α- bromo β-phenyl β -hydroxy propionic acid

(18) Benzaldehyde $\xrightarrow{\text{Pg 707}}$ Cinnamic acid $\xrightarrow{\text{Pg 624}}$ 2,3 -dibromo 3 - phenyl propionic acid $\xrightarrow{\text{Pg 631}}$ phenylene acetylene

(19) Benzaldehyde $\xrightarrow{\text{Pg 700}}$ benzalacetophenone $\xrightarrow{\text{Pg 623}}$ benzalacetophenone dibromide $\xrightarrow{\text{Pg 630}}$ α-bromo benzal acetophenone

(20) Benzaldehyde $\xrightarrow{\text{Pg 700}}$ Cinnamic acid $\xrightarrow{\text{Pg 691}}$ Styrene $\xrightarrow{\text{Pg 624}}$ 1,2- dibromo phenylethane

↓ 625

Cis Diol

(21) Benzaldehyde $\xrightarrow{\text{Pg 672}}$ m -nitro benzaldehyde $\xrightarrow{\text{Pg 653}}$ m - hydroxy benzaldehyde

(22) Benzaldehyde $\xrightarrow{\text{Pg 672}}$ m -nitro benzaldehyde $\xrightarrow{\text{Pg 707}}$ m- nitro cinnamic acid

(23) Benzaldehyde $\xrightarrow{\text{Pg 703}}$ Benzoin $\xrightarrow{\text{Pg 648}}$ Benzil $\xrightarrow{\text{Pg 646}}$ diphenyl acetylene

(24) Benzoin $\xrightarrow{\text{Pg 648}}$ Benzil $\xrightarrow{\text{Pg 656}}$ mesohydrobenzoin

(25) Benzoin $\xrightarrow{\text{Pg 648}}$ Benzil $\xrightarrow{\text{Pg 757}}$ 5,5- diphenyl hydantoin

(26) Benzoin $\xrightarrow{\text{Pg 648}}$ Benzil $\xrightarrow{\text{Pg 664}}$ Bibenzyl $\xrightarrow{\text{Pg 733}}$ stilbene dibromide

(27) Benzoin $\xrightarrow{\text{Pg 648}}$ Benzil $\xrightarrow{\text{Pg 720}}$ Benzilic acid $\xrightarrow{\text{Pg 658}}$ Diphenyl acetic acid

(28) Benzoin $\xrightarrow{\text{Pg 648}}$ Benzil $\xrightarrow{\text{Pg 616}}$ E and Z- stilbene diacetate

(29) Benzoin $\xrightarrow{\text{Pg 621}}$ Benzoin benzoate $\xrightarrow{\text{Pg 757}}$ 2,4,5 -triphenyl oxazole

(30) Benzoin → trans stilbene $\xrightarrow{\text{Pg 626}}$ stilbene dibromide $\xrightarrow{\text{Pg 629}}$ diphenyl acetylene (tolan)

(31) Benzaldehyde $\xrightarrow{\text{Pg 741}}$ Benzoic acid $\xrightarrow{\text{Pg 606}}$ methyl benzoate $\xrightarrow{\text{Pg 671}}$ methyl 3- nitro benzoate $\xrightarrow{\text{Pg 639}}$ methyl 3 -nitro benzoic acid

↓ 654

3- amino benzoic acid

(32) Benzoic acid $\xrightarrow{\text{Pg 671}}$ 3,5- dinitro benzoic acid $\xrightarrow{\text{Pg 727}}$ 3,5 -dinitroaniline

(33) Benzophenone $\xrightarrow{\text{Pg 717}}$ Benzophenone oxime $\xrightarrow{\text{Pg 717}}$ Benzanilide

(34) Benzophenone $\xrightarrow{\text{Pg 731}}$ Benzopinacol $\xrightarrow{\text{Pg 721}}$ Benzopinacolone

(35) Chlorobenzene $\xrightarrow{\text{Pg 677}}$ 2,4 - dinitro chlorobenzene $\xrightarrow{\text{Pg 781}}$ 2,4 -dinitrophenol

(36) Chlorobenzene $\xrightarrow{\text{Pg 677}}$ 2,4 -dinitro chlorobenzene $\xrightarrow{\text{Pg 696}}$ 2,4 -dinitro phenyl hydrazine

(37) Chlorobenzene $\xrightarrow{\text{Pg 677}}$ 2,4 -dinitro chlorobenzene $\xrightarrow{\text{Pg 780}}$ 2,4 -dinitro aniline $\xrightarrow{\text{Pg 660}}$ 1,2- diamino 4- nitrobenzene

(38) Fluorene $\xrightarrow{\text{Pg 674}}$ 2 -nitro fluorene $\xrightarrow{\text{Pg 652}}$ 2 -amino fluorene

(39) Fluorene →Fluorenone $\xrightarrow{\text{Pg 718}}$ Fluorenone oxime $\xrightarrow{\text{Pg 718}}$ 6-Phenanthridone

(40) Glycine $\xrightarrow{\text{Pg 616}}$ acetyl glycine $\xrightarrow{\text{Pg 754}}$ 4 -benzylidene 2- methyl oxazole $\xrightarrow{\text{Pg 754}}$ 2- acetamido cinnamic acid

(41) Glycine $\xrightarrow{\text{Pg 620}}$ benzoyl glycine $\xrightarrow{\text{Pg 755}}$ 4 -benzylidene 2-phenyl oxazole → DL- Phenylalanine

(42) Hydroquinone →hydroquinone dimethyl ether $\xrightarrow{\text{Pg 683}}$ 1,4 -di t butyl 2,5- dimethoxy benzene

(43) Hydroquinone $\xrightarrow{\text{Pg 643}}$ p- benzoquinone $\xrightarrow{\text{Pg 618}}$ 1,2,4-triacetoxy benzene

(44) Hyroquinone $\xrightarrow{\text{Pg 615}}$ hydroquinone diacetate $\xrightarrow{\text{Pg 723}}$ 2,5-dihydroxy acetophenone

(45) Isoborneol $\xrightarrow{\text{Pg 643}}$ Camphor $\xrightarrow{\text{Pg 668}}$ Camphor sulphonic acid

(46) p- Nitro toluene $\xrightarrow{\text{Pg 644}}$ p -nitro benzylidene diacetate $\xrightarrow{\text{Pg 639}}$ p- nitro benzaldehyde

(47) p-Nitro toluene $\xrightarrow{\text{Pg 645}}$ p- nitro benzoic acid $\xrightarrow{\text{Pg 655}}$ p- amino benzoic acid $\xrightarrow{\text{Pg 608}}$ ethyl p- amino benzoate

↓ 781 ↓ 777

p- nitrobenzoyl chloridep- iodo benzoic acid

(48) Malic acid $\xrightarrow{\text{Pg 779}}$ Coumalic acid $\xrightarrow{\text{Pg 608}}$ Methyl coumalate

(49) Malic acid $\xrightarrow{\text{Pg 779}}$ Maleic acid $\xrightarrow{\text{Pg 779}}$ Fumaric acid $\xrightarrow{\text{Pg 624}}$ α,β- dibromosuccinic acid

(50) Nitrobenzene $\xrightarrow{\text{Pg 670}}$ m dinitrobenzene $\xrightarrow{\text{Pg 657}}$ m nitroaniline $\xrightarrow{\text{Pg 637}}$ m nitrophenol

(51) Nitrobenzene $\xrightarrow{\text{Pg 670}}$ m- dinitrobenzene $\xrightarrow{\text{Pg 657}}$ m- nitroaniline $\xrightarrow{\text{Pg 739}}$ m- nitro chloro benzene

(52) Nitrobenzene $\xrightarrow{\text{Pg 670}}$ m -dinitrobenzene $\xrightarrow{\text{Pg 655}}$ m- phenylene diamine

(53) Nitrobenzene $\xrightarrow{\text{Pg 651}}$ N phenyl hydroxylamine $\xrightarrow{\text{Pg 729}}$ p- amino phenol $\xrightarrow{\text{Pg 615}}$ p- hydroxy acetanilide

(54) Nitrobenzene $\xrightarrow{\text{Pg 652}}$ hydrazobenzene $\xrightarrow{\text{Pg 726}}$ benzidine

(55) o- Nitro aniline $\xrightarrow{\text{Pg 663}}$ o- Phenylene diamine $\xrightarrow{\text{Pg 758}}$ 1,2,3-benzotriazole

(56) o- Nitro aniline $\xrightarrow{\text{Pg 663}}$ o- Phenylene diamine $\xrightarrow{\text{Pg 762}}$ Dibenzophenazine

(57) o- Nitro aniline $\xrightarrow{\text{Pg 663}}$ o- Phenylene diamine $\xrightarrow{\text{Pg 763}}$ Benzimidazole

(58) Phenol $\xrightarrow{\text{Pg 673}}$ o and p- nitro phenol→ 2- hydroxy 5- niitro benzyl chloride

(59) Phthalic anhydride $\xrightarrow{\text{Pg 659}}$ Phthalimide $\xrightarrow{\text{Pg 674}}$ 4-nitrophthalimide →4- nitro phthalic acid

(60) Phthalic anhydride $\xrightarrow{\text{Pg 659}}$ Phthalimide $\xrightarrow{\text{Pg 722}}$ Anthranilic acid $\xrightarrow{\text{Pg 738}}$ o- chloro benzoic acid $\xrightarrow{\text{Pg 712}}$ N- Phenyl anthranilic acid $\xrightarrow{\text{Pg 759}}$ 9- acridone

↓ 760

9-amino acridone

(61) Phthalic anhydride $\xrightarrow{\text{Pg 659}}$ Phthalimide $\xrightarrow{\text{Pg 722}}$ Anthranilic acid $\xrightarrow{\text{Pg 660}}$ Thiosalicylic acid

↓ 776

Biphenic acid

(62) Phthalic anhydride $\xrightarrow{\text{Pg 659}}$ Phthalimide $\xrightarrow{\text{Pg 722}}$ Anthranilic acid $\xrightarrow{\text{Pg 769}}$ Methyl red

(63) Phthalic anhydride $\xrightarrow{\text{Pg 659}}$ Phthalimide $\xrightarrow{\text{Pg 722}}$ Anthranilic acid $\xrightarrow{\text{Pg 758}}$ Benzoxazine

(64) Resorcinol $\xrightarrow{\text{Pg 752}}$ 4- methyl 7- hydroxy coumarin → 6 and 8 - nitro 4- methyl 7- hydroxy coumarin

(65) Resorcinol $\xrightarrow{\text{Pg 772}}$ fluorescein $\xrightarrow{\text{Pg 773}}$ eosin

(66) Salicylic acid $\xrightarrow{\text{Pg 612}}$ acetyl salicylic acid

(67) Salicylaldehyde $\xrightarrow{\text{Pg 708}}$ coumarin $\xrightarrow{\text{Pg 780}}$ cis methoxy cinnamic acid

(68) Tartaric acid $\xrightarrow{\text{Pg 763}}$ I-midazole 4,5- dicarboxylic acid $\xrightarrow{\text{Pg 691}}$ Imidazole

(69) p- Toluidine $\xrightarrow{\text{Pg 615}}$ p- aceto toluidide $\xrightarrow{\text{Pg 688}}$ 4 -acetamido 3 - bromo toluene $\xrightarrow{\text{Pg 638}}$ 4- amino 3 bromo toluene

Appendix

1. Organic Calculations

The chemical amount of a substance in moles or millimoles can be converted to its mass by multiplying with its molar mass. Whereas molar mass of a substance is obtained by appending the units gm/mol to its molecular weight. For example mass of 20 m mol of phenol (mol wt = 94)

$$20\text{m mol} \times \frac{1\,\text{mol}}{1000\,\text{m mol}} \times \frac{94}{1\,\text{mol}} = 1.88\ \text{gm}$$

Similarly the chemical amount of a liquid in moles or millimoles can be converted to its volume by multiplying the liquids molar mass with the inverse of its density. For example 30 m mol of acetic acid (d = 1.049 gm/mL, mol wt = 60)

$$30\ \text{m mol} \times \frac{1\ \text{mol}}{1000\,\text{m mol}} \times \frac{60\text{gm}}{1\,\text{mol}} \times \frac{1\,\text{ml}}{1.049\,\text{gm}} = 17.1\ \text{mL}$$

2.

Reagent	Approximate Strength
Glacial acetic acid	16N
Conc Hydrochloric acid	12N
Conc Nitric acid	16N
Conc Sulphuric acid	36N
Ammonium hydroxide	15N

3.

Reagent	*Approximate Strength*	*Preparation*
Acetic acid	5N	To 142.5ml of glacial acetic acid add 500ml of distilled water
Nitric acid	5N	To 155ml of conc acid add 500ml of distilled water

Sulphuric acid 5N	**To 70ml of conc acid add 250ml of distilled water**
Hydrochloric acid 5N	**To 215ml of conc acid add 500ml of water**
Ammonium hydroxide 5N	**5N To 167.5ml ammonia solution add 500ml distilled water**

INDEX